FROM SHORELINE TO ABYSS: CONTRIBUTIONS IN MARINE GEOLOGY IN HONOR OF FRANCIS PARKER SHEPARD

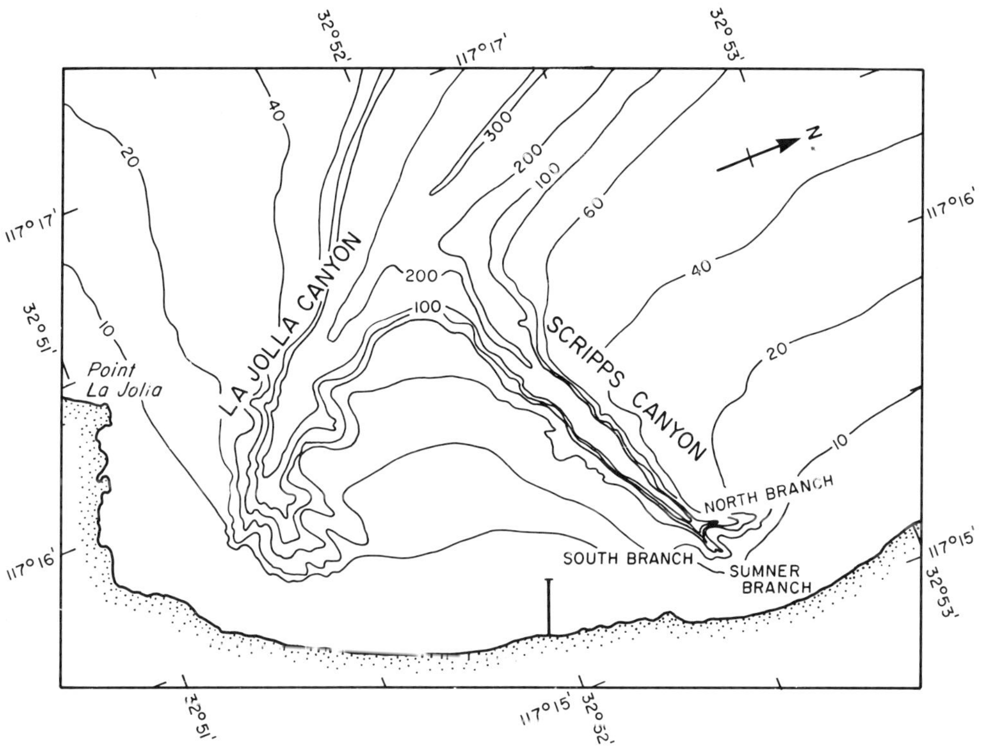

Edited by

Robert H. Osborne, Department of Geological Sciences, University of Southern California, Los Angeles, CA 90089-0740

Barbara H. Lidz, Editor of Special Publications
Special Publication No. 46

Tulsa, Oklahoma, U.S.A. *September 1991*

A PUBLICATION OF
SEPM (SOCIETY FOR SEDIMENTARY GEOLOGY)

ISBN # 0-918985-92-7

Printed in the United States of America

PREFACE

Although Francis P. Shepard passed away on April 25, 1985, he left a rich scientific legacy including more than 230 published papers and books primarily addressed to the study of submarine canyons and turbidity currents (101 publications), continental shelves and associated sediments (46), coastal processes and sediments (33), and marine physiography and tectonics (31). He is best remembered for his work on submarine canyons; however, his broad range of scientific interests and his remarkable ability to break new ground in each of these disciplines have served as a model for at least four generations of "Shepard" students. Upon his retirement from the Scripps Institution of Oceanography in 1964, about 30 of his students presented him with a book of their papers edited by R. L. Miller entitled Papers in Marine Geology: Shepard Commemorative Volume (MacMillan). Official retirement meant little to Shepard, as indicated by the fact that he authored almost another 100 publications, including seven books, from 1964 to 1985—a full research career for most mortals! The present volume was completed in honor of Shepard's innovative leadership in marine geology during a highly creative and productive professional life of 63 years. Inasmuch as the prestigious Francis P. Shepard Medal for Excellence in Marine Geology usually has been awarded annually by SEPM since the medal was established in 1966, it is most appropriate that this volume be included in the SEPM Special Publication series.

This volume is organized into five parts reflecting Shepard's major areas of scientific interest. With the exception of Part I, which contains only one paper, each section is organized principally from global- and regional-scale studies to more local ones. Where appropriate, secondary themes are presented in the introductory statements for each part of the volume. To personalize the Preface, the senior author of each paper was requested to submit a short paragraph relating the subject to Shepard and/or his research. Much of the Preface was extracted from these paragraphs.

Part I consists of a paper by Emery and Ross that sets the stage for the volume by providing an overview concerning the evolution of marine geology during the last 50 years. Some idea of the high rate of advance in marine geology is illustrated by the kinds of problems that were being addressed in Shepard's youth by him and others as compared with those being studied at the close of his career. In fact, present types of investigations were not possible earlier because of inadequate background information, data acquisition, methods of sampling and study, instrumentation, societal needs and funding. This history implies that we cannot now accurately forecast studies likely to occur in the year 2050, even though many attempts may be made.

Shepard was one of the pioneers in the modern study of beach processes, particularly along the Pacific coast of North America. Part II, which contains four papers, is dedicated to his coastal work, which spanned at least four decades; and included investigations of the classification of coastal systems, nearshore circulation, littoral compartmentilization, seasonal beach-sand transport, sea-cliff erosion, barrier-island development and late Quaternary sea-level rise. His book, Our Changing Coastlines, co-authored by Wanless, Sr., in 1971, has served as a major text dealing with the dynamics and evolution of coastal geology for the last two decades. Two themes that penetrate his work are the recognition of the coast as an interface between terrestrial and marine processes and the overlapping time scales associated with coastal processes. These themes continue to be central to ongoing investigations of shoreline response to climate, eustasy and tectonics.

Furthermore, Shepard stressed the concepts of littoral cells and their utility for understanding the input, transfer and outflow of sand within the shore zone. The identification of littoral cells and the computation of sediment budgets within such beach segments remain as fundamental elements in our understanding of coastal scientific and engineering problems. Peterson and others examine the areal distribution of active-beach widths, eolian-dune fields, coastal-terrace heights and beach grain-size data along a 1,000-km open coastline from Cape Mendocino in northern California to Cape Flattery in northernmost Washington. These data are used to identify the locations and dimensions of 42 continuous beach segments, each more than 2 km long, which occur along this convergent margin, and may serve as proxies for possible littoral cells. These segments are related to source and tectonic controls as well as net-littoral transport direction.

Best and Griggs identify the streams draining the Santa Cruz Mountains and 130 km of coastal bluffs as the primary sediment sources in the Santa Cruz Littoral Cell, which extends from San Francisco Bay to Monterey Submarine Canyon. The computed budget shows that sediment input to this cell is highly episodic in response to large and infrequent erosional events. Furthermore, bluff erosion and high-stream flow operate at different frequencies than associated longshore transport, and changes in the volumes of sand stored either on the beach or along the inner continental shelf may produce appreciable volumes of sand due to their large areal extents.

Although the littoral-cell concept is appealing and widely applied, it has never been demonstrated that foreshore sand from a given local source actually follows the suggested pathway along a substantive portion of a given cell to be trapped eventually either by a submarine-canyon head or by a point of land. Osborne and Yeh use Fourier analysis of detrital quartz grains to identify the local sources and grain-shape compositions within five divisions of the Oceanside Littoral Cell in southern California. Each subcell is compositionally distinct and, with one notable exception, appears to be internally homogeneous. Approximately 36 percent of the quartz sand appears to have been derived from adjacent inner continental-shelf deposits. It is interesting to note that grain-shape differences played an important role in the 1961 Shepard and Young paper.

Shepard's contribution to an understanding of marine sedimentary deposits led to a major advance in the interpretation of marine strata in the rock record. Much of what we know about ancient marine deposits has its origins in his field investigations of modern marine environments. The paper by Chiocci and Clifton is an example of the application of observations of modern sedimentary processes to the interpretation of Pleistocene nearshore deposits assigned to the Merced Formation, which are exposed in sea cliffs south of San Francisco, California. Gravel-filled gutter casts in the Merced Formation are oriented approximately normal to the inferred shoreline trend; accordingly, such structures may serve as useful indicators of shoreline trends in other gravelly nearshore deposits.

A considerable amount of Shepard's research was dedicated to a better understanding of the origin and evolution of continental shelves and slopes—the subject of Part III of this volume. For example, he published a synthesis of sediment texture on the continental shelves of the world in 1932. This pioneering work was a major step in the fledgling science of marine geology and served as a beacon for the growing cadre of scientists seeking answers from shipboard. His research was instrumental in focusing attention to key questions, such as the effects of sea-level changes, the origin and transport of sediments in shelf and slope settings, how shelves and slopes are modified by various geologic processes through time, and the role of these tectonic and sedimentologic elements as models for their ancient counterparts. Part II is divided into two sections. The first five papers are dominantly siliciclastic in character, whereas the last three papers concern carbonate strata.

The Lisitzin paper suggests that approximately 92 percent of the world's terrestrial sediment accumulates at two global levels; namely at river mouths and along continental rises. These two global levels are characterized by exceptionally high-sediment accumulation rates (avalanche sedimentation) as opposed to more dispersive processes within oceanic basins (normal sedimentation). Sediment accumulation at river mouths or along the continental rises is driven largely by sea-level changes; and accumulation at one of these global levels is temporally associated with the development of hiatuses at the other level.

Shepard's systematic study and contouring of unpublished U.S. Coast and Geodetic Survey hydrographic sheets of the Mississippi Delta region in the 1950s led to the discovery of shallow, discontinuous gullies off the distributary mouths. He suggested that erosion by density currents, especially turbidity currents, was the most likely cause. Continued research interest was low until 1969 when hurricane damage to offshore drilling platforms initiated another set of studies in the delta region. Using more modern data acquisition equipment, Bouma and co-authors show that the delta front is impacted by a variety of continuously active, subaqueous, mass movement processes. Differential loading, diapirism, and rotational lobes are but a few of the processes affecting delta-front morphology. Nevertheless, Shepard's earlier suggestions concerning mass movements provided a realistic starting point for their work.

Shepard's research revealed that shelf morphology and sediment are inherited largely from earlier times, when the shelf surface was the site of active fluvial, littoral and/or glacial processes. Field and Trincardi re-examine those precepts and build upon them to extend our understanding of how modern shelf surfaces evolved. By employing the recently developed concepts of sequence stratigraphy, they analyze the formation of shelf deposits during different stages of sea level on a variety of passive- and active-margin shelves. Their findings advance those of Shepard's initial work by identifying the conditions during falling and lowstand sea level that determine whether or not the coast will prograde and whether or not the associated deposits will be preserved.

The Yellow Sea was among the shelf areas studied by Shepard. In 1932, he identified the basic grain-size distribution on the seabed from nautical-chart notations, and showed the influence of the Yellow River sediment load and the occurrence of coarser grained material at the shelf edge. In 1949, Shepard and others described the area as analogous to classic epicontinental seas, and addressed possible effects of oceanic currents and storm waves. Booth and Winters discuss how storm waves might modify the morphology of the floor of the Yellow Sea and examine the susceptibility of the sea floor to erosion by wave-generated bottom currents. They also discuss what the implications of these possible modifications might be with regard to the sedimentologic record.

One of Shepard's longstanding research interests was in seafloor bathymetry. This interest is reflected in Gorsline's contribution concerning the slope morphology of the U.S. Pacific continental margin. Gorsline shows that profiles of slope-normal gradients of the continental slopes in this region can be related to large-scale structural elements and to secondary sedimentologic effects. The average gradient, steepest gradient, slope width and the relation of steepest slope segment to mid-slope depth were used to define three major slope provinces.

The clear, warm waters of the Bahamas were most attractive to Shepard. His pioneering work with J. E. Andrews and R. J. Hurley provided bathymetric maps and photographs of the Bahamian sea floor and also opened many new avenues of research. Although not widely recognized as a carbonate sedimentologist, Shepard's publications contain many new concepts that contributed greatly to our knowledge of tropical marine carbonates. Dill's paper concerning the subtidal stromatolites, ooids and crusted-lime muds of the southern Exuma chain exemplifies one of Shepard's greatest research legacies: namely, to go into the sea and understand the sediment and associated processes before you write about them.

In 1967, Shepard was the leader of the CARMARSEL expedition organized and executed by the Scripps Institution of Oceanography, which, with the help of the research vessel Horizon, investigated the lagoonal topography of the Caroline and Marshall Islands. The results of this expedition were published by Shepard in 1970 and provided an accurate representation of Truk, the finest almost-atoll known, as well as an improved knowledge of the geologic problems related to such structures. Guilcher's paper in the present volume discusses the origin and evolution of similar structures in French Polynesia. As such, this contribution adds new information for yet another area of the Pacific Ocean—much as Shepard did 20 years before.

In the third edition of Submarine Geology, Shepard included a description of the geology and carbonate sediments of the West Florida continental margin. Larry Doyle had the opportunity to discuss this area with him in 1978 and found that Shepard's knowledge of this area was remarkable and that he had many insightful comments and ideas concerning higher latitude carbonate systems and the sedimentologic processes associated with them. So, in a very real sense, Shepard contributed directly to the Brooks and Doyle paper concerning the Florida Middle Ground reef system, the northernmost living coral reef in the Gulf of Mexico.

Shepard is best remembered for his research concerning submarine-canyon and fan systems, which is the subject of Part IV of this volume. Normark and Piper discuss three case histories where turbidity currents were initiated by different processes: (1) storm surge on a carbonate-reef complex; (2) a seismically triggered failure within middle-slope sediment; and (3) a bedload delta displaying both hyperpycnal flow of fluvial bedload and failure of prodeltaic sediment. This paper emphasizes that the broad range of size, sediment distribution and depositional processes associated with turbidites reflects a wide variety of initiation processes. Normark and Piper's interests in submarine canyons date from their graduate work at the Scripps Institution of Oceanography during the middle and late 1960s.

As far back as 1938, Shepard promoted the role of tectonics in the development of submarine canyons. Although many of his papers were written before the advent of plate-tectonic theory, his concepts concerning the structural evolution of submarine-canyon systems can be accommodated by plate-tectonic concepts. Greene and others demonstrate that the location, morphology and sedimentary fill of many of the submarine canyons along the California continental borderland can be correlated with plate motion and plate-margin deformation largely associated with the San Andreas fault system.

Basin plain is a general term for ponds of sediment occurring at the outer margin of cross-continental-margin sediment transport systems. Such sediment is ponded by sills of other obstructions, which prevent further transport from the basin. Pilkey and Hokanson propose a classification of basin plains based on three attributes: (1) the extent to which sand layers cover the basin plain floor; (2) the regional-shape trends of sandbody distribution on the basin plain; and (3) the distribution around the basin plain margin of sediment from canyon mouth entry points (sand sources).

Inasmuch as Shepard was always seeking new areas or methods to map the sea floor, he would have been most receptive to the detailed analysis of the channel history of the Rhone Fan presented by O'Connell and others. Long-range (GLORIA) and high-resolution, deep-tow (Sea-MARC I) side-scan sonar imagery provide exciting details concerning the surface texture and morphology of a new thalweg exhibiting an unusual cuspate form rather than the more typical meandering or sinusoidal form.

Part V of this volume consists of two tectonically oriented papers. During and since Shepard's lifetime, increased knowledge of tectonism through plate-tectonic theory and of the age and distribution of seafloor sediments permits a broader view of global geomorphology than was possible even 40 years ago. This newly acquired knowledge is the basis for the Uchupi and Emery contribution concerning genetic global geomorphology.

The tectonic evolution of the California Continental Borderland includes each of the three major types of plate boundary interaction—subduction, rifting and transform faulting. This tectonic complexity, combined with its mostly submarine character, have hindered our understanding of this area. This volume concludes with a paper by Legg, who presents current models for the tectonic evolution of the borderland as well as uncertainties and controversial issues posed by these models.

I thank Robert F. Dill, K. O. Emery and William R. Normark for their encouragement during the formative stages of this volume, as well as Donn S. Gorsline for many helpful editorial suggestions. Gerald G. Kuhn assisted in the compilation of the initial list of potential authors and kindly submitted the photograph of Francis P. Shepard. All correspondence was typed by Susan I. Turnbow, Desser R. Moton and Cynthia H. Waite of the Department of Geological Sciences at the University of Southern California. The line drawing used for the cover and title page was drafted by Janet L. Dodds.

Robert H. Osborne

REVIEWERS

The scientific content as well as the editorial and illustrative aspects of this volume have benefited greatly by thoughtful and constructive reviews by the following individuals:

Kenneth A. Aalto
Mahlon M. Ball
Ron Boyd
William R. Bryant
Douglas W. Burbank
Thomas W. Dibblee, Jr.
John R. Dingler
Larry J. Doyle
William H. Easton
Robert Ehrlich
Michael E. Field
Peter Flemings
Reinhard E. Flick
Roger D. Flood
Eugene A. Fritsche
Louis E. Garrison
Donn S. Gorsline
Charles W. Holmes
Ralph E. Hunter
Herman A. Karl
Kim Kastens
Martin C. Kleinrock
Ronald L. Kolpack
Nicolas C. Krause
LaVerne D. Kulm
Homa J. Lee
Thomas Neurauter
Daniel G. Parrillo
R. Lawrence Philips
Erk Reimnitz
William C. Schwab
Douglas J. Sherman
Deborah K. Smith
Choule J. Sonu
Randolph P. Steinen
Charles E. Stelting
James P. M. Syvitski
Thomas A. Terich
Warren C. Thompson
Edward B. Thornton
Elazar Uchupi
John G. Vedder
Malcolm P. Weiss
Albert J. Williams, III
Robert F. Yerkes

CONTENTS

"Sit down before fact as a little child, be prepared to give up every preconceived notion, follow humbly wherever and to whatever abyss nature leads—or you shall learn nothing."

Thomas Henry Huxley

FRANCIS PARKER SHEPARD (1897–1985)

AFOOT AND AFLOAT ALONG THE EDGE: ADVENTURES OF AN INGENUOUS BEACHCOMBER— A TRIBUTE TO FRANCIS PARKER SHEPARD (1897–1985)

Francis Parker Shepard was born in Brookline, Massachusetts, on the 10th of May, 1897, to Thomas Hill and Edna (Parker) Shepard of the Shepard Shipping and Lumber Family. This was an auspicious year for geology because it was punctuated by the publication of Sir Archibald Geike's classic book on "the founders of geology" (Adams, 1954). Shepard was later to become one of the founders of marine geology for he would acquire the noble title of the Father of Marine Geology (Dietz and Emery, 1971).

Thomas Hill Shepard was of strong New England mariner's stock and provided well for his family, which included a daughter Katherine and son Francis. Both youngsters learned to love the sea from their parents, and spent many happy summer hours aboard personal pleasure yachts sailing the coastal waterways of the New England coast. The winters were devoted to classroom activities in both public and private schools.

After completion of his secondary-school education, F. P. Shepard entered Harvard University in 1916 with geology as a major. The next several years were filled with studies and summer field work in the Rocky Mountains with Professor J. B. Woodworth. Professor Wallace A. Atwood (Ph.D., University of Chicago, 1903) directed his thesis work and Shepard graduated with a B.A. at the end of the summer in 1919. Professor Atwood joined Harvard University as Professor of Physiography in 1913, succeeding W. M. Davis. He went on to become President of Clark University in 1920. At the fiftieth Anniversary Celebration of the University of Chicago in September 1941, W. W. Atwood was given a distinguished alumni medal by the Department of Geology. Two others received distinguished alumni medals as well: Professor K. F. Mather of Harvard, and W. E. Wrather, who became Director of the U.S. Geological Survey in 1943. At the same ceremony, Professor Reginald A. Daly was given an Honorary degree and delivered an address on "Glacial and Submarine Valleys."

Fran Shepard had already been introduced to the revolution of ideas and synthesis of earth and planetary phenomena that were being generated and published at the University of Chicago. Reginald A. Daly was the dynamic leader in the Harvard Geology Department and his stimulating comments often touched on the productivity at Chicago. Daly's enthusiasm spread throughout the entire department and was reflected in the teachings of J. B. Woodworth in stratigraphy and sedimentation and of Charles Palache in mineralogy. All touched on the work at the University of Chicago. Shepard decided on Chicago for his graduate studies, a second motivating catalyst was the fact that the future Elizabeth (Buchner) Shepard was just a little over an hour away by rail from Chicago to Milwaukee.

Francis P. Shepard, Esquire, arrived at the door of the University of Chicago Department of Geology, eager to start in the fall of 1919. Life was full of excitement as these new prospects unfolded in front of him. He plunged headlong into the work, motivated by the promise of devouring the wisdom of the masters and occasionally to promenade the environs between Chicago and Milwaukee.

Graduate studies at the University of Chicago were geared to the level of the professional geologist and, consequently, the students were expected to read and understand the literature, of which there was no shortage. This required dedicated research and critical evaluation during many hours in the library. Time passed quickly as the students gained a familiarity with the high-powered staff and the coveted collection of professional volumes, reprints and reports. The first year ended for Shepard with a joyful celebration of marriage to Elizabeth.

By this time, Shepard already had come under the influence of Rollin T. Chamberlain (son of the department founder Thomas Chrowder Chamberlain), who was then teaching structure and dynamic geology. Chamberlain was an enthusiastic mountain climber and through their mutual love of mountaineering they discussed possible problems for Fran's doctoral thesis. It was through these discussions that Shepard became fascinated by the great intermontane valley called the Rocky Mountain Trench, whose origin was very much a mystery. This unique, relatively straight valley is flat floored for most of its approximately 1,450-km length and is the birthplace of the great Columbia River, whose headwaters are known for their beauty and vastness.

He was especially fond of R. T. Chamberlain and Rollin D. Salisbury, who provided him with a great deal of research advice. T. C. Chamberlain had already retired (Emeritus, 1918, at age 75), but his principle of "The Multiple Working Hypothesis" remained, as well as remnants of his authoritarian teaching methods. T. C. Chamberlain seldom asked questions of his students in class lectures and he brooked no questions from them; the classroom time was fully allocated to lecture. Salisbury ("Saul"), Department Chairman at this time, was merciless in his questions and pity the student who was selected for interrogation on any one day. It was largely through these contrasting examples as well as through the published debates in the professional literature that Shepard's iconoclastic bent was fully atuned: namely, that the authorities were not always correct and differences of opinion did exist among experts. It was thus that Shepard later gained the reputation among his students that he would rather "count the teeth of the lion than to ask Aristotle" (Dietz and Emery, 1971, p. *A10*).

Shepard completed his Doctoral dissertation in 1922 on "Structure and Stratigraphy of Rocky Mountain Trench from Gateway to Golden," which today rests in the "stacks" of the Rosenwald Hall Geology Library. He concluded that the Rocky Mountain Trench was in part a graben and in part a horst. The oral defense completed, Shepard graduated in a class of 22 geology students with eight Ph.D. degrees given in 1922 (Fisher, 1963).

Francis Shepard was hired by the University of Illinois as an elementary geology instructor and his teaching career was underway by the fall of 1922. The year passed quickly

From Shoreline to Abyss, SEPM Special Publication No. 46

as he adjusted to the classroom regimen and the new faculty couple soon became familiar with the Illinois staff.

Shepard continued his interest in structural problems among these various classroom activities, and filling notebooks and three-by-five cards with ideas and possible research problems. This was to result in the publication of 11 papers devoted to structure and tectonics. As providence would have it, field work in the Rockies was out of the question the summer of 1923, because Elizabeth was with child and unable to participate in mountaineering. Thomas Hill II was born that spring, and the senior Shepards invited the junior Shepards and new grandson to Marblehead, Massachusetts, instead. There Fran was given the full use of the family 12-m (40-ft) "New Yorker" sloop to do geologic work to keep him busy. He had been intrigued that very last semester at Illinois by George Ekblaw's Master's thesis on the Hudson submarine canyon. This interest persuaded Shepard to sample the local Marblehead sea floor to satisfy his own curiosity about the so-called progression of sea-floor sediments from coarse sand to fine mud in moving from shore to deep water (a classic textbook theory of the time). He was surprised to find that the bottom-sediment pattern was not uniform or progressive, but seemingly very erratic in distribution. Thus began his career move into marine geology.

Subsequent examination of the nautical charts in the Boston offices of the Coast and Geodetic Survey and the U.S. Navy Hydrographic Office on bottom-sediment chart notations confirmed his own observations and excited him even more. Here were some "real" unstudied problems. He could hardly contain himself as he made plans for further sampling the following summer. This summer's experience did much to stimulate the three characteristics that drive a scientist with such passion and commitment to solve the puzzles of science (Kuhn, 1962, p. 37): 1) a desire to be useful; 2) the excitement of exploring new territory; and 3) the drive to test established knowledge. All of these motivating personality traits were highly developed in Francis P. Shepard. He was an early riser and hit the deck running with ideas. His field associates often found this attribute vexing, because they, in contrast, would require several cups of coffee to get moving, whereas Shepard was often well underway before breakfast.

The fall semester of 1923 and the spring semester of 1924 went quickly as the routine of teaching large classes became more firmly fixed. Shepard developed a pattern of visual aides consisting of lantern slides and charts that were to be his mainstay in teaching geology.

A departmental reorganization at the University of Illinois left a staff opening in Geology that soon was filled by one of the most illustrious teachers and scientists. That man was Harold R. Wanless, Sr., (Ph.D., Princeton, 1923), who, along with J. Marvin Weller (Ph.D., University of Chicago, 1927) and Raymond C. Moore (Ph.D., University of Chicago, 1916), brought new understanding to the Carboniferous of the United States. Wanless and Shepard soon became close personal friends and spent many friendly chats on problems of mutual concern (many years later, after both had retired to Emeritis status, they collaborated on a 579-page book entitled *Our Changing Coastlines*). Later, when William Shirley Bayley (Ph.D., Johns Hopkins, 1886, the first Doctoral degree given by that University; see Pettijohn, 1988) became the departmental chairman, his personal interest in the progress of his staff members helped to redirect the department. Bayley was instrumental in encouraging Shepard to place less emphasis on his multitude of ideas and to spend more time on gathering factual data supporting those ideas. Thus, the distinguished classical petrographer encouraged the budding marine geologist to return to the field to gather more data.

The summer of 1924 saw Shepard's return to Marblehead with George V. Cohee, a student assistant from the university. This time the sampling program was to include sediments from the Cape Cod region, along the Maine coast, and as far north as St. Andrews, New Brunswick, in the Bay of Fundy. Despite a few minor mishaps, the sampling was productive and a more comprehensive picture of the sea floor began to develop. Subsequent expeditions aboard his father's yachts provided evidence of glacial striations in the Gulf of St. Lawrence and suggested the role of glacial sculpture of this waterway. This latest finding was applicable directly to the working hypotheses that the deep gulfs along glaciated coasts may have been formed by glacial excavations during the Great Ice Age, rather than by crustal faulting as had been postulated by some geologists. That was to be one of the three geological "milepost" papers to his credit on glaciation. Up to this time, Shepard's research was privately subsidized by his father and by personal contributions from his university salary.

The first change in financing this marine research came in the form of a small grant from the University of Illinois graduate school dean. This grant allowed Shepard to obtain nautical charts from all over the world, and, with the help of interested geology students, he was able to prepare contour maps of geologic features and to classify these into conveniently identifiable categories. Shepard also compiled and delineated bottom notations of sediment character on these maps and was thereby able to develop an actual distribution of sediment types, rather than the hypothetical distributions based on armchair considerations. Several publications treated the classification of coastlines, bottom sediments, and submarine structures, but Shepard was the first to have physically examined the marine sedimentary environment and its geometries. That led to a revision and reclassification of coastal and submarine features that became the subject of several lively, published, discussions concerning the pros and cons of several classifications available during the 1930s. This early beginning in the study of continental-shelf sediments was to develop into some 46 separate publications during his entire career. "To him, geology was an observational and descriptive science. He was skeptical of highly sophisticated theoretical analysis and was impatient with oblique contemplation of the sea from ivory towers" (Emery and others, 1986, p. *332*).

The regular visits to the Boston office of the U.S. Coast and Geodetic Survey brought about a mutually beneficial cooperative effort for Shepard in that he was invited to participate in the Survey's hydrographic studies along the Atlantic Coast during the summer. His first trip with them was to Georges Bank in 1929, where he was able to observe submarine canyons off the bank's seaward edge and also to view those along the New England coast. The magnitude and complexity of the submarine canyons he had seen was to be a lifetime fascination that resulted in some 77 scientific publications by him.

Shepard was delighted by the newly devised echo-sounding apparatus and the remarkable data that were thus generated. On the return voyage from the Grand Banks, he recorded bottom reflections around the clock. The large volume of sounding data brought new resolution to the charting of bottom topography by revealing features that could only be imagined by the earlier hydrographic wireline sounding techniques. He also had an opportunity to accompany the hydrographer Paul A. Smith of the Coast and Geodetic Survey and A. C. Veatch of the U.S. Geological Survey aboard the *Corsair* as they gathered soundings to be published in 1939, as the first (somewhat stylized) contour map of the Atlantic submarine valleys of the United States.

Active interest in the study of the sea floor was being generated by several groups including the American Petroleum Institute, which in 1926 funded a world-wide study of the source sediments of petroleum by Parker D. Trask (1939). The earlier successful deep-sea expedition of the British *H.M.S. Challenger* (1872–1876) did much to stimulate the imaginations of the scientific community. The deep sea also came under investigation by the Germans with the *Meteor* Expedition to the South Atlantic, the Dutch *Snellius* Expedition to the East Indies, and the United States *Carnegie* Expedition to the Atlantic and the Pacific, all conducted in the middle to late 1920s. Henry C. Stetson of the Massachusetts Institute of Technology began his sediment studies along the Atlantic Coast shelf working out of Woods Hole Oceanographic Institution beginning in 1929.

Shepard was anxious to expand his studies of submarine canyons and was given a sabbatical leave from the University of Illinois to spend a year from 1933 to 1934 studying the submarine canyons of southern California. With a few small grants, mostly for supplies, he moved his whole family, which now consisted of Elizabeth, Thomas Hill II, and Anthony Lee (Tim), to Santa Monica, California, where several canyons just beyond the shoreline could be investigated by small boats.

Shepard quickly busied himself during this sabbatical leave studying previously published charts and acquainting himself with the individuals responsible for data acquisition at the Coast and Geodetic Survey. He was soon able to accompany hydrographic-survey crews from the Los Angeles office on surveys being conducted around the southern California Channel Islands as well as along several local submarine canyons. Shepard also made soundings of the Scripps Canyon with the helpful cooperation of the then Director T. Wayland Vaughan of coral reef fame. These experiences proved invaluable in the development of concepts concerning the origin of submarine canyons as well as interpreting the origin of the California islands. The year was considered to be scientifically profitable as well as very enjoyable for the family. Shepard was smitten by the potential geologic studies offered on the Pacific Coast; and, the whole family found it difficult to return to the midwest after this idyllic year.

T. Wayland Vaughan, formerly of the U.S. Geological Survey, was then Director of the Scripps Institution of Oceanography and was responsible for the official reorganization of the Scripps Institution of Biological Research to that of the Institution of Oceanography on October 14, 1925. Vaughan was a member of the National Academy of Sciences, Committee on Oceanography, at the time and was the chief consultant for planning the organization and facilities of the Woods Hole Oceanographic Institution established in 1930. With their mutual interest in geologic investigations and their familiarity with studies being conducted on both coasts, Vaughan was very much interested in encouraging Shepard to look for ways to bring needed research funding to the Institution. Roger R. Revelle, then a Ph.D. candidate from the Berkeley campus of the University of California, joined Vaughan in August 1931 as an assistant to study bottom sediments. Geology now made a strong showing at the former biological station. That influence was to play a major part in the rapid growth of the Institution. Moreover, when Vaughan was about to retire and a selection committee was appointed to find a replacement for him, it was headed by none other than the eminent geologist Professor George D. Louderback, an associate of the famous Professor Andrew Cowper Lawson of the geological "powerhouse" at the University of California, Berkeley.

The 1934–'35 academic year at Illinois actually passed very quickly with the busy teaching schedule and the processing of data for publication. The close of the year was punctuated by an invitation to Shepard from Captain F. S. Borden of the Coast and Geodetic Survey ship *Hydrographer,* out of the New Orleans district, to accompany the crew on an investigation off the Mississippi delta during the summer of 1935. Here was a chance to expand Shepard's knowledge of the United States coastal areas further to the Gulf of Mexico. The captain was particularly interested in getting Shepard's opinion about some anomolous round hills that were present in the delta area but were difficult to explain. The peculiar submerged circular features rose about 60 m above the surrounding sea floor and displayed a fairly flat top. Shepard was familiar with the salt domes extant on shore and concluded that the offshore circular features were of similar origin. As salt domes, these potential structural oil traps were of great importance to the petroleum industry. Many other deltaic features were observed by Shepard during this trip, which were to be the subject of later investigations that engendered 21 separate papers concerning shore and coastal sediments and 20 papers on submarine physiography.

The enthusiasm with which Professor Shepard conducted his animated geology lectures back at the University of Illinois so affected his students that they hardly missed a session. Two particularly bright undergraduates were quick to see the adventure in all this and swiftly decided that this was the stuff for them. Shepard had hinted at the prospects of having two openings for summer assistants for his next summer's field studies and Robert S. Dietz and Kenneth O. Emery instantly volunteered. Both were enterprising, independently motivated and were soon to be inventors of some unique geologic sampling devices. The times were very difficult during the depression years in the United States; so, to get to the University of Illinois from his home in New Jersey, Bob Dietz used his thumb and hitch-hiked the distance by auto. K. O. Emery was more direct and instead of waiting for a car to give him a ride, he grabbed the first Illinois-bound freight train he could find leaving from southern California, a common mode of travel in those days. It is no wonder that the two became fast friends.

As the year progressed, other events were taking place that were to have a profound influence on Shepard's future in marine geology. Dr. Charles Palache, Professor of Mineralogy at Harvard, became the newly elected President of the Geological Society of America. In this position, Shepard's former undergraduate mineralogy professor had as one of his responsibilities the granting of small amounts of money from the Penrose Fund to individuals conducting pioneering research. The amounts granted were hardly enough to conduct anything but the simplest of studies. Dr. Palache felt that a larger grant should be awarded to an individual who showed great promise in making a substantial contribution to the science. He was thoroughly familiar with the work of his former student on submarine canyons, so he suggested that Shepard submit a proposal for a substantial research grant. Shepard immediately communicated this news to T. Wayland Vaughan and his staff and the necessary expenses were totaled and combined into a proposal. Vaughan assured Shepard that the newly acquired E. W. Scripps research vessel would be available for this purpose. Scripps Institution of Oceanography was desperately in need of ship operating funds. The proposal was submitted and was granted by the Penrose Fund to the extent of $10,000. This was the largest grant given to any one individual at that time. The year was 1936.

In August 1936, the leadership of Scripps Institution of Oceanography passed from Vaughan to Professor Harald Ulrich Sverdrup of Norway. Vaughan had completed his restructuring of the Institution with added staff, reorganization, new buildings, an extensive library, the latest available oceanographic equipment, machine shops, and outstanding landscaping; he was anxious to return to Washington, D.C. Although the State of California had contributed some funds for the operation of the Institution, the bulk of the capital investments came from outright grants by Edward Wyllis Scripps, a bequest from the estate of George H. Scripps (E. W.'s brother), endowments from Ellen B. Scripps (E. W.'s sister), and later, grants from Robert P. Scripps (E. W.'s son). The Rockefeller Foundation made a substantial contribution for the latest in laboratory equipment in 1931 (Raitt and Moulton, 1967). T. Wayland Vaughan had carried the leadership "rod" passed to him by the first Director William E. Ritter very well. Now it was to be passed on to an eminent physical oceanographer Harald U. Sverdrup.

A new dynamic era of oceanographic activity followed Sverdrup's role of leadership as he attracted international interest and recruited researchers to the Institution. Shepard was one of those who was very productive and also capable of attracting funds for the continued growth of the Institution. The two worked well together and with other staff members to produce a great volume of scientific literature in this new science of oceanography, marine biology and marine geology. Emery and Dietz joined Shepared as paid assistants and now did much to advance the Institution's efforts by their enterprise and inventiveness (Dietz and Emery, 1976).

Shepard himself contributed 28 scientific papers from 1936 to the end of 1939 when he took part in the E. W. Scripps research expedition to the Gulf of California. Most of those papers dealt with submarine canyons, but others discussed a variety of natural phenomena, such as ocean-bottom currents, sand transport, geomorphology and coastal classification.

Shepard spent seven months of the year working out of Scripps and the rest of the year at his teaching post in Urbana, Illinois. In 1940, he and Roger Revelle received a followup grant from the Penrose Fund to continue work in the Gulf of California. When he joined the E. W. Scripps expedition to the Gulf of California in October 1940, he again brought with him his two graduate students Robert Dietz and K. O. Emery from the University of Illinois. The scientific party included C. A. Anderson, J. W. Durham, Roger Revelle, Shepard and his two graduate students. All of these individuals went on to become prominent in their respective fields. Anderson later became Chief Geologist for the U.S. Geological Survey; Durham gained world-wide distinction for his many paleontologic contributions; Revelle was to become the Director of Scripps and a world-class oceanographer; and, Shepard became the founder of modern marine geology. The contributions of Dietz and Emery stand alone as distinctive pivotal points in geologic thinking, but they are only a part of the multitude of students that were inspired and nurtured by Shepard to earn their own professional standing in universities, oil companies and governmental departments.

The results of the E. W. Scripps expedition to the Gulf of California were to become a milestone in the melding of land-based geology with submarine geology. However, publication of the results of this investigation was interrupted by more pressing world events as the United States became involved in World War II.

In 1942, Shepard joined the University of California Division of War Research, where he worked with the U.S. Navy in compiling continental-shelf sediment charts for use in submarine warfare. The construction of these charts covered the eastern and southeastern continental shelves of Asia from Kamchatka to the Bay of Bengal. Altogether, 43 sediment charts were constructed.

Among those involved in this war research were Eugene C. Lafond (the scientific coordinator on board the nuclear submarine *Skate* during its historical transit under the polar ice in 1958), Jeff Frautschy (who later became the Assistant Director at Scripps), K. O. Emery, Howard R. Gould (of Lake Mead fame), and others. Robert S. Dietz went into active duty with the Army Air Corps as a photo-reconnaisance B-25 bomber pilot conducting air-photo mapping for the duration of the war. By 1944, the need for sediment charts became less pressing and of the total 43 charts, only 25 were published, because by that time the enemy had retreated from those areas being studied (Schlee, 1973, p. 297, 311).

The war effort diminished the publication output of Shepard; but, with the end of the war, his output increased again to pre-war standards. In 1945, Shepard officially resigned his teaching position at the University of Illinois to devote full time to marine geology under a continuing Navy contract at Scripps. The seriousness of this decision was punctuated by the publication of the first text book in the field, *Submarine Geology,* in 1948. The first manuscript of this book was lost in a giant 1946 tsunami wave that breached the Hawaiian shoreline cottage of Francis and Elizabeth Shepard, and Fran almost lost his own life staying close to shore to photograph the waves. He subsequently wrote a significant paper on the nature of tsunamis in collaboration with Doak C. Cox and Gordon A. Macdonald of Hawaii.

New enthusiasm for research of all kinds was being generated by various organizations after the war to replenish the country's civilian needs. Among these was the massive effort of the several major oil companies to find new oil resources on a world-wide basis. Many of the oil fields had been recognized already by their structural configuration; now it was time to recognize stratigraphic traps and the facies changes associated with them. Some individual oil companies had their own research and development programs for this purpose, but the less fortunate ones had a great desire to obtain outside expertise in the recognition of stratigraphic traps and facies indicators.

As a result, the American Petroleum Institute, representing most of the oil industry, decided to fund a general unclassified research program to be conducted by a suitable public organization such as an active university organization. Enter Professor Shepard, who had done so much of the actual pioneering work in marine geology. Always searching for research funding with which to support his growing studies and his newly attracted students, Shepard made a very strong bid for and received the contract for the API Project 51 in 1951. The project, which he directed, was to last from 1951 to 1957, and resulted in many individually significant scientific publications and a compendium of original works in *Recent Sediments, Northwest Gulf of Mexico* edited by Francis P. Shepard, Fred B Phleger and Tjeerd H. van Andel.

That substantial body of new knowledge was followed by yet another major regional study designed to define the geological, geophysical, biological and oceanographic character of the Gulf of California beginning in 1958. Although Shepard himself did not direct this later investigation, he played a considerable part in the development and study of the Gulf and he involved a substantial number of his students in the study, many of whom had been associated with the original API Project 51 study. The Gulf of California program itself, under the directorship of Tj. H. van Andel, was now funded in part by the American Petroleum Institute, but also by the National Science Foundation, and in a large part by the Office of Naval Research; it also involved the cooperative efforts of the Seismological Laboratory, California Institute of Technology, and of the Geology Department, San Diego State University. Altogether about 12 students participated in that study, which was published in 1964 as the *Marine Geology of the Gulf of California* under the editorship of Tj. H. van Andel and G. G. Shor. Over the many years of Shepard's role as a teacher, more than three-score graduate students came under his tutelage and support, demonstrating that he devoted fully as much time to his students as to his own research (Miller, 1964).

The post-war years were perhaps the most dynamic of times for Shepard, for this period was heralded with the invention of the Aqua-Lung by Jacques Yves Cousteau and Captain Emile Gagnon of France. Although not mechanically inclined, Shepard was quick to recognize the usefulness of various devices capable of unraveling the puzzles of the sea floor. He spent many hours with snorkle and face mask before the Aqualung became available, but with the Aqualung, he had much greater freedom to examine more closely the underwater features that fascinated him so much. His students became especially adept at using this device and contributed many scientific papers on new findings from the sea floor. It was largely through the use of this remarkable tool that Shepard and his graduate student Robert F. Dill (who had been trained under K. O. Emery at the University of Southern California) were able to compile the necessary direct observations that went into the book entitled *Submarine Canyons and Other Sea Valleys* (1966).

One can well imagine the thrill it was for Fran when he was finally able to examine first hand the deep canyon walls of the Scripps Canyon in 1964 with Jacques Cousteau's diving saucer *SOUPCOUPE* and later in the Westinghouse *DEEPSTAR 4000* deep-sea research vehicle. Shepard describes the first dive in the *SOUPCOUPE* as, "It was one of the best days of my life. I felt I learned more about canyons that day than in the 30 years of remote-control study from the surface."

When satellite photography became available in our nation's space program, Shepard quickly realized its potential for coastal studies; after all, he had been remote sensing from land and the deck of ships for many years. These data were invaluable in the production of the book *Our Changing Coastlines,* which he published in collaboration with his old friend Harold R. Wanless, Sr. in 1971.

Shepard retired from scheduled teaching and administrative duties in 1967, but that in no way stunted his boundless energy and his interest in the events that surrounded him in marine geology and other natural phenomena. He continued his considerable output with the revision of two of his books, a book on *Geological Oceanography,* and three other books in collaboration with others. He also contributed 59 separate articles to a multitude of scientific journals and listings. He lost his eyesight to glaucoma the last three years of his life and was fortunate to have Gerry G. Kuhn act as his eyes and his literature researcher. Their teamwork was largely devoted to beach erosion studies and these flourished from 1975 until the end. Their studies led them to weather investigations as these related to beach erosion and coastal changes; a most important finding here was the correlation of volcanic activity with world-wide weather changes. Shepard was ever expanding his horizons and encouraged his students to do so.

His pioneering work and that of other early investigators came to fruition in our understanding of marine geology when the assemblage of magnetic profiling, seismic-reflection profiling, and detailed bathymetry surveys, plus the Joint Oceanographic Institution Deep Earth Sampling program of the Deep Sea Drilling Project were introduced into our thinking. The geologic dynamics resulting therefrom gave us new understanding with the formulation of the concept of plate tectonics by Robert S. Dietz in 1961 and also independently by Harry H. Hess about the same time. Dietz designated this concept by the descriptive term seafloor spreading. Although not directly involved with the deep ("blue-water") ocean, Shepard was very excited by the concept. Shepard was strictly a "green-water" geologist interested almost exclusively in the shore, shelf, slope and canyons along the continental edge. He did, however, become involved in the study of coral reef atolls from personal curiosity, the war effort, and in his search for evidence of a so-called highstand of post-glacial sea level some 5 ka. The hypotheses espoused by a few for this high stand could not be verified by Shepard and his team. Thus, the postulate could not be confirmed even in the same areas of the South Pacific used by the original investigators.

During his lifetime of scientific contributions, Fran and Elizabeth Shepard also generously funded a number of philanthropic endeavors such as marine geology libraries at the Scripps Institution of Oceanography, the Woods Hole Oceanographic Institution, and many other educational and social institutions through the Francis P. Shepard Foundation. He is also known to have helped several of his students financially during times of medical or other critical need. However, the principle witnesses of Shepard's contribution to marine geology are his many students and their followers, who continue as prominent teachers and researchers. "Most of his direct students have now retired, but their students are still turning out later generations of 'Shepard students' who share the enthusiasm for direct observation and precise data collection at sea that Shepard imbued in his direct followers" (Emery and others, 1986, p. 332).

Shepard was very active in the community of geologists and became a Fellow of the Geological Society of America and Geological Society of London, and a member of the American Association of Petroleum Geologists, the Society of Economic Paleontologists and Mineralogists, the Seismological Society of America, the American Geophysical Union, the American Association for the Advancement of Science, and Sima Xi. He also held Honorary Membership in the National History Society of Lausanne, Switzerland, the Netherlands Geological Society, and the Societé Vaudois des Sciences Naturalles of France.

In 1958, he was elected President (1958–1963) of the International Association of Sedimentologists. He was given an Honorary Membership in the Society of Economic Paleontologists and Mineralogists in 1966. He was given the prestigious Wollaston medal for outstanding contributions to geology at the April 1966 meeting of the Geological Society of London. Also in 1966, his colleagues honored him with the establishment of the Society of Economic Paleontologists and Mineralogists Francis P. Shepard Medal for Excellence in Marine Geology. The medal usually is awarded annually to an outstanding marine geologist for excellence in the fields of marine geology in which Shepard excelled. Partial funding for this award came from royalties set aside from the 1964 book entitled *Papers in Marine Geology, Shepard Commemorative volume,* dedicated to Shepard by about 30 of his former students at the time of his formal retirement.

In 1968, he was given an Honorary Doctor of Science degree from Beloit University and was named Man of Distinction by the City of San Diego for his many scientific contributions. The International Association of Sedimentologists bestowed its first issue of the highly acclaimed Sorby Medal to Francis Shepard in 1978 for his excellence in sedimentology. He was given an Honorary Doctor of Science Degree by the University of Southern California in January 1978 and was honored with a bronze statue of himself in Brest, France, by the French scientific community in recognition for his many contributions to marine geomorphology. In 1983 he was the recipient of the Lockheed Award for Ocean Science and Engineering presented by the Marine Technology Society. Francis Parker Shepard became Professor Emeritus of Submarine Geology, Scripps Institution of Oceanography, University of California, San Diego, upon his official retirement from administrative responsibilities in 1967.

Retirement did not mean an end to his productivity for in 1967 he published six scientific articles, in 1968 he contributed five additional papers; and, altogether between 1969 to 1985 he either wrote or collaborated with others to publish 54 separate articles, a revised third edition of *Submarine Geology,* and four separate books varying in size from 173 to 579 pages, a prodigious level of productivity.

Throughout his lifetime, Shepard continued to gain international fame and respect as a dynamic investigator for determining the truth of nature's secrets. He and his beloved partner Elizabeth travelled world-wide on many ocean voyages as well as by commercial means to many international meetings. He was frequently surprised, and privately delighted, to find himself a part of the revolutionary fabric surrounding him, not as an observer, but as a "shaker and mover" within the circle. Nowhere other than within the United States was Francis Shepard given greater regard and tribute than in France. Fran and Elizabeth Shepard visited regularly with their longtime friends Jacques Boucart and André Guilcher, who had similar research interests. The French made Fran an Honorary Member in their prestigious "Societé Vaudois des Sciences Naturalles," where only individuals of proven scientific truth and accuracy are admitted. With typical French flourish, they also dedicated a bronze statue of Shepard at the Université de Bretagne Occidentale at the seaport of Brest, France.

Professor Francis P. Shepard passed away April 25, 1985, and with his passing went the motivating force that encouraged many individuals to enter the field of marine geology. He was a creative scientist and a true gentleman, who lived through several major geological revolutions of thought and study. He was generous with his time and always gave considerate attention to his students with their questions and problems. Moreover, he was continually searching to obtain grants and assistantships for his students wherever these might be made available. As a result of this quest, he often enburdened himself with inordinate stress in satisfying the demands of those who insisted on regular reports and conferences with which to justify continued financial support for his students. He took particular satisfaction in seeing his students succeed in the practical world of scientific publications and professional standing.

Shepard left a legacy of more than 230 articles and books treating a large variety of geological phenomena he had studied personally. He did not fall into the category of scholars who fell compelled to add their own names to the products of their student's efforts, but chose instead to coauthor papers only in works to which he himself contributed an equal share as an associate researcher.

Francis Shepard was innately adapted at identifying logical inconsistencies in geological presentations; these seemed obvious to him, and he thereby questioned some ideas whose notions were often accepted as authoritative facts but were only interpreted observations. He was a natural iconoclast whose inquisitiveness was sharpened by his scientific training at Harvard and the University of Chicago. A very gentle man by nature, and of philanthropic sentiment, he was motivated by impulses of insight. This characteristic of his personality prevailed throughout his lifetime and provided the spark and vitality of his behavior. He sometimes antagonized his peers and irritated the prevailing authorities when he questioned their assumptions or the basis of their beliefs. His questions were without malice or intentional antagonism because he was always a complete gentlemen whose approach was genuinely honest curiosity rather than sarcasm. Shepard's drive and enthusiasm carried on right up to the end; he continued to see possible connections between seemingly unrelated natural phenomena.

In a tribute to Shepard, André Guilcher wrote of him "Pionnier dynamique, homme de terrain, équilibré, élégant, sportif, explorateur de domaines nouveaux, entraîneur gentil et persuasif d'enfants spirituels qui continuent sur sa lancée: tel a été « Fran », gloire de La Jolla où il repose face au Pacifique qu'il a tant aimé" (Guilcher, 1986, p. *94*). We all salute your memory, old friend and guide.

REFERENCES

ADAMS, F. D., 1954, The Birth and Development of the Geological Sciences: Dover Publications, Inc., New York, 506 p.

DIETZ, R. S., AND EMERY, K. O., 1971, Francis Shepard, portrait of a scientist: Earth-Science Reviews, v. 7, no. 1, p. A9–A15.

DIETZ, R. S., AND EMERY, K. O., 1976, Early days of marine geology: Oceanus, v. 19, no. 4, p. 19–22.

EMERY, K. O., DIETZ, R. S., KUHN, G. G., STEVENSON, R. E., AND CURRAY, J. R., 1986, Memorial: Francis Parker Shepard (1897–1985): American Association of Petroleum Geologists Bulletin, v. 70, p. 331–333.

FISHER, D. J., 1963, The Seventy Years of the Department of Geology, University of Chicago 1892–1961: The University of Chicago Press, Chicago, 147 p.

GUILCHER, ANDRE', 1986, Francis P. Shepard (1897–1985), peré de la géomorphologie marine: Annales de Géographie, v. 97, no. 527, p. 87–98.

KUHN, T. S., 1962, The Structure of Scientific Revolutions: University of Chicago Press, Chicago, 172 p.

MILLER, R. L., ed., 1964, Papers in Marine Geology; Shepard Commemorative Volume: MacMillan Company, New York, p. v–vii.

PETTIJOHN, F. P., 1988, A Century of Geology (1885–1985) at the Johns Hopkins University; Gateway Press, Inc., Baltimore, 316 p.

RAITT, HELEN, AND MOULTON, BEATRICE, 1967, Scripps Institution of Oceanography—First Fifty Years: The Ward Ritchie Press, Los Angeles, 217 p.

SCHLEE, SUSAN, 1973, The Edge of an Unfamiliar World; A History of Oceanography: Dutton, New York, 398 p.

TRASK, P. D., ed., 1939, Recent Marine Sediments: American Association of Petroleum Geologists, Tulsa, Oklahoma, 736 p.

GENE A. RUSNAK
2120 Pepper Tree Place
Escondido, California 92026

PART I
PERSPECTIVE

PAST AND FUTURE EVOLUTION OF MARINE GEOLOGY

K. O. EMERY AND DAVID A. ROSS

Department of Geology and Geophysics, Woods Hole Oceanographic Institution, Woods Hole, Massachusetts 02543

ABSTRACT: An attempt is made to understand some of the ways that marine geology developed during the past 50 years, essentially the working lifespan of an active but venerable scientist. This interpretation is aided by comparing marine geology with the development of land geology during a longer period, and by attempting to understand the relative roles of science and technology in the field of marine geology. Excursions from simple straight-line advance for all geology (and also for other fields) are provided by the unexpected appearances of broad generalizations, or paradigms, that commonly are developed by a few scientists and opposed by many, at least for a time. These sudden advances await the accumulation of critical masses of knowledge that, in turn, depend upon exceptional opportunities, partly in the form of adequate funding and partly by transfer of technology. These unusual circumstances make accurate prediction of future advances in maring geology (and in other scientific and technical fields) unreliable but still worthy of thought.

INTRODUCTION

This short essay is intended to provide a general statement concerning the development of marine geology during the past 50 years. This period lies within the time frame of our teacher and colleague, Francis P. Shepard, who worked and published in geology from 1922 until his death in 1985 at the age of 87. Much of the progress made in our field during this time was due to the development and use of new technology. Technology was not one of Frans' strong points; indeed, it generally was best to keep him away from the switches and buttons, but he always knew the value of new technology even though he hated to part with his beloved sextant.

SCIENCE AND TECHNOLOGY IN MARINE GEOLOGY

When one thinks of evolution or progress in a field of science, one should be careful to recognize the difference between technology and science. According to Webster's dictionary (Gove, 1976), a scientist is a person who has "accumulated and accepted knowledge that has been systematized and formulated with reference to the discovery of general truths or the operation of general laws"; has "ability to state problems, to frame questions, so that the technicians can make the machines yield facts that are significant." A technician is a person "who has learned the practical technical details and special techniques of an occupation."

Technology develops through invention and improvement of devices and/or techniques that enable a scientist to obtain new kinds or quantities of data that can be used to improve solutions of old problems or investigations of new ones. Among such devices that have become available and have been widely used by marine geologists since 1950 are satellites. Their passage provides accurate ship positions far from shore, their imagery yields color or absorption characteristics from which chlorophyll and suspended-sediment concentrations can be derived, and cloud cover from which weather maps can be created to guide shipboard operations. Radar altimetry from satellites can be used to measure sea-surface topography from which both gravity and ocean-floor topography can be inferred. Additionally, satellite measurements can be used for determination of precise geographical positions to permit direct measurements of movements of continents and islands by seafloor spreading. The rates and directions of movements so obtained should increase our knowledge of plate tectonics beyond the limits provided by geology alone.

New devices have been developed for use on surface ships to continuously record and produce real-time charts of the bottom topography by means of multiple sounding arrays and near-horizontal side-scanners to survey broad areas of the ocean floor. Topographic details on a finer scale can be provided by many improvements in underwater photography and television, and by direct human observation in submersibles, or by television in remotely operated vehicles (ROVs) capable of reaching any ocean depth. Regional surveys of general stratigraphy and structure of sediments and rocks beneath the ocean floor can be made through seismic reflection and refraction, magnetic signature and intensity, and gravity. More detailed information about materials beneath the ocean floor comes from drilling into the bottom to recover cores that are nearly continuous and have penetrations exceeding 2 km. New coring devices, such as a hydraulic piston corer used within the drill string, have increased sample recovery and reduced core disturbance. Drilling can be sited very accurately using dynamic positioning of the ship drilling platform, and very deep samples can be obtained by means of re-entering the bore holes after replacing worn bits. Boreholes can also be re-entered for insertion of seismic and other sensors.

There is a vague and uncertain boundary between the invention and use of these new technologies and the study of data obtained by new methods. In general, many devices and most methods have been adapted from previous usage for solving non-marine problems rather than being invented specifically for the ocean. Perhaps the best example of such methods is the use of computers to store, retrieve, process, and plot data of all kinds. Another method is the making and use of various radiometric measurements to determine the ages of sediments, fossils, and rocks from the ocean floor. Another is the analysis of trace elements and stable isotopes to determine sources, paths, and fates of water masses, organisms, sediments, and rocks. Other methods, also extrapolated from previous use on land, are techniques for plotting seismic stratigraphy, dating sediments and rocks by magnetic reversals, and determining their latitudes of deposition.

These and other new technologies and methods represent great advances over the relatively simple tools that were invented and built by oceanographers during earlier decades. Most are produced by skilled technicians whose in-

From Shoreline to Abyss, SEPM Special Publication No. 46

terests do not include interpretation of the results in terms of oceanic processes or history. The ideal match is a partnership between technicians having knowledge of materials, electronics, and computers and scientists having knowledge of the ocean and its waters, organisms, topography, or bottom and sub-bottom materials. Nevertheless, the devices, methods, and interpretations of recent results have reached new levels not even dreamed about 50 years ago. Scientists who did not live through these years of accelerating progress can scarcely visualize the change any more than a young child can imagine a world without television.

The main point here is that 50 years ago one could visualize and make only small improvements in then-existing corers, sediment traps, echo sounders, current meters, and other devices. We were unable to leap the gap between then-existing and present methods of measurements before the invention of devices for one field and their spread or adaptation to other fields. Geologists could and did imagine new kinds of information that they could use from application of geophysics (the big geologic hammer) and drilling of deep holes in the ocean floor, but probably it is beyond human mentality to conceive of quantum jumps as large as those within the past 50 years in terms of new concepts and geologic problems that came into focus from the new data. This limitation equally means that we probably cannot imagine the geologic problems that will be at the front 50 years from now anymore than we can visualize the devices and methods that will be available then to solve those problems.

PAST PROGRESS IN MARINE GEOLOGY

Some interesting comments about the state of geology existing in 1904 and its possible subsequent evolution came from lectures by prominent geologists and geophysicists at the International Congress of Arts and Science, St. Louis, and a farewell lecture by E. Suess in Austria the same year. The presidential address by I. C. Russell (1904) pointed out that previous studies of physiography had largely been exploratory and geographical in nature and that most terminology was inexact. He visualized that future studies would be concentrated in definite areas (such as the ice-covered Arctic and Antarctic regions) and directed toward investigations of specific problems (such as the work of wind, streams, glaciers, waves and currents). Oceanography was a newly developing branch of physiography in combination with geology and biology.

C. R. van Hise (1904) pointed out that chemistry and physics were the elementary sciences and that astronomy, biology and geology were applied (derivative) sciences. During previous decades and centuries, many interpretations of astronomy, biology and geology were laughable to contemporary chemists and physicists. The future progress of geology would require better grounding of geologists in chemistry and physics. G. F. Becker (1904), a geophysicist, briefly outlined past development of knowledge about the Earth's interior based upon now-obsolete hypotheses of origin and age. For improved understanding, he stated the need to concentrate upon better measurements of seismology, magnetics and thermal structure to improve knowledge of magmas, ore deposits, petroleum, glacial climates, weathering, erosion and geologic ages. Eduard Suess (1904) outlined the development of structural and other geological disciplines during his long professorship in Austria, pointing out that many problems remained to be investigated. These four eminent scientists with their many contributions to 19th-Century geology fully expected much further progress in understanding the origin, composition and history of the Earth during the 20th Century, but they had no way of knowing the many means of obtaining the necessary information that would be provided by technicians and the goals to be outlined by scientific generalizations during that century.

Interpretation of results in marine geology, as in all other fields of science, spans a wide range of levels. Fifty years ago marine geologists were more concerned with questions of what and where (descriptive) than with when and how (genetic). As a result, their reach was limited largely to areal distribution patterns of topography, rock types and sediment grain sizes. Later, their questions concerned dates of formation or deposition, geologic processes, changing structure and topography, and prediction of evolutionary changes.

About 1860, a major generalization, or paradigm, was developed by Charles Darwin on the basis of his observations of marine and land plants and animals—the evolution of life forms by natural selection. Only about 25 years ago, a similarly broad generalization was made by a few marine geologists—the concept of plate tectonics, a considerably modified form of the earlier continental-drift hypothesis. Each generalization aroused both opposition and broad approval by other scientists, with much effort spent upon testing, understanding, and extending the general concepts.

At present, perhaps the major concern of marine geologists is the relation of regional changes of topography and sediments to world-wide tectonics and their control of plate movements. Already, one can visualize that an important means for understanding both marine and land geology is via integration of regional geologic features with the plate-tectonic history of the world. We are only beginning to see that plate movements in one ocean are reflected in volcanism and mountain folding within other oceans, by rifting of distant continents, by differences in stratigraphy, and by changes of climates and, thus, changing environments for habitation and migration patterns of both marine and land plants and animals. On a more local scale, we can begin to see how plate movements have controlled the distributions of igneous, sedimentary and metamorphic rocks and their changes during Earth history. Plate movements also control topographic patterns of mountain ranges and flat plains on both ocean floor and continents as well as their subsequent alteration by erosion, burial and renewed tectonism. On a more immediately practical level, the plate movements yield information concerning the potential distributions of important mineral deposits and the processes that form such deposits.

Another paradigm, but one still in its early stages of testing, is that of asteroid or large meteorite impact on the Earth—notably one about 65 Ma probably following several earlier impacts. These impacts appear to have produced massive explosions that threw large volumes of dust into the atmosphere, blocking light from the sun, cooling the air, and considerably reducing photosynthesis or growth of plants. As a result, many species of plants and animals became extinct. It is these concurring extinctions that led to the impact concept. Broad generalizations such as these are the apex of scientific investigation, but they are few in number and are built upon many lesser generalizations and discoveries that, during decades or centuries, expanded and combined as knowledge increased. Such hard questions are being addressed by scientists using whatever tools can be adapted or developed by or in collaboration with qualified technicians.

FUNDING

The question immediately arises about what can be done to promote the synthesis of scientific observations and numerous small generalizations into the few and broader paradigms. It is easier to understand how to increase the invention of devices and methods: primarily by the provision of money and opportunity for study, experiment and information transfer from other fields of active study. A major source of such funding has been the military branches of governments through efforts to obtain strategic and tactical advantage. Frequently, the scientific gains have been indirect by-products. Examples are Captain James Cook's observations aboard H.M.S. *Endeavour* and *Resolution* during territorial surveys of 1769–1779, Darwin's time aboard H.M.S. *Beagle* during 1831–1836, and work by many scientists on material collected by H.M.S. *Challenger* during her epic voyage of 1873–1876. Important scientific results came not just from British naval ships but also were produced during voyages of naval vessels from France, Italy, Germany, Russia, Japan and a few others. In the United States and the Soviet Union, much was learned about the ocean during post-World War II expeditions aboard navy ships and aboard university ships funded by the Office of Naval Research and its Soviet parallel. To a certain extent, the value and timeliness of the findings aboard navy ships have been limited by some secrecy imposed in the name of military advantage.

Another major source of ship funds has been government agencies of non-military intent such as the U.S. National Science Foundation that, among other programs, sponsored the Deep Sea Drilling Program and now the Ocean Drilling Program, and extensive ocean-wide surveys conducted in the name of the International Decade of Ocean Exploration. The National Aeronautics and Space Administration and the National Oceanic and Atmospheric Administration enhanced the use and development of Earth-circling satellites and promoted shipboard and submersible operations. Universities have provided facilities and graduate students from their various departments, and oceanographic institutions are supported through Federal and some State agencies via research contracts and grants. Two Federal agencies have conducted considerable oceanic research using primarily their own staffs and ships: the U.S. Geological Survey and the U.S. Coast and Geodetic Survey (which later became part of the National Oceanic and Atmospheric Aministration).

Industry also had a large role in marine geology through its efforts to find and produce offshore oil, gas and other ocean-floor minerals. Several major oil companies made extensive geophysical surveys in all oceans aboard their own or chartered exploratory ships as well as conducting stratigraphic programs for their drilling ships and platforms. More local surveys have been made by companies for possible mining of sand and other minerals and by coastal cities for disposal of their sewage and other wastes. In contrast, the lesser developed countries of the world have provided few measurements (rarely even useful tide-guage recordings) and essentially no generalizations about the oceans because of inadequate education levels, other more pressing interests, and financial limitations. Education of selected individuals from lesser developed countries at universities within industrialized nations is an important step for enhancing interest and scientific contributions from those countries, but efforts can be lost when the trained individuals choose not to return to their own countries. Moreover, it appears that a disproportionate effort is made by lesser developed countries to train lawyers (rather than scientists) to obtain some benefits from increased knowledge of ocean resources.

The funds from navies and other other government agencies, university and industry sources in the industrialized nations of the world have provided billions of dollars for shipboard and laboratory studies of oceanic data. However, these sources appear to have been little used for finding and funding the few scientists who develop the broadest generalizations or paradigms. Perhaps this gap reflects the difficulty in identifying the scientists most likely to make such advances, but it may be due to the uncertainty of obtaining broad generalizations from those who are given funds. Perhaps those who are responsible for allocating research funds frequently feel safer if the funds are used for collection of new data. This suggests that a likely source of broad new concepts lies with those scientists who happen to have large, earned, inherited, or pension incomes—an uncertain and unreliable means of promoting science.

THE FUTURE

Accurate prediction of future scientific problems and of technological advances may be even more difficult to make now than in the past because the extent of present knowledge is so much broader. One may be less able now than earlier to identify successfully the parts of this knowledge that are most likely to serve as the bases for major further advances. It is relatively easy and certain to predict more and better methods of geophysical measurements to better understand the composition and structure of the Earth during the near future. But can we afterward successfully combine these measurements and abundant new dates of events into a comprehensive structural, stratigraphic and physiographic history of the Earth? What and how can we learn about the shapes and characteristics of previous generations of now-vanished oceans and their waters? The evolution of the present oceans has provided clues to the evolution and migration of both marine and land plants and animals; can we find similar clues to the more ancient past within geology and without a time machine? Before many decades have passed, we ought to be able to integrate the changes in Earth structure and stratigraphy with seafloor spreading during the past 200 million years; can we somehow extend that knowledge several billion years earlier?

A better understanding of these past relations should aid in predicting the future of the Earth, its climate, and its plant and animal populations. Will the glaciers remaining from the late Cenozoic finally melt and drown many coasts, or will a renewed advance of ice displace populations? Many answers relevent to humans are in the realm of politics, but cannot predictions from geology modify or supplant political forces? Examples could be in the form of increased coastal erosion, more severe earthquakes, more active volcanism, larger tsunamis, or more asteroid impacts—all of which are even now being studied and evaluated. Are humans more likely to annihilate themselves by bad practices in gathering, farming, hunting, herding, and mining on land and ocean than by military activity, especially during peak times of natural disaster?

Answers appear to be unclear, because great need for knowledge sometimes may inspire greater effort and greater success in obtaining and using knowledge. This need may produce unpredictable successes just as the steady accumulation of knowledge may reach a critical mass that allows rapid understanding and synthesis for a time. Such factors as urgent need, critical mass of knowledge, and unusual terrestrial and extraterrestrial events strew the path of a would-be predictor of science and technology with many unexpected hazards and opportunities. So, for the present, let's just agree that the future of marine geology is likely to be even more interesting and eventful than the past and not try to predict just how it will come about.

SUMMARY

Progress in marine geology during the past 50 years has been enormous. Emphasis is placed here upon the general trend from data collection, synthesis of data, and relation of results to other kinds of knowledge (from the adjacent land as well as from the entire Earth), and to final development of broad generalizations, or paradigms. A great debt is owed by marine geologists to the many sources of funding that have permitted the adaptation of devices from other fields and the invention of others to improve and speed the collection and interpretation of data that are basic to the later generalizations. Rapid development of these devices and methods during the past 50 years means that if we could visualize the devices and methods to be used fifty years hence, we probably would find them to be at least as different from present devices and methods as the present ones are different from those employed 50 years ago. New sources of data doubtlessly will provide the bases for new generalizations that are now unattainable or even totally unexpected. Clearly, technology and science are twin driving mechanisms, technology providing new kinds of data for science, and science providing needs for new kinds of technical data. This essentially is the same as the question of whether necessity is the mother of invention, or whether invention is the mother of necessity. The mix of scientists and technicians must be present in a culture to yield research and applications, perhaps helping to explain why science (in terms of numbers of publications and development of paradigms) has developed faster in nations that participated in the Industrial Revolution than in most other nations.

ACKNOWLEDGMENTS

We thank Dr. J. D. Milliman and Mr. Porter Hoagland III for their useful suggestions for improvement of this manuscript. This work is the result of research sponsored by NOAA Sea Grant National College Programs Office, Department of Commerce, under Grant No. NA86-AA-D-SG090, W. H. O. I. Sea Grant Project M/01. The U.S. government is authorized to produce and distribute reprints for governmental purposes not withstanding any copyright notation that may appear hereon. Woods Hole Oceanographic Institution Contribution No. 7174.

REFERENCES

BECKER, G. F., 1904, Present problems of geophysics: Science, v. 20, p. 545–556.

RUSSELL, I. C., 1904, Physiographic problems of today: Journal of Geology, v. 12, p. 524–550.

SUESS, EDUARD, 1904, Farewell lecture by Professor Eduard Suess on resigning his professorship (translation from German by Charles Schuchert): Journal of Geology, v. 12, p. 264–275.

VAN HISE, C. R., 1904, The problems of geology: Journal of Geology, v. 12, p. 589–616.

GOVE, P. B., editor in chief, 1976, Webster's Third New International Dictionary of the English Language, Unabridged: G & C Merriam Co., Springfield, MA, 2662 p.

PART II
COASTAL STUDIES

LITTORAL-CELL DEVELOPMENT IN THE CONVERGENT CASCADIA MARGIN OF THE PACIFIC NORTHWEST, USA

CURT D. PETERSON, MARK E. DARIENZO, AND DON J. PETTIT
Geology Department, Portland State University, Portland Oregon 97207
AND
PHILLIP L. JACKSON AND CHARLES L. ROSENFELD
Department of Geosciences, Oregon State University, Corvallis, Oregon 97731

ABSTRACT: The areal distributions of active-beach widths, eolian-dune fields, coastal-terrace heights and beach grain size are examined along a 1,000-km open coastline in the convergent Cascadia margin of the northeast Pacific. Beaches range from 10 to over 500 m in width in a total of 42 continuous beach segments, which are considered proxies for littoral cells at least 2 km long. Tectonic upwarping has exposed coastal-terrace deposits (1–30 m thick) and/or underlying bedrock to elevations as much as 120 m above mean sea level. The regional distributions of short-beach segments (average of 8–12 km long) in northern Washington and northernmost California—southern Oregon correspond to areas of maximum variation in terrace height, reflecting relatively rapid rates of tectonic vertical deformation. The three largest beach segments, 65 to 165 km long, are associated with major dune fields, and correspond to three of the largest river sources, namely the Eel, Columbia and Umpqua Rivers. Tectonic downwarping of inner-shelf sediment sinks might account for narrow, short-beach segments associated with the Klamath and Rogue Rivers. Six beach segments in central Oregon are isolated from either major river or high-terrace sand sources, and are assumed to be derived from transgressive shelf sands. However, the observed longshore variability in these beach deposits indicates that pre-existing distributions of offshore sands were not uniform along the coast. Finally, longshore trends in increasing beach width, increasing dune development and/or decreasing beach grain size to the north in about one-quarter of the analyzed beach segments suggest a small net northward transport. However, inconsistent trends of all three variables in the remaining segments confirm the occurrence of short-term conditions responsible for reversing longshore transport in littoral segments effectively bounded by prominant headlands.

INTRODUCTION

Highly segmented coastlines of active continental margins, i.e., collision coasts (Inman and Nordstrom, 1971) compartmentalize longshore sand distributions into littoral cells. Sand sources and sinks are likely to vary between active-margin cells due to localized sediment discharge from adjacent coastal ranges and longshore variability in the vertical displacements of Pleistocene coastal deposits and older bedrock. The existing reservoirs of sediments in modern littoral cells control beach-shoreline position and buffer sea cliffs from wave attack. The larger headlands likely influenced longshore transport during the late Holocene transgression, and possibly served as cell boundaries during previous highstands of sea level. However, little is reported concerning the cumulative effects of repeated littoral-cell development in "hardened" active-margin sites during successive marine highstands (Shepard and Wanless, 1971).

Interest in Quaternary coastal deposits in the convergent Cascadia margin of the U.S. Pacific Northwest (Fig. 1) has increased substantially during the last decade. Locally catastrophic beach erosion following the 1982–83 El Ninõ climatic event has been related to longshore displacements of beach sand in a number of Pacific Northwest cells (Komar, 1986; Komar, 1989; Peterson and others, 1990). Marine placer deposits in the region are found to be associated with coastal promontories and paleo-headlands (Peterson and Binney, 1988). Beach deposit cross sections demonstrate small sand reserves along some shorelines facing increasing development pressures (Pettit, 1990). Finally, knowledge of Quaternary shoreline morphology is important to neotectonic studies of margin deformation and related seismicity in this active subduction zone (Atwater, 1987; Kelsey and Carver, 1988; Darienzo and Peterson, 1990; Muhs and others, 1990).

Continuous beach segments between large headlands have been used by different investigators to define some of the larger littoral cells of the Washington and Oregon coasts (Terich and Schwartz, 1981; Galster, 1987). However, differences in the descriptive methodologies previously used to define the shorelines and their corresponding beach deposits prohibit their comparisons on a regional basis. In this paper we use 1) apparent beach widths, 2) continuous beach segments, 3) dune-field distributions, 4) estimated marine-terrace heights, and 5) beach-sediment grain size from established reference points to compare the open coastlines of this convergent margin. These shoreline data are used to document the distributions and dimensions of continuous beach segments, proxies for possible littoral cells, from Cape Flattery in northernmost Washington to Cape Mendocino in northern California (Fig. 1). Finally, we use longshore distributions of beach width, modern dune fields and beach sediment grain size in defined beach segments to examine directions of potential littoral drift in the study region.

BACKGROUND

Identification of Littoral Cells

Littoral cells are generally defined by zones of restricted, longshore sand transport. Length scales of littoral cells range from distances between adjacent rip currents (hundreds of meters) to distances between major structural features on the continental shelf (tens of kilometers) depending on the time scales of interest. For moderate time scales, i.e., tens to hundreds of years, longshore transport along the Pacific Northwest coast is thought to be restricted by prominent headlands (Fig. 2a-c), which terminate continuous beach segments (Terich and Schwartz, 1981). Establishing cell boundaries solely on the basis of beach-segment terminations is not strictly correct, as littoral sediments might bypass smaller headlands during extreme storm events. However, abrupt changes in longshore trends of beach width and sediment grain size on either side of existing headlands

From Shoreline to Abyss, SEPM Special Publication No. 46

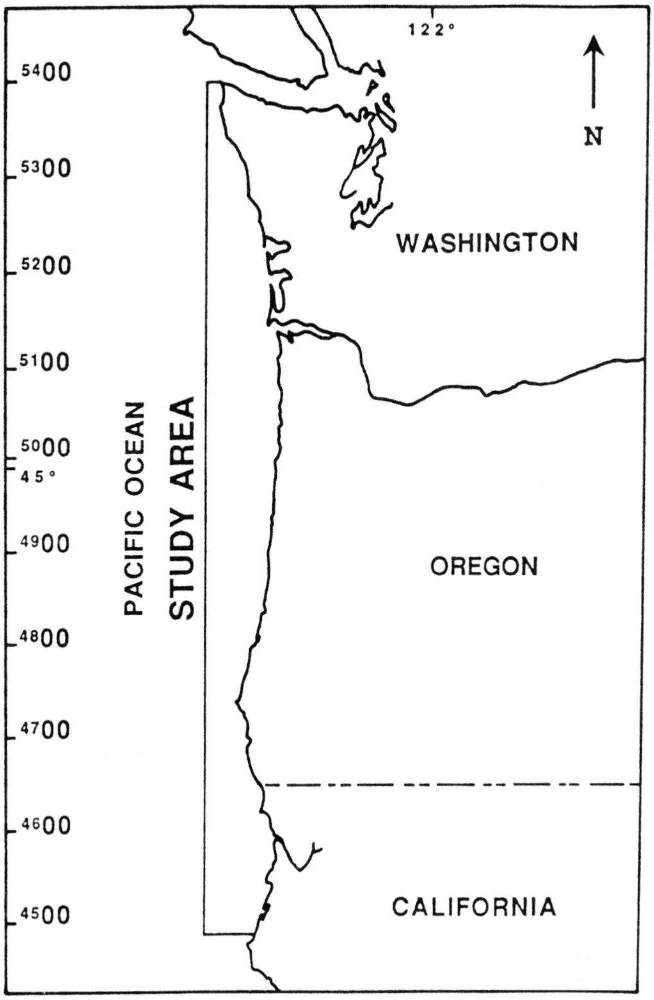

FIG. 1.—Map of study area, including 1,000-km-long open coastline from Cape Mendocino in northern California to Cape Flattery in northern Washington. UTM locations (metric coordinate distances: 5,400,000 to 4,500,000 m) and degrees latitude (45°) and longitude (122°) are shown at figure borders.

can indicate restricted zones of sediment supply and mixing, thereby identifying potential cell boundaries.

Cascadia Margin Coastal Tectonics

Irregularities of the Cascadia coastline that serve to isolate littoral deposits are derived from a variety of geologic structures and processes. Typically, resistant headlands and less resistant embayments in sea cliffs form by their differential retreat from wave attack during marine transgressions (Shepard and Grant, 1947). The resistant rock units can be juxtaposed longshore against less resistant units along faults, volcanic-flow boundaries and dipping sedimentary strata. In the Cascadia margin these structures have been inherited from accreted terranes including the Klamath, Coast, and Olympic Ranges (Cady, 1975; Irwin, 1981; Wells and others, 1984; Fig. 3) as well as formed by subsequent deposition and tectonic deformation in a subduction zone setting.

The southern end of the Cascadia subduction zone begins at the Mendocino triple junction in northernmost California (Fig. 3; Kelsey and Carver, 1988). The trench-to-coastline distance increases from 30 to 80 km in northernmost California to 120 km in northern Oregon, a zone of oblique convergence between the Juan de Fuca plate and the North American plate (Fig. 3). A curvature of the continental margin (about a 30° change in bearing) in northern Washington results in a complex convergence geometry, changing from oblique to normal convergence over a trench-parallel distance of about 100 km (Crosson and Owens, 1987). Finally, possible isostatic rebound (uplift) following post-Wisconsin deglaciation in the study area is limited to the northernmost Washington coast (Shipman, 1990).

Over short time scales (~500 yr), cycles of apparent coastal uplift and coseismic subsidence with 1 to 2 m vertical displacements have occurred along the Washington and northern Oregon coasts (Atwater, 1987; Darienzo and Peterson, 1990). These vertical displacements are thought to result from interplate coupling and dislocation in an active subduction zone. By comparison, short-term cycles of coseismic uplift and/or localized subsidence might be expected along the southern Oregon—northernmost California coast, which is closer to the trench and the deforming accretionary prism (Fig. 3). The possible effects of such episodic vertical displacements on beach deposits have yet to be addressed in this subduction zone margin, and are not further discussed in this paper.

Pacific Northwest Sand Supply and Wave Climate

Sand supply and wave climate are undoubtedly important factors in controlling the abundance and longshore distribution of littoral deposits. Numerous studies of beach-sand sources have been performed in the Pacific Northwest (Kulm and others, 1968; Scheidegger and others, 1971; Plopper, 1978; Bodin, 1982; Clemens and Komar, 1988). These studies establish that retreating sea cliffs, local rivers and possibly transgressive sands from the continental shelf have supplied sand to modern beaches. However, the relative contributions of these sources have not been quantified. Even less is known about sand sinks in this margin. Few quantitative estimates have been made of landward sand sinks including dune fields (Cooper, 1958, 1967) and tidal basins (Kulm and Byrne, 1966; Peterson and others, 1984a,b). None of the submarine-canyon heads in the study area reach nearshore depths at the present sea level (McManus, 1964; Kulm and others, 1975; Bodin, 1982). However, the potential for other offshore sinks of littoral sand has not been evaluated in the study area.

The Pacific Northwest has one of the highest wave energies in the Northern Hemisphere with winter, significant wave heights ($H_{1/3}$) commonly peaking at 5 to 7 m (Komar and others, 1976b). By comparison, the summer, significant wave heights rarely exceed 3 m. The annual variation in wave energy results in a seasonally reversing offshore (winter) to onshore (summer) sand-transport regime (Clifton and others, 1971; Komar and others, 1976b). In contrast to the well-documented across-shore transport cycle, the magnitude and direction of longshore transport in the

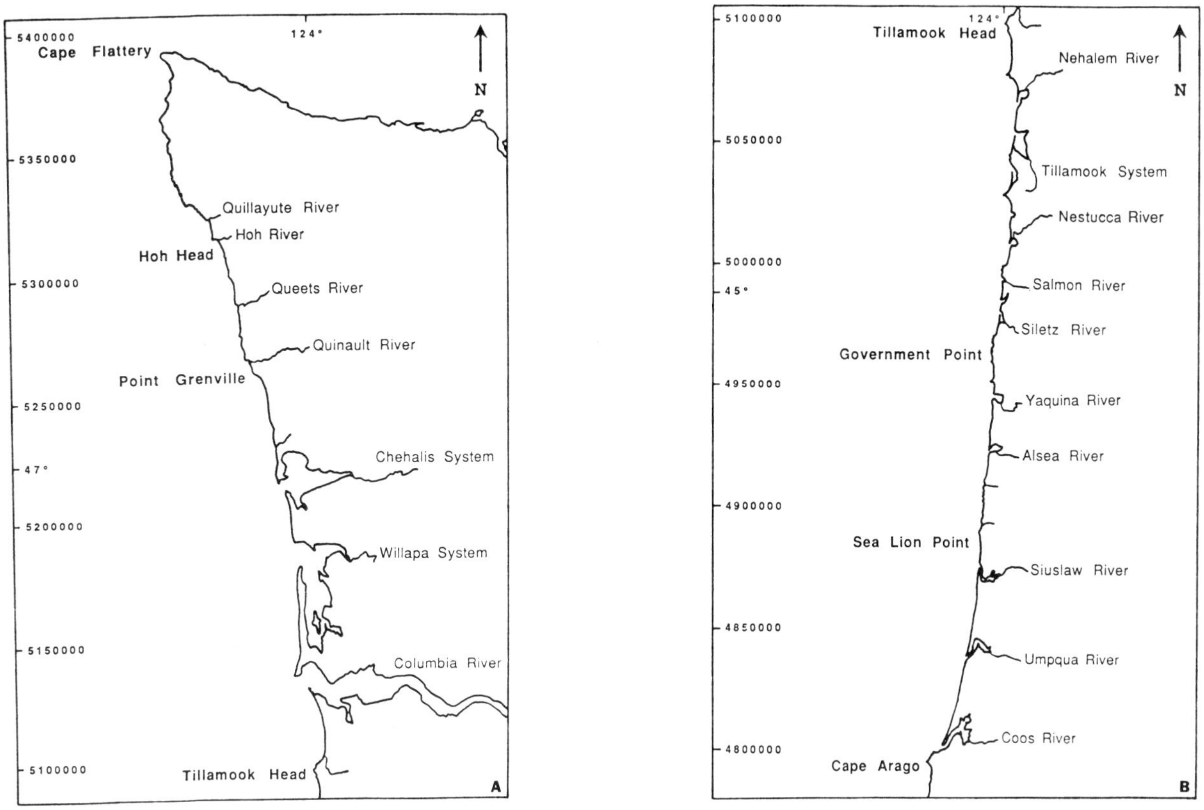

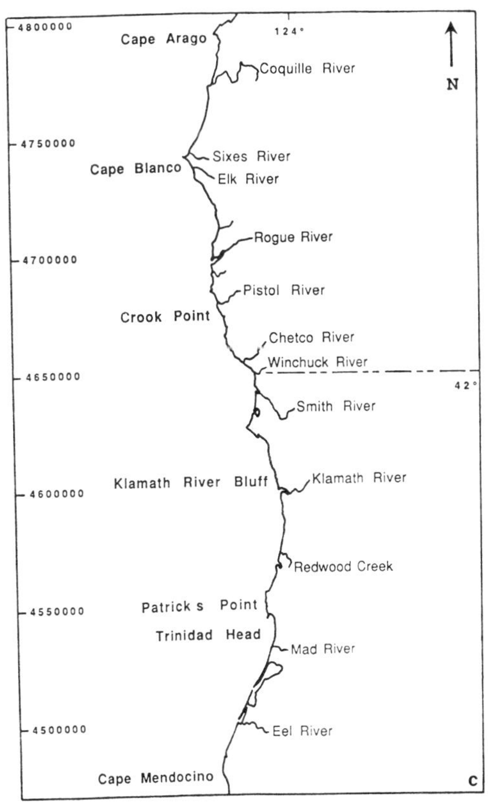

FIG. 2.—Map of headlands, river mouths and regional coastal morphology in study area including (a) from Cape Flattery to Tillamook Head, (b) Tillamook Head to Cape Arago, and (c) Cape Arago to Cape Mendocino. UTM coordinates (5,400,000-4,500,000 m) and latitude 47°, 45° and 42° positions are shown at figure borders.

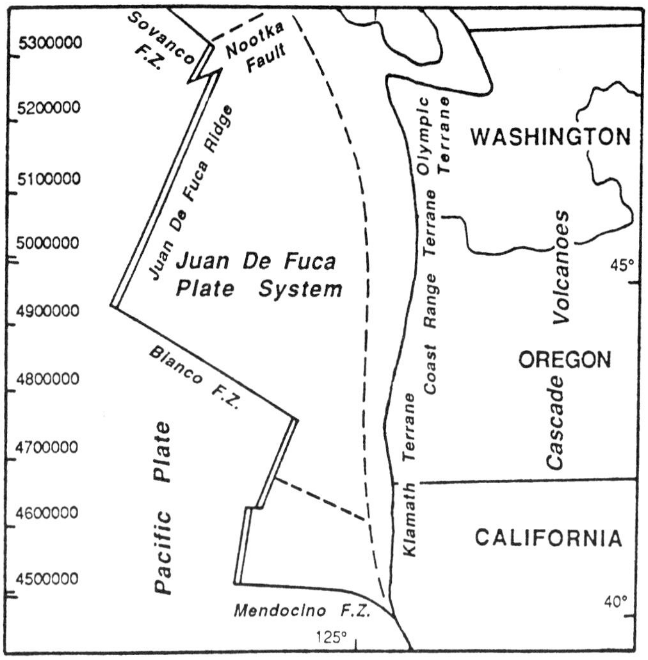

FIG. 3.—Tectonic map of Juan de Fuca Plate System, Cascadia subduction zone, and approximate boundaries of accreted terranes (Klamath, Coast Range and Olympics) along the U.S. Pacific Northwest coast. Dashed line represents subduction zone trench. Relative plate motions result in oblique convergence in northernmost California and Oregon, and normal convergence in northern Washington.

Pacific Northwest is open to debate. Annual reversals of longshore transport have been reported to follow seasonal reversals of dominant wave directions, e.g., winter storm surf from the southwest and fairweather swells from the northwest (Kulm and Byrne, 1966). However, recent sand displacements following the 1982–83 El Niño event indicate that reversing directions of longshore transport might instead be forced by interannual variations in winter wave climate alone (Peterson and others, 1990). For longer periods, Komar and others (1976a) convincingly argue for a condition of zero net transport along the Oregon coast, whereas investigations from either side of Oregon, i.e., Washington (Ballard, 1964; Plopper, 1978; Terich and Schwartz, 1981) and northernmost California (Bodin, 1982), suggest net northward littoral drift.

METHODS

To establish a baseline survey of beaches and associated shoreline conditions in the study area, we examined records of stereo aerial photographs (1:12,000 or 1:24,000 scale) from State and Federal agency archives in Oregon, Washington, and California. Continuous aerial photography of northernmost California was flown in May of 1978. By comparison, no single year provides continuous shoreline coverage in either Oregon or Washington. Our examinations indicate that the most complete shoreline coverage, prior to the anomalous erosion-accretion cycles following the 1982–83 El Niño, is for the years 1974–1981, with almost 70 percent of the coastal coverage in 1977–1978.

The aerial photographs used in this survey (Appendix) were chosen on the bases of 1) season, i.e., either summer or fall with exceptions of 11/79, 12/79, 1/80 and 4/80 in southern Oregon, 2) absence of cloud and fog cover, and 3) complete or nearly complete coverage of continuous beach segments between major headlands. There is little correspondence between flight times, e.g., approximately high noon ±1.5 hr, and a common tide level. Estimated tide levels for corresponding aerial photograph sequences ranged from +1.5 to −1.5 m mean tidal level (MTL; Appendix). However, the great majority of the flight sequences, representing 88 percent of total coastline distance, falls into a narrower range of ±1 m MTL. Assuming an average beach slope of 0.025 for the study area (Pettit, 1990), this range in tide level could result in a ±40 m range in observed beach width, relative to a given beach width at mean tidal level (0 m MTL).

A total of 2,024 aerial photograph reference sites have been established at intervals, i.e., approximately 0.5 km longshore, along the length of the 1000-km study area. The width of active-beach deposit at each reference site is estimated on the basis of the measured shore-normal distance of the exposed beach deposit from the mid-swash zone to the base of a sea cliff or to an established vegetation line. Potential bias of mid-swash zone positions from variable swash events is assumed to cancel when averaged over multi-kilometer distances. The established vegetation line is defined by the most seaward position of bushes or dense grass cover, typically located at the foredune crest.

The active-beach widths are thought to reflect the quantity of sand-size material in mobile beach deposits at reference sites during the time of aerial photography, i.e., generally during summer months. Local morphological features such as cusps, bars and troughs are known to effect beach profiles and corresponding beach widths. However, such factors should average out over multi-kilometer distances, yielding only regional trends in beach width over long distances, i.e., several kilometers. Depth constraints on beach sand cross sections and volumes in selected cells in the study area are discussed elsewhere (Pettit, 1990).

Dune widths are measured from the seaward line of established vegetation to the landward termination of the vegetated dune field or to a maximum shore-normal distance of 200 m. There is much uncertainty as to the ages of the dune fields, which might vary from hundreds to tens of thousands of years (Cooper, 1958, 1967; Rankin, 1983). Thus, maximum widths of the large dune fields (up to several kilometers) are not included in this survey. The widths of the modern dunes (maximum 500 m width cutoff) are thought to represent the sand reservoir above the reach of common high tides and storm surges. However, many foredunes do show signs of ongoing erosion and net shoreline retreat, their initial development having resulted from low topographic relief, and/or prior surplus sand supply. Nevertheless, extensive development of broad dune fields over regional scales (tens of kilometers longshore) must certainly reflect a long-term condition of excess-sand supply to corresponding beach sites.

Steep, boulder beaches and rocky coastlines are identified here as shorelines with zero beach-sand width. Longshore terminations of continuous beach segments by headlands are established where active-beach widths consistently fall below a 10-m cutoff over a distance of least 1 km. Potential headlands bounding the smallest pocket beaches (1–2 km in length) are not recognized in this study. Beach widths generally decrease at sites immediately adjacent to headlands due to wave refraction and the higher wave energy focused at the headlands (Shepard and Wanless, 1971). The maximum seaward projections of headlands that terminate continuous beach segments are measured at right angles from shore-parallel lines drawn across the base of each headland (Fig. 4). The shore-parallel lines are extended from relatively straight beaches, generally at least 1 km from the beach-segment termination. Tidal inlets are not considered to represent terminations of continuous beach segments in this study, as the shallow inlets are generally floored and flanked by beach sands (Dingler, pers. commun.). However, this does not preclude the potential for the largest inlets to serve as effective boundaries to longshore transport. Harbor jetties constructed at many of the larger tidal inlets (Grays Harbor, Columbia, Tillamook, Yaquina, Siuslaw, Umpqua, Coos, Humboldt Bays) might presently serve as modern cell boundaries, but such artificial structures are not addressed in this report.

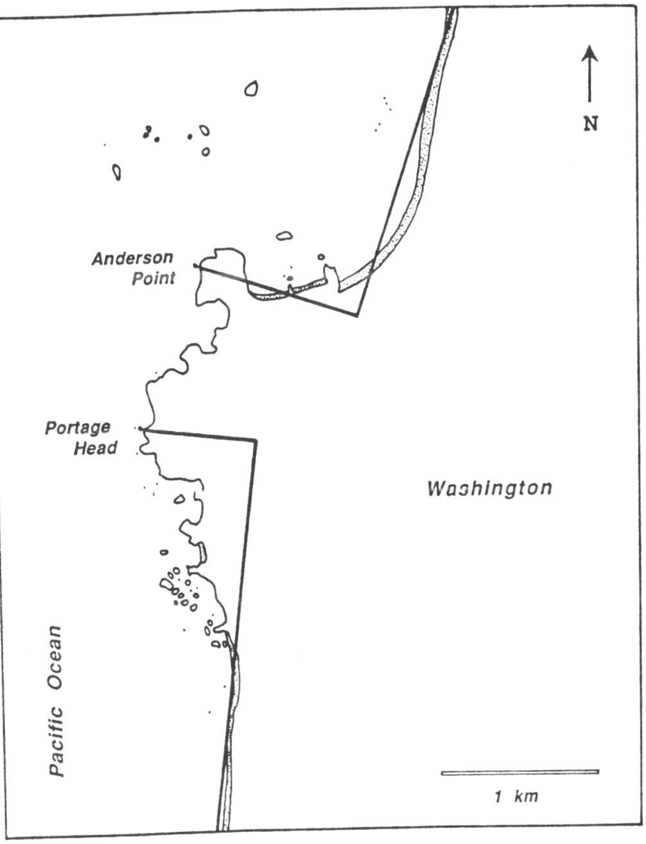

FIG. 4.—Scale drawing of coastal headlands in northern Washington and projection distances taken at right angles from shore-parallel lines drawn from adjacent straight beach segments. Dotted strips represent beach deposits adjacent to rocky shorelines at headlands. Offshore islands represent numerous sea stacks at this location just south of Cape Flattery (Fig. 2a; Table 1).

Coastal terraces at least 50 m wide that back either beach segments or rocky shorelines are identified from slope changes apparent in aerial photographs and corresponding U.S. Geological Survey 7.5-minute topographic maps. Marine-terrace deposits are important as potential sources of beach sand and as indicators of neotectonic coastal deformation. The elevations of the terraces are estimated from the first landward deflections in slope, i.e., the lowest terrace, as shown by map contour spacing and apparent slopes in corresponding aerial photographs. The heights of the first substantial slope deflections are estimated from the nearest elevation contours (6-m intervals). These terrace elevations are assumed to represent the youngest, preserved depositional surfaces at the present shoreline, and therefore reflect only very approximate paleosea-level positions. The terrace deposits have been dated in very few locations, but probably range from 80 to 120 ka (West and McCrumb, 1988; Muhs and others, 1990). Terrace-surface elevations of selected sites (about 50 in number) were field checked with an altimeter (accuracy ± 3 m) to verify mapped elevations.

Beach sands were collected from mid-beach faces at selected sites (133) for grain-size analysis. Gravel beaches and backshore gravel berms in otherwise sandy beaches were not sampled for grain-size analysis. Accessibility to wilderness beaches restricted sampling in some areas, but sample spacing generally averaged at least one sample site per 10 km longshore. Higher density sampling, on the order of 1 to 5-km spacing between sites, was employed for specific beach segments to evaluate grain-size variability over shorter longshore distances. At least 1 kg of sediment was scraped from 0 to 10-cm-depth intervals at mid-beach sites to provide sediment samples representative of fairweather beach deposits.

Exposed sand deposits in coastal terraces (44 sites) adjacent to beach sample sites were sampled for grain-size analysis. Composite samples were collected at terrace sites where grain size was observed to vary with depth in the section. The relative sparsity of analyzed terrace deposits is, in part, due to the absence of terrace deposits at many of the beach sample collection sites. Furthermore, terrace deposits composed of bay mud, colluvium and glacial drift were not sampled, due to their apparent low-sand content. Finally, exposed bedrock mudstones, volcanics and metasedimentary rocks, composing many non-terraced sea cliffs, were not sampled due to their low potential for beach-sand supply. Beach- and terrace-deposit sampling was performed between June 24 and September 5, 1989. Sample grain-size distributions were measured by sieving and settling tube (Pettit, 1990).

RESULTS

Apparent Beach Widths

The widths of active-beach faces are plotted as a function of position (UTM) in the study area (Fig. 5a-c). Even though estimated beach widths are taken from different tidal levels (generally ± 1 m MTL), it is evident that most of the longshore variability of apparent beach width does not correspond to breaks in flight date coverage (Fig. 5a-c, Appendix). The greatest changes in regional beach width occur

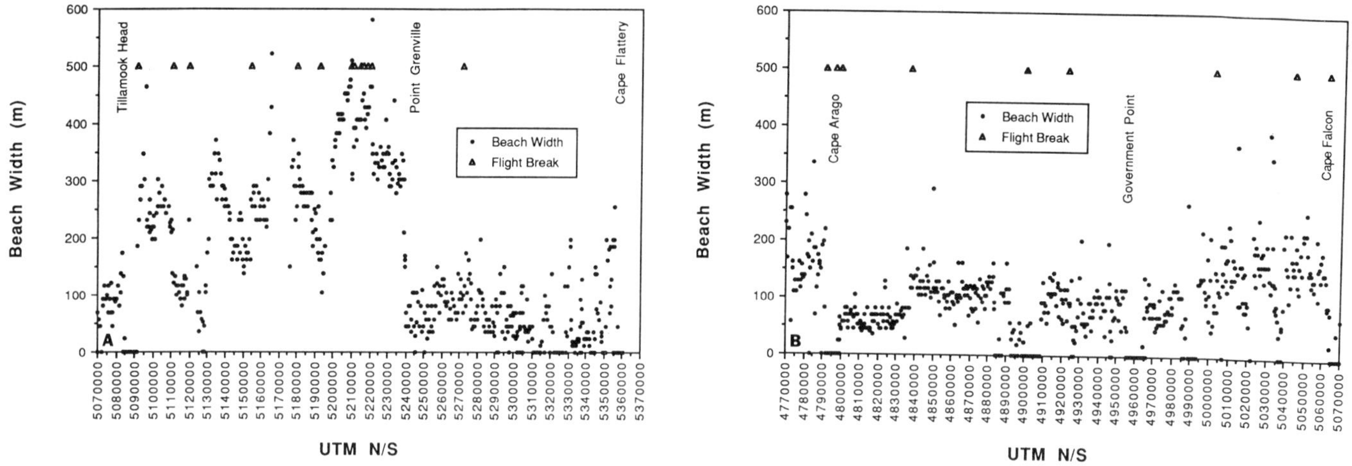

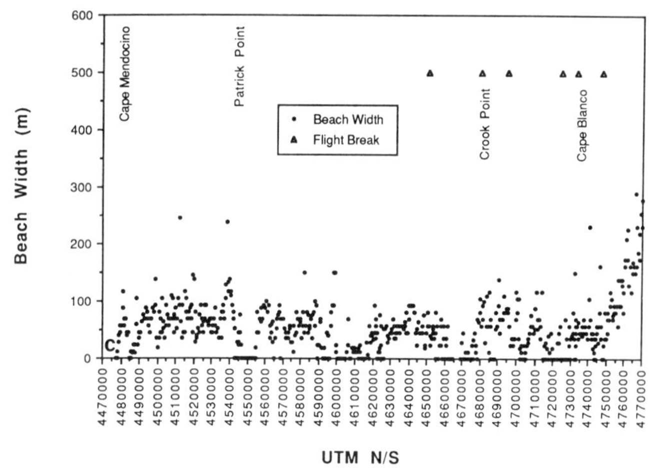

FIG. 5.—Plots of estimated, active-beach width in meters (solid circles) at reference sites spaced at approximately 0.5-km intervals longshore. Breaks in flight sequences (triangles) denoting changes in tidal level, season and year (Appendix) are also plotted as function of longshore position. Individual figures a–c are lettered from north to south including (a) Washington-northernmost Oregon, (b) Oregon, (c) southern Oregon-northern California. Active-beach deposits are from 0 to at least 600 m wide. Geographic place names are located in Fig. 2a–c.

within single flight-date sequences, rather than between flight-date sequences. These changes are generally associated with headlands or rarely with large tidal inlets.

The active-beach widths in the study area range from 0 m along rocky shorelines, unevenly distributed in all three states, to a maximum of 400 to 500 m at distances of 20 to 50 km south of Point Grenville in central Washington (Fig. 5a-c). The average apparent beach width between Tillamook Head and Point Grenville, on either side of the Columbia River (Fig. 2a), is just under 300 m (Fig. 5a). Rocky shorelines are absent in this region. By comparison, beaches to the north, between Cape Flattery and Point Grenville, average less than 100 m in width even though they correspond to flight periods of unusually low tidal elevation (Appendix). Rocky shorelines are common in this area, particularly north of Hoh Head (Fig. 2a), but account for less than 50 percent of the longshore distance north of Point Grenville (Fig. 5a).

In northern Oregon the beach widths between Tillamook Head and Government Point (Fig. 2a,b) are quite variable over short distances (1–5 km), commonly ranging from 50 to 250 m (Fig. 5a,b). By comparison, beaches in central Oregon, between Government Point and Cape Arago, demonstrate less longshore variability, and average about 100 m in width. Beaches in southern Oregon and northernmost California are among the narrowest found in the study area, averaging well under 100 m in width (Fig. 5c). Two notable exceptions include beaches averaging just under 200 m in width between Cape Arago and Cape Blanco in southern Oregon, and two isolated beaches exceeding 200 m in width, located 4.5 and 34.5 km south of Trinidad Head in northernmost California (Fig. 5b,c). A major jetty structure is located just north of the second wide beach area some 34 km south of Trinidad Head.

Continous Beach Segments

Continuous beach segments in the study area are established on the basis of minimum beach widths (10 m) and minimum lengths (2 km longshore) taken from the aerial photograph survey. The cutoff value for segment length (2 km) is based on the minimum number of reference sites (5) needed to constrain longshore trends of beach width. Shorter and narrower pocket beaches account for only a minor amount of the total coastline distance. The seaward projections of headlands bounding continuous beach segments are based on the first major promontory to the north or south of each beach segment termination (Table 1). The termi-

TABLE 1.—LITTORAL SEGMENTS AND BOUNDING HEADLANDS

Beach Segment Terminations Position UTM N-S	Segment Length km (Width Trend)*	Segment Boundaries Name, Position (UTM N-S)	Boundary** Projection Distance km
5355200-5350750	7.0 (0)	Waatch Point (5355200)	1.1
		Anderson Point (5350500)	1.1
5348000-5344950	4.5 (S)	Portage Head (5349500)	0.7
		Point of Arches (5344950)	0.5
5342400-5336500	6.5 (S)	Father and Son (5342800)	0.6
		Tskawahyah Island (5336700)	0.4
5336500-5331800	5.5 (0)	Tskawahyah Island (5336700)	0.4
		Sand Point (5331600)	0.4
5331450-5328750	3.5 (N)	Sand Point (5331600)	0.5
		North Yellow Banks (5328750)	0.4
5322050-5315050	7.0 (0)	North Kayostla (5323700)	0.4
		Cape Johnson (5313800)	0.5
5310950-5306150	4.5 (0)	Hole In The Wall (5311500)	0.4
		Quateata (5305500)	0.7
5305400-5304050	2.0 (0)	Quateata (5305500)	0.7
		Teahwit (5303300)	0.9
5302500-5298400	5.5 (S)	Taylor Point (5302500)	0.5
		Goodman (5297500)	0.7
5297750-5292400	5.5 (S)	Goodman (5297500)	0.6
		Hoh Head (5291400)	0.8
5291150-5251250	41.5 (0)	Hoh Head (5291400)	1.2
		Pratt Cliff (5250800)	0.2
5250400-5245500	5.0 (0)	Pratt Cliff (5250800)	0.3
		Cape Elizabeth (5245300)	0.2
5245100-5240200	5.5 (N)	Cape Elizabeth (5245300)	0.6
		Point Grenville (5239500)	0.3
5239650-5091700	163 (N)	Point Grenville (5239500)	0.6
		Tillamook Head (5090000)	2.5
5084150-5070000	14.5 (N)	Ecola Point (5085200)	0.6
		Cape Falcon (5068500)	1.1
5064450-5038100	25.5 (0)	Neahkanie Mt. (5065800)	1.3
		Cape Mears (5037700)	0.7
5036800-5022850	14.5 (0)	Cape Mears (5037700)	0.6
		Cape Lookout (5020800)	2.5
5020800-5008000	12.5 (0)	Cape Lookout (5020800)	2.7
		Cape Kiwanda (5007400)	0.6
5007400-4994100	13.0 (0)	Cape Kiwanda (5007400)	0.6
		Cascade Head (4991700)	1.3
4986150-4965350	21.0 (0)	Roads End (4986200)	0.4
		Government Point (4964200)	0.7
4955400-4947600	8.0 (N)	Otter Crest (4955800)	1.5
		Yaquina Head (4947500)	0.9
4947100-4927800	19.5 (0)	Yaquina Head (4947500)	1.5
		Seal Rock (4927500)	0.2
4926000-4909150	17.0 (0)	Seal Rock (4927500)	0.2
		Yachats (4906700)	0.4
4894850-4888350	6.5 (0)	Squaw Creek Bluff (4895300)	0.2
		Heceta Head (4887700)	0.5
4884300-4799150	88 (0)	Sea Lion Point (4885900)	0.4
		Gregory Point (4799700)	0.8
4791800-4773350	18.5 (0)	Cape Arago (4795900)	2.5
		Coquille Point (4774200)	0.3
4773350-4748550	26.5 (N)	Coquille Point (4774200)	0.3
		Blacklock Point (4748000)	0.5
4747400-4743800	4.5 (0)	Blacklock Point (4748000)	0.6
		Cape Blanco (4743000)	1.0
4743400-4732900	11.0 (0)	Cape Blanco (4743000)	0.6
		The Heads (4732700)	0.4
4732800-4727050	8.0 (N)	The Heads (4732700)	2.0
		Humbug Mountain (4725800)	0.8
4714650-4705900	9.0 (0)	Sisters Rock (4716500)	1.1
		Nesika Beach (4705700)	0.1
4702000-4688650	13.5 (0)	Otter Point (4702000)	0.5
		Cape Sebastian (4686500)	0.3
4686000-4678800	7.5 (0)	Cape Sebastian (4686500)	0.7
		Crook Point (4678500)	0.3
4655450-4626500	33.0 (0)	Chetco Point (4655300)	1.6
		Point St George (4626000)	0.6
4625000-4623400	5.0 (0)	Point St George (4626000)	0.8
		Battery Point (4622000)	0.2
4622500-4618400	5.0 (N)	Battery Point (4622000)	1.6
		Endberts Beach (4616500)	0.2
4611650-4608600	3.5 (0)	Midway Point (4612200)	0.5
		Footsteps Rock (4607600)	0.2
4600000-4594000	6.0 (0)	Klamath River Bluff (4602000)	1.5
		Split Rock (4594000)	0.2

TABLE 1.—*Continued*

Beach Segment Terminations Position UTM N-S	Segment Length km (Width Trend)*	Segment Boundaries Name, Position (UTM N-S)	Boundary** Projection Distance km
4591600-4589550	2.0 (0)	Johnson Creek Bluff (4592900)	0.4
		Ossagon Creek (4589300)	0.1
4589100-4554900	35.5 (0)	Ossagon Creek (4589300)	0.5
		Patricks Point (4554500)	1.4
4542700-4485700	65.5 (N)	Trinadad Head (4545000)	3.5
		False Cape (4485000)	0.3
4484400-4478200	6.5 (0)	False Cape (4485000)	0.7
		Cape Mendocino (4477000)	0.3

*Trend of increasing apparent beach width to the north (N), to the south (S); no clear trend of increasing beach width (0).
**Boundary seaward projections can vary between adjacent beach segments due to differences in beach offset and orientation on opposite sides of common headlands (Fig. 4).

nations of beach segments, defined at the points where beach widths fall under the 10-m-width minimum, are generally within 1 km of the major bounding promontories. Many beach segments also contain minor points, tidal inlets and/or other shoreline irregularities that might restrict longshore transport, particularly over short time scales.

Forty-two continuous beach segments, each greater than 2 km long, total some 767 km in longshore distance, or about three-quarters of the study area coastline (Table 1). The remaining one-quarter of the coastline is composed of rocky shorelines and smaller pocket beaches. The three longest beach segments include a 163-km segment between Point Grenville and Tillamook Head on either side of the Columbia River, an 88-km segment between Sea Lion Point and Gregory Point, in central Oregon, and a 65.5-km segment between Trinidad Head and False Cape in northern California (Table 1; Figs. 2a-c and 6). The average lengths of the remaining beach segments are as follows: 8 km for 13 segments in northern Washington, 15 km for 10 segments in northern Oregon, and 12 km for 16 segments in southern Oregon and northernmost California (Table 1; Fig. 6).

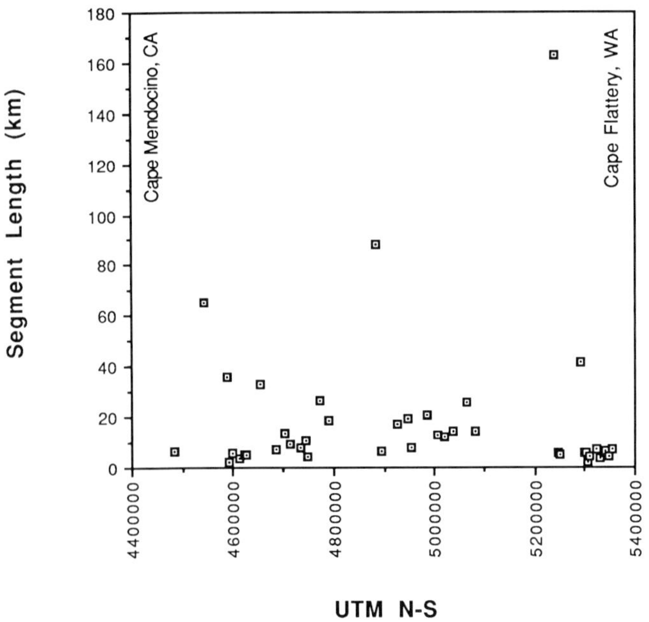

FIG. 6.—Regional distribution of continuous beach-segment lengths (Table 1) plotted as a function of location (UTM) along the study area coast from Cape Flattery, Washington, to Cape Mendocino, California. The segments are from 2 km (lower cutoff) to 165 km long, i.e., longshore extent of continuous beach deposits of at least 10 m width. Most beach segments in the study area are under 10 km long.

Headlands adjacent to terminations of the 42 beach segments range from 0.1 to 3.4 km in seaward projections, with average projection distances of 0.9 km for north-bounding headlands and 0.7 km for south-bounding headlands (Fig. 4; Table 1). The larger headlands, i.e., with seaward projections of at least 1.0 km, total 20 in number, and are relatively evenly distributed throughout the study area. The smallest promontories, i.e., 0.5 km and less in seaward projection, are particularly abundant, totaling at least 38 in number (Table 1). The abundance of these small headlands (45 percent of the 84 measured projections) demonstrates the relatively small amounts of shoreline offset between many of the resistant rocky headlands and less resistant, sand-fronted embayments of the Cascadia coastline.

Dune Fields

The presence and width (as much as 500 m) of dune fields in the study area are plotted as a function of location (UTM) along the coast in Figures 7a-c. The regional distribution of dune development is strikingly nonuniform. For example, major dune fields greater than 500 m wide are nearly continuous between Point Grenville and Tillamook Head and between Sea Lion Point and Gregory Point, just north of Cape Arago (Figs. 2a,b and 7a,b). Dune fields exceeding the 500-m-width cutoff are massed south of Cape Arago, north of Point St. George (4626850 UTM) and between Patricks Point/Trinidad Head and the Eel River (Figs. 2c and 7c). Narrower dune fields (50-150 m wide) with patchy longshore distributions are found interspersed between the major dune fields in Oregon and northern California. Dune fields are rarely observed north of Point Grenville, Washington, but where they do occur they average less than 50 m in width.

Continuous dune fields in the study area are established on the basis of minimum width (10 m) and minimum length (1 km) from the reference site dune widths reported earlier (Fig. 7a-c). Thirty distinct dune fields range from 1 km to 143 km long, and average 15.3 km in length (Table 2). The sum of the dune field lengths (458 km) is slightly less than half of the total coastline distance. Only seven of the dune

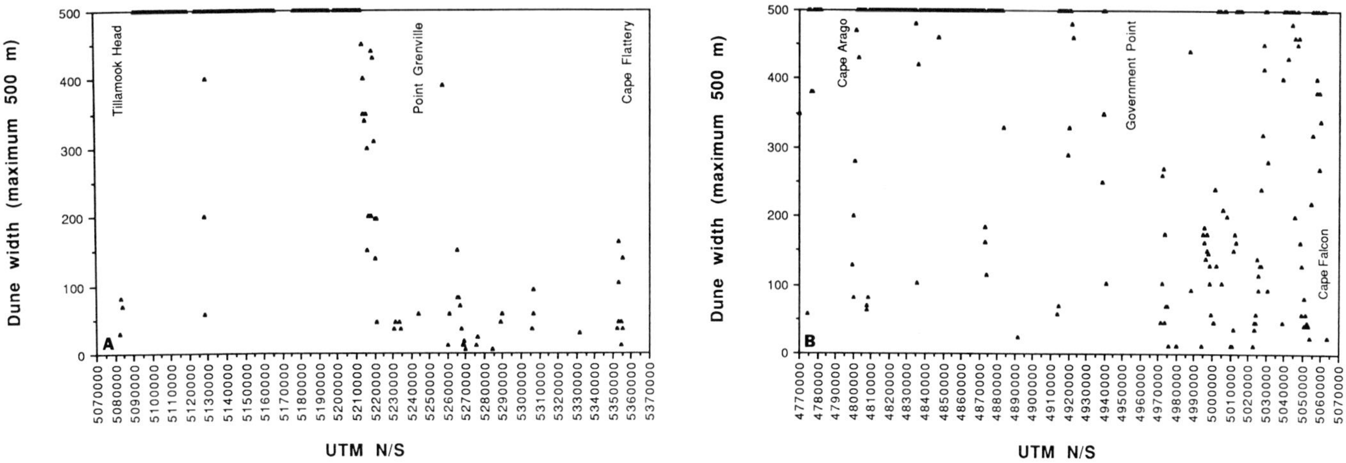

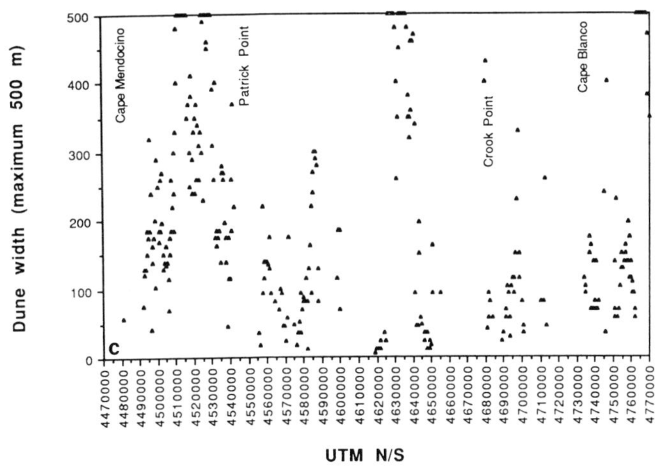

FIG. 7.—Plots of eolian dune-field widths (m) at reference sites spaced at approximately 0.5-km intervals longshore from (a) Washington to (b) Oregon to (c) southern Oregon-northern California. Modern dune-field widths are arbitrarily cut off at 500 m landward distance from the seaward edge of the established vegetation, i.e., usually the foredune crest.

fields are fully developed longshore, i.e., their lengths exceed 75 percent of the corresponding beach-segment lengths (Table 2). Thirteen of the dune fields are restricted to bay spits, whereas the 17 remaining dune fields cap prograding shorelines that are independent of bay spits or extend longshore beyond bay spits. Of the later group, a total of eight dune fields is clearly located at the northern end of corresponding beach segments, whereas only two dune fields occur at the southern ends of their respective beach segments (Tables 1 and 2).

Coastal-Terrace Heights

The elevations of the lowest, and presumably youngest, preserved surfaces of marine terraces are plotted as a function of location (UTM) in Figure 8a-c. These terrace surfaces range from less than 6 m to over 100 m above sea level in the study area. The development and preservation of marine-terrace surfaces along the Cascadia margin are extensive, backing some 40 percent of the beach- and sea-cliff-fronted coastline. Several large gaps in the longshore distributions of the marine terraces correspond to the presence of large dune fields (more than 200 m wide) that probably cover low terrace surfaces. Smaller gaps represent non-terraced sea cliffs (totaling 160 km longshore distance) and submerged river-cut valleys, presently bridged by barrier spits.

The terrace height distributions shown in Figure 8a–c represent sites of potential long-term sources of beach sand, but only in the most qualitative sense. For example, the marine-terrace deposits observed during our field studies ranged from 1 to 30 m in total thickness, often accounting for only a small fraction of the terraced seacliff exposure. Furthermore, examination of the terrace deposits shows them to range from variably weathered sand to silt and clay in composition. The coastal terraces frequently are covered by fluvial deposits and glacial drift in northern Washington, whereas marine terraces from study sites to the south commonly are capped by eolian deposits of fine sand. Where exposed above sea level, late Pleistocene wave-cut platforms were observed to cut into a variety of mudstones (recessed embayments) and volcanic rocks (resistant headlands) in Washington and northern Oregon, as well as variably metamorphosed sedimentary and igneous rocks in southernmost Oregon and northernmost California.

Longshore variability in terrace-surface elevations is apparent over both local and regional scales. For example, 10 to 30-m offsets in terrace elevations are common over longshore distances of 1 to 10 km throughout the study area (Fig. 8a-c). Some of the local variability in terrace-surface

TABLE 2.—DUNE FIELDS IN THE CASCADIA MARGIN

Location (UTM)	Nearest Feature	Length (km)	Position In Segment
5355350-5354400	Waatch	1.5	North
5353400-5352100	Bahobohosh Point	1.5	North
5306950-5305600	Quillayute River	1.5	Spit
5269550-5264600	Queets River	5.0	Spit
5221450-5091700	Point Grenville-Tillamook Head	143	Full
5083750-5082850	Ecola Point	1.0	North
5061900-5046600	Neahkanie Mountain	15.0	North
5046000-5038950	Tillamook Bay	7.0	Spit
5031150-5022850	Netarts Bay	8.0	Spit
5016350-5010400	Cape Lookout	5.5	North
5007600-5000950	Nestucca Bay	6.0	Spit
4999350-4994100	Neskowin	5.5	South
4975700-4971050	Siletz Bay	4.5	Spit
4940200-4938700	Yaquina Bay	1.5	Spit
4922550-4913900	Seal Rock	8.5	North
4883250-4799850	Heceta Head-Cape Arago	8.5	Full
4781650-4774400	Coquille Bay	8.0	Spit
4769900-4750500	Coquille Point-Blacklock Point	20.5	Full
4741900-4734400	Cape Blanco-The Heads	8.0	Full
4713150-4712100	Euchre Creek	1.0	Spit
4683450-4679400	Otter Point-Cape Sebastian	10.0	Full
4700950-4691100	Crook Point	4.0	South
4650900-4626850	Winchuck River-Point St George	25.5	Full
4621950-4619550	South Battery Point	3.0	North
4600000-4598450	Klamath River	1.5	Spit
4588250-4576800	Ossagon Creek	12.0	North
4571500-4569050	Redwood Creek	2.5	Spit
4567700-4566600	Freshwater Lagoon	1.0	Spit
4562150-4557450	Big Lagoon	4.5	Spit
4542000-4491900	Trinidad Head-Eel River	57	Full

*Segment Positions: North: dunes in northern quarter of defined beach segment (Table 1). South: dunes in southern quarter of beach segment. Full: dunes extending over three-quarters of beach-segment length. Spit: dunes limited only to spits fronting bays, rivers, lagoons.

elevations results from different thicknesses of eolian, fluvial and colluvial deposits above the marine deposits, particularly in northernmost Washington (Rau, 1973). Landslides and slumping associated with seacliff retreat locally have reduced apparent terrace-surface elevations or removed the youngest terrace deposits altogether.

However, larger scale variations in maximum terrace elevations, which occur over tens to hundreds of kilometers, must result from tectonic deformation. For example, maximum elevations of the lowest terrace surfaces (60–80 m MSL) are grouped at Cape Mendocino and Trinidad Head-Patricks Point in northernmost California, and at Cape Ferrelo-Crook Point and Cape Blanco in southern Oregon (Fig. 8c). By comparison, lowest terrace elevations barely exceed 40 m between Cape Arago and Tillamook Head in central and northern Oregon (Fig. 8a,b). The most extreme terrace-surface elevations are found in northern Washington (maximum elevations of 70–120 m MSL) with elevation peaks near Cape Elizabeth (5246900 UTM), Hoh Head (5291150 UTM), Taylor Point (5302400-5304600 UTM), Cape Johnson (5308950-5318650 UTM), Father and Son (5342200 UTM), and Watch Point (5357250 UTM), (Fig. 8a).

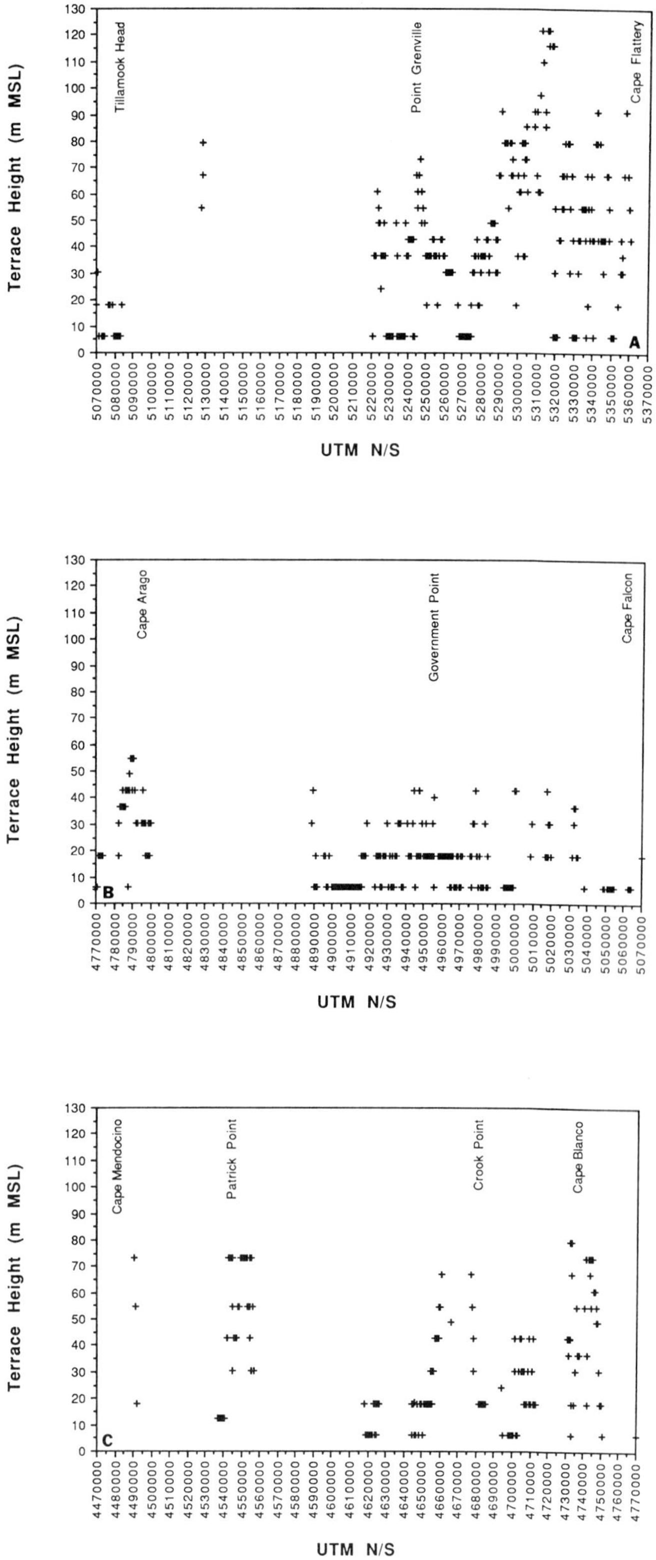

FIG. 8.—Plots of terrace heights (m) above mean sea level (MSL) at reference sites spaced at approximately 0.5-km intervals longshore from (a) Washington to (b) Oregon to (c) southern Oregon-northern California. Heights of youngest preserved terrace surfaces (at least 50 m wide) are estimated from elevations (6-m contour spacing) of lowest (first) breaks in slope observed on aerial photographs and U.S. Geological Survey topographic maps (1:24,000 scale).

Beach-Sediment Grain Size

Beach sands from 133 representative mid-beach sites in the study area range in mean grain size from 0.10 mm to 2.3 mm (Fig. 9a–c). Gravel-cobble beaches represent only

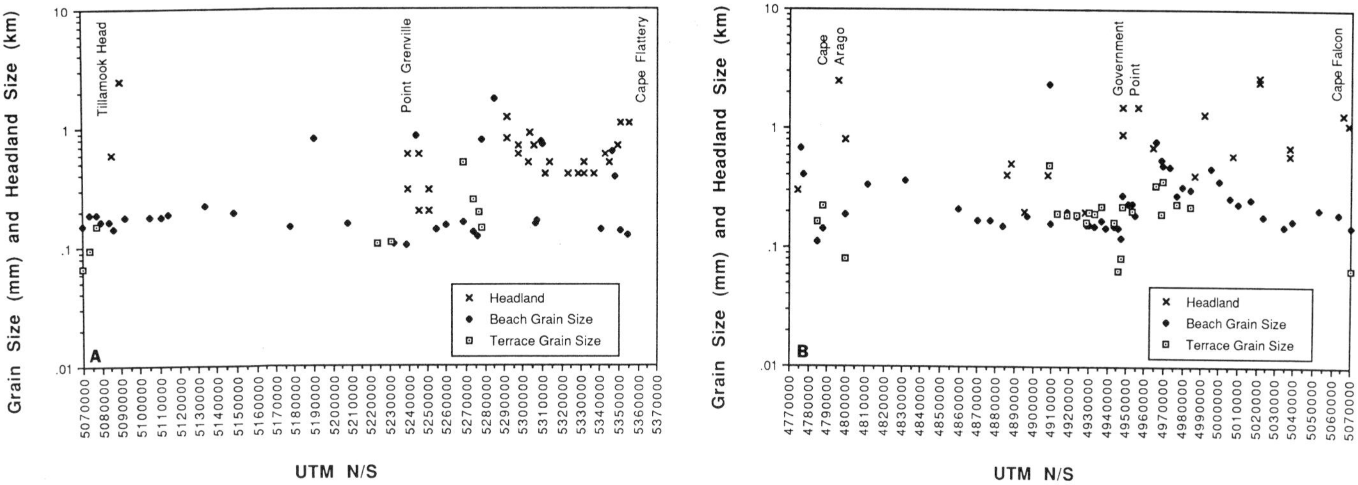

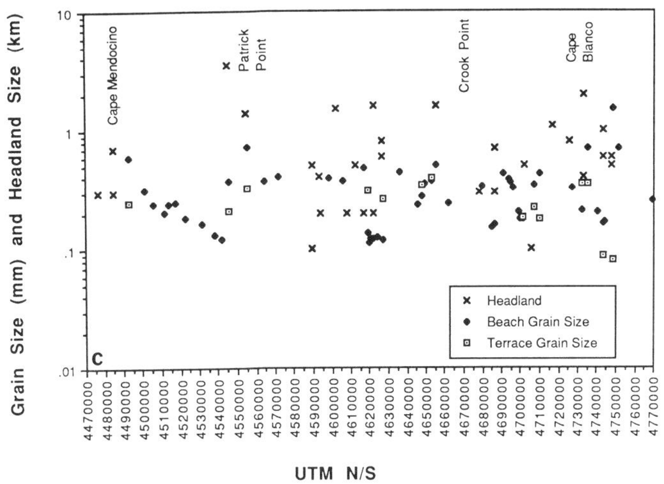

FIG. 9.—Plots of sample mean grain size (mm in log scale) from 133 mid-beach face sites (solid diamonds) and sandy coastal-terrace deposits (open squares) at representative locations, generally spaced 3 to 10-km distance longshore from (a) Washington to (b) Oregon to (c) southern Oregon-northern California. Headland locations (crosses) and corresponding size (projection distance in km in log scale) are also shown to identify segment boundaries.

a minor component of the study area coastline and were not sampled for sediment size analysis. Although the widest beaches are typically associated with the finest sands (Figs. 5a, 9a), a least squares correlation between beach mean grain size and corresponding beach width (correlation coefficient $r = 0.2$) shows little or no dependence between these two variables for most study area beaches. Finally, there appears to be little or no direct dependence between the size (projection distance) of bounding headlands and the mean grain size of adjacent beaches (Fig. 9a–c). Beach-sand size does not appear to be controlled by selective grain size bypassing around minor headlands.

Although not densely sampled, the beaches north of Point Grenville, Washington, are characterized by some of the coarsest grained deposits in the study area (average mean size of 17 samples is 0.43 mm). The finest grained beaches on average occur immediately to the south, between Point Grenville and Tillamook Head, yielding an average mean size of 0.17 mm for 12 samples. The northern Oregon beaches between Tillamook Head and Cape Arago demonstrate slightly coarser grained samples, i.e., an average mean size of 0.28 mm for 48 samples. In southern Oregon and northernmost California, the beach deposits show substantial variability in grain size, but average 0.30 mm mean grain size for 56 samples (Fig. 9a–c).

The mean grain size of analyzed samples from selected coastal-terrace deposits (44 sites) also are plotted as a function of location (UTM) in the study area in Figure 9a–c. Like the beach deposits, there is substantial longshore variation in the grain size of the analyzed terrace deposits, which range from 0.06 mm to 0.50 mm in mean grain size. Overall, the analyzed terrace-deposit samples are somewhat finer (average mean size of 0.22 mm) than the modern beach deposits (average mean size 0.30 mm). There appears to be little or no correspondence between measured mean grain-size values of adjacent beach and terrace deposits in southern Oregon and northernmost California (Fig. 9a–c). In some beach segments of north-central Oregon the mean grain sizes of adjacent beach and terrace deposits do show substantial similarities. Finally, the great abundance of fine beach sand together with the limited terrace exposure between Tillamook Head and Point Grenville (Figs. 8a and 9a) point to the large Columbia River source (Table 3) as the major control on beach grain size there.

TABLE 3.—POTENTIAL DRAINAGE-SEDIMENT SOURCES

River/System	Location UTM N-S	Suspended-Sediment Discharge* ($\times 10^3$ mt yr^{-1})	Hydrographic Factor (HF): Tidal/Fluvial Discharge**
Quillayute River	5307100	204	1 (c)
Hoh River	5289400	82	1 (c)
Queets River	5266600	144	1 (c)
Quinualt River (a)	5244600	86	1
Chehalis System	5196500	665	73 (d)
Willapa System	5169000	131	200 (d)
Columbia River	5121500	14,282	5
Nehalem River	5056100	216	10
Tillamook System	5046300	175	36
Nestucca	5000900	58	11
Salmon River (b)	4988500	24	3
Siletz River	4975200	65	8
Yaquina River	4940500	84	48
Alsea River	4919200	162	17
Siuslaw River	4874200	190	7
Umpqua River	4835600	3,243	11
Coos River	4800850	196	44
Coquille River	4775000	245	4
Sixes River	4745700	358	7
Elk River	4740300	85	8
Rogue River	4697300	4,688	1
Pistol River	4680500	96	6
Chetco River	4655250	246	1
Winchuck River (b)	4651000	70	1
Smith River	4644400	552	1
Klamath River	4599100	10,877	1
Redwood Creek	4571700	1,867	1
Mad River	4535200	2,518	1 (c)
Eel River	4499400	24,751	3

*Suspended-load transport data are from Karlin (1980) unless otherwise noted. Bedload transport is assumed to equal 10 to 30 percent of the total sediment transport in high-gradient streams of the Pacific Northwest (Tillamook Bay Task Force, 1978), or roughly one-quarter of the suspended-load transport, shown here.

**Hydraulic Factor (HF) = ratio of mean tidal-prism volume (MHW-MLW) to mean annual river discharge over 6-hr period (Peterson and others, 1984a; Clemens and Komar, 1988). HF data from Clemens and Komar (1988) unless otherwise noted. Rivers with HF = 10 to 15 likely bypass some sand to the littoral zone (Peterson and others, 1984a), whereas rivers approaching HF = 1 probably carry their entire sand-size bedload to the littoral zone.

(a) gauge at Quinault Lake, (b) Percy and others, 1974, (c) estimated from measured surface area and assumed tidal range of 2 m, (d) Knots and Barrick, 1976.

TABLE 4.—LITTORAL SEGMENT SEDIMENT SOURCES

Segment Boundaries	Average Beach-Plus Dune-Field Width (m)	Average Terrace Height (m MSL)	Estimated River-Sediment Discharge* ($\times 10^3$ mt yr^{-1})
5355200-5350750	195	15	—
5348000-5344950	80	45	—
5342400-5336500	45	50	—
5336500-5331800	20	45	—
5331450-5328750	114	20	—
5322050-5315050	56	60	—
5310950-5306150	61	50	51
5305400-5304050	104	75	—
5302500-5298400	55	55	—
5297750-5292400	79	75	—
5291150-5251250	94	30	57
5250400-5245500	71	50	—
5245100-5240200	73	40	21
5239650-5091700	671	00	3,769
5084150-5070000	100	10	—
5064450-5038100	432	00	98
5036800-5022850	294	05	—
5020800-5008000	302	25	—
5007400-4994100	367	15	15
4986150-4965350	114	15	16
4955400-4947600	88	25	—
4947100-4927800	133	20	21
4926000-4909150	334	10	41
4894850-4888350	90	15	—
4884300-4799150	558	00	907
4791800-4773350	363	25	61
4773350-4748550	334	20	89
4747400-4743800	81	65	21
4743400-4732900	136	45	—
4732800-4727050	45	45	—
4714650-4705900	91	15	—
4702000-4688650	147	10	1,172
4686000-4678800	177	20	24
4655450-4626500	294	15	217
4625000-4623400	43	10	—
4622500-4618400	65	05	—
4611650-4608600	16	NTSC	—
4600000-4594000	112	NTSC	2,719
4591600-4589550	40	NTSC	—
4589100-4554900	140	NTSC	467
4542700-4485700	321	00	6,817
4484400-4478200	64	NTSC	—

*Average of combined measurements of active-beach width and dune width (200 m maximum) from reference sites in corresponding beach segments.

**Average of measured terrace heights (Fig. 8) from reference sites in corresponding beach segments. Terraces and barrier spits covered by dunes are assigned terrace elevations of zero. Shorelines dominated by non-terraced sea cliffs are not evaluated for average terrace height.

***River-bedload sediment discharge based on 25% of corresponding suspended-sediment discharge from combined discharges of all significant rivers entering corresponding beach segments (Table 3). Sediment discharge from smaller creeks and direct-slope runoff are not evaluated here.

DISCUSSION

These results substantiate previous reports of great variability in the longshore distribution of modern littoral deposits in the Pacific Northwest region (Cooper, 1958; Dicken, 1961; Terich and Schwartz, 1981). Combined beach- and dune-width estimates, averaged for each beach segment (Table 4), are shown together with bounding headland projections in Figure 10a–c. The abrupt longshore changes in relative beach-sand abundance demonstrate the effectiveness of at least some of the existing headlands in restricting the longshore transport and redistribution of modern littoral deposits. However, there appears to be little or no correlation between headland size (projection distance) and beach-segment length or width (Figs. 6 and 10a–c; Table 4). Factors other than headland size must influence the distribution of beach deposits in this convergent-margin setting (see later discussion). Nevertheless, headlands do serve as primary segment boundaries in the study area. Their longshore spacing should control the length of beach segments in regions where low sand abundance does not permit beaches to prograde around coastal irregularities.

Coastal Tectonics

In convergent margins, coastal geologic structures can be inherited from accreted terranes or formed via deformation associated with ongoing convergence (Rau, 1973, 1980; Kelsey and Carver, 1988; Wells, 1990). Tectonic deformation differentially uplifts and/or juxtaposes variably resistant lithologic units longshore. Associated geologic structures ultimately control the positions of the differentially eroded embayments and headlands. Over the course of repeated marine transgressions and/or highstands, the focusing of wave energy on headlands serves to erode these protrusions preferentially, thereby smoothing the coast (Shepard and Kuhn, 1983). On the other hand, the catchment of littoral deposits along recessed shorelines fills these embayments (pocket beaches) with sediment buffers and

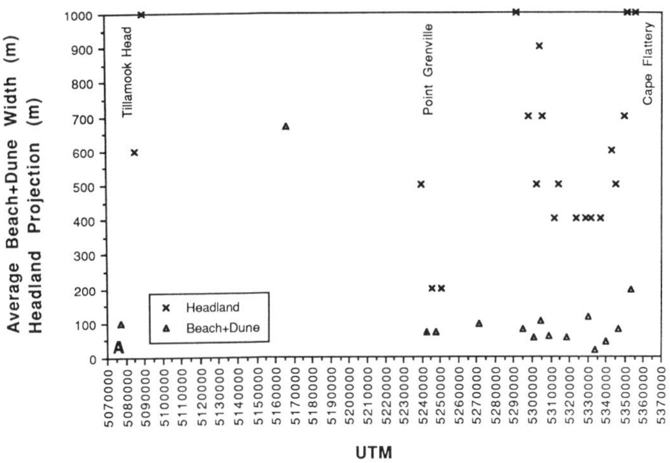

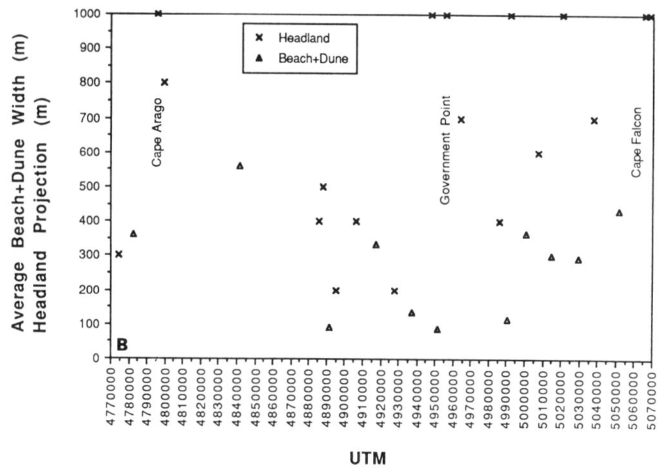

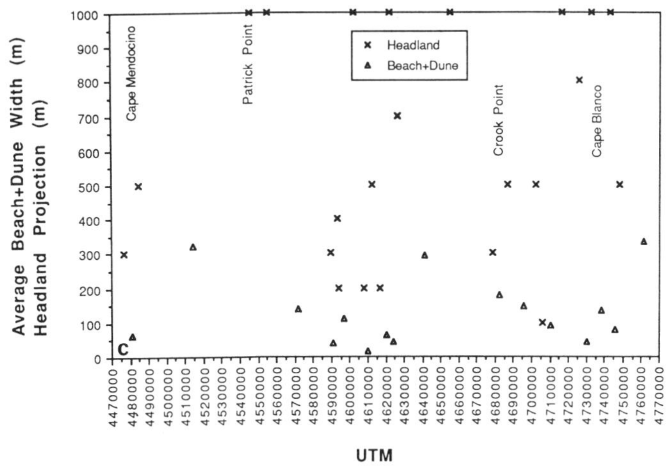

FIG. 10.—Plots of averaged beach + dune widths (triangles) from reference points within defined beach segments (Table 4) and corresponding headland projections (crosses, maximum 1 km shown here) as a function of position (UTM) along the study area from (a) Washington to (b) Oregon area to (c) southern Oregon-northern California.

protects them from further erosion during extreme storm events. Nevertheless, ongoing tectonism can overcome the erosional/depositional smoothing of the coast if the deformational rates exceed those of erosion.

On a regional scale, the Cascadia margin is characterized by relatively irregular shorelines in southern Oregon and northern California, and very irregular shorelines in northern Washington (Fig. 2a–c). Some of the regional coastal irregularities might be inherited from pre-existing tectonic structures, particularly in regard to relative resistances of lithologic units. However, there is a remarkable coincidence between the regional distribution of coastal irregularities and late Quaternary vertical deformation of coastal terraces. For example, the greatest heights and variations in terrace-surface elevations (Fig. 8a–c) occur in southern Oregon-northern California and northern Washington, substantiating previous reports of relatively rapid vertical deformation at the opposite ends of the study area (Rau, 1973, 1980; Kelsey and Carver, 1988; Muhs and others, 1990). The short-beach segments in northern Washington (average of 8 km for 13 segments) and southern Oregon-northernmost California (average of 12 km for 16 segments) possibly reflect these relatively high rates of Quaternary tectonic deformation.

In northernmost California, northwest-striking thrust faults and folds are associated with proximity to the Mendocino triple junction and the upper-plate deformation front, east of the trench (Fig. 3). Four major headlands, Cape Mendocino, False Cape, Trinidad Head and Patricks Point (Table 1), respectively, are associated with the northwest-striking faults, including Capetown, Russ, Trinidad and Bald Mountain faults (Kelsey and Carver; 1988). A minor headland, Table Bluff (4505800 UTM), is associated with another northwest-trending thrust fault, Little Salmon fault, located just north of the Eel River mouth. However, recent local progradation of this beach has at least temporarily reduced the effectiveness of this bluff as a boundary to longshore transport. Complex faulting and folding of the southern Oregon coast are also presumably related to this coastline's proximity to the actively deforming accretionary prism (Kelsey, 1990). Upwarping of resistant basement rocks associated with local anticlines has produced several major headlands, including Cape Arago, Coquille Point, and Cape Blanco (Table 1; Fig. 8b,c) in southern Oregon (Newton, 1980; Kelsey, 1990).

In northern Washington, a complex curvature of the subduction zone reportedly results in extraordinary interplate coupling and/or upper-plate deformation, as shown by the onshore exposure of the accretionary prism at a distance of 150 km from the trench (Crosson and Owens, 1987; Crosson, pers. commun.). Coastal upwarps associated with late Pleistocene folds and/or diapers have locally elevated resistant bedrock sections to produce the headlands between Point Grenville and Cape Elizabeth, at Hoh Head, Taylor Point, Teahwit, and Quateata among others (Table 1; Figs. 2a and 8a; Rau, 1973, 1980). Though less common, lateral offsets of differentially resistant units along strike-slip faults have also produced prominent headlands, such as Goodman (Table 1; Rau, 1980). The vertical deformation of variably resistant units from thrust faults, folds and piercement structures in the actively deforming accretionary prism is apparently responsible for the highly irregular coastline in northern Washington. The available sand supply to the area has simply not been sufficient to build out around bounding points, promontories, and tombolo-attached sea stacks, thereby resulting in the numerous, small, pocket beaches there.

By comparison, the least irregular shorelines in the study area occur in northern Oregon (Fig. 2b), where onshore and offshore tectonic deformation is characterized by strike-slip faulting and shore-oblique folds (Peterson and others, 1986; Snavely, 1987; Goldfinger and others, 1990; Wells, 1990). This neotectonic setting corresponds to the area of the lowest average terrace height and the least variation in terrace-surface elevations (Fig. 8b). The low-terrace heights in northernmost Oregon result in infrequent subaerial exposure of the underlying bedrock in many of the beach segments. Indeed, most of the major headlands in northern and central Oregon are composed of exotic basalt flows and invasions (Snavely and others, 1973; Beeson and Moran, 1979; Wells and others, 1989), rather than from vertical offsets of the regional bedrock, i.e., Tertiary mudstones and sandstones. Such basaltic headlands in northern Oregon include Sea Lion Point, Heceta, Squaw Creek Bluff, Yachats, Seal Rock, Yaquina Head, Otter Crest, Government Point, Cascade Head, Cape Lookout, Cape Mears, Neahkanie Mt., Cape Falcon, Ecola Point and Tillamook Head (Table 1; Peterson and others, 1986). In the absence of these unusual basaltic headlands, the littoral deposits of the northern Oregon coast might show even less segmentation (longer cells) than is now the case (Fig. 6).

Sand Sources

Of the several littoral sand sources identified in the Cascadia margin, two of the sources, i.e., eroding sea cliffs and rivers, are certain to vary longshore in their potential sand production. For example, coastal terraces are known to range from 0 to 120 m MSL in height over both local and regional scales (Fig. 8a–c). Surprisingly, many of the shortest and narrowest beach segments, generally located at the northern and southern ends of the study area, correspond to shorelines of maximum heights of terraces or non-terraced sea cliffs (Table 4; Fig. 11). The lack of a positive correlation between terrace/seacliff height and beach-sand accumulation probably results from local variability in terrace-deposit thickness and/or composition and lack of

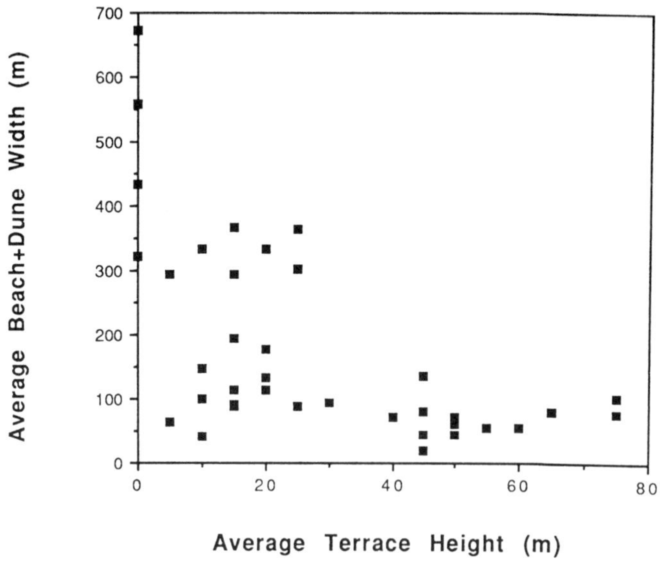

FIG. 11.—Relation between averaged beach + dune widths from reference points within defined beach segments (Table 4) and corresponding averaged terrace heights from the same defined beach segments.

sand production from exposed bedrock sections underlying the more highly elevated terraces, e.g., terrace heights greater than 30 m. Furthermore, several of the largest dune fields in the study area, e.g., south of Point Grenville, Heceta Head, Coquille Point, Winchuck River, and Trinidad Head (Table 2), either cover or have prograded seaward from any potential sources of sand from sea cliffs (Figs. 7a–c and 8a–c). In summary, the widest beaches and dune fields do not correspond to sites of extensive terrace-deposit exposure. Whereas eroding sea cliffs are currently supplying sediment to adjacent beach segments, their apparent sediment contributions cannot account for the great variability in littoral-sand abundance in the study area.

Rivers provide an alternative onshore source of littoral sands to the Pacific Northwest beaches. Unlike many passive-margin estuaries, most of the rivers in the Cascadia margin have largely filled their submerged river valleys with sediments, and are now likely bypassing sand to the littoral zone (Peterson and others, 1984a,b). Furthermore, the largest river systems are well entrenched into the uplifted coastal plain and foothills. As a result, these rivers have resupplied sediments to the same continental-shelf areas over successive marine highstands and regressions. First-order comparisons of potential river-sand sources (Tables 3 and 4) with appropriate littoral-deposit dimensions, such as averaged beach-dune width and segment length, are shown in Figure 12a and b. Not surprisingly, the largest littoral deposits are associated with major river-sediment sources, including the Columbia, Eel and Umpqua Rivers. However, two of the larger rivers, Klamath and Rogue Rivers, are associated with narrow, short, beach segments. Non-terraced sea cliffs and very low terraces adjacent to the Rogue and Klamath Rivers (Figs. 2c and 8c, Table 4) suggest possible sediment loss to downwarped nearshore and offshore

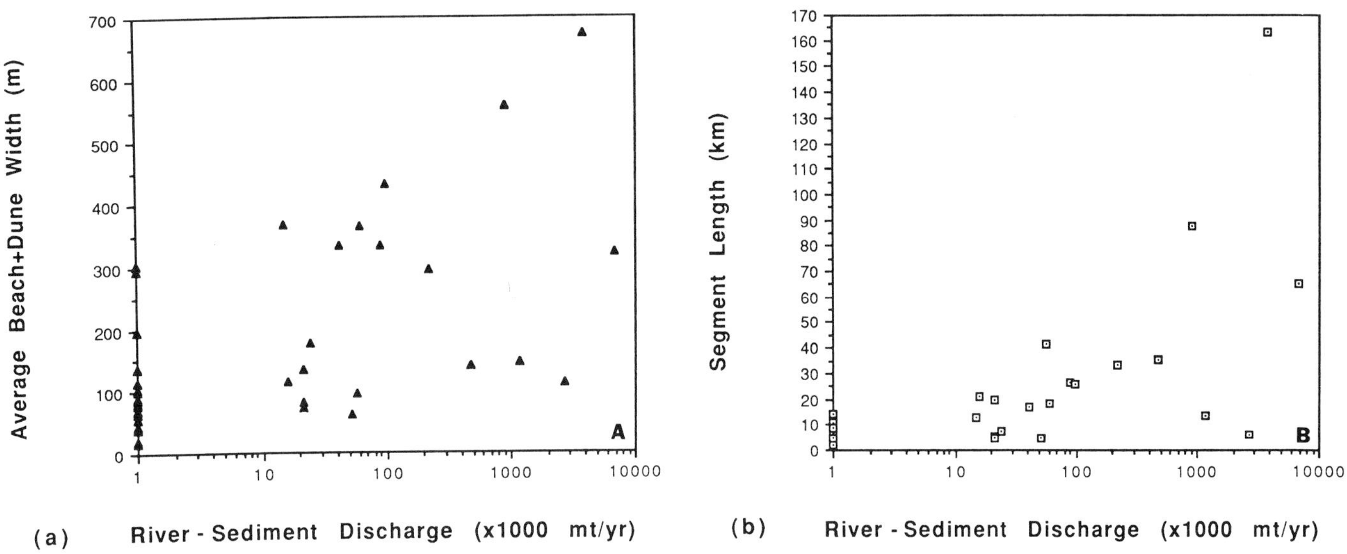

FIG. 12.—Plots of (a) averaged beach + dune width in defined beach segments versus total river-sediment discharge (mt/yr in log scale) in corresponding segments (Tables 3 and 4), and (b) beach-segment length versus total river-sediment discharge (mt/yr in log scale) in corresponding segments (Tables 1 and 4).

(inner-shelf) sinks in the vicinity of the Rogue and Klamath Rivers.

Finally, six of the 42 beach segments demonstrate beach-plus dune-field widths of at least 200 m, even though they are not associated with any large river systems (Table 4). These segments, located along the central Oregon coast, are generally backed by low, sandy terrace deposits (Figs. 8a–c and 9a–c). Transgressive sands from the inner shelf together with eroding sands from the onshore terrace deposits are the likely sources of littoral sand to these isolated beach segments, as suggested by Clemens and Komar (1988). Even the largest littoral deposits, clearly associated with major river-sediment sources (Fig. 12a,b), might be derived in part from locally recycled shelf sand. Such offshore deposits could be spatially restricted by point sources (entrenched river mouths) and major transport boundaries (paleoheadlands) now observed as submerged banks in the inner-middle shelf. Such spatially restricted sources of recycled sand along the nearshore and inner shelf are in apparent contradiction to regional northward drift of offshore sands on the middle-outer continental shelf in Pleistocene times (Scheidegger and others, 1971).

Littoral Transport

About one-third of the 42 beach segments analyzed demonstrate a clear longshore trend of increasing beach width (Table 1). Of these segments, nine show increasing beach widths to the north, whereas only four show increasing beach widths to the south. The latter four segments, located in northernmost Washington, are associated with very irregular coastlines that deviate from the general north-shoreline bearing in the study area (Fig. 2a). Local wave refraction around headlands and numerous sea stacks also might influence littoral transport in these small-beach segments of northernmost Washington. Significantly, two of the largest beach segments, i.e., south of Point Grenville and Trinidad Head, do contain relatively greater beach widths in their northern portions, possibly confirming a small net northward littoral transport in these large cells (Ballard, 1964; Terich and Schwartz, 1981; Bodin, 1982). The beach segment south of Trinidad Head is largely orientated northeast-southwest (Fig. 2c) in line with dominant southwest winter swells, and this might increase northward littoral transport there. The apparent variability of beach widths in the segment south of Point Grenville (Fig. 5a) implies the potential for additional transport boundaries, possibly including tidal inlets, in this very large littoral cell.

As previously noted, eight dune fields that are not restricted to barrier spits are located at the northern ends of corresponding beach segments, whereas only two are located at the southern ends of their beach segments (Table 2). In addition, several beach segments show decreasing grain-size trends to the north, including those south of Cape Kiwanda, Roads End, Cape Arago, Coquille Point, Otter Point, and Trinidad Head (Figs. 2b,c and 9b,c). By comparison, none of the analyzed beach segments demonstrate any clear decrease in beach grain size to the south. With the possible exception of the segment south of Roads End, the northward-fining trends of beach grain size do not appear to be related to longshore distributions of terrace-deposit grain size (Fig. 9a–c). Rather, these trends indicate a net northward littoral transport of the finest (most mobile) grain sizes relative to the coarser grain sizes. Northward littoral transport in these beach segments implies the presence of onshore and/or offshore sand sinks toward the northern end of the cells, or the bypassing of the finest sand fraction around northern headlands.

Neither measures of active-beach width, eolian-dune development or beach grain size demonstrate consistent longshore trends among the majority of analyzed beach segments in the Cascadia margin. The lack of such consistent trends in most of the beach segments supports previous reports of reversing littoral drift in the short 'pocket' beaches of the Pacific Northwest (Kulm and Byrne, 1966; Komar and others, 1976a). However, where consistent longshore trends of beach width, dune development and/or grain size do exist, they generally imply net northward littoral transport, as reported by other investigators (Ballard, 1964; Plopper, 1978; Terich and Schwartz 1981; Bodin, 1982). The apparent lack of net northward transport in many beach segments might not reflect an exact balance in reversing longshore-transport conditions (Kulm and Byrne, 1966), but might result from limited response to uneven transport conditions. Like water sloshing to and fro in a large tub, the beach sands in many of the small cells of this high-energy coastline have little or no memory of previous events, and therefore demonstrate only the most recent of littoral-transport conditions. In the absence of the numerous shoreline irregularities formed in this active-margin coast, the beaches might demonstrate a more consistent record of net littoral transport to the north.

CONCLUSIONS

Tectonics, sediment supply and wave climate are all important factors in controlling the distributions and dimensions of modern littoral deposits in the Cascadia margin. Geologic structures inherited from accreted terranes and/or formed by deformation in this active subduction zone provide the primary boundaries to longshore transport. Uplift or subsidence of coastal deposits relative to present sea level largely determine the composition and abundance of sediment supplied to modern beaches by seacliff retreat. However, the greatest regional variability of littoral-sand accumulation corresponds to the spatial distribution of major river sources, including the Columbia, Eel, and Umpqua Rivers. The presence of substantial beach deposits in segments lacking either major rivers or extensive coastal-terrace deposits, requires an offshore sand supply, i.e., transgressive sands from the inner shelf. Finally, variable longshore trends of beach width, dune-field development and beach-sediment grain size within individual beach segments indicate the potential for net northward littoral drift, in the absence of the numerous headlands that presently segment the Cascadia margin shoreline.

ACKNOWLEDGMENTS

This research was supported by the NOAA Interstate Coastal Zone Management program, administered by the National Coastal Resources Research and Development Institute, under grants No. 2 5632 03 (1988 89) and No. CZ17.90-5635-01 (1990–91). We thank Charles Clough and Mike Boyer for performing the grain-size analyses, Tara Karnes, Bob Carson, and Margot Truini for assisting with aerial-photograph data reduction, and finally Mike Hammond and Doann Hamilton for completing the figure drafting. Thomas Terich and Gary Carver provided valuable information on local beach conditions, site access and tectonic framework in Washington and northern California, respectively. Aerial photographs were made available and/or loaned to the project from archives maintained by the California Coastal Commission, Oregon Department of Transportation, U.S. Army Corps of Engineers, and the Washington Department of Natural Resources.

APPENDIX.—AERIAL PHOTOGRAPHY FLIGHT DATA

UTM Location	Date	Time	Estimated Tidal Level (m) MTL*
Washington			
5360450-5359150	09/11/77	12:48–13:38	+0.4
5358750-5310050	07/18/77	10:05–12:00	−0.4
5309100-5305500	09/11/77	12:48–13:38	+0.3
5305400-5295950	07/18/77	10:05–12:00	−0.8
5295500-5271650	07/22/77	11:00–12:05	−1.0
5271200-5221000	06/06/77	10:07–12:54	−1.4
5220500-5218500	08/06/81	11:00	−1.0
5218000-5216000	06/06/77	10:07–12:54	−1.4
5215650-5215250	08/06/81	11:00	−1.0
5214550-5211100	06/06/77	10:07–12:54	−1.4
5210750-5209750	05/28/81	10:10	−0.1
5209300-5192650	06/06/77	10:07–12:54	−1.5
5192350-5180200	07/19/77	10:43–10:45	−0.8
5179800-5154300	06/12/74	12:50–12:55	−0.8
5153700-5124150	06/17/74	13:40	−0.2
Oregon			
5119200-5110900	10/06/78	11:16–11:20	+0.0
5110350-5091700	10/12/78	12:34–13:55	−0.3
5091600-5062500	10/02/78	13:11–13:30	+1.5
5061900-5043900	10/25/78	10:07–11:12	+0.2
5043300-5002700	10/12/78	12:34–13:55	−0.1
5002150-4923100	10/25/78	10:07–11:12	+0.2
4922550-4899600	10/31/78	11:33–11:53	+1.2
4899400-4837550	09/28/79	11:31–14:09	+0.2
4837000-4799450	11/06/79	10:35–11:18	+0.9
4799300-4797050	09/28/79	11:31–14:09	+0.1
4796550-4791800	07/17/80	10:35–11:18	−0.9
4791600-4747400	04/11/80	13:20–13:56	−0.9
4746800-4732550	04/02/80	11:16–13:36	+0.7
4732350-4724000	04/11/80	13:20–13:56	−1.0
4723650-4695500	04/02/80	11:16–13:36	+0.5
4694900-4669150	12/08/79	11:36–12:11	+0.5
4664350-4651600	01/23/80	11:16–13:36	0.0
California			
4650900-4477000	05/05/78	10:36–12:58	+0.4

*Estimated from tide tables (National Oceanic Atmospheric Administration, 1974–1981).

REFERENCES

ATWATER, B. F., 1987, Evidence for great Holocene earthquakes along the outer coast of Washington State: Science, v. 236, p. 942–944.

BALLARD, R. L., 1964, Distribution of beach sediment near the Columbia River: University of Washington Department of Oceanography, Technical Report No. 98, 82 p.

BEESON, M. H., AND MORAN, M. R., 1979, Columbia River Basalt Group stratigraphy in western Oregon: Oregon Geology, v. 41, p. 11–14.

BODIN, P., 1982, Longshore and seasonal variations in beach sand, Humboldt Country, California: Implications for bulk longshore transport direction. [Unpublished M.S. Thesis]: Humboldt State University, Arcata, California, 64 p.

CADY, W. M., 1975, Tectonic setting of the Tertiary volcanic rocks of the Olympic Peninsula, Washington: U.S. Geological Survey Journal of Research, v. 3, p. 537–582.

CLEMENS, K. E., AND KOMAR, P. D., 1988, Oregon beach-sand compositions produced by the mixing of sediments under a transgressive sea: Journal of Sedimentary Petrology, v. 58, p. 519–529.

CLIFTON, H. E., HUNTER, R. E., AND PHILLIPS, R. L., 1971, Depositional structures and processes in the non-barred high-energy nearshore: Journal of Sedimentary Petrology, v. 41, p. 651–670.

COOPER, W. S., 1958, Coastal sand dunes of Oregon and Washington:

Geological Society of America Memoir 72, 169 p.

COOPER, W. S., 1967, Coastal dunes of California: Geological Society of America Memoir 104, 131 p.

CROSSON, R. S., AND OWENS, T. J., 1987, Slab geometry of the Cascadia subduction zone beneath Washington from earthquake hypocenters and teleseismic converted waves: Geophysical Research Letters, v. 14, p. 824–827.

DARIENZO, M. E., AND PETERSON, C. D., 1990, Episodic tectonic subsidence of late Holocene salt marshes, northern Oregon, Central Cascadia margin: Tectonics, v. 9, p. 1–22.

DICKEN, S. N., 1961, Some recent physical changes of the Oregon coast: U.S. Department of Navy, Office of Naval Research, Project Report NR 388-062, 151 p.

GALSTER, R. W., 1987, A survey of coastal engineering geology in the Pacific Northwest: Bulletin of the Association of Engineering Geologists, v. 24, p. 161–197.

GOLDFINGER, CHRIS, MACKAY, M. E., KULM, L. D., AND YEATS, R. S., 1990, Neotectonics and possible segmentation of the Juan de Fuca plate and Cascadia Subduction zone off central Oregon (abst.): Transactions, American Geophysical Union, v. 71, p. 1580.

INMAN, D. L., AND NORDSTROM, C. E., 1971, On the tectonic and morphologic classification of coasts: Journal of Geology, v. 79, p. 1–21.

IRWIN, W. P., 1981, Tectonic accretion of the Klamath Mountains, *in* Ernst, W. G., ed., The Geotectonic Evolution of California, Rubey Volume 1: Prentice-Hall, Englewood Cliffs, New Jersey, p. 29–49.

KARLIN, ROBERT, 1980, Sediment sources and clay mineral distributions off the Oregon coast: Journal of Sedimentary Petrology, v. 50, p. 543–560.

KELSEY, H. M., AND CARVER, G. A., 1988, Late Neogene and Quaternary tectonics associated with the northward growth of the San Andreas transform fault, northern California: Journal of Geophysics, v. 93, p. 4797–4819.

KELSEY, H. M., 1990, Late Quaternary deformation of marine terraces on the Cascadia subduction zone near Cape Blanco: Tectonics, v. 9, p. 983–1014.

KNOTS, N. P., AND BARRICK, R. C., 1976, Hydrodynamics of Grays Harbor estuary, Washington. Appendix A: Maintenance Dredging and the Environment of Grays Harbor Washington: U.S. Army Corps of Engineers, Seattle, Washington, 95 p.

KOMAR, P. D., LIZARRAGA, J. R., AND TERICH, T. A., 1976a, Oregon coast shoreline changes due to jetties: Journal of Waterways, Harbors and Coastal Engineering Division, American Society of Civil Engineers, v. 102, p. 13–30.

KOMAR, P. D., QUINN, W., CLAYTON, C., REA, C. C., AND LIZZARRAGA-ARCINIEGA, J. R., 1976b, Wave conditions and beach erosion on the Oregon coast: The Ore Bin, v. 38, p. 103–112.

KOMAR, P. D., 1986, The 1982–83 El Niño and erosion on the coast of Oregon: Shore and Beach, v. 54, p. 3–12.

KOMAR, P. D., 1989, Erosion of Netarts Spit, Oregon: Continued impacts of the 1982–83 El Niño: Shore and Beach, v. 57, p. 11–19.

KULM, L. D., AND BYRNE, J. V., 1966, Sedimentary response to hydrography in an Oregon estuary: Marine Geology, v. 4, p. 85–118.

KULM, L. D., ROUSH, R. C., HARLETT, J. C., NEUDECK, R. H., CHAMBERS, D. M., AND RUNGE, E. J., 1975, Oregon continental shelf sedimentation: Interrelationships of facies distribution and sedimentary processes: Journal of Geology, v. 83, p. 145–175.

KULM, L. D., SCHEIDEGGER, K. F., BYRNE, J. V., AND SPIGAI, J. J., 1968, A preliminary investigation of the heavy mineral suites of the coastal rivers and beaches of Oregon and northern California: The Ore Bin, v. 30, p. 165–184.

MCMANUS, D. A., 1964, Major bathymetric features near the coast of Oregon, Washington and Vancouver Island: Northwest Science, v. 38, p. 65–82.

MUHS, D. R., KELSEY, H. M., MILLER, G. H., KENNEDY, G. L., WHELAN, J. F., AND MCINELLY, G. W., 1990, Age estimates and uplift rates for late Pleistocene marine terraces: Southern Oregon portion of the Cascadia forearc: Journal of Geophysical Research, v. 95, p. 6685–6698.

NATIONAL OCEANIC AND ATMOSPHERIC ADMINISTRATION, 1974–1981, Tide Tables, High and Low Water Predictions, West Coast of North and South America Including the Hawaiian Islands: U.S. Department of Commerce, NOAA, National Ocean Service, Riverdale, Maryland.

NEWTON, V. C., 1980, Prospects for oil and gas in the Coos Basin, Western Coos, Douglas and Lane Counties, Oregon: State of Oregon Department of Geology and Mineral Industries, Oil and Gas Investigation No. 6, 74 p.

PETERSON, C. D., AND BINNEY, S. E., 1988, Compositional variations of coastal placers in the Pacific Northwest, USA: Marine Mining, v. 7, p. 397–416.

PETERSON, C. D., JACKSON, P. L., O'NEIL, D. J., ROSENFELD, C. L., AND KIMERLING, A. J., 1990, Littoral cell response to interannual climatic forcing 1983–1987 on the central Oregon coast, USA: Journal of Coastal Research, v. 6, p. 87–110.

PETERSON, C. D., SCHEIDEGGER, K. F., AND KOMAR, P. D., 1984a, Sediment composition and hydrography in six high-gradient estuaries of the northwestern United States: Journal of Sedimentary Petrology, v. 54, p. 86–97.

PETERSON, C. D., SCHEIDEGGER, K. F., AND SCHRADER, H. J., 1984b, Holocene depositional evolution of a small active-margin estuary of the northwestern United States: Marine Geology, v. 59, p. 51–83.

PETERSON, C. P., KULM, L. D., AND GRAY, J. J., 1986, Geologic map of the ocean floor off Oregon and the adjacent continental margin: Oregon Department of Geology and Mineral Industries, Geologic Map Series, Map GMS-42.

PETTIT, D. J., 1990, Distribution of sand within selected littoral cells of the Pacific Northwest. [Unpublished M.S. Thesis]: Portland State University, Portland, Oregon, 249 p.

PERCY, K. L., SUTTERLIN, C., BELLA, D. A., AND KLINGEMAN, P. C., 1974, Descriptions and Information Sources for Oregon Estuaries: Sea Grant College Program, Oregon State University, Corvallis, Oregon, 294 p.

PLOPPER, C. S., 1978, Hydraulic sorting and longshore transport of beach sand, Pacific Coast of Washington: [Unpublished Ph.D. Thesis]: Syracuse University, New York, 184 p.

RANKIN, D. K., 1983, Holocene geologic history of the Clatsop Plains foredune ridge complex: [Unpublished M.S. Thesis]: Portland State University, Portland, Oregon, 176 p.

RAU, W. W., 1973, Geology of the Washington coast between Point Grenville and the Hoh River: Washington Department of Natural Resources, Division of Geology and Earth Resources Bulletin No. 66, 58 p.

RAU, W. W., 1980, Washington coastal geology between the Hoh and Quillayute Rivers: Washington Department of Natural Resources, Division of Geology and Earth Resources Bulletin No. 72, 57 p.

SCHEIDEGGER, K. F., KULM, L. D., AND RUNGE, E. J., 1971, Sediment sources and dispersal patterns of Oregon continental shelf sands: Journal of Sedimentary Petrology, v. 41, p. 1112–1120.

SHEPARD, F. P., AND GRANT, U. S., 1947, Wave erosion along the southern California coast: Geological Society of America Bulletin, v. 58, p. 916–926.

SHEPARD, F. P., AND KUHN, G. C., 1983, History of sea arches and remnant stacks of La Jolla, California, and their bearing on similar features elsewhere: Marine Geology, v. 51, p. 139–161.

SHEPARD, F. P., AND WANLESS, H. R., Sr., 1971, Our Changing Coastlines: McGraw-Hill, New York, 597 p.

SHIPMAN, HUGH, 1990, Vertical land movement in coastal Washington: Washington Department of Natural Resources, Washington Geologic Newsletter, v. 18, p. 26–32.

SNAVELY, P. D., Jr., 1987, Tertiary geologic framework, neotectonics, and petroleum potential of the Oregon-Washington continental margin, in Scholl, D. W., Grantz, A., and Vedder, J. G., eds., Geology and Resource Potential of the Continental Margin of Western North America and Adjacent Ocean Basins: Houston, Texas, Circum-Pacific Council for Energy and Mineral Resources, Earth Science Series, v. 6, p. 305–335.

SNAVELY, P. D., Jr., MACLEOD, N. S., AND WAGNER, H. C., 1973, Miocene tholeiitic basalts of coastal Oregon and Washington and their relations to coeval basalts of the Columbia Plateau: Geological Society America Bulletin, v. 84, p. 387–424.

TERICH, T. A., AND SCHWARTZ, M. L., 1981. A geomorphic classification of Washington State's Pacific Coast: Shore and Beach, v. 49, p. 21–27.

Tillamook Bay Task Force, 1978, Tillamook Bay drainage basin erosion and sediment study, Oregon, Main Report: A cooperative study by the Tillamook Bay Task Force, the Oregon State Water Resources Department, and the U.S. Department of Agriculture Soil Conservation Service, Portland, Oregon, 203 p.

WELLS, R. E., ENGEBRETSON, D. C., SNAVELY, P. D., Jr., AND COE, R. S., 1984, Cenozoic plate motions and the volcano-tectonic evolution of western Oregon and Washington: Tectonics, v. 3, p. 275–294.

WELLS, R. E., SIMPSON, R. W., BENTLEY, R. D., BEESON, M. J., MANGAN, M. T., AND WRIGHT, T. L., 1989, Correlation of Miocene flows of the Columbia River Basalt Group from the central Columbia River Plateau to the coast of Oregon and Washington: Geological Society of America, Special Paper 239, p. 113–130.

WELLS, R. E., 1990, Paleomagnetic rotations and the Cenozoic tectonics of the Cascade arc, Washington, Oregon and California: Journal of Geophysical Research, v. 90, p. 19, 409–19, 417.

WEST, D. O., AND MCCRUMB, D. R., 1988, Coastline uplift in Oregon and Washington, and the nature of Cascadia subduction-zone tectonics: Geology, v. 16, p. 169–172.

A SEDIMENT BUDGET FOR THE SANTA CRUZ LITTORAL CELL, CALIFORNIA

TIM C. BEST
Foxx-Nielsen & Associates, 501 Mission Street, Santa Cruz, CA 95060
AND
GARY B. GRIGGS
Department of Earth Sciences and Institute of Marine Sciences, University of California, Santa Cruz, CA, 95064

ABSTRACT: Beach compartments or littoral cells form the framework for our understanding of the sources, transport, sinks, and storage of sand in the nearshore zone. In general, along the California coast, beach sand is derived from rivers or bluff erosion, moves alongshore under the influence of prevailing wave conditions, and ultimately is lost either to a submarine canyon or dune field.

The Santa Cruz Littoral Cell appears to extend as far north as San Francisco Bay and terminates downcoast at Monterey Submarine Canyon. Northwesterly waves drive littoral drift at a rate of about 200,000 to 250,000 m^3/yr at the Santa Cruz Harbor.

The major sources of sand within the Santa Cruz cell are coastal streams draining the Santa Cruz Mountains and 130 km of coastal bluffs. On the basis of the grain-size distribution of beach and nearshore sediments, a littoral cutoff diameter (0.18 mm) was established in calculating a sediment budget. Fluvial sediment transport measurements combined with flow-duration curves and grain-size distributions were employed to calculate input from coastal streams to the cell. Cliff height and length, sand content, and average long-term (50 to 75 years of aerial-photograph coverage) erosion rates were utilized to determine the littoral contribution from seacliff and bluff retreat. Coastal streams supply about 75 percent of the total littoral-sand input to the cell, bluff erosion contributes about 20 percent and the remaining 5 percent is from gully erosion and sand-dune deflation.

Sediment input to the cell is highly episodic in response to large and infrequent erosional events. The processes that deliver sand to the cell (principally bluff erosion and high stream flow) may operate at different frequencies than those that move sand through the cell (longshore transport). An additional complexity arises from potential changes in sand storage, either on the beach or along the inner shelf, which are capable of producing significant volumes of sand due to the large areas involved.

INTRODUCTION

A sediment budget is a quantitative statement of the relations between the rate of sediment production, transport, storage and discharge. For a littoral cell, it is the conceptual simplification of the interaction of processes that deliver sand into the littoral zone and transport it along the shoreline. Discrete beach compartments or littoral cells form a framework for our understanding of sources, transport, and outputs of sand in the nearshore zone. In a typical beach compartment, littoral transport begins at a rocky headland or section of coast where the upcoast supply of sand or littoral drift is restricted or minimal. Sediments enter the littoral cell primarily from coastal streams and bluff erosion, and are transported alongshore under the influence of the prevailing wave conditions (Inman and Frautschy, 1966). Ultimately, the sand is lost from the system or cell through either a submarine canyon, a coastal-dune field, or in some cases, direct removal through sand mining. In theory, each cell exists as a distinct entity with little or no transport of sediment between cells.

The use of sediment budgets has become common practice among fluvial geomorphologists; however, its application to the littoral zone is still in its infancy. Bowen and Inman (1966), working on the central California coast, were the first to develop a sediment budget for a littoral cell. In subsequent years, other littoral-budget studies have been completed by Hales (1978), Weggel and Clark (1983), Inman and Jenkins (1983), Oradiwe (1986), Dolan and others (1987), Morelock (1987), and Osborne and others (1989). These reports often focus on longshore transport rates or sediment input volumes and only qualitatively address the issue of sediment routing. Due in part to a lack of data, contributions from individual sources have often been evaluated either in very general terms, or using unverified assumptions or extrapolations.

Our lack of a quantitative understanding of littoral cells and sand budgets has become all too obvious along the California coast (Griggs, 1987a). The problems and costs associated with harbor dredging, where jetties or breakwaters have been constructed in the middle or downcoast ends of littoral cells with high-drift rates, on one hand, and the reduction of sand delivery to beaches due to impoundment of sediment behind dams in the coastal watersheds (Norris, 1964; Brownlie and Taylor, 1981), on the other, stem directly from the failure to incorporate this type of information in the land-use decision-making process. The application of a sediment budget to the nearshore zone is a useful tool in coastal land-use management, and is an essential step in understanding the importance of sediment routing along the coast. On the central and northern California coastline, a large gap exists in our present state of knowledge regarding both littoral-cell boundaries and production, transport, storage and loss of littoral sediment within these cells. This investigation was directed toward quantitatively evaluating sediment routing of littoral sands within a portion of the Santa Cruz Littoral Cell. A secondary goal was to evaluate the methodology used in developing a littoral-sediment budget.

PHYSICAL SETTING

The study area extends along the central California coast from Tunitas Creek in San Mateo County in the north to Twin Lakes Beach in Santa Cruz County in the south, and inland to the crest of the Santa Cruz Mountains (Fig. 1). The northern boundary of the study area was initially assumed to correspond to the upcoast boundary of the Santa Cruz Littoral Cell. Although a regional study by Inman (1976) suggested that this cell extended as far north as the entrance to San Francisco Bay, more recent studies by Weber

From Shoreline to Abyss, SEPM Special Publication No. 46

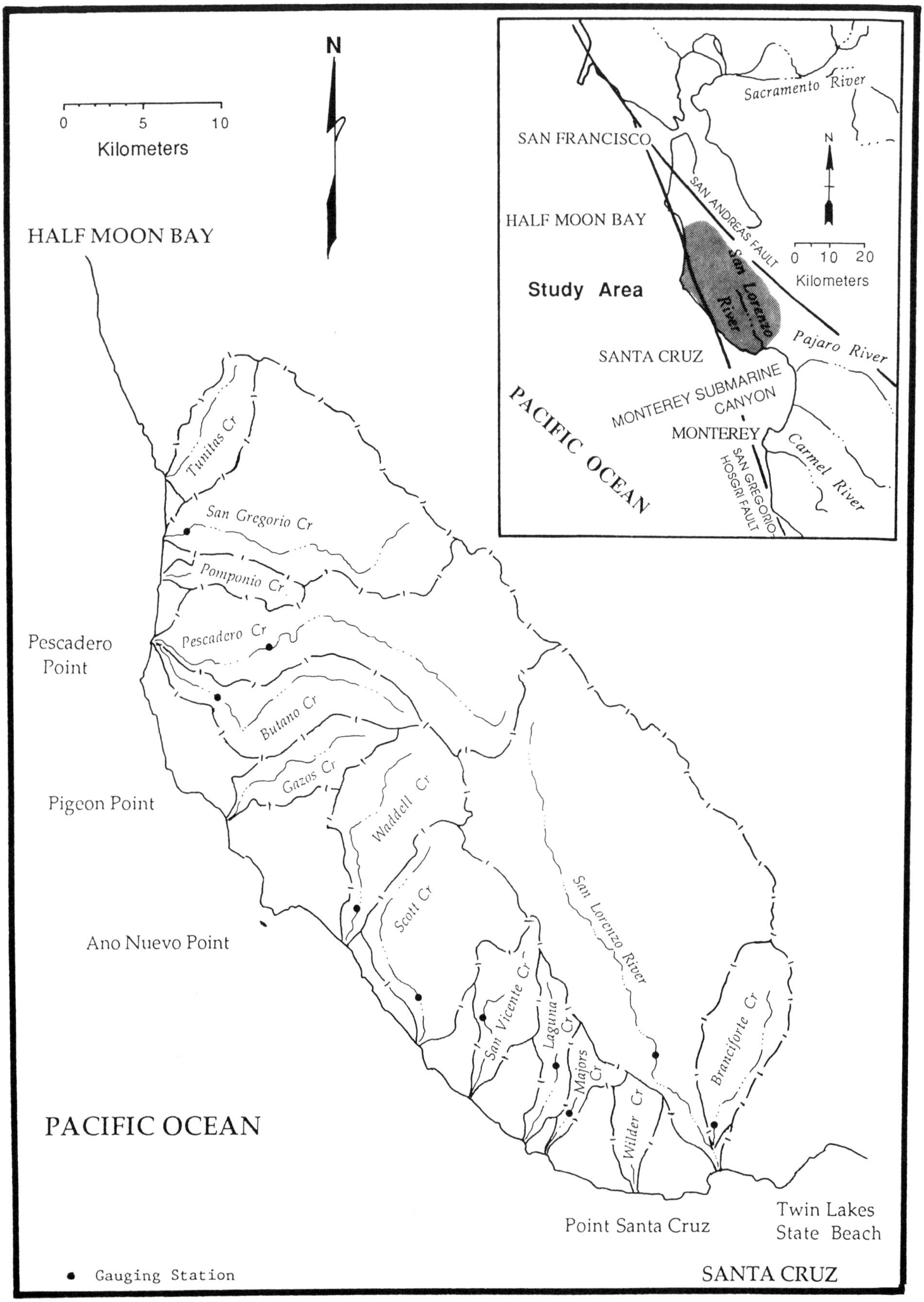

FIG. 1.—Study area and location of principal drainage basins.

(1979) and Griggs (1987a) based on heavy-mineral provinces identified by Yancey and Lee (1972) delineated the northern boundary of the cell at Tunitas Creek, which was the northern boundary used in this study. Habel and Armstrong (1978) in a state-wide overview, terminated the Santa Cruz cell at the northern edge of Monterey Bay. It became apparent at the end of this study that the Santa Cruz Littoral Cell extends at least as far north as the entrance to San Francisco Bay.

The Santa Cruz Littoral Cell terminates downcoast at Monterey Submarine Canyon (Fig. 1). The border of the study area, however, was chosen at Twin Lakes State Beach (the Santa Cruz Small Craft Harbor), a point within the cell where rates of littoral transport are reasonably well known. Net transport within the cell is from northwest to southeast.

The 80 km of littoral-cell coastline north of Monterey Bay are characterized by steep, 5- to 50-m-high coastal bluffs, occasional pocket beaches and coastal dunes. To the east, the Santa Cruz Mountains are drained by a number of small, steep coastal streams, the mouths of which have been drowned by the Holocene rise in sea level. Their lower reaches are typically marked by low-gradient flood plains, coastal lagoons and marshes.

The study area is located within the Salinian Block, an elongate, northwest-trending tectonic sliver bounded on the northeast by the San Andreas fault and on the southwest by the San Gregorio-Hosgri fault system (Fig. 1). In general, Paleozoic metamorphic rocks (marbles and schists) and Cretaceous granitic rocks form the basement complex of this portion of the Santa Cruz Mountains. These rocks are overlain by a sequence of more than 10,000 m of Tertiary marine conglomerates, sandstones, siltstones and mudstones (Nielsen and Brabb, 1979; Clark, 1981).

A series of up to five prominent marine terraces indents the seaward slope of the Santa Cruz Mountains along the length of the littoral cell (Fig. 2). Overlying these terraces is a thin (usually 1.5 to 6 m), nearly continuous mantle of poorly consolidated marine and non-marine gravels, sands and silts. Quaternary sand dunes are found at the mouths of Pescadero, San Gregorio, Waddell and Scott Creeks, and locally overlie the marine-terrace deposits at Ano Nuevo Point, Franklin Point and Sand Hill Bluff.

The Santa Cruz Littoral Cell is characterized by a Mediterranean climate with 90 percent of the rainfall occurring between November and March. The relatively abrupt rise of the Santa Cruz Mountains produces pronounced orographic effects with mean annual rainfall ranging from 45 to 70 cm along the coast and increasing to 150 cm along the crest of the range (Rantz, 1971). Storms are typically of short duration but often of high intensity.

FIG. 2.—Coastal marine terraces and actively eroding sea cliffs 11 km north of Santa Cruz. Tertiary Santa Cruz Mudstone capped by unconsolidated marine and non-marine Quaternary terrace deposits form the eroding sea cliffs.

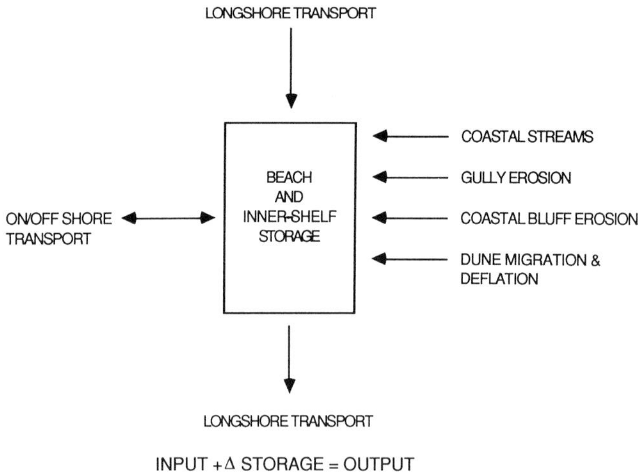

FIG. 3.—Conceptual diagram of sediment routing in the Santa Cruz littoral cell.

APPROACH

Construction of any sediment budget requires 1) identification of source, storage, and output elements, 2) recognition and quantification of erosional and transport processes, including an understanding of the relative importance of persistent versus episodic events, and 3) an understanding of grain size and composition of material in storage and transport (Dietrich and others, 1982; Lehre, 1981).

Sediment routing in the Santa Cruz Littoral Cell can be represented by a conceptual model (Fig. 3). To develop the budget, we will systematically evaluate the importance of each source and output element and attempt to evaluate a range of potential error. In this investigation we are concerned only with material coarse enough to remain on the beach face or in the nearshore zone in appreciable quantities. Fine-grained material (i.e., fine sands, silts and clays) that is lost permanently from the cell will not be included in any budget calculations.

GRAIN-SIZE DISTRIBUTION OF MATERIAL IN LONGSHORE TRANSPORT

The rate of longshore transport is greatest immediately shoreward of the breaker zone (Komar, 1971, 1976) and decreases offshore. It is reasonable to assume that the areal extent of longshore transport is limited in a shore-normal direction by available wave energy and nearshore bathymetry. Because of the direct relation between wave energy (near-bottom current velocity) and particle diameter entrained in transport (Silvester and Mogridge, 1971; Komar and Miller, 1973, 1975), a threshold velocity exists for sediment movement. Assuming a steady-state environment, some minimum grain size should exist below which particles will not remain within the active zone of littoral transport in any appreciable quantity. Material smaller than this size (designated the littoral cutoff diameter) will be lost permanently offshore.

The cutoff diameter for the Santa Cruz cell was determined by comparing the grain-size distribution of 42 composite beach-face samples taken throughout the study area, offshore bottom sediments, and dredge spoil collected at the Santa Cruz Harbor.

Beach samples are all well sorted with an abrupt cut off to a fine-grained tail. The maximum curvature of the fine tail is interpreted to represent the local cutoff diameter of particles in transport. The D_{10} is the particle size for which 10 percent of the sample by weight is finer, and closely approximates the maximum curvature of the fine tail. The mean D_{10} or cutoff diameter within the study area is 0.22 mm but decreases downcoast from an average of 0.45 mm near Tunitas Creek to 0.18 mm at Twin Lakes Beach (Fig. 4). We interpret this alongshore reduction to be a product of both selective sorting as well as differences in available wave energy. Giving more weight to the finer-grained samples, and also to those obtained geographically closer to the downcoast end of the study area, we chose the littoral cutoff diameter of 0.18 mm for the cell, a value equal to the mean plus one standard deviation of all samples. Hicks (1985) calculated a cutoff diameter from Santa Cruz Harbor dredge spoils at 0.18 mm using D_{16} (the average particle size plus one standard deviation).

Grain-size distributions are known for offshore samples collected at depths from 25 to 200 m (Lee and others, 1970; Yancey and others, 1970). In general, mean size decreases offshore, and nearly all sediment is finer than 0.18 mm (of the 44 offshore samples, only four had D_{90} values greater than 0.18 mm), which offers additional support for the selection of 0.18 mm for the littoral cutoff diameter within the Santa Cruz Cell.

SEDIMENT INPUT

Sediment input to the Santa Cruz Littoral Cell is principally from coastal streams, gullies, seacliff erosion, sand dunes, and longshore transport. Onshore transport from the inner shelf appears to constitute only a minor contribution. Biogenous and hydrogenous inputs are unimportant to the budget as well.

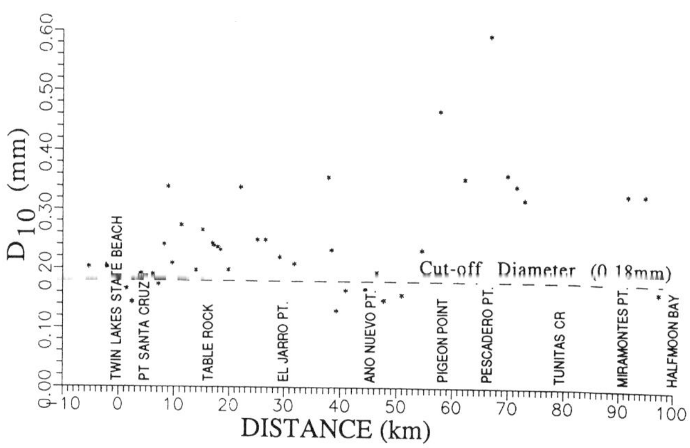

FIG. 4.—D_{10} and cutoff diameter of beach sand collected along the coastline from Santa Cruz to Half Moon Bay. Stars represent D_{10} values for individual beach-sand samples; dashed line represents cutoff diameter for material in littoral transport.

Coastal Streams

Coarse-grained sediment delivery by coastal streams is usually the dominant source of sand to littoral cells along the coastline of California (Komar, 1976; Griggs, 1987a; Stow and Chang, 1987). The coastal watersheds draining the west side of the Santa Cruz Mountains are limited in areal extent but have relatively large sediment yield. To evaluate the littoral-sediment yield requires: 1) estimating total sediment yields, 2) adjusting for the coarse-grained fraction (greater than 0.18 mm), and 3) adjusting the total yield for any changes in floodplain storage.

Thirteen streams with basin areas greater than 12 km^2 drain 85 percent of the 1140-km^2 study area (Fig. 1). In general these basins are narrow; streams are deeply incised and drain steep terrain. The mouths of these streams have been drowned by the Holocene rise in sea level, and thus have low-gradient flood plains or coastal lagoons that serve as temporary sediment storage sites. Although the streams are perennial, sand bars commonly block their mouths during dry summer months. The mouths of all of the coastal streams between Wilder and San Vicente Creeks have been blocked by highway and railroad fill, leading to sediment impoundment during the past century.

The study area is underlain predominantly by Tertiary clastic marine sediments with local exposures of granitic basement. Overlying the bedrock is a thin veneer of colluvium and soil. The shallow depth of colluvium on hillslopes, combined with lack of large sediment bars and overbank deposits along the stream channels, suggests that storage of sediment in the upper portions of the watersheds is minor.

A large portion of the sediment produced within these basins is believed to be due to mass movements (primarily shallow-debris flows), which occur during winters of intense rainfall (Griggs, 1982; Smith and others, 1982; Ellen and Wieczorek, 1988). Lehre (1981) determined that over 80 percent of all sediment mobilized and 55 to 65 percent of all sediment reaching the main channel in a small 1.4-km^2 drainage basin northwest of San Francisco was due to shallow landsliding. Thus, sediment input to stream channels should be very episodic in response to very large storms.

Vegetation in these basins consists of redwoods, oak and madrone on side slopes and valley bottoms, and chaparral and grasslands on ridge crests. Whereas the San Lorenzo River basin has been considerably altered by human activity (principally logging, quarrying and grading associated with construction activities), which has increased erosion rates and sediment yields, most of the smaller coastal basins have remained relatively undisturbed since the redwood logging of the past century.

Several previous studies have addressed sediment yield of these stream basins. Brown (1973), Macy (1976), Griggs and Paris (1982), Hicks (1985), and Nolan and Marron (1988) have estimated sediment yield for portions of the San Lorenzo River, whereas Curry and others (1986) and Williams (pers. commun. 1990) have addressed the sediment yield from Butano Creek. The majority of these studies, however, are based on limited data and do not necessarily reflect the long-term sediment yield of the basin.

Daily stream flow has been monitored on 10 coastal streams, which drain 68 percent of the total study area (Table 1). Sediment transport has been monitored at five of these stations (U.S. Geological Survey, 1931–1960, 1963–1988). Only the San Lorenzo River and Pescadero Creek have over one year of sediment-transport data, however (Table 1). Stream flow and sediment transport at Waddell Creek were monitored during the winter of 1986 as part of this study. The focus of this portion of the investigation was to evaluate the sediment yield at ungauged basins or at basins where sediment data are limited.

Sediment yield can be computed by developing a number of relations that integrate measurements of stream discharge and sediment transport. By normalizing characteristic variables between watersheds (i.e., basin area, precipitation, geology, etc.) and developing relations between stream discharge and sediment load, sediment yields from gauged basins can be used to predict the yield at ungauged basins.

A flow-duration curve relates discharge and percent of time that this discharge is exceeded. At ungauged basins or basins where flow record is short, an average flow-duration curve can be generated from nearby gauged basins having similar terrain by normalizing to mean annual discharge (Fig. 5). Mean annual discharge at an ungauged stream, in turn, can be estimated from rainfall-runoff relations (Fig. 6).

Sediment-transport curves, relating water discharge to sediment discharge, can be developed from measurements of suspended sediment, bedload, or any size fraction for which data are available (Colby, 1956). Where sediment data are unavailable, an average sediment-rating curve based

TABLE 1.—SUMMARY OF GAUGING RECORDS FOR STREAMS IN NORTH SANTA CRUZ COUNTY AND SOUTH SAN MATEO COUNTY

Stream	Drainage Area at Gauging Sta. (km^2)	Stream-Flow Measurements	Suspended-Sediment Measurement	Bedload Measurement
San Gregorio Creek near San Gregorio	131.2	1969–present	1986	1986
Pescadero Creek near Pescadero	118.9	1951–present	1971, 73, 75, 86	1986
Butano Creek near Pescadero	47.4	1962–1974	—	—
Waddell Creek near ranger station[a]	59.8	1986	1986	1986
Scott Creek near Big Creek	65.0	1958–1973	—	—
San Vicente Creek near Davenport	15.8	1970–1988	—	—
Laguna Creek near Davenport	8.0	1970–1976	—	—
Majors Creek near Santa Cruz	9.8	1970–1976	—	—
San Lorenzo River at Big Trees	274.5	1937–present	1972–82, 86[b]	1972–82[b]
Branciforte Creek at Santa Cruz	44.8	1940–43, 1951–68	—	—

[a]Gauged for the purpose of this study. All other gauging stations established by USGS.
[b]Mean daily sediment discharge values available from 1972–1982.

FIG. 5.—San Lorenzo River at Santa Cruz during the flood of January 4, 1982.

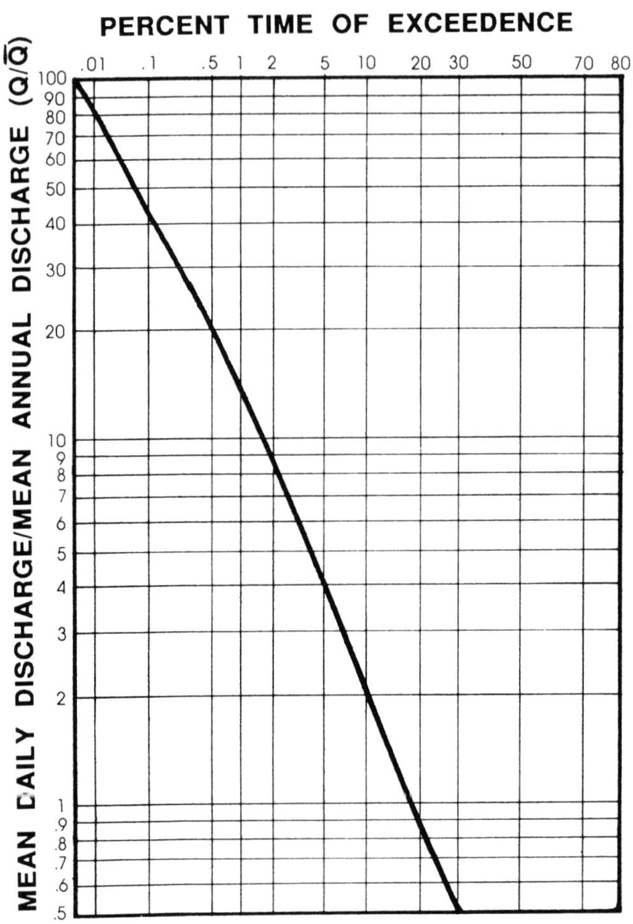

FIG. 6.—Unit flow-duration curve for the coastal streams in north Santa Cruz County and south San Mateo County for the period of record 1937–1988. Discharges have been normalized by mean annual discharge. Flow-duration curve is based on U.S. Geological Survey gauged streams, summarized in Table 1.

on measurements obtained from similar basins in an area can be used to approximate the rates of sediment transport at the ungauged stream (Figs. 7 and 8). Combining the flow-duration curve with the sediment-transport relation allows us to compute sediment yield from a basin. Sediment-transport values for the San Lorenzo River, the largest basin within the study area, have been reported for the period 1972–1982 (U.S. Geological Survey, 1963–1988) and provide the most accurate measure of actual sediment yield.

The sediment yields for the major streams draining the study area are summarized in Table 2. For the period of available data, the littoral-size fraction comprised 16 percent of the total load, of which 40 percent was transported as bedload. The relation between stream flow and the percentage of littoral sediment in transport indicates that at high flows a larger fraction of the total load is coarse-grained sediment. As flow increases, a larger portion of the total load is contributed to the cell.

The sediment delivery from coastal watersheds is very episodic. Forty-six percent of the total load of the San Lorenzo River for the period of sediment discharge record (1972–1982) was transported during five individual storm events, with 32 percent of the total suspended load carried during the January 4–5, 1982 flood, a 30-year event (Griggs, 1987b; Fig. 9). Suspended load for 1982 accounts for 52 percent of the total load for the entire 11 years of sediment record. Hicks (1985) suggests that 70 percent of the littoral-size load of the San Lorenzo River for the period of record 1972–1983 was delivered by the high flows during the major storms of 1982 and 1983. The infrequent but high-intensity rainfall events and subsequent stream discharge from these steep short basins can clearly dominate sediment input to the littoral cell.

The average annual total sediment yield calculated for the entire study area amounts to 365 tons/km^2, with the San Lorenzo River (at the Big Trees gauging station) having a slightly higher yield of 553 tons/km^2. This value for the San Lorenzo River is about half of that presented in the U.S. Geological Survey reports for the period of record using mean daily flow values (931 tons/km^2). Sediment yield determined from the mean daily method is normally more accurate than that generated from flow-duration curves. Unfortunately, mean daily sediment-discharge values are only available for the San Lorenzo River. Because this study relies on sediment-data collection at lower flows and over a shorter time frame than those of the U.S. Geological Survey, the values reported here likely underestimate actual "long-term" conditions. Based on this comparison, we believe that this underestimate is no more than 50 percent.

The average annual littoral-sediment yield (material larger than the cutoff diameter) for the study area is therefore in the range of 61 to 122 tons/km^2. Applying this value to the remaining basins (those larger than 12 km^2) in the 1140-km^2 study area and assuming a bulk density of 1600 kg/m^3 produces an average of 43,000 to 87,000 m^3 of littoral material annually from all stream sources.

Aggradation or an increase in sediment storage in coastal lagoons and flood plains due to sea-level rise could reduce these littoral-sediment yield values. Few studies address recent stream-channel aggradation in the study area, although

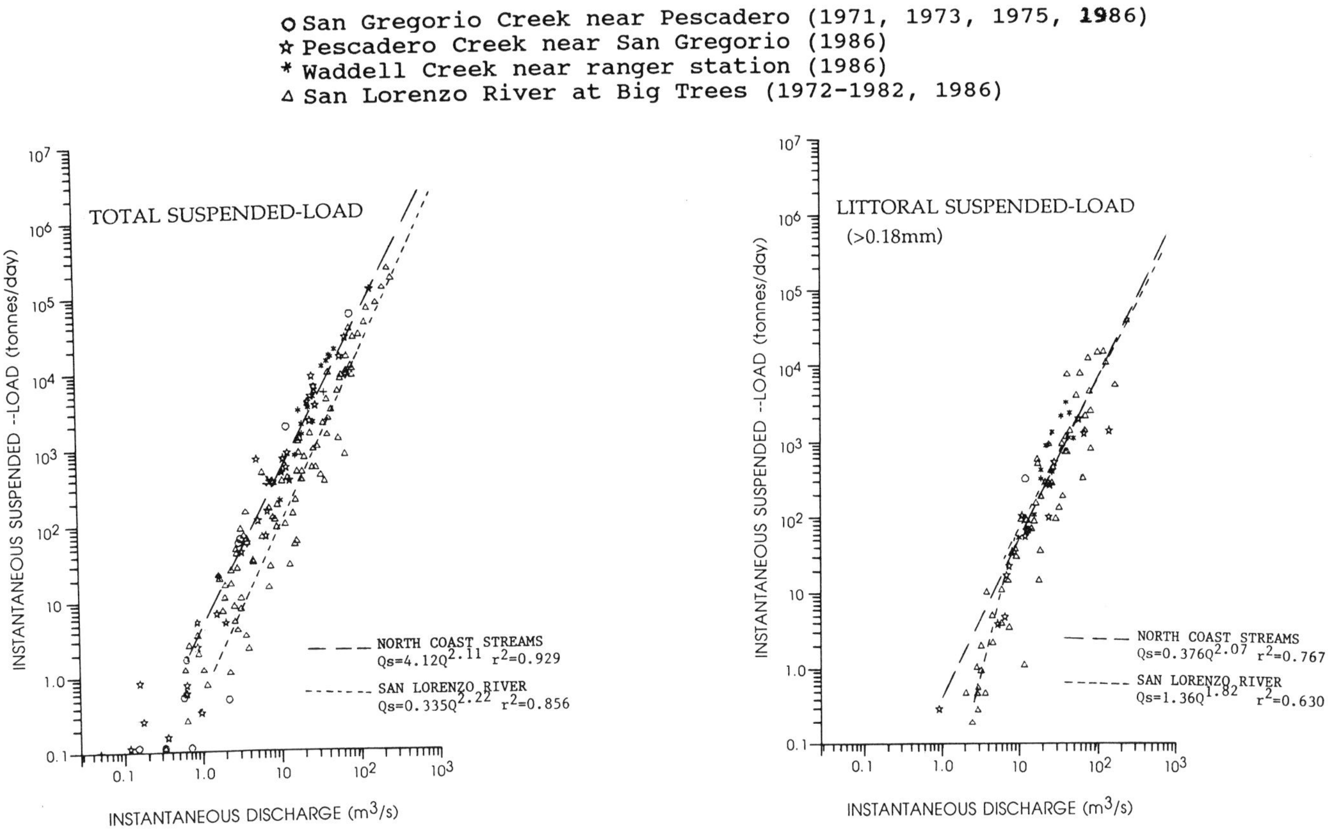

FIG. 7.—Total and littoral suspended-load rating curve for the coastal watersheds between Santa Cruz and Half Moon Bay.

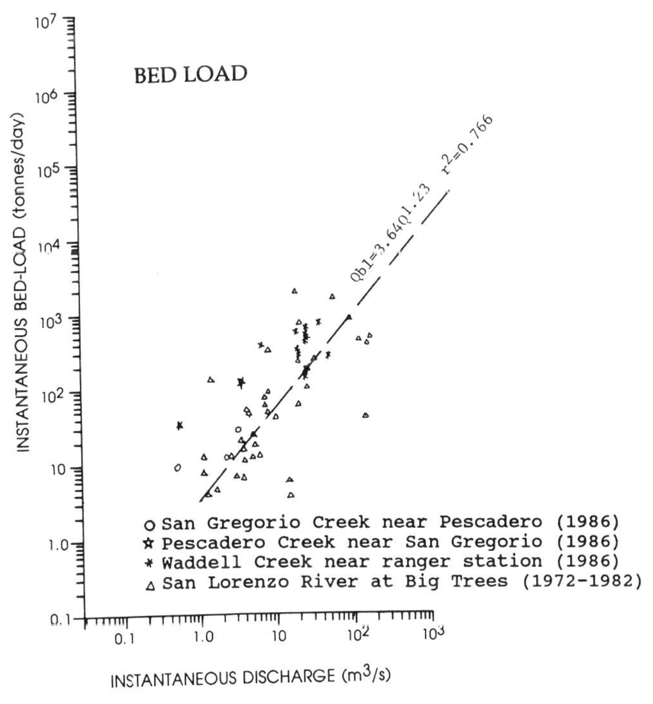

FIG. 8.—Bedload-rating curve for the coastal watersheds between Santa Cruz and Half Moon Bay.

Griggs and Paris (1982) do report on the aggradation of the lower reaches of the San Lorenzo River in response to dredging below natural grade during construction of a flood-control project in 1957–1959. Aggradation expectedly followed dredging until an equilibrium gradient was attained. Scour continues to occur during high flows followed by aggradation during low-flow periods. Aggradation of the Butano Creek flood plain, which has been partially modified by levees, has been addressed by both Curry and others (1986) and Williams (pers. commun., 1990). Both workers examined floodplain morphology through a comparison of topographic maps and aerial photographs dating back over 100 years. Williams concluded that net current yield is relatively small; Curry suggested that localized aggradation followed the 1986 storms. Based on all available information, we believe that the amount of net aggradation in the lower reaches of the coastal streams over the past 50 to 100 years is relatively small with respect to the overall sediment yield of the individual basins. However, we recognize that additional work is required in this area.

Gully Erosion

Extensive gullying characterizes the coastal hills along the northern portion of the study area (Fig. 10). The majority of gully erosion occurs along a 4-km-wide coastal

TABLE 2.—SEDIMENT LOAD FOR COASTAL STREAM BASINS IN NORTH SANTA CRUZ AND SOUTH SAN MATEO COUNTY (in tons/yr)

Stream	Drainage Area (km^2)	Suspended Load		Bedload		Total Load	
		Total Load	Littoral Load	Total Load	Littoral Load	Total Load	Littoral Load
Tunitas Creek at beach	30.1	1,290	105	299	227	1,590	332
San Gregorio Creek near San Gregorio	131.2	47,800	3,620	2,450	2,030	5,025	5,650
Pomponio Creek at beach	18.6	679	55	206	163	885	219
Pescadero Creek near Pescadero	118.9	49,700	3,760	2,510	1,930	52,210	5,690
Butano Creek near Pescadero	47.4	11,000	857	1,040	807	12,040	1,670
Gazos Creek at beach	29.8	2,700	218	462	299	3,160	517
Waddell Creek near Ranger Station	59.8	19,600	1,506	1,460	953	21,060	2,460
Scott Creek at beach	76.6	36,100	2,750	2,090	1,900	38,190	4,740
San Vicente Creek at beach	29.1	3,090	247	234	209	3,324	455
Laguna Creek at beach	20.8	1,290	105	299	236	1,580	341
Major Creek at beach	12.1	372	29	140	109	512	138
Wilder Creek at beach	14.6	366	30	143	109	509	139
San Lorenzo River at Big Trees	274.5	142,000	21,900	9,890	9,250	140,100	31,100
Branciforte Creek at Santa Cruz[a]	44.8	1,500	606	877	816	2,377	1,422
Total metric tons/yr		318,000	35,600	22,300	19,200	294,000	54,900

[a]Tributary of the San Lorenzo River.

strip between Pigeon Point and Half Moon Bay, with the basins between San Gregorio and Pescadero Creeks being the most extensively affected. Over the past 50 years there has been a dramatic increase in both the size and number of gullies (Swanson, 1983). Gully networks are preferentially confined to the marine-terrace deposits and alluvial and colluvial materials overlying the Purisima Formation. Gullying results from subsurface soil piping through more permeable horizons, and channelized overland flow (Swanson, 1983).

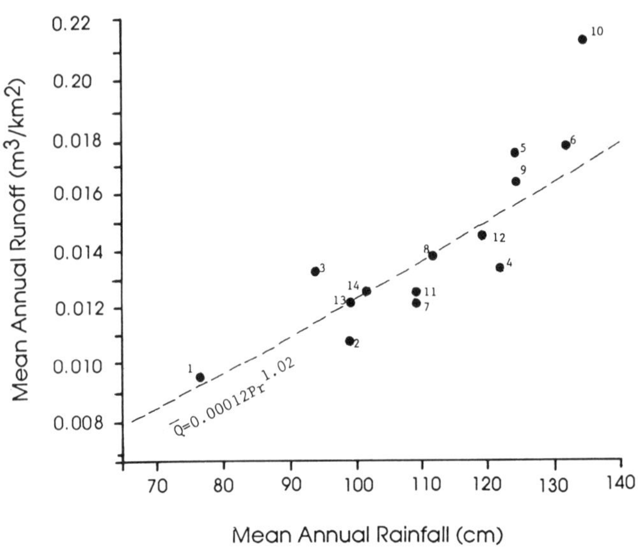

NORTH COAST BASINS
1 San Gregorio Creek near San Gregorio
2 Pescadero Creek near Pescadero
3 Butano Creek near Pescadero
4 Scott Creek near Big Creek
5 San Vicente Creek near Davenport
6 Laguna Creek near Davenport
7 Majors Creek near Santa Cruz

SAN LORENZO RIVER BASINS
8 San Lorenzo River near Boulder Creek[b]
9 Boulder Creek near Boulder Creek[b]
10 Bear Creek at Boulder Creek[b]
11 Zayante Creek at Zayante[b]
12 San Lorenzo River at Big Trees
13 Branciforte Creek at Santa Cruz

SOQUEL CREEK BASINS
14 Soquel Creek at Soquel[ab]

a: Outside of study area
b: Not used in regression analysis

FIG. 9.—Mean annual rainfall-runoff relation for the coastal Santa Cruz mountains region. Mean annual rainfall from Rantz (1971); mean annual runoff from U.S. Geological Survey (1931–1960, 1963–1988).

Sediment yield was estimated at the 1.2-km^2 Long Gulch sub-basin located 1 km south of Pomponio Creek (Figs. 1 and 10) that represents approximately 17 percent of the 7.4-km^2 area currently undergoing active gullying. We determined the sediment yield from Long Gulch by integrating measurements of individual gully lengths and widths (determined from aerial photographs) with the field relations between top width and cross-sectional area. The littoral fraction of the total load was determined from sieve analysis of materials exposed in the gully walls. We then applied the sediment yield from Long Gulch to the other gullied areas to determine the total and littoral input from coastal gullying to the cell.

Average sediment yield from the coastal gullies is increasing as new gullies form and existing gullies enlarge. Between 1928 and 1982 the average sediment yield from the gullies of this coastal area was about 1,660 m^3/km^2. For the period of stream sediment record, however (1972 to 1982), average annual yield was roughly twice this value, about 3,900 m^3/km^2. This represents a total sediment load of 29,000 m^3/yr for the gullied area.

Based on grain-size analysis of the materials exposed in the gully walls, only about 10 to 20 percent of the total load has a particle diameter greater than the littoral cutoff size. The majority of littoral-size sediment originates from the overlying marine-terrace and alluvial deposits. Although gullying is evident in the Purisima Formation, the sediments are too fine grained to be a significant source of littoral sediment. The littoral input to the cell from gullying is currently about 2,900 to 5,800 m^3/yr but is expected to increase slightly.

Seacliff Erosion

The shoreline throughout the study area consists predominantly of an uplifted sequence of marine terraces fronted by actively eroding, near-vertical sea cliffs. The cliffs are

FIG. 10.—Actively eroding gullies in Long Gulch, between Pomponio Creek and Pescadero Creek, San Mateo County coast. Note road in foreground for scale.

interrupted by relatively small pocket beaches that have formed at the mouths of the coastal streams (Fig. 2). Terraces are well developed due primarily to the nearly flat-lying or gentle seaward dip of the underlying well-bedded sedimentary rocks. Sea cliffs range in height from 5 to 50 m, have an abrupt cliff edge, and typically erode to form an irregular coastline. Lithology and structure are the dominant factors influencing seacliff morphology.

The Santa Cruz Mudstone dominates the sea cliffs throughout the southern portion of the study area and is exposed between Pt. Santa Cruz and Ano Nuevo Creek. The Pigeon Point Formation (well indurated silts, sands and gravels) crops out from White House Creek to Pescadero Creek. The Purisima Formation (mudstone, siltstone and very fine-grained sandstone) is exposed both north of Pescadero Creek and along the cliffs within the city of Santa Cruz. Quaternary dunes, terrace deposits and alluvium comprise the low bluffs in the Ano Nuevo area.

The production of littoral sand from a segment of shoreline through seacliff erosion (Qs) is the product of the cross-sectional area of sea cliff (A, equal to cliff length × height), the average annual rate of cliff retreat (dx/dt), and the percentage of that material which is greater than the cutoff diameter (percent CD), or:

$$Qs = A(\text{percent CD})dx/dt.$$

Rates of seacliff retreat are governed by the ability of large storm waves to attack the base of the cliff and the relative ease with which cliff material can be dislodged, either directly by wave attack, or through subaerial processes such as rock falls (Fig. 11). Episodic and locally variable rates of cliff retreat result from a combination of 1) alongshore differences in the strength of cliff materials (Griggs and Johnson, 1979; Sunamura, 1983), 2) the infrequent occurrence of high tides and extreme storm waves capable of producing significant erosion and removing debris from the base of the cliff, 3) concentration of wave energy due to local bathymetry, and 4) the presence or absence of a protective beach.

Rates of bluff retreat also may be influenced by large earthquakes, such as the 7.1-magnitude October 17, 1989 Loma Prieta earthquake. This was the largest earthquake to affect the central coast since the 1906 San Francisco earthquake, and it produced isolated cliff failure throughout the entire study area (Plant and Griggs, 1990). For the most part, these failures consisted of isolated rock falls or soil slips resulting in up to 1 m of bluff retreat. Similar cliff failures were noted during the 1906 event (Lawson, 1908). In addition to the instantaneous bluff failure during the 1989 earthquake, tension cracks formed as much as several meters landward of the cliff edge in many locations along the shoreline of the study area. These are expected to lead to

FIG. 11.—Block fall landsliding along the coastal seacliffs south of Santa Cruz. Failure occurred along joint sets due to undercutting by surf action. Note person on side of photo for scale.

additional failure in subsequent wet winter months. It is clear that a large-magnitude earthquake may play an important but infrequent role in cliff recession.

Rates of seacliff retreat over the past 50 to 100 years were computed from a comparison of time-sequential aerial photographs, ground photographs and historic coastal surveys and maps. Methods used involved measurements of the position of the seacliff edge at individual locations during the time covered by available maps or photographs.

Lithology and structure of the cliff-forming materials exert the dominant control on rates of seacliff retreat. Within the 50- to 100-year data base used, the well-cemented sandstones and conglomerates of the Pigeon Point Formation show negligible erosion (Figs. 12 and 13). In contrast, cliffs containing the Purisima Formation and Santa Cruz Mudstone erode at average rates of 0 to 20 cm/yr, with the Purisima eroding more rapidly by virtue of its general lack of cementation and induration. The mudstone is extensively jointed, which allows for hydraulic quarrying or block removal as well as rock falls from the cliff. The common presence of an intertidal shore platform in the mudstone throughout much of the study area (Fig. 2) buffers the base of the cliff from direct wave attack, thus allowing for only moderate rates of erosion.

The most rapid rates of cliff retreat occur in the Ano Nuevo Point area and typically exceed 30 cm/year. These high-erosion rates are due to a combination of 1) the relatively low height (less than 10 m) of the marine terrace and sea cliff at this location, 2) the material exposed in the bluff being dominantly unconsolidated and therefore, very erodible terrace deposits and dune sands, and 3) the direct exposure of the point to severe storm waves. This small section of the cell is a potentially significant source of sand for the budget due to a combination of the rapid-erosion rates and availability of sand-size material.

The area of sea cliff that provides potential sediment to the littoral budget was determined from measurements of cliff length and height. Height was measured from the bluff top to a depth of about 4 m below sea level, the approximate zone within which active erosion of the bedrock is concentrated. The thicknesses of the different stratigraphic units exposed in the sea cliff were determined from a combination of direct field measurements, near-horizontal aerial photographs taken at cliff-top altitude from offshore, topographic maps, and vertical aerial photographs using a parallax bar.

Based on both a visual evaluation of the seacliff materials and a number of grain-size analyses, it is evident that the

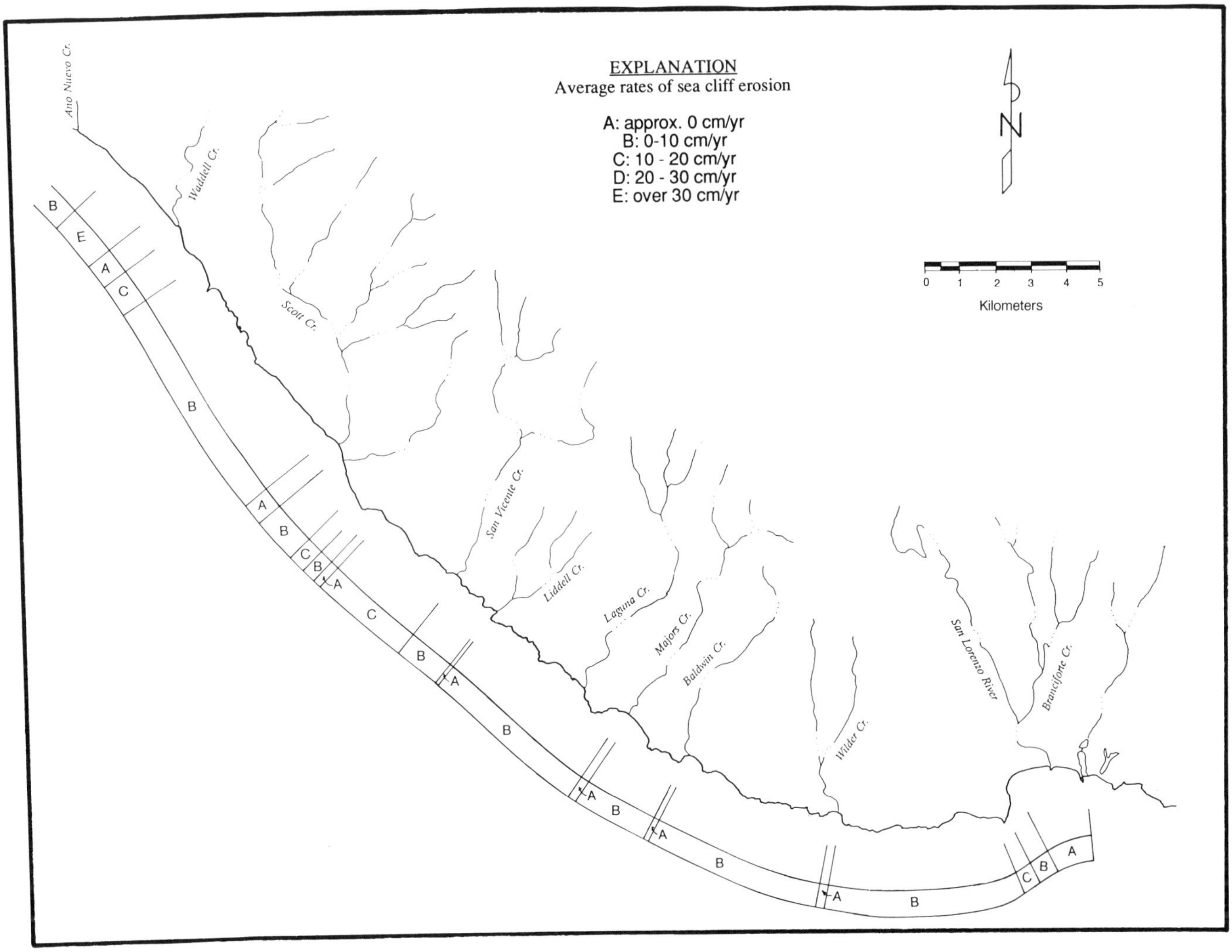

FIG. 12.—Average rates of seacliff retreat—Santa Cruz Small Craft Harbor to Ano Nuevo Creek.

exposed bedrock materials do not contain or provide a significant amount of littoral sand. The Santa Cruz Mudstone and Purisima Formation are too fine grained. The Pigeon Point Formation is sufficiently coarse grained, but is eroding very slowly. Terrace deposits, on the other hand, are relatively coarse grained. Between Santa Cruz and Ano Nuevo Creek, the amount of littoral sand (greater than 0.18 mm) contained in the terrace deposits ranges from 16 to 78 percent, with an average of 52 percent. From Ano Nuevo Creek to Tunitas Creek the range is from 1 to 28 percent, with an average of 19 percent. In the Ano Nuevo area, 77 percent of the dune sand is larger than the cutoff diameter.

For the portion of the littoral cell updrift of the Santa Cruz Small Craft Harbor, the contributions of the individual geologic units to the littoral budget from seacliff retreat are summarized in Table 3. The unconsolidated, relatively coarse-grained Quaternary deposits, which cap the bedrock terrace along the entire length of the cell, supply 52 percent of the littoral sand produced from seacliff retreat. Although the Purisima Formation and Santa Cruz Mudstone form the sea cliffs for most of the study area, they are too fine grained to contribute significant amounts of littoral-size material. The short stretch of recent dunes at Ano Nuevo is the source for most of the remainder of the sand provided by seacliff retreat to the cell.

Sand Dunes

Modern sand dunes overlie the marine-terrace deposits at Ano Nuevo Point, Franklin Point, and Sand Hill Bluff, and also occur locally at the mouths of Pescadero, San Gregorio, Waddell and Scott Creeks. Evidence has been put forth that in the recent past (.2–.25 ka) the coastline at Ano Nuevo Point extended seaward to include Ano Nuevo Island (Weber, 1979). The existence of a receptive windward-facing beach and a low-relief peninsula allowed for a broad beach and dune field to develop. Sand slowly moved across the point in a series of barchan dunes and eventually cascaded off the cliff along the south shore. The opening of the channel between the point and the island in the mid-1700s re-

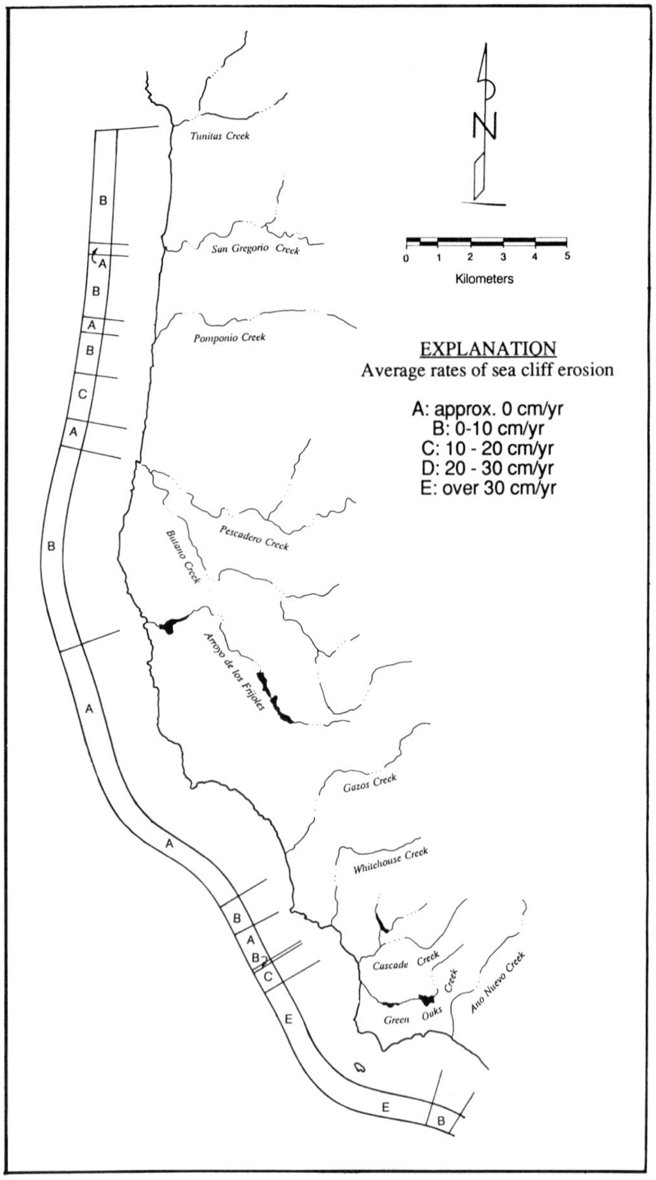

FIG. 13.—Average rates of seacliff retreat—Ano Nuevo Creek to Tunitas Creek.

duced the sediment supply to the dune field and the dunes have subsequently stabilized and deflated. Currently, only a minor amount of sand is blown onto the point along the north shore between Table Rock and North Point.

An evaluation of the significance of this source involving time-sequential aerial photographs combined with field measurements of dune frontage and height, as well as grain-size distribution, indicates present contribution of littoral-size sand from the dunes is about 1,700 m^3/yr, and is expected to decline as the dune field stabilizes further.

Because of the relatively small size of the dune fields at Franklin Point, Sand Hill Bluff, and locally at the stream mouths, the amount of sand transferred between these storage sites and the littoral cell is minor.

TABLE 3.—SUMMARY OF AVERAGE ANNUAL SEDIMENT INPUT OF BEACH-SIZE SAND TO THE SANTA CRUZ LITTORAL CELL BY SEACLIFF EROSION BETWEEN SANTA CRUZ AND TUNITAS CREEK

Geologic Unit	Average Sediment Input (m^3/yr)	Percent of Cliff Budget
Purisima Formation	250– 360	2
Quaternary marine terrace	6,880–12,100	58
Quaternary dunes	5,040– 7,500	40
TOTAL	12,170–19,960	100%

Onshore Transport

A number of processes have been discussed in recent years as potentially significant in transporting littoral sediments across the shelf, either onshore or offshore (Wright, 1987). Whether or not this type of transport takes place is of considerable importance to this study. To date there have been no specific investigations within the study area addressing the on/offshore transport of coarse-grained material.

Significant storage of littoral sediment on the inner shelf is a difficult component in any budget to evaluate (Bowen and Inman, 1966). Due to the large area involved, a small increase in thickness of sediment cover can produce a very large volume. Typical offshore-survey methods utilized do not allow for the resolution in sediment depth or thickness required to make precise estimates of this component.

Extensive offshore sampling across the shelf of the study area (Lee and others, 1970; Yancey and others, 1970) indicates that at depths greater than 25 m nearly all of the sediment is finer than 0.18 mm (the littoral cutoff diameter), suggesting that neither significant onshore nor offshore transport of littoral sand takes place in the cell. This storage component is potentially significant, however, and needs better resolution.

Longshore Transport

If the boundaries of a littoral cell are selected correctly, and there is no leakage or transfer of sediment between cells, then littoral transport is not a budget input to consider. As this study progressed, however, it became evident that the heavy mineral-province boundary identified by Yancey and Lee (1972) did not constitute a littoral-cell boundary. Although Habel and Armstrong (1978) delineated three littoral cells between Monterey Bay and San Francisco, there are no major sinks along this entire coastline capable of forming the termination of a cell. It is now evident that the Santa Cruz cell extends upcoast at least to the entrance of San Francisco Bay.

Because the sediment sources north of Tunitas Creek were not systematically evaluated, and because there have been no studies on longshore-transport rates in that area, sediment input to the littoral cell from upcoast longshore transport is a largely unknown quantity at present.

The majority of coarse-grained sediment entering the San Francisco Bay from either the Sacramento or San Joaquin Rivers is trapped within the bay (Griggs and Hein, 1980). The D_{50} of bottom sediments collected off the mouth of the San Francisco Bay range between 0.10 and 0.33 mm, fining offshore (Thompson, 1981), suggesting that there may be some very fine-grained sand that does leave the bay entrance.

Some information regarding sediment transport does exist for the area immediately offshore of the mouth of San Francisco Bay (Doug Pirie, Corps of Engineers, South Pacific Division, pers. commun., 1990). About 380,000 m^3 of material is dredged each year on average from the shipping channel at the entrance to San Francisco Bay and placed on the outer end of the bar, which lies immediately to the south of the shipping channel (average annual dredging volumes range from about 225,000 to 750,000 m^3). The north side of the channel experiences the greatest shoaling each year following dredging, whereas the south bar does not exhibit any net accretion despite the dumping of dredged material, suggesting that sediment transport is dominantly from north to south, or downcoast. The dredged material has a D_{50} of about 0.2 mm, or is very fine grained. Some of the material appears to move onshore and accumulates in the sand dunes along the northern San Francisco peninsula.

The ultimate depositional site for the remaining material is not completely understood at present. There is no evidence of large accumulations of this material on the immediate downcoast beaches nor is there evidence of large littoral-drift rates or volumes progressing southward through the littoral cell in the Half Moon Bay area. Some of this large volume of relatively fine-grained dredged material undoubtedly is transported alongshore to the south, and may in fact support the submerged fine-grained portion of the beach as well as contribute to the exposed beach face. The volumes or importance of these materials are unknown at present, however.

SEDIMENT OUTPUT

The principal output of coarse-grained material from the littoral cell is by longshore transport. Due to channel filling and the annual dredging associated with the maintenance of the Santa Cruz Small Craft Harbor, a number of studies have been directed toward understanding the littoral-drift processes and rates at this location within the cell (Table 4).

Potential littoral-transport rates have been calculated in the vicinity of the harbor from both wave hindcast data (Anderson, 1971; Walker and others, 1978) and from wave data measured directly offshore from a slope array (Seymour and others, 1980). The calculation of the potential rate of littoral transport is a relatively common approach in coastal-engineering investigations and involves the conversion of incident wave energy into a downcoast littoral component. This quantity depends on the longshore component of wave power, which is a function of wave energy and the angular approach of the wave to the shoreline. There are some difficulties and limitations with this approach, which include the transformation of deep-water wave spectra to the shoreline in areas of significant refraction, as well as the very close dependence on nearshore bathymetry.

It is worth comparing the potential littoral transport values determined at the harbor area using this approach (Table 4), specifically those of Seymour and others (1980) and Walker and others (1978), which vary by a factor of six. One partial explanation for this difference is that Seymour only used one year of wave data, and this was a year of only moderate wave conditions. Another more significant variable in this approach, however, is the extreme sensitivity to the amount of wave refraction. The lack of very good bathymetric control, for example, can significantly affect the calculations of potential transport. Oradiwe (1986), for example, in calculating potential littoral transport around the shoreline of Monterey Bay, noted that an angular difference of 1° between the angle of wave approach and the bottom contours at the break point produced annual drift rates varying from 152,000 m^3 upcoast to 160,000 m^3 downcoast. A 2° angular difference produced a difference in potential transport rate of nearly 600,000 m^3 annually.

Moore (1972) and Walker and others (1978) estimated net annual longshore transport rates of 191,000 and 230,000 m^3, respectively, based on accretion rates against the upcoast jetty of the Santa Cruz harbor following its construction in 1963–1965. Walker and Williams (1980) suggest that 134,000 to 288,000 m^3 of sand annually bypass the harbor mouth. Seymour and others (1980) conclude that the amount of bypassing is minimal. An analysis of aerial photographs from the harbor mouth and the beach shape on opposite sides of the harbor suggests some bypassing is taking place, although its volume is unknown.

Griggs and Johnson (1976), Walker and Williams (1980) and Griggs (1987a) used dredging data from the Santa Cruz Harbor to obtain minimum values for rates of littoral transport. The more recent dredging volumes (120,000 to 145,000 m^3 annually) are more representative than the earlier data due to the continuing impoundment of drift against the upcoast jetty in the first 15 years following construction. Dredge figures are minima for two reasons. First, they do not account for any sand bypassing the harbor entrance and secondly, the figures are for "pay yardage," which is the volume contractors actually bid to remove based on required channel depths. Periodic comparison of pay yardage with actual dredged yardage indicates that the latter volumes may be significantly larger, in part due to continued infilling of the channel during the weeks of dredging.

Based on this evaluation of all previous calculations of rates of littoral transport and potential littoral transport at the harbor, we feel that a reasonable value at this point in the cell is from 200,000 to 250,000 m^3 annually.

The principal sink for sand within the Santa Cruz Littoral Cell at the present time is Monterey Submarine Canyon (Fig.

TABLE 4.—SUMMARY OF LONGSHORE-TRANSPORT RATES IN THE VICINITY OF THE SANTA CRUZ SMALL CRAFT HARBOR

Reference	Transport Rate (m^3/yr)
Moore (1972)	191,000
Anderson (1971)	270,000 (max)
Walker and Dunham (1978)	230,000–380,000
Seymour and others (1980)	47,000
Walker and Williams (1980)	114,000–380,000
Griggs (1987a)	120,000–145,000 (min)

1). The canyon heads into the shoreline in the center of Monterey Bay at Moss Landing. Three branches of the canyon extend to within about 100 m of the shoreline, where they appear to be effective in removing most of the sand moving downcoast within the cell.

DISCUSSION OF SEDIMENT BUDGET

Input and output volumes for the portion of the Santa Cruz Littoral Cell within the study area are summarized in Table 5. Principal sources of littoral sand are the coastal streams, bluff erosion, and probable longshore transport from upcoast. Annual input volumes from the streams and bluffs range from about 60,000 to 114,000 m^3, of which about 72 to 76 percent is from coastal streams and 18 to 20 percent is from bluff erosion. Output from the cell through longshore transport at the Santa Cruz Small Craft Harbor averages about 200,000 to 250,000 m^3 annually.

Excluding longshore transport from upcoast into the study area, the computed average annual input volumes total only 24 to 57 percent of the longshore transport near the downcoast end of the cell; 86,000 to 190,000 m^3 of material transported out of the cell annually are unaccounted for by our measured inputs. There are three possible explanations for this discrepancy: 1) the cell boundary initially selected is incorrect and there is significant input from sources upcoast of the cell boundary; 2) a reduction of storage either along the beaches or on the inner shelf; and 3) temporal variations in sediment delivery (e.g., stream input during major floods) that have not been adequately accounted for due to the relatively short data base.

In retrospect, the northern limit of the cell as delineated on the basis of heavy-mineralogy provinces (Yancey and Lee, 1972; Weber, 1979; Griggs, 1987a) is incorrect. Although there is a mineralogical change that takes place at Tunitas Creek, it does not appear that this is a result of a littoral-cell boundary, rather a change in the bedrock in the coastal drainage basins. Based on coastal geomorphology, the presence of beaches and known bathymetry, we now believe that the northern boundary of the Santa Cruz Littoral Cell extends at least to the mouth of San Francisco Bay, about 50 km upcoast (Fig. 1).

There are no major streams that drain this additional portion of coastline, but the entire area consists of near-vertical, high, eroding cliffs of the Purisima Formation to the south, granite at Montara Point, and young erodible Pleistocene sediments of the Merced Formation, as well as unconsolidated recent dunes in the San Francisco peninsula area. Smaller drainage basins, more subdued topography, and lower annual rainfall suggest that sediment yield from the streams north of the study area is less than those within the study area. Although some areas of this coastline are relatively stable, many sections are undergoing rapid retreat (Tinsley, 1972; Griggs and Savoy, 1985); this high-erosion rate combined with some large deep-seated coastal landslides suggests that coastal-bluff erosion plays a more important role in sediment input to the cell than the bluffs within the original study area. In addition, the gullies, which are prominent in southern San Mateo County south of Tunitas Creek, also extend northward to Half Moon Bay.

Preliminary estimates of additional littoral sediment input from coastal streams, gullies, and bluff erosion from Tunitas Creek to the mouth of San Francisco Bay amount to 20,000 to 45,000 m^3/yr (Table 5). This excludes sediment entering the cell either through or across the mouth of San Francisco Bay from the north. These are both unknown quantities at present.

As discussed earlier, on the basis of information from the Corps of Engineers dredging history (Doug Pirie, pers. commun., 1990), there is considerable fine-grained material (380,000 m^3 with a D_{50} of 0.2 mm) dredged from the shipping channel each year that appears to be moving into the northern end of the cell in the San Francisco peninsula area. Some of this material ends up in the dunes of the peninsula and some significant quantity may move southward alongshore under the prevailing longshore currents. This relatively fine-grained material may add significant volume to either the subaerial or offshore portion of the beach prism of the San Mateo and Santa Cruz County coastlines. There are locations along this coastline where large inputs of littoral-size sand should be apparent, however, and are not, which still leaves considerable uncertainty regarding the ultimate depositional site of this large volume of material. Detailed mineralogical and grain-size studies would be required in order to resolve this component of the budget.

Changes in storage may account for some of the remaining deficit. Excess sand in littoral transport may be stored temporarily along the inner shelf and also subaerially on the beach. A thinning of the beach and/or nearshore bottom-sediment accumulation can generate significant quantities of sediment. Quantifying these changes in storage is difficult without precise long-term surveys. A comparison of pocket beach elevations northwest of Santa Cruz indicates that the resolution of the time-sequential aerial photographs is inadequate to address fully the subtle changes in beach volume. Long-term changes in storage may also be masked by the seasonal on/offshore sediment exchange between the beach and nearshore bottom. This potential discrepancy, or deficit, can potentially be significant due to the large volume of material that can be generated by a relatively small change in either beach height or nearshore-sediment thickness over a large area. For example, the re-

TABLE 5.—SUMMARY OF THE AVERAGE ANNUAL SEDIMENT INPUT AND OUTPUT OVER THE PAST 50 YEARS TO THE SANTA CRUZ LITTORAL CELL BETWEEN THE SAN LORENZO RIVER AND TUNITAS CREEK

	Sediment Input (m^3/yr)
Coastal streams	43,000–87,000
Gully erosion	2,900–5,800
Bluff erosion	12,000–20,000
Dune deflation	1,700
On/offshore	0
Longshore transport from upcoast	
Streams, bluffs, gullies	20,000–45,000
San Francisco bar and bay	?
TOTAL	79,600–159,500
	Sediment Output (m^3/yr)
Longshore transport	200,000–250,000
TOTAL	200,000–250,000

moval of 1 cm from a stretch of shoreline 100 m wide and 130 km long (the total distance from the entrance of San Francisco Bay to the Santa Cruz Harbor) could generate 130,000 m^3 of sediment.

In addition, thinning of the beaches and a reduction in the amount of littoral sand in storage, in part, may have taken place in response to major changes in the Ano Nuevo Point area over the past 100 to 200 years. Weber (1979) suggests that prior to the 18th century, Ano Nuevo Island was attached to the mainland as a peninsula. This peninsula provided a large upcoast trap for sediment and is believed to have accumulated a 6- to 9-million-m^3 reservoir of beach and dune sand. The opening of the channel between the island and the present point some time in the mid-1700s allowed the trapped sand to move downcoast, creating temporarily wider beaches south of Ano Nuevo. There is some evidence in beach-width and shoreline-erosion trends (Weber, 1979) to suggest that this hypothesized pulse of sand now may be nearly gone, with subsequent thinning of the beaches. Thus, potential contribution from this large volume of sand in storage may have amounted to as much as 30,000 to 45,000 m^3/yr for this 200-year period.

CONCLUSIONS

1) Based on an analysis of coastal and nearshore geomorphology as well as the sediment budget, it is now apparent that the Santa Cruz Cell extends upcoast at least as far as the entrance to San Francisco Bay. Mineralogical differences in beach sand are not adequate in themselves to define the cell boundary. In the Santa Cruz Cell, sediment enters the littoral zone principally by coastal streams and seacliff erosion. Material is reworked in the surf zone; finer grained sediment is transported offshore and the coarser grained material (greater than the littoral cutoff diameter) remains in littoral transport. Sand is temporarily stored on beaches and along the inner shelf before being ultimately lost to Monterey Submarine Canyon.

2) A careful look at the grain size of both beach and offshore sand in the littoral cell reinforces the significance of a littoral cutoff diameter, a grain size below which the sediment cannot remain on the beach, but moves offshore. The littoral cutoff diameter within the Santa Cruz cell is about 0.18 mm.

3) Sediment input to the cell is highly episodic in response to large and infrequent erosional events. Considerable caution must be exercised in calculating littoral sand inputs from individual sources due to both temporal variations (e.g., the extremely episodic nature of stream sediment delivery) and spatial variations (e.g., the alongshore variations in cliff height, stratigraphy, grain-size distribution, and erosion rates).

4) Current littoral drift rates at the downcoast end of the study area are relatively well documented and range from 200,000 to 250,000 m^3/yr.

5) Calculated input volumes from all sources that supply the Santa Cruz Littoral Cell range from 80,000 to 160,000 m^3/yr. Additional material may enter the cell from the area off the entrance to San Francisco Bay.

6) Coastal streams supply about 75 percent of the total littoral sand to the cell, whereas bluff erosion contributes about 20 percent.

7) Those processes that deliver sand to a cell (bluff erosion and high-stream flow, for example) may operate at different frequencies than those that move sand through a cell (longshore transport).

8) Changes in sand storage, either on the beach or along the inner shelf, can potentially produce significant volumes of sand due to the large areas involved. The lack of precision in the methods of investigation makes these quantities elusive to determine, however. In the Santa Cruz Cell it appears that a significant portion of the current littoral drift may have been due to a reduction of sediment storage. Historically, this is apparent at Ano Nuevo Point, although a general thinning of all beaches may also be occurring.

ACKNOWLEDGMENTS

The California State Department of Boating and Waterways provided the funding for this project. The authors are grateful to Mark Reid and Robert Osborne, whose constructive comments greatly improved an early version of the manuscript.

REFERENCES

ANDERSON, R. G., 1971, Sand budget for Capitola Beach, California: [Unpublished M.S. Thesis], Naval Postgraduate School, Monterey, California, 57 p.

BOWEN, A. J., AND INMAN, D. L., 1966, Budget of littoral sands in the vicinity of Point Arguello, California: U.S. Army Coastal Engineering Research Center Technical Memorandum No. 19, 56 p.

BROWN, W. M., 1973, Erosion processes, fluvial-sediment transport, and reservoir sedimentation in part of Newell and Zayante Creek basins, Santa Cruz County, California: U.S. Geological Survey Open-File Report, 31 p.

BROWNLIE, W. R., AND TAYLOR, B. D., 1981, Coastal sediment delivery by major rivers in southern California: Sediment Management for Southern California Mountains, Coastal Plains and Shoreline, Part D, 314 p.

COLBY, B. R., 1956, Relationship of sediment discharge to streamflow: U.S. Geological Survey Open-File Report, 170 p.

CLARK, J. C., 1981, Stratigraphy, paleontology, and geology of the central Santa Cruz Mountains, California Coast Ranges: U.S. Geological Survey Professional Paper 1168, 51 p.

CURRY, ROBERT, HOUGHTON, ROBERT, KIDWELL, T., AND TANG, P., 1986, Sedimentologic and hydrologic analysis of the Pescadero marsh and its watershed: Unpublished Report for the State Department of Parks and Recreation, 42 p.

DIETRICH, W. E., DUNNE, TOM, HUMPHREY, N. F., AND REID, L. M., 1982, Construction of sediment budgets for drainage basin, *in* Sediment Budgets and Routing in Forested Drainage Basins: U.S. Forest Service, Pacific Northwest Forest and Range Experiment Station General Technical Report PNW-141, p. 5–23.

DOLAN, T. J., CASTENS, P. G., SONU, C. J., AND EGENSE, A. K., 1987, Review of sediment budget methology: Oceanside Littoral Cell, California, *in* Coastal Sediments '87: American Society of Civil Engineers, p. 1289–1304.

ELLEN, S. D., AND WIECZOREK, G. F., 1988, Landslides, floods and marine effects of the storm of January 3–5, 1982 in the San Francisco Bay region, California: U.S. Geological Survey Professional Paper 1434, 310 p.

GRIGGS, G. B., 1982, Flooding and slope failure during the January 1982 storm, Santa Cruz County, California: California Geology, v. 35, p. 158–163.

GRIGGS, G. B., 1987a, Littoral cells and harbor dredging along the California coast: Environmental Geology and Water Science, v. 10, no. 1, p. 7–20.

GRIGGS, G. B., 1987b, The production, transport, and delivery of coarse-grained sediment by California's coastal streams, *in* Coastal Sediments '87, American Society of Civil Engineers, p. 1825–1838.

GRIGGS, G. B., AND HEIN, J. R., 1980, Sources, dispersal, and clay mineral composition of fine-grained sediment off central and northern California: Journal of Geology, v. 88, p. 541–566.

GRIGGS, G. B., AND JOHNSON, R. E., 1976, The effect of the Santa Cruz small craft harbor on coastal processes in northern Monterey Bay, California: Environmental Geology, v. 1, p. 299–312.

GRIGGS, G. B., AND JOHNSON, R. E., 1979, Coastline erosion, Santa Cruz County: California Geology, v. 4, p. 67–76.

GRIGGS, G. B., AND PARIS, LANCE, 1982, Failure of flood control on the San Lorenzo River, Santa Cruz County, California: Environmental Management, v. 6, p. 407–419.

GRIGGS, G. B., AND SAVOY, L. E., eds., 1985, Living with the California Coast: Duke University Press, Durham, North Carolina, 393 p.

HABEL, J. S., AND ARMSTRONG, G. A., 1978, Assessment and atlas of shoreline erosion along the California coast: State of California, Department of Navigation and Ocean Development, 277 p.

HALES, L. Z., 1978, Coastal processes study of the Oceanside, California, Littoral Cell: U.S. Army Waterways Experiment Station, Miscellaneous Paper H-78-8, Vicksburg, Mississippi.

HICKS, D. M., 1985, Sand dispersion from an ephemeral delta on a wave-dominated coast: [Unpublished Ph.D. Dissertation], Earth Science Department, University of California, Santa Cruz, California, 210 p.

INMAN, D. L., 1976, Man's impact on the coastal zone: California Department of Navigation and Ocean Development, 150 p.

INMAN, D. L., AND FRAUTSCHY, J. D., 1966, Littoral processes and the development of shorelines, *in* Santa Barbara Specialty Conference on Coastal Engineering, p. 511–536.

INMAN, D. L., AND JENKINS, S. A., 1983, Oceanographic report for the Oceanside Beach Facilities: prepared for the City of Oceanside, California.

KOMAR, P. D., 1971, The mechanics of sand transport on beaches: Journal of Geophysical Research, v. 76, no. 3, p. 713–721.

KOMAR, P. D., 1976, Beach Processes and Sedimentation: Prentice-Hall, Inc., Englewood Cliffs, New Jersey, 429 p.

KOMAR, P. D., AND MILLER, M. C., 1973, The threshold of sediment movement under oscillatory water waves: Journal of Sedimentary Petrology, v. 43, p. 1101–1110.

KOMAR, P. D., AND MILLER, M. C., 1975, Sediment threshold under oscillatory waves: Proceedings, 14th Conference on Coastal Engineering, p. 756–775.

LAWSON, A. C., ed., 1908, California Earthquake of April 18, 1906: Report of the State Earthquake Investigation Commission, Carnegie Institution of Washington, Publication No. 87, v. I, part I, 451 p.

LEE, J., YANCEY, T. E., AND WILDE, P., 1970, Recent sediments of central California continental shelf—Pigeon Point to Sand Hill Bluffs: Part A. Introduction and grain size data: University of California, Berkeley, Hydraulic Engineering Laboratory, Technical Report HEL 2-28, 62 p.

LEHRE, A. K., 1981, Sediment budget of a small California Coast Range drainage basin near San Francisco: International Association of Hydrological Sciences Publication 132 (Christchurch, 1981), p. 123–139.

MACY, J. S., 1976, Sediment yield of the upper San Lorenzo watershed using quantitative geomorphic and sediment sampling approaches: [Unpublished M.S. Thesis], San Jose State University, San Jose, California, 127 p.

MOORE, J. T., 1972, A case history of Santa Cruz Harbor, California: University of California, Berkeley, Hydraulic Engineering Laboratory, Technical Report HEL-24-14, 42 p.

MORELOCK, JACK, 1987, Beach sand budget for western Puerto Rico, *in* Coastal Sediments '87, American Society of Civil Engineers, p. 1333–1345.

NILSEN, T. H., AND BRABB, E. E., eds., 1979, Geology of the Santa Cruz Mountains, California: Field Trip Guidebook for the Geological Society of America Cordilleran Section, San Jose, California, 97 p.

NOLAN, K. M., AND MARRON, D. C., 1988, Stream channel response to the storm in the Santa Cruz mountains, *in* Landslides, Floods and Marine Effects of the Storm of January 3–5, 1982 in the San Francisco Bay region, California: U.S. Geological Survey Professional Paper 1434, p. 245–264.

NORRIS, R. M., 1964, Dams and beach sand supply in southern California, *in* Miller, R. L., ed., Papers in Marine Geology: Macmillan, New York, p. 154–171.

ORADIWE, E. N., 1986, Sediment budget for Monterey Bay: [Unpublished M.S. Thesis], Naval Postgraduate School, Monterey, California, 110 p.

OSBORNE, R. H., FOGARTY, T. M., AND KUHN, G. G., 1989, A quantitative comparison of coarse-grained sediment yield from contributing cliffs and associated rivers: southern Orange and San Diego Counties, California (abst.): Geological Society of America Annual Meeting Program.

PLANT, N., AND GRIGGS, G. B., 1990, Coastal landslides caused by the October 17, 1989 earthquake, Santa Cruz County, California: California Geology, v. 43, no. 4, p. 73–96.

RANTZ, S. E., 1971, Precipitation depth-duration-frequency relations for the San Francisco Bay Region, California: U.S. Geological Survey Basic Data Contribution 25, 23 p.

SEYMOUR, R. J., DOMURAT, G. W., AND PIRIE, D. M., 1980, A sediment trapping experiment at Santa Cruz, California: Proceedings, 17th International Conference of Coastal Engineering, American Society of Civil Engineers, p. 1416–1435.

SILVESTER, ROBERT, AND MOGRIDGE, G. R., 1971, Reach of waves to the bed of the continental shelf: Proceedings, 12th Conference on Coastal Engineering, p. 651–667.

SMITH, T. C., HART, E. W., BALDWIN, J. E., AND RODRIGUES, R. J., 1982, Landslides and related storm damage, January 1982, San Francisco Bay region, California: California Geology, v. 35, p. 139–152.

STOW, D. A., AND CHANG, H. H., 1987, Coarse sediment delivery by coastal streams to the Oceanside Littoral Cell, California: Shore and Beach, v. 55, p. 30–40.

SUNAMURA, T., 1983, Processes of sea cliff and platform erosion, *in* Komar, P. D., ed., CRC Handbook of Coastal Processes and Erosion: CRC Press, Inc., Boca Raton, Florida.

SWANSON, M. L., 1983, Soil piping and gully erosion along coastal San Mateo County, California: [Unpublished M.S. Thesis], University of California, Santa Cruz, 141 p.

THOMPSON, J. K., 1981, Sediment grain size distribution in San Francisco Bay, California, January/February and August 1973: U.S. Geological Survey Open-File Report 81-1332, 33 p.

TINSLEY, J. C., 1972, Sea cliff retreat as a measure of coastal erosion, San Mateo County, California: [Unpublished Ph.D. Dissertation], Department of Geology, Stanford University, California, 88 p.

U.S. GEOLOGICAL SURVEY, 1931–1960, Surface water supply in the United States: Department of Interior, Water Supply Paper, published annually.

U.S. GEOLOGICAL SURVEY, 1963–1988, Water resource data for California, vol. 2: Department of Interior, U.S. Geological Survey Water-Data Report Series, CA-82 through 88-2, published annually.

WALKER, J. R., AND DUNHAM, J. W., 1978, Santa Cruz Harbor shoaling study: Report for San Francisco District, U.S. Army Corps of Engineers (pages not consecutively numbered).

WALKER, J. R., AND WILLIAMS, P. J., 1980, A phased-dredging program for Santa Cruz Harbor: Proceedings, 17th Conference on Coastal Engineering, American Society of Civil Engineers, p. 1493–1511.

WEBER, G. E., 1979, Vertical crustal movements near Point Año Nuevo, San Mateo County, possible cause for the recent stabilization of a dune field, *in* Weber, G. E., Lajoie, K. R., and Griggs, G. B., eds., Coastal Tectonics and Coastal Geologic Hazards in Santa Cruz and San Mateo Counties, California; Geological Society of America Cordilleran Section Field Trip Guide, 75th annual meeting, 187 p.

WEGGEL, J. R., AND CLARK, G. R., 1983, Sediment budget calculations—Oceanside, California: U.S. Army Waterways Experiment Station, Miscellaneous Paper CERC-83-7, Vicksburg, Mississippi, p.

WRIGHT, L. D., 1987, Shelf-surf zone coupling: Cross-shore transport mechanisms on the shore face: Coastal Sediments '87, American Society of Civil Engineers, p. 25–40.

YANCEY, T. E., ISSELHARDT, C., OSUCH, L., LEE, J., AND WILDE, P., 1970, Recent sediments of the central California shelf—Pillar Point to Pigeon Point: Part A. Introduction and grain size data: University of California, Berkeley, Hydraulic Engineering Laboratory, Technical Report HEL 2-26, 64 p.

YANCEY, T. E., AND LEE, J. W., 1972, Major heavy mineral assemblages and heavy mineral provinces of the central California coast region: Geological Society of America Bulletin, v. 83, p. 2099–2104.

FOURIER GRAIN-SHAPE ANALYSIS OF COASTAL AND INNER CONTINENTAL-SHELF SAND SAMPLES: OCEANSIDE LITTORAL CELL, SOUTHERN ORANGE AND SAN DIEGO COUNTIES, SOUTHERN CALIFORNIA

ROBERT H. OSBORNE AND CHIA-CHEN YEH

Department of Geological Sciences, University of Southern California, Los Angeles, California 90089-0740

ABSTRACT: The identification of sand sources and the volumetric contribution from each during a given time interval is essential to our understanding of coastal processes and the computation of meaningful sediment budgets for the littoral zone. Fourier grain-shape analysis (FGSA) of detrital quartz from the medium-sand fraction of 83 fluvial, seacliff, beach and inner continental-shelf samples was performed to identify and quantify the contributions from local sand sources within the Oceanside Littoral Cell, which occupies the coastal reach between Dana Point and Point La Jolla in southern Orange and San Diego Counties.

Differences in shoreline orientation, physiography and the occurrence of Oceanside Harbor may be used to divide the Oceanside Littoral Cell informally into five segments. These are, from northwest to southeast, the Capistrano subcell (Dana Point to San Mateo Point), the San Onofre subcell (San Mateo Point to the north breakwater at Oceanside Harbor), the Oceanside-Carlsbad subcell (south jetty at Oceanside Harbor to the onshore projection of Carlsbad Submarine Canyon), the Encinitas subcell (Carlsbad Submarine Canyon to La Jolla Submarine Canyon), and the La Jolla Cove subcell (La Jolla Submarine Canyon to Point La Jolla). FGSA demonstrates that each of these subcells is compositionally distinct with regard to grain-shape populations, which, in part, reflects a characteristic set of local sources supplying each subcell.

Although shape-compositional differences occur between and among the five subcells, the shape composition within each cell is rather uniform with the possible exception of the Santa Margarita River sand in the San Onofre subcell. This homogenization may reflect sediment mixing as a result of seasonally related, bidirectional, longshore and onshore transport during a relatively recent but undefined time interval.

With the probable exception between the San Onofre and Oceanside-Carlsbad subcells since 1942, an appreciable amount of sand has been exchanged naturally between each pair of adjacent subcells. Although the reason(s) is (are) not clear, the northernmost two subcells are dominated by downcoast and southern shelf sources associated with each subcell, the central subcell appears somewhat transitional with regard to upcoast and downcoast sources, and the southernmost two subcells are dominated by upcoast and northern shelf sources. Continued monitoring is required to estimate the volumes involved and the rates of exchange, which will determine the effectiveness of the subcell boundaries as barriers to littoral-sand transport.

An average of approximately 36 percent of the quartz in the medium-sand fraction of the foreshore samples was derived from the inner continental shelf. Although it seems likely that storm activity, perhaps coupled with shoreward transport through ridge and runnel systems, may be involved, fundamental questions concerning the mechanism, volumetric aspects, rates of transport and transport pathways remain unanswered.

Weighted local source estimates supplied by FGSA, grain-size and mineralogic analyses must be integrated with mass-balance computations to improve sediment budget analysis and to constrain sand-transport pathways to and within the littoral zone.

INTRODUCTION

Emery (1960) and Inman and Chamberlain (1960) identified a series of littoral cells along the southern California coast. These cells are based on the concept of longshore transport of dominantly fluvially derived sediment, which is entrapped either by submarine-canyon heads or by points of land that extend seaward from the general position of the coastline. The 86.3-km-long Oceanside Littoral Cell extends from Dana Point south to Point La Jolla (Fig. 1). This is a complex coastal reach backed by cliffs and bluffs in the San Onofre, Camp Pendleton and La Jolla areas (Fig. 2). It contains several major rivers (San Juan Creek, San Mateo Creek, San Onofre Creek, Santa Margarita River, San Luis Rey River and San Dieguito River), numerous coastal streams, and several major lagoons (Buena Vista, Agua Hedionda, Batiquitos, San Elijo, and Los Penasquitos). The Del Mar Boat Basin and Oceanside Harbor constitute a major small-craft harbor in the Oceanside area (Fig. 2).

This coastal reach has experienced a long history of shoreline erosion (Shaw, 1980, p. 22–36) and suffered $116 million in damages during the severe winter storms of 1982–1983. The 100- to 200-year storm of January 16–18, 1988 (Seymour, 1989a, b) resulted in damages and clean-up costs of approximately $6,700,000 in San Diego County (Armstrong and Flick, 1989). Much of this expense was for damage in the Oceanside Littoral Cell. In addition to such natural disasters, the harbor facility at Oceanside has altered the natural sediment-transport system significantly since its initial construction in 1942 (Hales, 1978, p. 17–21). Recently, concern has been expressed regarding the impact of the nuclear generating station at San Onofre on the distribution of sediment on the adjacent coast and inner continental shelf (Schroeter and others, 1989; Sediment Dynamics Group, 1989). In addition, significant subaerial erosion has occurred along the coastal cliffs and bluffs at San Onofre, Camp Pendleton, Carlsbad, Leucadia, Cardiff-by-the-Sea, Solana Beach, Del Mar and La Jolla (Kuhn and Shepard, 1984; Kuhn and others, 1987; Osborne and others, 1988). There also are several areas, particularly near Del Mar, with a history of major landslide activity (Kuhn and Shepard, 1984).

It is clear that an accurate sediment budget, based on a comprehensive, long-term study of shoreline changes and the quantification of the factors responsible for such changes, is required to provide a basis for sound scientific, engineering and management decisions affecting the Oceanside Littoral Cell. A sediment budget is a conceptual framework that mathematically describes the transport of sand and coarser grained sediment into, within, and from a defined coastal reach during a specified time interval.

Natural and artificial sediment sources and sinks for the littoral zone are shown schematically in Figure 3. Natural

From Shoreline to Abyss, SEPM Special Publication No. 46

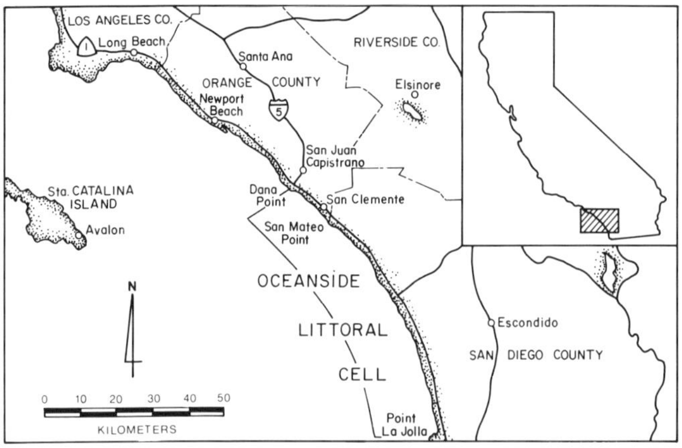

FIG. 1.—Map showing the location of the Oceanside Littoral Cell.

sediment sources include river discharge; cliff, bluff and dune erosion; wind and overwash transport; longshore transport; and onshore transport from the inner continental shelf. Natural sediment sinks include losses to submarine canyons and offshore transport to the inner shelf. Beach nourishment (artificial source) and sand mining (artificial sink) may be important locally. Factors affecting sediment transport within a defined coastal reach include the nearshore wave climate, nearshore currents, tides, sediment characteristics, bathymetry, wind and wave overwash, engineered structures, and sea-level fluctuations.

The principal objective of a sediment budget is to predict quantitatively how the previously identified factors will interact to alter the quantity and distribution of sediment within a given coastal reach (the control volume) from both short- and long-term perspectives. Theoretically, the use of an appropriate budget would make it possible to assess the potential impacts of short-term episodic events such as major storms and floods, of intermediate events such as dam and harbor construction, and of long-term events such as sea-level fluctuations and tectonism.

A sediment budget normally is constructed from an analysis of historical records of shoreline position and associated bathymetric changes, an analysis of historical records of flooding and sediment delivery to the shorezone, and an analysis of the interaction of the shoreline and forces acting on it. Historical data are used to establish long-term trends to determine associations between shoreline change and the factors affecting the shoreline, and to calibrate an appropriate mathematical model. Usually, short-term field observations are integrated into such an analysis to confirm and/or constrain relation suggested or established by historical analysis, as well as to define more accurately conditions at the time the budget was formulated.

The essence of sediment budget analysis is to account for each source that increases and each sink that decreases the volume of sediment in transport or temporary storage within the defined coastal reach (the control volume) during a specified time interval, which is usually measured in decades. The sediment contributions and losses are considered so that the mass-conservation equation is satisfied:

Sum of volumetric gains from sources
= Sum of volumetric losses to sinks

A topographic feature of the beach or nearshore within the defined coastal reach may serve as an "interior" source or sink, depending upon whether it is eroding or accreting, respectively. Such "interior" terms, as well as "exterior" terms (outside of the control volume of the defined coastal reach), must be considered to balance the preceding equation.

Although the objectives of sediment budget analysis, including numerical modeling, are of the utmost importance in coastal research, state-of-the-art estimates of the sediment volumes and transport rates associated with each source and sink, as well as the use of different computational procedures to derive such estimates, may produce quite different results. For example, after comparing the estimates for fluvial coarse-grained sediment delivery in the Oceanside area reported in 12 recent studies, Stow and Chang (1987) concluded that the results are far from conclusive. They attributed this condition to the sparsity of estimates, assumptions and inaccuracies associated with the methodologies employed, as well as a general failure to account for river mouth/deltaic processes following major runoff events.

Likewise, after reviewing existing data, literature, sediment budget methodologies and five recent budgets for the Oceanside area (Nordstrom and Inman, 1973; Inman, 1976; Hales, 1978; Inman and Jenkins, 1983; and Weggel and Clark, 1983), as well as proposing a new preliminary budget, Tekmarine, Inc. (1987) concluded that at present, it is impossible to compute a definitive budget due to uncertainties surrounding key elements of the budget. The most recent sediment budget for the Oceanside Littoral Cell was computed by Moffatt & Nichol, Engineers (1989). Appendix L in Moffat & Nichol, Engineers (1989) summarizes each of the previous budgets for the Oceanside Harbor area and Oceanside Littoral Cell.

There are several key problems in computing accurate sediment budgets that require resolution. One is the fact that conventional monitoring of changes in seafloor topography (profiling) in the littoral zone does not in itself serve to define uniquely the volumetric contributions of the sources or the pathways of the sand in transport. Another key problem is accurately estimating the volume of sediment delivered and the exchange rate associated with the cross-shore vector. A third is the difficulty of incorporating the effects of intense, natural episodic events such as major storms and floods, which contrast markedly with the relative consistency of fairweather processes. Paradoxically, in southern California the large storm waves, which may cause significant coastal erosion and structural damage, often are associated with large discharges of sediment and water to the shorezone. Such events are likely to cause major adjustments within an affected coastal reach. Thus, regardless of the conceptual complexity and mathematical sophistication of existing sediment budget and numerical-modeling procedures, fundamental questions remain concerning the vol-

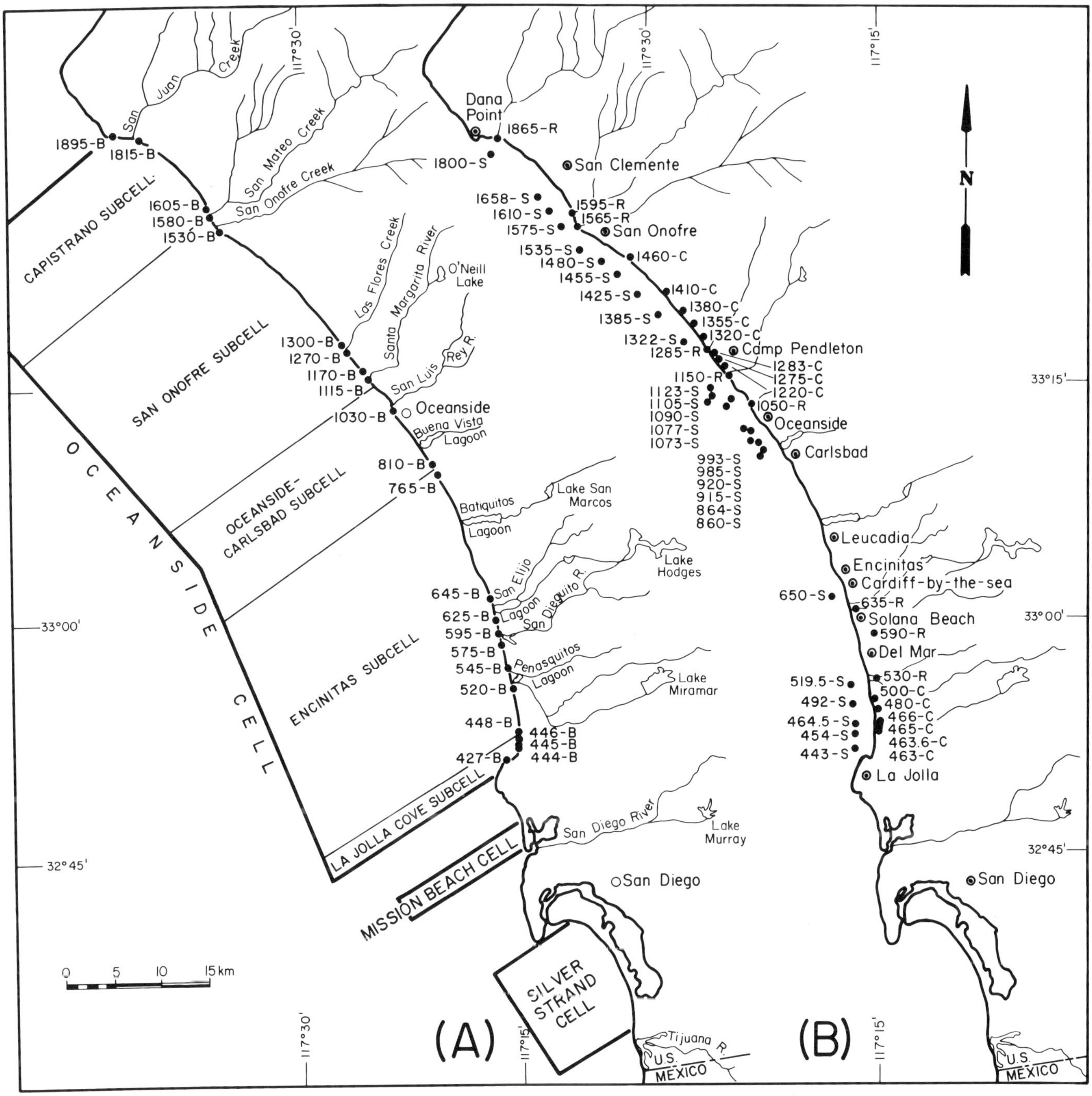

FIG. 2.—Maps showing the principal geographic features, littoral subdivisions and sample locations from Dana Point to the United States-Mexico border. The three- to four-digit number associated with each sample locality refers to a range line system defined by the U.S. Army Corps of Engineers. The range line values represent the distance in miles along the coastline north of the United States-Mexico border. A decimal point is understood to precede the last digit so that range line 1580 is 158.0 miles north of the border. The letter S, B, R and C following each range line refers to an inner continental shelf, beach, river or cliff sample, respectively. (A) Locations of beach samples; (B) locations of shelf, river and cliff samples.

umetric aspects, transport pathways and rates associated with littoral-sediment sources and sinks. Perhaps such questions may be addressed best by observational data derived from relatively long-term monitoring experiments.

The purpose of this paper is to investigate the shape composition of detrital-quartz grains from the medium-sand fraction of beach samples from the Oceanside Littoral Cell to test for the possible existence of subdivisions (subcells) within this cell, to estimate the volume percent of medium sand derived from each local source for each subcell, and to estimate the volume of sand exchanged between adjacent subcells. Such information will greatly improve the accu-

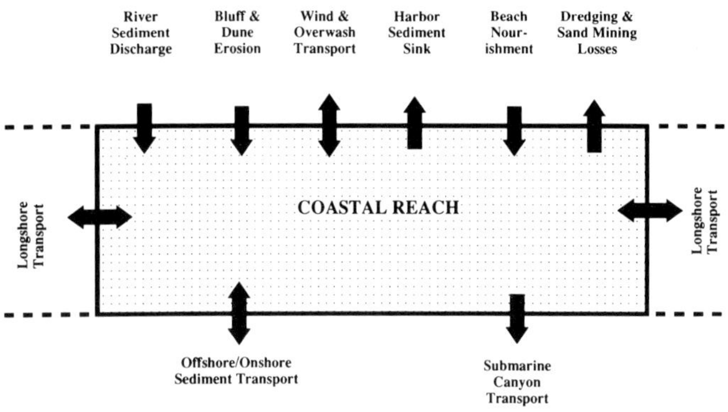

FIG. 3.—Schematic diagram showing the conceptual sediment sources and sinks as well as idealized transport pathways for beach sand in a defined coastal reach (after Fig. 8, U.S. Army Corps of Engineers, 1987, p. 7).

racy of future sediment budgets for the Oceanside Littoral Cell.

SEDIMENT SOURCES

Ultimate Sediment Sources

The results of mineralogic analyses reported by Osborne and others (1983), Darigo (1984), Osborne (1985), Darigo and Osborne (1986), and Osborne and others (1990) indicate that the siliciclastic fraction of the river, cliff, beach and shelf samples from the Oceanside Littoral Cell ultimately were derived from the Cretaceous Woodson Mountain Granodiorite and Bonsall Tonalite of the Southern California batholith (Larsen, 1948) and the Catalina Schist terrane (potassium-argon ages from 95–109 Ma; Suppe and Armstrong, 1972). Inasmuch as the major rivers entering the Oceanside Littoral Cell head in the rocks of the Southern California batholith, it is hardly surprising to identify this pluton as the major source terrane.

Although chlorite may be derived from rocks assigned to the Southern California batholith, the occurrence of glaucophane schist, glaucophane and actinolite-tremolite in the sand fraction of the observed samples reflects ultimate derivation from the Catalina Schist terrane. Although compositionally diverse, exposures of this terrane at Santa Catalina Island and Palos Verdes Hills consist dominantly of albite-rich greenschist, amphibolite and altered mafic plutonic rocks that are associated with a glaucophane-rich, blueschist terrane (Sorensen, 1984). The principal source for such grains in the Oceanside Littoral Cell is the early to middle Miocene San Onofre Breccia (Stuart, 1979), which crops out as a continuous ridge about 3 km wide from approximately 0.75 to 3 km inland of the shoreline from San Mateo Creek to the San Luis Rey River. The San Onofre Breccia contains pebbles, cobbles and boulders derived from the Catalina Schist terrane and was deposited as a sequence of alluvial-fan, fluvial and transitional marine units in this area (Stuart, 1979). Blueschist grains from the San Onofre Breccia also have been resedimented into younger formations and terraces along this coastal strip (Osborne and others, 1990) and may be delivered to the shorezone by gullying and sheetwash where no river drainage is established. Although there is no doubt that some part of the sand fraction in the observed samples was derived from the Catalina Schist terrane via the San Onofre Breccia, it is of interest to determine whether or not a significant volume of the medium-sand size quartz grains, on which this study is based, was derived from this source rock.

This question may be addressed using the detrital quartz-grain classification proposed by Basu and others (1975). Using this classification, Osborne and Pipkin (1982) indicated that secondary strain increased the percentage of undulose monocrystalline quartz, which caused the 51 samples examined to plot erroneously in the low-rank metamorphic source rock field. There is, however, another way to use these data. The strained and unstrained monocrystalline quartz may be combined to form a new variable, total monocrystalline quartz. The percentages of total monocrystalline quartz, total polycrystalline quartz with 2 to 3 subunits and total polycrystalline quartz with >3 subunits may be compared with the average percentages for acid plutonic, high-rank metamorphic and low-rank metamorphic source rocks provided in the histogram of Basu and others (1975, their Fig. 8). The results of this procedure are shown in Table 1, which clearly shows the close affinity of the Oceanside sample set with the acid plutonic source rock field. Therefore, it appears that little medium-sand size quartz was derived from rocks of the Catalina Schist terrane.

Local Sediment Sources

Although information concerning ultimate source rocks is sedimentologically important, the thrust of this paper is to estimate the percentage of medium sand contributed from potential local sources to the foreshore of a set of subcells defined within the Oceanside Littoral Cell. The subcells considered are presented in the next section of this paper. Potential local sources for the subcells include upcoast and downcoast beaches (longshore contributions), major rivers, sea cliffs and bluffs, and the inner continental shelf (cross-shore contribution).

INITIAL DIVISION OF THE OCEANSIDE LITTORAL CELL

Although the littoral-cell concept is appealing, it has never been demonstrated that sand from a given local source actually follows the suggested pathway through a substantive part of a given cell. This, in turn, gives rise to the possibility of littoral-cell segmentation. The distribution of sed-

TABLE 1.—AVERAGE PERCENTAGES OF DETRITAL QUARTZ TYPES FROM MAJOR SOURCE TERRANES (BASU AND OTHERS, 1975) AND AN OCEANSIDE SAMPLE SET (OSBORNE AND PIPKIN, 1982)

Type of Detrital Quartz Graian	Acid Plutonic (%)	High-rank Metamorphic (%)	Low-rank Metamorphic (%)	Oceanside Sample Set (%)
Total Monocrystalline	89	73	45	90
Nonundulatory	84	56	23	27
Undulatory	4	16	22	63
Total 2–3 Unit Polycrystalline	8	22	30	5
Total > 3 Unit Polycrystalline	4	5	25	5

Underlined values may be used for source rock identificataion.

iment and energy impinging along a given shoreline may be affected by, among other factors, the orientation and physiography of that coastal reach as well as by the presence of engineered structures. From northwest (upcoast) to southeast (downcoast), the occurrence of Dana Point, San Mateo Point, the north breakwater at Oceanside Harbor, Carlsbad Submarine Canyon, Scripps Submarine Canyon and Point La Jolla makes it possible to divide the Oceanside Littoral Cell initially into five subcells (Fig. 2). These are named the Capistrano subcell (Dana Point to San Mateo Point), the San Onofre subcell (San Mateo Point to the north breakwater at Oceanside Harbor), the Oceanside-Carlsbad subcell (Oceanside Harbor to the onshore projection of Carlsbad Submarine Canyon), the Encinitas subcell (from the onshore projection of Carlsbad Submarine Canyon to La Jolla Submarine Canyon), and the La Jolla Cove subcell (La Jolla Submarine Canyon to Point La Jolla) (Osborne and others, 1990). The average shoreline orientation and average slope from 0 to −18 m (−60 ft) MLLW for each of the five subcells considered are listed in Table 2.

FOURIER GRAIN-SHAPE ANALYSIS OF SAND SAMPLES

Introduction

Since its introduction by Ehrlich and Weinberg in 1970, Fourier analysis of quartz-grain shapes has been applied to a wide variety of natural tracer and petrofacies problems. The interested reader is referred to the following papers: Brown and others (1980), Ehrlich and others (1974), Ehrlich and Chin (1980), Ehrlich and others (1980), Hudson and Ehrlich (1980), Mazzullo and Ehrlich (1980, 1983), Mrakovitch and others (1976), Riester and others (1982), Van Nieuwenhuise and others (1978), and Mazzullo and Magenheimer (1987). Fourier grain-shape studies completed in California include Ehrlich and others (1974), Porter and others (1979), Bloom (1979), Clark and Osborne (1982), Gaynor (1984), Osborne and others (1985), Bomer (1985), Ahlschwede (1988), Broadhead (1988), Cho (1989), and Osborne and others (1990).

Ehrlich and Weinberg (1970) describe a closed-form Fourier method to analyze the observed variation of two-dimensional, maximum-projection, grain-shape area. Grain shape may be estimated by an expansion of the periphery radius as a function of angle about the grain's center of gravity by a Fourier series. In Fourier analysis, a series of sine and cosine curves with periods equal to fundamental harmonics is fit to the observed data by a least-squares technique. Fundamental harmonics are the prime fractions (1/2, 1/3, 1/4, . . . 1/*N*), where *N* equals half the number of digitized points used to define the periphery of a grain. As the number of fundamental harmonics is increased, the computed curve converges with the observed data. The highest frequency that can be estimated is the Nyquist frequency, which is equal to twice the distance between successive observations. If the Nyquist frequency is exceeded, error may be introduced by the incorporation of irresolvable high frequencies into lower frequencies (aliasing). The radius is given by:

$$R(\theta) = Ro + \sum_{n=1}^{\infty} Rn \cos(n\theta - \varnothing n)$$

where θ is the polar angle measured from an arbitrary reference line. The first term in the series *Ro* is equivalent to the average radius of the grain in the maximum projection orientation. For the remainder of the terms, *n* is the harmonic order, *Rn* is the harmonic amplitude, and ∅*n* is the phase angle. The phase angle appears to provide little additional grain-shape information, and therefore is not considered further. It is important to note that the *n*'th harmonic contributes to the explanation of the observed shape variation as a figure with *n* "bumps." For example, the "zeroth" harmonic is a centered circle with an area equal to that of the maximum projection; the first harmonic is an off-centered circle; the second is a figure eight; the third is a trefoil; etc. The center of gravity of the maximum-projection shape is used as the origin of the radius expansion to simplify interpretation of the Fourier series. Coordinates of points along the periphery of the maximum-projection outline are required for the Fourier expansion. At least twice the number of such points must be known as the number of the highest desired harmonic. The initial origin of the periphery points may be arbitrary, because a later transformation places the origin at the center of gravity of the maximum grain-projection area. If a harmonic or periodic function exists within the data, the amplitude of the sine and cosine curves with periods close to the natural harmonic will be considerably larger than the amplitudes of other harmonics in the sequence.

Although conceptually similar to the closed-form methodology described by Ehrlich and Weinberg (1970), the methodology employed in this report makes use of the newer and more widely used Fast Fourier Transform (FFT). This procedure involves the calculation of many values of the line spectrum using the FFT computer algorithm to produce a smoothed estimate of the continuous spectrum. The FFT algorithm, as its name implies, is extremely rapid and requires only $n \cdot \log_2 n$ arithmetic operations rather than the n^2 operations as do alternative methods. The reader is referred to Brigham (1974), Bloomfield (1976), and Bendat and Piersol (1971) for extensive treatments of the mathematically complex FFT.

TABLE 2.—AVERAGE SHORELINE ORIENTATION AND AVERAGE SLOPE FROM 0 TO −18 M (−60 FT) MLLW FOR EACH OF THE FIVE SUBCELLS CONSIDERED

Subcell	Trend	Average Slope (0 to −18 m MLLW) m/km	Degrees
Capistrano	N42°W	5.9	0.34
San Onofre	N45°W	6.0	0.34
Oceanside-Carlsbad	N39°W	11.6	0.66
Encinitas	N12°W	18.0	1.03
La Jolla Cove	N28°E	40.2	2.30

Sample Collection

This study is based on the Fourier grain-shape analysis (FGSA) of 83 sand samples collected from the major rivers, sea cliffs and beaches in the Oceanside Littoral Cell as well as the adjacent inner continental shelf. The location of each sample is shown in Figure 2. All of the river and virtually all of the cliff samples used were collected during April 3 to 5 and June 1986, respectively. The river samples were obtained from bars in active channels, and were collected at least 2 km upstream of the river mouths to avoid the possible effects of tidal currents on grain size, grain shape and mineral composition. The tidal effects on the Santa Margarita River are so far reaching that it was necessary to collect this sample about 5 km upstream of the mouth. The exact geographic and stratigraphic positions of the seacliff and bluff samples collected in the San Onofre, Camp Pendleton and Torrey Pines areas (Fig. 2) are given in Kuhn and others (1987). The first beach-sample set was collected from April 3 to 5, 1986, and the second set on October 8, 1986. The April set approximates the end of the oceanographic winter in southern California, and the October set approximates the end of the oceanographic summer. The beach samples were collected at MLLW at locations about 1 km northwest and southeast of each major river mouth, as well as at other critical locations along the shoreline. Beach samples 765-B and 810-B were collected in February 1989 to provide additional information concerning downcoast transport for the Oceanside-Carlsbad subcell and upcoast transport for the Encinitas subcell, respectively. The La Jolla Cove subcell was defined after the primary sampling program in 1986; therefore, the beach samples used for this segment were collected in September 1989. The beach samples were obtained by inserting a tin can normal to the shoreface to a depth of about 15 cm, which produced a small core with a volume of approximately 500 cm^3.

Inner continental-shelf samples were included in this study to understand the role of cross-shore sand exchange better. The 17 samples with range lines from 443-S through 1123-S (Fig. 2) were obtained from vibracores recovered during January and February 1981 at water depths ranging from -11 to -21 m (-37 to -70 ft) MLLW. Core descriptions as well as the results of grain-size, petrographic and Fourier grain-shape analyses on samples obtained from these cores are discussed by Osborne and others (1983), Darigo (1984), Darigo and Osborne (1986), Cho (1989), Osborne and Cho (1989), and Osborne and others (1990). The 10 shelf samples with range lines from 1322-S through 1800-S (Fig. 2) were collected north of Oceanside Harbor in February 1989 at a water depth of -9 m (-30 ft) MLLW. Depending upon the closure depth used for the Oceanside Littoral Cell, these samples may be considered either lowermost foreshore or innermost shelf.

Sample Preparation

Quartz grains from the 0.25- to 0.50-mm (1.0–2.0 phi) or medium-sand fraction from each sample were used for the Fourier grain-shape analysis. Samples were first wet-sieved to obtain the desired size fraction to minimize attrition of the quartz grains during sieving. Sieving was followed by drying in convection ovens. After drying, all strongly magnetic minerals were removed from the samples using a hand magnet. The samples were boiled in a solution of 10 percent hydrochloric acid with stannous chloride crystals for 20 min to remove iron-oxide coatings and carbonate cement (Carver, 1971). The hydrochloric-acid residue was removed by rinsing the samples in de-ionized water. The quartz grains were etched clean by washing in a hydrofluoric-acid bath for 1 min. Such etching did not significantly alter the shape of the grains (Schultz, 1980). All samples were then rinsed in de-ionized water and dried thoroughly.

Samples were examined under a petrographic microscope using a reflecting light source. Quartz grains were picked by using a very small brush and were dry mounted on a glass slide, which enhanced the contrast between the grain and its background (Ehrlich and Weinberg, 1970). Adequate contrast was essential in the operation of the Fourier grain-shape digitizing program.

Fourier Methodology

Digitizing of the individual quartz grains was accomplished using an optical digitizer and the EDGE Fortran program. In this process, a black-and-white camera projected the maximum two-dimensional image of the quartz grain from a standard binocular petrographic microscope onto a video screen. With lighting from below, the image of the grain appeared dark in relation to the surrounding background. A maximum contrast between grain edge and background allowed the computer to scan the pixels of the video image and record all boundary points of the grain as Cartesian coordinates (X–Y coordinates). A pixel was determined as a boundary point if it was dark and one of its neighbors was light. While the Cartesian coordinates were computed, the centroid of the projected grain also was computed. The data were stored on floppy disk and transferred from the microcomputer to a Digital Equipment Corporation VAX-11/750 computer.

To prepare the data for the Fourier transform, the Cartesian coordinates were converted to polar coordinates (radius versus angle). The radius is the distance from a boundary point to the centroid; the angle used is the angle from the horizontal line bisecting the centroid to the boundary point. In order to examine the entire wave-number amplitude spectrum (each wave number is analogous to one harmonic) of a grain during statistical analysis, the Fast Fourier Transform (Brigham, 1974) was used. One requirement of the Fast Fourier Transform is that the data be evenly spaced. Since the boundary points of a grain are not evenly spaced, the data must be linearly interpolated. Here, the boundary points were interpolated to 128 evenly spaced points. The Fast Fourier Transform also requires that the data set be centered around zero. To center the data, the mean radius is subtracted from the radius versus angle data. This radius is later placed in the final amplitude spectrum set.

Upon completion of the Fast Fourier Transform, the amplitude of each wave number was calculated by the equation:

$$SQRT(\text{Real}^{**}2 + \text{Imaginary}^{**}2)$$

where *SQRT* = square root, Real = real number, Imaginary = imaginary number, and **2 = square of the number (Brigham, 1974). This computation allows the amplitude spectrum of an individual grain to be compared to the amplitude spectrum of other grains. This comparison is the first step in the analysis process and directs the path the statistical analyses will follow. Since the input data are real rather than imaginary, half of the wave numbers are reflections and are not used in the statistical analyses. This potentially leaves 64 wave numbers for analysis. Each wave number is represented by a vector and the occurrence of 64 wave numbers results in the analysis of 64 dimensions in space.

Discriminant-Function Analysis

Discriminant-function analysis (DFA) is employed as the most appropriate multivariate statistical procedure to estimate the percentage of each beach sample derived from various sand sources within each littoral subcell. DFA is one of the most powerful statistical procedures available to assign or partition samples into previously defined populations. In the present study, interest is focused on assigning each of the 215 digitized grains in each beach sample to one of the sand sources identified for a given littoral subcell.

Mathematically, DFA consists of computing a transform, which gives the minimum ratio of the difference between or among a set of multivariate means to the multivariate dispersion (variance) of the set of previously defined groups. For example, if one considers only two groups as consisting of two clusters of points in multivariate space, DFA computes the equation of the one orientation along which the two clusters have the greatest spatial separation, while simultaneously each cluster has the least dispersion (variance). In this manner, the spatial separation of the two clusters is maximized, and is, therefore, most distinct. Once the transform is computed, new samples easily may be classified into one of these two groups. Davis (1986, p. 478–502) provides a most readable summary of the mathematics associated with DFA.

In the present study, the International Mathematical Subroutine Library DFA subroutine was used to compute the required transform for the harmonic-amplitude values of samples assigned to each sand source within a given littoral division. Thus, the multivariate spatial separation among the sand sources within a division is maximized, while the dispersion for each source is minimized. Once the discriminatory equations were computed, the harmonic-amplitude values for each of the 215 grains per beach sample within that cell or subcell were analyzed, and each grain was assigned to one of the defined sources. The number of grains in each beach sample assigned to a given source within that cell or subcell was computed and expressed as a percentage.

When DFA is used initially to assign the percentages of grains in a beach sample to a given source, the assignment assumes that each potential source is 100 percent unique with respect to grain shape. In reality, each potential source is a mixture of grain-shape components; therefore, the initial assignment must be recast to reflect the unique grain-shape components characteristic of that source. All resultant end-member sets are stochastically distinct.

RESULTS AND DISCUSSION

Capistrano Subcell

The results of FGSA for the three April 1986 beach samples examined in the Capistrano subcell are listed in Table 3A and illustrated in Figures 4A and 5. The average sand contributions from the five local sources supplying the Capistrano subcell are as follows: San Juan Creek, 12 percent; northern shelf, 8 percent; southern shelf, 12 percent; San Mateo Creek, 28 percent; and downcoast beach (San Onofre subcell), 40 percent.

The results, shown in Figure 4A, indicate a very high degree of compositional consistency with regard to grain shape within this subcell. The values from 26 to 31 percent, indicating the volumetric contribution from San Mateo Creek, show the greatest difference, which is only 5 percent. The 1- to 2-percent differences observed for the four other local sand sources are well within the range of sampling and analytical error.

These results indicate three major sedimentologic relations. First, the compositional homogeneity within this subcell suggests thorough mixing of the sand at each sample site derived from each of the five local sources. Such thorough mixing probably results from cross-shore and bidirectional longshore transport within the Capistrano subcell during a relatively recent, but unknown, period of time.

The second major relation is the compositional dominance of sand derived from the downcoast portions of the subcell. The observed dominance of sand from the downcoast portions of this subcell reflects the net effect of incremental, relatively long-term sand accumulation. Any bedload discharge from San Juan Creek (Simon & Associates, Inc., 1988) has been greatly diluted by and thoroughly mixed with sand from downcoast sand sources, which suggests either 1) little sand has been derived from San Juan Creek during the residence time represented by the beach samples analyzed, or 2) a tremendous volume of downcoast sand was transported northward during this time interval. The lack of a delta at the mouth of San Juan Creek and the fact that Dana Point Harbor requires little maintenance dredging suggest that little sand has been contributed from San Juan Creek in the recent past. Therefore, it is more likely that the sand derived from the downcoast portions of the subcell has been added incrementally. The dominance of sand from the downcoast portions of this subcell, coupled with probable net longshore transport to the south (Tekmarine, Inc., 1987), argues for a relatively small bedload contribution from San Juan Creek.

The third major relation is the occurrence of approximately 20 percent inner-shelf sand in the beach samples. It is clear that some net onshore transport has occurred at both the northern and southern parts of this subcell. It is interesting to note that the average sand contribution from the southern shelf (12 percent) is greater than that of the northern shelf (8 percent), which supports the concept of a

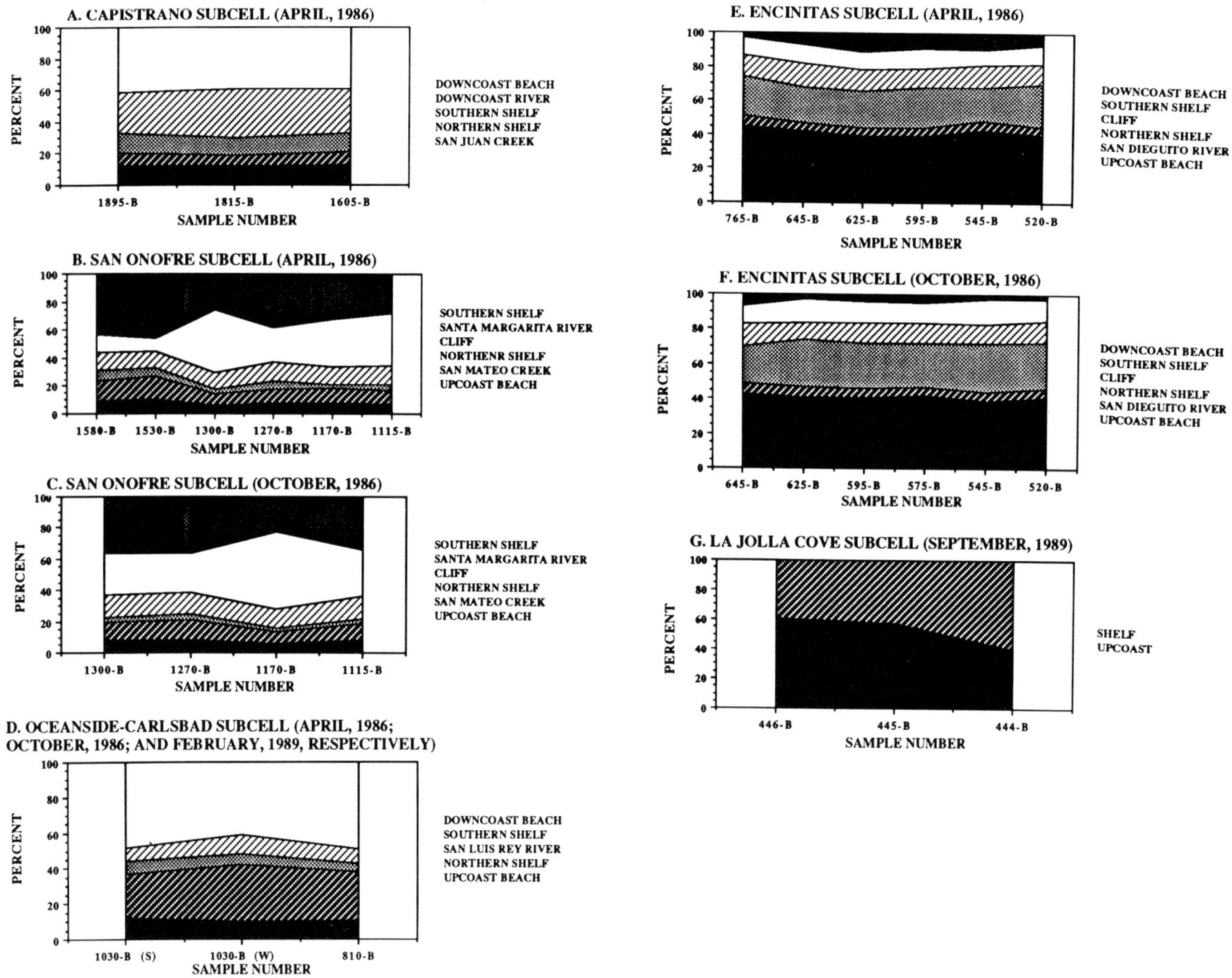

FIG. 4.—Area plots based on the data listed in Table 3. The patterns in each diagram are in the same stratigraphic order as the sources listed along the right margin. For example, in (A) the percentage of medium-grained quartz from San Juan Creek is shown in solid black, whereas the percentage from the downcoast beach is shown in solid white.

downcoast source for most of the sand in the Capistrano subcell.

San Onofre Subcell

April 1986 Beach Samples.—

The grain-shape composition for the six April 1986 beach samples from the San Onofre subcell displays more grain-shape variation than was observed in the Capistrano subcell (Table 3B and Figs. 4B and 5). The average sand contributions from the six local sources supplying the San Onofre subcell are as follows: upcoast beach (Capistrano subcell), 8 percent; San Mateo Creek, 12 percent; northern shelf, 5 percent; cliff, 13 percent; Santa Margarita River, 27 percent; and southern shelf, 35 percent.

The range of values for the upcoast beaches (4 percent), northern shelf (5 percent) and cliff (2 percent) contributions (Fig. 4B) is quite similar; therefore, little genetic significance can be assigned to the observed longshore variations for these sand sources. However, interesting longshore trends do occur for the contributions from San Mateo Creek, Santa Margarita River and the southern shelf.

The sand contribution from San Mateo Creek averages 16 percent for station 1850-B and 1530-B. This decrease to the south may represent either downcoast transport or dilution of San Mateo River sand by upcoast sand transport from the Santa Margarita River (station 1150) or southern shelf. Table 3B and Figure 4B indicate that approximately 62 percent of the beach sand in the San Onofre subcell is derived from the Santa Margarita River (27 percent) and southern shelf (35 percent). Ignoring the prominent peak at station 1300-B for the moment, the contribution of Santa Margarita River sand decreases from approximately 32 percent south of 1300-B to 11 percent north of 1300-B, which implies net upcoast transport. The peak of Santa Margarita River-like sand at 1300-B is intriguing and may represent bedload contribution from Las Flores Creek or, more likely,

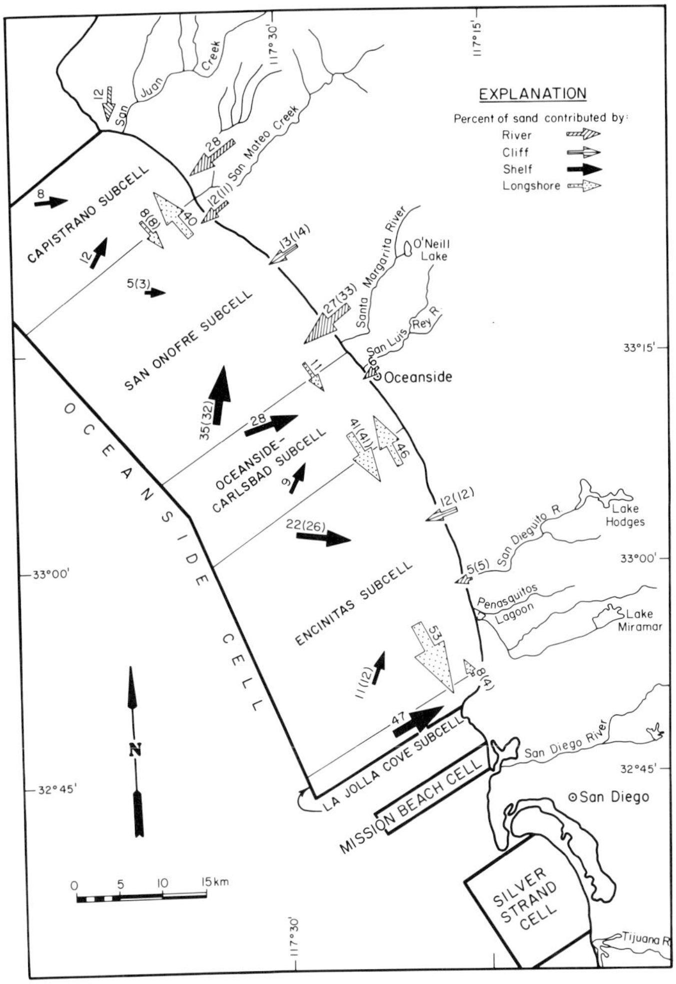

FIG. 5.—Map showing the percentage of medium sand-size quartz derived from sources contributing to each of the five subcells of the Oceanside Littoral Cell. The average percentage of sand contributed from each source is indicated along the shaft of each arrow. Where one value is given, it represents the composition of the winter-sample set. Where two percentages are present, e.g., 22(26), the first value is for the winter- and the second for the summer-sample set.

a pulse of sand discharged from the Santa Margarita River itself. Although speculative, the likelihood of the Santa Margarita River as opposed to Las Flores Creek as being the source for the sand in this peak is greater, because of much higher average annual bedload-discharge rates for the Santa Margarita River. Simon & Associates, Inc. (1988) estimate the average annual sand and gravel yield for the Santa Margarita River as 25,500 tons as compared with 3,610 tons for Las Flores Creek. There is no record of substantive discharge from the Santa Margarita River during the winter of 1985–1986; therefore, the observed peak must be the result of an earlier discharge event. The last major discharge of the Santa Margarita River was in November-December 1984. During this period, 7.21 inches of rain fell at Lindberg Field, and this was recorded as the third wettest November and December in 134 years (G. G. Kuhn, pers. commun.). Consequently, the observed peak must be at least 1.25 years old. Although the sand contribution from San Mateo Creek obviously was transported downcoast, the observed decrease in the amount of San Mateo River sand to the south may represent dilution by Santa Margarita River and southern shelf sand rather than distance from the San Mateo River. The percentage of San Mateo River sand in the San Onofre subcell may be rather constant, and the observed decrease south of 1530-B may reflect dilution by Santa Margarita sand transported upcoast.

The contribution of southern shelf sand (35 percent) argues for considerable onshore as well as upcoast sand transport. Disregarding the effect of the peak of Santa Margarita River-like sand at 1300-B, the amount of southern shelf sand increases from approximately 33 percent south of 1300-B to 44 percent north of this station. Inasmuch as the shape characteristics of the southern shelf sand are distinct from that of the northern shelf, this substantive upcoast increase must reflect a relatively long-term accumulation of incremental amounts of southern shelf sand. This, in turn, may imply a rather continuous supply of shelf sand to the southern beaches and net upcoast transport.

Although a more temporally controlled experiment is desirable, it appears that rather thorough sediment mixing has produced a relatively consistent grain-shape composition for the San Onofre subcell with respect to the upcoast beach (Capistrano subcell), San Mateo Creek, northern shelf and cliff sources. Southern shelf sand apparently is being added rather continuously through onshore and upcoast transport. The principal source of variation observed in this sample set appears to be sand input from the Santa Margarita River, which apparently is delivering sand to the upcoast beaches in an episodic manner, presumably related to high-discharge events.

October 1986 Beach Samples.—

The grain-shape composition for the four end-of-summer beach samples from the San Onofre subcell is shown in Table 3C and Figures 4C and 5. The average sand contributions from the six local sources considered are as follows: upcoast beach (Capistrano subcell), 8 percent; San Mateo Creek, 11 percent; northern shelf, 3 percent; cliff, 14 percent; Santa Margarita River, 33 percent; and southern shelf, 32 percent.

The compositional variation for upcoast beach (Capistrano subcell; 2 percent), San Mateo River (5 percent), northern shelf (2 percent) and cliff (2 percent) is insufficient with respect to the number of beach samples analyzed to assign any genetic significance to the observed longshore trends. As for the April sample set, the greatest range of observed values is associated with the Santa Margarita River (24 percent) and the southern shelf (13 percent) sources.

There is an average increase of approximately 6 percent in Santa Margarita grain shapes in the San Onofre subcell from April to October, 1986. This clearly demonstrates net upcoast transport during this six-month interval. It appears that the April 1986 peak at station 1300-B migrated upcoast approximately 24 km from the mouth of the Santa Margarita River and was homogenized into a more uniform percentage of Santa Margarita sand. The October 1986 peak at 1170-B occurs 3.2 km north of the river mouth and represents a distinct discharge event. Inasmuch as there is no

TABLE 3.—RESULTS OF FOURIER GRAIN-SHAPE ANALYSIS FOR BEACH SAMPLES COLLECTED AT THE TIMES INDICATED WITHIN EACH OF THE FIVE INITIAL SUBCELLS. VALUES LISTED FOR EACH SAMPLE ARE THE PERCENTAGES OF MEDIUM-SAND SIZE QUARTZ DERIVED FROM EACH OF THE LOCAL SOURCES INDICATED FOR THAT SUBCELL. SAND SOURCES CONSIDERED FOR EACH SUBCELL AND ASSOCIATED SAMPLES ARE LISTED BELOW EACH MATRIX

A. Capistrano Subcell (April, 1986): Dana Point to San Mateo Point

	San Juan Creek	Northern Shelf	Southern Shelf	San Mateo Creek	Downcoast Beach
1895-B	12	8	13	26	41
1815-B	12	7	11	31	39
1605-B	13	8	12	28	39

River sample: 1865-R, San Juan Creek
Northern Shelf Sample. 1800-S, Offshore of Dana Point Harbor
Southern Shelf Samples. 1658-S, 1610-S, Offshore of San Clemente
Downcoast River Sample: 1595-R, San Mateo Creek
Downcoast Beach Sample: 1580-B

B. San Onofre Subcell (April, 1986): San Mateo Point to North Breakwater at Oceanside Harbor

	Upcoast Beach	San Mateo Creek	Northern Shelf	Cliff	Santa Margarita River	Southern Shelf
1580-B	9	15	7	13	13	43
1530-B	10	17	6	12	9	46
1300-B	6	8	4	12	45	25
1270-B	8	10	6	13	25	38
1170-B	6	11	2	13	34	32
1115-B	7	10	4	14	37	28

Upcoast Beach Sample: 1605-B
Northern River Samples: 1595-R, 1565-R, San Mateo Creek
Northern Shelf Samples: 1575-S, 1535-S, 1480-S, 1455-S, 1425-S, 1385-S, 1322-S, Offshore of San Mateo Creek
Cliff Samples: 1380-C, 1460-C, 1410-C, 1355-C, 1320-C, 1283-C, 1275-C, 1220-C, San Onofre and Camp Pendleton Areas
Southern River Samples: 1150-R, 1285-R, Santa Margarita River
Southern Shelf Samples: 1123-S, 1105-S, 1090-S, 1077-S, Offshore Oceanside

C. San Onofre Subcell (October, 1986): San Mateo Point to North Breakwater at Oceanside Harbor

	Upcoast Beach	San Mateo Creek	Northern Shelf	Cliff	Santa Margarita River	Southern Shelf
1300-B	8	12	3	14	27	36
1270-B	8	13	4	14	25	36
1170-B	6	8	2	12	49	23
1115-B	8	11	3	14	30	34

Upcoast Beach Sample: 1605-B
Northern River Samples: 1595-R, 1565-R, San Mateo Creek
Northern Shelf Samples: 1575-S, 1535-S, 1480-S, 1455-S, 1425-S, 1385-S, 1322-S, Offshore of San Mateo Creek
Cliff Samples: 1380-C, 1460-C, 1410-C, 1355-C, 1320-C, 1283-C, 1275-C, 1220-C, San Onofre and Camp Pendleton Areas
Southern River Samples: 1150-R, 1285-R, Santa Margarita River
Southern Shelf Samples: 1123-S, 123-S, 1105-S, 1090-S, 1077-S, Offshore of Oceanside

D. Oceanside-Carlsbad Subcell (April, 1986; October, 1986; and February, 1989, respectively): Oceanside Harbor to Onshore Projection of Carlsbad Submarine Canyon

	Upcoast Beach	Northern Shelf	San Luis Rey River	Southern Shelf	Downcoast Beach
1030-B	12	25	7	8	48
1030-B	10	32	6	11	41
810-B	11	27	6	8	49

Upcoast Beach Sample: 1115-B
Northern Shelf Sample: 1073-S, Offshore of Oceanside
River Sample: 1050-R, San Luis Rey River
Southern Shelf Samples: 985-S, 993-S, 920-S, 915-S, 864-S, 860-S, Offshore of Carlsbad
Downcoast Beach Sample: 765-B

TABLE 3.—*Continued*

E. Encinitas Subcell (April, 1986): Onshore Projection of Carlsbad Submarine Canyon to La Jolla Submarine Canyon

	Upcoast Beach	San Dieguito River	Northern Shelf	Cliff	Southern Shelf	Downcoast Beach
765-B	45	6	23	13	10	3
645-B	42	5	21	14	11	7
625-B	39	5	22	12	11	11
595-B	39	5	24	11	12	9
545-B	43	5	20	13	9	10
520-B	40	5	25	12	11	7

Upcoast Beach Sample: 810-B
River Samples: 590-R, 635-R, 530-R, San Dieguito River
Northern Shelf Samples: 519.5-S, 650-S, Offshore of Encinitas
Cliff Samples: 500-C, 466-C, 465-C, 463.6-C, 463-C, Torrey Pines Area
Southern Shelf Samples: 454-S, 492-S, 464.5-S, Offshore of La Jolla
Downcoast Beach Sample: 427-B

F. Encinitas Subcell (October, 1986): Onshore Projection of Carlsbad Submarine Canyon to La Jolla Submarine Canyon

	Upcoast Beach	San Dieguito River	Northern Shelf	Cliff	Southern Shelf	Downcoast Beach
645-B	43	6	21	13	10	7
625-B	41	6	27	10	13	3
595-B	41	5	26	12	12	4
575-B	42	5	25	12	11	5
545-B	39	5	28	11	14	3
520-B	41	5	27	12	12	3

Upcoast Beach Sample: 810-B
River Samples: 590-R, 635-R, 530-R, San Dieguito River
Northern Shelf Samples: 519.5-S, 650-S, Offshore of Encinitas
Cliff Samples: 500-C, 466-C, 465-C, 463.6-C, 463-C, Torrey Pines Area
Southern Shelf Samples: 454-S, 492-S, 464.5-S, Offshore of La Jolla
Downcoast Beach Sample: 427-B

G. La Jolla Cove Subcell (September, 1989): La Jolla Submarine Canyon to La Jolla Point

	Upcoast Beach	Shelf
446-B	61	39
445-B	58	42
444-B	40	60

Upcoast Beach Sample: 448-B
Shelf Sample: 443-S, Offshore of Point La Jolla

evidence for a high-discharge event from April to October, 1986, the exact age of this event is not known. It is important to note that Fourier grain-shape analysis has the resolution to identify and trace such pulses of beach sand and would be especially useful as a tracer following a known high-discharge event.

There is an average reduction of 3 percent southern shelf sand in the San Onofre subcell from April to October, 1986. For the October sample set, the southern shelf contribution ranges from 34 to 36 percent, except at station 1170-B, where it apparently is diluted to 23 percent by the addition of Santa Margarita sand.

Overall, the grain-shape composition of the San Onofre subcell remained much the same from April to October, 1986. The only significant change is the upcoast transport of Santa Margarita River sand, which altered the grain-shape composition of associated beach samples.

Oceanside-Carlsbad Subcell

The results of FGSA for the three beach samples from the Oceanside-Carlsbad subcell are shown in Table 3D and Figures 4D and 5. The average sand contributions from the five local sources supplying this subcell are: upcoast beach (San Onofre subcell), 11 percent; northern shelf, 28 percent; San Luis Rey River, 6 percent; southern shelf, 9 percent; and downcoast beach (Encinitas subcell), 46 percent.

The range of variation for the upcoast-beach contribution (San Onofre subcell; 2 percent), San Luis Rey River (2 percent) and southern shelf (3 percent) is quantitatively unimportant; however, the variation for the northern shelf (7 percent) and downcoast-beach (Encinitas subcell; 8 percent) contributions appears to be important from both a seasonal and a longer term sense. Interestingly enough, the variation observed between the April and October 1986 beach samples from station 1030-B and the February 1989 beach sample from station 810-B appears to be on the same order of magnitude as the seasonal variation between the April and October beach samples for 1986. Although the sample size for this subcell is small, it appears that the beach-sand samples display a high degree of compositional homogeneity (Fig. 4D).

The most interesting aspects of the grain-shape composition of the beach samples in this sample set are the apparent compositional homogeneity and the dominance of sand shapes derived from the northern shelf (28 percent) and the downcoast beach (46 percent). One might expect a substantive contribution from the San Luis Rey River (Simon & Associates, Inc., 1988); however, bedload discharge from the San Luis Rey River may have been greatly reduced by the Pacific Street crossing. In any case, it is clear that any such contribution has been greatly diluted by sand from other local sources. Inasmuch as the last major rainfall occurred in November and December 1984, any bedload contribution from the San Luis Rey River was greatly diluted during the following 1.25 or more years. The dominance of sand shapes characteristic of the northern shelf and downcoast beaches indicates the importance of onshore and upcoast longshore transport, respectively. The compositional homogeneity reflects rather thorough mixing during an unknown time interval.

Encinitas Subcell

April 1986 Beach Samples.—

Table 3E and Figures 4E and 5 show the results of FGSA for the six end-of-winter samples from the Encinitas subcell. The average sand contribution from each source is as follows: upcoast beach (Oceanside-Carlsbad subcell), 41 percent; San Dieguito River, 5 percent; northern shelf, 22 percent; cliff, 12 percent; downcoast beach (La Jolla Cove subcell), 8 percent; and southern shelf, 11 percent.

The range of variation for the contributions from the San Dieguito River (1 percent), cliff (3 percent) and southern shelf (3 percent) is statistically insignificant; however, the observed range of variation for the upcoast beach (Oceanside-Carlsbad subcell; 6 percent), northern shelf (5 percent) and particularly the downcoast beach (La Jolla Cove subcell; 8 percent) may require explanation. It is useful to present the results obtained for the end-of-summer samples before discussing the sources of compositional variation for the Encinitas subcell.

October 1986 Beach Samples.—

The grain-shape results for the six end-of-summer beach samples from the Encinitas subcell are listed in Table 3F and illustrated in Figures 4F and 5. The average sand contribution from each local source is as follows: upcoast beach (Oceanside-Carlsbad subcell), 41 percent; San Dieguito River, 5 percent; northern shelf, 26 percent; cliff, 12 percent; downcoast beach (La Jolla Cove subcell), 4 percent; and southern shelf, 12 percent.

The range of variation observed for the grain-shape contributions from the upcoast beach (Oceanside-Carlsbad subcell; 4 percent), San Dieguito River (1 percent), cliff (3 percent), downcoast beach (La Jolla Cove subcell; 4 percent) and southern shelf is insignificant. The variation associated with the northern shelf (6 percent) may be important.

The April and October 1986 values for the average sand contributions are equal for the upcoast beach (41 percent), San Dieguito River (5 percent) and cliff (12 percent), and vary by only 1 percent for the southern shelf (11 and 12 percent), respectively. Thus, at least 70 percent of the shape composition for the Encinitas subcell appears to be seasonally consistent for the observed sample set.

The end-of-summer beach samples contain an average of 4 percent more northern shelf sand and 4 percent less downcoast-beach sand. Inasmuch as these 4-percent differences are close to the limits of methodological resolution, the differences between the means of the end-of-winter and end-of-summer sample sets for the northern-shelf and for the downcoast-beach sources were examined using Student t-tests. In both cases, the observed differences are not significant at the 5-percent significance level. Therefore, the beach samples from the Encinitas subcell must be considered seasonally homogeneous with respect to the 1986 sample sets.

The results of FGSA show the dominance of the upcoast beach and northern shelf in supplying sand to the beach of the Encinitas subcell. This indicates onshore transport of inner-shelf sand in the northern part of the subcell and net downcoast transport of upcoast-beach and northern shelf sand. The small volumetric contribution of the San Dieguito River agrees well with the small annual-discharge rates computed by Simon & Associates, Inc. (1988). The relatively small contributions from the downcoast beach and southern shelf indicate the secondary importance of upcoast transport in the Encinitas subcell. In terms of sediment budget considerations, it is important to note that the beach samples in this subcell contain an average of 33 percent inner-shelf sand and 12 percent cliff sand. Again, the compositional homogeneity of the grain shapes in the beach-sample set suggests thorough mixing during an unknown residence time.

La Jolla Cove Subcell

Although contributing cliffs occur along the shore of the La Jolla Cove subcell, they consist dominantly of fine-grained sandstone and clay shale assigned to the Cretaceous Point Loma Formation. Inasmuch as these strata are not important sources for medium-grained sand, no major rivers enter this subcell, and upcoast longshore transport of sand around Point La Jolla is unlikely, only upcoast-beach and inner continental-shelf sand are considered as medium-grained sand sources for this subcell. The results of FGSA for the three beach samples collected in September 1989 are listed in Table 3G and illustrated in Figures 4G and 5. The average grain-shape composition for these samples is as follows: upcoast beach (Encinitas subcell), 53 percent and shelf, 47 percent.

Figure 4G shows a 3-percent decrease in upcoast-beach sand from station 446-B to 445-B, followed by an 18-percent decrease from 445-B to 444-B. This southward decrease in upcoast-beach sand may simply reflect the condition that a pulse of southward-directed upcoast-beach sand that has reached stations 446-B and 445-B has not yet reached 444-B.

The fact that inner-shelf sand comprises from 39 to 60 percent of the grain-shape composition in the La Jolla Cove subcell reflects the paucity of local sources and a reduction in the degree of dilution by upcoast-beach sand, particularly from station 445-B to 444-B.

Although the identification and quantification of local sand sources for each subcell are important scientific contributions and potentially lead to more accurate budget calculations, the results obtained bear on three other substantive coastal issues. These are: 1) the degree of mixing of littoral sand within each subcell; 2) the sedimentologic reality of the initial set of five littoral subcells; and 3) the role of the inner continental shelf as a source of naturally delivered beach sand.

Compositional Mixing Within Each Subcell

With the important exception(s) of the preserved peak(s) of Santa Margarita River sand in the San Onofre subcell, each of the subcells considered appears to be rather internally homogeneous with respect to grain-shape populations (Table 3 and Fig. 4). Assuming temperate drought conditions, which are temporally dominant in historical records for southern California (Inman, 1981), winter waves generally have a net energy-flux component to the south due to their generation by northern Pacific storms passing close to southern California, whereas summer waves often show a net energy flux to the north due to their generation from more distant storms, either southwest of Acapulco, Mexico, or from Antarctica. As such, longshore sediment transport is commonly bidirectional in the short term. It is likely that such bidirectional longshore and onshore transport, which may vary both in intensity and duration and which affect virtually all of the Oceanside Littoral Cell, may be largely responsible for the observed shape-compositional uniformity. Unfortunately, the exact mechanism and residence time required to evolve such compositional uniformity are not known. Monitoring of the peak(s) of Santa Margarita River sand in the San Onofre subcell (Fig. 4A, B) is in progress to understand these sedimentologic and temporal factors better.

Sand Exchange Between Adjacent Subcells

The five subcells considered were defined largely on the basis of shoreline orientation, natural physiographic features and engineered structures associated with Oceanside Harbor. Table 3 and Figures 4 and 5, as well as much of the preceding discussion, indicate that each of the five subcells identified is compositionally distinct with regard to quartz grain-shape populations, which is due, in part, to the occurrence of a different set of local sand sources for each subcell. However, it also is important to consider the amount of sand exchanged between adjacent pairs of subcells.

Only 8 percent of the beach sand from the Capistrano subcell is transported downcoast around San Mateo Point to the San Onofre subcell, and approximately 40 percent of the beach sand is transported upcoast from San Mateo Point to the Capistrano subcell. The relatively small volume transported downcoast must partially reflect the relatively small volume of sand delivered from San Juan Creek in recent years, because the wave flux in this area indicates net downcoast transport (Osborne and others, 1990). Therefore, San Mateo Point is only a partially effective barrier to longshore transport.

The north breakwater and south jetty at Oceanside Harbor also serve as partially effective barriers to longshore transport between the San Onofre and Oceanside-Carlsbad subcells, and the harbor itself is a very effective sink for sand. Therefore, there can be little debate regarding the overall effectiveness of Oceanside Harbor as a barrier separating these two subcells. The upcoast-beach sand (11 percent) in the Oceanside-Carlsbad subcell reflects either a pre-harbor natural exchange and/or a post-harbor beach nourishment contribution.

Approximately 41 percent of the beach sand from the Oceanside-Carlsbad subcell is transported downcoast past the onshore projection of the Carlsbad Submarine Canyon to the Encinitas subcell, and about 46 percent of the beach sand from the Encinitas subcell is transported into the Oceanside-Carlsbad subcell. Likewise, approximately 53 percent of the beach sand from the Encinitas subcell is transported downcoast past the head of La Jolla Submarine Canyon, whereas only 6 percent of the sand from the La Jolla Cove subcell is transported upcoast into the Encinitas subcell. There appears to be a great deal of longshore sand exchange across the boundaries separating both the Oceanside-Carlsbad and Encinitas subcells as well as the Encinitas and La Jolla Cove subcells.

It is interesting to note that downcoast and associated southern shelf sources within each subcell constitute an average of 80 percent in the Capistrano subcell and from 62 percent (April, 1986) to 65 percent (October, 1986) in the San Onofre subcell. Although downcoast and southern shelf sources are predominant (55 percent) in the Oceanside-Carlsbad subcell, upcoast and northern shelf sources also are volumetrically significant (39 percent). The Encinitas subcell is dominated by upcoast and northern shelf sources (63 percent for April 1986; and 67 percent for October 1986), as is the La Jolla Cove subcell (at least 53 percent from upcoast beaches alone). It appears that the northernmost two subcells are dominated by downcoast and associated southern shelf sources within each subcell, the central subcell appears somewhat transitional with regard to upcoast and downcoast sources, and the southernmost two subcells are dominated by upcoast and northern shelf sources. It is tempting to explain these observations in terms of the potential sheltering effects of Dana Point and Point La Jolla, but it is not clear that deep-water wave statistics support this contention. The possible roles of storms and nearshore topography in producing the observed compositional trends require further work.

Although each of the five subcells can be defined in terms of distinct sets of grain shapes representing the local sources supplying a given coastal reach, it is clear that an appreciable amount of sand has been exchanged naturally between adjacent pairs of subcells, with the probable exception of the San Onofre and Oceanside-Carlsbad subcells after 1942, during a relatively recent but unknown time interval. Continued monitoring is required to determine the volumes involved and rate of exchange between adjacent pairs of subcells to determine the effectiveness of the subcell boundaries as barriers to littoral-sand transport as well as to compute more accurate and refined sediment budgets.

The Inner Continental Shelf as a Source for Naturally Delivered Beach Sand

An average of approximately 36 percent of the quartz in the medium-sand fraction of the beach samples examined was derived from the inner continental shelf (Table 3 and Figs. 4 and 5). The Capistrano subcell contains an average of 20 percent shelf sand; the San Onofre, Oceanside-Carlsbad and Encinitas subcells contain from 35 to 38 percent; and the La Jolla Cove subcell contains an average of 47 percent. The fact that approximately 36 percent of the foreshore beach sand in this area was derived from the inner shelf clearly points to the need for a greater understanding of the cross-shore transport vector (Fig. 3).

Using Fourier analysis, Riester and others (1982) describe quartz grain-shape populations from beach to outer-shelf depths off Long Island. Of interest to this study is that more abraded beach sand, which is formed at water depths from 0 to −5 m MLW, is mixed with inner-shelf sand, which occurs at depths from −20 to −10 m MLW, within a bar complex at water depths from −5 to −10 m MLW. They suggest that during storms, wave base may migrate to water depths deeper than the bar, resulting in the mobilization and landward transport of inner-shelf sand to the bar complex, where it is mixed with sand eroded from the beach. Riester and others (1982) did not discuss transport or mixing rates, nor was an explicit mechanism proposed to explain the shoreward transport of inner-shelf sand.

Bowen and Inman (1966), Inman and others (1966) and Clemens and Komar (1988) have documented net onshore transport of shelf sand at widely spaced localities along the Pacific coast: namely, at Sebastian Vizcaino Bay, Mexico; Point Arguello, California; and selected beaches north of Cape Blanco, Oregon, respectively.

Dayton and others (1989) describe erosion in San Diego County, which was associated with the exceptionally intense storm of January 17–18, 1988. Seymour (1989a, b) suggested this may have been a 100- to 200-year event. Dayton and others (1989) report that two concrete-filled scuba tanks were transported from a water depth of −12 m to a point inshore of −8 m—a distance of at least 0.5 km from their original location. Likewise, reef-derived boulders were transported from a depth of −15 m as much as 10 m shoreward, and a rock reef projecting about 2.5 m above the substrate at a depth of −18 m was reduced to rubble, and much of this was projected shoreward. Clearly, there is ample evidence of shoreward transport of even boulder-size sediment at inner-shelf depths during intense storms.

Dean (1987) discusses several lines of evidence as well as mechanisms concerning onshore sediment transport from continental-shelf depths as well as "in situ" biogenic sediment production for several portions of the Florida coast. Of importance to the present study is that using stream function (nonlinear) wave theory, Dean (1987) shows that the average, nondimensional bottom shear stress would tend to cause shoreward sediment transport across the continental shelf. The shoreward forces require considerable time to transport large volumes of sediment landward; however, the average bottom shear stress increases substantially for shallower water and higher waves, which would accelerate associated transport rates. In contrast to the Bruun Rule, Dean (1987) also concludes that following a sea-level rise, landward sediment transport of continental-shelf sediment is required to reachieve an equilibrium profile.

These studies, among others, clearly demonstrate the mobilization and shoreward transport of sand and coarser grained sediment at inner-shelf depths; however, the fundamental questions concerning the exact mechanism, volume of sediment involved, rates of transport, transport pathways and residence times remain unanswered. It seems likely, however, that storm activity, perhaps coupled with shoreward transport through ridge and runnel systems, may be involved.

CONCLUSIONS

FGSA of detrital quartz has been used to identify local sources for medium-grained sand in a series of five subcells within the Oceanside Littoral Cell. The subcells initially were defined by the orientation and physiography of that coastal reach as well as by the presence of engineered structures associated with Oceanside Harbor. From northwest to southeast, these are named the Capistrano subcell (Dana Point to San Mateo Point), the San Onofre subcell (San Mateo Point to the north breakwater at Oceanside Harbor), the Oceanside-Carlsbad subcell (from the south jetty at Oceanside Harbor to the onshore projection of Carlsbad Submarine Canyon), the Encinitas subcell (from the onshore projection of Carlsbad Submarine Canyon to La Jolla Submarine Canyon), and the La Jolla Cove subcell (from La Jolla Submarine Canyon to Point La Jolla). Each of these five subcells is compositionally distinct with regard to quartz grain-shape populations, which, in part, reflects the occurrence of a characteristic set of local sand sources supplying each subcell.

With the exception(s) of the preserved peak(s) of Santa Margarita River sand in the San Onofre subcell, each of the subcells considered appears to be rather internally homogeneous with respect to grain-shape populations. Although the exact mechanism and residence time required to evolve such compositional homogeneity are not known, it appears that seasonally related, bidirectional longshore transport may be a major factor.

Although each of the five subcells is distinct in terms of grain-shape composition, it is clear that an appreciable amount of beach sand has been exchanged naturally between all pairs of adjacent subcells, with the probable exception of natural exchange between the San Onofre and Oceanside-Carlsbad subcells since 1942. Although the reason(s) is (are) not clear, the northernmost two subcells are dominated by downcoast and southern shelf sources associated with each subcell, the central subcell appears somewhat transitional with regard to upcoast and downcoast sources, and the southernmost two subcells are dominated by upcoast and northern shelf sources. Continued monitoring is required to estimate the volumes involved and rates of exchange to determine the effectiveness of the subcell boundaries as barriers to littoral-sand transport.

Approximately 36 percent of the quartz in the medium-sand fraction of the foreshore samples was derived from the inner continental shelf. Although it is likely that storm activity, perhaps coupled with shoreward transport through ridge and runnel systems, may be involved, the exact mechanisms, volumetric aspects, rates of transport, transport pathways and residence times are unknown.

FGSA clearly can be used to identify the local sand sources within the subcells considered and is useful to estimate the percentage contributed from each source. The utility of this method may be extended by computing compositional estimates weighted with respect to the percentage of medium sand and the percentage of quartz in each sample within a given coastal reach. However, to be most useful for sediment budget analysis, the weighted local source estimates supplied by FGSA, grain-size and mineralogic analyses must be integrated with mass-balance computations to obtain estimates for the volumes contributed from each local source, as well as to define or constrain sand-transport rates and paths. Such work is now in progress for the Oceanside Littoral Cell.

ACKNOWLEDGMENTS

This work is based partially on sand samples and Fourier grain-shape data collected by Kelly S. Ahlschwede, Sean D. Broadhead, Eleanor A. Compton and Kyung H. Cho of the Sedimentary Petrology Laboratory at the University of Southern California. Several additional samples were collected by Gerald G. Kuhn. The vibracores were recovered by Ocean Surveys, Inc., under contract to the U.S. Army Coastal Engineering Research Center, on board the *RV Diamond Lady*. Tim Fogarty wrote the EDGE Fortran Program for digitizing quartz grains. Anthony B. Price, Los Angeles District, U.S. Army Corps of Engineers, contributed several samples and participated in informative discussions regarding the experimental design employed and the interpretation of the results. However, the authors take full responsibility for the content of this paper. The manuscript was typed by Desser Moton and Sue Turnbow. Figures 1 and 2 were drafted by Janet Dodds.

Partial funding for this research was provided by the Los Angeles District, U.S. Army Corps of Engineers Contract No. DACW09-87-D-0017.

REFERENCES

AHLSCHWEDE, K. S., 1988, Sources and net littoral transport of sand in San Diego and southern Orange Counties, southern California: Fourier grain-shape analysis: Unpublished M.S. Thesis, University of Southern California, Los Angeles, California, 135 p.

ARMSTRONG, G. A., AND FLICK, R. E., 1989, Storm damage assessment for the January 1988 storm along the southern California shoreline: Shore & Beach, v. 57, no. 4, p. 18–23.

BASU, ABHIJIT, YOUNG, S. W., SUTTNER, L. J., JAMES, W. C., AND MACK, G. H., 1975, Re-evaluation of the use of undulatory extinction and polycrystallinity in detrital quartz for provenance interpretation: Journal of Sedimentary Petrology, v. 54, p. 873–882.

BENDAT, J. S., AND PIERSOL, A. G., 1971, Random data: Analysis and Measurement Procedures: Wiley-Interscience, Inc., New York, 407 p.

BLOOM, LAURIE, 1979, The relationships among river, beach and submarine canyon sands in the southern Santa Barbara littoral cell, Ventura County, California: Fourier grain-shape analysis: Unpublished M.S. Thesis, University of Southern California, Los Angeles, California, 115 p.

BLOOMFIELD, PETER, 1976, Fourier analysis of time series: An introduction: Wiley-Interscience Inc., New York, 258 p.

BOMER, E. J., 1985, Quartz grain-shape modification in high-gradient fluvial environments: Fourier grain-shape analysis: Unpublished M.S. Thesis, University of Southern California, Los Angeles, California, 139 p.

BOWEN, A. J., AND INMAN, D. L., 1966, Budget of littoral sands in the vicinity of Point Arguello: Coastal Engineering Research Center, U.S. Army Corps of Engineers, Technical Memorandum Number 19, 41 p.

BRIGHAM, E. O., 1974, The Fast Fourier Transform: Prentice-Hall, Inc., Englewood Cliffs, New Jersey, 252 p.

BROADHEAD, S. D., 1988, Fourier grain-shape analysis of the sources and littoral transport of sand, San Elijo Lagoon to Point La Jolla: Unpublished M.S. Thesis, University of Southern California, Los Angeles, California, 235 p.

BROWN, P. J., EHRLICH, ROBERT, AND COLQUHOUN, D. J., 1980, Origin of patterns of quartz sand types on the southeastern United States continental shelf and implications on contemporary shelf sedimentation—Fourier grain shape analysis: Journal of Sedimentary Petrology, v. 50, p. 1095–1100.

CARVER, R. E., 1971, Procedures in Sedimentary Petrology: Wiley-Interscience, Inc., New York, 653 p.

CHO, K. H., 1989, Sedimentology of a composite inner-shelf sand body resulting from the resuspension of nearshore sediment by episodic, storm-generated currents, Oceanside California: Unpublished M.S. Thesis, University of Southern California, Los Angeles, California, 104 p.

CLARK, R. A., AND OSBORNE, R. H., 1982, Contribution of Salinas River sand to the beaches of Monterey Bay, California, during the 1978 flood period: Fourier grain-shape analysis: Journal of Sedimentary Petrology, v. 52, p. 807–822.

CLEMENS, K. E., AND KOMAR, P. D., 1988, Oregon beach-sand compositions produced by mixing of sediments under a transgressing sea: Journal of Sedimentary Petrology, v. 59, p. 519–529.

DARIGO, N. J., 1984, Quarternary stratigraphy and sedimentation of the mainland shelf of San Diego County, California: Unpublished M.S. Thesis, University of Southern California, Los Angeles, California, 2 vols., 447 p.

DARIGO, N. J., AND OSBORNE, R. H., 1986, Quaternary stratigraphy and sedimentation of the inner continental shelf, San Diego, California, *in* Knight, R. J., and McLean, J. R., eds., Shelf Sands and Sandstones: Canadian Society of Petroleum Geologists, Memoir 11, p. 73–98.

DAVIS, J. C., 1986, Statistics and Data Analysis in Geology: John Wiley & Sons, Inc., New York, 646 p.

DAYTON, P. K., SEYMOUR, R. J., PARNELL, P. E., AND TEGNER, M. J., 1989, Unusual marine erosion in San Diego County from a single storm: Estuarine, Coastal and Shelf Science, v. 29, p. 151–160.

DEAN, R. G., 1987, Additional sediment input to the nearshore region: Shore & Beach, v. 55, nos. 3–4, p. 76–81.

EHRLICH, ROBERT, AND CHIN, MAUREEN, 1980, Fourier grain-shape analysis: A new tool for sourcing and tracking abyssal silts: Marine Geology, v. 38, p. 219–232.

EHRLICH, ROBERT, AND WEINBERG, BERNHARD, 1970, An exact method for characterization of grain shape: Journal of Sedimentary Petrology, v. 40, p. 205–212.

EHRLICH, ROBERT, ORZECK, J. J., AND WEINBERG, BERNHARD, 1974, Detrital quartz as a natural tracer—Fourier grain shape analysis: Journal of Sedimentary Petrology, v. 44, p. 145–150.

EHRLICH, ROBERT, BROWN, P. J., YARUS, J. M., AND PRZYGOCKI, R. S., 1980, The origin of shape frequency distributions and the relationship between size and shape: Journal of Sedimentary Petrology, v. 50, p. 475–484.

EMERY, K. O., 1960, The Sea off Southern California: A Modern Habitat of Petroleum: John Wiley and Sons, New York, 366 p.

GAYNOR, J. M., 1984, Sources and transport of sand in the littoral zone of Lake Tahoe, California and Nevada: Fourier grain-shape analysis: Unpublished M.S. Thesis, University of Southern California, Los Angeles, California, 112 p.

HALES, L. Z., 1978, Coastal processes study of the Oceanside, California littoral cell: U.S. Army Corps of Engineers, Miscellaneous Paper H-78-8, 60 p.

HUDSON, C. B., AND EHRLICH, ROBERT, 1980, Determination of relative provenance contributions in samples of quartz sand using Q-mode factor analysis of Fourier grain shape data: Journal of Sedimentary Petrology, v. 50, p. 1101–1110.

INMAN, D. L., 1976, Summary report of man's impact on the California coastal zone: State of California, Department of Navigation and Ocean Development, Sacramento, California, 150 p. (Reissued by State of California, Department of Boating and Waterways, June 1980).

INMAN, D. L., 1981, Evaluation of the budget of waves and sand, Oceanside Littoral Cell and recommendations for monitoring and sand bypassing: Los Angeles District, U.S. Army Corps of Engineers, 24 p.

INMAN, D. L., AND CHAMBERLAIN, T. K., 1960, Littoral sand budget along the southern California coast: Report of the 21st International Geological Congress, Copenhagen, Volume of Abstracts, p. 245–246.

INMAN, D. L., EWING, G. C., AND CORLISS, J. B., 1966, Coastal sand dunes of Guerrero Negro, Baja California: Geological Society of America Bulletin, v. 77, p. 787–802.

INMAN, D. L., AND JENKINS, S. A., 1983, Oceanographic report for Oceanside beach facilities: City of Oceanside, California, 149 p.

KUHN, G. G., AND SHEPARD, F. P., 1984, Sea Cliffs, Beaches and Coastal Valleys of San Diego County: University of California Press, Berkeley, 193 p.

KUHN, G. G., OSBORNE, R. H., AHLSCHWEDE, K. S., AND COMPTON, E. A., 1987, Processes, locations and rates of coastal cliff erosion from 1947 to present, Dana Point to the United States-Mexico border and the stratigraphy of contributing cliffs and bluffs at San Onofre, Camp Pendleton, and Torrey Pines: Los Angeles District, U.S. Army Corps of Engineers, Technical Report CCSTWS 87-2, 298 p.

LARSEN, E. S., 1948, Batholith of Corona, Elsinore and San Luis Rey quadrangles, southern California: Geological Society of America, Memoir 29, 182 p.

MAZZULLO, J. M., AND EHRLICH, ROBERT, 1980, A vertical variation in the St. Peter Sandstone—Fourier grain shape analysis: Journal of Sedimentary Petrology, v. 50, p. 63–70.

MAZZULLO, J. M., AND EHRLICH, ROBERT, 1983, Grain shape variation in the St. Peter Sandstone: A record of eolian and fluvial sedimentation of an early Paleozoic cratonic sheet sand: Journal of Sedimentary Petrology, v. 53, p. 105–119.

MAZZULLO, J. M., AND MAGENHEIMER, STEWART, 1987, The original shapes of quartz sand grains: Journal of Sedimentary Petrology, v. 57, p. 479–487.

MOFFATT & NICHOL, ENGINEERS, 1989, Sediment budget report Oceanside Littoral Cell (draft): Los Angeles District, U.S. Army Corps of Engineers, Los Angeles, California, 112 p.

MRAKOVITCH, JOHN, EHRLICH, ROBERT, AND WEINBERG, BERNHARD, 1976, New technique for stratigraphic analysis and correlation—Fourier grain-shape analysis, Louisiana offshore Pliocene: Journal of Sedimentary Petrology, v. 46, p. 226–233.

NORDSTROM, C. E., AND INMAN, D. L., 1973, Beach and cliff erosion in San Diego County, California, *in* Ross, A., and Dowlen, R. J., eds., Studies in the Geology and Geologic Hazards of the Greater San Diego Area: San Diego Association of Geologists, San Diego, California, p. 125–131.

OSBORNE, R. H., 1985, Coast of California storm and tidal wave study: Littoral zone sediments: Los Angeles District, U.S. Army Corps of Engineers, Technical Report CCSTWS 85-11, 148 p.

OSBORNE, R. H., AND CHO, K. H., 1989, Sedimentology of a composite inner-shelf sand body resulting from the resuspension of nearshore sediment by episodic, storm-generated currents: Proceedings Volume, Coastal Zone '89, American Society of Civil Engineers, p. 4391–4405.

OSBORNE, R. H., AND PIPKIN, B. W., 1982, Geomorphic and sedimentologic analysis for the Oceanside Project: Phase II: Los Angeles District, U.S. Army Corps of Engineers, 121 p.

OSBORNE, R. H., DARIGO, N. J., AND SCHEIDEMANN, R. C., JR., 1983, Potential offshore sand and gravel resources of the inner continental shelf of southern California: State of California, Department of Boating and Waterways, 302 p.

OSBORNE, R. H., EDELMAN, M. C., GAYNOR, J. M., AND WALDRON, J. M., 1985, Sedimentology of the littoral zone in Lake Tahoe, California-Nevada: California State Lands Commission, 88 p.

OSBORNE, R. H., KUHN, G. G., COMPTON, E. A., AND FOGARTY, TIMOTHY, 1988, Processes, locations and rates of coastal cliff erosion from 1887 to 1947, Dana Point to the Mexican border: Los Angeles District, U.S. Army Corps of Engineers, Technical Report CCSTWS 88-8, 300 p.

OSBORNE, R. H., AHLSCHWEDE, K. S., BROADHEAD, S. D., CHO, K. H., COMPTON, E. A., AND YEH, C.-C., 1990, Fourier grain-shape and mineralogic analyses of coastal and inner shelf sand samples: Dana Point to the United States-Mexico border: Los Angeles District, U.S. Army Corps of Engineers, Technical Report CCSTWS 89-1, 116 p.

PORTER, G. A., EHRLICH, ROBERT, OSBORNE, R. H., AND COMBELLICK, R. A., 1979, Sources and nonsources of beach sand along southern Monterey Bay, California—Fourier shape analysis: Journal of Sedimentary Petrology, v. 49, p. 727–732.

RIESTER, D. D., SHIPP, R. C., AND EHRLICH, ROBERT, 1982, Patterns of quartz sand shape variation, Long Island littoral and shelf: Journal of Sedimentary Petrology, v. 52, p. 1307–1314.

SCHROETER, S. C., REITZEL, JOHN, DIXON, JOHN, ELWANY, HANY, AND SMITH, R. O., 1989, Anomalous sediments in the San Onofre kelp forest: Marine Review Committee, Inc., Department of Biological Sciences, University of California at Santa Barbara, Santa Barbara, California, 180 p.

SCHULTZ, D. J., 1980, The effects of hydrofluoric acid on quartz grain shape—Fourier grain shape analysis: Journal of Sedimentary Petrology, v. 50, p. 644–645.

SEDIMENT DYNAMICS GROUP, 1989, Sediment dynamics of the area off San Onofre, California: Environmental Research, Southern California Edison Company, Rosemead, California, 294 p.

SEYMOUR, R. J., 1989a, Unusual damage from a California storm: Shore & Beach, v. 57, no. 3, p. 31.

SEYMOUR, R. J., 1989b, Wave observations in the storm of 17–18 January, 1988: Shore & Beach, v. 57, no. 4, p. 14–17.

SHAW, M. J., 1980, Artificial sediment transport and structures in coastal southern California: Scripps Institution of Oceanography, SIO Reference No. 80-41, 109 p.

SIMON, LI & ASSOCIATES, INC., 1988, River sediment discharge study San Diego region: Los Angeles District, U.S. Army Corps of Engineers, Technical Report CCSTWS 88-3, 95 p.

SORENSON, S. S., 1984, Petrology of basement rocks of the California Continental Borderland and the Los Angeles Basin: Unpublished Ph.D. Dissertation, University of California, Los Angeles, California, 423 p.

STOW, D. A., AND CHANG, H. H., 1987, Coarse sediment delivery by coastal streams to the Oceanside Littoral Cell, California: Shore & Beach, v. 55, no. 1, p. 30–40.

STUART, C. J., 1979, Lithofacies and origin of the San Onofre Breccia, coastal southern California, *in* STUART, C. J., ed., A Guidebook to Miocene Lithofacies and Depositional Environments, Southern California and Northwestern Baja California: Los Angeles, Pacific Section of the Society of Economic Paleontologists and Mineralogists, p. 25–42.

SUPPE, JOHN, AND ARMSTRONG, R. L., 1972, Potassium-argon dating of Franciscan metamorphic rocks: American Journal of Science, v. 272, p. 217–233.

TEKMARINE, INC., 1987, Oceanside Littoral Cell Preliminary Sediment Budget Report: Los Angeles District, U.S. Army Corps of Engineers, Technical Report CCSTWS 87-4, 158 p.

U.S. ARMY CORPS OF ENGINEERS, 1987, Consolidated Plan of Study Coast of California Storm and Tidal Wave Study: Los Angeles District, Technical Report CCSTWS 87-1, 369 p.

VAN NIEUWENHUISE, D. S., YARUS, J. M., PRZXYGOCKI, R. S., AND EHRLICH, ROBERT, 1978, Sources of shoaling in Charleston Harbor: Fourier grain shape analysis: Journal of Sedimentary Petrology, v. 48, p. 373–384.

WEGGEL, J. R., AND CLARK, G. R., 1983, Sediment budget calculations, Oceanside, California: Los Angeles District, U.S. Army Corps of Engineers, Coastal Engineering Research Center, Vicksburg, Mississippi, Coastal Engineering Research Center Miscellaneous Paper 83-7, 55 p.

GRAVEL-FILLED GUTTER CASTS IN NEARSHORE FACIES—INDICATORS OF ANCIENT SHORELINE TREND

FRANCESCO L. CHIOCCI
Consiglio Nazionale delle Ricerche, C. S. Geologia Tecnica, Roma, Italia
AND
H. EDWARD CLIFTON
U.S. Geological Survey, 345 Middlesfield Road, MS 99, Menlo Park, California 94025

ABSTRACT: Pleistocene nearshore deposits (part of the Merced Formation) exposed in sea cliffs south of San Francisco, California, contain elongate, gravel-filled casts at the bases of some gravel or sand beds. The size and shape of these linear features resemble those of gutter casts of sand protruding into shale in other deposits. The gravel-filled gutter casts of the Merced Formation show a strongly preferred east to northeast orientation, approximately normal to the shoreline trend inferred on the basis of the modern shoreline trend and the regional structural grain. Truncation and absence of deformation in the lamina of the sand into which the gutter casts extend indicate that the features are erosional rather than constructional (due to loading), and their orientation implies that they were cut by oscillatory currents beneath shoaling waves. As such, they provide a useful indication of the shoreline trends in other ancient gravelly nearshore deposits. In the Merced Formation, high-angle cross-bedding is abundant and dips primarily in a longshore or obliquely offshore direction. The variability of the cross-bedding makes it a less reliable indicator of shoreline trends. Steep, in some cases vertical or overhanging, walls to the gutter casts and the presence of clasts of sandstone in the casts that are indistinguishable from the underlying material indicate that the structures are formed by rapid cutting and filling of nearshore sand that is unexpectedly cohesive.

INTRODUCTION

The only commonly occurring directional feature within ancient clastic nearshore deposits is trough or tabular cross-bedding (Leckie and Krystinik, 1989). Unfortunately, the relation between cross-bedding dip direction and shoreline trend is highly variable. Asymmetric oscillatory currents generated by shoaling waves drive bedforms shoreward, producing onshore-dipping cross-strata (Clifton and others, 1971; Clifton, 1981). Longshore currents produce shore-parallel cross-bedding and rip currents generate cross-strata that dip directly or obliquely offshore (Clifton, 1981; Dupré, 1984). As a consequence, very little generalization can be drawn regarding the orientation of the shoreline from cross-bedding in the nearshore environment.

Another, generally unrecognized, directional structure occurs in some gravelly nearshore deposits. This structure consists of elongate gravel-filled scours that, in size and shape, resemble the sandy gutter casts described by others; it may be a substantially more reliable indicator of shoreline trend than is cross-bedding. Where it occurs in conjunction with cross-bedding, this structure provides a basis for interpreting the origin of the cross-stratification.

In this paper we describe gravel-filled gutter casts in progradational nearshore sequences of Pleistocene age within the Merced Formation south of San Francisco, California. The depositional environment and shoreline trend for these young deposits are reasonably well constrained and provide a basis for analyzing the orientation relative to the shoreline and the origin and paleogeographic significance of the gravel-filled gutter casts. The features also indicate that the nearshore sand into which they were cut was surprisingly cohesive.

GUTTER CASTS

The term "gutter cast" was introduced by Whitaker (1973) to describe elongate protuberances in the underside of sedimentary strata, some meters long and centimeters to a few decimeters wide and deep. As defined, the structures may be symmetric or asymmetric in cross section and can have vertical or even overhanging sides. The trend of a gutter cast on the base of a bed ranges from very straight to somewhat sinuous.

Most of the gutter casts described in the literature are sand-filled depressions cut into underlying lutitic sediment (Goldring, 1971; Whitaker, 1973; Aigner and Fütterer, 1978; Allen 1982; Collinson and Thompson, 1982; Goldring and Aigner, 1982). A few contain gravel, mostly of intraformational or biogenic origin (Goldring, 1971; Whitaker, 1973). Leithold and Bourgeois (1984) noted the dimensional similarity of gutter casts to small steep-walled sand- and gravel-filled scours cut into shallow-water wave-dominated sandstone of Miocene age in Oregon.

Although gutter casts have been referred to an intertidal setting, they are more likely formed in a shallow submarine environment (Aigner and Fütterer, 1978). Whitaker (1973) noted the occurrence of gutter casts in turbidite successions and concluded that the structure has little value as a paleobathymetric indicator. Nearly all authors agree that gutter casts originate where moving water, with or without tools, scours a small, elongate channel into the sediment underlying the sediment-water interface (Allen, 1982). The currents responsible have been thought to be either completely unidirectional or tidal (Aigner and Fütterer, 1978), or generated by passing waves (Goldring, 1971). Although sandy gutter casts are likely to be useful paleocurrent indicators, their orientation relative to the nearest shoreline may differ substantially. In some cases they trend parallel to inferred shorelines (Aigner and Fütterer, 1978); in others, especially where waves are involved, they may lie normal to an inferred shoreline (Goldring, 1971).

NEARSHORE FACIES OF THE MERCED FORMATION

The Merced Formation is a late Cenozoic unit that is superbly exposed in tilted section in sea cliffs south of San Francisco (Fig. 1). The formation, in its exposure north of the San Andreas fault, comprises more than 1,750 m of

From Shoreline to Abyss, SEPM Special Publication No. 46

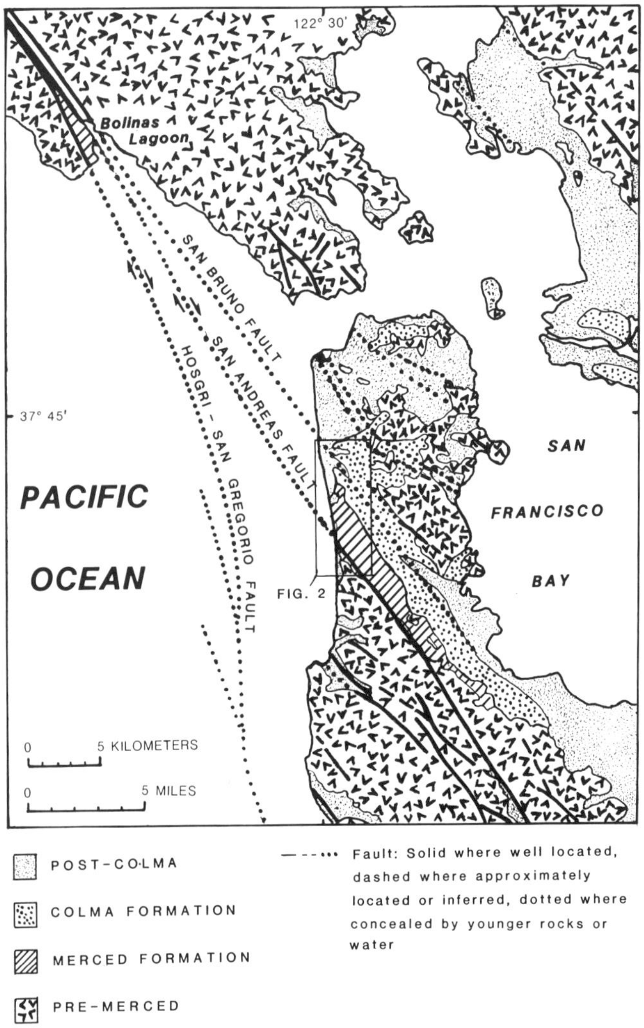

FIG. 1.—Location (inset box) of the seacliff exposures of the Merced Formation near San Francisco, California.

section that appears to be entirely of Pleistocene age (Clifton and others, 1988). The succession consists of at least 30 transgressive-regressive cycles (Fig. 2), most of which probably resulted, at least partly, from Pleistocene glacio-eustatic fluctuations of sea level. The regressive intervals are characterized by shallowing-upward progradational sequences, and 24 of these sequences contain cross-bedded gravelly facies inferred to have originated within a nearshore environment (Hunter and others, 1984).

Sixteen of the nearshore units lie within a continuous sequence that extends from inner-shelf facies below to beach-foreshore facies above. The rest are either truncated at the top by a transgressive erosional surface or begin with such a sequence at the base. The nearshore units range in thickness from less than 1 to 9 m. The nearshore facies consists of interstratified coarse- to fine-grained sand and gravel. High- and low-angle, medium-scale cross-stratification is common in the sandy beds. Shells and shell fragments are present in most of the units, and their absence in others probably is due to subsequent chemical leaching. Evidence of faunal burrowing exists in many of the units.

The general oceanographic and paleogeographic setting for these Pleistocene deposits almost certainly resembled that of today. Most if not all of the nearshore facies originated along an open oceanic coastline dominated by large (1- to 2-m-high) waves under normal conditions, punctuated by storm waves in excess of 10 m at intervals of tens to hundreds of years. The Pleistocene shorelines probably trended somewhere between the present coastal orientation (north) and that of the Coast Range faults (north-northwest) that probably bounded the Merced basin.

The nearshore facies of the Merced Formation reflects deposition in the zone adjacent to the shoreline, where asymmetric currents of shoaling waves and wave-generated rip and longshore currents predominate. The depths of this zone are not precisely established. Rip currents typically extend shortly beyond the breaker line, which depends on the height of the waves. The water depth at the point of the breaking waves is commonly considered to be about 0.78 of the wave height (Iverson, 1951). Accordingly, under 10-m-high waves, the nearshore would extend through a breaker zone to depths of 12 to 15 m. Although the preservation potential of such storm deposits is not clear, it is probable that the Merced nearshore facies developed in water depths that ranged from zero to 10 ± 5 m (Clifton, 1988).

GRAVEL-FILLED GUTTER CASTS IN NEARSHORE FACIES IN THE MERCED FORMATION

The nearshore successions within the Merced Formation are marked by numerous erosional surfaces. Most of these are flat or gently undulatory, but in nearly every sequence that contains abundant gravel, a few erosional surfaces are marked by distinct irregularities. Some of these irregularities result from infaunal burrowing, but others take a cross-sectional form of pockets of gravel some centimeters across that protrude a few centimeters into underlying sand (Fig. 3). Most of these protuberances lie at the base of gravel beds, but a few extend beneath erosional contacts at the base of sand beds. Upon excavation, the protrusions can be seen to be elongate and relatively straight (Figs. 4 and 5). These linear features are herein referred to as "gravel-filled gutter casts."

The width of the gutter casts in the Merced gravel beds ranges from a few centimeters to a few decimeters; the average is on the order of 10 cm. The nature of exposure makes it difficult to ascertain their length; a few have been excavated for more than a meter into the sea cliff.

The shape of the casts in transverse section generally is "U-shaped," although "V-shaped" cross sections sometimes occur. The side walls of the casts can be very steep, and a few are vertical or even overhanging. The cross-sectional shape may change longitudinally. The ends of gutter casts are rarely found; where they are seen, the gutter casts lose their relief and gradually disappear in the longitudinal direction (Fig. 5). Where two or more gutter casts are closely spaced, the sand between the structures seems to stand in positive relief above the base of the adjacent gravel bed. Multiple gutter casts at the base of an individual bed are approximately parallel to one another.

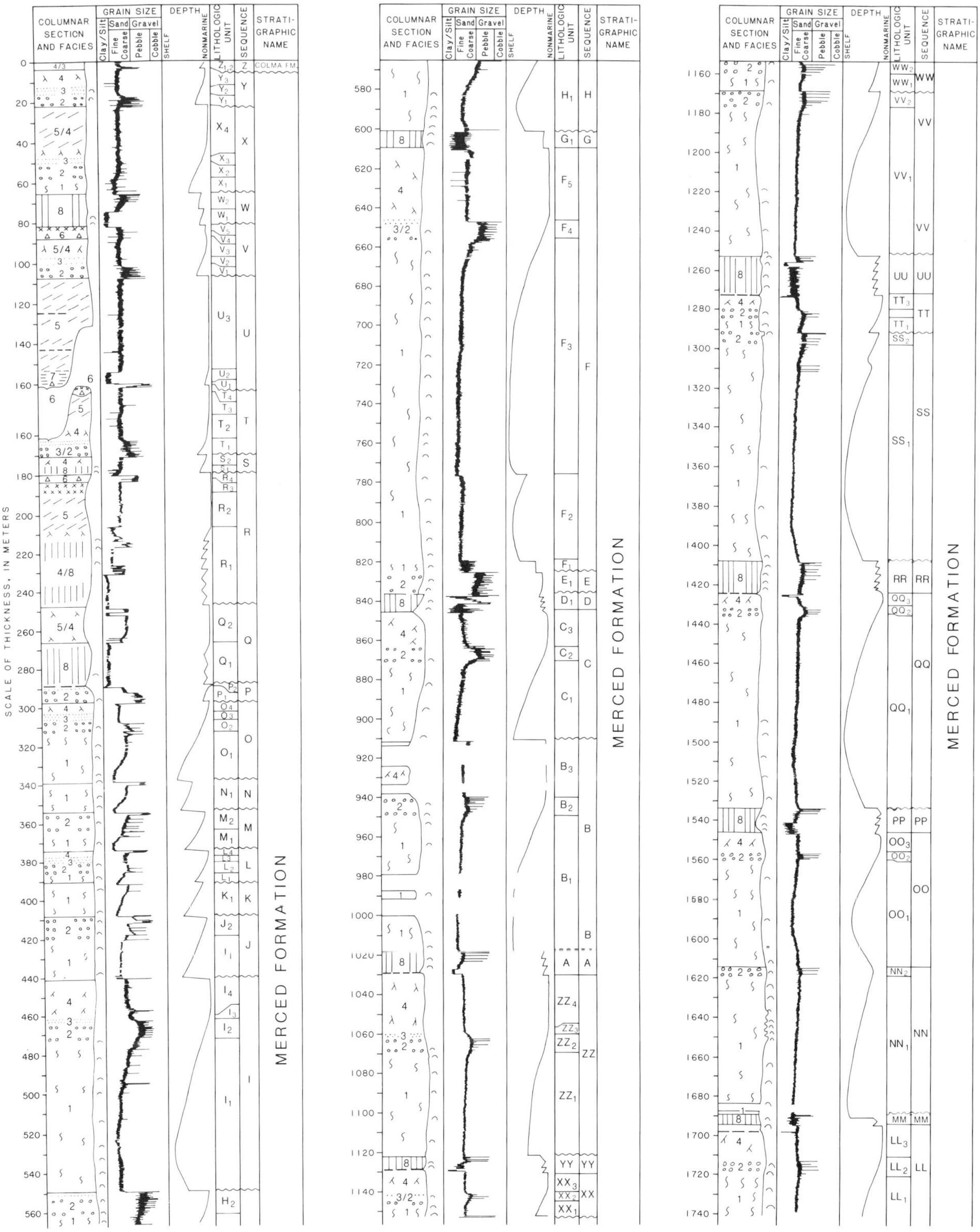

FIG. 2.—Stratigraphic succession and depositional facies of the Merced Formation in seacliff exposures south of San Francisco, California (after Clifton and others, 1988). Curve to the right of the stratigraphic column indicates inferred depth of deposition. Sequences as designated in Hunter and others (1984). Depositional facies as follows: 1) shelf, 2) nearshore, 3) foreshore, 4) backshore, 5) eolian dune, 6) alluvial, 7) swamp/marsh/pond, and 8) embayment.

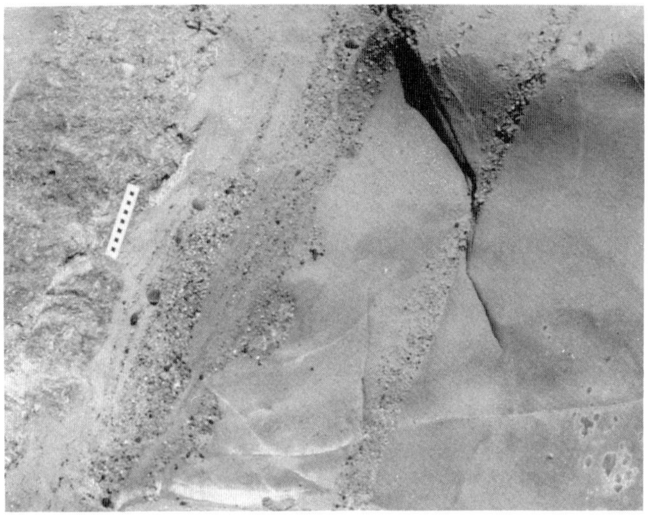

FIG. 3.—Gravel-filled irregularities at the base of a gravel bed in nearshore facies of the Merced Formation. Features are typical cross sections of gravel-filled gutter casts. Scale in centimeters.

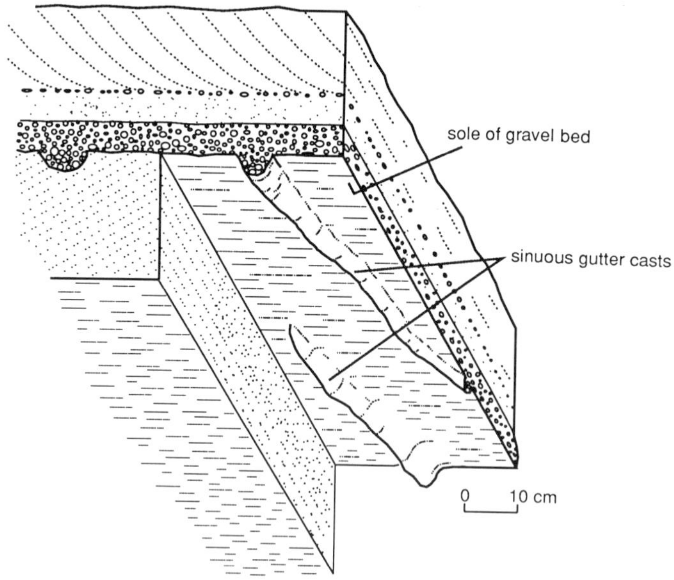

FIG. 5.—Field sketch of slightly sinuous gravel-filled gutter casts at the base of a gravel bed in nearshore facies of the Merced Formation.

Most of the gravel filling these gutter casts consists of pebbles 1 cm or less in diameter. One gutter cast contained a cobble 14.5 × 10 × 6 cm. A few gutter casts contain clasts, several centimeters across, of sand that strongly resembles the wall material (Fig. 6). Typically, the gravel within a gutter cast is indistinguishable from that of the overlying gravel bed, although in some cases the clasts in the cast are somewhat larger and the proportion of sand may be somewhat less. In some gutter casts, the fill is stratified and reflects a filling from one side or the other of the gutter. No relation seems to exist between the size of gravel-filled gutter casts and the thickness of the overlying bed.

The gutters clearly are eroded into the subjacent sand rather than produced by loading or injection. Where lamination can be seen in the sand adjacent to a cast, it is invariably truncated at the margin of the cast without evidence of deformation (Fig. 7).

DIRECTIONAL RELATIONS

The generally low degree of induration of the Pleistocene deposits of the Merced Formation permits sufficient excavation of the sea cliff to establish the trend of the gravel-filled gutter casts. Although most of the sequences yielded only one or two measurable gutter casts, two (sequences *P* and *V* in Fig. 8) were sufficiently rich in gravel and exposed over a broad enough area that gutter casts could be measured at 7 and 11 intervals, respectively. Most of the gutter casts exhumed within the Merced Formation, when rotated to horizontal around the strike of the enclosing strata, trend

FIG. 4.—Partly exhumed gravel-filled gutter cast at the base of a gravel bed showing linear character of the cast. Note absence of deformation of the lamina adjacent to the cast.

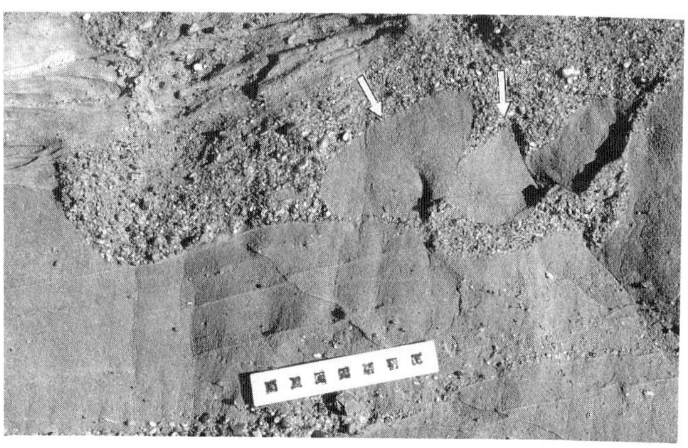

FIG. 6.—Clasts of sandstone (arrows), within a gravel-filled gutter cast, that are indistinguishable from the sand into which the cast extends. Scale in centimeters.

FIG. 7.—Cross section of gravel-filled gutter cast (to left of centimeter scale) and gravel-filled burrow (to right of scale). Note truncation of faint lamination against the gutter cast.

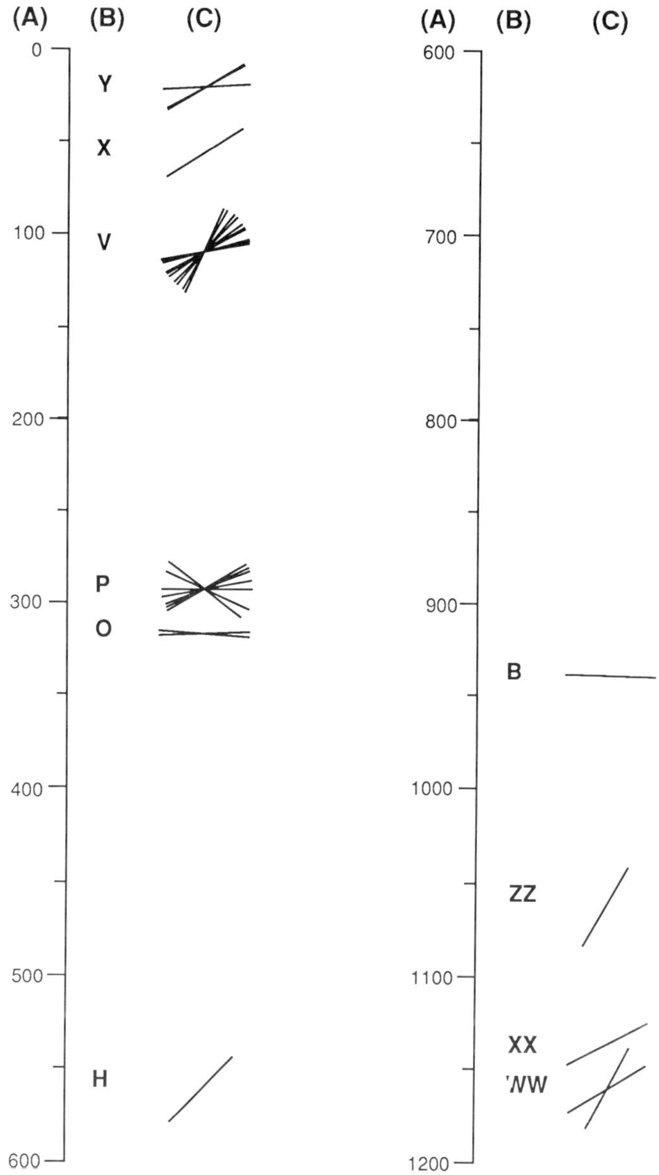

FIG. 8.—Thickness (in meters) (A) of the interval of the Merced Formation and sequences (B) in which gravel-filled gutter casts were exhumed, and the trends of these casts (C). Trends corrected for tectonic rotation.

from east to northeast (Fig. 8). The measurements, which encompass nearly 1,200 m of section, suggest a possible stratigraphic variability in the orientations. Gutter casts in the interval of 350 to 1,200 m below the top of the section are sparse, but generally trend northeast. Those in the interval of 270 to 320 m below the top (sequences *O* and *P*) trend more to the east. Those near the top of the section (sequences *V*, *X* and *Y*) have a northeast trend.

The other prominent directional structure in the Merced nearshore facies is high-angle cross-stratification. Forest dip directions were measured in 149 cross-bedded units in these deposits, encompassing more than 1,700 m of section. The foresets, after rotation of the units to horizontal around the strike of the enclosing strata, dip predominantly toward the south and south-southeast (Fig. 9). The orientation of the gutter casts over this same general stratigraphic interval appears to be neither parallel nor normal to the foreset directions.

The lack of any approximate rectilinear relation of gutter-cast and foreset orientation also is apparent in the summary of measurements from individual sequences (*O* + *P* and *V*) where enough gutter-cast and cross-bedding data exist to draw a useful comparison (Figs. 10 and 11). In both of these sequences, the gutter casts trend obliquely to the dominant direction of foreset inclination. A number of gravel-filled gutter casts in the Merced Formation lies at the base of cross-bedded pebbly sand (Fig. 12). In every case observed, the gutter cast was oriented with a general east trend beneath foresets that generally were inclined to the south.

ORIGIN OF THE GRAVEL-FILLED GUTTER CASTS

The gravel-filled gutter casts in the nearshore facies of the Merced Formation clearly formed in an energetic environment that included both oscillatory flow generated beneath passing waves and unidirectional flow of wave-induced longshore and rip currents. Most of the largest pebbles within the gutter casts, about 1 cm long, require a velocity on the order of 1 m/s to reach the threshold for grain transport under oceanic waves (Komar and Miller, 1975).

The vertical to overhanging walls of the casts could result either from erosion of the underlying sand or from subsequent deformation by loading. The absence of sand injections (flame structures) into the overlying gravel, the abrupt truncation of undeformed stratification in the underlying sand at the margin of some of the casts, and the presence of

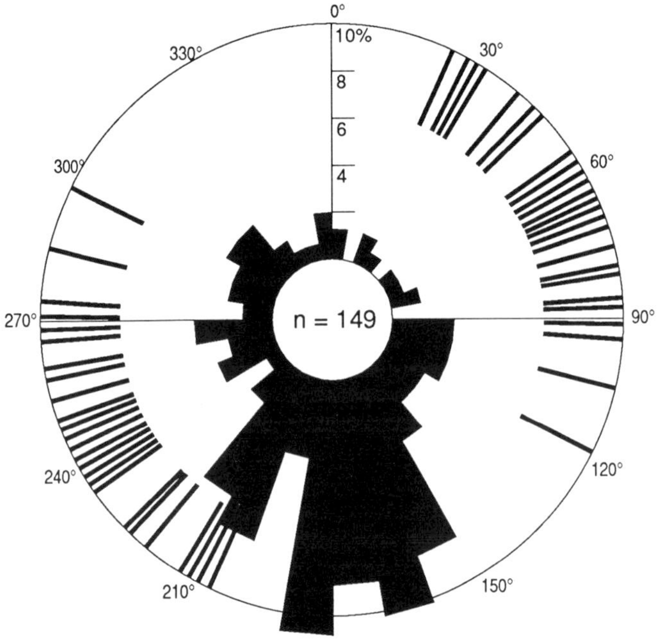

FIG. 9.—Summary of cross-bedding dip directions (rose diagram) and gutter-cast trends (on outer circle) within the Merced Formation in its seacliff exposures south of San Francisco. All data displayed after enclosing strata have been rotated to horizontal around the strike of the beds. Number in center of rose diagram indicates number of cross-bedding measurements.

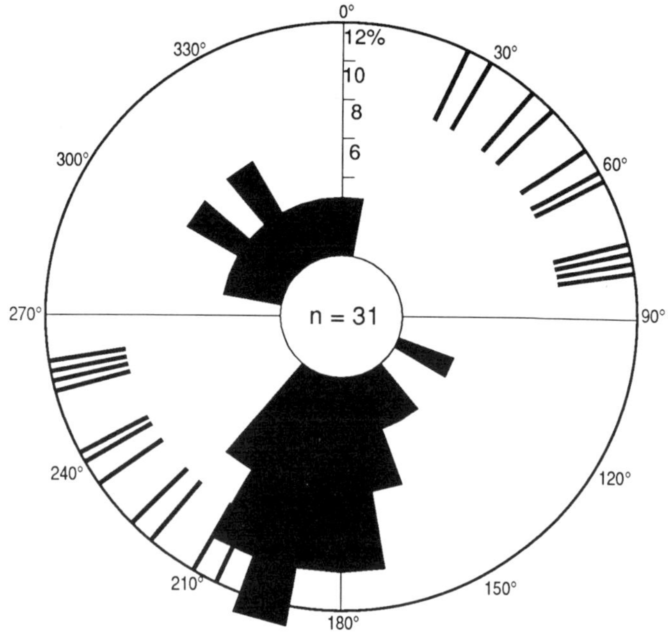

FIG. 11.—Cross-bedding dip directions (rose diagram) and gutter-cast trends (on outer circle) in sequence *V*. All data displayed after enclosing strata have been rotated to horizontal around the strike of the beds. Number in center of rose diagram indicates number of cross-bedding measurements.

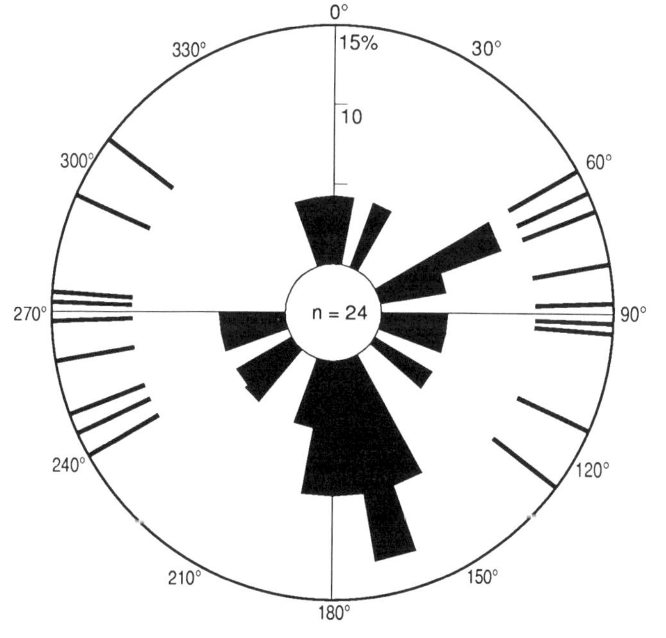

FIG. 10.—Cross-bedding dip directions (rose diagram) and gutter-cast trends (on outer circle) in sequences *O* and *P* (combined because they are stratigraphically separated by only a few meters). All data displayed after enclosing strata have been rotated to horizontal around the strike of the beds. Number in center of rose diagram indicates number of cross-bedding measurements.

clasts of the underlying sand within a few casts argue against oversteepening by deformation. The production and maintenance of such steep walls during an episode of erosion implies rapid scouring and immediate filling of the structures.

The orientation of the gutter casts in the Merced nearshore facies is approximately normal to the inferred shoreline. The structures therefore probably were produced either by the oscillatory currents generated by the shoaling waves and/or by rip currents. We consider a rip-current origin less likely for two reasons. First, the overlying bed typically does not contain offshore-directed cross-bedding; the gravel

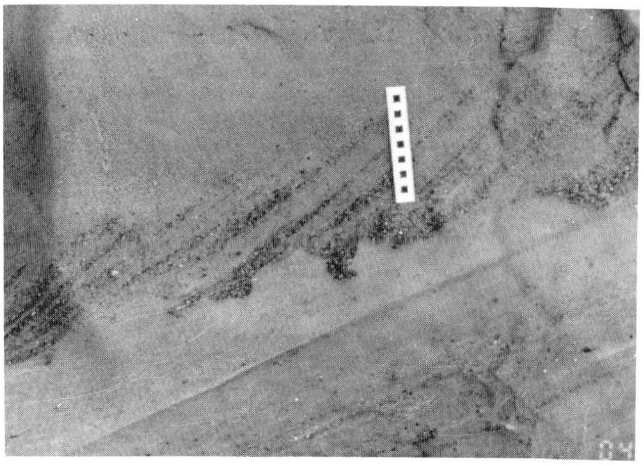

FIG. 12.—Gravel-filled gutter casts at the base of a cross-bedded sand and gravel bed in sequence *P*, Merced Formation. Scale in centimeters.

beds accordingly seem more appropriately interpreted as storm lags than as concentrations on the floor of rip channels. Second, if unidirectional flow were important in the creation of the casts, we would expect a significant number to be produced by longshore currents (given the prevalence of generally shore-parallel cross-stratification in the nearshore facies). In contrast, gutter casts with an approximately shore-normal trend commonly occur at the base of beds that show shore-parallel cross-bedding. Shore-parallel gutter casts seem nearly nonexistent in the Merced (possibly in part because a bias exists against recognition of gutter casts oriented parallel to the trend of the outcrop).

The more-or-less isolated occurrence of the gutter casts at the base of any bed suggests that a special set of circumstances may have been required for their initiation. It is likely that some, if not all, were initiated by turbulence associated with the presence of burrows, as suggested by Leithold and Bourgeois (1984). Burrow structures are not uncommon in the Merced nearshore deposits (Figs. 13 and 14) and may have either provided localized bottom irregularity or specific sites of biogenically reworked and therefore less cohesive sand that was preferentially susceptible to scouring. Alternatively, preferential scouring may have begun around an isolated large clast, as suggested by Aigner and Fütterer (1978). In either case, pebbles are likely to have accumulated in a depression caused by the initial scouring, where they would then provide an additional source of turbulence or act as abrasive tools for extending the gutter into the underlying sand.

COHESIVENESS OF THE SAND CUT BY GRAVEL-FILLED GUTTERS

The gutters in the Merced nearshore facies were almost certainly cut into sand that was no more consolidated than that which presently underlies the sea floor of the open Pacific Coast nearshore zone. Yet, the steep-to-overhanging walls of the casts and, particularly, the presence of clasts of sandstone identical to the wall material within the gutter fill imply some cohesiveness.

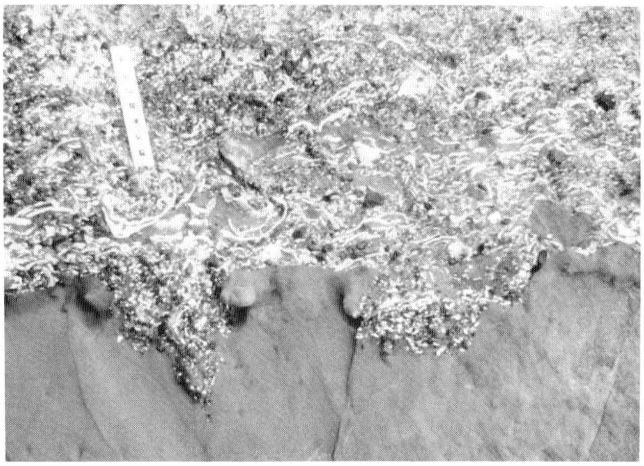

FIG. 14.—Cross sections of probable burrows at the base of a gravel bed that resemble cross sections of gutter casts. Sequence *M*, Merced Formation. Scale in centimeters.

It is tempting to assume that the sand into which the gutters were cut would not sustain slopes greater than the angle of repose for loose sand. Simple digging into natural water-saturated sand on the sea floor indicates that this is not the case. Nearly vertical walls persisted for at least a minute in the margins of holes dug rapidly into poorly sorted carbonate sand in Truk Lagoon (Fig. 15), and for nearly 20 seconds or so in holes dug into well-sorted, fine quartzose sand in a small tidal channel in a Pacific Coast embayment. Sand beneath the wave-swept nearshore environment is likely to be more tightly packed and therefore more cohesive, and very rapid filling of the gutters might preserve temporarily oversteepened walls. Binding by biogenic processes such as addition of mud by filter feeders, digestive packing by small infaunal deposit feeders (Clifton and Thompson, 1978), or algal growth may provide further mechanisms for increasing cohesiveness. Possibly, gravel within the gutter

FIG. 13.—Large irregularity at the base of a gravel bed in the nearshore deposits of the Merced Formation, probably created by a feeding ray. Scale in centimeters.

FIG. 15.—Underwater excavation in poorly sorted carbonate sand, Moen Island, Truk Lagoon. Scale is 15 cm protractor with round side facing camera. Excavation was made about 30 seconds before photograph was taken; walls continue to exceed substantially the angle of repose.

provided support for the walls as they were cut; this, however, would still not adequately explain the presence of clasts of wall sediment within the gutter. Conceivably, a relatively high degree of substrate cohesiveness is required to form gutter casts. Otherwise, the scour could be less localized and erosion would proceed to about the same degree over the entire bed, without producing distinct, separate gutters.

PALEOGEOGRAPHIC SIGNIFICANCE OF GRAVEL-FILLED GUTTER CASTS IN NEARSHORE FACIES

The gravel-filled gutter casts in the Merced Formation appear to have been formed by currents associated with shoaling waves. As such, these structures are potentially important sources of paleogeographic information. Waves entering shallow water along a straight coast refract until their crests are nearly parallel to the shoreline (Komar, 1976, p. 110). Gravel-filled gutter casts in nearshore deposits can therefore be inferred to be nearly normal to the paleoshoreline orientation.

In the Merced Formation, variation in the orientation of gutter casts within the stratigraphic column may reflect subtle changes in shoreline trend. For example, during deposition of sequences *O* and *P* (Fig. 8), the shoreline may have trended more north in contrast to the rest of the succession, where it appears to have been more northwest. Such a difference is not detectable in the direction of cross-bedding foresets in these units.

The predominantly south-dipping cross-bedding in these nearshore deposits (Fig. 9) indicates a strong influence of south-flowing longshore currents. Within individual sequences, the relation between foreset dip direction and shoreline appears to be more complex. For example, in sequence *V*, the primary mode of cross-bedding inclinations is to the south-southwest, obliquely offshore from the northwest-trending shoreline indicated by the gutter casts, which implies a strong rip-current influence. A secondary cross-bedding mode toward the northwest suggests episodes of longshore flow in that direction. The cross-bedding dip directions in sequences *O* and *P* are more scattered, but the mode toward the south-southeast (Fig. 10) denotes a prevalence of longshore flow in this direction. Such a flow would be generated if waves approaching from the west (as indicated by the gutter-cast trends) were not completely refracted against a north-northwest—trending shoreline. Such subtleties of interpretation are not evident in the summary directional diagram (Fig. 9).

Where gutter casts occur at the base of cross-bedded sand and gravel beds, it seems probable that both gutter casts and cross-stratification were generated within such a short period of time that the orientation of approaching waves would be unchanged. The presence of general shore-normal gutter casts at the bases of beds with generally shore-parallel cross-bedding suggests sets of waves that generated both the gutter casts and the longshore currents that formed the cross-bedding. Combining gutter-cast trends and foreset inclinations in such cases may provide for further refinement of shoreline trend.

SIMILAR FEATURES IN OTHER SHALLOW-WATER DEPOSITS

Structures that resemble the gravel-filled gutter casts of the Merced Formation are present in shallow-water conglomeratic sandstone of Miocene age on the southern Oregon coast. Leithold and Bourgeois (1984) describe these features in detail, which they term "high angle scours." Like the gutter casts of the Merced, they are cut into sand, have steep and, in some cases, overhanging walls, and many are filled with gravel (Fig. 16). Leithold and Bourgeois note their similarity to gutter casts as defined by Whitaker (1973).

These "high angle scours" differ, however, in a number of potentially significant ways from those in the Merced Formation. Although the nature of the exposure made the three-dimensional geometry of the scours difficult to discern, Leithold and Bourgeois (1984) felt that they were more scoop- or flute-shaped than channel-shaped. The features are also larger than those in the Merced; their width and depth is measured in decimeters rather than in centimeters. The high-angle scours typically are asymmetric in vertical cross section, and many exhibit multiple episodes of filling (Fig. 17). Although Leithold and Bourgeois (1984) infer development of the features in water depths (0–5 m) similar to those envisioned for the nearshore facies of the Merced, the Miocene scours occur in sediment characterized by hummocky-stratified and parallel-bedded organic-rich sandstone, an association different from the cross-bedded sand and gravel of the Merced occurrence. Leithold and Bourgeois (1984) attribute the Miocene scours largely to unidirectional currents, owing to the asymmetry of the features.

Despite these differences, both the gravel-filled gutter casts of the Merced and the Miocene high-angle scours represent the effects of localized scouring on a shallow sandy sea floor. Leithold and Bourgeois (1984) noted the requirement of substantial cohesion of the eroded material to explain the steep walls of the scours and the presence of intraclasts of wall material in the scour fill. They suggested that this co-

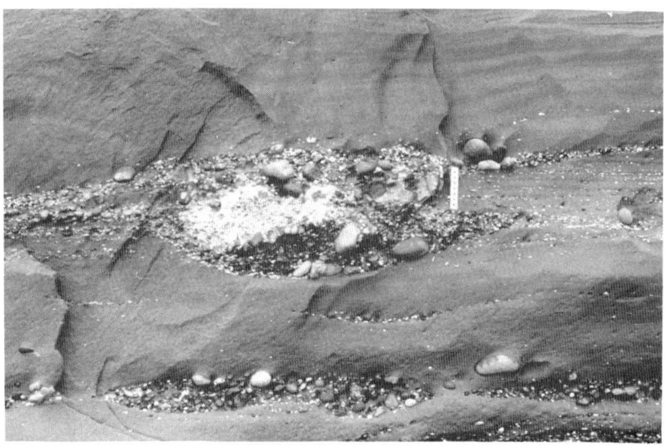

FIG. 16.—"High angle scour" (Leithold and Bourgeois, 1984) in shallow-water conglomerate bed of Miocene age on the southern Oregon coast. Note similarity to gravel-filled gutter casts in the Merced Formation. Scale in centimeters.

FIG. 17.—Large "high angle scour" in shallow-water conglomerate bed of Miocene age in southern Oregon that shows evidence of multiple episodes of filling. Note overhanging wall on the upper right margin of the "scour."

hesiveness may result from compaction and early diagenesis of the sand, facilitated by the presence of volcanic clays and organic matter (which are generally absent in the nearshore facies of the Merced).

The gravel-filled gutter casts of the Merced and the high-angle scours of the Miocene rocks in southwestern Oregon almost certainly are members of the same family of features. The gutter casts in the Merced appear to be more linear and are better attributed to scouring by oscillatory flow under passing waves. The Miocene features are larger, apparently less linear, and may be better attributed to scouring by unidirectional flow. Both sets of structures imply an unexpected degree of cohesiveness in the eroded bed, although it seems unlikely that the time span between events in the dynamic nearshore facies of the Merced would be sufficient to allow for the compaction and early diagenesis postulated for the Miocene rocks.

CONCLUSIONS

Small linear scours cut into underlying sand are common features at the bases of gravel beds in Pleistocene nearshore facies of the Merced Formation, San Francisco, California. In approximately transverse vertical section, these structures may be mistaken for burrow fills; in approximately longitudinal section, they may be seen only as a broad shallow scour. Exposure of the sole of the bed (by excavation, if necessary) is generally required to establish the linear geometry of the features.

The gravel-filled gutter casts of the Merced exhibit a strongly preferred orientation that is approximately normal to the inferred shoreline. The structures are best attributed here to scouring by oscillatory currents under shoaling storm waves, although a rip-current origin cannot be completely discounted.

Gravel-filled gutter casts provide a useful indicator of paleoshoreline trend. Although the structures are presumed specifically to reflect the direction of wave approach, their occurrence close to the beach is in an area where refraction is likely to have turned the approach nearly normal to the shore. Gutter casts in a nearshore setting, with their consistently shore-normal orientation, provide a more reliable indicator of shoreline trend than cross-bedding dip direction, which may be either longshore, offshore, or onshore.

Gravel-filled gutter casts and similar high-angle scours in shallow marine sandstone may require a high level of cohesiveness of the substrate for their formation. A variety of factors, such as tight packing, organic binding, compaction and early diagenesis, may contribute to this cohesiveness. It is likely that the sand that exists today beneath high-energy surf zones is more cohesive than previously thought.

ACKNOWLEDGMENTS

Funding for the senior author's field research was provided by The Departimento di Scienze delle Terra, Universita delgi Studi, "La Sapienza," Roma, Italia.

REFERENCES

AIGNER, THOMAS, AND FÜTTERER, E., 1978, Kolk-Töpfe und-Rinnen (pot and gutter casts) im Muschelkalk-Anzeiger für Wattenmeer?: Neues Jahrbuch für Geologie und Paläontologie Abhandlungen, v. 156, p. 285–304.

ALLEN, J. R. L., 1982, Sedimentary Structures, Their Character and Physical Basis, Volume II: Developments in Sedimentology 30B, Elsevier Scientific Publishing Company, New York, 663 p.

CLIFTON, H. E., 1981, Progradational sequences in Miocene shoreline deposits, southeastern Caliente Range, California: Journal of Sedimentary Petrology, v. 41, p. 651–670.

CLIFTON, H. E., 1988, Sedimentologic approaches to paleobathymetry, with applications to the Merced Formation of central California: PALAIOS, v. 3, p. 507–522.

CLIFTON, H. E., AND THOMPSON, J. K., 1978, *Macaronichnus segregatis*, a feeding structure of shallow marine polychaetes: Journal of Sedimentary Petrology, v. 48, p. 1293–1302.

CLIFTON, H. E., HUNTER, R. E., AND GARDNER, J. V., 1988, Analysis of eustatic, tectonic, and sedimentologic influences on transgressive and regressive cycles in the late Cenozoic Merced Formation, San Francisco, California, *in* Paola, C., and Kleinspehn, K. L., eds., New Perspectives in Basin Analysis: Springer-Verlag, New York, p. 109–128.

CLIFTON, H. E., HUNTER, R. E., AND PHILLIPS, R. L., 1971, Depositional structures and processes in the nonbarred high-energy nearshore: Journal of Sedimentary Petrology, v. 41, p. 651–670.

COLLINSON, J. D., AND THOMPSON, D. B., 1982, Sedimentary Structures: George Allen and Unwin, Inc., Boston, 194 p.

DUPRÉ, W. R., 1984, Reconstruction of paleo-wave conditions during the late Pleistocene from marine terrace deposits, Monterey Bay, California: Marine Geology, v. 60, p. 435–454.

GOLDRING, ROLAND, 1971, Shallow-water sedimentation as illustrated in the Upper Devonian Baggy Beds: Memoir, Geological Society of London, v. 5, p. 1–80.

GOLDRING, ROLAND, AND AIGNER, THOMAS, 1982, Scour and fill: The significance of event separation, *in* Einsele, G., and Seilacher, A. eds., Cyclic and Event Stratification: Springer-Verlag, Berlin, p. 354–362.

HUNTER, R. E., CLIFTON, H. E., HALL, N. T., CSÁSZÁR, GEZA, RICHMOND, B. M., AND CHIN, J. L., 1984, Pliocene and Pleistocene coastal and shelf deposits of the Merced Formation and associated beds, northwestern San Francisco Peninsula, California: Society of Economic Paleontologists and Mineralogists Field Trip Guidebook No. 3, 1984 Midyear Meeting, p. 1–29.

IVERSON, H. W., 1951, Studies of wave transformation in shoaling water, including breaking *in* Gravity Waves: National Bureau of Standards Circular 521, p. 9–32.

KOMAR, P. D., 1976, Beach Processes and Sedimentation: Prentice-Hall, Inc., Englewood Cliffs, New Jersey, 417 p.

KOMAR, P. D., AND MILLER, M. C., 1975, The initiation of oscillatory ripple marks and the development of plane-bed at high shear stresses under waves: Journal of Sedimentary Petrology, v. 45, p. 697–703.

LECKIE, D. A., AND KRYSTINIK, L. F., 1989, Is there evidence for geostrophic currents preserved in the sedimentary record of inner to middle shelf deposits?: Journal of Sedimentary Petrology, v. 59, p. 862–870.

LEITHOLD, E. L., AND BOURGEOIS, JOANNE, 1984, Characteristics of coarse-grained sequences deposited in nearshore, wave-dominated environments—examples from the Miocene of southwest Oregon: Sedimentology, v. 31, p. 749–775.

WHITAKER, J. H. MCD., 1973, 'Gutter casts', a new name for scour-and-fill structures, with examples from the Llandoverian of Ringerik and Malmöya, southern Norway: Norsk Geologisk Tidsskrift, v. 53, p. 403–417.

PART III
CONTINENTAL-SHELF AND -SLOPE STUDIES

CHARACTERISTICS OF GLOBAL-SEDIMENT FORMATION IN REGIONS OF HIGH- AND SUPERHIGH-ACCUMULATION RATES

ALEXANDER P. LISITZIN
Institute of Oceanology, USSR Academy of Sciences, Krasikova 23, Moscow 117218, USSR

ABSTRACT: Sedimentation in regions of high (100–1,000 Bubnoff units [B]) and superhigh accumulation rates (more than 1,000 B) differs drastically from that characterized by normal rates (less than 100 B), in terms of distribution, mass, composition and structure. One Bubnoff unit is 1 mm/1 ka. High-accumulation rates result in high-water content, high-sediment mobility, and high organic-carbon content. The ability of sediments to slide and flow into depressions increases, and the main transport processes are gravitational flows. Sedimentation associated with avalanche rates is the principal sedimentation process on Earth. The thickness of normal deposits in the pelagic realm does not exceed 1 km, but in the avalanche-sedimentation regions, thickness reaches as much as 10 to 15 km. Normal sedimentation is associated with dispersion of sedimentary material, whereas avalanche sedimentation is associated with the concentration and formation of thick, autonomously developing sedimentary bodies. Avalanche sedimentation occurs where suspension concentration exceeds 5 mg/l, where sediment accumulation rates of more than 100 B occur, and absolute masses of more than 5 mg/cm^2/1 ka accumulate.

According to the hypographic curve of the Earth, three global levels are revealed: 1) river mouths (river-sea border), 2) continental rises, and 3) deep-sea troughs. They are each separated by 3 to 6 km of topographic relief, but are related closely by sedimentary source material. Sediment translocation occurs with hiatuses and is a function of sea-level changes. With an increase in sea level, deposition occurs on the first level of concentration, and/or hiatuses form on the second and 3rd levels; with a decrease in sea level, hiatuses form on the first level and concentration occurs on the second and third levels. These three vertical levels of sediment deposition, related to each other in space and time, are unified by sediment composition (including organic carbon). These sedimentary systems (each system includes collection, transportation and accumulation of sediment) include subaerial and subaqueous aspects. Concentration of organic-rich material with isostatic crustal downwarping takes place in these zones, and in some cases results in oil and gas formation. This is the principal zone of oil and gas formation. Capture of the main supply, more than 90 percent of the river-suspension load on the first level, results in a deficiency of pelagic material. Precipitation of the dissolved load forms sediment by bioprocesses ($CaCO_3$, SiO_2 amorphous, C organic, etc.). However, river-suspension residues (7–8 percent), absorption, eolian, ice and endogenous-material input are most important. The concept of avalanche sedimentation requires new approaches to the questions of sedimentation, geochemistry and the formation of mineral deposits.

CHARACTERISTICS OF SEDIMENT FORMATION IN REGIONS OF HIGH AND SUPERHIGH ACCUMULATION RATES

As studies of suspended-matter concentration in water, rates of modern-sediment accumulation, absolute masses, thickness and volumes of sediment deposits illustrate, the distribution of sedimentary material on the sea and ocean floors is extremely uneven. Most sedimentary material (more than 90 percent) accumulates close to the location of its input, in regions of high and superhigh sediment accumulation rates. If the major part of the ocean floor is characterized by sedimentation rates of less than 1 to 10 mm/1 ka (Fig. 1), regions of high-sedimentation rates are characterized by 100 to 1,000 mm/1 ka, and regions of superhigh sedimentation by more than 1,000 mm/1 ka. One Bubnoff unit equals 1 mm/1 ka.

High-sedimentation rates result in significant changes in the associated sediments, e.g., in an increase of their water contents, which, in turn, results in instability on slopes, and in the addition of significant organic-matter content, which also decreases stability (gas-sapping, etc.) and results in the formation of special geochemical structures, textures, etc.

The processes associated with high- and superhigh sedimentation rates were named *avalanche sedimentation* by Lisitzin (1982; 1983; 1984a,b; 1985a,b; 1986, 1988), because they can be compared with the avalanche accumulation of snow in mountains (Seemann, 1978), avalanche electron processes, or storm concentrations of elements in geochemistry. Where sediment accumulates with normal rates (<100 mm/1 ka), the main process is "particle by particle" deposition. In the regions of avalanche sedimentation, the distribution of sediment is determined mainly by the process of gravitational translocation from shallow hypsometric levels to deeper ones, i.e., relief is very significant in the formation of the resultant sedimentary bodies.

Investigations carried out during the last 10 years have shown that most sediment accumulates not on the vast territories of the World Ocean floor, but on small areas along the periphery of the continents, the basins of avalanche sedimentation. In these places the thicknesses of sediment often exceeds 10 to 15 km, while at the same time it reaches only 1 km in the pelagic regions of the deep ocean (Fig. 2).

The purpose of this paper is to compare the genetic charateristics of avalanche and normal pelagic sediments in the World Ocean, thus revealing the characteristic features of accumulation of the major mass and volume of sediment on the contemporary Earth and in past time.

The extremely uneven character of sediment distribution in the World Ocean with its concentrations in particular places results in the deficiency of sediment elsewhere and may result in the formation of hiatuses. Because the hypsometric position of the depositional site has key importance for avalanche sedimentation, it is possible to delineate three major global levels of sedimentation that correspond to inflections of the Earth's hypsometric curve, i.e., to the places, where the power of gravitational fluxes are lost. These levels are: 1) at river mouths, 2) at the base of the continental slope, and 3) in deep-sea trenches of active continental margins. Although related to each other in vertical section, these differ in depth over a range from 4 to 10 km and form the sedimentary system of earth. Transport of sediment from one level to another and the appearance of large hiatuses in one level and of avalanche sedimentation in the other are related temporally at a global scale, i.e., they are synchronous. These relations are determined mainly by variations

From Shoreline to Abyss, SEPM Special Publication No. 46

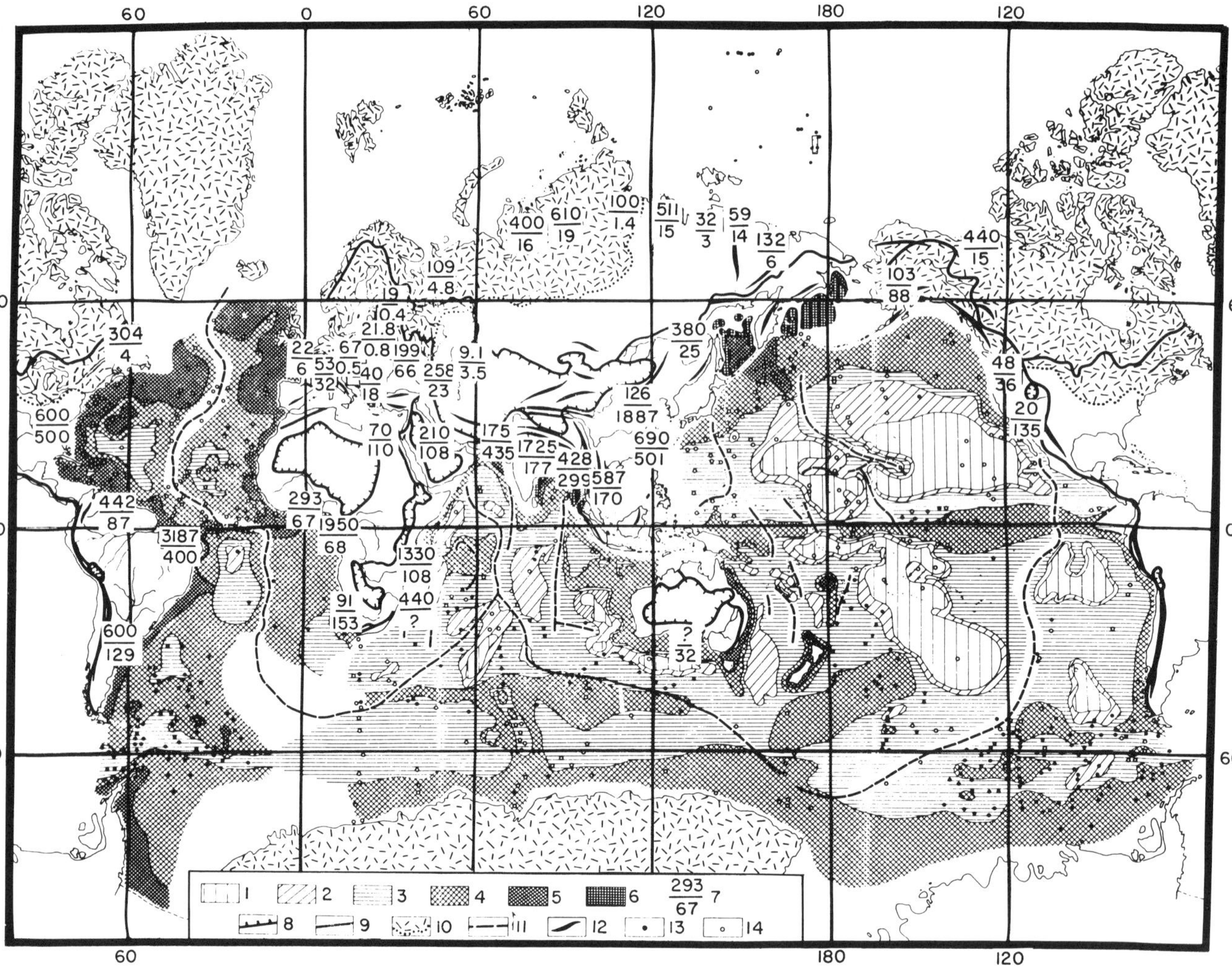

FIG. 1.—Sediment accumulation rates in the World Ocean for the regions of normal (particle by particle) sedimentation. Sediment accumulation rates are given in mm/1 ka: (1) <1; (2) from 1 to 3; (3) from 3 to 10; (4) from 10 to 30; (5) from 30 to 100; and (6) >100. Other important factors are: (7) the flow rate of the largest rivers (numerator is the water discharge in km^3/yr; denominator is particulate matter in millions of tons/yr); (8) the border of the regions with internal drainage; (9) borders of the water-collection basins of the ocean; (10) largest regions of influence by Quaternary glaciation; (11) ridges on the ocean floor; (12) main ridges and mountain systems on land; (13) stations with measurements of sedimentation rate using paleomagnetic methods; and (14) estimates of sedimentation rate using other methods (Lisitzin, 1974).

of sea level with time, and are the recorders of these variations. Sea-level fluctuations are the natural regulators of avalanche sedimentation in translocation from the first level (river mouths) to the second (slope base) on a global scale.

In scale, these events, which take place underwater and are invisible to observers, significantly exceed known events for the continents. With potential change of hypsometric position on the slope by as much as 4 to 5 km, underwater avalanches acquire enormous dynamic power. The associated sediment volume often exceeds 10 to 30 km^3, and they may have a weight of tens of billions of tons. Transport of these masses from higher levels produces various types of gravitites, and the travel distances of these concentrated and water-diluted fluxes of sediment often reach 2,000 to 3,000 km.

The continental slope, extending for about 350,000 km, is much like a giant "factory" that produces gravitites of all forms. The largest mapped landslides have volumes of 10 to 100 times more than those previously mentioned. Thus, the Storegya landslide on the continental slope of Norway has a volume of about 3,900 km^3, and the distance of translocation of material in it reaches 500 km. The volume of this landslide alone exceeds the annual supply of the sedimentary material of all the rivers of the world by more than 300 times (Bugge, 1989).

The upper level of sediment accumulation (at river mouths) produces a global network of sediment traps that operate with an efficiency of about 90 percent. According to the author's data, only 7 percent of the particulate river discharge passes through this trap (i.e., 0.8 km^3 of the 12 km^3 of the annual supply of suspended sediment to the river mouths).

The upper limit of the upper level of avalanche sedimentation corresponds to modern sea level, which is high and still rising at a rate of 1 to 2 mm annually. As sea level decreases eustatically for the whole planet simultaneously,

FIG. 2.—Principal regions of avalanche sedimentation, which are concentrated along the periphery of the World Ocean. Thickness (in km) of the sedimentary deposits is based on geophysical and deep-sea drilling data: (1) <1; (2) from 1 to 3; and (3) >3. (4) Contour lines are isopachs in km (Lisitzin, 1980).

sediment is discharged from places of temporary accumulation at river mouths to the middle and lower level (slope base), where most of it remains in permanent storage. These massive shifts result in the formation of numerous hiatuses and discontinuities, which are revealed in the composition of river-mouth deposits with the help of seismostratigraphy and drilling. The appearance of breaks in the record, even on a global scale, does not mean that the sediment accumulation has totally stopped: it has simply been transferred from one level to another (contraphase principle).

Hiatuses on one hypsometric level of sedimentation occur at the same time as avalanche sedimentation on another: these are the two primary phases, replacing one another in the process of cyclic (two-phase) sedimentation. Finally, the processes of avalanche sedimentation are significant because they form the largest sedimentary bodies on Earth; they form the most abundant types of sedimentary deposits with specific features, and also are related to the largest accumulations of mineral resources, primarily oil and gas.

In this paper only one type of avalanche sedimentation is considered, the terrigenous one, which plays the major role in the formation of the Earth's surface. Other types of avalanche sedimentation exist: biogenic (upwelling, coral reefs, etc.), chemogenic (evaporite formation), and volcanogenic (active margins) (Lisitzin, 1982).

MATERIALS AND METHODS

Methods for quantitative studies of sediment have improved rapidly. The following methods have been used here: 1) data concerning the concentration of suspended sediment in the water column; 2) suspended-matter fluxes, measured with sedimentation traps installed at various depths and with isotopic methods; 3) sedimentation rates; 4) absolute masses or accumulation rates (burial); 5) the distribution of thickness of sedimentary deposits defined with the help of seismic studies; 6) drilling records, and 7) the areal and spatial distribution of volumes and masses of the sediment.

These studies, developing rapidly in the last 10 years, are usually carried out separately for various regions of the world. By generalization from many such studies, it is possible to obtain a quantitative estimate concerning the distribution of sediment on the ocean floor at the present time. These data are supplemented by studies of the composition of the various types of sediment. In the course of expeditions to various parts of the seas and oceans, the author has been collecting the material for this generalization for about 35 years (Lisitzin, 1974, 1978, 1988).

Special attention was paid to terrigenous input, the major component by mass of the sedimentary material delivered from continents by rivers. To obtain the following data using a standard system, the Institute of Oceanology carried out numerous expeditions to the mouths of the largest rivers of the USSR, which are little known to the scientists in the West (Lisitzin and Gordeev, 1974; Demina and others, 1978; Morozov, 1979; Artem'ev and others, 1981, Gordeev, 1983), and a major oceanographic expedition to the Amazon River (Monin and Gordeev, 1988). Studies of the regions of avalanche sedimentation of the largest rivers of the world also were carried out by Curray and Moore, 1971; Folger, 1972; Handa, 1972; Lisitzin, 1974, 1979, 1982; Broussard, 1975; Carbonnel and Meybech, 1975; Milliman and Boyle, 1975; Milliman and others, 1975, 1985, 1986; Subramanian, 1979; Gibbs, 1981; Berg, 1982; Hu and others, 1982; Gilbert, 1983; Stallard and Edmond, 1983, 1987; Li and others, 1984; Chough, 1985; Ren, 1986; Kolla and Coumes, 1987). The data concerning the mean composition of the world river-suspension loads, obtained by these authors, were confirmed by Martin and Meybeck (1979). On the basis of particular examples, the vast majority of the suspension load was shown to be deposited at the river-sea boundary. The predominance of the particulate matter typical of river water was shown to shift at the barrier to the predominance of dissolved load typical of open ocean.

Beyond the river mouths, large-scale quantitative studies of sediment suspended in sea water, but also in aerosols as well as in cryosols in the Antarctic, also were carried out. Along a series of profiles, extending from the mouths of the largest rivers (discharge glaciers in polar areas) to the pelagic realm, we managed to show the distribution of sediment, and to compare it with the distribution of suspensates in river water, and of aerosols and cryosols. These quantitative studies of sediment at the time of its preparation and translocation (in the river and sea suspensions) were supplemented by the analysis of many thousands of stratigraphic sections and drilling cores obtained for the measurement of modern sedimentation rates. For the first time, the direct quantitative comparison of sediment concentrations during formation, transportation and accumulation was considered.

Quantitative studies of sediment for even longer periods were carried out using cores obtained by the *Glomar Challenger* (more than 1,000 locations in various parts of the ocean) combined with seismic data, which permits the comparison of modern and ancient sedimentation rates (back to 150–160 million years). Given the global scale of this study, some aspects of the data base are not complete. However, the actual material considered seems reliable enough to draw large-scale conclusions. Contemporary methods of studying sediment concentrations in the water column and on the bottom permit one to characterize the rate of processes at various temporal scales, from an almost synoptic view of collection of suspended sediment in water during periods of days and weeks to months and years. Sediment fluxes can be estimated using sediment traps and even more significantly, by the study of sedimentation rates and absolute masses of the bottom sediments over periods of hundreds and thousands of years. Finally, for temporal scales of hundreds of thousands to millions of years, one can approach the study in terms of the thickness of deposits as determined by cores and geophysical data. Such data permit one to draw broad conclusions concerning the regularities of high and superhigh sediment accumulation rates both in time and space.

MAIN CONCEPTS AND TERMS

The processes associated with high and superhigh (avalanche) sedimentation are defined as those that start to appear when the suspension content in water is more than 10 mg/l, when the sedimentation rate exceeds 100 mm/1 ka, and when sediment accumulation rates exceed 5 g/cm^2/1 ka. High sediment accumulation rates result not only in the formation of specific patterns of sediment deposition (mass, granulometric composition, texture, etc.), but also in the isostatic compensation of the Earth's crust, which, in turn, results in the development of thermobarometric conditions favorable for oil and gas formation. The term avalanche sedimentation itself is envisioned in terms of a combination of a number of criteria. Thus, the accumulation of material in a small basin during a short time interval without isostatic compensation is not avalanche sedimentation, because these high rates turn out to be ephemeral, and the accumulation of the sediment is not fixed in the geological record.

To initiate the transport of sediment prepared by surficial weathering, bathymetric relief is required. Under the influence of gravity, the transport of sediments from shallow to deeper hypsometric levels takes place. The greater the relief, the more energetic the transportive processes. The World Ocean is a global basin. The average depth of the ocean is 3,794 m, and the average elevation of the land is +840 m, i.e., the total relief is about 5 km. Most of the transported sediment is usually deposited quickly in the depocenters. The competence of the flux decreases abruptly at the slope base where the transition to the deep-ocean plain occurs.

The hypsographical curve of the Earth (Fig. 3) shows that there are three transitions or minima at the largest scale: 1) close to sea level, where the fluxes enter the ocean and lose transport power; 2) at the base of the continental slope, where fluxes originating from the slope pass to the plain; and 3) deep-sea trench, where the deepest part of the ocean floor occurs. These are the first, second and third levels of avalanche sedimentation, respectively. For preservation of sediment in the geologic record, it is important that the

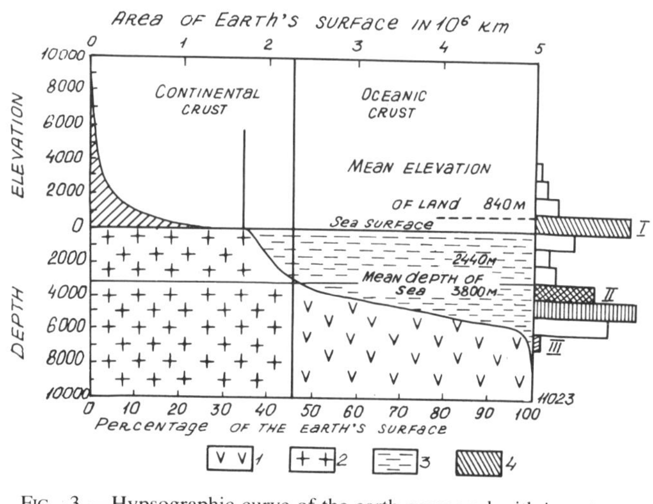

FIG. 3.—Hypsographic curve of the earth compared with its principal structural units (continental and oceanic crust): (1) oceanic crust; (2) continental crust; (3) ocean water; and (4) continents. A histogram showing the distribution of areas corresponding to a given depth interval is shown at right: I, II and III are the global levels characterized by avalanche sedimentation.

boundary between sialic and oceanic crust lies parallel to the slope at a depth of approximately 300 m. The sediments deposited on continental crust have fates determined by the evolution of the sialic crust, whereas the evolution of ocean crust follows its own laws. Continental areas (according to the type of crust) are about two times larger than in the geographical sense (based on the sea-level line between ocean and land). The floor of deep-sea trenches reaches a maximum depth of −11,023 m. The maximum global difference in relief is from +8,843 m (Mount Everest) to −11,023 m (Mariana Trench), i.e., about 20 km. This is the upper limit of the potential energy reserve for gravitational transport on the planet.

In the text that follows, suspension concentration in water is given in mg/l, flux of suspended matter in FU units (flux units), FU: = 1 $mg/cm^2/1$ ka. For the bottom sediments of the pelagic area of the ocean, the unit of sedimentation rate, the Bubnoff unit (B) is used (1B = 1 mm/1 ka). Sediment accumulation rates in general, and of its components individually, are expressed in Strachov units (S), proposed by the author (1 $g/cm^2/1$ ka). The relation between the units of sediment flux in water to the sediment accumulation rates on the bottom (burial rates) is given by the equation 1,000 FU = 1 S.

FIRST GLOBAL LEVEL OF HIGH AND SUPERHIGH RATES—ESTUARIES AND DELTAS OF THE RIVERS OF THE WORLD

Proof of rapid sediment deposition on the river-sea border (first global level of avalanche sedimentation) can be obtained by the comparison of concentrations of river suspensate with the suspension concentrations beyond the "river-sea" barrier. Table 1 shows that the mean fluvial-suspension load is usually in the range of 100 to 10,000 mg/l and more, but beyond that barrier it ranges from 0.1 to 1 mg/l. Therefore, at the barrier, the reduction of the sediment content in water changes by a factor of 100 to 1,000 and more takes place, i.e., avalanche sedimentation of river suspensates occurs, which only slightly extends beyond the mixing zone. The sediment supply to the mixing zone is determined by the average elevation of the drainage basin, intensity of weathering, geologic peculiarities, etc.

Considering the total input of sediment from land (about 18 billion tons annually), about 75 percent (14.4 billion tons) is from the rivers of Asia. This yield is determined by continent elevation and weathering intensity in associated drainage areas. The average elevation of Asia is +960 m, which is a maximum for continents (except for Antarctica), which provides significant driving power to the transportive fluxes. Erosion is intensified there by both tropical and equatorial climates with high chemical-weathering rates. In the equatorial zone, and close to orogenic areas as well, 1 km^2 of water-collecting area yields more than 500 to 1,000 tons of sediment. In the cold parts of the moderate zones, yields are less than 5 tons/km^2, i.e., 200 times less. The probability of avalanche sedimentation is higher, respectively, in the equatorial zones and close to the orogenic zones.

From the rivers of the USSR, the maximum turbidity is from the Kalaus River, which drains unconsolidated deposits. Its turbidity is 48,000 mg/l. For the mountain rivers of the Caucasus, turbidity values of 10,000 to 25,000 mg/l are characteristic. Using the example of the rivers of the USSR, which occupy one-sixth of the Earth's surface, it is possible to show that in both cold and moderate zones, river turbidity is closely related to climatic zones. The minimum concentration characteristic for the rivers of the tundra and the taiga is less than 50 mg/l. For the forest-steppe zone, it is 50 to 150 mg/l; for the steppe zone, 150 to 500 mg/l; and for the regions of abrupt relief changes (mountains and plateaus), it is more than 500 mg/l. Together with the zonal factors (climate and elevation), azonal factors also are very important (geological composition of the water-collecting basins, resistance to weathering, etc.). Abnormally high-concentration values for the corresponding zones are typical for rivers, whose drainages include large area of loess and other unlithified sedimentary deposits (e.g., China, Middle Asia, and USA).

TABLE 1.—INDICES OF AVALANCHE SEDIMENTATON AT THE FIRST GLOBAL LEVEL: DELTAS AND ESTUARIES OF THE LARGEST RIVERS

Rivers 1	Suspended-Matter Discharge (10^6 tons/yr) 2	Drainage Area (10^6 km^2) 3	Water Discharge (km^3/yr) 4	Turbidity of river water (mg/l) 5	Water Turbidity at the River-Sea Barrier 6	Sedimentation Rate in Delta and Fan (B) 7	Delta-Sediment Thickness, Maximum (km) 8
1. Ganges and Brahmaputra	1,670	1.48	971	600–1,700	1.0–0.1	100–1,000	16
2. Huanghe	1,080[1]	0.077	49	14,000–15,000		5,000	
3. Amazon	900	6.15	6,300–3,187	142–156	0.5–0.1	100–1,000	12
4. Changjing	478[2]	0.194	900			100,000–500,000	
5. Mississippi	210/620	3.27	580	362–800	0.1	1,000	to 15–18
6. Indus	100	0.97	238	420	0.1–1.0	100–1,000	10
7. Irrawaddy	299	0.429	428	600–700			
8. Mekong	170	0.810	387	400–500			
9. Orange	153	1.020	91	1,600–1,700			
10. Colorado	153	0.637	20	6,660			
11. Red	130	0.119	—	—			
12. Nile	110 (recent 0)	2.96	30–70	1,800–3,600		400	>4
13. Niger	40	1.21	192	208	1	200	9–12
14. Congo	43	3.82	1,250	34	1–2 and 1	400	6

1) Lee and Chough, 1989
2) Milliman and Meade, 1983
B = Bubnoff units

Direct observations of the suspended content in the mouths of many rivers also are confirmed by data concerning sediment fluxes in river mouths and deltas, as studied using sediment traps, measurements of sedimentation rates and thicknesses of the sedimentary deposits. Values for sedimentation rates in deltas, bays and fjords, i.e., on the river-sea border, not only reach avalanche values (more than 100 B), but exceed them significantly. For example, in the Hwang Ho River mouth, the rates reach 52,800 B, and values of more than 1,000 B, which correspond to the superfast sediment accumulation rate level, are common.

It is notable, that in the zone of river- and sea-water mixing, suspended content and sedimentation rates are much higher than in the river, and, of course, much higher than on the oceanic side of the mixing zone. These conditions are determined by a complex of physical, physiochemical and biological processes. Deposition of suspended sediment is related not only to reduction of the flux-transport power of a river as it flows toward the sea, where it is dammed, but also to the physiochemical environmental change, which can be quantitatively represented by the salinity shift. In the freshwater part of an estuary (salinity less than 1‰), the normal river suspended content is characteristic, where particulate forms of transported elements form 65 to 95 percent of the total. High contents of humic and fulvo acids, as well as iron compounds, are typical.

In brackish conditions (salinity from 1 to 30‰), deposition of most of the river suspended load takes place, as well as recharging of colloidal particles by the precipitation of clay mineral flocules, humic and fulvo acids as well as iron oxides (80–95 percent). It is especially here that a unique condition appears, where the suspended concentration is much higher than in the river, the so-called "silt-clay barrier," which most often appears in the salinity range of 2 to 7‰. During the process of flocculation of the humic substances and iron, many of the metals are precipitated from water (as much as 50–90 percent), and that is why the "element-organic barrier" is situated here. The characteristic precipitation here is of metals determined by their affinity to humic acids, iron and clay minerals.

Farther toward the sea in the zone (salinity more than 30‰), another region of high suspended concentration appears. This is related to plankton, a "biological barrier," with the magnificent development of phyto- and zooplankton, including filtrator organisms, which are capable of filtering the entire volume of coastal waters in several days. The margins of this barrier move depending on the season and other factors. In the region of this biological barrier, the transformation of dissolved forms of elements into protoplasm, tissues and tests, organic-matter formation and biosorption takes place. Large-scale biofiltration also operates here, e.g., the accumulation of particles and floccules into pellets.

Therefore, estuary-mouth zones are not only the regions of deposition of the predominant part of terrigenous suspended load brought by rivers, but also of mobilization and the transformation of a significant (as much as 40–60 percent) part of the dissolved elements into suspension load. This is a unique region for the whole sedimentation process!

The mass flocculation of the river suspended load and its fast deposition may reach high and superhigh sediment accumulation rates as long as sediment is not washed from the mixing zones by the influence of waves and tides, as takes place, for example, at the Amazon River mouth (Fig. 4). A significant part of the suspended load is washed from the mouth of the Amazon River by strong, coast-parallel currents, which form a suspended flux as well as depositing of clay shoals along the coast. These shoals move at a rate of about 900 m/yr (Nedeco, 1968; Froidefond and others, 1988). Transport takes place within a band about 20 to 40 km wide with an associated suspension-load transfer of about 150 million m^3 annually, and if the silt shoals are considered, about 100 million m^3 annually. The transport distance is about 1,600 km to the west from the river mouth. The dispersion of sediment to the neighboring regions also occurs in the estuaries, where influenced by strong, local wave action and variations of tidal level. Therefore, the sediment is not always deposited at a river mouth, although there is the possibility for it to be concentrated here. Under the influence of waves and tides, it may be transported along the coast downslope, passing into the region of avalanche sedimentation of the second level. There are many intermediate stages between the main types of sediment distributed at river mouths. Considering such intermediate stages, and independently of the mouth type (deltas, estuaries, bays, fjords, etc.), the mixing of river and sea waters is the main factor for sediment concentration. It is here where the material, gathered from vast water-collecting areas, is mobilized in the form of the sediment within small geographical areas. This is where deposition of the terrigenous matter from the suspended load of rivers takes place, as well as transformation of the dissolved forms of matter to suspended load by organisms and absorption (Fig. 5). Thus, the formation of large sediment accumulations on the first level is possible by the combination of its voluminous supply and its insignificant removal from this zone (Broussard, 1975; Galloway, 1975; Berg, 1982). Alterations of the ocean level, which significantly determine the preservation of sedimen-

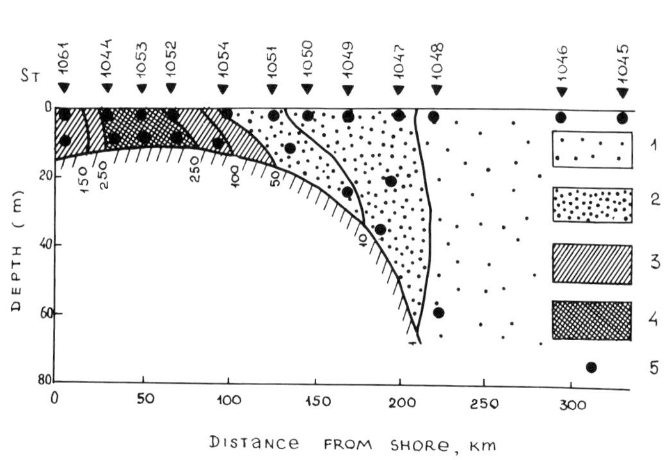

FIG. 4.—Distribution of suspended load associated with the mouth of the Amazon River in mg/l: (1) <1; (2) from 1 to 50; (3) from 50 to 250; and (4) >250 ("silt barrier") (Monin and Gordeev, 1988).

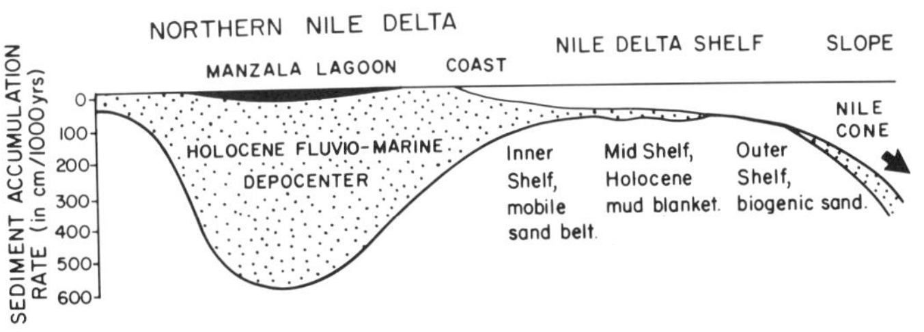

FIG. 5.—Avalanche-sedimentation rates in the delta and submarine fan of the Nile River. Maximum rates in the Holocene deltaic depocenter are as much as 6,000 B and as much as 500 B in the subaqueous cone (Stanley, 1988).

tary bodies over time, must be added to this model. The limits of this paper restrict us to the consideration of sediment accumulation characteristics using a few examples of deltas, which are most important in terms of sediment discharge (Table 2). Tables 1 and 2 show that the main part of the Earth's surface is drained by just 12 giant rivers with a suspension yield exceeding 100 million tons (including two super-giant rivers with discharges exceeding 1,000 million tons per year). Therefore, about 6 billion tons of suspended matter are concentrated in 12 locations on the Earth's surface, i.e., more than 30 percent of the total terrigenous matter of the first sedimentation level accumulates here.

The zonal characteristics of the distribution of these largest avalanche-sedimentation basins are characteristic. Most often they occur in humid zones and on the border of humid and arid zones, but do not occur in glacial and arid zones. Depocenters of modern avalanche sedimentation are situated along the shores of Asia, extending from India to China. It is clear that the study of the characteristics of sediment accumulations in these particular regions is most important.

The Ganges and Brahmaputra Rivers, united by a common delta, comprise the largest sedimentary system of the world. According to different data sets, they supply from 1,670 to 2,177 million tons of suspended sediment per year, i.e., more than two-thirds of the river flow into the Indian Ocean. This is more than the supply of all the rivers entering the Atlantic basin. Turbidity of these river waters, which drain the Himalayas, is also high, from 600 to 1,700 mg/l, i.e., 2 to 6 times higher than the average turbidity of rivers. Tidal variations of sea level reach 7 m in the mouth of these rivers, which results in the removal of a significant part of the material from the area of the river-sea barrier into the Bay of Bengal. A giant underwater prolongation of the delta occurs here, a cone about 3,000 km long and 1,000 km wide. The total cone area is about 2 million km^2, which is 25 times more than the area of the subaerial portion of the delta. With thickness as much as 16 km, the volume of the sediment here reaches 5 million km^3, which is 10 times more than the volume of the whole modern Black Sea. The avalanche-sedimentation basin is not a unique one in terms of its scale: the volume of the ancient West-Siberian basin is equal to about 7 million km^3. Nowadays, the main part of the sediment accumulates close to these river mouths (in the range of the 5-km isopach). The suspended concentration close to the mouth is reduced to 1 to 2.5 mg/l, i.e., by 700 to 2,000 times as compared to the river waters, though traces of very fine-grained material can be identified in the bay water column nearly to the Equator (Lisitzin, 1988).

Another good example may be observed close to the mouth of another big river of Southeast Asia, the Irrawaddy River. The turbidity of its waters is 600 to 700 mg/l, but is reduced quickly from the mixing zone to 1 mg/l and less. Deep-sea drilling in the underwater cone of the Ganges and Brahmaputra (the maximum core length is 2.5 km, i.e., one sixth part of the thickness is penetrated) showed that the cone formation has been taking place since shortly before and after collision of the Indian subcontinent with the Asian plates about 55 ma. Subaqueous cones of the Ganges-Brahmaputra and the Indus contain correlative thicknesses that may be related to the uplift and erosion of the Himalayas. This giant oceanic basin did not experience uplift, and during all of its history has been situated below sea level. With the maximum depth of the Bay of Bengal being 5,258 m, the accumulated sediment could have filled it three times over, if not for isostatic crustal adjustment.

The second important modern sedimentary system is the Hwang Ho River, which has a comparatively small water-collecting basin, composed of loess, which results in its comparatively rapid removal, and enormous turbidity values (up to 15,000 mg/l, i.e., the border between dense suspension and bottom-carried sediments is lost (Table 3). The Yellow Sea is a vast mixing zone, a constant depocenter, where the sediment from the Chinese and Korean rivers accumulates. The Yellow Sea obtained its name because of the very high suspended-particle content (more than 100 mg/l in summer and more than 500 mg/l in winter), which mostly is the product of loess erosion (Lee and Chough, 1989). The sedimentation rate in Bochay Bay averages about 5,000 B, which qualifies this as a region of superhigh sedimentation rates. About 70 percent of the suspended matter is deposited here (Milliman and others, 1986). The supply of suspended sediment to the river mouth is so significant that changes in the position of the mouth can be easily identified. It reached the modern position in 1,128 to 1,855 A.D. Sedimentation rates at the Bochay Bay mouth decreased rapidly to 1,000 B and less, i.e., superhigh sediment accumulation decreases into the high zone south of this peninsula.

The avalanche-sedimentation region of the Hwang Ho River on the Yellow Sea floor merges with the avalanche-sedimentation region of the Yantze River, which transports 478 million tons of suspended sediment annually to its mouth

TABLE 2.—DEPOSITION OF SUSPENDED LOAD AT THE RIVER-SEA BORDER. VALUES ARE IN PERCENT OF THE TOTAL AMOUNT OF DELIVERED SUSPENDED LOAD

River	Loss (%)	Reference
Amazon	95	Milliman and others, 1975
Mississippi	90	Trefrey and Presley, 1976
Saint Lawrence	93	Bewers and Yeats, 1977
Zair	95	Eisma and others, 1978
Shelda	92	Duinker and others, 1979
Rivers of the Black and Azov Seas	83	Demina and others, 1978
Kura	90–95	Ivanov, 1955

(Table 3). The sedimentation rate at this river mouth reaches enormous values of 300,000 B. That is why even in the central parts of the Yellow Sea avalanche rates of about 3,000 to 4,000 B can still be observed. About one half of the suspended sediment from the Yantze River is deposited north of 30° latitude, the sedimentation rate here reaching 528,000 B (DeMaster and others, 1985).

On the basis of water discharge, the Amazon River is the biggest river of the world. As Table 1 shows, the turbidity of this river's waters is hundreds of times lower than that of the rivers of China and southeastern Asia and is related to the influence of forest vegetation, anchoring the soft deposits and weathering products. New data on sediment accumulation in the Amazon River mouth were recently obtained by Americn and Soviet expeditions (Milliman and others, 1975; Milliman and Boyle, 1975; Meade and others, 1979; Monin and Gordeev, 1988). The mean content of suspended load in river water is 156 mg/l here; and beyond the river-sea barrier, it decreases by 100 to 150 times (Fig. 4). The maximum sedimentation rates, about 1,700 B (Gibbs, 1981), are related to the mixing zone (salinity interval from 2 to 10 percent) and the location of the modern depocenter.

The input of sediment to the subaqueous Amazon cone has not been observed; however, the canyons are filled with sediment, as in the case of the cones of the Ganges, Indus and many other rivers, which is related to the modern high sea-level stand. The thickness of the sediment in the Amazon cone exceeds 10 km, the sedimentation rate during the last 2.2 ma varied from 50 to 1,150 B. The bed has been depressed under the influence of isostatic loading. Similar data can be obtained for many other rivers (Niger-Irrawaddy, Indus, Irrawaddy; Fig. 6) (Litsitzin, 1974, 1978, 1988).

Comparison of the data concerning the distribution of suspended sediment, sediment fluxes in the water column, as well as on the modern and ancient sediment accumulation rates at the river-sea boundary permits one to make several general conclusions. 1) In river waters, almost all the sediment is particulate; for hydrolyzate elements it constitutes 97.5 to 99.9 percent. Only alkaline and alkaline-earth elements are carried from rivers mainly in solution. 2) At the river-sea boundary, 70 to 90 percent of the suspended load present in the river water is deposited in a small area, namely, where flocculation of dissolved organic substances and iron occurs. Co-sedimentation of many of the elements having an affinity to organic matter and iron (Sholkovitz, 1976, 1978; Romankevich, 1977; Morozov,

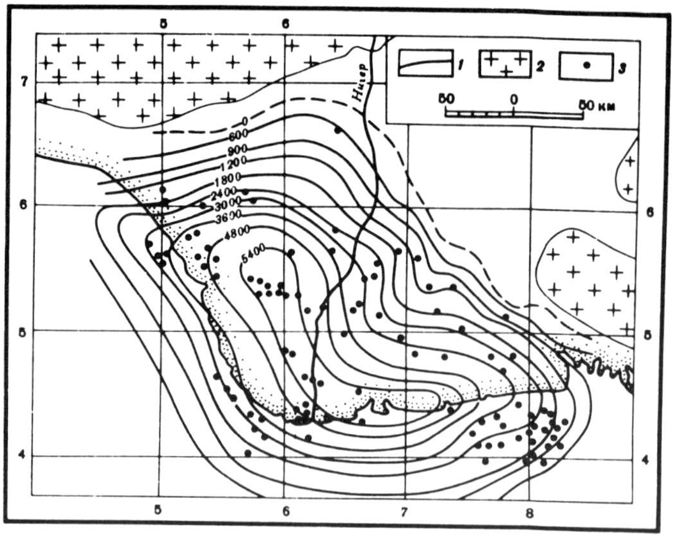

FIG. 6.—Thickness of sedimentary deposits in the Niger delta: (1) isopachs in m; (2) crystalline shield rocks; and (3) location of drill holes (Avbovbo, 1978).

1979) and biological extraction also take place, i.e., mass transformation of dissolved-element forms to suspended particulates occurs. 3) From 70 to 90 percent of the sediment accumulates at high and superhigh rates. It is soft, has high-water content (Terzagi, 1956) and depresses the Earth's crust under the associated load. 4) As drilling in deltas and seismostratigraphy show, numerous hiatuses and disconformities are characteristic of avalanche sedimentation at the first level (Vail and others, 1977; Vail and Hardenbol, 1979; Weimer, 1990). Thus, most of the sediment is not preserved in the great thicknesses associated with deltaic deposits, but has been truncated and removed from the margins of these deposits with changes in sea level, and has been transported to the second global level by avalanche sedimentation.

THE SECOND AVALANCHE-SEDIMENTATION GLOBAL LEVEL: THE CONTINENTAL RISE

The continental slope of the oceans has a total length of 350,000 km and a relief from 130 to 200 m (the shelf edge) to about 4 km. The width of the slope is most often from 20 to 100 km; therefore, the gradient is very high (an average of about 4°). The combination of significant depth changes with the slope steepness provides a huge reserve

TABLE 3.—AVERAGE VALUES FOR THE PHYSICAL PROPERTIES AND ACCUMULATION RATES FOR FINE-GRAINED SEDIMENT

Province	Grain Size (%)			Water Content (%)	Wet Bulk Density (g/cm^3)	Undrained Shear Strength (kPa)	Accumulation Rate (B)
	Sand	Silt	Clay				
Changjiang Estuary[1,3]	6	49	45	78	1.58	2.94	300,000
South of Cheju[1,3]	8	41	51	118	1.41	0.98	3,000
Central Yellow Sea[1,4]	6	47	87	87	1.52	3.55	3,000
Southestern Yellow Sea/Mud belt[2,4]	2	62	36	71	1.60	5.46	1,000

Mass physical properties after [1]Butenko and others (1985) and [2]Lee and others (1987).
Accumulation rates after [3]DeMaster and others (1985) and [4]Kim and others (1988).

of potential energy. The continental slope is a global "spring-board" on which masses of the sediment are moved downward from the shelf and accelerated to significant rates (absorbing water at the same time). Mass failures of such sediment have occurred periodically and simultaneously for the World Ocean with sea-level decreases, as demonstrated by seismostratigraphy and studies of delta deposits (Vail and others, 1977; Vail and Hardenbol, 1979).

The slope of this giant precipice, situated on the ocean margins, is cut by systems of underwater canyons and valleys (Shepard, 1973). A subaqueous drainage system, similar to the land system, is being formed here. In contrast to the land-drainage system, this one is determined not by the transfer of water, but by the transfer of sediment in various degrees of dilution with water. When the sediment is slightly diluted, it can be transported as solid blocks (underwater falls) or as subaqueous landslides. Therefore, all the types of slope processes exist here, as well as drainage sediment-collecting systems. The feeding area is situated on the shelf and in the upper parts of the slope. The transport area is the upper and the middle parts of the slope, and the accumulation area is at the slope base within a band equal to or exceeding 500 m wide. A remarkable peculiarity of sedimentaton at the slope base is that it is cyclic, because of its conjugation with the processes of the first level, and because almost no natural primary sediment occurs on the slope.

When sea level is high (e.g., as in the modern phase), the main part of the sediment accumulates on the first level, in the river mouths; when sea level is low, the material is discharged downslope and accumulates at the base. This is confirmed by comparison of the data on large accumulations at the slope base, where they are always associated with huge rivers. Confirmation also comes from the comparison of the stratigraphy with mass-deposit cones at the slope base, mineralogic studies, the studies of organic matter, etc. Underwater slope-failure cone deposits are mainly upper global-level deposits, partially changed in the process of gravitational transport downslope.

The continental slope is a complex area, where steep regions interchange with gently sloping ones. When a soft sedimentary block reaches the critical mass, it starts moving downslope, most often, first in the form of the whole block (rock fall). Then with acceleration, more and more dilution of the block takes place, and other blocks are incorporated into the movement. Many of them fall from steep regions, breaking at the base, being further saturated with water. These processes of gravitational transport of the sediment do not result in significant changes in the deposits' compositions; new matter is neither added, nor extracted. In gravitational translocation processes, a definite sequence may be revealed, a succession from falls to fine turbidites (Lowe, 1976).

The first phase of sedimentation, preparation of the sediment, is related strongly to the first avalanche-sedimentation level, which, in turn, is related to climatic zone and other factors previously mentioned. At this phase, the differentiation of sediment is very weak, as it is transported in masses, differentially saturated with water, streaming downslope. The amount of the material available for transport and its composition are determined by the processes at the upper level.

The second phase, transportation, occurs in two forms: 1) diffusion along the slopes without the concentration of sediment into channels; and 2) concentration into channels (underwater valleys, canyons). The latter form has major importance, which is confirmed by the association of the upper parts of the fans with the lower parts of associated fans and valleys. With the channelled (drainage) transportation system, it is usually possible to define the upper sediment-collecting fan, the middle transporting part of the canyon and the lower sediment-collecting fan (Gorsline, 1978, 1984; Normark, 1978; Nardin and others, 1979; Bouma and others, 1985; Pilkey and Cleary, 1986). It is possible to isolate in fans the rudimentary differentiation of material, transportation from the first level: the coarsest material accumulates in the upper parts of the cones and particle size progressively decreases to the generally finer sizes in the distal direction.

The sizes of underwater transportation systems and slump cones are surprising. The canyon-cut depth commonly reaches 200 to 500 m, and 800 to 1,000 m and more for the largest ones. They have a characteristic Y-shape from the upper part, and trough-shape form in the lower part. The Ganges and Brahmaputra subaqueous cone is about 3,000 km long and 1,000 km wide, with an area of about 2 million km^2. The area of the Indus subaqueous slump cone is even bigger, about 1,500 km long and 960 km wide, with an area of 1.1 million km^2 (Nani and Kolla, 1982; Kolla and Coumes, 1987).

Even the comparatively small Saint Lawrence River has a subaqueous slump cone about 600 km long and 150 to 300 km wide with maximum thickness as much as 120 km (Uchupi and Austin, 1979). Seismic studies of recent years show sedimentary accumulations at some locations along the continental slope base also have thicknesses in excess of 10 km (16–18 km for the Ganges and Brahmaputra; more than 10 km for the Indus and others). In the "small oceans," sea reservoirs with oceanic crust, thicknesses of the sediment at the slope base reach 9 to 10 km in the South China Sea, 4 to 5 km in the Sea of Japan, 5 to 6 km in the Sea of Okhotsk and more than 9 to 10 km in the Bering Sea (Fig. 7).

If we imagine a final discharge reservoir of circular shape, from 70 to 90 percent of the sediment would be concentrated along its periphery as a sedimentary toroid, flattened at the top. The width and length of the toroid increase close to big river mouths and decrease in areas of minimum sediment input. This is a special type of sedimentation, differing from the normal particle-by-particle accumulation, a concentrated form of preparation, transportation and deposition.

The toroidal sedimentary body often possesses breaks, dividing it into autonomous sedimentary basins, but the general shape of the toroid usually can be observed. It appears that the sedimentary toroid is composed of several constructional modules, subaqueous slump cones, which are complexly changing in time and space. The slope sedimentary formations, which seem to be unified, are therefore like a "cone store" of modules of various size and shape.

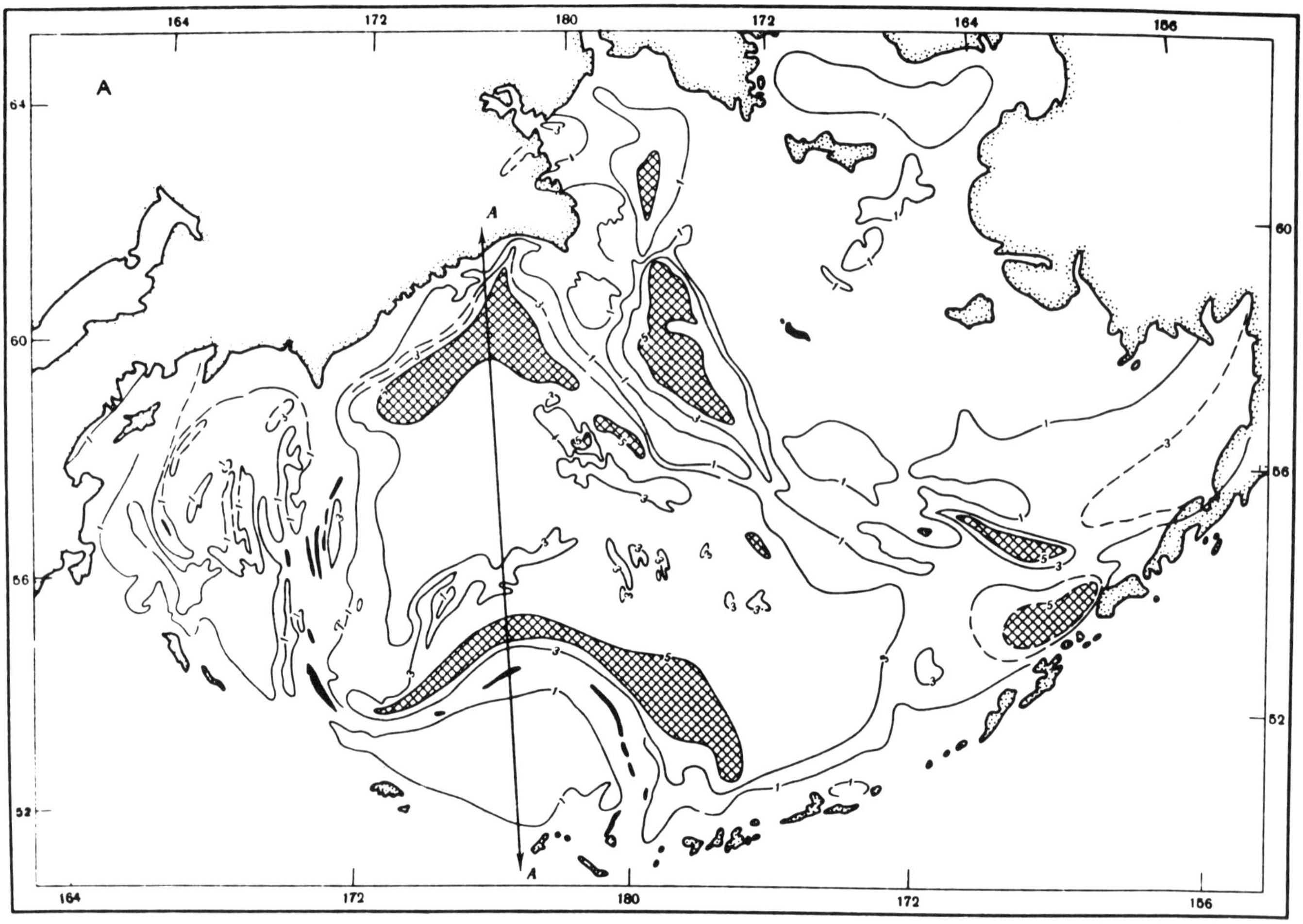

FIG. 7.—Sedimentary basins of the Bering Sea with the thickness of the sedimentary cover in km. The 1-, 3-, and 5-km isopachs are shown, and the thickness of the sedimentary cover exceeding 5 km (maximum of more than 10 km) is shaded (Gnibidenko, 1961).

The total length of the sedimentary toroid is about 350,000 km; the width ranges from 100 to 1,000 km and in some places exceeds 1,500 km, with thickness reaching 10 km and more, i.e., these are the strata corresponding to miogeosynclinal deposits identified on land. The formation of particularly large sedimentary bodies is related to avalanche- (high and superhigh) sedimentation rates, which are stable for long periods. Such conditions often occur close to large mountain systems; and the avalanche-sedimentation sites are associated with sediment derived from areas characterized by high-denudation rates. Moreover, in some cases, the volume of these correlative deposits significantly exceeds the volume of the mountains feeding them. This is possible in the case where a long period of uplift and denudation of the mountain range has occurred.

The giant sedimentary basins of the Ganges and the Indus River mouths are related to the Himalayas uplift with rates of 100 to 200 mm/1 ka, and in some stages up to 750 mm/1 ka (Zeitler and others, 1980). Similar conditions exist in the Alps and the neighboring sedimentary basins. For the formation of the sedimentary deposits existing in the basins, it was necessary not only to erode the Alps substantially (the modern height is 4.8 km), but also to create a depression in their place as deep as 12 km (Menard, 1964; Clark and Jager, 1969).

The phases of gravitite accumulation in toroids are interchanged with phases of normal sedimentation, when most of the sediment coming from the land accumulates at river mouths (modern phase). The avalanche transport by gravitites associated with low sea-level stands enhances organic-matter preservation, due the possibility of its transport to great depths (as much as 5,000 m). This permits us to consider the second major global avalanche-sedimentation zone as highly promising with respect to the occurrence of gas and oil deposits.

DEPOSITS OF THE REGIONS OF HIGH AND SUPERHIGH SEDIMENT ACCUMULATION RATES: REGIONS OF SEDIMENT DEFICIENCY AND HIATUSES, A GLOBAL CORRELATION

The logic of sedimentation shows that, with extraction in the modern cycle of 90 to 91 percent of the sediment entering the river mouths on the river-sea border, sediment deficiency for vast floor areas outside of this region must take place, i.e., hiatuses must exist.

One of the discoveries of recent years is the wide distribution of hiatuses in the distal deep-sea regions of the ocean, which were studied by deep-sea drilling. Thorough analysis of the hiatuses shows that two of their global levels (for unlithified sediments) could be distinguished where the most widely developed and long hiatuses are concentrated. The upper one is associated with the level of avalanche sedimentation. The hiatuses of this level truncate soft sediments with a period equal to sea-level fluctuations. The position of these hiatuses corresponds to the phases of global downslope sediment failure. Hiatuses of the lower level are not always observed in the continental slope where avalanche sedimentation occurs, but they often are associated here with the stages of normal sediment accumulation. In contrast, they are mostly developed in the pelagic ocean area, the region of a constant sediment deficiency. The lower the sediment accumulation rate, the less the change must be to lower the rate to zero or even to turn it to negative values, i.e., so that erosion of the bottom would occur.

Hiatuses are widespread in modern sediment, but they are even more clearly seen in the pelagic sediment of the ocean for the longer geologic periods, when significant sea-level changes took place. Most widespread are hiatuses 10 million years long. The hiatuses in the pelagic areas were most numereous during the Paleogene. The highest position of sea level occurred at that time; therefore, the most complete extraction of the sediment into first-level deposits occurred, and, consequently, the most marked deficiency in pelagic sediments also took place. Dramatic sea-level changes also were related to the Quaternary ice ages. These were accompanied by the formation of cold water, which flowed near the sea floor and transported sediment.

Thus, for the avalanche sediments of the first and second levels, two global levels can be distinguished for the hiatuses. Like unlithified strata, hiatuses of the upper and lower levels are closely related to each other in time and phase: the maximum development of the hiatuses on one level corresponds to the maximum development on the other. According to distribution by area, they can be local, regional and global. The last notion does not mean that sedimentation has been interrupted everywhere at the same time, rather, that synchronous hiatuses can be observed in all the parts of the World Ocean, which usually correspond to synchronous sea-level changes.

Therefore, a close relation exists between regions of high and superhigh sedimentation and concentration of the main portions of the sediment, and the deficient regions, where only the residual part of the matter balance takes place (7–10 percent at modern stage). These appear to be various parts of the same mechanism of pumping matter from the upper global level to the lower one with dispersion of just a small part of the sediment (about 10 percent) on the huge ocean floors by "normal" sedimentary processes. The doctrine of the hiatuses is closely related to the doctrine of avalanche and normal sedimentation (Lisitzin, 1984a, 1986, 1988).

SEDIMENTARY SYSTEMS OF THE EARTH

The revelation of widespread and regular accumulations of huge masses of sedimentary material in some places, with simultaneous deficiency and hiatus formation in others, changes our ideas concerning the general character of sedimentation drastically. Most of the sediment accumulates with high and superhigh rates in local basins, which develop autonomously. Sedimentation in these regions goes on in the avalanche manner and is interrupted by hiatuses and erosional phases. These basins are situated mainly on two different hypsometeric levels, and are related to each other by a direct downslope line; mass transport only goes downslope. The main mechanism controlling the transport of the sediment from the upper into the lower basins is sea-level change.

The doctrine of local basins is significantly expanded, if we consider not only the basin itself, but the whole system as well. Usually for the basins of the upper and lower levels, the sedimentary system includes three general parts: 1) sediment collecting (water collecting for subaerial), 2) transportational, and 3) accumulation (sedimentary basin itself).

In the plan view, a water-collecting (sediment-collecting) basin usually has the shape of a triangle, pointed from the top in the direction of sediment transport (downslope). The triangle top turns into the transport bed (river, canyon). The upper and the lower parts of the bed are divided vertically (hypsometrically), which provides for the flow of water and suspended sediment, or the movement of gravitational flows. The lower part of the transport bed turns into an accumulation triangle (cone), which in its area is hundreds and thousands times less than the sediment-collecting triangle. This is where the first concentration of sediment occurs. Thus, the subaerial and subaqueous sedimentary systems are similarly built in principle (Figs. 8, 9). The area of sediment-collecting triangles for many rivers may exceed 1 million km^2 (7.18 million km^2 for the Amazon River).

Subaerial and subaqueous systems are characterized by close relations in the vertical sense: sediment transport goes in one direction, from upper hypsometric levels to lower ones, i.e., from subaerial sedimentary system to the subaqueous. This also determines the similarity of the mass composition (Fig. 9). The largest mass-transport events are related to global sea-level changes. The sediment transport time is determined by the potential energy differences (relief) between the upper and the lower triangles. The mechanism of the sediment transport from the upper level to the lower one is determined by sea-level change. This, in turn, provides for the cyclic character of the sedimentologic process in the upper and associated lower basins, as well as the contraphase movement of the process in them. The duration of many of the sedimentary basins, hidden by the ocean waters (on the slope base), reaches 150 to 160 million years.

The importance of the largest sediment accumulations of the first and second levels is determined by the significant content of organic matter in them (usually 1–2 percent, at some places more than 10 percent, the usual content in pelagic sediments being 0.25 percent). This in combination with huge thicknesses, lithological peculiarities and thermobarometric conditions results in the formation of oil and gas accumulations at some locations. Even turbidites, which are the most water-diluted parts of these sediment masses, usually maintain an internal reducing environment as well as

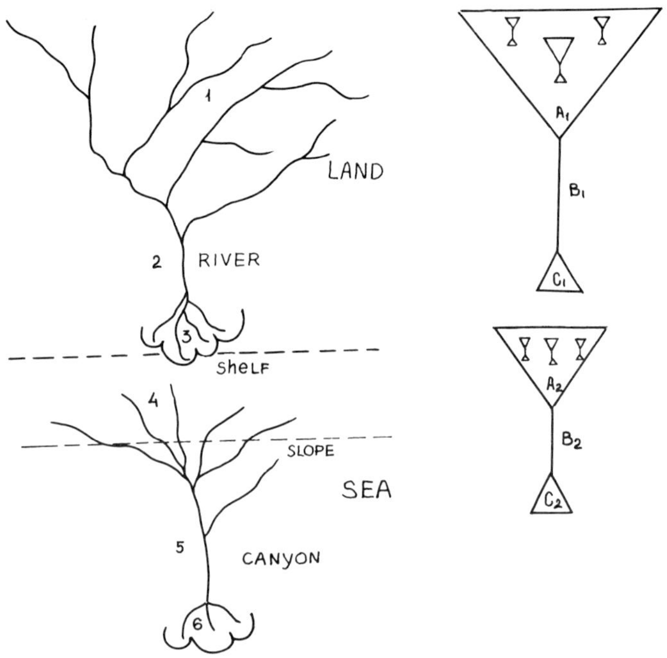

FIG. 8.—Schematic diagram showing the vertically interrelated sedimentary systems of the Earth: (1) subaerial drainage basin; (2) transport bed; and (3) subaerial and subaqueous delta. This system continues oceanward as: (4) the sediment-collecting cone (submarine valleys and canyons); (5) transportaton bed; and (6) submarine fan. To the right is a schematic diagram of the operative sedimentary system at a global as well as regional and local scales. Each system is composed of (A) a sediment-collecting triangle, (B) a transportation bed and (C) an accumulation triangle. The general tendency of this process is to collect sediment from large areas and to concentrate it in small sedimentary basins.

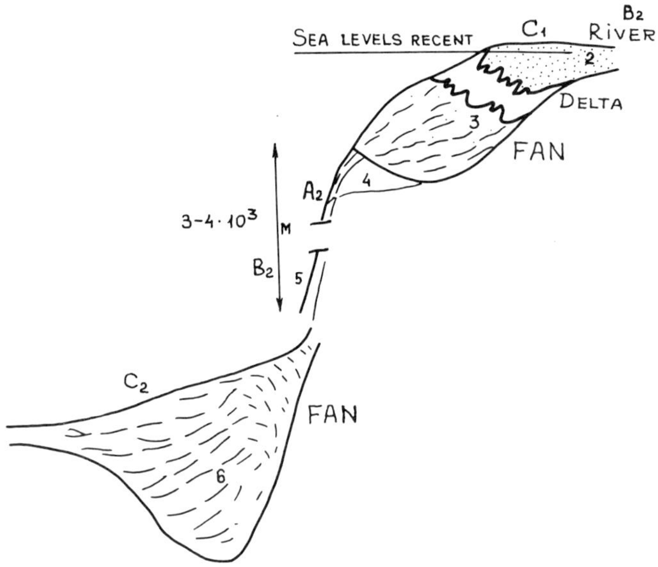

FIG. 9.—Schematic diagram showing the sedimentation systems in a vertical transect of the first and second levels of avalanche sedimentation. Lowering of sea level results in massive sediment transport from the upper to the lower levels. Numbers and letters correspond to those in Figure 8.

a high content of $C_{org.}$ and uranium (Colley and others, 1969), which may be successfully used for the correlation of stratigraphic columns at distances of 100 km from each other.

QUALITATIVE COMPOSITION OF SUSPENDED SEDIMENT AND SEDIMENTS FROM REGIONS OF AVALANCHE SEDIMENTATION

Granulometric Composition and the Volume of Sedimentary Bodies on the First Level of Avalanche Sedimentation

Studies of granulometric composition of the suspended load of rivers have been conducted in the USSR for more than 50 years, and maps showing the granulometric composition of such loads were developed. For the USSR and for the rivers of the world, the influence of climatic zones and water-collecting basin relief on the granulometry of river suspended load and sediment of the first level of avalanche sedimentation can be seen clearly. Mountain rivers transport larger amounts of material to associated estuaries than rivers draining low-relief areas. The higher rate of chemical weathering in the water-collecting basin, the more the original rocks turn into clay, and therefore, the higher the content of the pelitic fraction in the associated rivers. As can be seen from Table 4, based on direct studies of weathering on land and observations on river flow (Lisitzin, 1978), sharply different amounts of sediment are transported from various climatic zones to the ocean in the present period. Avalanche sedimentation in river estuaries of the equatorial zone is represented by large sedimentary bodies (a total of 19 billion tons is supplied annually here, but 92 percent, i.e., 17.5 billion tons, is sedimented) composed of the finest pelitic material; about 80 percent of the fraction is less than 0.01 mm in diameter. The sedimentary bodies in regions of avalanche sedimentation for the two global moderate humid zones by their volume and weight are much smaller than those in equatorial zones (only 3 billion tons are supplied annually, but 92 percent, i.e., 2.76 billion tons, is sedimented), which is six times less than in sedimentary bodies in the equatorial zone. By their granulometric composition, they differ from equatorial sedimentary bodies in the predominance of sandy silt (mainly silt) fractions. The supply of sedimentary material into the two global ice and two arid zones (northern and southern) is even less. Together they receive an amount of sediment equal to that which is carried into the moderate humid zones (3 billion tons). The formation of any large basins of avalanche sedimentation is unlikely to occur here, except through flowing rivers, which transit through arid zones, e.g., the Nile. In glacial zones, physical weathering and the supply of the material into the ocean mainly by ice are most important. Coarse material predominates; pelite content is usually from 3 to 5 percent, but sometimes is as much as 20 percent. In the moderate humid zones, sand and silt are predominant in the suspended load of rivers; the main fraction is 0.05 to 0.02 mm in diameter. The pelitic fraction (less than 0.01 mm in diameter) increases to as much as 20 to 30 percent. In arid zones, the pelitic-material content in river suspensions is as much as 50 to 60 percent, and coarse fractions (more than 1 mm in diameter) are absent.

TABLE 4.—GRANULOMETRIC COMPOSITION IN PERCENT (NUMERATOR) AND ABSOLUTE MASSES OF VARIOUS GRAIN-SIZE FRACTIONS IN BILLIONS OF TONS PER YEAR (DENOMINATOR) OF FLUVIAL SEDIMENT SUPPLIED TO THE OCEAN FROM DRAINAGE BASINS IN VARIOUS CLIMATIC ZONES

Fraction of Terrigenous Material	Zones				
	Glacial	Moderate Humid	Arid	Equatorial Humid	Total Supply to the World Ocean
Rough desintegrated	50/0.75	30/0.9	0/0	0/0	6.6/1.65
Sandy alevrite	45/0.68	50/1.5	40/0.64	20/3.8	26.4/6.62
Pelite	3–5/0.08	20/0.6	60/0.96	80/15.2	67.0/16.84
TOTAL	100/1.5	100/3.0	100/1.6	100/19.0	100/25.1

The maximum output of the pelitic fraction (about 80 percent) occurs in the equatorial zone. With the maximum development of weathering, this zone also discharges the maximum amount of sediment into rivers, nearly 19 billion tons from the total supply of 25.1 billion tons. From the total of 16.8 billion tons of the pelitic fraction delivered to the estuaries of the rivers of the world, the rivers of this zone transport 15.2 billion tons (90.5 percent).

Therefore, the granulometric composition of the suspended-load rivers and consequently, the composition of associated estuarine sediments, where more than 90 percent of the suspended load is deposited, is determined by the combination of climate and relief (mountain ridges) and by the composition of the rocks of water-collecting basins, e.g., wide distribution of forests and clay rocks.

The granulometric composition of the sediment of the continental-slope base (underwater fans, etc.) is determined by the composition of the original material, i.e., sediment of the first level. That is because, in general, it is close to the sediment of the underwater parts of the estuaries. It should be noted, however, that gravitational flows produce partial differentiation of the estuarine sediment. In general, the strata of the continental-slope base are characterized mainly by sediment with pelitic contents from 60 to 80 percent; of sandy silt fraction from 10 to 20 percent; and of the larger material from 5 to 10 percent. The greater the distance from the continental-slope base, the more differentiation of original matter by size occurs. The largest sand content is situated where the highest dynamic activity of gravitational flows (channels) occurs, and the largest pelitic content occurs in calmer locations on the periphery.

Mineral Composition

Sediment composition is changed from rock composition of the water-collecting basin as a function of the intensity of chemical weathering (the stronger the weathering, the larger the fraction of minerals of the rocks that turn into clay minerals), by relief and by other less important factors. In glacial conditions, physical weathering predominates; this is the region where disintegrated broken minerals predominate in the material of avalanche sedimentation, even the minerals non-resistant to weathering are preserved. Physical weathering also predominates in moderate zones. Here disintegrated minerals are widespread, although the most unstable of them are removed by chemical weathering and are transformed into clay minerals. The clay mineral complex illite and chlorite predominates. In the equatorial zone, practically all of the minerals, excluding the most stable ones, turn into clay (kaolinite, montmorillonite and various other secondary minerals). The zonal characteristics of the distribution of clay minerals (Rateev and others, 1968, Lisitzin, 1978), in general, correspond with climatic-zone characteristics, and predetermine the mineral composition of the sediments of avalanche basins of the first and the second levels. Additional information is given by Lisitzin (1978, 1988).

Geochemistry of Sedimentary Bodies, Macro- and Trace Elements

Similarities in the average chemical composition of the suspended loads of rivers were obtained from studies of the chemical composition for more than 100 rivers, including many of the rivers in the USSR. In total, they cover about 18 percent of the solid-river flow (Lisitzin and others, 1982). Table 5 lists the results of average geometric composition of the suspended loads (where data are available). Data were obtained independently by Martin and Meybeck (1979) for 20 of the largest rivers in the world (about 15 percent of the world's solid flow). These authors give the average weighted values for macro-elements of the suspended load and median values for trace elements. The data of Martin and Meybek (1979) cover mainly tropical rivers, whereas our data cover predominantly the rivers of the moderate humid zone. It can be seen from Table 5 that the data concerning the majority of the elements are in relatively close agreement; however, there are some exceptions, e.g., lithium.

Inasmuch as more than 90 percent of river suspended load is sedimented on the first level of avalanche sedimentation, its chemical composition determines the geochemical image of the contemporary regions of avalanche sedimentations. That means that the geochemistry of these sediments can be determined from the compositions of the suspended load of associated rivers. The average composition of ancient clays and argillaceous schists of the continents can be used to estimate the composition of ancient sediments in regions of avalanche sedimentation, because in the regions of avalanche sedimentation, these rocks are the main ulti-

TABLE 5.—AVERAGE COMPOSITION OF FLUVIAL SUSPENDED LOADS DERIVED FROM CLAYS AND PELITIC SCHISTS COMPARED TO THAT OF BOTTOM SEDIMENTS IN REGIONS CHARACTERIZED BY AVALANCHE SEDIMENTATION, AS WELL AS A COMPARISON OF THE COMPOSITIONS OF SUSPENDED AND BOTTOM LOADS IN OCEANIC REGIONS CHARACTERIZED BY PELAGIC SEDIMENTATION

	Suspended and Bottom Sediments of Rivers in Areas of Avalanche Sedimentation			Suspended and Bottom Sediments of the Ocean Pelagic Zone	
Elements Oxides	Average for Suspended Load of the World's Rivers	Average Compositon of Clays and Argillaceous Schists		Average Composition of the Ocean's Suspended Load	Average Composition of Bottom Sediments of the Oceanic Pelagic Zone
SiO_2	54.80[1]	60.30[2]	55.09[3]	6.41[4]	42.08[4]
TiO_2	0.67	0.97	0.86	0.03	0.57
Al_2O_3	15.65	11.98	16.30	1.03	10.11
Fe_2O_3	7.28	7.36	4.17	0.91	5.41
FeO	—	—	1.87	—	—
MnO	0.14	0.16	0.05	0.02	0.37
MgO	2.07	1.82	2.46	—	2.36
CaO	3.52	4.08	4.75	—	19.05
Na_2O	1.35	0.96	0.75	—	2.45
K_2O	1.81	2.53	3.01	—	1.60
P_2O_5	0.14	0.29	0.11	0.69	—
C_{org}	2.5	—	—	11.0	0.48
SiO_2/Al_2O_3	3.48	—	3.38	6.22	4.16
Al_2O_3/TiO_2	18.23	—	18.95	34.33	17.74
Fe_2O_3/Al_2O_3	0.46	—	0.26	0.88	0.54
$\frac{Fe + Mn}{Ti}$	13.03	—	8.45	32.25	12.0
CaO/MgO	1.69	—	1.93	—	.07
Fe/Mn	46.36	—	74.75	42	13.06

[1]Lisitzin and Gordeev (1974); [2]Martin and Meybeck (1979); [3]Ronov and Yaroshevsky (1976); [4]Lisitzin (1978)

mate sources, and more than 90 perecent of the clay is deposited in this manner. Comparison of the average composition of river suspended load and average composition of clays and schists indicates that their macro-element composition is very similar. This is related to the fact that these elements are transported in mineral crystal lattices as suspended sediment, i.e., in chemically inert forms.

For Cd, Zn, Pb, Ag, and Mo, enrichment relative to the composition of the river suspended load is characteristic, which is related to absorption of these elements on fine pelitic particles as well as to the anthropogenic factor. Impoverishment of river suspended load with respect to the composition of sedimentary rocks is observed for U, Ta, Cs, Hf, and Ba. A high degree of similarity in the composition of river suspended load and sedimentary rocks is characteristic mainly for rivers of the moderate zone. It is less in the equatorial zone, where the chemical composition of original rocks is much less stable. Hydrogenic forms of many of the materials constitutes from 80 to 99 percent in the pelagic realm, which means that the soluble form predominates for practically all the elements in the ocean, as contrasted to the suspended load. The transformation of the elements from solution into suspension is conducted by organisms and absorption processes.

When comparing average content of elements in the river suspended loads and pelagic ocean sediments, three groups of elements can be distinguished: 1) The majority of macro-elements (Si, Al, Fe, Mg, Ti, and K) and low-mobile elements (REE, Se, Yb, Y, etc.) are found in pelagic ocean sediments and river suspended loads nearly in the same amounts. This is because these minerals are principally detrital (terrigenous matrix), which means that this group of elements cannot be used for the differentiation of avalanche and normal pelagic sediments in the deposits of the past. 2) The second group of elements (Tl, Cd, Zn, Rb, and Cr) exist in river suspended loads in concentrations considerably higher than in pelagic sediments. This enrichment is partially related to the influence of man. 3) The most interesting contributions are the elements of the third group (Ba, Ca, B, As, Co, Cu, Mn, Sr, and P), which are enriched in the pelagic sediments as compared to river suspended load and sediments of the avalanche regions. Accumulation of these elements is related to biogenic processes as well as to absorption.

For the particulate forms of elements, the main process at the river-sea barrier is mechanical differentiation (mainly at the limit of 2 μm) with sedimentation and coagulation of the fine fraction (less than 2 μm). For the soluble forms, the main role belongs to the formation of iron-organic flocculi, absorption and desorption co-sedimentation, and biological extraction. The geochemistry of solubilized forms is determined here mainly by the affinity of the elements for humic substances and to iron hydroxides.

Not only is the decrease in the elements' total content in soluble form and mainly in the particulate form significant (Fig. 10), which results in the predominance of solubilized (and mobile) forms of the elements in the ocean, but the changes of their forms are significant as well. In this process, the main point is that at the river-sea barrier the number of passive-element forms (incorporated into crystal lat-

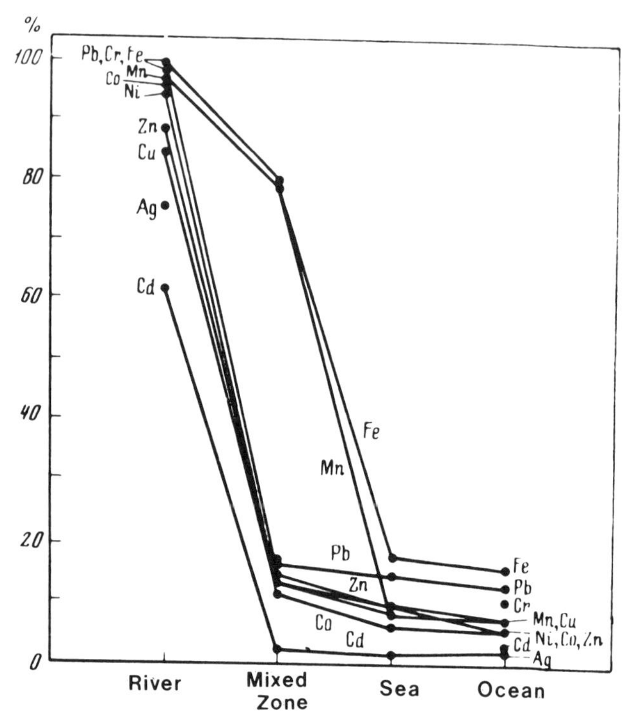

FIG. 10.—Reduction of particulate elements at the river-sea barrier (Gordeev, 1983).

tices of disintegrated minerals) sharply decreases and the number of mobile forms (solubilized, co-sedimented, absorbed) sharply increases.

The new details, which shed light on the dynamics of the sedimentation process in regions of avalanche sedimentation (average composition of the suspended load of rivers of the world as compared to the average composition of sedimentary rocks of the continents for the regions of avalanche sedimentation) and in the regions of ocean pelagic sediments (the average of 1,500 measurements) can be distinguished by normalization of macro- and of the most representative trace-element contents to aluminium (Table 6). Table 6 shows that: 1) the average composition of the river suspended load is very close to that of the sedimentary rocks of the continents, concerning mainly the hydrolizate elements and Fe with Mn, i.e., the composition of river suspended load corresponds to continental sedimentary material. 2) In the sedimentary rocks much more of the elemental composition is related to biogenic extraction from water (Si, Ca, Mg, Sr). 3) Ocean pelagic sediments correspond by element hydrolizates to the river suspended load and to the mean sedimentary rocks of the continents; however, the content of titanium is much higher in them. 4) The most significant differences can be noted (as well as in sedimentary rocks) for the biogenically cycled elements. Silica, which is precipitated in the form of opal, as well as Ca, Mg, Sr and mainly Ba, whose ratio in pelagic sediments is higher than the Ba/Al ratio in the river suspended load and sedimentary rocks, are enriched by an order of magnitude in pelagic sediment. 5) The Fe/Al and Mn/Al ratios are much higher in pelagic sediments (by 4–5 times for Mn). 6) Therefore, for pelagic ocean sediments, the most remarkable peculiarity is the high impact of biogenic elements (Ca, Si, Ba, and Sn), as well as Fe and Mn. If the excess of biogenic components is explained by the transformation of these elements by bios from solutions (mainly by plankton), then for Fe and Mn, the input of endogenic substances from hydrothermal areas of the mid-ocean ridges is significant.

TABLE 6.—A COMPARISON OF SELECTED ELEMENTAL COMPOSITION OF FLUVIAL SUSPENDED LOADS, CONTINENTAL SEDIMENTARY ROCKS AND OCEANIC SEDIMENT (NORMALIZED TO ALUMINUM)

Elements	River Suspension	Continental Sedimentary Rocks	Ocean Sediments
Al	1	1	1
Si	3.08	3.40	3.67
Ti	0.048	0.043	0.064
Fe	0.61	0.62	0.71
Mn	0.012	0.012	0.052
Mg	0.15	0.27	0.27
Ca	0.169	0.66	1.42
Ba	0.0045	0.0043	0.048
Sr	0.0032	0.0043	0.014
Cu	0.00096	0.00054	0.0044
Zn	0.0037	0.00076	0.0024
Pb	0.0019	0.00019	0.00097
Ga	0.00022	0.00028	0.00024
Zr	0.0024	0.0019	0.0023
Hf	0.00003	0.0005	0.00007
Th	0.00011	0.00010	0.00014
Y	0.00032	0.00029	0.00099
La	0.00047	0.00038	—

Oceanic Matter

The most important peculiarity of the sediments of the regions of high and superhigh rates of sedimentaton is a very significant content of organic matter in them. About 90 percent of the organic matter of the World Ocean is concentrated in the regions characterized by avalanche sedimentation. The combination of high contents of organic matter in these sediments (1–10 percent and >10 percent $C_{org.}$) with significant thickness, size and long development of the sedimentary basins incorporating them results in oil and gas accumulation. At present, about 33 percent of the total oil and gas production is obtained from avalanche sedimentation zones, and by the end of the century, the major part will be produced there. Thus, the regions of avalanche sedimentation are, at the same time, the regions of avalanche accumulation of organic matter, the principal site of origin for oil and gas.

Gravitites are one of the four environments in which about 90 percent of the known world oil and gas deposits have been discovered (Yarborough, 1971). This is the predominant type of oil-bearing deposit in California (Curray and others, 1971; Gardett, 1971; Nagle and Parker, 1971). In particular, they were found in deep-sea environments (avalanche sedimentation of the second level) in the Gulf of Mexico (Hoyt, 1959; Paine, 1966), where a significant $C_{org.}$ content is combined with high and superhigh sedimentaton rates and with the presence of sandy oil and gas accumulation layers.

One of the most remarkable peculiarities of estuaries and deltas is a very high primary production (phytoplankton and phytobenthos) reaching the highest known levels, which results in high content of organic matter in the bottom sediments. According to the roughest estimations, the content of $C_{org.}$ in these sediments is one to two orders higher than in pelagic deposits. The high $C_{org.}$ content in the sediments, together with avalanche rates of their accumulation, results in fast burial of the organic matter, and in the appearance of conditions favorable for oil and gas formation. The data concerning primary production in coastal waters and estuarics arc shown in Table 7.

The primary production of macrophytes reaches 30 kg/m^2 of that production in moderate latitude zones. The majority of macrophytes are macroalgae and sea grasses, and mangroves in tropical latitudes. The average values of the marshes and swamps production can be considered to be 400 to 1,400 gC/m^2year. The input of allochtonous organic matter (solubilized and particulate) from river waters is also great. Mean $C_{org.}$ content in pelagic sediments is 0.25 to 0.30 percent, and the mean $C_{org.}$ content in avalanche sediments (of the continental slope) is about 1.3 to 1.5 percent, i.e., is higher by five times and more.

Figures obtained by Hunt (1972) are close to these with an amount of 0.99 percent of $C_{org.}$ for shelves and slopes and 0.22 percent of $C_{org.}$ for pelagic sediments. Trask (1939) considered the mean values for coastal sediments to be 1 to 3 percent (sometimes up to 8 percent), and for deep-sea sediments to be 0.3 to 1.5 percent. High values could be related to the slope sediments (Trask, 1939). Mean $C_{org.}$ content in estuaries and deltas is usually much higher, and

TABLE 7.—A COMPARISON OF PRIMARY PRODUCTION IN ESTUARIES AND FJORDS (HEAD, 1976) AND THE OCEANIC PELAGIC REALM ($g\ c/m^2/yr$)

Location	
	Estuaries and sea-estuaries
	1. Plankton
Long Island	470
Tropical	135
Azov Sea	200–300
Baltic Sea	130
	2. Macrophytes
	500–2000
	Fjords
British Columbia	480
Sweden	100
Scotland	70
	Pelagic regions of the ocean
Oligotrophic waters	25.6
Transitional	51.1
Eutrophic (regions of equatorial divergence and Antarctic regions)	73.0
Coastal	124

reaches 2 to 3, sometimes 5 percent (in some cases higher). It must be noted that mean $C_{org.}$ content in shelf sediment is about 0.7 percent, i.e., shelf sediment occupies the transition position between the deposits of the first sedimentation level (estuaries and deltas), situated above, and the second-level deposits, at the slope base.

Spatial localization of the deposits, rich in organic matter, of the regions of high and superhigh sedimentation rates, their close quantitative relation on the first and the second levels, is supplemented by the influence of climatic-zone characteristics. The $C_{org.}$ content in deposits of deltas of the glacial zone was established long ago to be much lower (Trask, 1939) than that of the zones with moderate and tropical climate (Subba Rao, 1960; Naidu, 1966).

According to the calculations made recently for the Atlantic Ocean (the data from more than 60 scientific cruises of Soviet vessels), 26.3 billion tons of $C_{org.}$ are buried in 1,000 years in the bottom sediments (Emelyanov and Romankevitch, 1979) with primary production of 5.8 billion tons of $C_{org.}$ Maximum $C_{org.}$ accumulation is observed on the periphery of the ocean (Fig. 11), where 22.6 billion tons (87 percent of the total) are sedimented. At the same time, on the vast areas of the ocean bed, just 3.4 billion tons of $C_{org.}$ are sedimented in 1,000 years (13 percent). Maximum $C_{org.}$ content in the Atlantic Ocean sediments is observed up to 9.3 percent in deposits of bays, fjords and upwelling zones and up to 16 percent in the upwelling zone of southern West Africa.

Leaving aside important details, we may conclude that regions of avalanche sedimentation of the first and the second levels constitute the principal bodies of sediment on Earth and the principal (avalanche) accumulations of organic matter, which make them the main source of hydrocarbons. Prospecting for hydrocarbons for industrial purposes is, in fact, prospecting for ancient regions of avalanche sedimentation.

The data presented show that the most characteristic differences in the major (chemical) composition between the

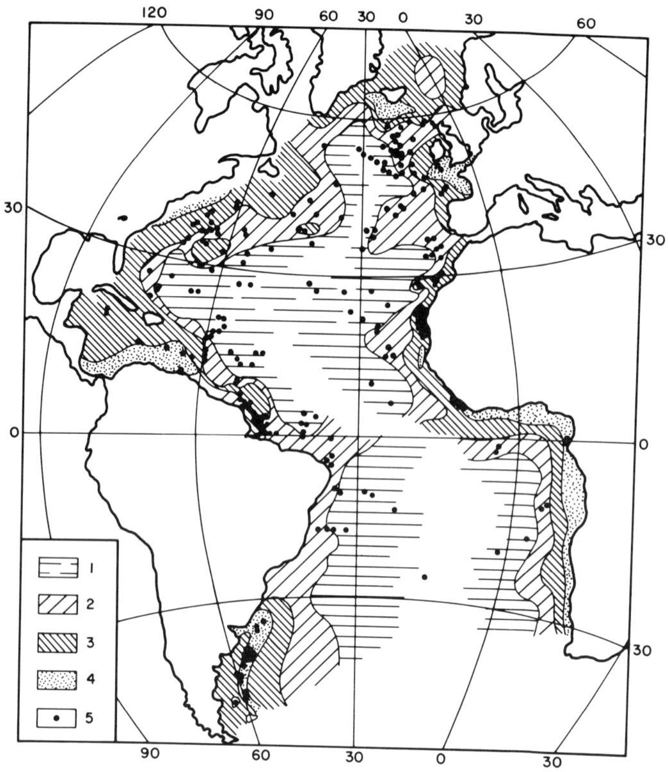

FIG. 11.—Distribution of $C_{org.}$ in the upper layer (0.5 cm) of Atlantic Ocean sediment in percent of dry matter: (1) <0.25; (2) 0.25–0.5; (3) 0.5–1.0; and (4) >1.0. Dots indicate sampling locations (Emelyanov and Romankevitch, 1979).

sediments of avalanche-sedimentation regions and the normal pelagic ocean sediments are the following: 1) avalanche sediments are enriched with organic matter by five to 10 times (and more), as compared to the deep-water sediments, and 2) higher contents of $CaCO_3$ (two times and more), manganese (three times), barium (seven times), copper, cobalt and arsenic (three to four times) are characteristic of the pelagic sediments. Thus, qualitative (granulometric, mineral and chemical) composition of the sediments of deltas and corresponding deposits of the second avalanche-sedimentation level (deep-sea fans and other gravitite deposits) are strongly dependent on the composition of the river suspended load. The greatest influence is produced by climate, because it determines the rate and depth of protolith processing by chemical weathering.

TRANSFORMATION OF THE SEDIMENTARY MATTER ON THE RIVER-SEA BARRIER AND ITS KEY IMPORTANCE FOR AVALANCHE AND NORMAL PELAGIC SEDIMENTATION: GLOBAL SIGNIFICANCE OF AVALANCHE SEDIMENTATION

Thus, on the river-sea barrier, drastic changes of the original sediment of river water (particulate and dissolved) take place, passing from a supremacy of particulate forms of elements (freshwater geochemistry of sediment) to the predominance of dissolved forms of elements (sea and ocean geochemistry of sediment). There is a group of elements in river suspension for which the dissolved form is more significant than the particulate form (Cl, S, I, F, Br, Ca, Na,

Sb). For many years this extremely important process was not considered, which leads to serious mistakes in understanding the whole sedimentary process and its geochemistry.

On a vast scale, the transformation of sediment and rebuilding of its structure takes place. Two types of sediment evolve, which differ sharply by quantity and composition: 1) sediment of avalanche-sedimentation zones (about 90 to 92 percent of the total); and 2) sediment of the ocean pelagic area (8 to 10 percent of the total). In a general way, this is obvious from the following calculations. Above the barrier, the particulate river flow is equal to 18.5 billion tons of the particulate forms sedimented at the barrier. The input of elements in dissolved form to the barrier is 4.4 billion tons annually. Thus, rivers annually yield 18.5 + 4.4 = 22.9 billion tons of solution and suspension load. If we accept that, then an average of about 30 percent of dissolved forms is sedimented at the barrier, or about 1.32 billion tons. Thus, in the region of avalanche sedimentation of the first level, 17.02 + 1.31 = 16.34 billion tons of sediment are deposited annually, i.e., 80 percent of its total (22.9 billion tons) input. Here the main part of the ocean sediment budget is extracted and the mass-accumulation potential for pelagic sedimentation is limited to material "which bypassed" the barrier (1.43 billion tons of particulate and 3.08 billion tons of dissolved matter). Thus, in total, 4.56 billion tons, i.e., about 25 percent from avalanche regions, or 19.9 percent, pass from the total input to the barrier. If the particulate element form is predominant in the river sediment, it exceeds dissolved forms here by four times, then the particulate form in oceanic sediment contributes only 0.48, i.e., two times less, than the dissolved form.

The formation of sedimentary bodies in the avalanche-sedimentation regions is characterized by a marked excess of particulate (disintegrated) material. Particulates are 13 times (17.02/1.32 = 12.8) more abundant here than the dissolved load. The typical ratio is quite different for the sedimentary bodies formed in the ocean pelagic areas. The ratio of particulate forms/dissolved forms is 1.48/3.08 = 0.48. That is, for pelagic sediment accumulation, the mechanisms of transforming dissolved load into suspended load and then into sediments, play the major role, as well as the addition of 7 to 8 percent of terrigenous material, which passed through the river-sea barrier.

Therefore, the doctrine of avalanche sedimentation drastically changes the image of the pattern of sedimentation processes in seas and oceans not only in the regions of avalanche sedimentation itself, but also in the pelagic-sedimentation regions. That is why the main role in the pelagic realm is played not by the process of mechanical fractionation of river suspended loads (supplied into the ocean), but by a more complex process of biogeochemical removal of element dissolved forms, and their transformation into suspension and further sedimentation. That is why geochemistry of the sedimentation process in seas and oceans is biogeochemistry (Lisitzin and others, 1983; Lisitzin, 1988). Biotransformation and biological differentiation includes the processes of bioassimilation, biofiltration, bioabsorption and biotransformation, i.e., in pelagic areas it is the bios that determines the processes of preparation, transportation and deposition of sediment and its differentiation. On the river-sea border the change of lithogenic forms of elements to hydrogenic ones, available to the bios, takes place. Elemental matter, transformed into suspended load by organisms (mainly by phytoplankton) and terrigenous matter, pass through the barrier, enter the pelagic area not "particle by particle," but by biofiltration associated with zooplankton and benthonic organisms.

ACKNOWLEDGMENTS

The author expresses sincere thanks to Professors Robert H. Osborne and Donn S. Gorsline for their time and interest in editing the translation of this paper. Although the original text was improved by valuable corrections, possible shortcomings are the responsibility of the author. The manuscript was typed by Sue Turnbow and Desser Moton.

REFERENCES

Artem'ev, V. E., Bobylyeva, N. U., and Lutsarev, C. V., 1981, Dissolved and particulate organic matter in the zone of river and sea waters mixing (Riga Bay): Oceanology, v. 21, no. 2, p. 286–292.

Avbovbo, A. A., 1978, Tertiary lithostratigraphy of Niger delta: American Association of Petroleum Geologists Bulletin, v. 62, no. 2, p. 235–240.

Berg, O. R., 1982, Seismic detection and evaluation of delta and turbidite sequences: Their application to exploration for the subtle trap: American Association of Petroleum Geologists Bulletin, v. 66, no. 9, p. 1271–1288.

Bewers, I. M., and Yeats, P. A., 1977, Oceanic residence times of trace elements: Nature, v. 268, no. 5621, p. 595–598.

Bouma, A. H., Normark, W. R., and Barnes, N. E., eds., 1985, Submarine Fans and Related Turbidite Systems: Springer-Verlag, New York, 351 p.

Broussard, M. L., ed., 1975, Deltas, Models for Exploration: Houston Geological Society, 555 p.

Bugge, T., 1989, Submarine slides of the Norwegian continental margin, with special emphasis in the Storegya area: Continental Shelf Institut, Norway, Publication No. 110, p. 1–152.

Butenko, J., Milliman, J. D., and Ye, Y. C., 1985, Geomorphology, shallow structure and geological hazards in the East China Sea: Continental Shelf Research, v. 4, p. 121–142.

Carbonnel, J. P., and Meybech, M., 1975, Quality variations of the Mekong River at Pnom Peuh, Cambodia and chemical transport in the Mekong Basin: Journal of Hydrology, v. 27, p. 249–265.

Chough, S. K., 1985, Further evidence of fine-grained sediment dispersal of the southern Yellow Sea: Sedimentary Geology, v. 41, p. 152–172.

Clark, J. G., and Jager, E., 1969, Denudation rate in the Alps from geochronologic and heat flow data: American Journal of Science, v. 267, p. 1143–1160.

Colley, S., Thomson, J., and Toole, J., 1969, Uranium relocations and derivation of quasi-isochrons for turbidite/pelagic sequence in the Northeast Atlantic: Geochimica et Cosmochimica Acta, v. 53, p. 1223–1234.

Curray, J. R., Hall, K. B., and Herron, R. F., 1971, Geology, oil fields and future petroleum potential of Santa Barbara Channel area, California, *in* Cram, I. H., ed., Future Petroleum Provinces of the United States—Their Geology and Potential: American Association of Petroleum Geologists Memoir 15, p. 192–211.

Curray, J. R., and Moore, D. G., 1971, Growth of the Bengal deep-sea fan and denudation in the Himalayas: Geological Society of America Bulletin, v. 82, p. 563–572.

DeMaster, D. L., McKee, B. A., Nillroner, B. A., Qi Al, C. A., and Cher, J. C., 1985, Rates of sediment accumulation and particle removing based on radiochemical accumulation measurements from continental shelf deposits in the East China Sea: Continental Shelf Research, v. 4, p. 143–158.

DEMINA, L. L., GORDEEV, V. V., AND FOMINA, L. S., 1978, Fe, Mn, Zn, Cu forms in river water and suspensions and their changes in the river and sea waters mixing zone (on the example of the Black, Azov and Kaspian Seas): Geochemistry, no. 8, p. 1211–1229.

DUINKER, J. C., WOLLAST, R. G., AND BILLEN, G., 1979, Manganese in the Rhine and Scheldt estuaries: Part 2, Geochemical Cycling Estuarine Coast: Marine Science, v. 9, no. 7, p. 727–738.

EISMA, D., KALF, J., AND KAAST, J., VON DER, 1978, Suspended matter in the Zaire estuary and the adjacent Atlantic Ocean: Netherlands Journal of Sea Research, v. 12, nos. 3-4, p. 382–406.

EMELYANOV, E. M., AND ROMANKEVITCH, E. A., 1979, Geochemistry of the Atlantic Ocean: Organic compounds and phosphorus: Nauka, Moscow, 220 p.

FOLGER, D. W., 1972, Characteristics of estuarine sediments of the United States: U.S. Geological Survey Professional Paper 72, Government Printing Office, Washington, D.C., p. 1–94.

FROIDEFOND, J. M., PUJOS, M., AND ANDRE, Y., 1988, Migration of mud banks and changing coastline in French Guiana: Marine Geology, v. 84, p. 19–30.

GALLOWAY, W. E., 1975, Evolution of deltaic systems, *in* Broussard, M. L., ed. Deltas, Models for Exploration: Houston Geological Society, p. 87–89.

GARDETT, P. H., 1971, Petroleum potential of Los Angeles, California, *in* Cram, I. H., ed., Future Petroleum Provinces of the United States—Their Geology and Potential: American Association of Petroleum Geologists Memoir 15, p. 296–308.

GIBBS, R. I. 1981, Sites of river-derived sedimentations in the ocean: Geology, v. 9, no. 1, p. 77–80.

GILBERT, R., 1983, Sedimentary processes of Canadian arctic fjords: Sedimentary Geology, v. 36, p. 147–176.

GNIBIDENKO, G. C., 1961, The Bering Sea tectonics, *in* Geological History of the Bering Sea: Far Eastern Scientific Centre, USSR Academy of Sciences, Vladivostok, p. 5–26.

GORDEEV, V. V., 1983, River Flow to the Oceans and the Patterns of its Geochemistry: Nauka, Moscow, 180 p.

GORSLINE, D. S., 1978, Anatomy of margin basins: Journal Sedimentary Petrology, v. 48, p. 1055–1068.

GORSLINE, D. S., 1984, A review of fine-grained sediments, origins, characteristics, transport and deposition, *in* Stow, D. A. V., and Piner, D. J. W., eds., Fine-Grained Sediments: Deep-Water Processes and Facies: The Geological Society, Blackwell Scientific Publication, Oxford, London, p. 17–34.

HANDA, B. K., 1972, Geochemistry of the Ganges River: Geohydrology, v. 2, p. 71–78.

HEAD, P. C., 1976, Organic processes in estuaries, *in* Burton, J. D., and Liss, P. S., eds., Estuarine Chemistry: Academic Press, London, p. 53–91.

HOYT, W. V., 1959, Eorsional channel in the middle Wilcox near Yoakum, Lavaca County, Texas: Gulf Coast Association of Geological Societies, Transactions, v. 9, p. 41–50.

HU, M., STALLARD, R. F. AND EDMOND, J. M., 1982, Major ion chemistry of some large Chinese rivers: Nature, v. 298, p. 550–553.

HUNT, J.M., 1972, Distribution of carbon in crust of Earth: American Association of Peteroleum Geologists Bulletin, v. 56, p. 2273–2277.

IVANOV, K. L., 1955, On suspension sedimentation on the estuary seaside of Kura: Proceedings of the State Oceanographic Institute, issue 28, no. 40, p. 131–136.

KIM, S. W., LEE, Y. O., KONG, Y. S., PARK, Y. S., LEE, C. W., AND MIN, G. H., 1988, Marine geological study of the continental shelf off southern Korea: Energy Research Report, KR-87-26, p. 1–64.

KOLLA, V., AND COUMES, F., 1987, Morphology, interval structure, seismic stratigraphy and sedimentation in Indus fan: American Association of Petroleum Geologists Bulletin, v. 71, no. 6, p. 650–677.

LEE, H. J., AND CHOUGH, S. K., 1989, Sediment distribution, dispersal and budget in the Yellow Sea: Marine Geology, v. 87, p. 195–205.

LEE, H. J., CHOUGH, S. K., JEONG, K. S. AND HAN, S.-J., 1987, Geotechnical properties of sediment cores from the southern Yellow Sea: Effects of depositional processes: Marine Geotechnology, v. 7, p. 37–52.

LI, Y., TERSAKA, H., YANG, T. S., AND CHEN, J. S., 1984, The elemental composition of suspended particles from the Yellow and Yangtze River: Geochimica et Cosmochimica Acta, v. 48, p. 1561–1564.

LISITZIN, A. P., 1974, Sedimentation in the Oceans: Nauka, Moscow, 438 p.

LISITZIN, A. P., 1978, Processes of Ocean Sedimentation: Nauka, Moscow, 392 p.

LISITZIN, A. P., 1979, Global zones of sedimentogenesis, *in* Brechovskich, M. L., ed., Soviet Oceanology Success: Nauka, Moscow, p. 118–135.

LISITZIN, A. P., 1980, General regularities of the structure of sedimentary thickness of the ocean, *in* Monin, A. S., and Lisitzin, A. P., eds., Geological History of the Ocean: Nauka, Moscow, p. 36–103.

LISITZIN, A. P., 1982, Avalanche sedimentation, *in* Lisitzin, A. P., ed., Avalanche Sedimentation in the Ocean: Rostov University Publishers, Rostov-na-Donu, p. 3–59.

LISITZIN, A. P., 1983, Avalanche sedimentation in the ocean. Paper 1: Lithology and Minerals Resources, no. 6, p. 3–27.

LISITZIN, A. P., 1984a, Avalanche sedimentation in the ocean. Paper 2: Lithology and Mineral Resources, no. 5, p. 3–29.

LISITZIN, A. P., 1984b, Sedimentary ocean bodies, *in* Bogdanov, N. A., and Pucharovsky, Y. M., eds., Geology of the Floor According to the Data of Deep-Sea Drilling: Nauka, Moscow, p. 12–62.

LISITZIN, A. P., 1985a, Avalanche sedimentation in the ocean. Paper 3: Lithology and Mineral Resources, no. 4, p. 3–25.

LISITZIN, A. P., 1985b, Avalanche sedimentation in the ocean. Paper 4: Lithology and Mineral Resources, no. 6, p. 3–20.

LISITZIN, A. P., 1986, Avalanche sedimentation in the ocean. Paper 5: Lithology and Mineral Resources, no. 4, p. 3–29.

LISITZIN, A. P., 1988, Avalanche Sedimentation and Hiatuses in Sediment Accumulations in Seas and Oceans: Nauka, Moscow, 306 p.

LISITZIN, A. P., DYEMINA, L. L., AND GORDEEV, V. V., 1983, Biosystem of sea water, *in* Monin, A. S., and Lisitzin, A. P., eds., The Biogeochemistry of the Ocean: Nauka, Moscow, p. 32–60.

LISITZIN, A. P., AND GORDEEV, V. V., 1974, On chemical composition of suspension and ocean waters: Lithology and Mineral Resources, no. 3, p. 38–58.

LISITZIN, A. P., LUKASHIN, V. N., AND GURVICH, E. G., 1982, On the ratio of elements output by rivers and their accumulation in bottom sediments of the ocean: Geochemistry, no. 1, p. 106–113.

LOWE, D. R., 1976, Subaqueous liquified and fluidized sediment flows and their deposits: Sedimentology, v. 23, p. 285–308.

MARTIN, J. M., AND MEYBECK, M., 1979, Elemental mass-balance of material carried by major world rivers: Marine Chemistry, v. 7, p. 173–206.

MEADE, R. K., NORDIN, C. F., CURTIS, W., RODRIGUES, F. M. C., DEVALE, C. M., AND EDMOND, J. M., 1979, Sediment loads in the Amazon River: Nature, v. 278, p. 161–163.

MENARD, H. W., 1964, Marine Geology of the Pacific: McGraw-Hill, New York, 271 p.

MILLIMAN, J. D., AND BOYLE, E. M., 1975, Biological uptake of dissolved silica in the Amazon River estuary: Science, v. 183, p. 395–397.

MILLIMAN, J. D., LI, F., ZHAO, Y. Y., ZHENG, T. M., AND LIMEBURNER, R., 1986, Suspended matter regime in the Yellow Sea: Progress in Oceanography, v. 17, p. 215–227.

MILLIMAN, J. D., AND MEADE, R. H., 1983, World-wide delivery of river sediment to the ocean: Journal of Geology, v. 91, p. 34–42.

MILLIMAN, J. D., HUANG-TING, S., ZUO-SHENG, Y., AND MEADE, R. H., 1985, Transport and deposition of river sediment in the Chomgjiang estuary and adjacent continental shelf: Continental Shelf Research, v. 4, p. 37–46.

MILLIMAN, J. D., SUMMERHAYES, C. P., AND BARETTO, H. T., 1975, Oceanography and suspended matter off Amazon River, February–March, 1973: Journal of Sedimentary Petrology, v. 45, no. 1, p. 189–206.

MONIN, A. S., AND GORDEEV, V. V., 1988, The Amazonia: Nauka, Moscow, 214 p.

MOROZOV, N. P., 1979, On the ratio of migration forms of trace elements in waters of rivers, bays, seas and oceans: Geochemistry, no. 8, p. 784–795.

NAGLE, H. E., AND PARKER, E. S., 1971, Future oil and gas potential of onshore Ventura Basin, California, *in* Cram, I. H., ed., Future Petroleum Provinces of the United States—Their Geology and Potential: American Association of Petroleum Geologists Memoir 15, p. 254–297.

NAIDU, A. S., 1966, Lithological and chemical facies changes in the recent deltaic sediments of the Godavari River, India, *in* Shirley, M. L., and Ragsdale, J. A., eds., Deltas in their Geological Framework: Houston Geological Society, p. 125–157.

NANI, B. R., AND KOLLA, V., 1982, Acoustic character and thickness of sediments of the Indus fan and the continental margin of western India: Marine Geology, v. 47, nos. 3–4, p. 181–195.

NARDIN, T. R., HEIN, F. J., GORSLINE, D. S., AND EDWARDS, B. D., 1979, A review of mass movement processes, sediment and acoustic characteristics, and contrast in slope and base-of-slope systems versus canyon-fan-basin floor systems: Society of Economic Paleontologists and Mineralogists Special Publication 27, p. 61–73.

NEDECO, 1968, Submarine Transportation Study. Report on Hydrualic Investigations: Nedeco Engineering Consultants, The Hague, 293 p.

NORMARK, W. R., 1978, Fan valleys, channels and depositional lobes on modern submarine fans: Characteristics for recognition of sandy turbidite environments: American Association Petroleum Geologists Bulletin, v. 62, p. 912–931.

PAINE, W. R., 1966, Stratigraphy and sedimentation of subsurface beds of southwestern Louisiana: Gulf Coast Association of Geological Societies, Transactions, v. 16, p. 161–174.

PILKEY, O. H., AND CLEARY, W. J., 1986, Turbidite sedimentation in the northwestern Atlantic Ocean basin, *in* Vogt, P. R., and Tucholke, B. E., eds., The Western North Atlantic Region: Geological Society of America, Boulder, Colorado, p. 437–450.

RATEEV, M. A., GORBUNOVA, Z. N., LISITZIN, A. P., AND NOSOV, G. L., 1968, Climatic zonal regularities of clay mineral distribution in sediments of the World Ocean: Oceanologic Studies No. 18, Nauka, Moscow, p. 282–310.

REN, MEI-E., ed., 1986, Modern Sedimentation in the Coastal and Nearshore Zones of China: Springer-Verlag, Berlin, 466 p.

ROMANKEVICH, E. A., 1977, Geochemistry of Organic Substances in the Ocean: Nauka, Moscow, 234 p.

RONOV, A. B., AND YAROSHEVSKIY, A. A., 1976, A new model for the chemical structure of the Earth's crust: Geochima, no. 12, p. 1761–1795.

SEEMANN, U., 1978, Snow sedimentation—diagenesis and avalanches; a correlation with sedimentary rocks: Sedimentary Geology, v. 21, p. 189–201.

SHEPARD, F. P., 1973, Submarine Geology (3rd ed): Harper and Row, New York, 517 p.

SHOLKOVITZ, E. R., 1976, Flocculation of dissolved organic and inorganic matter during the mixing of river water and seawater: Geochimica et Cosmochimica Acta, v. 40, p. 831–846.

SHOLKOVITZ, E. R., 1978, Flocculation of dissolved Fe, Mn, Al, Ca, Ni, Ce and Cd during estuarine mixing: Earth and Planetary Science Letters, v. 41, no. 1, p. 77–86.

STALLARD, R. F., AND EDMOND, J. M., 1983, Geochemistry of the Amazon, the influence of geology and weathering environment on the dissolved load: Journal of Geophysical Research, v. 88, p. 9671–9688.

STALLARD, R. F., AND EDMOND, J. M., 1987, Geochemistry of the Amazon, weathering chemistry and limits to dissolved inputs: Journal of Geophysical Research, v. 92, p. 8293–8302.

STANLEY, D. J., 1988, Low sediment accumulation rates and erosion on the middle and outer Nile delta shelf off Egypt: Marine Geology, v. 84, p. 111–117.

SUBBA RAO, M., 1960, Organic matter in marine sediments off east coat of India: American Association of Petroleum Geologists Bulletin, v. 44, p. 1705–1713.

SUBRAMANIAN, Y., 1979, Chemical and suspended sediment charcteristics of rivers of India: Journal of Hydrology, v. 44, p. 37–55.

TERZAGI, K., 1956, Varieties of submarine slope failures: University of Texas, Bureau of Engineering Research Special Publication 23, 41 p.

TRASK, P. D., 1939, Organic content in recent marine sediments, *in* Trask, P. D., ed., Recent Marine Sediments: Thomas Marby and Company, London, p. 428–453.

TREFRY, J. H., AND PRESLEY, B. J., 1976, Trace metal transport from the Mississippi River in the Gulf of Mexico, *in* Windom, H. L., and Duce, R. A., eds., Marine Pollutant Transfer: Lexington Books, Toronto, p. 39–76.

UCHUPI, ELAZAR, AND AUSTIN, J. A., 1979, The stratigraphy and structure of the Laurentian cone region: Canadian Journal of Earth Sciences, v. 16, p. 1726–1752.

VAIL, P. R., AND HARDENBOL, J., 1979, Sea level change during the Tertiary: Oceanus, v. 22, p. 71–79.

VAIL, P. R., MITCHUM, R. M., JR, AND THOMPSON, S., III, 1977, Seismic stratigraphy and global changes of sea level. 4. Global cycles of relative changes of sea level, *in* Payton, C. E., ed., Seismic Stratigraphy—Application to Hydrocarbon Exploration: American Association of Petroleum Geologists Memoir 26, p. 83–97.

WEIMER, PAUL, 1990, Sequence stratigraphy, facies geometries and depositional history of the Mississippi fan, Gulf of Mexico: American Association of Petroleum Geologists Bulletin, v. 74, no. 4, p. 425–453.

YARBOROUGH, H., 1971, Sedimentary environments and the occurrence of major hydrocarbon accumulations: Gulf Coast Association of Geological Societies, Transactions, v. 21, 82 p.

ZEITLER, P. K., TAHIRUHELL, K. A., AND NAESER, C. W., 1980, Uplift of the lesser Himalayas, Northern Pakistan, as inferred from fusion track ages of sphene, apatite and zircon: Fission Trace Dating Workshop, Pisa, 239 p.

DELTA-FRONT GULLIES AS PART OF MASS-MOVEMENT PHENOMENA: MISSISSIPPI RIVER DELTA FRONT

A. H. BOUMA,[1] H. H. ROBERTS,[1] J. M. COLEMAN,[1] AND D. B. PRIOR[2]

[1]*School of Geosciences, Louisiana State University, Baton Rouge, Louisiana 70803; and* [2]*Bedford Institute of Oceanography, Dartmouth, Nova Scotia, B2Y 4A2, Canada*

ABSTRACT: Shepard's classic paper on delta-front gullies on the Mississippi delta slope initiated a host of modern studies that revealed the complexity and importance of subaqueous mass-movement processes. Differential weighting and diapirism, rotational slumps, retrogressive slides and their gullies and depositional lobes are but a few of the processes that mold the morphology of the delta-front area. These instabilities form on extremely low-angle slopes (often less than 0.5°) and are responsible for the transport of large volumes of shallow-water sediment to deeper environments.

INTRODUCTION

Modern subaqueous-slope mass-movement processes have been known since the deployment and subsequent breaking of submarine telephone cables in the 19th century (Coleman and Prior, 1988). Submarine slides have been positively identified as a mechanism to explain submarine cable breaks, anomalous sediment properties, sudden bathymetric changes, and complex submarine-slope morphology (Archanguelsky, 1930; Shepard, 1932).

Based on the classical studies concerning physiography and sediment distribution in the Mississippi River delta complex (e.g., Trowbridge, 1930; Russell, 1936; Fisk and others, 1954), a study was funded by the American Petroleum Institute (API Project 51) to obtain more information on the Mississippi delta area and the northern Gulf of Mexico shelf (Shepard and others, 1960). The API study revived interest among Louisiana investigators; their work led to the Mississippi delta being recognized as a classic locality for deltaic research (Coleman and Gagliano 1964; Coleman, 1988).

In his comparison of subaqueous morphologic features, Shepard (1955) was the first investigator to study and contour systematically the unpublished hydrographic sheets in the Mississippi delta region prepared by the U.S. Coast and Geodetic Survey. He delineated shallow, discontinuous valleys on the continental shelf and adjacent upper continental slope, specifically in front of the distributaries, and compared the gullies to those cutting the submerged delta fronts of lakes in Switzerland. Daly (1936) already had attributed delta-front gullies to erosion by turbidity currents, and Shepard wondered if such a process was applicable to the ones he observed in front of the Mississippi delta.

TYPES OF SUBAQUEOUS-MASS WASTING

Various authors have proposed classification systems for subaqueous-mass movements. Dott (1963) initially suggested four main categories: submarine falls, slides or slumps, flows, and turbidity flows. Moore (1977) omitted turbidity flows and subdivided the other categories on the basis of type of transported material. Middleton and Hampton (1976) proposed the term "sediment gravity flows" and suggested that landslides may evolve into different types of gravity flows. More recently, Prior and Coleman (1984) suggested a subdivision of submarine slides and flows based on the morphology and geometry of the features, showing a continuum from slides to flows. Modern acoustical techniques and concepts of marine-soil mechanics made it possible to expand on previous classifications and to show the relation between processes and forms (Fig. 1).

MISSISSIPPI DELTA-FRONT SEDIMENT INSTABILITY

By the late 1960s, the offshore petroleum industry was drilling and constructing bottom-fixed platforms in water as deep as 100 to 150 m. In the aftermath of a major hurricane (Camille) in 1969, several platforms in water depths of 110 m were severely damaged or lost. Post-hurricane analysis revealed that the damage was not the result of hurricane-driven wind, wave or current forces. Instead, massive slides, resulting from sediment instability, were generated on slopes of less than 0.3° and were responsible for the platform damage (Bea, 1971). Interestingly, the type of failure that caused this structural damage had been discussed some 15 years earlier in a classic paper by Shepard (1955). It required some 20 years to prove that Shepard had been correct in his original assessment that the complex bathymetry observed in the offshore Mississippi River delta was created by subaqueous-mass movements. Figure 2 illustrates the bathymetric variability of the upper-delta front resulting from the mass-movement processes.

More recent detailed marine geological investigations, using modern high-resolution seismic-reflection profiles and sidescan sonar imagery on the subaqueous parts of the continental shelf and upper slope seaward of the Mississippi delta, showed that contemporary recurrent subaqueous gravity-induced mass movements are common phenomena (Coleman and others, 1974; Coleman and Garrison, 1977; Prior and Coleman, 1978a, b; Coleman and Prior, 1980; Roberts and others, 1976, 1980). These subaqueous-sediment instabilities are an integral component of normal deltaic and offshore marine sediment-transport processes. Such sediment instabilities generally occur on very low slopes (often less than 0.5°), where large volumes of sediment may be transported by a variety of processes from shallow to deeper marine locations along well-defined transport paths.

The major factors that influence stability of bottom sediments are: 1) rapid sediment accumulation at the river mouth, resulting in widespread sedimentary loading on the upper delta-front slope (Fig. 3); 2) deposition of coarser grained sands and silts (distributary-mouth bar) on underconsolidated prodelta clays; 3) rapid biochemical degradation of organic material, leading to the formation of large volumes

From Shoreline to Abyss, SEPM Special Publication No. 46

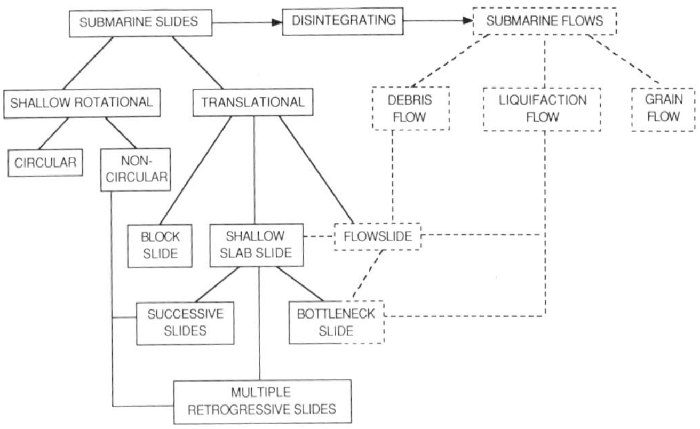

FIG. 1.—Diagram illustrating types of submarine mass-movement features and associated processes and gradations between slides and flows (redrawn from Prior and Coleman, 1984).

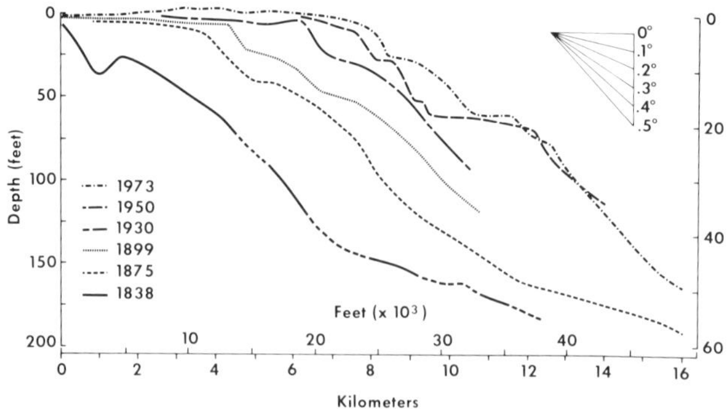

FIG. 3.—Bathymetric profiles of the Mississippi River delta front illustrating progradation from 1838 to 1973. Based on published hydrographic map of South Pass, Mississippi River.

of methane gas. In bubble phase, this gas contributes to the generation of excess pore pressure within the sediments (Whelan and others, 1975); 4) cyclic wave loading of the sea floor caused by winter storms and hurricanes. Variations of wave-induced seafloor pressures impart downslope stresses and contribute to excess porewater-pressure generation.

The main types of subaqueous-slope instabilities presented in Figure 1 can be observed in water depths ranging from 5 to 200 m. In the immediate vicinity of the river mouth, differential weighting caused by implacement of

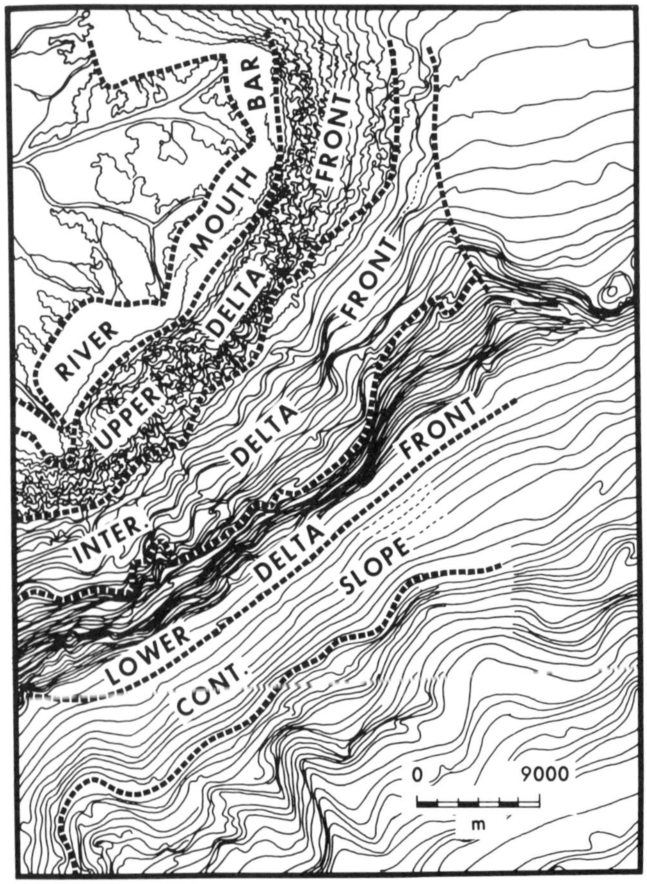

FIG. 2.—Bathymetry of the Mississippi River delta front illustrating complex patterns of bottom contours in the upper- and lower-delta front, which are related to numerous mudflow gully systems.

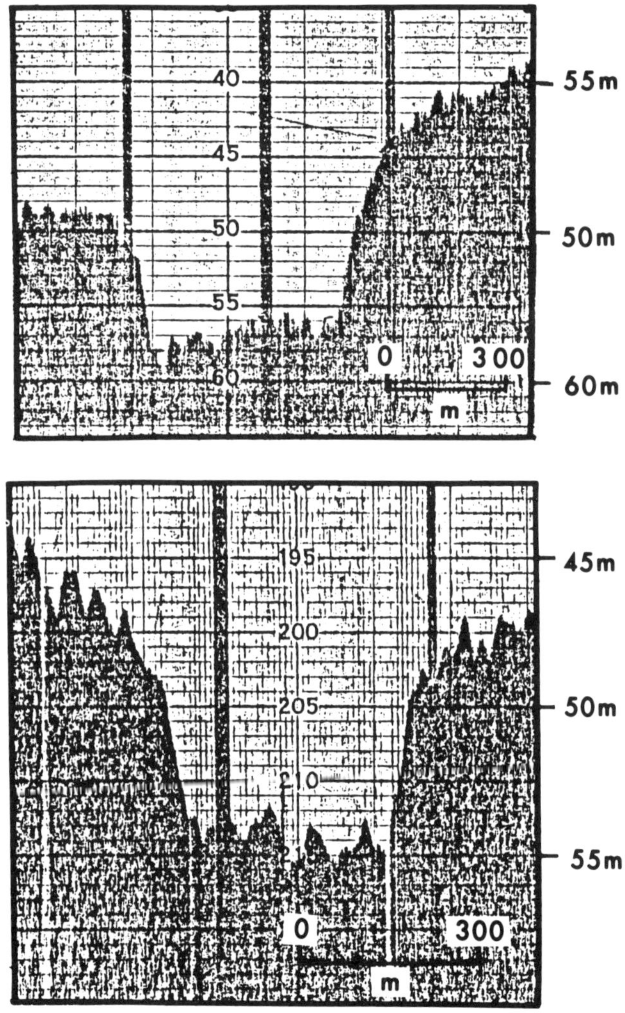

FIG. 4.—Echo-sounder profiles oriented essentially normal to mudflow gullies of the upper-delta front. Note the U-shaped gully cross sections and flat floors.

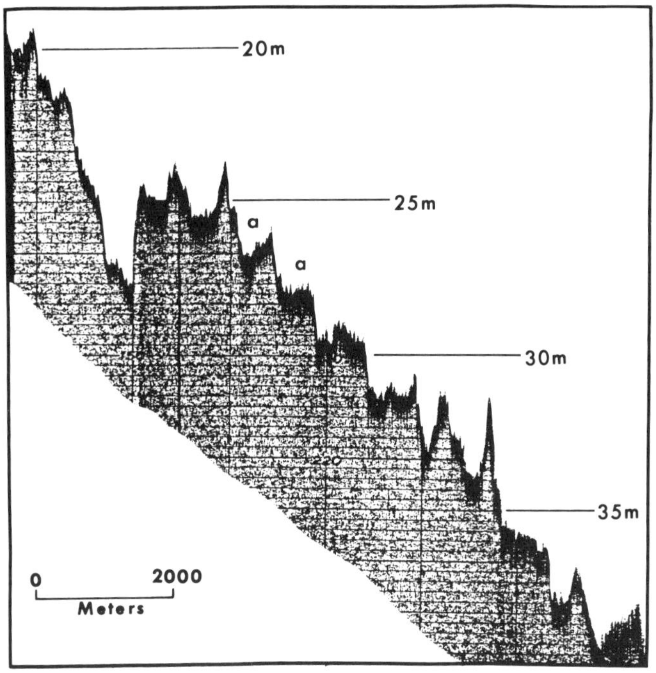

FIG. 5.—Echo-sounder profile run down the upper-delta front (essentially a dip profile) showing mudflow gullies and small peripheral slumps. This erratic seafloor topography is typical of the delta-front province.

dense distributary mouth-bar sands, causes vertical diapiric uplift and the formation of clay diapirs (mudlumps). In the adjacent shallow-water bays, a variety of small-scale features, such as collapse depressions and bottleneck slides, is formed (Prior and Coleman, 1978a; Coleman and Prior 1980). On the peripheral edge of the distributary-mouth bar, one observes a wide variety of arcuate peripheral slumps. Immediately seaward of the bar, commencing in water depths of 10 m or less, bottom topography becomes very erratic, reflecting numerous channel-like features generally radiating from the major distributary mouths. Bathymetric profiles (Fig. 4) across these channel-like features display relative flat bottoms and steep sides, revealing a classic U-shaped profile. A dip profile across the upper-delta front illustrates the complexity of the delta-front morphology (Fig. 5). Sidescan sonar imagery shows that these gullies are the updip parts of elongate retrogressive slides (Coleman, 1988).

Rotational Slumps

Sediment mass movement begins high on the upper delta-front slope near the distributary-mouth bar, where bottom slopes range from 0.2 to 1.0° (Coleman and others, 1974). The surface expressions of the shear planes associated with these rotational slumps have a distinctive curve or shape in plan view and are concave upward in the subsurface. Often they merge into bedding planes with depth. Scarp heights

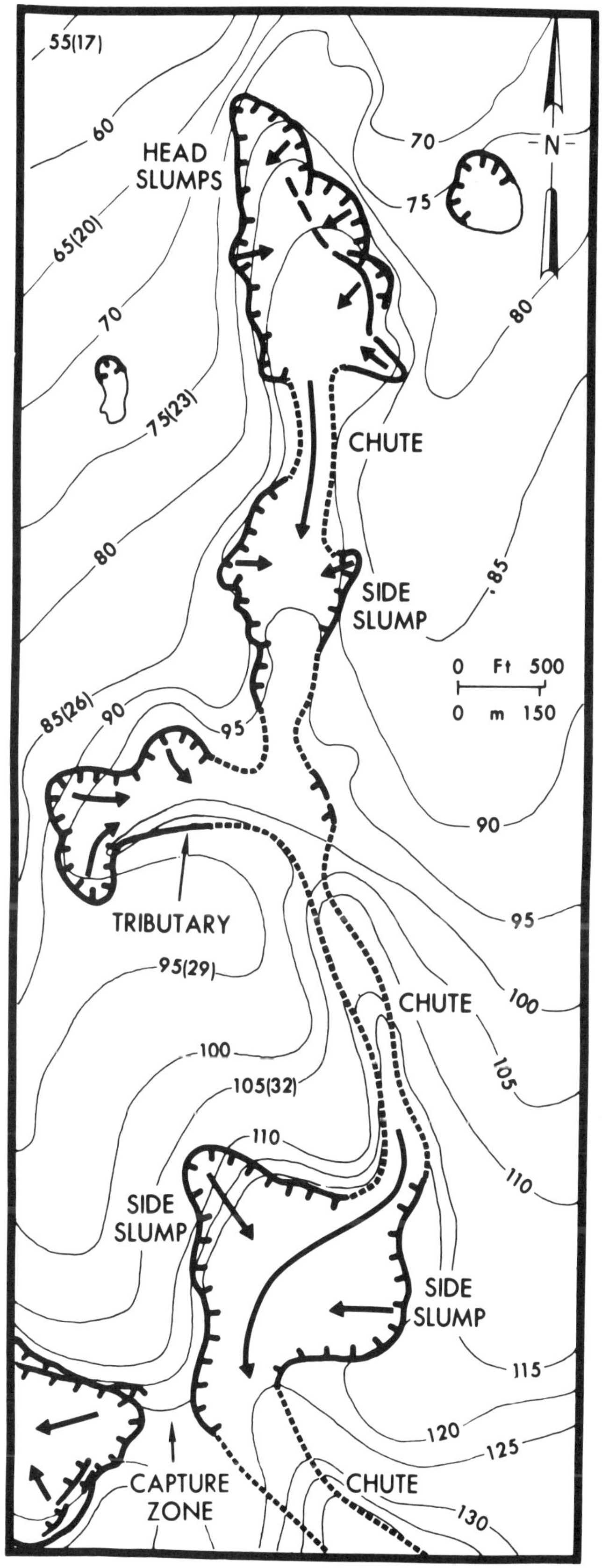

FIG. 6.—Illustration of the components of the upslope end of a complex mudflow gully system. The bowl-shaped depressions serve as source areas for blocks of sediment that are degraded in size as they are transported downslope.

vary from 3 to 7 m, and the shear planes cut to depths of 30 to 35 m. The surface of the slump block is rotated upslope into the fault plane, giving a stair-stepped appearance to the delta front (Fig. 5). Irregular and hummocky bottom topography and displaced clasts of sediment occur downslope of the scarps. This process of slumping moves large quantities of highly fractured and partially remodeled sediment downslope from delta-front and distributary-mouth bar environments into deeper waters at the shelf edge or onto the upper continental slope.

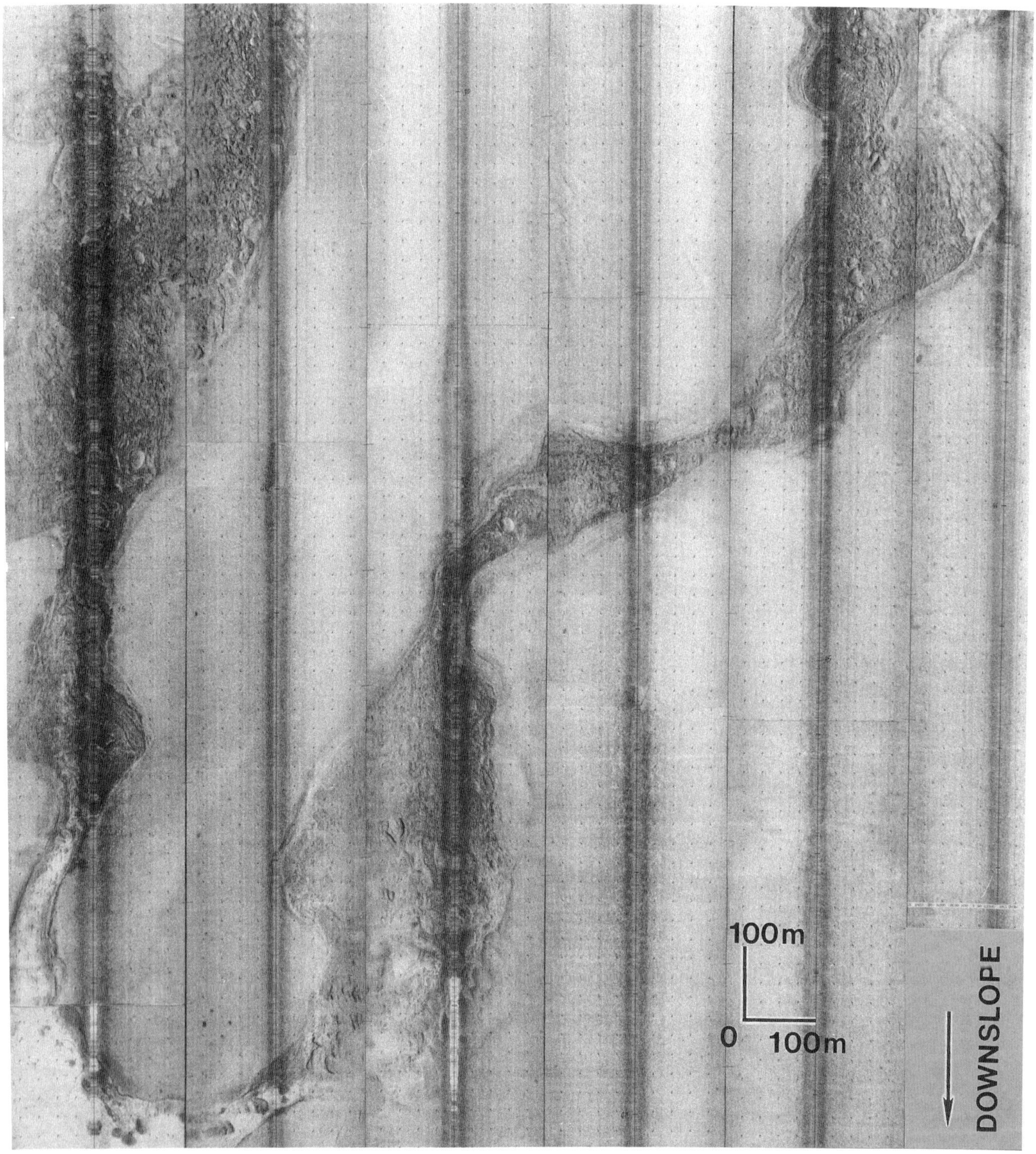

FIG. 7.—A sidescan sonar mosaic illustrating two major gully systems that originate on the upper-delta front. Note the large sediment blocks in the gullies and the contrasting smooth, undisturbed sea floor outside the gullies.

Mudflow Gullies

Major elongate systems of gullies, initially described as delta-front gullies by Shepard (1955), extend radially from each distributary (Fig. 2). High-resolution seismic data and drill hole information indicate that these "gullies" represent a type of mass movement that is generally referred to as retrogressive slides (Prior and Coleman, 1978a, b). The major characteristics of an elongate retrogressive slide (also commonly known as mudflow gullies) are illustrated in Figure 6. Gullies emerge from an extremely disturbed area of slump topography. Each gully has a clearly recognizable area of rotational instability or shear slumps at its upslope margin, and a long, often sinuous, narrow chute that links a depressed hummocky source area on the upslope end to composite overlapping depositional lobes. The instability is bounded by a bowl-shaped depression that serves as the source area for the blocky debris (Fig. 6). Multiple head scarps (head slumps) and crown cracks are present, indicating upslope retrogression. Within the source area, hummocky to irregular but nevertheless distinctive blocks of various sizes and arrangements can be identified. Downslope, an essentially elongate, narrow chute is bounded by sharp linear escarpments and numerous sidewall rotational slumps (Fig. 6). The gully floors are incised from 3 to 20 m into the bottom. The gullies extend downslope nearly perpendicular to the depth contours for distances as much as 7 to 10 km.

The sidescan sonar image in Figure 7 illustrates two gully systems, both commencing high on the delta front. This figure is composed of adjacent sidescan sonar images that have been merged to form a mosaic. The mosaic sharply contrasts the smooth undisturbed sea floor and the chaotic bottom topography within the mudflow gullies. The narrow gullies are relatively deeply incised, and evidence of sidewall instability is indicated by small slumps along the gully

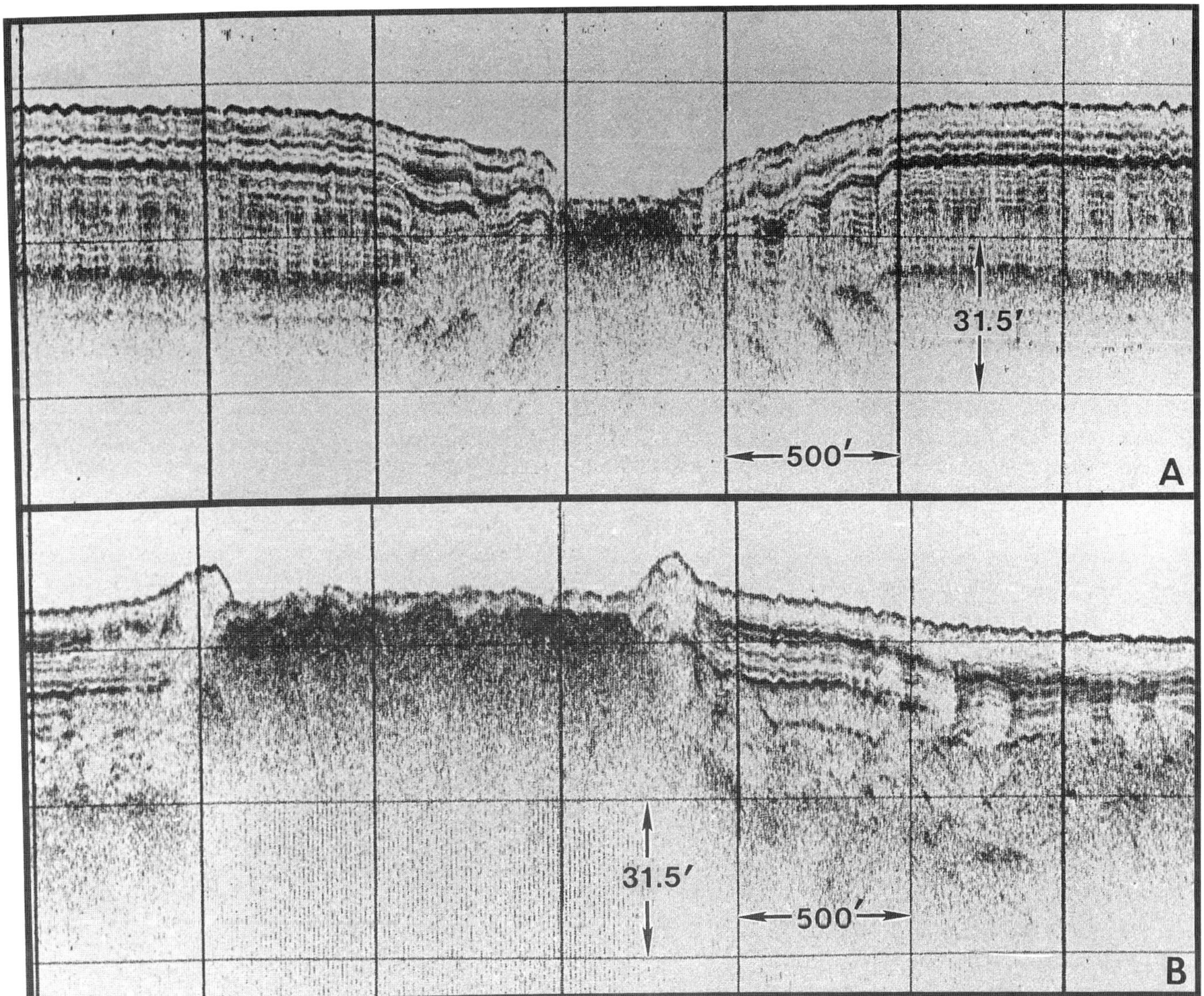

FIG. 8.—High-resolution seismic profiles across the upslope portion of a mudflow gully (A), and one across the downslope extension (B). Note the instability features in the gully wall in (A) and depositional levees associated with the downslope counterpart (B).

margins and by the alternations of bulbous and restricted regions (Fig. 7). The widths of the individual gullies range from 18 to 150 m at the narrow points to as much as from 370 to 550 m at the widest. The floors of the gullies are characterized by large erratic blocks of different sizes (Fig. 7). High-resolution seismic profiles across the gullies show that the side walls are subject to major instability (Fig. 8A); this process is responsible for the widening of the gully in a seaward direction (Fig. 8). Farther downslope, these gullies become wider, more depositional and develop levees and overbank deposits (Fig. 8B).

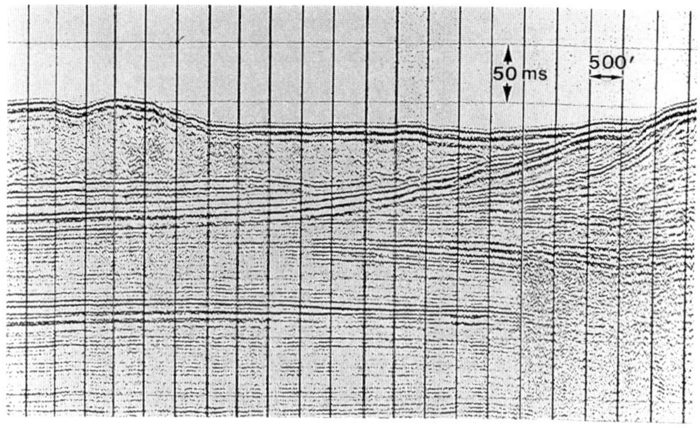

FIG. 10.—A high-resolution seismic (sparker) profile across the lower-delta front, showing overlapping lobes of sediment discharged from different mudflow gully systems. Strike section of South Pass, Mississippi delta.

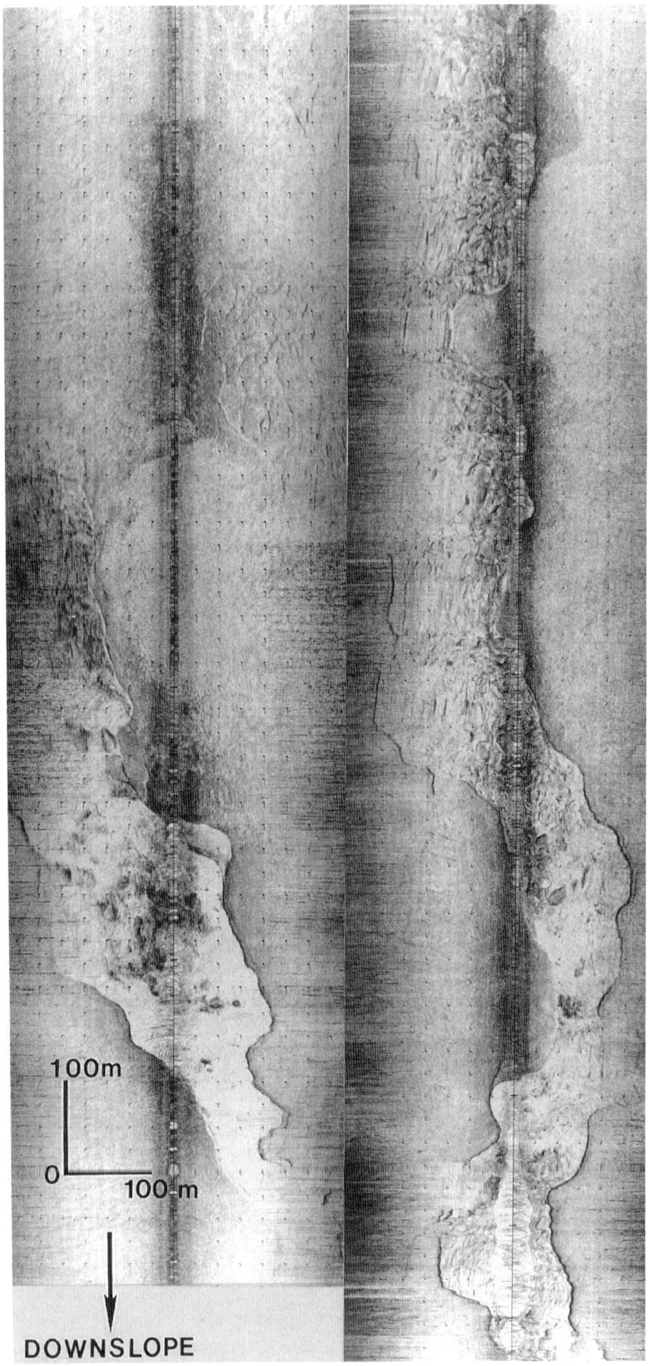

FIG. 9.—A sidescan sonar mosaic illustrating the distal ends of two mudflow gullies, where sediments are being discharged onto the outer continental shelf.

Extensive areas of irregular bottom topography, composed of discharged, blocky, disturbed debris, characterize the depositional lobe at the downslope ends of the mudflow gullies (Fig. 9). Large erratic blocks may be as much as 1 km wide. The transported debris often plows into the underlying undisturbed marine sediments, and a series of pressure ridges form in front of the advancing lobe. Figure 10 shows a high-resolution seismic profile across the intermediate- to lower-delta front, where gullies deposited their debris in overlapping lobes of acoustically opaque mudflow material.

Shepard (1955) observed no distinct difference in sediment grain size from cores collected from the bottom of the gullies or from the intergully (ridge) areas. However, Roberts and others (1980), using X-ray radiography, found that gully sediments were commonly sheared, inclined, fractured, and remolded. They also found that stratification and textures, considered typical of upper delta-front sediments were found at middle delta-front water depths. In a few cases, Shepard (1955) observed that the bottom sediments in the gullies were slightly more consolidated than those on the areas between the gullies. The sediments can be described as silty clays and clayey silts with a trace of sand. All cores collected during API Project 51 contained typical, thin, interbedded, sand and silt laminae, characteristic of delta-front deposits (Shepard and others, 1960).

In contrast, the sediments of the mudflow lobes are poorly stratified and show the effects of consolidation, as revealed by dewatering and degassing structures. Results of these consolidation processes can be observed both as small-scale sedimentary structures in cores (Roberts and others, 1980) and as large surficial features such as mud volcanoes.

CONCLUSIONS

Shepard (1955) suggested that the gullies radiating from the main Mississippi River delta distributaries are the result of mass-movement processes. Later studies, using more sophisticated equipment, have shown that the delta front is impacted by a variety of continuously active mass-movement processes. The delta-front gullies of Shepard proved to be only part of the complex of mass-movement phenomena, rather than isolated features as simple depth soundings would suggest. These slides start updip with the collapse of sediment resulting from underconsolidation due to rapid deposition and/or rapid porewater-pressure changes caused by major storm waves. Part of the sediment involved in that collapse can be transported downdip like a debris flow with masses of undisturbed sediment floating in a "fluid" of remolded clay-rich matrix. Farther downdip, the gully functions as a conduit through which sediment is transported to one or more depositional lobes that form a complex relief that protrudes above the surrounding undisturbed seafloor. The net result of this process is the transport and redeposition of sediments downslope. In most cases, this type of mass movement takes place in several steps, which involve deposition followed by transportive reactivation.

Although processes related to mass movements on the Mississippi River delta front operate over short periods of geologic time, ranging from a few years to a few thousand years, they result in deposition of significant volumes of sediment over extensive areas. Continued accumulation develops a thick sequence of cyclic deposits with complex vertical and lateral facies changes. Knowledge of the rapidity of changes and processes in this modern-delta setting aids in the interpretation and understanding of other river-dominated modern and ancient sedimentary sequences. Shepard's classic paper of 1955, based only on bathymetric measurements, not only proved the validity of mass movements in front of the Mississippi delta, but also initiated a host of modern studies that showed the complexity of the area and the importance of understanding slope stability on nearly horizontal shelves.

REFERENCES

ARCHANGUELSKY, A. D., 1930, Slides of sediments on the Black Sea bottom and the importance of this phenomenon for geology: Bulletin of the Society of Nature, Moscow, v. 38, p. 38 80.

BEA, R. G., 1971, How sea floor slides affect offshore structures: Oil and Gas Journal (November), p. 88–92.

COLEMAN, J. M., 1988, Dynamic changes and processes in the Mississippi River delta: Geological Society of America Bulletin, v. 100, p. 999–1015.

COLEMAN, J. M., AND GAGLIANO, S. M., 1964, Cyclic sedimentation in the Mississippi River deltaic plain: Gulf Coast Association of Geological Societies, Transactions, v. 14, p. 67–80.

COLEMAN, J. M., AND GARRISON, L. E., 1977, Geological aspects of marine slope stability, northwestern Gulf of Mexico: Marine Geotechnology, v. 2, p. 9–44.

COLEMAN, J. M., AND PRIOR, D. B., 1980, Deltaic Sand Bodies: Continuing Education Course Note Series No. 15: American Association of Petroleum Geologists, Tulsa, Oklahoma, 171 p.

COLEMAN, J. M., AND PRIOR, D. B., 1988, Mass wasting on continental margins: Annual Review of Earth and Planetary Science, v. 16, p. 101–119.

COLEMAN, J. M., SUHAYDA, J. N., WHELAN, THOMAS, III, AND WRIGHT, L. D., 1974, Mass movements of Mississippi River delta sediments: Gulf Coast Association of Geological Societies, Transactions, v. 24, p. 49–68.

DALY, R. A., 1936, Origin of submarine "canyons": American Journal of Science, v. 31, p. 401–420.

DOTT, R. H., Jr., 1963, Dynamics of subaqueous gravity depositional processes: American Association of Petroleum Geologists Bulletin, v. 47, p. 104–128.

FISK, H. N., MCFARLAN, EDWARD, JR., KOLB, C. R., AND WILBERT, L. J., JR., 1954, Sedimentary framework of the modern Mississippi delta: Journal of Sedimentary Petrology, v. 24, p. 79–99.

MIDDLETON, G. V., AND HAMPTON, M. A., 1976, Subaqueous sediment transport and deposition of sediment gravity flows, *in* Stanley, D. J., and Swift, D. J. P., eds., Marine Sediment Transport and Environment Management: Wiley, New York, p. 197–218.

MOORE, D. G., 1977, Submarine slides, *in* Voight, B., ed., Rockslides and Avalanches: Developments in Geotechnical Engineering 1, Elsevier, Amsterdam, p. 563–604.

PRIOR, D. B., AND COLEMAN, J. M., 1978a, Disintegrating retrogressive landslides on very low-angle subaqueous slopes, Mississippi delta: Marine Geotechnology, v. 3, p. 37–60.

PRIOR, D. B., AND COLEMAN, J. M., 1978b, Submarine landslides on the Mississippi River delta-front slope: Geoscience and Man: School of Geoscience, Louisiana State University, Baton Rouge, v. 19, p. 41–53.

PRIOR, D. B., AND COLEMAN, J. M., 1984, Submarine slope instability, *in* Brunsden, D., and Prior, D. B., eds., Slope Instability: Wiley, New York, p. 419–455.

ROBERTS, H. H., CRATSLEY, D., AND WHELAN, THOMAS, III, 1976, Stability of Mississippi delta sediments as evaluated by analysis of structural features in sediment borings: 8th Offshore Technology Conference, Houston, Texas, p. 9–28.

ROBERTS, H. H., SUHAYDA, J. N., AND COLEMAN, J. M., 1980, Sediment deformation and transportation on low-angle slopes: Mississippi River delta, *in* Coates, D. R., and Vitek, J. D., eds., Thresholds in Geomorphology: George Allen and Unwin, London, p. 131–167.

RUSSELL, R. J., 1936, Physiography of the lower Mississippi River delta, *in* Reports on the Geology of Plaquemines and St. Bernard Parishes: Louisiana Department of Conservation Geological Bulletin 8, p. 3–193.

SHEPARD, F. P., 1932, Landslide modifications of submarine valleys: Transactions, American Geophysical Union, v. 13, p. 226–230.

SHEPARD, F. P., 1955, Delta front valleys bordering the Mississippi distributaries: Geological Society of America Bulletin, v. 66, p. 1489–1498.

SHEPARD, F. P., PHLEGER, F. B., AND VAN ANDEL, Tj., eds., 1960, Recent Sediments, Northwest Gulf of Mexico: American Association of Petroleum Geologists, Tulsa, Oklahoma, 394 p.

TROWBRIDGE, A. C., 1930, Building of Mississippi delta: American Association of Petroleum Geologists Bulletin, v. 14, p. 867–901.

WHELAN, THOMAS, III, COLEMAN, J. M., SUHAYDA, J. N., AND GARRISON, L. E., 1975, The geochemistry of Recent Mississippi River delta sediments: Gas concentration and sediment stability: 7th Offshore Technology Conference, Houston, Texas, p. 71–84.

REGRESSIVE COASTAL DEPOSITS ON QUATERNARY CONTINENTAL SHELVES: PRESERVATION AND LEGACY

MICHAEL E. FIELD
U.S. Geological Survey, 345 Middlefield Rd., Menlo Park, CA 94025
AND
FABIO TRINCARDI
Istituto per la Geologia Marina, via Zamboni 65, 40127 Bologna, Italy

ABSTRACT: Regressive coastal deposits containing internal downlapping surfaces are common on continental shelves of the world. Through theoretical considerations and evaluation of examples from the literature and our own studies in California and Italy, we have examined the conditions that lead both to the formation and preservation of these deposits.

Coastal downlapping deposits form by progradation of coastal and deltaic lithosomes during stable and falling sea level. In addition to the rate and direction of sea-level change, the major controls governing the development of downlapping deposits are sediment availability and shelf morphology (gradient, surface irregularity, and depth of shelf break). Continental margins receiving a large, continuous supply of sediment commonly have vertical stacks of thick, laterally extensive deposits; those fed by relatively small coastal streams may have well-developed shelf-margin deposits if accommodation space was available. In the absence of feeder streams, shelf-margin downlapping deposits can form from locally derived sediment.

Regressive coastal deposits are preserved in a variety of settings, but they are *least* likely to be preserved on broad, low-gradient (coastal plain) shelves, where small drops in relative sea level were accompanied by large seaward shifts of the shoreline. Even on high-gradient tectonic margins, downlapping deposits commonly are not preserved on the midshelf. The process of shoreface erosion during a transgression is efficient at planing off deposits from previous depositional cycles. Where the deposits are thin, the deposit is partially or wholly reworked; where they are thick, the basal part may be preserved.

Regressive deposits are *most* likely to be preserved along the shelf margin, where relatively thick sequences form at or below the position of the lowstand sea level. On many continental shelves, the coarse sandy texture of outer-shelf sediment largely reflects original deposition in a regressive coastal environment during a fall and lowstand of relative sea level.

INTRODUCTION: SEDIMENT FACIES ON MODERN SHELVES

Early Insights

In 1932, F. P. Shepard published a paper that would forever change the way geologists viewed modern continental shelves. Equally important, he also set the standard for how marine geology was to be conducted: by observation and interpretation of field data. Shepard (1932, p. 1017–1018) begins with the observation that:

> "Many geologists have stated that marine sediments vary from coarse to fine directly as the distance from shore and the depth of water increase. Such gradation seems so reasonable in view of our knowledge of waves and currents that what is practically a geologic axiom has come into existence. Most standard textbooks assume this principal applies to the sediments on the present continental shelves."

In this pioneering paper, Shepard systematically reviewed the distribution of particle sizes from continental shelves, embayments and estuaries around the world. His results clearly demonstrated what is now common knowledge to several generations of marine geologists: that present gradations of particle sizes, from coarse near shore to fine on the outer shelf, rarely exist (Fig. 1). More important than these crucial observations were the interpretations based on those observations. Shepard recognized that present shelves had a surficial-sediment facies resulting from a complex history and concluded that: 1) the lack of outwardly decreasing gradation in grain size was due to the presence of coarse sediment on the outer shelf that formed from stream transport and local erosion during lowstands of sea level; 2) the coarse sediment on the outer shelf remains uncovered by fine particles because of the action of currents; 3) the abundance of coarse sediment on shelves off glaciated coasts supports the idea that the adjacent shelves had been glaciated; and 4) gravel patches far from shore and rocky areas indicate that the shelves are only thinly veneered by sediment.

Although the concept of continental shelves being glaciated or exposed to subaerial erosion and deposition had been discussed previously by a number of geologists, Shepard synthesized observations of surface-sediment types on a global scale. After firmly establishing the impact of sea-level fluctuations on shelf sediment with his 1932 paper, he conducted, with colleagues, detailed studies of the shelves off the U.S. Atlantic coast (Shepard and Cohee, 1936), Gulf of Mexico coast (e.g., Shepard and Moore, 1955) and Pacific coast (Emery and others, 1952).

Shepard's leadership in continental-shelf research was followed by colleagues and students, many of whom sought to explain the recent history of today's continental shelves. A landmark paper by J. R. Curray in 1964 explained for the first time shelf facies as a composite result of two major variables: rate of deposition and sea-level movement (Fig. 2). Curray demonstrated, with examples from the Gulf of Mexico and Pacific shelves, that coastal recession or progradation depends not only on whether sea level is rising or falling but on its relative *rate* of change and on the availability of sediment to offset or augment a sea-level change. Thus, a slow rise in sea level in areas of high rates of sedimentation might lead to coastal progradation rather than coastal retreat (Fig. 2). Many of the ideas of Curray (1964) form the underpinnings of concepts developed later by others on shelf facies and sequences. Studies by many workers, for example Milliman and Emery (1968) and Pilkey and others (1969), began to establish the details of the age and inherited properties of shelf sediment. In 1968, a landmark paper by K. O. Emery describing relict sediment on

From Shoreline to Abyss, SEPM Special Publication No. 46

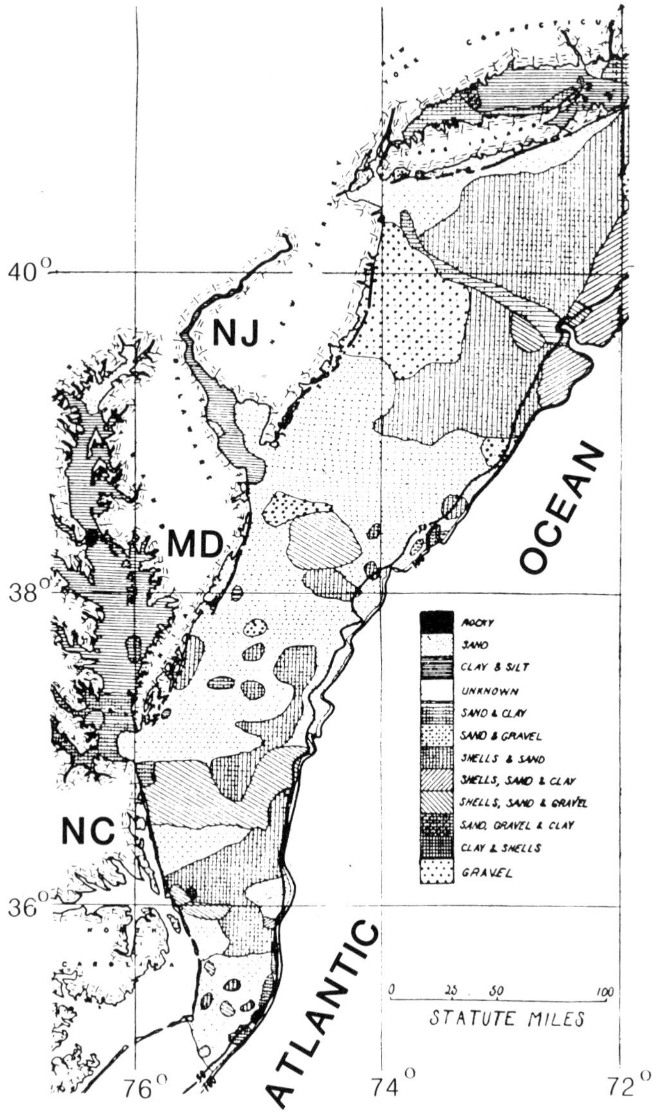

FIG. 1.—Map of sediment distribution on the continental shelf off northeastern United States (Shepard, 1932). Note the abundance of sand and gravel on the shelf, a characteristic which led Shepard to infer that changing sea levels had a world-wide influence on shelf-sediment patterns.

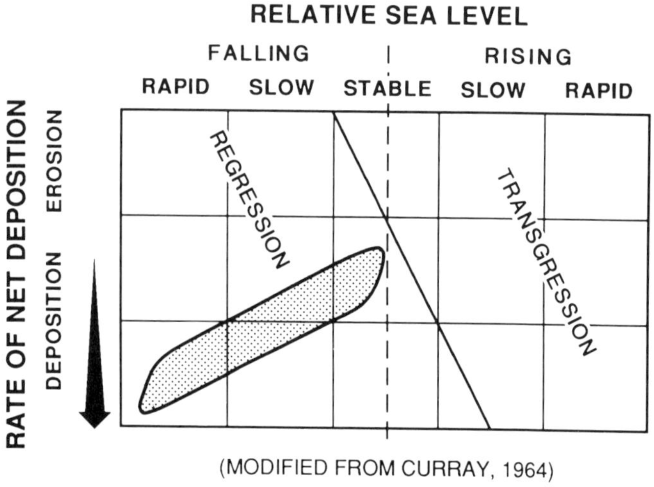

FIG. 2.—Diagram showing the effects on lateral migration of the shoreline (in a direction perpendicular to the shoreline) due to the rate of change of relative sea level and the local net rate of deposition (from Curray, 1964). Curray's diagram shows that the tendency of a shoreline to retreat or advance seaward depends on *both* parameters. The added field (shaded in gray) defines schematically the conditions for coastal progradation. The addition of new sediment is required for the coast to prograde, and as the rate of sea-level fall increases, the amount of sediment required to maintain progradation increases. The increasing size of the arrow represents increasing rates of deposition.

continental shelves of the world brought together what Shepard had initiated 36 years earlier by establishing the antiquity of the shelf surface.

As the shelf sample collections grew and as new and improved measurements of currents and storm deposits were made, the collective thinking about the shelf surface grew another step. In the United States, D. J. P. Swift and colleagues (Swift and others, 1971), and in Europe, A. Stride and colleagues (Stride and others, 1972) observed that relict-shelf sediment was actually in a modified equilibrium state: that is, it responded to and was shaped by active shelf processes. By the early 1970s, the surficial-sediment facies of modern continental shelves was reasonably well understood. In a simplified view, shelves adjacent to large rivers are veneered by shore-parallel belts and seaward-thinning wedges of Holocene terrigenous silt deposited under highstand conditions. Most shelves, however, are not dominated by stream deposition, and the dominant shelf facies is a relatively coarse deposit that formed under lower sea-level conditions and is now continually modified by storms, tides, waves, and animals.

Purpose

In this paper we examine the nature of regressive coastal deposits on Quaternary continental shelves. Because our interpretation is a product of the concepts and observations on continental-shelf sedimentation and sea-level history that have developed over roughly the past 30 years, we thought it important to review those observations. Most studies have emphasized the effects of the last transgression on shelf-surface deposits. Our study is a departure from that pattern in that we focus on the roles of sediment supply and shelf geometry during *falling* and *lowstand* sea levels in forming and preserving diagnostic progradational sediment packages. Because we are interpreting sediment sequences that formed over time spans of several thousand to tens of thousands of years, the effect of basin subsidence, normally a major factor in sequence stratigraphy, can be disregarded. Throughout this paper we use the term *downlapping* in the context of internal downlapping stratal surfaces within shallow-marine deposits; there is no intent to include *all* downlapping deposits in this discussion. Finally, the following discussions on sequence stratigraphy and on coastal responses to transgressions and regressions are provided as a prologue that we think is important to our synthesis of the geometry, growth, and preservation of prograding coastal-sediment packages.

Sequence Stratigraphy and Late Quaternary Shelf Deposits

With the advent of sequence stratigraphy during the 1980s, siliciclastic sequences and their bounding surfaces came to be studied within a time/space framework as a function of eustasy, subsidence (thermal cooling plus loading), and sediment input (Jervey, 1988). Accommodation, the space in which sediment can accumulate, is produced as a function of both eustasy and subsidence (Posamentier and others, 1988). Sea-level fluctuations combine with subsidence to control changes in accommodation through time, and it is the *rate* of subsidence that controls the delay in time until maximum accommodation is achieved following a maximum highstand of sea level. This interaction between subsidence and sea-level also implies that, in an extreme case, no decrease in accommodation would occur during a falling eustatic sea level.

Facies architecture and sequence boundaries develop in response to the interaction between variations of accommodation and sediment input. During an increase in accommodation, the volume of available sediment will affect the style of sedimentation. For example, high sediment input results in shoreline aggradation and progradation with a basinward migration of coastal facies (depositional regression). Where sediment input is capable of balancing the increase in accommodation, the shoreline facies will remain stationary through time. On the other hand, low sediment input can induce either retrogradation (aggradation on the shelf and maintenance of depositional slope) or transgression (landward movement of the coastline accompanied by non-deposition and reworking on the shelf).

Concepts and terminology derived from sequence stratigraphy are now widely applied to modern shelf and coastal systems (Nummedal and Swift, 1987; Suter and others, 1987). This approach has proved to be constructive in investigating the nature of late Quaternary deposits, particularly in areas characterized by high rates of deposition and minor tectonic influence (Boyd and others, 1989). On a broader scale, Galloway (1989a,b) has recently described in full the association and range of genetic stratigraphic sequences that typically occur in large basins, such as the Gulf of Mexico.

Modern continental margins in tectonically active areas (rift, convergent, or borderland margins) show significant lateral changes in physiography. In such areas, the change in physiography is an important parameter controlling lateral variability of deposits forming in response to changes in accommodation through time. In areas with variable shelf gradients, a drop in sea level can result in deposition at one locale and non-deposition at an adjacent locale. During a rapid fall in relative sea level, shelf-perched deposits document the rapid seaward migration of the shoreline (Posamentier and Vail, 1988).

Coastal Response to a Transgression

The last major event to affect shelf sedimentation on a global scale was the late Pleistocene and Holocene rise in sea level. Sea level stood approximately 120 m below its present position at about 18 ka (Aharon, 1983). Its rise was relatively rapid (on the order of 1 m/ky), and as it transgressed the shelf, all of the surficial sediment passed through a nearshore setting. Because such a history should be recognizable, much effort went into identifying the influence that the coastal phase had on particles residing on the shelf today (e.g., Pilkey and others, 1969). As a result of these studies, criteria were established for recognition of still-stand deposits. In 1968, Swift introduced the concept of shoreface erosion to explain the recycling of coastal facies into inner-shelf strata. As coastal lithosomes retreat by erosion and recycling, the nearshore-sediment wedge partially migrates landward and is partially left behind as a thin carpet on the shoreface erosional surface. Thus, as Fischer (1961) pointed out, all, some or none of the coastal-barrier and lagoonal deposits laid down in front of a rising sea may be preserved.

Because the last major event on the shelf was a rise in sea level, most studies of shelf sediment have emphasized the effects of transgressions. Curray (1964) suggested that a continuum exists between rapid erosional transgressions and depositional transgressions, although some of his examples of depositional transgressions may actually be pre-transgressive (lowstand) or post-transgressive (highstand). One of the main effects of a rising sea is to produce a transgressive surface of erosion that obliterates most or all of the record of the preceding depositional event or events. Gradual erosional retreat of the shoreface produces abrupt vertical changes in facies (Demarest and Kraft, 1987). The preservation of older fluvial and paralic deposits is a function of the topographic relief, areal extent, and distance from the shoreline of those depositional settings (Demarest and Kraft, 1987). Recycled sediment accumulates on the inner shelf seaward of the landward-advancing zone of nearshore erosion (Field, 1980). The shelf and coastal deposits that are cycled through the shoreface are left behind the landward-migrating coast as a veneer of reworked material or, in some settings, molded into a series of sand ridges (Swift and others, 1972; Duane and others, 1972). Rarely has a transgression produced a thick, continuous, sediment sequence. Most thick transgressive deposits seem to occur in areas of high levels of sedimentation (Boyd and others, 1989), rapid basinal subsidence (Bourgeois, 1980), or rapid load compaction and subsidence (Penland and others, 1988).

Sets of parasequences backstepping onto the basin margin develop during relative sea-level rise if enough sediment is available (Van Wagoner and others, 1988). Such parasequences show onlap and downlap terminations in the landward and basinward directions, respectively. Condensed sections develop in basin or shelf areas away from major sediment-entry points during times of relative sea-level rise and early highstand (Coleman and Roberts, 1988). As the rate of relative sea-level rise decreases, the set of parasequences changes from retrogradational (backstepped) to aggradational and then to progradational (development onto the downlap surface).

On many continental margins not receiving sediment from a major fluvial source, late Quaternary transgressive deposits are absent or underdeveloped and replaced by a thin veneer of hemipelagic sediment (Suter and others, 1987;

Kindinger, 1988; Tesson and others, 1990; Trincardi and Field, 1991). In these cases, the sequence-boundary, ravinement surface, and maximum-flooding surface usually coincide.

Coastal Response to a Regression

The sedimentary sequences that develop in response to a falling sea level have received relatively little attention in the literature. Curray (1964) considered most regressive deposits on the U.S. Gulf and Pacific shelves to be buried beneath younger transgressive deposits; he also thought them to be thin and subject to erosion immediately following deposition. In their study of the morphology of the U.S. Atlantic shelf, Swift and others (1972) inferred that most morphological features were formed during the last transgression on an inherited erosional surface. Their study emphasized surficial features only, however, and did not address the nature of subsurface deposits. Other workers showed that in areas of locally high rates of sedimentation, as for example the Costa de Nayarit (Curray and Moore, 1964), prograding delta lobes can be built and preserved on the shelf.

One reason for the inattention to regressive deposits on modern shelves has been the lack of good-quality three-dimensional data (i.e., high-resolution seismic-reflection profiles) that illustrate subsurface facies and relation between units. In recent years there has been an improvement in acoustic techniques and a growing collection of shelf data (e.g., Coleman and Roberts, 1988). With the advent of the concepts of seismic stratigraphy, increased emphasis has been placed on deposits that form during a drop in sea level (Vail and others, 1977). For example, Suter and Berryhill (1985) and Matteucci and Hine (1987) present information on midshelf and shelf-margin deltas that formed during falling and low sea levels on the Texas and North Carolina shelves. More recently, Trincardi and Field (1991) provided an example of downlapping sediment bodies formed by coastal progradation on the Tyrrhenian Sea shelf of Italy.

Posamentier and Vail (1988) and Posamentier and others, (1988) discuss from a conceptual standpoint the nature of regressive facies. Simply stated, regressive deposits form as a result of a seaward shift in the shoreline. The seaward shift might be caused by rapid sediment accumulation and progradation during steady sea-level conditions, or even during a rising sea level, provided the sediment accumulation rate is sufficient to counterbalance the rate of shoreline loss due to a rising sea (e.g., the Mississippi delta). In areas of high rates of coastal-sediment accumulation, the shoreline progrades to the shelf edge as part of the highstand systems tract. Along most shelf segments, however, sediment delivery is insufficient to allow the coast to prograde significantly during stillstand conditions. In such areas regressive facies are formed solely in response to a falling sea level and seaward growth of the coastal lithosome, a condition referred to as forced regression by Posamentier and others (1990). Regressive progradational wedges deposited on the shelf edge adjacent to the Rhône delta (Tesson and others, 1990) are taken to represent a Quaternary example of "forced regression."

Transgressive Regressive Cycles

The record of cyclic coastal erosion and deposition in response to Quaternary glacio-eustatic sea-level fluctuations is complex and incompletely documented thus far. Three studies that document the depositional geometry and facies relations associated with major and minor Quaternary sea-level changes are those from the terraces of New Guinea (Aharon, 1983), the continental shelf off Louisiana (Coleman and Roberts, 1988), and the uplifted coastal deposits along central California (Clifton, 1989). The focus of the remainder of this paper is on the depositional geometry of coastal-sediment bodies formed during stable and falling relative sea levels.

PROGRADATION DURING HIGHSTAND CONDITIONS

During a relative rise of sea level not compensated by sediment input, the shelf surface becomes a surface of nondeposition. A condensed section, typified by increased content of calcareous macrofossils, planktonic foraminifera, and authigenic minerals, develops during this period on the outer shelf (Loutit and others, 1988). Sediment from upland-source regions begins to fill stream valleys as a condensed section develops on the inner shelf. Once the valleys are filled, coastal progradation is initiated onto the shelf. Progradation develops inversely with the rate of accommodation increase (Jervey, 1988). That is, if accommodation space increases rapidly, progradation will slow. Sediment input, coastal energy, and shelf geometry control whether the coast will prograde during high sea-level conditions and determine what form the deposits will take. The geometry of sediment bodies built during a highstand will differ from those built during a sea-level fall (Okamura, 1989).

Progradation of a coastline during high sea-level conditions requires a steady supply of sediment. This is in contrast to falling sea-level, when seaward migration of the coast occurs *de facto*. The high sea level provides accommodation space for material delivered to the coast by streams or local erosion; as deposition occurs, however, the accommodation space for the depositional profile becomes increasingly larger. That is, progradation into increasing water depths requires increasing sediment loads or longer time periods to maintain progradation (Fig. 3). From a volumetric standpoint, it is clear that progradation and complete infilling of the available space can be achieved only with a major supply of sediment. In first-, second-, or third-order cycles of sea-level fluctuation, time alone may be sufficient to achieve highstand progradation. During Quaternary glacio-eustatic cycles, however, when highstand conditions are short-lived, a large sediment source is required for progradation. The Mississippi River is an example of such a source; in the time since sea level reached its present position, it has infilled its downcut valley and prograded the shoreline significant distances onto the shelf (Boyd and others, 1989).

The level of coastal energy controls how, or if, a section of coast can prograde (Thorne and Swift, 1991). For example, stream-dominated deltas, such as the Mississippi,

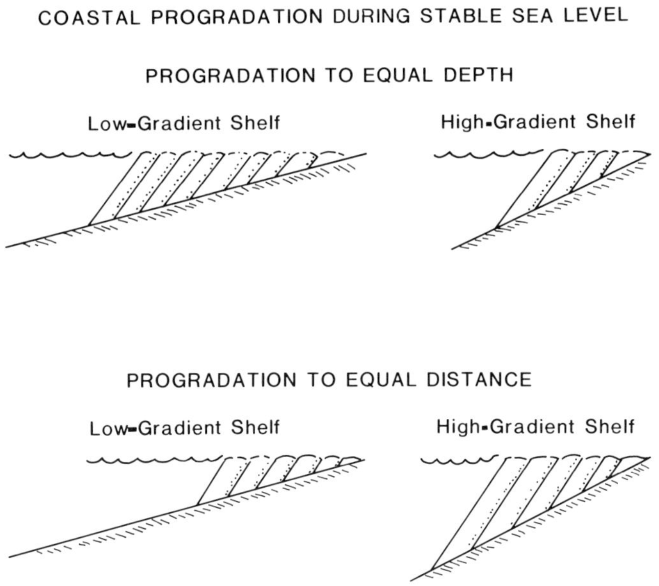

FIG. 3.—Schematic diagram comparing coastal progradation during stable highstand sea level on high- and low-gradient shelves. Clearly, more sediment or more time is required for a coastal deposit to prograde the same distance on a steep shelf as on a gentle shelf. During stable sea-level conditions, progradation is self-limiting and will markedly decrease as accommodation increases, unless the rate of sediment input increases.

that experience low to moderate wave and current conditions, develop in a series of overlapping lobes that potentially will prograde across the shelf to the shelf edge. Sediment depocenters that experience high levels of coastal energy are reshaped into a different depositional style. The Columbia River, which discharges onto the high-energy shelf of the U.S. Pacific Northwest, produces a deposit that is distributed along the midshelf for more than 100 km (Nittrouer and Sternberg, 1981). Another example is the shelf deposit formed by the Amazon River, which has one of the highest sediment-discharge rates in the world (Milliman and Meade, 1983). The Amazon shelf has a thick, modern, sediment wedge that extends seaward to a water depth of 60 m, but beyond that the shelf is an uncovered surface of transgression (Kuehl and others, 1882). The Amazon shelf has high-velocity tidal flows (Cacchione and others, 1990), which redistribute exiting sediment along the coast for hundreds of kilometers (Rine and Ginsburg, 1985). Clearly, the regressive sequences developed in such a setting will differ markedly from ones formed in areas of moderate coastal energy.

Shelf geometry is a key parameter controlling development of the regressive facies, even during a highstand of sea level. The greater the accommodation created, the thicker the aggradational deposit required before progradation takes place. Perhaps the best way to view the effect of shelf gradient is by comparison of a low-gradient ("coastal plain") shelf with a high-gradient ("Pacific-style") shelf (Fig. 3). The two shelf types differ in gradient by an order of magnitude. Coastal-plain shelves are typically more gentle than 1:1,000 (0.06°), whereas Pacific-style shelves commonly are steeper than 1:500 (0.12°), and sometimes as steep as 1:100 (0.6°). The effect that gradient has on progradation during highstand conditions is illustrated in Figure 4. On Pacific-style shelves, substantially less sediment is required to build seaward to a given depth than to the same depth on a coastal-plain shelf (Fig. 4). Although somewhat intuitive, this point is helpful for explaining the steep constructional shorefaces that existed shortly after sea level reached its present position. The comparison between the shelves also shows that for progradation to extend a given distance, far more material is required for a Pacific-style shelf. Coastal progradation occurred in many embayments along the New South Wales coast of Australia following the end of the transgression at about 6 ka (Thom and others, 1978; Chapman and others, 1982). The shelf off southeastern Australia is relatively steep and narrow, and Figure 4 illustrates one explanation as to why progradation ended: increasingly *larger* amounts of sediment were needed to maintain progradation.

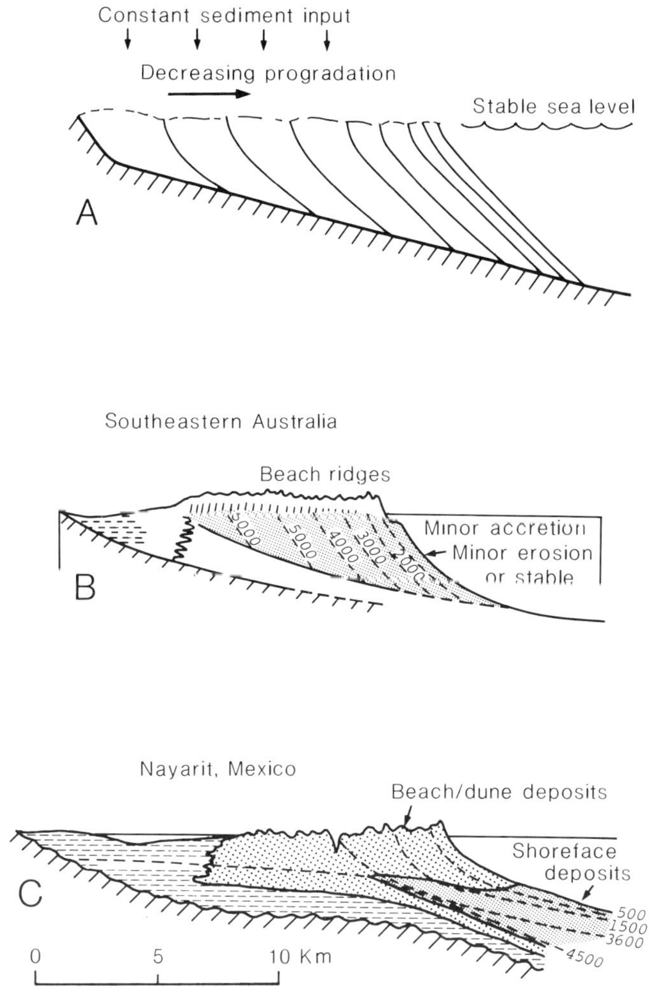

FIG. 4.—Sketch (A) shows decreasing seaward growth with distance offshore when sediment input is held constant. Diagrams (B) (modified from Chapman and others, 1982) and (C) (modified from Curray and others, 1969) show examples of seaward decreases in progradation rate because of increasing accommodation space. The numbers in (B) and (C) represent time datums.

Above all else, coastal progradation during highstands of sea level requires a large and continual influx of sediment and/or a long time period. Given the brevity of glacio-eustatic highstands (Fig. 5), the amount of sediment required for continual coastal progradation is large, probably on the order of that delivered only by the major streams of the world—about 10^8 tons of sediment per year. Given an adequate source, the influence of shelf geometry and local coastal energy are minimal. *However,* if the sediment supply is relatively small, then shelf geometry, including steepness, and the strength of coastal currents become more significant in determining whether or not progradation occurs.

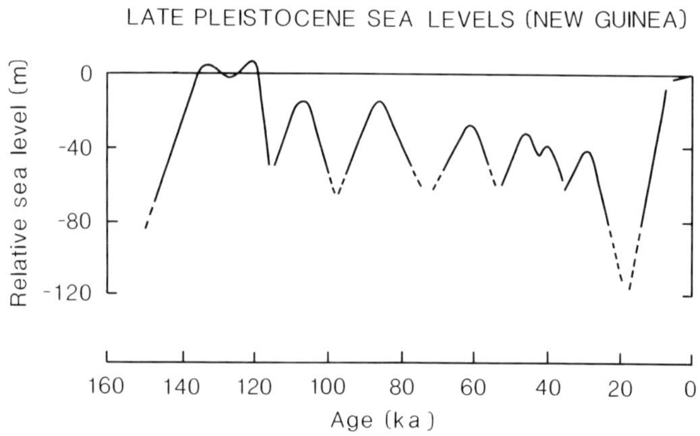

FIG. 5.—Modified diagram from Aharon (1983) showing sea-level changes, based on oxygen isotope records, for the past 140,000 years in New Guinea.

PROGRADATION DURING FALLING SEA-LEVEL AND LOWSTAND CONDITIONS

Sedimentation on continental shelves during falling and lowstand sea levels is controlled largely by two factors: sediment supply and shelf geometry. Sediment supply is the most important factor; without an adequate supply of sediment, coastal lithosomes cannot build seaward. Similarly, no matter how unfavorable conditions may be for coastal progradation, growth will occur if enough sediment is available.

Effect of Shelf Morphology

Shelf Gradient and Irregularity.—If a very large supply of sediment continually reaches the shoreline, the gradient does not affect development of the lithosome. However, given a fixed rate of supply of coastal sediment, the local gradient of the shelf is a primary control on whether or not a thick regressive deposit can be constructed. Shelves having steeper gradients are more likely to have coastal lithosomes prograding in an uninterrupted manner. Figure 6A illustrates this point schematically by comparing deposition on a low-gradient coastal-plain shelf and a high-gradient active-margin shelf. For one unit of sea-level change, the coastal sequence has to prograde farther on the low-gradient shelf (Fig. 6A) than on the high-gradient shelf. If the rate of sediment supply is the same, lower gradients will yield thinner deposits and increase the likelihood that progradation will not occur. In the extreme condition, a nearly horizontal shelf will require a shoreline shift of tens of kilometers for a sea-level drop of 1 m, a condition that eliminates coastal progradation in practically all areas except for those fed by major streams.

All shelves, particularly those on tectonically active margins, have local changes in gradient. Figure 6B, which is more typical of most shelves than the constant gradient depicted in Figure 6A, illustrates such an example. Assuming a steady-state influx of sediment, a prograding coastal wedge would be a laterally varying deposit that thickened, thinned, or disappeared as a result of local changes in gradient.

Depth of Shelf Break.—The depth of the shelf break was recently identified by Trincardi and Field (1991) as crucial for development of downlapping deposits on the eastern Tyrrhenian Sea shelf margin of Italy. The shelf on this margin varies markedly in width, orientation and depth of the shelf break over short lateral distances due to abrupt changes in subsurface geologic structure (Fig. 7). On the shelf, remnants of late Pleistocene prograded coastal deposits are found on both the middle shelf and the shelf margin (Figs. 7, 8). Trincardi and Field (1991) found that in shelf areas where the shelfbreak was *shallower* than the lowstand shoreline (120 m), sediment bypassed the coastal zone to the slope and no downlapping regressive deposits formed (Fig. 8A).

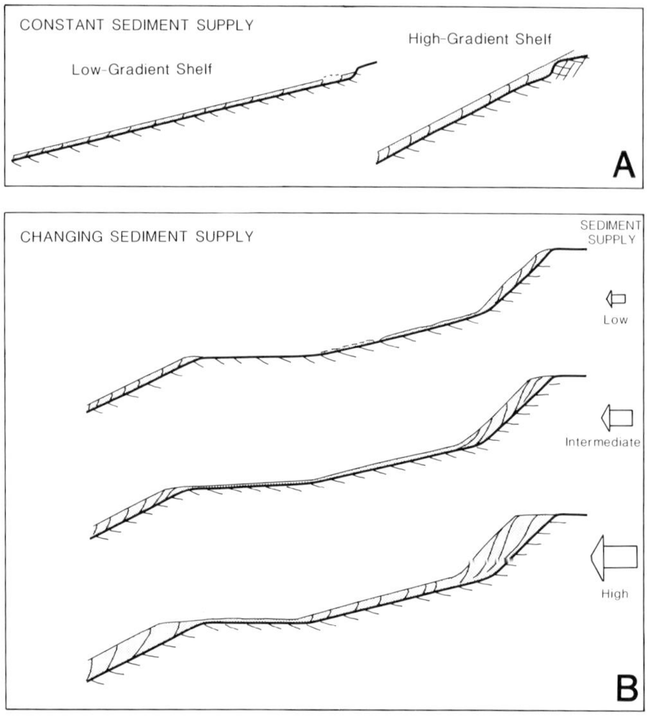

FIG. 6.—Schematic diagram showing the effect of shelf gradient and sediment supply on formation of regressive deposits. (A) Given a constant rate of sediment influx during a fall in relative sea level, regressive deposits will be thicker (and more likely to be preserved) on a high-gradient shelf than on a low-gradient shelf. (B) On shelves with variations in gradient, low-gradient sections will have thinner regressive deposits relative to the higher gradient sections. If sediment input is low, regressive deposits will not form on the flatter sections; as sediment supply increases, the thickness of the regressive deposits will increase accordingly.

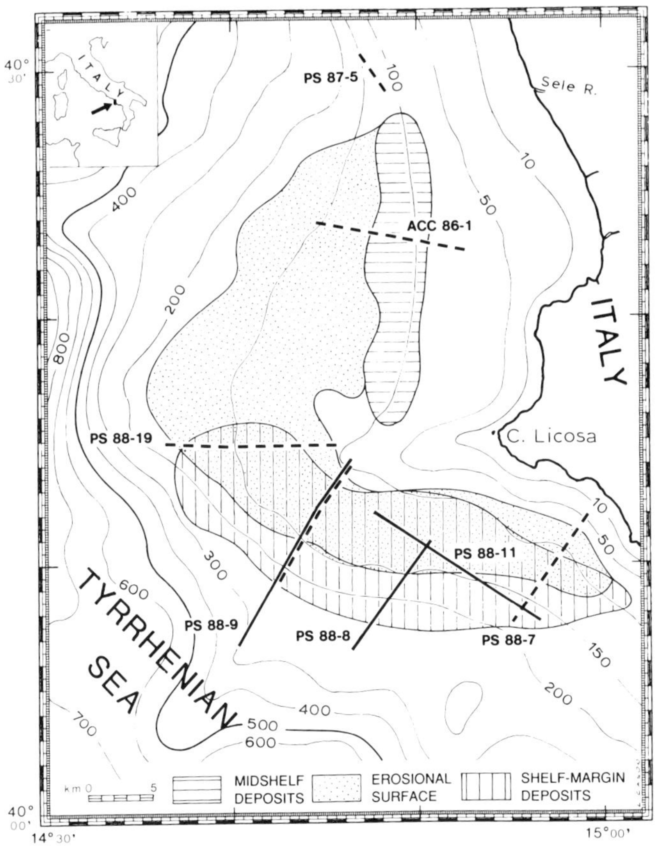

FIG. 7.—Map of a section of the continental shelf off southwestern Italy showing the location of preserved regressive downlapping deposits on the midshelf and shelf margin. Also shown is a broad erosional surface thought to have formed seaward of the lowstand sea level by shoreface erosion. Labeled profiles are ones best illustrating the character of downlapped regressive deposits. Profile PS 88-9 is shown in Figure 8; profile ACC 86-1 in Figure 13. From Trincardi and Field (1991); depths are in meters.

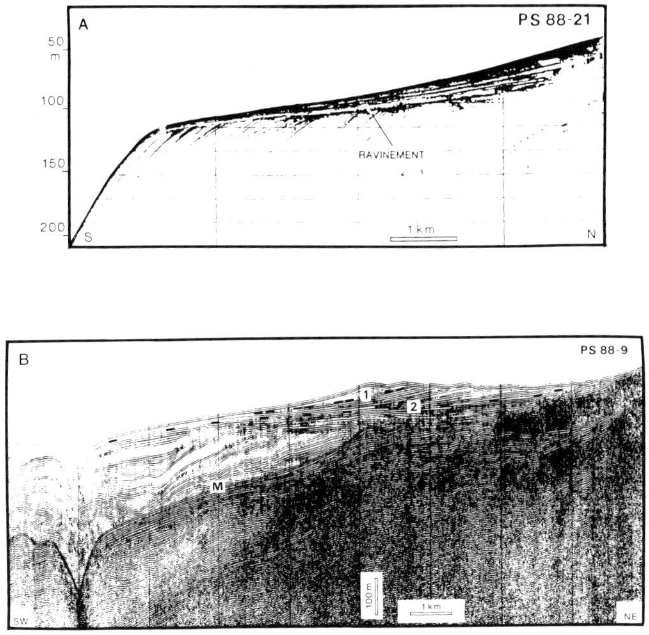

FIG. 8.—(A) Profile PS88-21 from the shelf off southwestern Italy showing the *absence* of a shelf-margin deposit. The lowstand shoreline was seaward of the shelf break and sediment was deposited on the slope rather than in a coastal sequence. This profile was collected north of the map area of Figure 7. (B) Profile PS 88-9 from the shelf off southwestern Italy; location of the profile is shown in Figure 7. This profile demonstrates development of a shelf-margin downlapping deposit; two distinct downlap surfaces are labeled. Both profiles are from Trincardi and Field (1991).

In areas where the shelfbreak was *deeper* than the lowstand shoreline, coastal deposits prograded seaward across the remaining low-angle shelf and onto the upper slope (Fig. 8B). In their evaluation, the position of the shelf break relative to the lowstand shoreline is a first-order control on the development of shelf-margin downlapping deposits.

The coast and shelf are fed by small coastal streams draining uplifted regions of southern Italy. Although shelf-margin deposits only occur where the shelf break lies seaward of the lowstand shoreline, their width and thickness show lateral variation, which can be related to variations in sediment supply (Trincardi and Field, 1991). This relationship between sediment supply and deposit width (Fig. 9) can be controlled either by input from coastal streams or by the processes, such as storms and shore-parallel currents, which distribute the material along the prograding coasts. Because the deposit progrades onto the upper slope, it was inferred that outbuilding continued during lowstand conditions, but that once sediment supply decreased or abated, truncation of the upper surface of the prograded sequence was initiated by shoreface retreat.

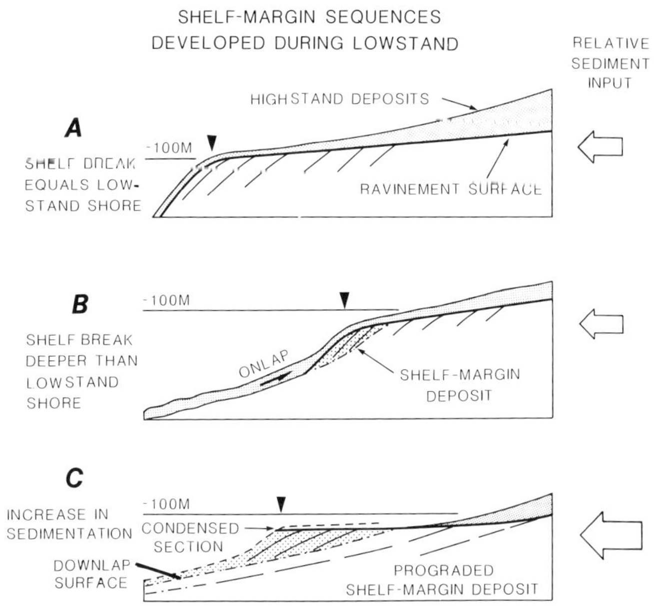

FIG. 9.—Shelf-margin sequences off southwestern Italy developed during a lowstand of sea level. (A) Where the shelf break is equal to or shallower than the lowstand shoreline position, as in Figure 8A, shelf-margin deposits do not form. (B) and (C) Where shelf-margin deposits form, their geometry is controlled by the shelf gradient and availability of sediment. Modified from Trincardi and Field (1991).

Effect of Sediment Supply

There can be no coastal progradation without introduction of new sediment. During a transgression, the coastal facies is recycled by shoreface erosion and overturning (recycling) of the barrier complex (Field and Duane, 1976; Swift, 1976). During a regression, however, the coastal lithosome is abandoned by the seaward-advancing shoreline; new material must be introduced into the system to maintain beach accretion and seaward growth. Increased stream efficiency due to downcutting and steeper gradients may achieve this in large measure. Whether the material forms a prograding deltaic complex or is redistributed along the shoreline into a series of prograding beach ridges probably depends on the size of the stream and the efficiency of the coastal-transport system. Swift and others (1991) discuss more fully the distinction between deltaic and nondeltaic progradation. As they point out, the distinction is more one of semantics than one of substance.

Sediment supply is important to the regressive system both as the shoreline crosses the shelf and at the lowstand position (Thorne and Swift, 1991). The effect during a *sea-level* fall is illustrated in Figure 6. With minimal sediment available, prograding coastal bodies will develop only where the shoreline remains essentially in place for long time periods, i.e., where the local gradient is steep. With increased sediment supply, regressive coastal deposits can develop in many shelf areas, and with abundant sediment, downlapping deposits may form on all but the flattest section of the shelf. During lowstand conditions, the sediment supply is equal in importance to the depth of the shelf break. A continuous supply of sediment from a stream functioning at peak efficiency can allow continuous progradation into deep water until the depth of the water becomes too great to allow significant outbuilding.

Shelf margins receiving a large, continuous supply of sediment commonly have thick, laterally extensive deposits that are stacked one on top of another (Fig. 10). Stacked sequences are common off major streams, such as the Mississippi River (Fig. 10A; Suter and Berryhill, 1985), and also off medium-size streams such as the Rio Grande de Santiago, along the Costa de Nayarit (Fig. 10B; Curray and Moore, 1964) and the Ventura and Santa Clara Rivers in southern California (Fig. 10C; Field and others, 1983; Dahlen and others, 1990). Each regressive phase is represented by a thick, prograded, shelf-margin deposit, which overlies a thin transgressive deposit lying on the erosional unconformity that defines the sequence boundary. The accommodation space required for the stacking of these cycles is made available in part by the large sediment load, which causes subsidence and compaction through time. The Eocene Coaledo Formation in Oregon contains stacked delta lobes (Chan and Dott, 1986) and may represent an ancient analogue for the stream-fed shelf-margin deposits discussed above.

Continental margins fed by relatively small coastal streams also show development of shelf-margin deposits. The eastern Tyrrhenian Sea margin of southern Italy is one such example. The shelf and coast are fed by small coastal streams, and regressive sequences of the type shown in Figure 8B have no detectable direct link to a fluvial source of sediment. As discussed above, Trincardi and Field (1991) found that the thickness and width of the downlapping deposit were variable, and they interpreted the variation to result from changes in the supply of sediment to the lowstand coast (Fig. 9).

Other examples of shelf-margin deposits in areas fed by relatively small coastal streams are found on the North Carolina shelf (Fig. 11A; Matteucci and Hine, 1987), the Portuguese shelf near Lisbon (Fig. 11B, C; Mougenot and others, 1983), the Texas shelf of the Gulf of Mexico (Morton and Price, 1987), and the Alboran Sea shelf of Morocco (Gensous and others, 1986). Matteucci and Hine (1987) describe the deposit off Cape Fear, North Carolina, as a shelf-edge delta, largely because internal stratal surfaces prograde away from a point source. The distinction between deltas and prograded coastal deposits is minor in such a setting as this, where a sandy medium-energy coast is fed by a small low-gradient stream. The point reinforced by the work of Matteucci and Hine is that shelf-margin deposits are preserved on coastal-plain shelves, but only where an increased gradient (accommodation space) exists and a local source of sediment is available. Belknap and Kraft (1981) also demonstrate that the extent and preservation of paralic lowstand and transgressive deposits on the U.S. Atlantic margin are influenced by pre-existing topography.

Similarly, Mougenot and others (1983) document prograding shelf-margin deposits along the Iberian and French Mediterranean margins. A comparison of the profiles presented by those authors again demonstrates the requirement for accommodation space on the shelf margin and availability of sediment. The margin off southern Portugal has been tectonically tilted, providing space over a steep outer-shelf surface (Fig. 11B,C); oblique downlapping beds, presumably from a local sediment source, have prograded over the shelf edge. In areas of increased sediment availability, such as the Gulf of Genoa, the outer shelf is composed of a stacked sequence of multiple downlapping deposits separated by thin transgressive deposits (Mougenot and others, 1983), producing a composite section similar to those illustrated in Figure 10.

Shelf-margin deposits can form in the absence of feeder streams, by lowstand redistribution and progradation of coastal sediment. The capability of coastal processes to construct prograded sediment terraces at the shelf edge *without* a significant fluvial source is best illustrated from

FIG. 10.—Profiles and interpreted line drawings of stacked regressive deposits off large fluvial-sediment sources. (A) Line drawing from the southwestern Louisiana shelf of the Mississippi River; from Suter and Berryhill (1985). Circled numbers refer to significant horizons or deposits. (B) Composite structural cross section of the shelf off Costa de Nayarit, Mexico; from Curray and Moore (1964). (C) Profile and interpretation from the shelf off Hueneme, California, near the mouths of the Ventura and Santa Clara Rivers; from Field and others (1983). The U identifies the unconformity defined by the downlap surface.

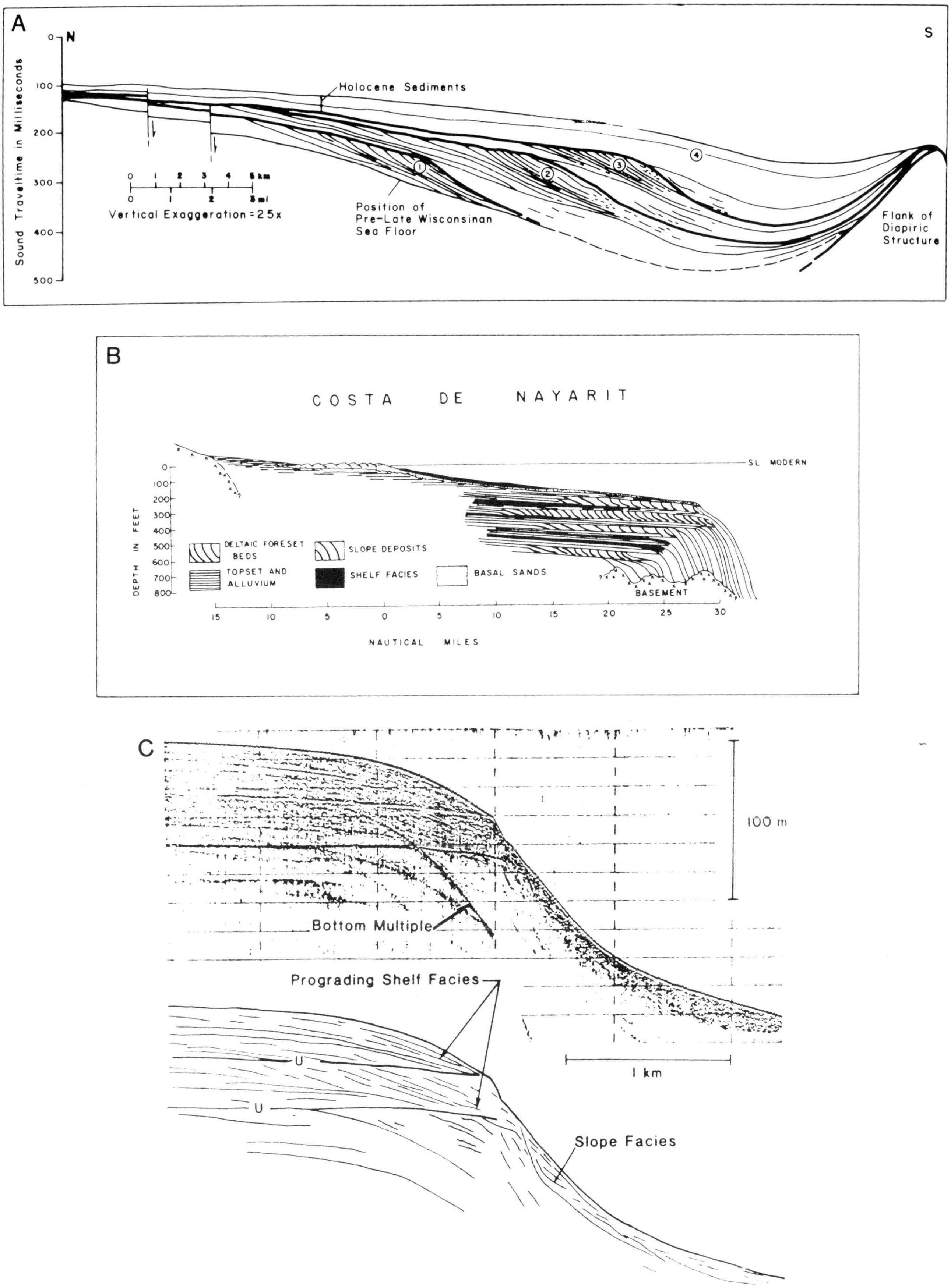
A
N
S
Sound Traveltime in Milliseconds
0
100
200
300
400
500
Holocene Sediments
0 1 2 3 4 5 km
0 1 2 3 mi
Vertical Exaggeration = 25x
Position of
Pre-Late Wisconsinan
Sea Floor
Flank of
Diapiric
Structure
B
COSTA DE NAYARIT
SL MODERN
DEPTH IN FEET
DELTAIC FORESET BEDS
SLOPE DEPOSITS
TOPSET AND ALLUVIUM
SHELF FACIES
BASAL SANDS
BASEMENT
NAUTICAL MILES
C
100 m
Bottom Multiple
Prograding Shelf Facies
1 km
U
U
Slope Facies

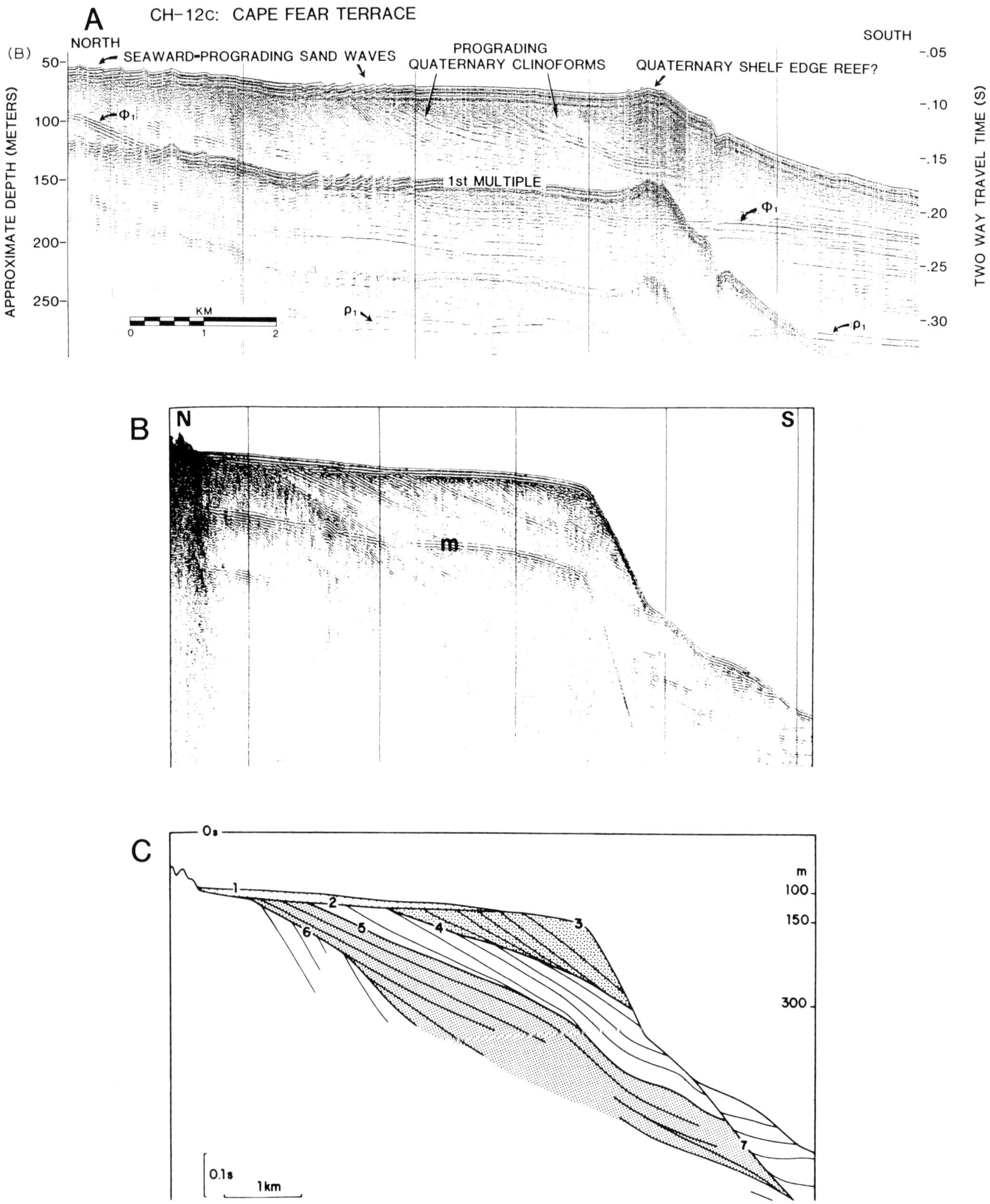

FIG. 11.—(A) Profile showing the progradational nature of shelf-margin deposits off Cape Fear, North Carolina, on the U.S. Atlantic margin; from Matteucci and Hine (1987). The Greek letters identify major stratigraphic boundaries. (B) Seismic-reflection profile with interpretation (C) of a well-developed shelf-margin deposit off Portugal; from Mougenot and others (1983). The numbers refer to significant horizons and deposits. Holocene highstand deposits (1) overlie the Pleistocene surface of erosion (2). The shelf break (3) formed by shelf-margin progradation following tilting of the margin (identified by unconformity 4).

the banks and island platforms of the southern California continental borderland. The island platforms and ridges are without significant sediment supply except for very small coastal streams and recycling of material previously eroded from pre-Quaternary outcrops. On some of the banks, such as Tanner and Cortes, the bank tops are completely submerged to depths of 80 to 120 m and are completely isolated from terrestrial regions. Thus, lowstand shelf-margin deposits (Fig. 12) had to be shaped from unconsolidated sediment by coastal and shelf processes; only very minor fluvial sources existed. The shelf-margin deposits on Cortes Bank shown in Figure 12 display the variation in downlap deposits that occurs in this type of setting. Each profile shows cycles of deposition, with each cycle being generated by erosion of the bank top. All three profiles presumably have the same amount of sediment available, yet the resulting deposits are significantly different because the local gradients are different. In Figure 12A, the gradient of the downlap surface is steep, and the prograded deposit quickly becomes stabilized; that is, its prograding slope is the same gradient as the downlap surface and enormous quantities of sediment are required to continue progradation (Fig. 4A). In contrast, Figure 12B shows a low-gradient downlap surface that constrains the shape of the shelf-margin deposit, resulting in a thin and relatively broad sequence. Where the gradient of the downlap surface is complex, as in Figure 12C, the geometry of the shelf-margin deposit is complex. The underlying surface changes from low gradient, to slightly higher gradient, to nearly flat. The result is a prograded deposit that is both thick and laterally extensive.

Other examples of shelf-margin deposits not associated with fluvial feeder systems come from Elba Ridge in the northern Tyrrhenian Sea (Aiello and others, 1979), the eastern Tyrrhenian Sea margin off Calabria (Trincardi and others, 1987), and Adventure Bank in northwestern Sicily Channel (Colantoni and others, 1985). On both Elba Ridge and Adventure Bank, shelf-margin deposits are constructed of biogenic and lithic sediment derived from local erosion and planation of structural highs. On the Calabrian margin, wave-cut terraces (presently found at about 135 m water depth and characterized by erosional steps as high as 20 m) formed during the late Pleistocene lowstand. Their geographic distribution indicates an origin related to the focusing of storm waves against coastal headlands. Seaward of the erosional scarps are small-scale sediment deposits characterized by internal downlapping reflectors that become steeper in deeper water. These ancestral shoreface deposits, fed only by sediment eroded from the scarps, provide another example of progradation in the absence of deltaic deposition. Other examples of prograded downlapping deposits constructed from locally derived sediment come from shelves enriched in carbonate sediment, such as the west Florida shelf (Brooks and Holmes, 1990).

PRESERVATION OF REGRESSIVE DEPOSITS

Midshelf Downlapping Deposits

Downlapping regressive deposits are not widely reported from continental shelves. Many of the places where they have been recognized—California, Italy, Australia, Iberia, Gulf of Mexico, North Carolina—are single examples from large shelves. Most of the examples discussed and illustrated in the preceding sections are from shelf margins and not from midshelf areas, where they seem to be rare. Their apparent scarcity in midshelf regions may simply reflect poor identification or interpretation; however, it is more likely that, although they are commonly formed during falling sea levels, their potential for preservation on the shelf is limited. As Davis and Clifton (1987) note, the preservation of coastal sequences is complex and conceptually requires the right intermix of sea-level change, sediment input and wave or tidal energy. The process of shoreface erosion during a transgression is efficient at planing off deposits from previous cycles. The shelf off northern California (Fig. 13A) is characterized by a prominent erosional surface separating underlying deformed strata from late Quaternary deposits. Overlying the erosion surface on the inner shelf is a remnant of downlapping deposits, only 10 m thick, capped by Holocene highstand shelf deposits. Apparently, on the outer shelf all traces of a regressive coastal deposit have been eliminated (see right side of profile in Fig. 13A). The same concept applies to the eastern Tyrrhenian Sea shelf, where Trincardi and Field (1991) report that midshelf downlapping deposits typically are not preserved (Fig. 13C). Their presence corresponds to a deepening (relative to adjacent areas) of the underlying surface that was infilled by seaward-advancing coastal deposits. Thus, prograding coastal deposits are best preserved in shelf areas where they are thicker and more likely to survive reworking by shoreface erosion. Such a scenario is illustrated in Figure 14, where prograded coastal sequences are deposited unevenly across an irregular shelf surface during a fall in relative sea level; another example is given by Nardin and others (1981) from the tectonically controlled shelf off Santa Monica, California. Shoreface erosion during a rise in relative sea level obliterates the character of the upper part of the prograded deposits. Where the deposits are thin, the entire deposit is reworked; where they are thick, the basal part is preserved, as in Figure 14C. A contrast to this is exhibited by recent seismic-reflection evidence obtained from southeast Australia that shows a relatively thick, continuous downlapping deposit preserved across the midshelf (P. Roy, written commun., 1990). Other likely settings for preservation are narrow embayments or estuaries, where sediment is channeled into elongate seaward-thickening deposits (e.g., Syvitski and others, 1988).

Shelf-Margin Downlapping Deposits

It appears from available evidence that the shelf margin is the site of greatest potential for preservation. Where shelf-margin downlapping deposits have been observed off southwest Italy, southern California, Portugal, North Carolina, and other locations, they are not connected to midshelf downlapping deposits, which are thin or non-existent. There are several reasons why shelf-margin deposits are preferentially preserved.

First, the outer shelf and upper slope commonly have a steeper gradient, which provides for increased accommodation space. Under stable sea-level conditions during a lowstand, the increase in gradient provides for thicker wedges

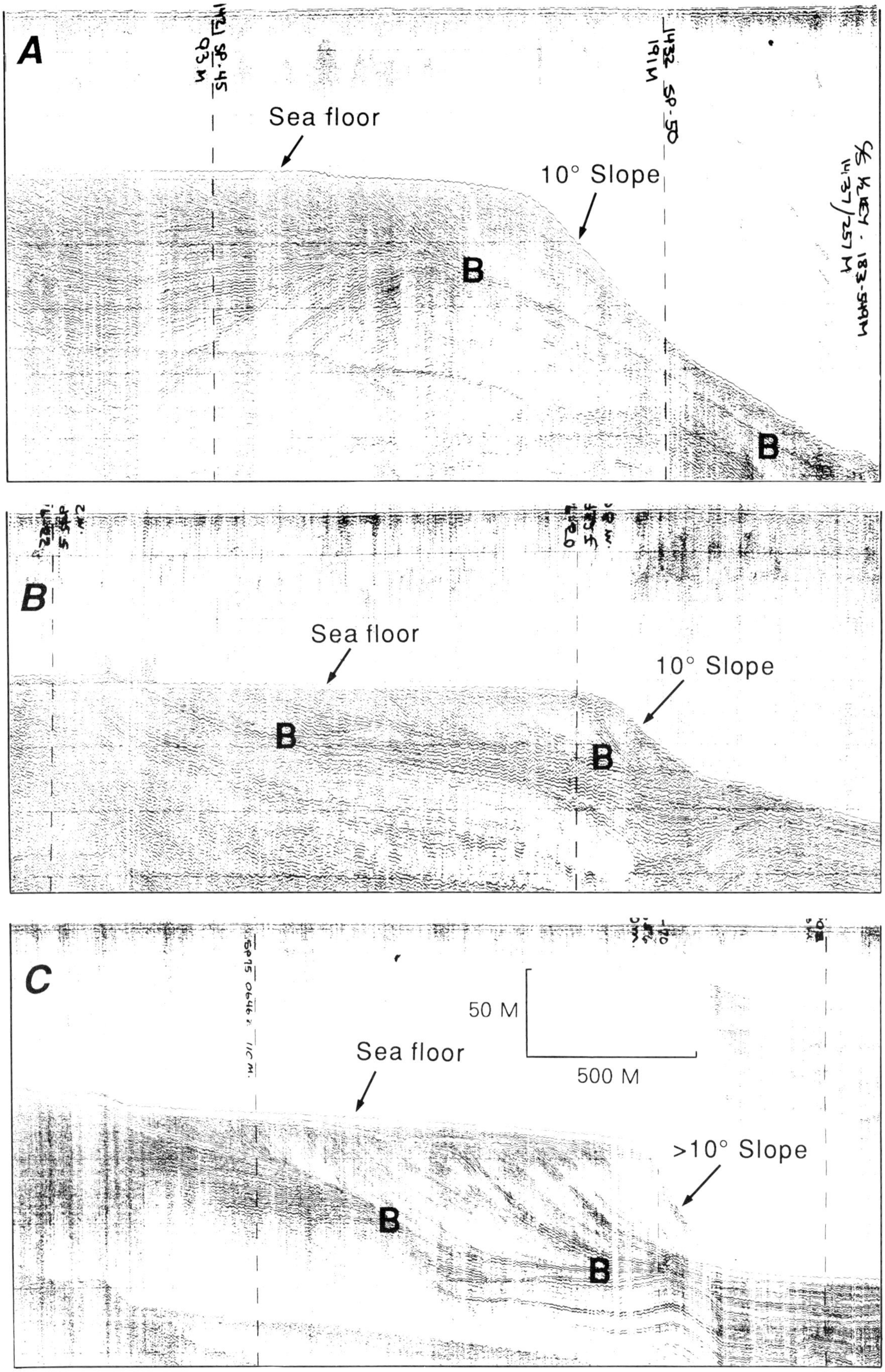
A
Sea floor
10° Slope
B
B
B
Sea floor
10° Slope
B
B
C
50 M
500 M
Sea floor
>10° Slope
B
B

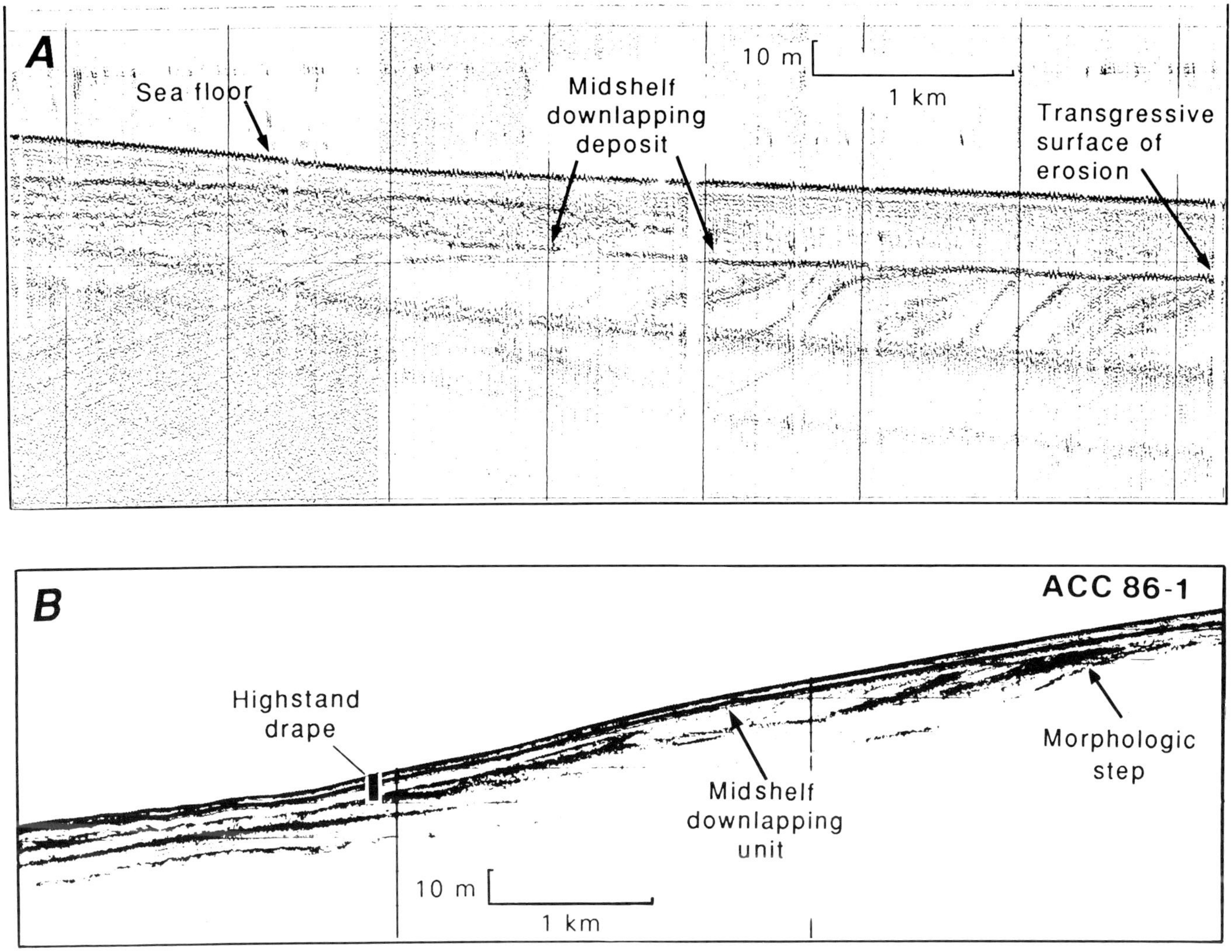

FIG. 13.—Examples of preserved midshelf regressive deposits. (A) Northern California shelf off the Klamath River. (B) Shelf off Licosa, southwestern Italy; from Trincardi and Field (1991); location shown in Figure 7.

of prograding sediment to be deposited (Fig. 3). As with any sedimentary deposit, greater thicknesses increase the likelihood that at least some of the deposit will be preserved.

A second reason for high-preservation potential relates to relative sea level and sediment availability. As relative sea level reaches its lowstand position, shoreline migration is slowed or stalled, and more time is available for the development of coastal-sediment bodies. Sediment is contributed by stillstand erosion and by coastal streams that have reached their maximum gradient and are transporting material at peak efficiency.

A third factor favoring preservation of shelf-margin deposits is their formation seaward of the lowstand shoreline. As relative sea level begins to rise, the coastline retreats by the process of shoreface erosion. The tops of the lowstand shelf-margin deposits are reworked by the transgressing sea, but as most of the deposits lie deeper than the lowstand shoreline, erosion is minimal. As sea-level rise continues, stream base levels change and sediment is trapped in stream valleys. The lack of sediment feeding the coast increases the likelihood of erosion and reworking of the thin downlapping units that extend across the shelf.

FIG. 12.—Shelf-margin prograded deposits from Cortes Bank in the southern California Continental Borderland. Cortes Bank is isolated from the mainland by deep basins and receives no fluvial input; these deposits were formed solely by reworking of sediment from the bank top. These three profiles illustrate some of the principles defined in the text and earlier figures; the "B" marks the surface of downlap. Where the downlap surface is steep, as in (A), progradation is slowed because of the increasing volume of sediment required to maintain growth. Where the downlap surface is relatively flat and shallow, as in (B), the shelf-margin prograded deposit is thin and laterally extensive. Profile (C) shows a complex downlap surface, which results in a thick and extensive deposit. Profiles are from Greene and others (1975). The terrace surfaces are roughly 100 m below sea level.

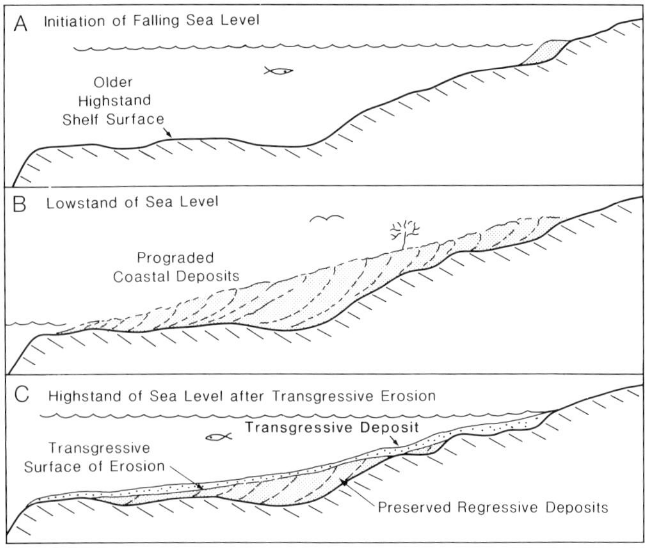

FIG. 14.—Diagram illustrating the sequence of events leading to the deposition and preservation of regressive deposits on the midshelf. Prograded units are deposited unevenly across the continental shelf during a relative fall in sea level, leaving thicker deposits in depressions and on shelf sections having steep gradients. Erosion of the shelf surface during a relative rise in sea level obliterates the entire regressive deposit where it was initially thin. In areas where the regressive beds are thick, only the top is reworked and the basal part of the deposit is preserved.

As Shepard (1932) first demonstrated from his systematic compilation of sediment grain-size data from around the world, most continental shelves are mantled by fine to coarse sand and gravel. Since Shepard's initial observations, most investigators have concurred that the coarse shelf sediment resulted from coastal and fluvial deposition at times of lower sea level. The coarsest deposits commonly occur on the outer shelf and near the shelf break. This has been interpreted by some as a result of winnowing during present relative highstand conditions by active phenomena, such as breaking internal waves. The distribution and character of outer-shelf sands become more understandable, however, in light of the style of regressive coastal deposition.

During the maximum lowstand, streams were at their maximum gradient and peak efficiency; deposition at the shelf margin formed a thick and sometimes laterally extensive deposit (Figs. 12B,C) of coastal and shoreface sand (Morton and Price, 1987), much of it relatively coarse. The coarse sandy facies of the thick lowstand coastal lithosome was the most likely to survive transgressive erosion. Deposits inshore of the lowstand shoreline, however, commonly were truncated and reworked, leading to preservation of the fine-sand texture typical of the shoreface environment (Fig. 15). One effect of this may be that on many continental shelves, outer-shelf sediment texture may largely reflect their original deposition in a regressive coastal environment, despite 10,000 to 20,000 years of marine reworking.

THE LEGACY OF REGRESSIVE COASTAL DEPOSITS

In the foregoing discussion, we have emphasized the formation and preservation potential of downlapping regressive deposits on continental shelves. This section focuses on the legacy of those deposits on Holocene shelves—the record they have left in the surface distribution of sediment.

Regressive downlapping coastal deposits have been shown to be common on many types of shelves around the world. The scientific literature contains a number of examples, and doubtless many others exist. They have received little attention until relatively recently, because little was known about their origin or their significance for sequence-stratigraphic interpretation. Downlapping deposits are preserved in a variety of settings. They are most likely to be preserved along the shelf margin, where thick sequences formed at or below the position of the lowstand sea level, and they are least likely to be preserved on broad low-gradient (coastal plain) shelves, where small drops in relative sea level were accompanied by large seaward shifts of the shoreline. Even on high-gradient tectonic margins, downlapping deposits commonly are not preserved on the midshelf. The factors that control the presence, geometry and preservation of these regressive deposits are: sediment availability and shelf gradient, irregularity, and depth of the shelf break. On most continental shelves, conditions were probably satisfactory for the formation of downlapping deposits across the shelf. These deposits, along with the thicker and better developed deposits along the shelf margin, formed the grist for the transgressive erosion that accompanied the return of sea level to highstand conditions.

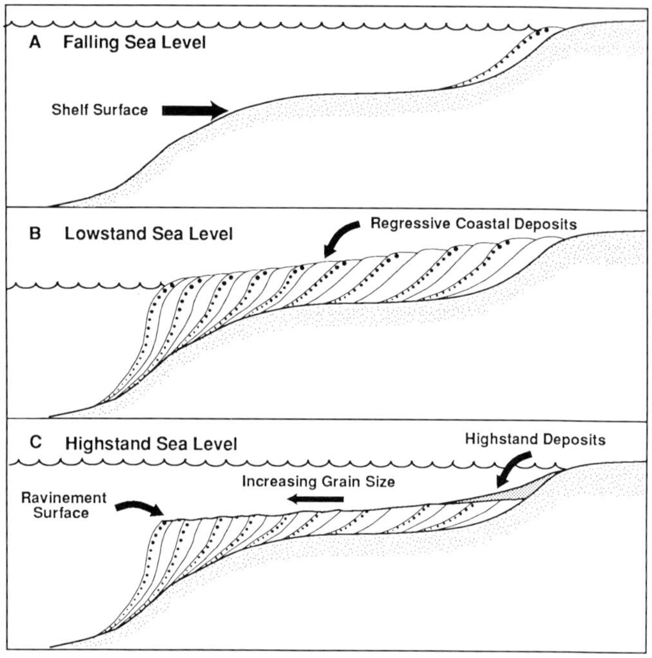

FIG. 15.—Diagram illustrating one mechanism for the origin of coarse sediment on the outer shelf. Prograded coastal units are truncated by shoreface erosion during transgression. Erosion into the prograded deposits will expose increasingly finer sediment (lower-shoreface facies) in a landward direction, particularly with respect to the outer shelf, where only beach and dune sands are available for reworking. The result is a remnant veneer of coarse sediment (sand) on the outer shelf, especially near the shelf break.

ACKNOWLEDGMENTS

The ideas we have expressed here have benefited greatly from discussions and reviews by colleagues. In particular, we thank Ed Clifton, Ralph Hunter, and Larry Phillips of the U.S. Geological Survey and Ron Boyd of Dalhousie University for their time and thoughtful suggestions that immeasurably improved this manuscript. Much of the data presented here was collected through the marine programs of the U.S. Geological Survey and the Istituto per la Geologia Marina. We gratefully acknowledge the use of colleagues' results reproduced in Figures 4, 5, 10, 11, 12, and 13 in this paper. This is Contribution No. 794 of the Istituto per la Geologia Marina.

REFERENCES

AIELLO, EROS, BARTOLINI, CARLO, GABBANI, GIULANO, ROSSI, SERGIO, VALLERI, GILIOLA, CERTINI, LUCIANO, CLERICI, CARLO, AND LENAZ, RENZO, 1979, Studio della piattaforma continentale medio-tirrenica per la ricerca delle sabbie metallifere: 1. da Capo Linaro a Monte Argentario: Bollettino Societá Geologica Italiana, v. 97, p. 495–525.

AHARON, P., 1983, 140,000-yr isotope climatic record from raised coral reefs in New Guinea: Nature, v. 304, p. 720–723.

BELKNAP, D. F., AND KRAFT, J. C., 1981, Preservation potential of transgressive coastal lithosomes on the U.S. Atlantic coast: Marine Geology, v. 42, p. 429–442.

BOURGEOIS, JOANNE, 1980, A transgressive shelf sequence exhibiting hummocky stratification: The Cape Sebastian Sandstone (Upper Cretaceous), southwest Oregon: Journal of Sedimentary Petrology, v. 50, p. 681–702.

BOYD, RON, SUTER, JOHN, AND PENLAND, SHEA, 1989, Relation of sequence stratigraphy to modern sedimentary environments: Geology, v. 17, p. 926–929.

BROOKS, G. R., AND HOLMES, C. W., 1990, Modern slope configuration of the southwest Florida carbonate slope: Development by shelf margin progradation: Marine Geology, v. 94, v. 301–315.

CACCHIONE, D. A., DRAKE, D. E., KINEKE, G. C., AND STERNBERG, R. W., 1990, GEOPROBE measurements of strong bottom currents and rapid sedimentation on the Amazon prodelta: EOS Transactions, American Geophysical Union, v. 71, p. 1377.

CHAN, M. A., AND DOTT, R. H., JR., 1986, Depositional facies and progradational sequences in Eocene wave-dominated deltaic complexes, southwestern Oregon: American Association of Petroleum Geologists Bulletin, v. 70. p. 415–429.

CHAPMAN, D. M., GEARY, M., ROY, P. S., AND THOM, B. G., 1982, Coastal Evolution and Coastal Erosion in New South Wales: D. West Government Printer, Sydney, Australia, 341 p.

CLIFTON, H. E., 1989, Stratigraphic sequence response to glacio-eustatic fluctuations in a rapidly subsiding late Cenozoic Basin, San Francisco, California, *in* Morton, R. A., and Nummedal, D., eds., Gulf Coast Section, SEPM Foundation Seventh Annual Research Conference Proceedings, p. 203–211.

COLANTONI, PAOLO, CREMONA, GIUSEPPE, LIGI, MARCO, BORSETTI, A. M., AND CATI, FRANCO, 1985, The Adventure Bank (off Southwestern Sicily): a present-day example of carbonate shelf sedimentation: Giornale di Geologia, v. 47, p. 165–180.

COLEMAN, J. M., AND ROBERTS, H. H., 1988, Sedimentary development of the Louisiana continental shelf related to sea-level cycles: Part I—sedimentary sequences: Geo-Marine Letters, v. 8, p. 63–108.

CURRAY, J. R., 1964, Transgressions and regressions, *in* Miller, R. L., ed., Papers in Marine Geology: Macmillan, New York, p. 175–203.

CURRAY, J. R., AND MOORE, D. G., 1964, Pleistocene deltaic progradation of continental terrace, Costa de Nayarit, Mexico, *in* van Andel, Tj. H. and Shor, G. G., Jr., eds., Marine Geology of the Gulf of California: American Association of Petroleum Geologists Memoir 3, Tulsa, Oklahoma, p. 193–215.

CURRY, J. R., EMMEL, F. J., AND CRAMPTON, P. J. S., 1969, Holocene history of a strand plain, lagoonal coast, Nayarit, Mexico, *in* Castanaras, A. A., and Phleger, F. B., eds., Coastal Lagoons—A Symposium: Universidad Nacional Autónoma de México/UNESCO, Mexico City, p. 63–100.

DAHLEN, M. Z., OSBORNE, R. H., AND GORSLINE, D. S., 1990, Late Quaternary history of the Ventura mainland shelf, California: Marine Geology, v. 94, p. 317–340.

DAVIS, R. A., JR., AND CLIFTON, H. E., 1987, Sea-level change and the preservation potential of wave-dominated and tide-dominated coastal sequences, *in* Nummedal, Dag, Pilkey, O. H., and Howard, J. D., eds., Sea-Level Fluctuation and Coastal Evolution: Society of Economic Paleontologists and Mineralogists Special Publication 41, Tulsa, Oklahoma, p. 164–180.

DEMAREST, J. M., AND KRAFT, J. C., 1987, Stratigraphic record of Quaternary sea levels: Implications for more ancient strata, *in* Nummedal, Dag, Pilkey, O. H., and Howard, J. D., eds., Sea-Level Fluctuation and Coastal Evolution: Society of Economic Paleontologists and Mineralogists Special Publication 41, Tulsa, Oklahoma, p. 223–240.

DUANE, D. B., FIELD, M. E., MEISBURGER, E. P., SWIFT, D. J. P., AND WILLIAMS, S. J., 1972, Linear shoals on the Atlantic inner continental shelf, Florida to Long Island, *in* Swift, D. J. P., Duane, D. B., and Pilkey, O. H., eds., Shelf Sediment Transport: Process and Pattern: Dowden, Hutchinson and Ross, Stroudsburg, Pennsylvania, 656 p.

EMERY, K. O., 1968, Relict sediments on continental shelves of the world: American Association of Petroleum Geologists Bulletin, v. 52, p. 445–464.

EMERY, K. O., BUTCHER, W. S., GOULD, H. R., AND SHEPARD, F. P., 1952, Submarine geology off San Diego, California: Journal of Geology, v. 60, p. 511–548.

FIELD, M. E., 1980, Sand bodies on continental shelves: Holocene record of the U.S. Atlantic inner shelf off Maryland: Journal of Sedimentary Petrology, v. 50, p. 505–528.

FIELD, M. E., AND DUANE, D. B., 1976, Post-Pleistocene history of the United States inner continental shelf: Significance to origin of barrier islands: Geological Society of America Bulletin, v. 87, p. 691–702.

FIELD, M. E., CARLSON, P. R., AND HALL, R. K., 1983, Seismic facies of shelf-edge deposits, U.S. Pacific continental margin, *in* Stanley, D. J., and Moore, G. T., eds., The Shelfbreak: Critical Interface on Continental Margins: Society of Economic Paleontologists and Mineralogists Special Publication 33, Tulsa, Oklahoma, p. 299–314.

FISCHER, A. G., 1961, Stratigraphic record of transgressing seas in light of sedimentation on Atlantic coast of New Jersey: American Association of Petroleum Geologists Bulletin, v. 45, p. 1651–1666.

GALLOWAY, W. E., 1989a, Genetic stratigraphic sequences in basin analysis I: Architecture and genesis of flooding-surface bounded depositional units: American Association of Petroleum Geologists Bulletin, v. 73. p. 125–142.

GALLOWAY, W. E., 1989b, Genetic stratigraphic sequences in basin analysis II: Application to northwest Gulf of Mexico Cenozoic basin. American Association of Petroleum Geologists Bulletin, v. 73. p. 143–154.

GENSOUS, B., TESSON, M., AND WINNOCK, E., 1986, La marge méridionale de la Mer d'Alboran: caracteres structuro-sédimentaires et évolution récente: Marine Geology, v. 72, p. 341–370.

GREENE, H. G., CLARKE, S. H., JR., FIELD, M. E., LINKER, F. I., AND WAGNER, H. C., 1975, Preliminary report on the environmental geology of selected areas of the southern California continental borderland: U.S. Geological Survey Open-File Report 75-596, Menlo Park, California, 70 p.

JERVEY, M. T., 1988, Quantitative geological modeling of siliciclastic rock sequences and their seismic expression, *in* Wilgus, C. K., Hastings, B. S., Kendall, C. G. St. C., Posamentier, H. W., Ross, C. A., and Van Wagoner, J. C., eds., Sea-Level Changes—An Integrated Approach: Society of Economic Paleontologists and Mineralogists Special Publication 42, Tulsa, Oklahoma, p. 47–70.

KINDINGER, J. L., 1988, Seismic stratigraphy of the Mississippi-Alabama shelf and upper continental slope: Marine Geology, v. 83, p. 79–94.

KUEHL, S. A., NITTROUER, C. A., AND DEMASTER, D. J., 1982, Modern sediment accumulation and strata formation on the Amazon continental shelf: Marine Geology, v. 49, p. 279–300.

LOUTIT, T. S., HARDENBOL, JAN, VAIL, P. R., AND BAUM, G. R., 1988, Condensed sections, the key to age determination and correlation of continental margin sequences, *in* Wilgus, C. K., Hastings, B. S., Kendall, C. G. St. C., Posamentier, H. W., Ross, C. A., and Van Wagoner, J. C., eds., Sea-Level Changes—An Integrated Approach: Society of Economic Paleontologists and Mineralogists Special Publication 42, Tulsa Oklahoma, p. 183–213.

MATTEUCCI, T. D., AND HINE, A. C., 1987, Evolution of Cape Fear terrace; A complex interaction between the Gulf Stream and a paleo shelf edge delta: Marine Geology, v. 77, p. 185–205.

MILLIMAN, J. D., AND EMERY, K. O., 1968, Sea levels during the past 35,000 years: Science, v. 162, p. 1121–1123.

MILLIMAN, J. D., AND MEADE, R. H., 1983, World-wide delivery of sediment to the oceans: Journal of Geology, v. 91, p. 1–21.

MORTON, R. A., AND PRICE, W. A., 1987, Late Quaternary sea-level fluctuations and sedimentary phases of the Texas coastal plain and shelf, *in* Nummedal, Dag, Pilkey, O. H., and Howard, J. D., eds., Sea-Level Fluctuation and Coastal Evolution: Society of Economic Paleontologists and Mineralogists Special Publication 41, Tulsa, Oklahoma, p. 181–198.

MOUGENOT, DENIS, BOILLOT, GILBERT, AND REHAULT, J. P., 1983, Prograding shelfbreak types on passive margins: Some European examples, *in* Stanley, D. J., and Moore, G. T., eds., The Shelfbreak: Critical Interface on Continental Margins: Society of Economic Paleontologists and Mineralogists Special Publication 33, Tulsa, Oklahoma, p. 61–77.

NARDIN, T. R., OSBORNE, R. H., BOTTJER, D. J., AND SCHEIDEMANN, R. C., JR., 1981, Holocene sea-level curves for Santa Monica shelf, California continental borderland: Science, v. 213, p. 331–333.

NITTROUER, C. A., AND STERNBERG, R. W., 1981, The formation of sedimentary strata in an allochthonous shelf environment: The Washington continental shelf, *in* Nittrouer, C. A., ed., Sedimentary Dynamics of Continental Shelves: Developments in Sedimentology 32, Elsevier Scientific Publishing Company, Amsterdam, p. 201–232.

NUMMEDAL, DAG, AND SWIFT, D. J. P., 1987, Transgressive stratigraphy at sequence-bounding unconformities: Some principles derived from Holocene and Cretaceous examples, *in* Nummedal, Dag, Pilkey, O. H., and Howard, J. D., eds., Sea-Level Fluctuation and Coastal Evolution: Society of Economic Paleontologists and Mineralogists Special Publication 41, Tulsa, Oklahoma, p. 241–260.

OKAMURA, YUKINOBU, 1989, Multi-layered progradational sequences in the shelf and slope of the southwest Japan forearc, *in* Taira, A., and Masuda, F., eds., Sedimentary Facies in the Active Plate Margin: Terra Scientific Publishing Company, Tokyo, p. 295–317.

PENLAND, SHEA, BOYD, RON, AND SUTER, J. R., 1988, Transgressive depositional systems of the Mississippi delta plain: A model for barrier shoreline and shelf sand development: Journal of Sedimentary Petrology, v. 58, p. 932–949.

PILKEY, O. H., BLACKWELDER, B. W., DOYLE, L. J., ESTES, E., AND TERLECKY, P. M., 1969, Aspects of carbonate sedimentation on the Atlantic continental shelf off the southern United States: Journal of Sedimentary Petrology, v. 39, p. 744–768.

POSAMENTIER, H. W., JERVEY, M. T., AND VAIL, P. R., 1988, Eustatic control on clastic deposition, I. Conceptual framework *in* Wilgus, C. K., Hastings, B. S., Kendall, C. G. St. C., Posamentier, H. W., Ross, C. R., and Van Wagoner, J. C., eds., Sea Level Changes—An Integrated Approach: Society of Economic Paleontologists and Mineralogists Special Publication 42, Tulsa, Oklahoma, p. 109–124.

POSAMENTIER, H. W., AND VAIL, P. R., 1988, Eustatic control on clastic deposition, II. Sequence and systems tract models, *in* Wilgus, C. K., Hastings, B. S., Kendall, C. G. St. C., Posamentier, H. W., Ross, C. R., and Van Wagoner, J. C., eds., Sea Level Changes—An Integrated Approach: Society of Economic Paleontologists and Mineralogists Special Publication 42, p. 125–154.

POSAMENTIER, H. W., JAMES, D. P., AND ALLEN, G. P., 1990, Aspects of sequence stratigraphy: Recent and ancient examples of forced regressions (abst.): American Association of Petroleum Geologists Bulletin, v. 74, p. 742.

RINE, J. M., AND GINSBURG, R. N., 1985, Depositional facies of a mud shoreface in Suriname, South America—a mud analogue to sandy shallow-marine deposits: Journal of Sedimentary Petrology, v. 55, p. 633–652.

SHEPARD, F. P., 1932, Sediments of the continental shelves: Geological Society of America Bulletin, v. 43, p. 1017–1040.

SHEPARD, F. P., AND COHEE, G. V., 1936, Continental shelf sediments off the mid-Atlantic states: Geological Society of America Bulletin, v. 47, p. 441–458.

SHEPARD, F. P., AND MOORE, D. G., 1955, Central Texas coast sedimentation: Characteristics of sedimentary environment, recent history, and diagenesis: American Association of Petroleum Geologists Bulletin, v. 39, p. 1463–1593.

STRIDE, A. H., BELDERSON, R. H., AND KENYON, N. H., 1972, Longitudinal furrows and depositional sand bodies of the English Channel: Memoire Bureau Recherches Geologique et Minieres, no. 79, Paris, p. 233–244.

SUTER, J. R., AND BERRYHILL, H. L., 1985, Late Quaternary shelf-margin deltas, northwest Gulf of Mexico: American Association of Petroleum Geologists Bulletin, v. 69, p. 77–91.

SUTER, J. R., BERRYHILL, H. L., AND PENLAND, SHEA, 1987, Late Quaternary sea-level fluctuations and depositional sequences, southwest Louisiana continental shelf, *in* Nummedal, Dag, Pilkey, O. H., and Howard, J. D., eds., Sea-Level Fluctuation and Coastal Evolution: Society of Economic Paleontologists and Mineralogists Special Publication 41, Tulsa, Oklahoma, p. 199–219.

SWIFT, D. J. P., 1968, Coastal erosion and transgressive stratigraphy: Journal of Geology, v. 76, p. 444–456.

SWIFT, D. J. P., 1976, Coastal sedimentation, *in* Stanley, D. J., and Swift, D. J. P., eds., Marine Sediment Transport and Environmental Management: Wiley-Interscience, New York, p. 255–310.

SWIFT, D. J. P., STANLEY, D. J., AND CURRAY, J. R., 1971, Relict sediments on continental shelves: A reconsideration: Journal of Geology, v. 79, p. 322–346.

SWIFT D. J. P., DOFOED, J. W., SAULSBURY, F. P., AND SEARS, PHILLIP, 1972, Holocene evolution of the shelf surface, central and southern Atlantic shelf of North America, *in* Swift, D. J. P., Duane, D. B., and Pilkey, O. H., eds., Shelf Sediment Transport: Process and Pattern: Dowden, Hutchinson and Ross, Stroudsburg, Pennsylvania, p. 499–574.

SHEPARD, F. P., PHILLIPS, SANDRA, AND THORNE, J. A., 1991, Sedimentation on continental margins. Part IV: Lithofacies and depositional systems, *in* Swift, D. J. P., Tillman, R. W., and Oertel, G. F., eds., Shelf Sands and Sand Bodies: Special Publication Number 12, International Association of Sedimentologists.

SYVITSKI, J. P. M., SMITH, J. N., CALABRESE, E. A., AND BOUDREAU, B. P., 1988, Basin sedimentation and the growth of prograding deltas: Journal of Geophysical Research, v. 93, p. 6895–6908.

TESSON, M., GENSOUS, B., ALLEN, G. P., AND RAVENNE, CH., 1990, Late Quaternary deltaic lowstand wedges on the Rhône continental shelf: Marine Geology, v. 91, p. 325–332.

THOM, B. G., POLACH, H. A., AND BOWMAN, G. M., 1978, Holocene age structure of coastal sand barriers in New South Wales, Australia: Report of the Department of Geography, Faculty of Military Studies, University of New South Wales, Duntroon, Australia, 86 p.

THORNE, J. A., AND SWIFT, D. J. P., 1991, Sedimentation on continental margins, Part VI: A regime model for depositional sequences, their component systems tracts, and bounding surfaces, *in* Swift, D. J. P., Tillman, R. W., and Oertel, G. F., eds., Shelf Sands and Sand Bodies: Special Publication Number 12, International Association of Sedimentologists.

TRINCARDI, FABIO, CIPOLLI, MARCO, FERRETTI, PAOLO, LA MORGIA, JASEMINE, LIGI, MARCO, MAROZZI, GABRIELE, PALUMBO, VITO, TAVIANI, MARCO, AND ZITELLINI, NEVIO, 1987, Slope basin evolution on the Eastern Tyrrhenian Margin: Preliminary report: Giornale di Geologia, v. 49, p. 1–9.

TRINCARDI, FABIO, AND FIELD, M. E., 1991, Geometry, lateral variability, and preservation of downlapped regressive shelf deposits: eastern Tyrrhenian margin, Italy: Journal of Sedimentary Petrology, v. 61, p. 0000.

VAIL, P. R., MITCHUM, R. M., JR., AND THOMPSON, S., 1977, Seismic stratigraphy and global changes of sea level, Part 3: Relative changes of sea level from coastal onlap, *in* Payton, C. E., ed., Seismic Stratigraphy—Applications to Hydrocarbon Exploration: American Association of Petroleum Geologists Memoir 26, Tulsa, Oklahoma, p. 63–82.

VAN WAGONER, J. C., POSAMENTIER, H. W., MITCHUM, R. M., JR., VAIL, P. R., SARG, J. F., LOUTIT, L. S., AND HARDENBOL, JAN, 1988, An overview of the fundamentals of sequence stratigraphy and key definitions, *in* Wilgus, C. K., Hastings, B. S., Kendall, C. G. St. C., Posamentier, H. W., Ross, C. A., and Van Wagoner, J. C., eds., Sea Level Changes—An Integrated Approach: Society of Economic Paleontologists and Mineralogists Special Publication 42, Tulsa, Oklahoma, p. 39–46.

WAVE PROCESSES AND GEOLOGIC RESPONSES ON THE FLOOR OF THE YELLOW SEA

JAMES S. BOOTH AND WILLIAM J. WINTERS

U.S. Geological Survey, Quissett Campus, Woods Hole, Massachusetts 02543

ABSTRACT: The floor of the Yellow Sea is a geologically mundane surface: it is nearly horizontal, lacks relief, and, with few exceptions, is devoid of conspicuous geomorphologic features. However, it is the principal repository for the prodigious sediment load of the Huanghe (Yellow River); and, due to its inherent shallowness (average depth is 40 m), it is frequently stressed by waves generated by winter storms and typhoons. Analyses of mass physical properties of cores representing the upper few meters of sediment in the central and north-central Yellow Sea (near the Shandong Peninsula), in conjunction with analyses of slope stability, failure modes, and erodibility, permit an assessment of the likelihood and effect of dynamic, transient geologic events on the seabed.

Vane shear-strength profiles along with consolidation test data indicate that the present surface of the seabed is in a depositional mode and is compacting normally. In addition, liquid-limit profiles imply that in the study area these neritic sediments have been accumulating in an environment that probably has not been modified significantly since sea level reached its current level. There is no geotechnical evidence in the nine cores recovered that slope failures have occurred, and clasts, sand lenses or other manifestations of mass movements, including flows, also are absent. These observations support previous interpretations of seismic records. Moreover, slope stability analysis for static conditions shows that the sea floor is quite stable.

Regardless, shear-stress levels generated by cyclic loading during major storms may approach the sediment shear strengths, and, when coupled with concomitant excess pore pressures, could cause slope failure. Unless the failed beds collapsed or flowed, however, there probably would be little conspicuous evidence of such a failure. In fact, evaluation of the potential of these sediments for disintegrative behavior suggests that they are not prone to either collapse or flow.

Storm waves also generate oscillatory bottom currents that may erode the seabed. Whether the sediment is considered as cohesionless or cohesive, typhoons could have the potential to erode at all water depths within the Yellow Sea (i.e., to 90 m), and winter storms to water depths of 60 m or more. However, in the case of cohesive behavior, it could be that the effect of winter storms and most typhoons is generally less extreme. If the sea floor is repeatedly scoured, it is likely limited to the top few centimeters.

Despite the fact that storm waves may cause slope failure and are certainly responsible for frequent scouring, they probably leave only a subtle sedimentologic imprint on the seabed.

INTRODUCTION

The Yellow Sea is a shallow body of water on the continental platform between the Chinese mainland and the Korean Peninsula (Fig. 1). It is a classic epicontinental sea. Its characteristics include an average water depth of about 40 m, and a sea floor that is generally flat and of low relief. Regional gradients vary from 0.2° to less than 0.01°, and average less than 0.1°. Moreover, major geomorphic features are scarce. Among the more prominent of such features are the Shandong mud wedge, a 40-m-thick mound of sediment located south of the tip of the Shandong Peninsula (Milliman and others, 1985), the delta built by the Huanghe (Yellow River) in the western Yellow Sea prior to 1855 A.D., and linear sand bodies derived from the delta that have as much as 15 m of relief locally (Liu and others, 1989).

The shallow stratigraphy is marked by an erosional surface 10 to 40 m below the present sea floor (Milliman and others, 1989). This surface was cut by channels, suggesting that the Huanghe may have carried sediment across the exposed shelf during the last lowstand of the sea and delivered it to the nearby Okinawa Trough, which lies southeast of the Yellow Sea. After a period of channel infilling and accumulation of deltaic sediments, the ensuing transgression produced a facies consisting of peats and muddy sands. Milliman and others (1989) show that the neritic Holocene silts and clays that top this section are on the order of 2 to 5 m thick throughout much of the area.

The most salient characteristic of the Yellow Sea, which accounts for the fine-grained sediment veneer on its floor, is that it receives the sediment load of the Huanghe, which carries the second largest sediment load among world rivers (Milliman and Meade, 1983). Most of the load is deposited in the Gulf of Bohai near the present mouth of the river (Fig. 1), where land accretion is occurring at a rate of approximately 24 km^2/yr (Wang and Aubrey, 1987). About one-tenth of the total load, however, is re-introduced into the water column, most likely as a consequence of winter storms, and eventually exits the Gulf of Bohai. It is then transported around the tip of the Shandong Peninsula and into the main body of the Yellow Sea by the southward-moving Yellow Sea Cold Current (Milliman and others, 1989). Additional fine sediment is derived from the 1855 delta, again by erosion during winter storms, and carried southward along the coast (Milliman and others, 1985; Graber and others, 1989); this sediment may leave, then return to the Yellow Sea on the Korean side by entrainment into the Yellow Sea Warm Current (Graber and others, 1989). In the area proximal to and south of the tip of the Shandong Peninsula into the central region of the Yellow Sea, present rates of sediment accumulation have been estimated at near, but less than, 4 mm/yr, except for a ramp (Shandong mud-wedge "foreset beds") facing south from the peninsula. There rates may approach 9 mm/yr (Clark Alexander, pers. commun.).

Evidence that the surface texture of the sediment may be episodically governed by bottom currents generated from winter-storm waves has been presented by Graber and others (1989). Other vigorous processes that affect the sea floor have been discussed by Milliman and others (1985, 1989), including faulting just south of the Shandong Peninsula and some possible outgassing in that same area. Because notable seismic events are essentially unknown in the area, the faulting may be related to a subsidence phenomenon. The gas, possibly originating in the peaty sands in the transgressive facies beneath the Holocene neritic sediments, may locally increase sediment pore pressures, but apparently it

From Shoreline to Abyss, SEPM Special Publication No. 46

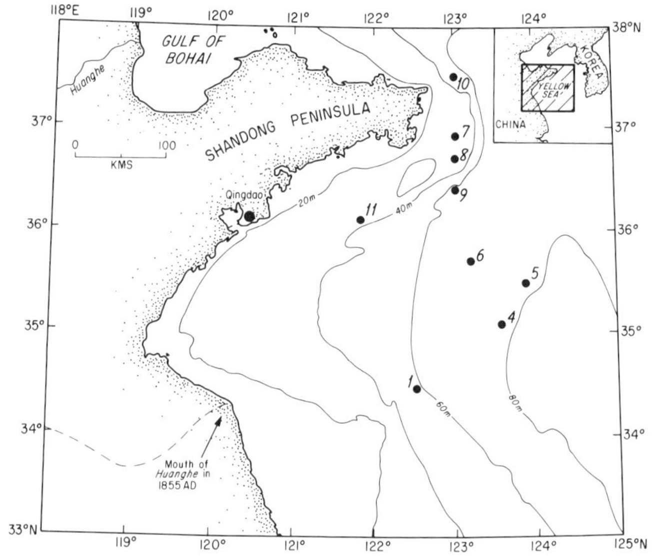

FIG. 1—Station locations and reference points. Sites 7, 8, and 9 are on the Shandong mud wedge.

has not left a regionally significant imprint on the seafloor morphology or geology. Despite the faulting and outgassing, Milliman and others (1985, 1989) found no evidence of slope failures or mass movements near the Shandong Peninsula, or, in fact, elsewhere in the Yellow Sea south of the peninsula.

Given the shallow, nearly horizontal sea floor, the tectonic stability; the limited evidence of faulting, and the apparently restricted extent of gas-charged sediment, it appears that the most consequential regional dynamic process may be the wave action generated by major winter storms and typhoons. Thus, within the contemporary oceanographic and sedimentologic framework, it was our general purpose to assess the potential of these wave forces to perform geologic work, both in terms of establishing and modifying the recent neritic-sediment veneer, and in initiating large-scale, transient, geologic events.

METHODS

Field

The sampling program was carried out aboard the *RV Science 1,* a research vessel operated by the Institute of Oceanology for Academia Sinica. Eleven Kasten cores and eight box cores were recovered from a total of nine stations (Fig. 1). The length of the Kasten cores was characteristically <3 m, the box cores <0.7 m, thereby mostly restricting the study to Holocene neritic sediments. The Kasten cores were tested for vane shear strength at sea and subsampled for later, onshore laboratory index property testing (e.g., water content, liquid limit, grain-specific gravity); subcores from box cores were stored for later triaxial and consolidation testing. Core and site information are provided in Figure 1.

Undrained shear strength (S_u) was measured at sea with a motorized vane shear device. Measurements were made at 25-cm intervals on split cores beginning at 50 cm below the top of the core (the upper 50 cm were removed for radiometric dating). Although the Kasten corer (Kuehl and others, 1985) minimizes mechanical disturbance, some disturbance probably took place during coring due to the low strength of the sediment. Thus, the strength values reported herein are probably less than the *in-situ* values. The inserted vane was rotated 360° after the initial failure and the remolded strength (S_{ur}) was measured; sensitivity (S_t) which is the "initial" shear strength to remolded shear-strength ratio, was then calculated.

Laboratory

The geotechnical index property tests were conducted as follows: water content (w) was measured in accordance with standard test method D2216-80 of the American Society for Testing and Materials (ASTM, 1985), except that the samples were dried at 50°C. Liquid limit (w_L) values were determined by the cone penetrometer method (standard test BS 1377 of the British Standards Institution [Head, 1980]). Plastic limit (w_P) was measured in accordance with standard test method D4318-84 (ASTM, 1985). Grain-specific gravity (G_s) measurements were made with an air-comparison pycnometer, in which the sample chamber was put under a vacuum, then purged with helium. The index property data were salt-corrected by assuming a salinity of 35 ‰. From these data, bulk density (ρ_t), which was used for overburden estimates (i.e., effective vertical stress, σ'_{vo}), was calculated. Estimates of the maximum past stress experienced by the sample (σ'_{vm}) and the coefficient of consolidation (c_v) were determined from constant rate-of-strain consolidation tests that were performed at the top, middle, and bottom of each subcore. Overconsolidation ratios, OCR($\sigma'_{vm}/\sigma'_{vo}$) were derived from these results. Standard test method D4186-82 (ASTM, 1985) was used. Consolidated, undrained triaxial tests were conducted on selected subcores. The method was based on the ASTM (1985) standard test for unconsolidated, undrained compressive strength (D2850-82). The effective friction angle (ϕ') was determined from the triaxial test data. Details of all testing and handling procedures are given in Winters (1988) and Booth and Winters (1989).

CONTEMPORARY GEOLOGIC ENVIRONMENT OF THE SEA FLOOR

The floor of the Yellow Sea is comprised of weak, soupy sediments. Even 0.5 m below the mud line, shear strengths were below threshold values for measurement (i.e., <0.2 kPa) of the vane shear apparatus in about one-fifth of the 68 cases. As shown in Figure 2, however, strength values are scattered about those that would be predicted, based on a typical increase in vertical stress (overburden) with subbottom depth, for a sediment undergoing normal compaction (the scatter is attributed in part to the technique and the disturbance during coring). Consequently, there is no compelling indication from the strength data that these sediments are in a state of underconsolidation (existing effective overburden stress calculated assuming no excess pore pressure is greater than maximum past effective stress, i.e., the sediment has yet to fully consolidate under the weight of the existing overburden) due to excess pore pressure from

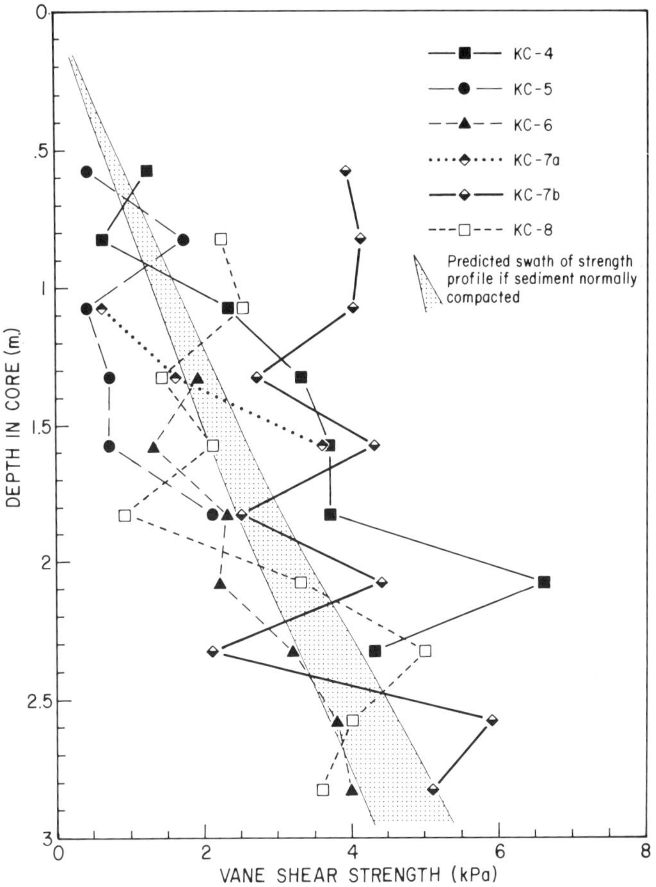

FIG. 2—Vane shear-strength profiles in the Holocene neritic sediments. Swath profile is based on assumption that strength increase with increasing overburden in a normally consolidated sediment is in the ratio of 0.24 to 0.30 (Ladd and others, 1977).

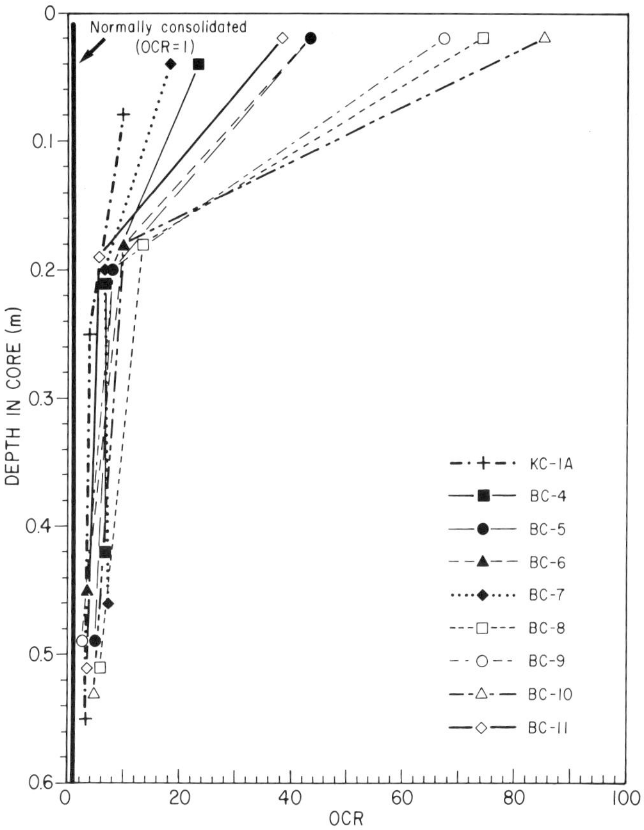

FIG. 3—OCR (ratio of maximum past effective-overburden stress to present effective-overburden stress) profiles in the upper 0.5 m. Note how profiles converge toward line representing normal compaction.

rapid deposition or the presence of gas. In fact, X-radiographs exhibit no indication of bubble structures, voids, or expansion. Similarly, a state of overconsolidation (maximum past effective stress greater than existing effective stress), which may imply, for example, that an erosional surface is present or that cements are contributing to the shear strength, is not apparent, except possibly for core 7b, which, seemingly, is only so in the upper 2 m. In general, a depositional sedimentary environment, or at least one of non-erosion, is suggested.

The consolidation data support the assertion that the seabed is fundamentally a depositional surface and is compacting normally. The overconsolidation ratios are high (10–85) at the surface, which is in accord with values typically reported for marine sediment (Silva and Jordan, 1984; Booth and others, 1989). However, they decrease rapidly and approach a value of unity (normal consolidation) within the upper 50 cm (Fig. 3).

As inferred by the grain-size liquid-limit profile overlays (Fig. 4a), liquid limit is a fairly sensitive index to modest changes in texture in these sediments. The relatively straight, vertical profiles of liquid limits (Fig. 4b) thus suggest that sediment texture is relatively invariant downcore. Accordingly, we surmise that the existing or similar local depositional environments may have been sustained at these core sites throughout recent geologic time, at least insofar as texture is an indicator. This implies that since the Holocene transgression, the depositional environment has remained stable—at least in terms of texture, despite fluctuating sedimentation rates and shifting delta sites. However, as also shown in Figure 4a, not all the cores exhibit a nearly straight vertical-texture profile. The bottom of core KC-4 is much coarser than the top, with a gradual transition in the middle. It is conceivable that the fine-grained neritic sediment may be only a veneer over parts of this region. The Holocene transgressive sequence (Milliman and others, 1989) may be present in the section that core KC-4 represents. The textural significance of liquid limit is underscored in Figure 4 in that the cores clearly fall into two groups, excluding KC-4. The profiles reflecting the lower w_L values are comprised of ≥65 percent silts and sands; the profiles with the higher w_L values are comprised of ≥50 percent clay. Also, the silty clay samples are all from water depths exceeding 60 m, whereas none of the coarser samples were recovered from these depths.

DYNAMIC SEAFLOOR PROCESSES

Slope Stability

Classic, telltale signs of mass movement, such as scarps, gouges, local unconfirmities (e.g., as created by deposition

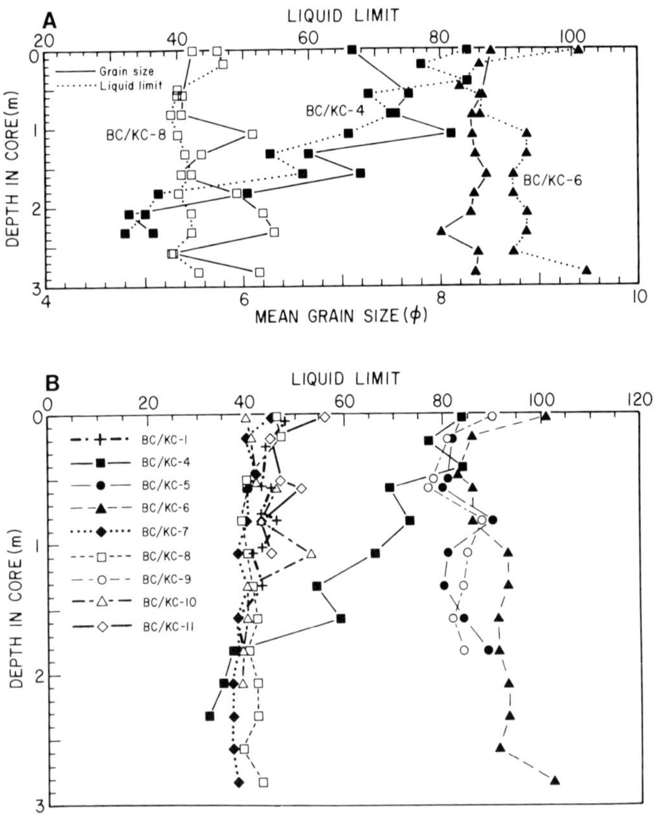

FIG. 4—Liquid-limit and grain-size profiles. The close correlation between liquid limit and grain size in these sediments is shown in the combination plot (A) for stations 4, 6, and 8. Two distinct sediment groups are evident in the liquid-limit profiles (B): the higher w_L values of the right-hand group represent finer grain sizes. Note the progressive change in BC/KC-4.

over an exposed glide "plane"), or irregular topography (e.g., a hummocky surface or lobate mass) were not observed in seismic-reflection profiles (Milliman and others, 1985; 1989) or sidescan sonar images (Milliman and others, 1985). The absence of identifiable local unconformities in seismic profiles in the Holocene neritic sediment (Milliman and others, 1989) does not necessarily imply that none are present, because changes in acoustic impedance within a thin, low-density bed may not be easily detected. Still, no unconformities were noted by the remote sensing. In addition, results of limited sidescan coverage indicate the presence of an apparently smooth sea floor (J. D. Milliman, pers. commun.), which also suggests that mass movements have not occurred. Despite the explicit imagery of mass-movement features north of the Shandong Peninsula off the Huanghe Delta presented by Prior and others (1986), it is feasible that in a nearly horizontal seabed that is comprised of weak sediment, geomorphic evidence may be difficult to detect in some cases. If true, mass movement may have occurred, but gone undetected.

Core descriptions and analyses reveal none of the typical manifestations of mass deposition either. As shown in Figures 2–4, profiles of the upper 3 m demonstrate a certain vertical homogeniety or predictability: major, abrupt, downcore changes in mass physical properties that could be attributed to coring through scarps or glide 'planes' were not encountered, nor were allochthonous clasts, sand lenses, or structureless sediments. Sensitivity values, which can be an index of fabric or structure, showed no trends or major changes in profile and averaged >3.6 (normal for marine sediments). With the exception of KC-4, texturally graded beds were not observed either. However, as above, given the rather uniform, watery sediments in conjunction with extremely low-slope angles, it may be difficult to note evidence of mass movement in the cores, particularly on a macroscale.

Has slope failure occurred? Although this cannot be addressed directly, we can examine the likelihood of slope failure in this environment. Factors that bear on slope stability and conditions that favor slope failure have been presented and discussed elsewhere (e.g., Morgenstern and Sangrey, 1978; Booth and others, 1985; Lee and Edwards, 1986) and thus will not be presented in detail here. The primary factors that would control slope stability in this particular case are slope angle, pore pressure (in excess of hydrostatic), and dynamic loading. Regional gradients within the study area are, as previously noted, extremely low, varying from about 0.2° to 0.01°, and locally steep slopes are unknown. Factor of Safety (FS) calculations (static shear resistance/gravitational shear stress) based on these slope angles, as well as on the average angle of internal friction determined from triaxial tests (≈35°) for the drained case, and on the premise of normal consolidation (assumed strength-overburden ratio, S_u/σ'_v, ≈ 0.25) for the undrained case, result in values ≥40, where a value less than unity indicates the potential for slope failure.

Extremely low seafloor gradients also typify the subaqueous portion of the Mississippi and Huanghe deltas, as well as other regions that have experienced slope failure (Prior and Coleman, 1978; Prior and others, 1986), so this fact alone does not intrinsically indicate a stable sea floor. Nonetheless, the absence of bubble structures in the cores and general absence of evidence of gas in seismic records (Milliman and others, 1989) suggest that excess pore pressure due to gas does not contribute to seabed instability. Moreover, although rates of sediment accumulation of nearly 9 mm/yr have been reported for part of the study area, this is not rapid enough for this type of sediment, which is, relative to this particular rate of deposition, too permeable and/or incompressible (coefficient of consolidation, c_v, $\approx 10^{-3}$) to permit generation of a significantly overpressured, and hence more failure-prone section (Gibson, 1958; Morgenstern, 1967). On the other hand, outside forces that can increase shear stress and elevate pore pressures, thus promoting instability, are common in the shallow Yellow Sea in the form of winter-storm- or typhoon-generated waves.

Larger scale waves have a demonstrable effect on seabed stability. Similar to earthquakes, which are rare in this stable platform, they impart a cyclic shear stress to the sediment column that may approach or exceed the original (prestorm) sediment shear resistance and, simultaneously, may degrade the inherent strength of the sediment by increasing pore pressures (thereby reducing effective stresses). Thorough investigations in the Mississippi delta region (Doyle,

TABLE 1—Hypothetical Data for Analyzing Erodibility

Hypothetical data set of wave and wave-induced bottom-current characteristics and bed stresses constructed from empirical relations between assumed wind fields and wave characteristics, linear wave theory, the Grand-Madsen wave-current bottom boundary-layer theories (Grant and Madsen, 1979; 1982; 1986), and approximations of grain-size data. Values shown for τ_b, τ_c, β_m, β_c were determined using a computer program (see Glenn [1983] for theoretical basis and program elements).

Case	d (m)	L (m)	U_b (cm/s)	τ_b (Pa)	τ_c (Pa)	β_m	β_c	Grain size
winter storm	20	121	128	4.06	.18	2.16	.09	fine to very fine sand (125 μm)
T = 10 sec	40	146	58	1.00	.18	.56	.09	fine to very fine sand (125 μm)
H_s = 5 m	60	154	27	.16	.06	.81	.19	medium silt (20 μm)
	80	156	12	.03	.06	.23	.19	medium silt (20 μm)
10-yr typhoon	25	216	240	11.51	.18	5.93	.09	fine to very fine sand (125 μm)
T = 15 sec	45	273	153	3.60	.06	12.16	.19	medium silt (20 μm)
H_s = 9 m	65	305	106	1.84	.06	6.57	.19	medium silt (20 μm)
	85	326	76	1.01	.06	3.82	.19	medium silt (20 μm)
100-yr typhoon	30	326	334	20.01	.18	10.11	.09	fine to very fine sand (125 μm)
T = 20 sec	50	407	239	7.90	.06	25.43	.19	medium silt (20 μm)
H_s = 13 m	70	461	186	4.96	.06	16.39	.19	medium silt (20 μm)
	90	506	149	3.30	.06	11.18	.19	medium silt (20 μm)

Wave attributes: T = wave period, H_s = significant wave height, L = wave length
Seabed attribute: d = water depth
Current attributes: U_b = current velocity at top of boundary layer (in free stream), τ_b = maximum bed shear stress
Shields parameters β_m = maximum value of the Shields parameter, τ_c = critical bed shear stress (to initiate movement of grain size indicated), β_c = critical value of Shields parameter (to initiate movement of grain size indicated)

1973) have shown the potential of storm waves to affect the ocean bottom, and numerous studies that deal with the theoretical aspects of the problem have been published (Seed and Rahman, 1978; Madsen, 1978). In the Yellow Sea case, we note that for different-size storms, whose wave characteristics are shown in Table 1, the moment associated with the trough-crest pressure differential may cause shear stresses that approach the vane shear strengths in the upper 2 m (Fig. 5): the winter storm may develop shear stresses of about 50 percent of the observed vane shear strength; the larger typhoon approaches 70 percent. As indicated in Figure 5, even at a representative thickness for these Holocene sediments (about 5 m, as shown by Milliman and others, 1989), the shear stresses are an appreciable percentage (>50 percent for the two types of typhoons) of the vane shear strength. Furthermore, the associated increase in pore pressure combined with the strain that results from cyclic loading can degrade shear resistance considerably. This is especially true in these relatively impervious sediments, assuming that the duration of the storm is several hours (>1,000 cycles). Estimating the amount of strength degradation is tenuous. Strength actually may increase in some sediments (Lee and Edwards, 1986). More likely, strength may decrease from 10 to 50 percent or more (Andersen, 1976; Lee and Focht, 1976; Lee and Edwards, 1986). Accordingly, we suggest that slope failure may occur in these sediments under certain storm conditions.

On virtually level terrain composed of weak sediments, as heretofore discussed, such failures may be almost imperceptible in acoustic records in some cases, or even in cores, if the failed sediment mass remains intact. Actual movement of the seabed may be nil (a parautochthonous slide), and surface scars, due primarily to the presence of the relatively plastic, soupy sediments, may be "healed" or somewhat masked in a relatively short period of time. If the sediment liquified or disintegrated upon failure, however, the resulting flows or collapse features might be more conspicuous, such as those on the subaqueous Huanghe delta,

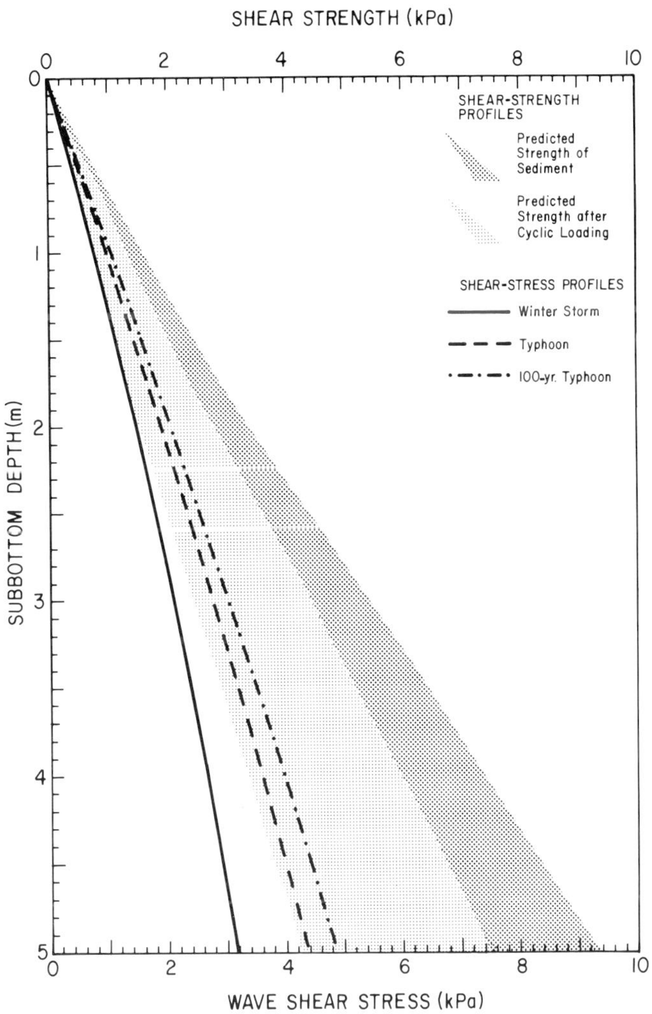

FIG. 5—Storm wave-induced subbottom shear-stress profiles (τ_w), swath profile of predicted strengths, and swath profile of reduced strength (after cyclic loading) in the upper 5 m of sediment.

which, in fact, may have been triggered by storm waves (Prior and others, 1986).

Based on concepts and methods presented in Poulos (1981), Poulos and others (1985) and Whitman (1985), the near-surface sediment in the Yellow Sea does not appear to be susceptible to liquefaction (herein liquefaction refers to the process of reduction in shear strength of a sediment mass and the consequent fluidity of the failed mass). One of Whitman's criteria for this disintegration (i.e., liquefaction) is that the sediment behaves as what he labels as a State I sediment. This type of behavior is defined (from triaxial tests) on the basis of a stress path in p′-q (mean effective normal stress-shear stress) space (Fig. 6). A sediment in this state, in which it would be if its stress path were to continue to bend to the left (into the more darkly shaded area), will exhibit a loss in shearing resistance once the peak (fundamental) shear strength has been mobilized and can continue to deform ("fail") as long as the driving shear stress is greater than the resulting, diminished shear strength (steady-state shear strength). According to Poulos and others (1985), a sediment in State I is clearly contractive under shear stresses that exceed shear resistance. All 19 triaxial tests conducted on these sediments indicate the presence of a marginally contractive to dilative material. They are in State II (also defined in Fig. 6). A sediment in State II continues to retain or increase its strength with increasing stress even though large strains (deformation) may be developing (in Fig. 6, the stress path would move into the more lightly shaded area). Unlike State I, therefore, a reduction in stress would result in a cessation of strain after the removal of the cyclic loading, and disintegration would not take place. Representative results of the triaxial tests are plotted on Figure 6. As indicated, some of the samples,

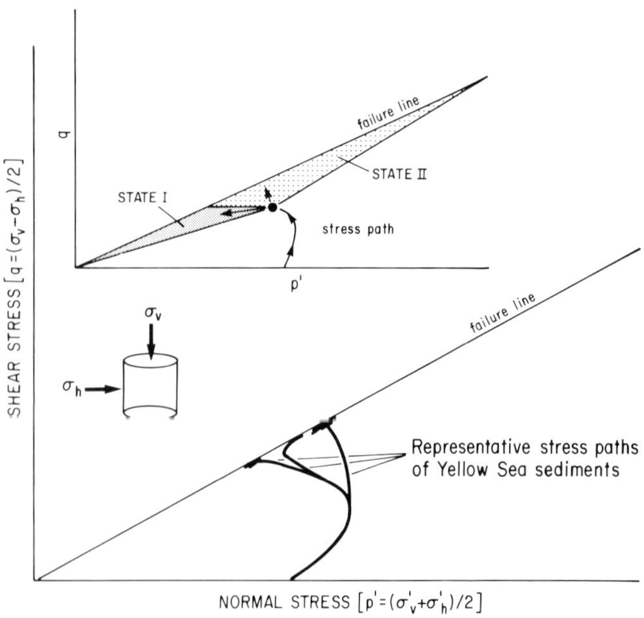

FIG. 6—Representative stress paths from consolidated isotropic undrained triaxial tests. Shear stress $[(\sigma'_v-\sigma'_h)/2]$ continued to increase or remain nearly constant throughout the duration of the test in all cases. σ'_v and σ'_h refer to vertical and horizontal effective stress, respectively. INSERT: Definitions of Whitman's (1985) State I and State II soil behavior.

although all are in State II, are close to the state I-II boundary. Estimating the position of this state boundary is somewhat subjective in that *in-situ* lateral stresses are unknown. It is evident, however, that none of the samples is unequivocally contractive and thus potentially susceptible to collapse and flow.

This nondisintegrative behavior is significant because cyclic loading through wave action is commonplace on the floor of the Yellow Sea and, as noted, may well cause slope failure (Fig. 5). After the storm passes, however, stress, as would be the case before a storm, would be imparted by gravity alone. Assuming that the sediment did not convert to State I during the cyclic loading (a phenomenon which may have happened in the case of some Gulf of Alaska shelf sediments [Schwab and Lee; 1988]), this gravity-induced stress would be unable to sustain the deformation of an initial slope failure: a disintegrative slide, which may manifest itself as a depression or collapse structure, or possibly as a perceptible flow feature, would likely not occur and slope failure, if occurring, may remain somewhat inconspicuous. It is therefore likely that although the wave-induced shear stresses and concomitant reduction in strength may have a profound effect on the Holocene section, there is apparently no profound effect on the geologic record.

Erosion

Waves associated with winter storms and typhoons not only generate cyclic shear stresses within the sediment that can promote slope failure, but also induce currents that can erode the bed surface as well. Graber and others (1989) have shown that the passage of a typical cold front (15 m/s winds) in the vicinity of the Yellow Sea can produce waves with periods up to 10 s and heights up to 5 m. Based on Shields criterion, which is used to relate current velocity to the threshold of grain movement, they determined that the wave-induced bed shear stresses from a storm of this intensity would be sufficient to entrain and suspend fine sand and coarse silt over a wide area; the latter to a water depth of 50 m. Because cold fronts are frequent throughout the winter, Graber and others (1989) suggest that they may be an effective agent in southward transport, although suspended sediment may eventually be returned to the Yellow Sea by the Taiwan Warm Current.

In addition to the likely frequent attacks by winter storms, the seafloor sediments also can be agitated in response to typhoon-generated waves. Tracks of recent tropical storms (wind speeds from 35–64 kts [18–33 m/s]) and typhoons ($>$65 kts [34m/s]) in the region of the Yellow Sea are shown in Figure 7. These types of storms, particularly if duration and fetch are ample, are capable of generating seas that influence the bottom sediments. Table 1 is a hypothetical data set that has been developed to estimate such effects. The table represents the "neutral case" in that the values of the maximum bed shear stress (τ_b) and the maximum value of the Shields parameter (β_m) assume an undisturbed, smooth, and sterile (no biological effects) bed, i.e., no suspended sediment, fluid-mud layer, or change in bed characteristics due to increasing velocities as their maxima are approached. Accordingly, the maximum values presented

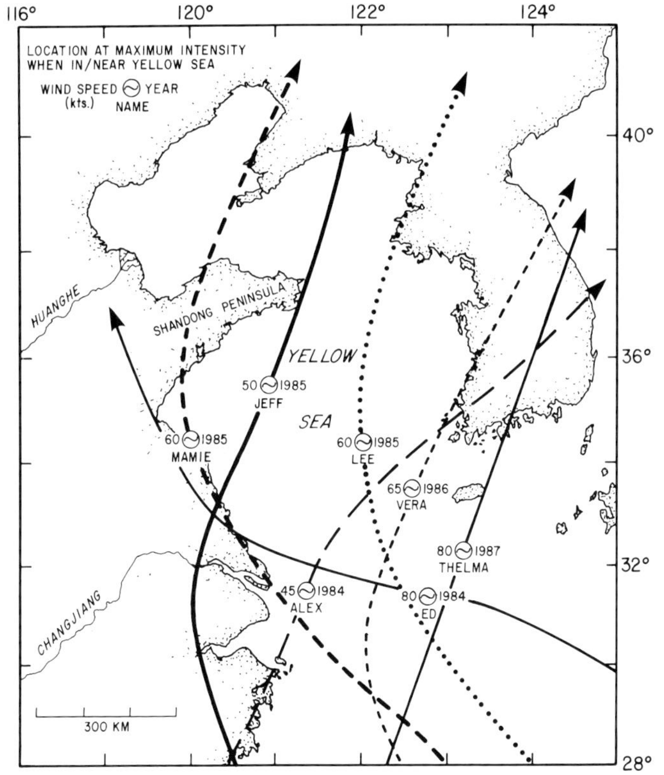

FIG. 7—Tracks of selected, recent tropical storms and typhoons in the Yellow Sea (data from Joint Typhoon Warning Center, 1984–1988).

should be considered first-order approximations only. The bottom velocities (i.e., in the free stream immediately above the turbulent boundary layer) given in Table 1 were computed from linear wave theory using the following equation:

$$U_b = \frac{H_s g T}{2L} \cdot \frac{1}{\cosh(2\pi d/L)}$$

where U_b is the wave-induced current velocity at the sea floor, H_s is the significant wave height, g is the acceleration due to gravity, T is the wave period, L is the wave length, and d is the water depth. Bed shear stresses are derived from bottom velocities, and were determined from:

$$\tau_b = \frac{1}{2} \rho_w f U_b^2$$

where ρ_w is the density of sea water (taken as 1.025 g/cm^3) and f is the wave-friction factor, which depends only on the ratio of the physical bottom roughness to the near-bottom wave excursion (see Grant and Madsen, 1982).

Cohesionless sediment.—The values of bed shear stress may be used with respect to both cohesionless and cohesive sediment. For cohesionless sediment they may be applied in the form of the Shields parameter:

$$\beta = \frac{\tau_b}{(S - 1)\rho_w g D}$$

where S is the ratio between grain and seawater density, and D is grain diameter.

As discussed by Mantz (1977), the Shields criterion may be valid for silts down to a size of perhaps 15 μm, as well as coarser grains. Accordingly, the application of the Shields criterion was extended to include the 20-μm grain size, which approximates a prominent textural mode in Yellow Sea sediments. A mode at 125 μm, also prominent in Yellow Sea sediments, especially in shallower water, was also selected for analysis. As shown in Table 1, the values of β_m exceed β_c (the critical value for erosion, or threshold of grain movement) for the winter-storm case at all water depths, although below 60 m movement of bottom sediment may be problematical because of the approximate nature of the calculations and uncertainties associated with the Shields model. During typhoons, however, the resuspension threshold is readily exceeded over the entire range of depths. The data from Table 1 are presented graphically in Figure 8a.

Cohesive sediment.—The erodibility of the seabed, particularly at water depths where silts and clays predominate (≥40 m), may be controlled by cohesion: the Shields criterion would not apply. Specifically, if the fine fraction behaves as a cohesive material because a significant fraction of the sediment is finer than the grain-size limit for noncohesive behavior (15 μm) experimentally determined by Mantz (1977), or because the grains are fundamentally more

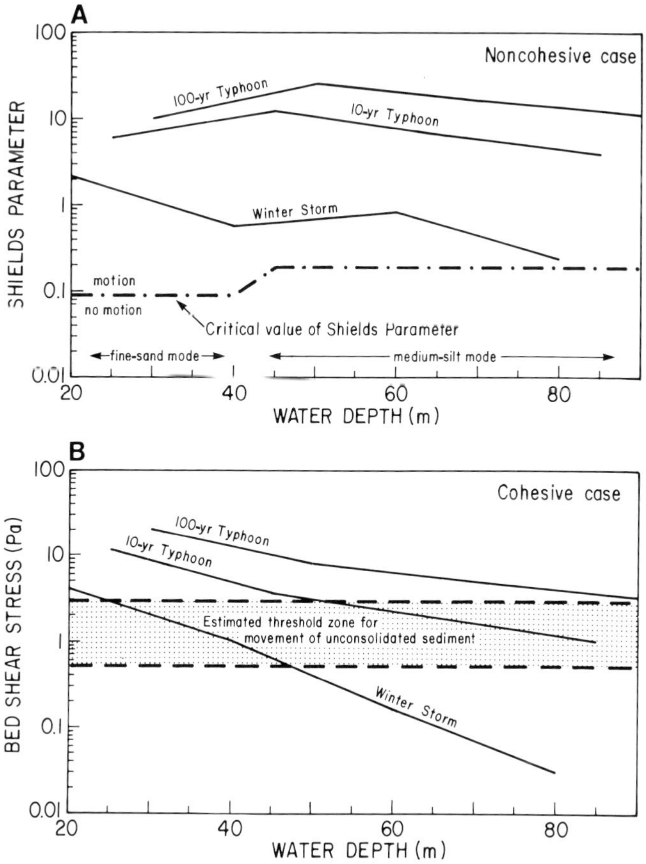

FIG. 8—Erodibility of Yellow Sea bottom sediments as a function of storm type (i.e., wave characteristics) and water depth (see Table 1 for data). Values of the Shields parameter and bed shear stress associated with the different storms and depths are considered rough estimates.

cohesive than the relatively inert grains used by Mantz, the erodibility of the sediment must be examined in terms of cohesion. Viewing erodibility in an analogous manner to factor of safety, we suggest that cohesive sediments erode as a function of the ratio between S_e (shear strength with respect to erosion) and τ_b; if the bed shear stress exceeds its shearing resistance with respect to erosion ($S_e/\tau_b < 1$), erosion may take place. Although we can estimate τ_b in our study area (Table 1), we cannot estimate S_e. However, based on numerous flume experiments, fines within the upper few millimeters of sediment were shown to erode at bed shear stresses $\approx$ 0.5 Pa (Parchure and Mehta, 1985; Partheniades, 1986; Maa and Mehta, 1987). The value for S_e tended to increase with salinity, consolidation time, and cohesiveness (i.e., increased quantities of more 'active' clay minerals), but decrease under wave action. Furthermore, an indication of bed shear stresses required to erode sediment below the initial mud-line has been provided by Kamphuis and Hall (1983), who consolidated samples to the equivalent pressure of 5 to 10 m of overburden ($\sigma'_v \approx 50$ kPa) or more and experimentally showed that the erodibility of cohesive sediments in unidirectional flow is linearly related to shear resistance. At the lowest consolidation states, they showed that erosion took place at bed shear stresses of $\approx$5 to 10 Pa, and suggested that the value may be somewhat less at zero consolidation pressure (gravity and cohesion still must be overcome). On the basis of the cited flume studies, we estimate that the threshold of erosion for surficial cohesive sediment (S_e) generally lies in the range of 0.5 to 3 Pa. If this estimate were applied to the Yellow Sea sediments (the shaded zone in Figure 8b), it would appear that, for all but water depths below 45 to 50 m in the winter-storm case, the uppermost layers (cm scale) of sediment may be susceptible to erosion and that the potential for erosion may extend much beyond the upper few centimeters. On the other hand, in the winter-storm case the sediments may only be in motion at water depths <25 m (Fig. 8b), and the wave-induced bottom currents associated with the 10-yr typhoon only may be eroding down to depths of 50 m (Fig. 8b).

Despite the fact that the upper few centimeters may readily erode under many of the aforementioned circumstances, core evidence does not suggest that the most recent episode of erosion (and redeposition) could have affected the seabed on a scale of more than a few centimeters. Bioturbation, particularly in the form of burrows, is extensive in these sediments and extends throughout the length of the cores. Accordingly, although recolonization times, and hence, an indication of the time required to re-burrow (homogenize) the sediment, are unknown, a recent, major episode of erosion (of several centimeters or more) and redeposition would seem unlikely. The fact that bioturbation may obliterate the effects of less recent events compounds the problem of recognition in general. The cores are apparently normally consolidated, have normal sensitivities, and show no pattern or noteworthy textural differences between their surfaces and subsurfaces that would indicate emplacement of (a) storm deposit(s), which also suggests that no recent event has taken place. The sedimentation rates, of 9 mm/yr or less, belie any suggestion of recent, *en masse* deposition as a product, for example, of mass movement.

If there has been no major downcutting event, it may be because the rate of erosion is not great enough to allow a single storm, having a duration of hours, to remove a vertically significant amount of material. Partheniades (1965, 1986), Mehta and Partheniades (1979), Parchure and Mehta (1985), Sheng (1986), and Maa and Mehta (1987) have reported a range of erosion rates from 10^{-4} to 10^{-6} g/cm^2/s for specific experiments on fine-grained sediments at bed shear stresses that were <3 Pa. Assuming that these rates are useful as order-of-magnitude approximations for our case, we speculate that a few centimeters, at most, would be removed by an individual storm in the Yellow Sea.

If the storm waves are somewhat limited in their ability to cut vertically into the seabed, even in the presence of the continuous, unidirectional currents in the Yellow Sea that can carry entrained material away from a specific site, then it would seem possible that an environment of net erosion may be as much a function of the number of storms as it is individual storm intensities. We surmise, given that 1) most Huanghe sediment is trapped in the Yellow Sea (Milliman and others, 1987), 2) there are positive accumulation rates, and 3) geotechnical data indicate a depositional, normally compacted seabed, that the possibly severe erosion associated with storms is transient. Redeposition follows immediately. The effects of the event are not preserved.

SUMMARY

The seabed near and south of the Shandong Peninsula is a depositional environment. It is comprised of weak, yet normally compacted, sediment that has apparently been deposited under the same or similar marine conditions throughout most of the Holocene. The floor of the Yellow Sea, being virtually flat (regional slopes of 0.2°–.01°), is not inherently prone to slope failure. This is supported by field evidence, which is characterized by an absence of hummocky terrain, lobate features, scarps, or unconformities in the acoustic records (Milliman and others, 1985, 1989); laboratory data, which showed that the upper 3 m are vertically homogeneous or predictable and that sharp, downcore discontinuities in mass physical properties that could be indicative of scarps or glide planes were absent, as were allochthonous clasts, sand lenses, or other possible manifestations of slides or flows; and static slope-stability analysis, which shows, even in the worst case, that slope failure is highly improbable.

Waves are the dominant geologic agent in the Yellow Sea. Shear stresses generated by large waves during major winter storms and typhoons may reach 50 to 70 percent or more of the sediment shear strength, and, coupled with their capability to degrade shear resistance by generating excess porewater pressure, could possibly produce slope failures. If the failed beds remained intact, however, there would probably be little evidence of such a failure, particularly on a macroscale. If the sediments tended to liquify or disintegrate, collapse features, or possibly some minor flows, would be evident. These sediments do not appear to be in State I, though, which is a requisite for disintegration (Whitman, 1985).

Storm waves also can generate high-velocity, oscillatory bottom currents and, therefore, bed shear stresses capable of eroding the seabed. Analyses of erosion potential for both the cohesionless sediment (Shields criterion) and cohesive sediment cases suggest that typhoons could erode the surface sediments at all water depths and winter storms to water depths of 45 to 60 m. However, if the sediment is fundamentally cohesive, in the winter-storm case the bed shear stress only may exceed its strength with respect to erosion in water depths <25 m; and in the 10-yr typhoon case, it may be eroding to depths <50 m. Because winter storms and typhoons are frequent, we assume that part, if not most, of the sea floor is repeatedly scoured. Also, it is likely that despite this frequent dynamic interaction between bottom currents and sediments, there is little net effect on the geological attributes of the seabed, except for a constant reworking of the top few centimeters.

In terms of event stratigraphy, as inferred by Dott (1983), a 100-yr typhoon is not a rare event, but even events much rarer and more cataclysmic may leave only a subtle geologic imprint of their passing. Furthermore, we believe that the chance for preservation of that imprint is in jeopardy: the record we see only represents geologic time since the last rise in sea level.

ACKNOWLEDGMENTS

This investigation received its primary support from the National Science Foundation and the Academia Sinica of the Peoples Republic of China through the Woods Hole Oceanographic Institution (John D. Milliman, Principal Investigator) and the Institute of Oceanology at Qingdao (Professor Yun-shan Qin, Chief Scientist).

We most gratefully acknowledge the invitation, support and cooperation of John Milliman. We also acknowledge the assistance and cooperation of Charles A. Nittrouer (State University of New York—Stoney Brook) and David J. DeMaster (North Carolina State University), as well as the field assistance of Ross Elliott. Clark Alexander (Skidaway Institute of Oceanography) graciously provided core X-radiographs and grain-size data. David E. Drake (U.S. Geological Survey, Menlo Park) was responsible for the computer analysis that led to the data shown in Table 1, and contributed to the interpretation of those data.

REFERENCES

American Society for Testing and Materials, 1985, 1985 Annual book of ASTM standards, v. 4.08: Soil and rock; Building stones: Philadelphia, ASTM, 1078 p.

Andersen, K. H., 1976, Behavior of clays subjected to undrained cyclic loading: Proceedings, First International Conference, Behavior of Off-Shore Structures, Trondheim, Norway, p. 392–433.

Booth, J. S., Sangrey, D. A., and Fugate, J. K., 1985, A nomogram for interpreting slope stability of fine-grained deposits in modern and ancient marine environments: Journal of Sedimentary Petrology, v. 55, p. 29–36.

Booth, J. S., and Winters, W. J., 1989, Geotechnical description of Yellow Sea sediments: U.S. Geological Survey Open-File Report 89-149, 265 p.

Booth, J. S., and Winters, W. J., Poppe, L. J. Neiheisel, James, and Dyer, R. S., 1989, Geotechnical, geological, and selected radionuclide retention characteristics of the radioactive waste disposal site near the Farallon Islands: Marine Geotechnology, v. 8, p. 111–132.

Dott, R. H., Jr., 1983, The 1982 Presidential Address: Episodic sedimentation—How normal is average? How rare is rare? Does it matter?: Journal of Sedimentary Petrology, v. 53, p. 5–23.

Doyle, E. H., 1973, Soil-wave tank studies of marine soil instability: Proceedings, Fifth Offshore Technology Conference, v. 2, p. 753–766.

Glenn, S. M., 1983, A continental shelf bottom boundary layer model: The effects of waves, currents, and a moveable bed: Unpublished Sc.D. Thesis, Massachusetts Institute of Technology, Cambridge, Massachusetts, 275 p.

Gibson, R. E., 1958, The progress of consolidation in a clay layer increasing in thickness with time: Geotechnique, v. 8, p. 171–182.

Graber, H. C., Beardsley, R. C., and Grant, W. D., 1989, Storm-generated surface waves and sediment resuspension in the East China and Yellow Seas: Journal of Physical Oceanography, v. 19, p. 1039–1059.

Grant, W. D., and Madsen, O. S., 1979, Combined wave and current interaction with a rough bottom: Journal of Geophysical Research, v. 84, no. C4, p. 1797–1808.

Grant, W. D., and Madsen, O. S., 1982, Movable bed roughness in unsteady oscillatory flow: Journal of Geophysical Research, v. 87, no. C1, p. 469–481.

Grant, W. D., and Madsen, O. S., 1986, The continental-shelf bottom boundary layer: Annual Review of Fluid Mechanics, v. 18, p. 265–305.

Head, K. H., 1980, Manual of Soil Laboratory Testing, V. 1: Soil classification and compaction tests: Pentech Press, London, p. 69–73.

Joint Typhoon Warning Center, 1984–1988, Annual tropical cyclone report, 1984 through 1988: U.S. Naval Oceanography Command Center, Guam.

Kamphuis, J. W., and Hall, K. R., 1983, Cohesive material erosion by unidirectional current: Journal of Hydraulic Engineering, American Society of Civil Engineers, v. 109, p. 49–61.

Keuhl, S. A., Nittrouer, C. A., DeMaster, D. J., and Curtin, T. B., 1985, A long, square-barrel gravity corer for sedimentological and geochemical investigation of fine-grained sediments: Marine Geology, v. 62, p. 365–370.

Ladd, C. C., Foott, R., Ishihara, K., Schlosser, F., and Poulos, H. G., 1977, Stress-deformation and strength characteristics: State-of-the-Art Reports, Ninth Annual International Congress on Soil Mechanics and Foundation Engineering, Tokyo, p. 421–494.

Lee, H. J., and Edwards, B. D., 1986, Regional method to assess offshore slope stability: Journal of the Geotechnical Engineering Division, American Society of Civil Engineers, v. 112, p. 489–509.

Lee, K. L., and Focht, J. A., Jr., 1976, Strength of clay subjected to cyclic loading: Marine Geotechnology, v. 1, p. 165–185.

Liu, Zhenxia, Huang Yichang, and Zhang Qinian, 1989, Tidal current ridges in the southwestern Yellow Sea: Journal of Sedimentary Petrology, v. 59, p. 432–437.

Maa, P.-Y., and Mehta, A. J., 1987, Mud erosion by waves: A laboratory study: Continental Shelf Research, v. 7, p. 1269–1284.

Madsen, O. S., 1978, Wave-induced pore pressures and effective stresses in a porous bed: Géotechnique, v. 28, p. 377–393.

Mantz, P. A., 1977, Incipient transport of fine grains and flakes by fluids: Journal of the Hydraulics Division, American Society of Civil Engineers, v. 103, p. 601–615.

Mehta, A. H., and Partheniades, Emmanuel, 1979, Kaolinite resuspension properties: Journal of the Hydraulics Division, American Society of Civil Engineers, v. 105, p. 411–416.

Milliman, J. D., and Meade, R. H., 1983, World-wide delivery of river sediment to the oceans: Journal of Geology, v. 91, p. 1–21.

Milliman, J. D., Beardsley, R. C., Yang, Z. S., and Limeburner, Richard, 1985, Modern Haunghe-derived muds on the outer shelf of the East China Sea: Identification aand potential transport mechanisms: Continental Shelf Research, v. 4, p. 175–188.

Milliman, J. D., Qin, Y. S., and Butenko, J., 1985, Geohazards in the

Yellow Sea and East China Sea: Proceedings, 17th Offshore Technology Conference, p. 73–81.

MILLIMAN, J. D., QIN, Y. S., REN, M. E., AND SAITO, YOSHIKI, 1987, Man's influence on the erosion and transport of sediment by Asian rivers: The Yellow River (Huanghe) example: Journal of Geology, v. 95, p. 751–762.

MILLIMAN, J. D., QIN, Y. S., AND PARK, Y. A., 1989, Sediments and sedimentary processes in the Yellow and East China Seas, *in* Taira, A., and Masuda, F., eds., Sedimentary Facies in the Active Plate Margin: Terra Scientific Publishing Company, Tokyo, p. 233–249.

MORGENSTERN, N. R., 1967, Submarine slumping and the initiation of turbidity currents, *in* Richards, A. F., ed., Marine Geotechnique: University of Illinois Press, Urbana, p. 189–220.

MORGENSTERN, N. E., AND SANGREY, D. A., 1978, Methods of stability analysis, *in* Landslides: Analysis and Control; National Research Council, Transportation Research Board, Washington, D.C., p. 155–171.

PARCHURE, T. M., AND MEHTA, A. J., 1985, Erosion of soft cohesive sediment deposits: Journal of Hydraulic Engineering, American Society of Civil Engineers, v. 111, p. 1308–1326.

PARTHENIADES, EMANUEL, 1965, Erosion and deposition of cohesive soils: Journal of the Hydraulics Division, American Society of Civil Engineers, v. 91, p. 105–139.

PARTHENIADES, EMANUEL, 1986, A fundamental framework for cohesive sediment dynamics, *in* Mehta, A. J., ed., Estuarine Cohesive Sediment Dynamics: Springer-Verlag, New York, p. 219–250.

POULOS, S. J., 1981, The steady state of deformation: Journal of the Geotechnical Engineering Division, American Society of Civil Engineers, v. 107, p. 553–562.

POULOS, S. J., CASTRO, GONZALO, AND FRANCE, J. W., 1985, Liquefaction evaluation procedure: Journal of the Geotechnical Engineering Division, American Society of Civil Engineers, v. 111, p. 772–791.

PRIOR, D. B., AND COLEMAN, J. M., 1978, Disintegrating retrogressive landslides on very low-angle subaqueous slopes: Marine Geotechnology, v. 3, p. 37–60.

PRIOR, D. B., YANG, Z.-S., BORNHOLD, B. D., KELLER, G. H., LU, N. Z., WISEMAN, W. J., JR., WRIGHT, L. D., AND ZHANG, J., 1986, Active slope failure, sediment collapse, and silt flows on the modern subaqueous Haunghe (Yellow River) delta: Geo-Marine Letters, v. 6, p. 85–95.

SCHWAB, W. C., AND LEE, H. J., 1988, Causes of two slope-failure types in continental-shelf sediment, northeastern Gulf of Alaska: Journal of Sedimentary Petrology, v. 58, p. 1–11.

SEED, H. B., AND RAHMAN, M. S., 1978, Wave-induced pore pressure in relation to ocean floor stability of cohesionless soils: Marine Geotechnology, v. 3, p. 123–150.

SHENG, Y. P., 1986, Modeling bottom boundary layer and cohesive sediment dynamics in estuarine and coastal waters, *in* Mehta, A. J., ed., Estuarine Cohesive Sediment Dynamics: Springer-Verlag, New York, p. 43–65.

SILVA, A. J., AND JORDAN, S. A., 1984, Consolidation properties and stress history of some deep sea sediments, *in* Denness, Bruce, ed., Seabed Mechanics: International Union of Theoretical and Applied Mechanics, London, p. 25–40.

WANG, YIN, AND AUBREY, D. G., 1987, The characteristics of the China coastline: Continental Shelf Research, v. 7, p. 329–349.

WHITMAN, R. V., 1985, On liquefaction: Proceedings, 11th International Conference on Soil Mechanics and Foundation Engineering, San Francisco, v. 4, p. 1923–1926.

WINTERS, W. J., 1988, Geotechnical testing of marine sediment: U.S. Geological Survey Open-File Report 88-36, 52 p.

SLOPE MORPHOLOGY OF THE U.S. PACIFIC CONTINENTAL MARGIN

Donn S. Gorsline
Department of Geological Sciences, University of Southern California, Los Angeles, California 90089-0740

ABSTRACT: Profiles of slope-normal gradients of continental slopes of the U.S. Pacific margin can be related to the large-scale structural framework of the margin, and to the secondary effects of sedimentation. Tabulations of average gradient, steepest gradient, slope width and the relation of steepest segment to midslope depth define three major slope provinces.

The first two regions form the area extending from the Agua Blanca Fault Zone and its offshore extensions off northern Baja California to the Mendocino Fracture Zone. The boundary between region 1 and region 2 has been taken at Point Conception. The boundary separates 1) the outer steep-slope province off the California Borderland, the Patton Escarpment, and 2) a northern province with gentler declivities, usually because of modification by deposition. Both provinces show possible effects of the intersection of oceanic fracture zones with the margin. The northern province (region 3), from Cape Mendocino to Juan de Fuca Strait, can be subdivided into two subregions approximately at Heceta Bank, Oregon.

The major provinces correspond on the first-order level to the two main modes of margin interaction with the adjacent ocean plates, underthrusting and transform motion. Second-order factors are the location of fracture zones, and deposition.

Within the California Continental Borderland, the shallower base-of-slope depths and narrower shelves produce slopes that have lower relief and are generally steeper than the main continental slopes. These are secondarily modified by depositional processes and mass failure. There is a general increase in depth of slope base and in steepness to the south, illustrating the regional depression to the south, as well as the decrease in sediment supply in that same direction.

It is evident that regional characteristics of the morphology of slopes is a function of the structural setting and of the local sediment supply.

INTRODUCTION

Objectives

Professor F. P. Shepard, throughout his career, was interested in the morphology of the sea floor. He documented the general character of the morphology of regions and the global ocean in the various editions of *Submarine Geology* (e.g., 1973) and such frequently cited monographs as the study of the margin off southern California (Shepard and Emery, 1941). His book with Dill on submarine canyons (Shepard and Dill, 1966) is a later example. His data on average gradients of slopes, width of shelves, and so forth have been widely cited.

Like Professor Shepard, my own interest in margin morphology reflects the starting point of all marine geological studies: an adequate morphologic framework into which deposits, processes and development can be fitted. Examples include Shepard's (1948) careful description of the morphology of the various world provinces, and the work by Heezen and others (1959) on North Atlantic Ocean morphology and development of a global bathymetric classification scheme (Inman and Nordstrom, 1971).

Sources of Data

The National Ocean Service (NOS) of the National Ocean and Atmosphere Administration (NOAA), and the U. S. Geological Survey (USGS) have joined forces to produce an excellent series of metric bathymetric maps of the United States margins at scales of 1:250,000 and 1:1,000,000, using contour intervals of 50 m. Table 1 shows the charts used in this study. The next generation of NOAA bathymetric maps, which are just beginning to appear, is based on swath mapping using the SEABEAM system of multi-transducer acoustic arrays.

As is true of all marine charts, the data for any given chart may come from surveys conducted over a wide range of dates and with equipment of several levels of precision and accuracy. In addition, the contouring interval selected and the scale of the chart will limit the representation of very steep slopes (closely spaced contour lines). The contours are subject to the biases or protocols involved in the chart construction and these may introduce additional sources of error (e.g., the classic study of Veatch and Smith, 1939). These factors apply to new multi-beam methods as well as to the older single-beam sounding methods. Chart scale is the limiting factor in any contour presentation.

For this study, 120 profile transects were arbitrarily positioned along the outer slope of the U.S. Pacific margin from about the United States-Mexico border to the United States-Canada border. This yielded an average spacing of about 15 km along the roughly 2,200-km length of the slope. The transects were purposely arranged to avoid obvious submarine canyons, mass-failure scars, and sharp slope reentrants produced by possible tectonic offset. The primary purpose was to define the morphology of the undissected slope.

Profile locations are shown in Figure 1, and the general profile data have been tabulated in Table 2. Off southern California, the margin is broad and is the product of a complex structural history (Gorsline and Teng, 1989). The main profile set was extended along the outer margin of the Borderland at the Patton Escarpment, the true edge of the continental margin.

For comparative purposes, a second series of profiles was compiled for Borderland basin slopes within the outer slope (Patton Escarpment) noted above. Forty-three profiles were measured and constructed along the mainland shelf and slope (19 profiles), and in the central Borderland to include slopes of banks and islands (24 profiles). The locations are shown in Figure 2, and a summary of the profile data is given in Table 2.

This paper deals only with the U.S. Pacific margin slope. Emery and Uchupi (1984) have written a masterful analysis of the general margin morphology for the Atlantic region. Heezen and others (1959) included in their illustrations and discussion several profiles of Atlantic slopes that are parts of transocean transects.

From Shoreline to Abyss, SEPM Special Publication No. 46

TABLE 1.—REFERENCE BATHYMETRIC MAPS

Map Title	Number	Scale
PACIFIC MARGIN:		
The Rampart	NOS/NH 11-4, NOAA/USGS	1:250,000
Bushnell Knoll	NOS/NH 11:1, NOAA/USGS	1:250,000
Patton Ridge	NOS/NI 10-12, NOAA/USGS	1:250,000
San Clemente Island	NOS/NI 11-10, NOAA/USGS	1:250,000
San Diego	NOS/NI 11-11, NOAA/USGS	1:250,000
Santa Ana	NOS/NI 11-8, NOAA/USGS	1:250,000
Long Beach	NOS/NI 11-7, NOAA/USGS	1:250,000
Los Angeles	NOS/NI 11-4, NOAA/USGS	1:250,000
Santa Rosa Island	NOS/NI 10-9, NOAA/USGS	1:250,000
Santa Maria	NOS/NI 10-6, NOAA/USGS	1:250,000
San Louis Obispo	NOS/NI 10-3, NOAA/USGS	1:250,000
Sur Canyon	NOS/NI 10-2, NOAA/USGS	1:250,000
San Francisco	NOS/NJ 10-8, NOAA/USGS	1:250,000
Arena Canyon	NOS/NJ 10-4, NOAA/USGS	1:250,000
Noyo Canyon	NOS/NJ 10-1, NOAA/USGS	1:250,000
Santa Rosa	NOS/NJ 10-5, NOAA/USGS	1:250,000
Ukiah	NOS/NJ 10-2, NOAA/USGS	1:250,000
Eureka	NOS/NK 10-10, NOAA/ USGS	1:250,000
Crescent City	NOS/NK 10-7, NOAA/USGS	1:250,000
Monterey	NOS/NJ 10-12, NOAA/USGS	1:250,000
Smith River	NOS/NK 10-7, NOAA/USGS	1:250,000
Coos Bay	NOS/NK 10-4, NOAA/USGS	1:250,000
Cape Disappointment	NOS/NL 10-4, NOAA/USGS	1:250,000
Tillamook Bay	NOS/NL 10-7, NOAA/USGS	1:250,000
Newport	NOS/NL 10-10, NOAA/USGS	1:250,000
Cape Flattery	NOS/NM 10-10, NOAA/ USGS	1:250,000
Umpqua River/Cape Ferrelo	NOS/1308N-17	1:250,000
Pt. St. George/Pt. Delgada	NOS/1308N-12	1:250,000
Santa Barbara/Huntington Beach	NOS/1206N-15	1:250,000
Tillamook Head/Heceta Head	NOS/1308N-22	1:250,000
Pt. Reyes/Tolo Bank	NOS/1307N-18	1:250,000
Pt. Conception/Cape San Martin	NOS/1306N-20	1:250,000
Huntington Beach/Punta Sal Si Puedes	NOS/1206N-16	1:250,000
Santa Rosa Island	NOS/1306N-19	1:250,000
Central California	NOS/1307N-11	1:250,000

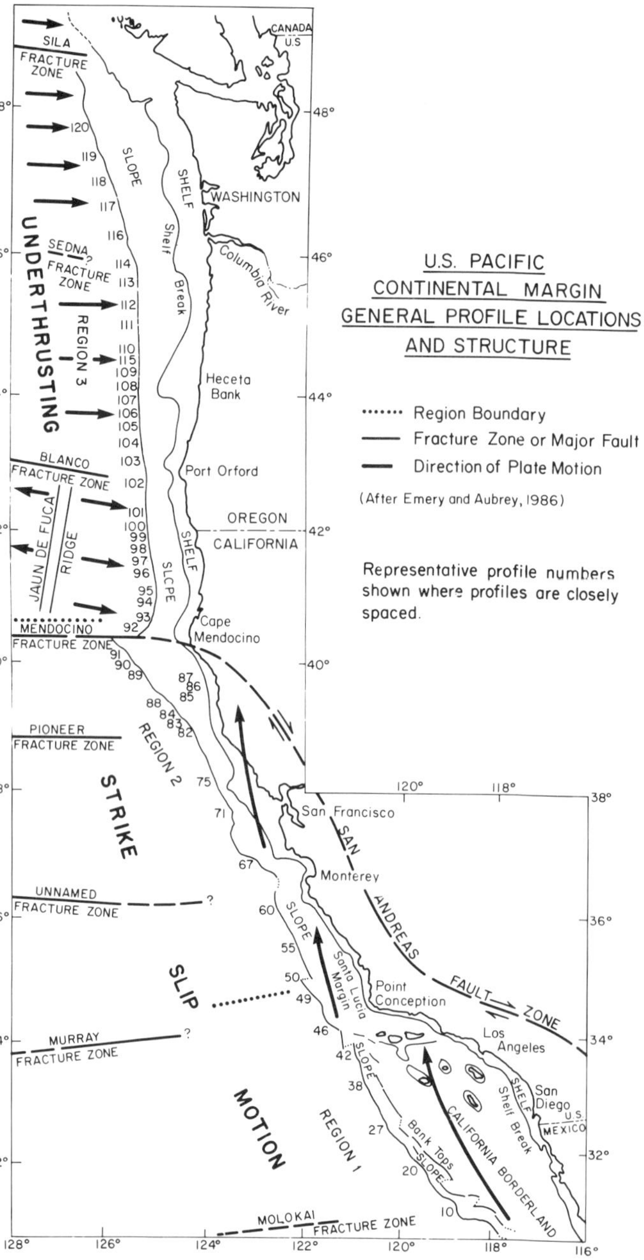

FIG. 1.—U.S. Pacific continental margin, profile locations and boundaries of slope regions defined in text. Due to close spacing in some areas, not all profile locations are shown, but representative numbers give general location. Major structural elements are also shown with plate motions (Emery and Aubrey, 1986).

Methods

The method used was to construct slope-normal gradient profiles rather than the apparent-gradient profiles drawn when a simple straight-line profile is plotted down an irregular slope surface. Data were obtained by measuring the minimum horizontal distance normal to successive contours at 250-m-depth intervals along a transect path (see Fig. 3). This produced a set of standard-depth-interval, straight-line-slope segments that were combined to form the complete profile. Width of the adjacent shelf was also measured and the approximate shelfbreak depth estimated.

The angle of declivity for each segment was calculated, and the mean of all segment gradients provided the average slope gradient for the profile. Depths to top of slope and base of slope were tabulated, the angle and depth of steepest segment, and the mid-depth of the slope profile were recorded. The five basal segments of each profile were averaged to obtain the average gradient of the slope base. This yielded a measure related to presence or absence of a sediment apron at the base of slope, or indicated where lateral transport and deposition along the base of slope produced a sharp sediment level like a bathtub ring.

Given the depth of shelf break and the width of shelf, the simple declivity of the adjacent shelf can be calculated. This average "straight-line" simplification yields a value that is larger than the actual midshelf gradient, given that most shelf profiles are actually convex-up profiles. However, it gives at least a regional indication of changes in shelf morphology and gradient.

The position of the depth of steepest declivity relative to the midslope depth can be visualized in three simple profile forms. Where the steepest depth coincides with midslope,

TABLE 2.—U.S. PACIFIC CONTINENTAL-MARGIN DATA

Dimension	Regions 1	2	3	Total
Shelf Width, km	13.0	20.0	35.0	24.0
Slope Width, km	13.8	55.3	48.1	36.2
Mean Average Gradient, degrees	11.9	5.2	6.9	8.4
Mean Steepest Slope Gradient, degrees	18.6	11.6	16.4	15.7
Length, km	—	—	—	2,200
Area, km^2	—	—	—	132,500
Mean Depth, Slope Base, m	—	—	—	3,200
Southern California Borderland Margin:	Mainland		Offshore	All
Shelf Width, km	—		—	—
Slope Width, km	9.9		9.2	9.5
Mean Average Gradient, degrees	12.1		20.9	17.0
Mean Steepest Gradient, degrees	21.8		33.3	28.2
Length, km				—
Area, km^2				40,000
Mean Depth, Slope Base, m				—

the profile is symmetrical or linear; where the steepest segment is deeper than midslope, the simple profile is concave up; where the steepest segment is shallower than midslope, the profile is convex up. Examples of each type taken from real profiles are shown in Figure 4. This relation can be quantified in terms of the depth difference between the midslope depth (M) and the depth to the steepest segment (S). Where S is greater than M, the slope steepens to base and is convex up. Where S is shallower than M, the slope gradient decreases with depth and is concave up. Where $M = S$, the slope gradient is a symmetrical "S" curve or is a straight-line gradient profile.

GENERAL PACIFIC SLOPE CHARACTERISTICS

General Statement

The Pacific U.S. margin extends for almost 2,200 km from the Mexican border to the Juan de Fuca Strait between the United States and Canada. Along this trend the slope width averages 36 km (range of 15–110 km), comprises an area of about 130,000 km^2 (includes 40,000 km^2 of the Continental Borderland), and generally has a base-of-slope depth of about 3,200 m. Data for major subdivisions of the area are shown in Table 2.

The adjacent continental shelf (exclusive of the Continental Borderland) has an average width of 24 km (range from 2–70 km), an area of about 50,000 km^2, and a mean simplified gradient of 0.4° (7 m/km). The mean shelfbreak depth is 140 m. The outline of the two provinces is shown in Figure 1.

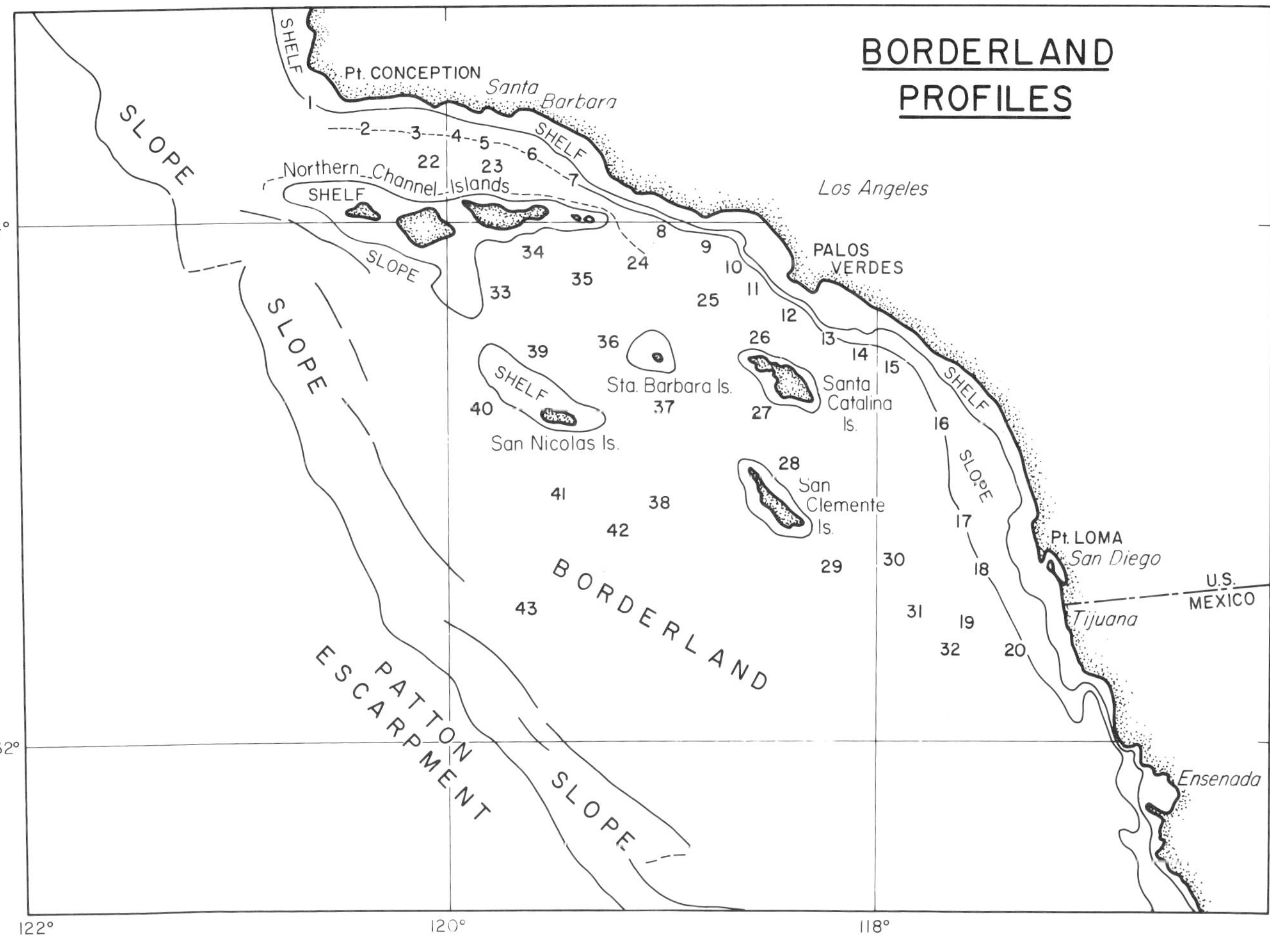

FIG. 2.—California Continental Borderland slope-profile locations and boundaries of slope and shelf provinces. Location numbers are as in borderland figures later in paper.

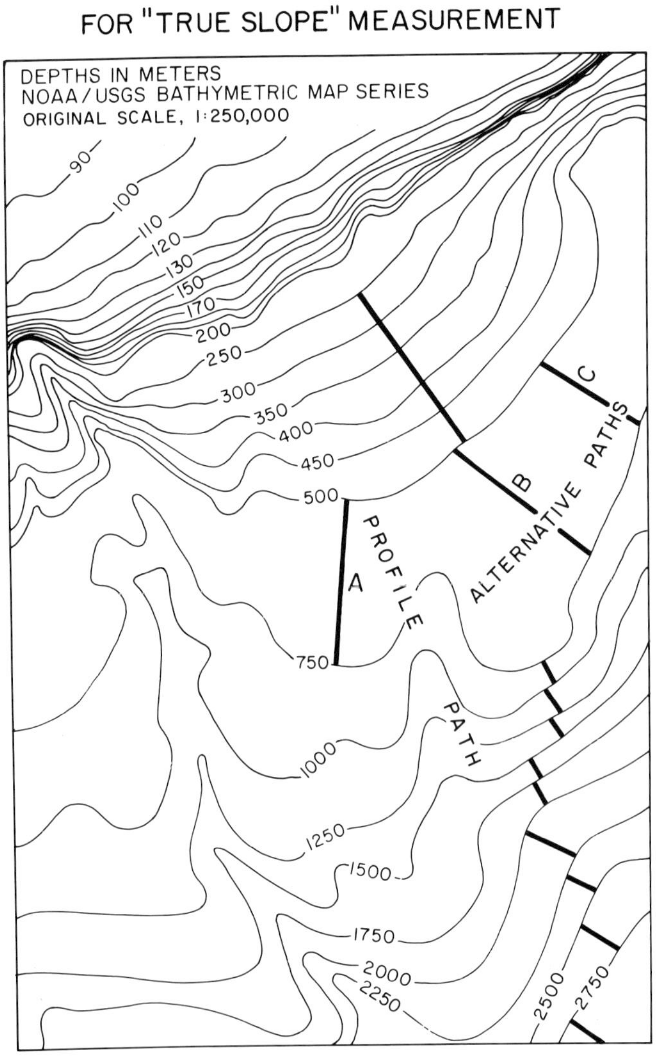

FIG. 3.—Graphic method of true slope-profile generation used in this paper. Where alternative paths are possible, one closest to general line (B) is taken.

Pacific Outer-Slope Characteristics

The general outline and gradient of the Pacific Continental Slope changes in response to such factors as tectonic activity and style, deposition and erosion. The changes from province to province are striking and obvious, as will be seen in the discussion later, and figures.

Slope-depth characteristics—

Figure 5 illustrates the depth to top, midslope, and base of slope. The figure is essentially a shoreward-looking view of the aspect of the slope along the U.S. Pacific margin.

South of Point Conception (approximately profile 50 in Fig. 1), the top of the Outer Continental Slope is actually the top of the outermost bank of the California Continental Borderland, hence, the great variation in depth that can be seen along the crest of the outer slope, the Patton Escarpment. It is also evident that the depth to the base of the slope along the escarpment is generally deeper than that to

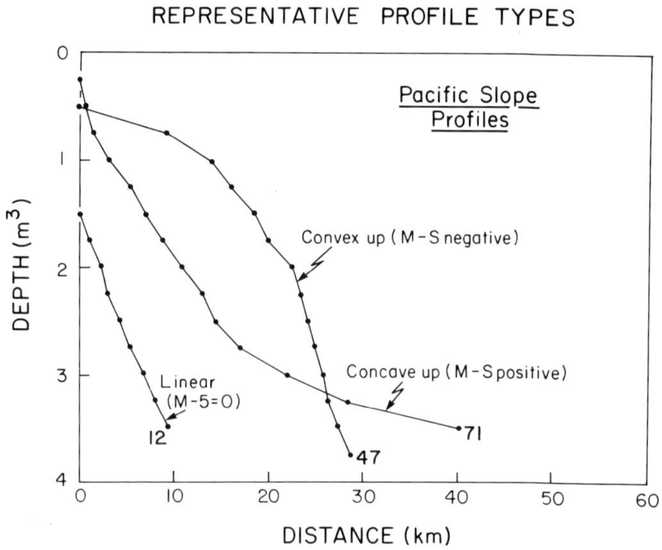

FIG. 4.—Representative slope-profile types based on difference between midslope depth and depth of steepest segment (M-S). Linear form is also termed steep form in text.

the north of Point Conception. This reflects relatively slower rates of sediment accumulation at the base of the area, far from the continental-sediment sources (Uchupi, 1962), and also probably is an indication of a deeper initial seafloor level (profiles 27–35) as compared to the region north of the Murray Fracture Zone. The latter region is probably beyond the area influenced by large deep-sea fan deposition (Monterey and Arguello Fans).

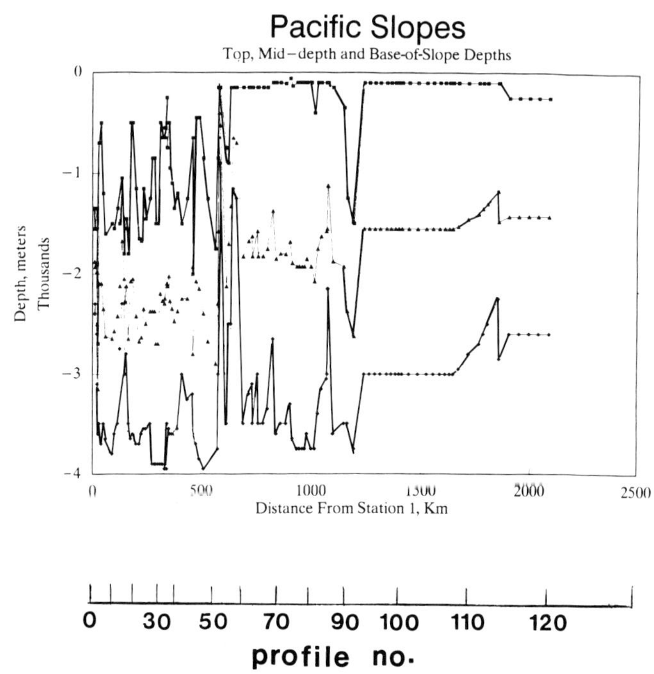

FIG. 5.—Depth to top, base and midslope along the Pacific margin. Lower scale indicates profile number along the distance scale. Mid-depths are shown with light lne where numerous points might confuse top- and bottom-depth points.

South of profile 27, the base-of-slope depth is highly variable and represents the effects of irregular topography and local deposition at the mouths of gaps in the Patton Escarpment. The top and the base become deeper from about profile 40 to profile 50, corresponding to the San Miguel Gap in the Patton Escarpment and ending at the southern end of the Santa Lucia Escarpment.

Another indication of the depth of the offshore Pacific Plate may be seen between profiles 74 and 82, a region that does not appear to be influenced by fan deposition. Thus, the base of slope here represents the intersection of the escarpment with the oceanic sea floor. South of profile 74, the base of slope depths are primarily a function of the deposition of deep marine-fan sediments (Monterey Fan, Arguello Fan, etc.).

Another break in the depth of base of slope occurs from about profile 85 to profile 91. The break corresponds to the broadening, and structural offset, of the slope from just south of and up to the Mendocino Ridge, where the Mendocino Fracture Zone intercepts the continental margin. Figure 1 shows the relations of the major Eastern Pacific structural elements to the U.S. Pacific margin.

In summary, the slope aspect shows highly variable top and bottom depths south of Point Conception, and a more uniform aspect north of the cape. The cape region is affected by transform motions of plates and changes in the accessibility of sediment sources. The boundary at Point Conception is the point at which the outer slope separates from the mainland shelf and is the boundary between regions 1 and 2 (Fig. 1). This boundary separates the broad Borderland from the narrow mainland shelf north of Point Conception.

North of Cape Mendocino, slopes are affected by underthrusting in a regime of collision-margin tectonics (Kulm and others, 1984; Emery and Aubrey 1986). The ridge defines the boundary between regions 3 and 2 (Fig. 1).

Note in Figure 5 that the general depth of the base of slope markedly decreases north of the fracture zone as a result of the offset of the oceanic floor along the transcurrent fault. A similar effect is seen at about the latitude at which the Sedna Fracture Zone possibly intercepts the margin (Fig. 1). North of that point, the general seafloor depth is again shallower. The Juan de Fuca Spreading Center is nearing the margin progressively to the north, and hence, the shallower depth of the plate segment north of the transcurrent fault.

The local shallowing of the base of slope, seen progressively from about profile 108 to profile 114, represents the surface of the Astoria Fan built out from Astoria Canyon (Kulm and others, 1984). The distal plain associated with the fan forms the base level of the slope south of Profile 108.

South of Mendocino, there are at least four fracture zones, Pioneer, an unnamed fracture, Murray, and Molokai (Fig. 1), that may interact with the margin. These intercept the margin roughly at profiles 75 to 82, 60 to 67, 46 to 50 and 1 to 10, respectively. In these regions the extrapolated intersection points cannot be closely defined, except for Murray, which probably comes in at the Point Conception area and is possibly the cause of slope morphology changes in that area that define the boundary between regions 1 and 2 (Fig. 1).

The Pioneer Fracture Zone is a possible cause of the shelf-width bulge just north of San Francisco Bay; the unnamed fracture zone off Monterey Canyon may be a factor in the position of that canyon; the swing of Patton Escarpment to the east at profile 10 appears to be closely related to the Molokai Fracture Zone and corresponds to the major Agua Blanca Fault Zone passing seaward across Baja California.

Slope width.—

Figure 6 shows slope width along the Pacific margin, and tabulations of average values are listed in Table 2. Although locally variable, there is a clear shift in general width from narrow south of profile 45 (a mean of about 15 km), a progressive increase to profile 69 and then an almost cyclic variation north of that locale, but with generally wider slopes as the norm (regional means between 45 and 55 km). The shift matches slope-gradient data (see next section) and is a logical result; steep gradients to roughly common basal depths will produce narrower slope width. If we eliminate the Mendocino shift at profiles 91 and 92, then slope widths north of San Francisco oscillate from about 20 to 60 km (mean width = 48 km). The primary factor is probably structural offset, but no direct data are available to fix a good explanation.

Profile form.—

As was noted earlier, the general form of each slope profile can be defined as symmetrical or straight, concave up or convex up. This relation can be quantitatively shown in terms of the difference between the midslope depth (M) and

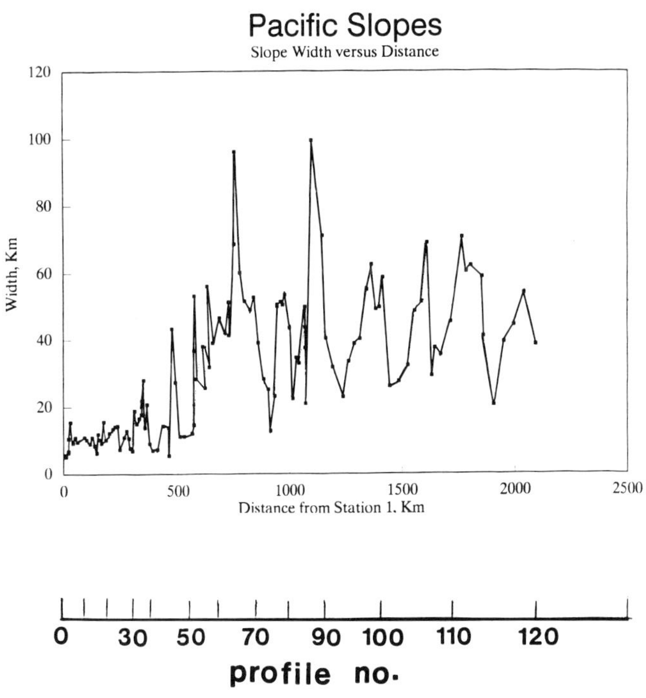

FIG. 6.—Width of the slope along the Pacific margin. Profile location shown in lower scale versus distance scale.

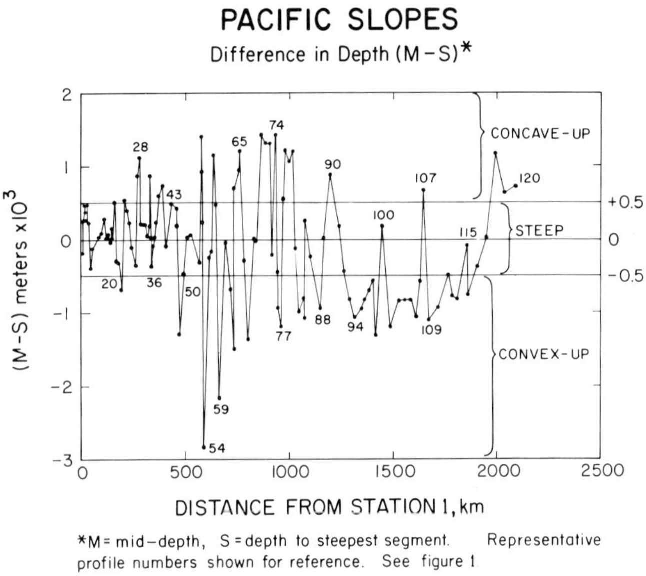

FIG. 7.—Difference between midslope depth and depth of steepest segment (M-S) for the Pacific margin slope profiles. From south to north.

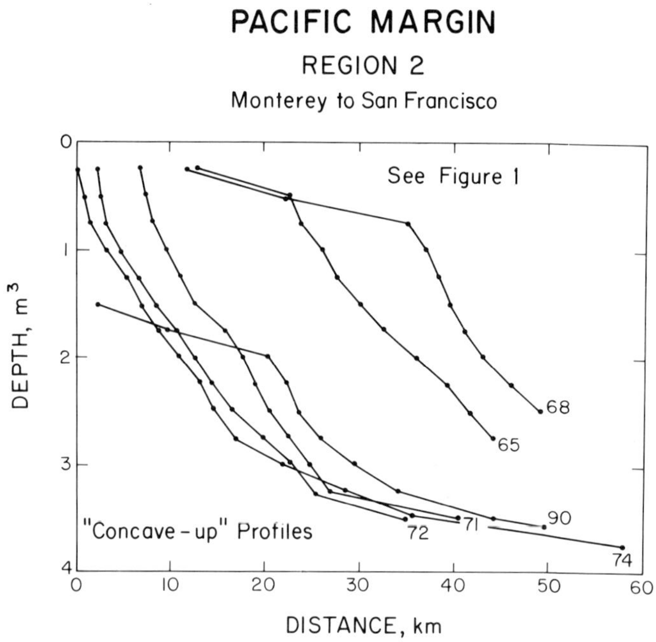

FIG. 9.—Examples of concave up profiles, some terraced. (M-S) for these is positive.

depth to the steepest segment (S). Figure 7 shows the difference between the depth of midslope and that of the steepest 250-m-depth segment (M-S). Where M-S is 0 or close to 0 ($\pm$500 m), then the slope profiles are linear or "steep" (Fig. 8). Where M-S is positive (M deeper than S), then the slope profile tends to be concave up (Fig. 9). Where M-S is negative (M shallower than S) then the simple profile is convex up (Fig. 10).

In Figure 8, it is evident that profiles up to profile 20 are virtually all steep, near-straight-line gradients; from profile 20 to profile 50 they tend to be somewhat more variable, with 76 percent near-linear (steep), 17 percent concave up and 7 percent convex up. From profile 50 to profile 90 (Mendocino), the majority, 68 percent, are either strongly

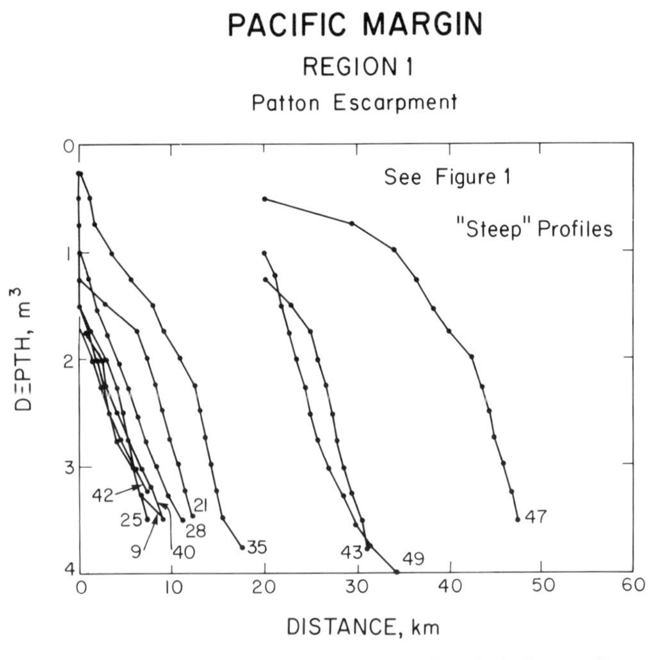

FIG. 8.—Examples of steep, linear slope profiles, including profile 47 showing typical convex up for comparison. Profile 49 is slightly concave up. For steep and slightly curved profiles, (M-S) is within ±500 m or equals zero.

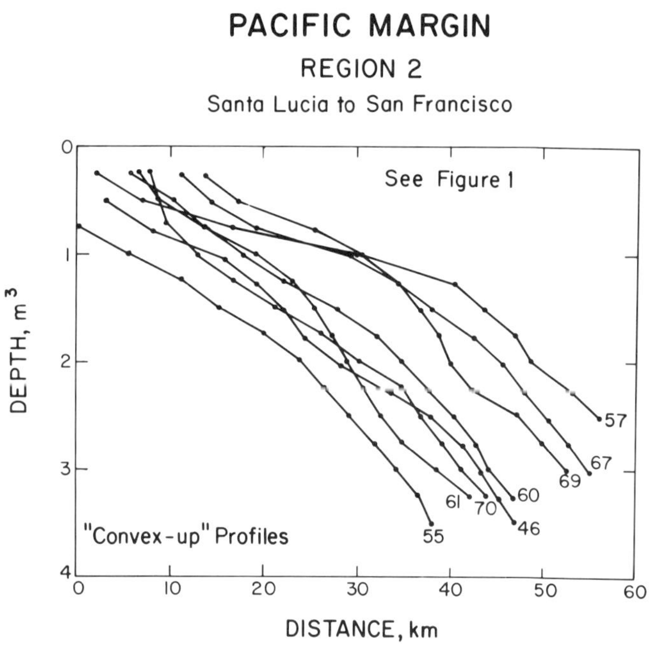

FIG. 10.—Profiles for the portion of region 2 north of Point Conception to San Francisco showing generally convex-up form typical of this region.

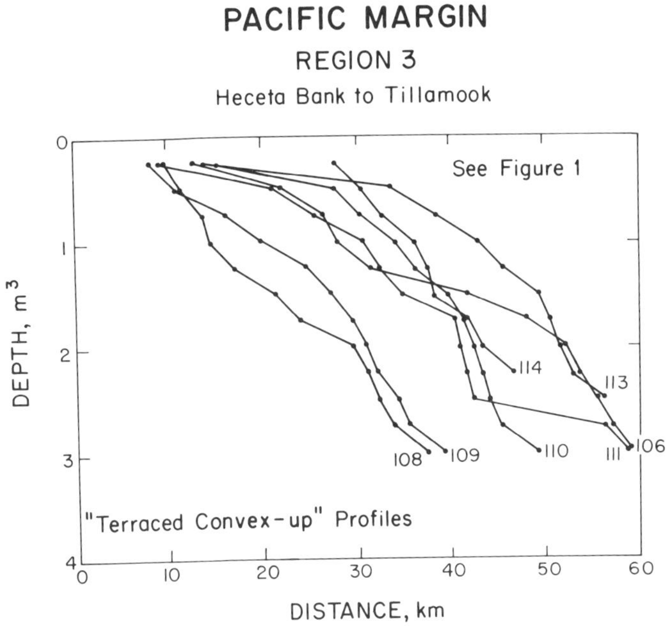

FIG. 11.—Profiles north of Cape Mendocino with typical convex up profiles. This is in the area of underthrusting of the ocean plate.

concave up (38 percent), or convex up (30 percent). Profiles are more evenly distributed between positive and negative differences in the *M-S* factor and tend to be broader irregular profiles.

North of Mendocino the common form is convex up; 60 percent of the profiles are in the negative range of *M-S* (Figs. 11, 12). Only 13 percent of profiles 90 to 120 are typically

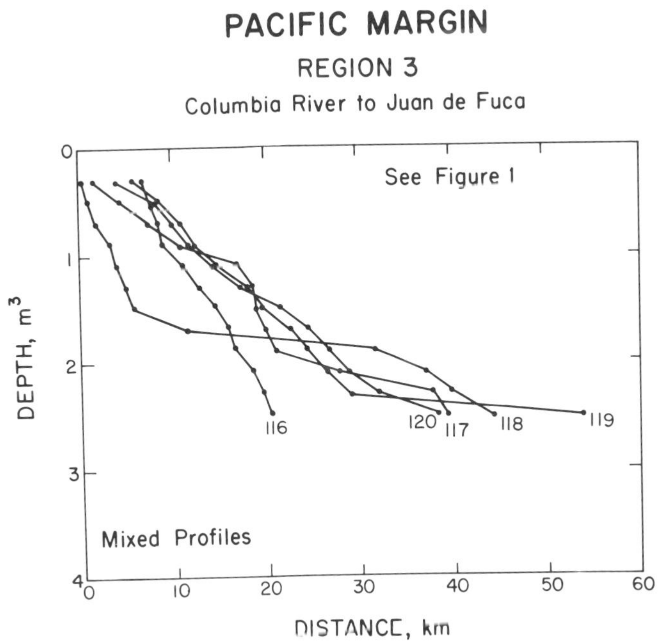

FIG. 12.—Profiles north of the general point of possible intersection of the Sedna Fracture Zone and close approach of the Juan de Fuca Ridge. These are highly irregular slopes in detail with terraced and convex-up profiles.

concave up; only 20 percent would fall in the steep category of Figure 8.

It is evident that the strongest differences in *M-S* occur north of Point Conception, whereas most differences south of the cape (along the Patton Escarpment at the outer edge of the California Borderland) are generally not more than ±500 m.

The borderland outer scarps may have minor aprons at the base, but in general the majority of the profiles is close to steep, linear profiles. Concave up profiles can be visualized as steep escarpments with base-of-slope aprons. Convex-up profiles can be visualized as steep scarps with a shelfbreak (bank-edge) roll off and no sediment aprons at their bases. Of the profiles north of Conception to Cape Mendocino, where the margin is the traditional shelf-slope type, more have aprons at their bases than not. From Mendocino north to Tillamook (Fig. 11), a majority of the slope profiles is convex up with no apron and is dominated by underthrusting and steep basal-slope gradients. From that general area to Juan de Fuca, profiles are mixed and show some basal-apron (submarine-fan) modification (roughly from profile 114 to 120, Fig. 12).

Profile gradient.—

Figures 13 and 14 illustrate average slope gradient and mean gradient of the basal 1,500 m of slope, respectively. A plot of steepest gradients (not used here) essentially parallels the data in Figure 14.

In Figure 13, it is evident that the Patton Escarpment is generally much steeper than the slopes north of Point Conception. The average gradient of profiles 1 to 50 (Mexican border to Point Conception) is 11.9° ± 4.5°, whereas the average gradient north of Point Conception is 5.9° ± 2.3°. Note that the standard deviation of values about the southern mean is ±4.5°, whereas for the northern gradients it is ±2.3°, illustrating the much more regular nature of the northern slope. A linear regression of the slope gradients for the southern area is a very weak +0.12, and much the same for the northern values, suggesting a very weak tendency in both regions for increasing gradients.

Mean slope gradient rapidly steepens between profiles 1 and 10 along the southward-facing escarpment at the southern end of the Patton Escarpment, where the line of the scarp swings east. The south face is normal to the direction of regional plate motion (Fig. 1). The scarp north of profile 10 is increasingly parallel to the regional direction of right-lateral shear. The combination of plate motion and margin-edge orientation may be exerting the subtle effect seen in the weak northward increase in average gradient (Fig. 13). As was mentioned in an earlier section, the Molokai Fracture Zone may intercept the margin at this latitude and that may well be a controlling factor.

Emery (1960) and Doyle and Gorsline (1977) have both shown that there is a progressive deepening of basin sills, bank tops and basin floors to the south in the Borderland, and this may be the cause of the very weak trend of deepening outer-slope crests to the south from profile 45 to profile 12 in Figure 5. It also may account for the weakly defined decrease in slope gradient to the south in that region related to decreasing relief of the scarp (Fig. 13). The ap-

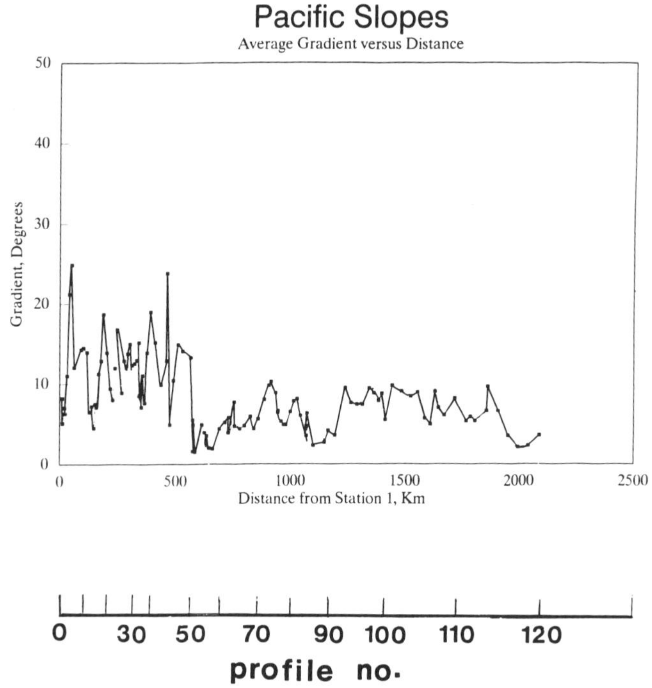

FIG. 13.—Average slope gradient of Pacific margin. Each profile value is the mean of the segment gradients in that profile.

parent decrease in gradient also is influenced by the several gaps in the Patton Escarpment, where tectonic breaks have interrupted the slopes.

To the north, from profile 51 to profile 91, the general configuration of the slope shows little variability in average and basal gradients. From profiles 92 to 103, along the margin between Mendocino and the Blanco Fracture Zone, the average steepness abruptly increases. As previously noted, base of slope depth is roughly constant to the latitude of Heceta Bank (profile 108), after which the base-of-slope depth shallows to about profile 115 at an abrupt widening of the adjacent shelf. This may represent an effect of the Sedna Fracture Zone, which can be roughly extrapolated from its offshore location into a possible margin intersection in this vicinity. Heceta Bank is a horst-and-graben structure oriented North Northwest-South Southeast, and these local structures may reflect shears or compression between larger, regional, structural elements.

The zone of shallowing base-of-slope depth from profile 108 to 115 (Fig. 1) along the margin from Heceta Bank north must represent a structural offset of the margin. This could be related to impingement of minor fracture zones upon the margin block, but no offshore fracture zones appear to be present in this area (Kulm and others, 1984). The Sila Fracture Zone can be extended to meet the straight, east-west flank of the Olympic Peninsula on the south side of the Juan de Fuca Strait north of profile 120. Between profile 115 and 120, a sharp change in basal-slope steepness occurs, which may indicate some structural interactions. Figure 14 illustrates that this factor, and sharp jumps in basal steepness, occur near the possible positions of the Blanco and Sedna Fracture Zones, depending on whether these extend into the margin block and also upon the precise location of such intersections. Data are absent to answer this question. Base-of-slope depth certainly is significantly higher north of profile 115.

Depth to base of slope, as noted earlier, shows large oscillations from about profile 50 to profile 60 that may reflect possible interactions between the Murray and the unnamed fracture zones (Fig. 1). Depth to base of slope then

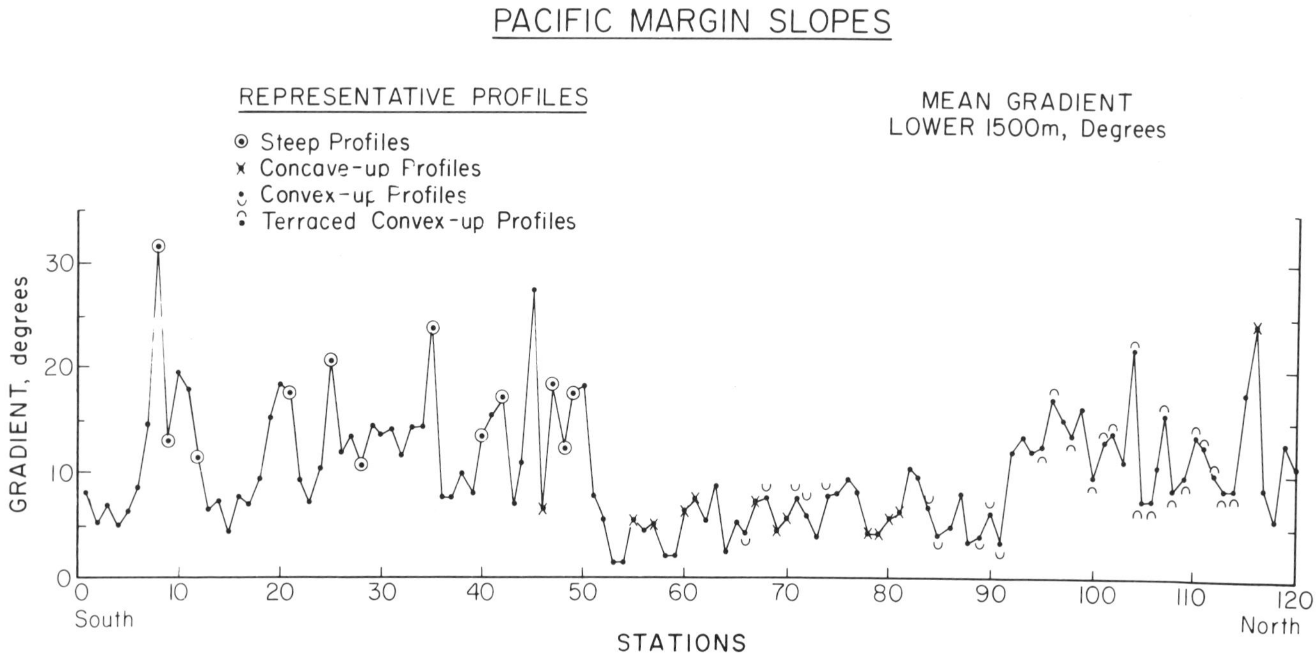

FIG. 14.—Basal 1500-m gradient along the Pacific margin. Where this value is high, the profiles are generally either linear or convex up. The value is the average of the five deepest segments of each profile.

becomes less variable to about profile 82, at which point variation is large and generally shallowing to the vicinity of the Mendocino Escarpment, after which the base-of-slope depth is significantly shallower. The Pioneer Fracture Zone, if extended to the margin, would intersect the slope at about profiles 81 to 82.

Slope-morphology regions (Fig. 1) can be established from inspection of Figures 13 and 14, which show mean slope gradient and basal-slope gradient. A major change in mean gradient occurs at about profile 50 to 52, and a second change (less marked) occurs in basal gradient at about profile 92. As we have seen, slope width (Fig. 6) changes progressively from about profile 45 to 46 to about profile 70, and then varies around a markedly increased average width from there north (ca. 50 km versus 15 km).

Slope provinces.—

In summary, the slope-depth boundaries are highly variable south of Cape Mendocino and quite uniform north of the cape. This is the divide between transform motions of plates south of the ridge area and underthrusting collision-margin tectonics north of the ridge (Kulm and others, 1984). The divide defines a major slope-morphology province boundary (region 3 to 2, Fig. 1).

Three regions from south to north have been defined based upon morphological characteristics alone. Region 1 is the Patton Escarpment and the Santa Lucia Escarpment from Point Conception south to below the Mexican border; the steep, narrow slope is the outer slope of the California Continental Borderland. Region 2 includes the margin from Point Conception north to the Mendocino Fracture Zone and is characterized by variable widths, including the widest elements of the entire margins, variable slope forms, lowest average gradients and variable relief. Region 3 is the margin north of Mendocino and is the most regular in its character of the regions defined here. A change in depth-of-slope base and profile form can be seen in the northern portion of this region and could define another subregion.

In terms of structural relation, regions 1 and 2 are transform margins. Region 1 probably terminates south of the United States-Mexico border at the Agua Blanca Fault Zone. The trend of this major fault can be extended along the eastward-swinging southern end of the patton Escarpment at about profiles 1 to 10 and merges with the Patton Escarpment, which then swings to the northwest. South of this area, the general structure of the Borderland of Baja California changes, probably in response to large-scale shifts of margin blocks in Neogene time that have been rifted north along the outer edge of the old margin (Teng, 1985; Teng and Gorsline, 1989).

Acoustic-reflection seismic-profile data.—

Additional information can be gained from an examination of the U.S. Geological Survey acoustic seismic profiles of the Extended Economic Zone along the Pacific Margin (EEZ Scientific Staff, 1988). Figure 15 shows the transects of that survey and, in bold, the segments that represent slope crossings. Approximately 30 near-normal crossings can be examined along the margin that yield information on the structure and sediment cover on the slopes. Figure 15 also shows the general location of matching profile locations from this study.

The southern acoustic seismic-reflection profiles corresponding to slope profiles 1 to 40 (region 1) typically show steep, rugged escarpments with little or no evident sediment cover. Sediments are often seen ponded at the base of slope, but these appear to have been deposited by laterally moving turbidity currents along a shallow base-of-slope trough. Contacts with the adjacent slope are sharp and abrupt (Fig. 16).

North of Point Conception, the seismic profiles of region 2 show sediment cover, slumps and base-of-slope aprons that evidence the higher rates of sediment contribution from the adjacent coast (Fig. 17).

Figure 18 shows typical profiles from region 3 north of the Mendocino Fracture Zone. These profiles show the effects of underthrusting in a convergent margin, modified by extensive submarine-fan deposition at the slope bases away from the feeder submarine canyons. Profiles are typically convex up.

CALIFORNIA BORDERLAND SLOPES

General Statement

Although the main thrust of this paper is on the Outer Continental Slope of the U.S. Pacific margin of the North American Plate, a selected set of profiles from within the California Borderland are also included. All of these slopes are actually the margins of the tectonically developed basins that comprise the Borderland. They are therefore quite different in scale and environment from those of the main continental slope. This region encompasses about 40,000 km^2 of the total 130,000 km^2 of slope provinces discussed here, and the significant portion of the "slope" area deserves mention for that reason alone.

Within the California Borderland, the slopes are generally of much lower relief than the continental-margin slopes, having their basal levels in the adjacent borderland basins at depths that range from 600 to over 2,000 m. Because of the limited relief, these profiles were constructed by the same method described earlier, but using a 100-m contour interval. General morphological data for these profiles are listed in Table 2 and locations of profiles are shown in Figure 2.

As was done for the larger region of the U.S. Pacific margin, the width of the adjacent shelves and the estimated depth to shelf break were tabulated for each profile. Along the mainland margin, the width of the shelf averages about 8 km, ranging from 1 to 20 km and with 70 percent between 2.5 and 12.5 km. Depth to the mainland shelf break is about 100 m. Shelf gradient as a simple angle from shore to shelf break is 0.7°, but the range is from less than 0.3° to 4° and in all instances is much more than the true slope of the concave-up surfaces in actual shelf profiles.

Borderland-Slope Characteristics

Slope-depth characteristics.—

Figure 19 (locations 1–20) shows the changes in relief of the mainland profiles from west to south from Point Con-

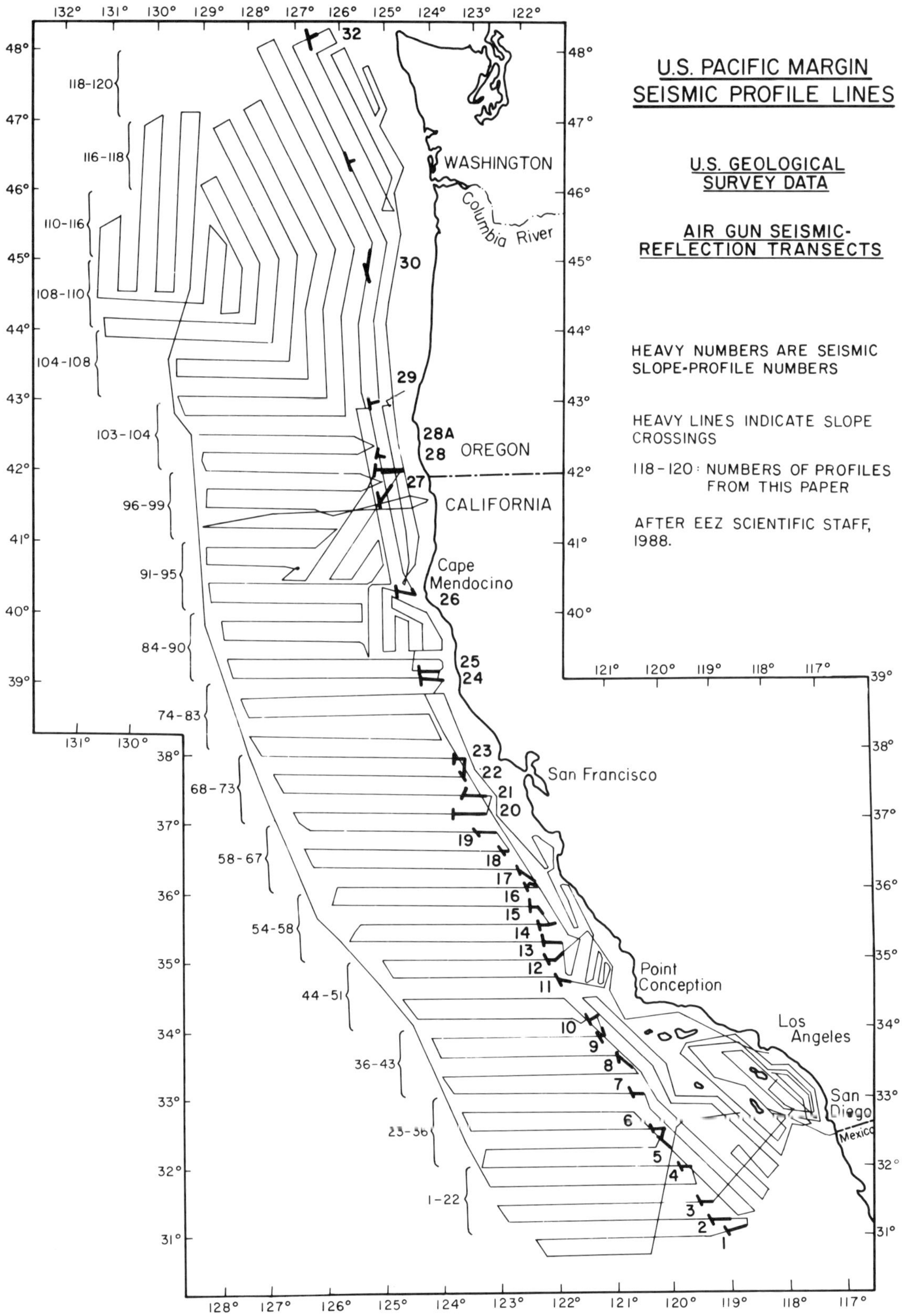

FIG. 15.—Locations of U.S. Geological Survey "GLORIA" transects and acoustic seismic profiling. Locations of slope crossings and seismic-profile number are shown in bold. For reference, the slope-profile locations for each set are also shown.

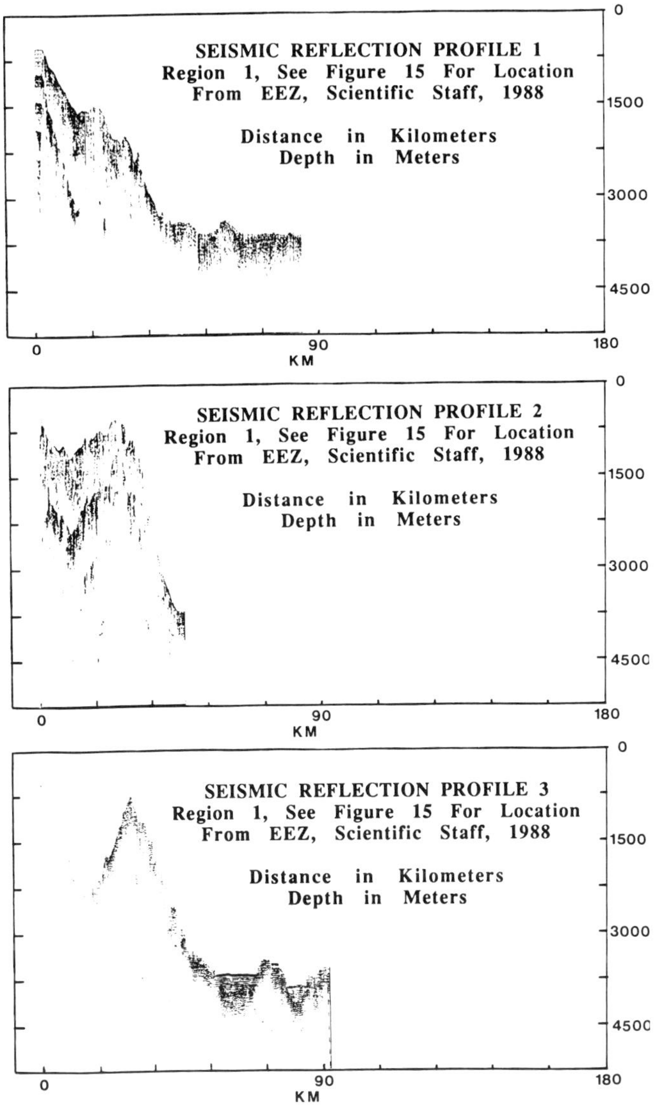

FIG. 16.—Representative seismic-reflection profiles from region 1 (see Fig. 15) showing linear, steep-profile forms with little or no sediment cover and sharp abutment with adjacent oceanic sediments.

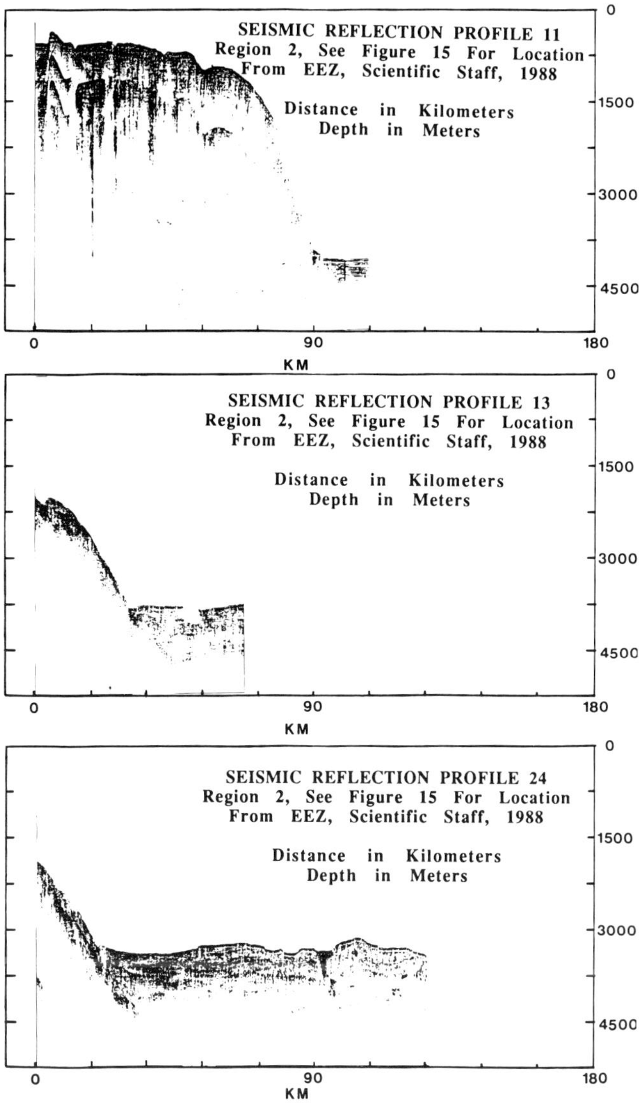

FIG. 17.—Representative seismic-reflection profiles from region 2 (see Fig. 15) showing linear to slightly concave-up profiles with marginal aprons and slumps. Sediment-draped profiles.

ception to the United States-Mexico border. This trend follows the offshore topography of a sequence of inner basins gradually deepening to the south, as noted by Emery (1960). The regional downwarping to the southwest is a factor in all of the Borderland-slope data. Downwarping is partly due to cooling of the margin crust following the overwhelming of the East Pacific Rise, commencing in Oligocene time and continuing to the recent, between Mendocino and the Gulf of California mouth (Atwater, 1970; Teng, 1985). Isostatic adjustment and compression also have contributed to the downwarping, and progressively thicker basin deposition to the north, nearer the principal sediment sources, has filled the northern basins and decreased their basin-floor depths (Emery, 1960; Gorsline and Teng, 1989).

Figure 19 also shows slope relief for all of the Borderland profiles. Note that the offshore profiles (32–43) are all generally from deeper bank edges or insular-shelf edges to deeper basin floors, following the general basin-floor deepening offshore and south.

Slope width.—

Figure 20 shows width of slope for the mainland slope (profiles 1–20), and for central Borderland bank and insular slopes (profiles 32–43). The mainland slope ranges in width from 1 to 15 km with an average width of 8 km.

Within each mainland coastal cell (Emery, 1960; Inman and Frautschy, 1966), the shelf widths increase to the central embayment, where the shelf width is greatest, and then decrease to the next promontory. The next peak in width corresponds to central Santa Monica Shelf and so forth. Slope widths show no such simple relation. Except for profiles 2 and 6 (the fifth and ninth points in Fig. 21), there is a weak general tendency for slope widths to increase south to San Diego. Regionally, the east-trending northern coast

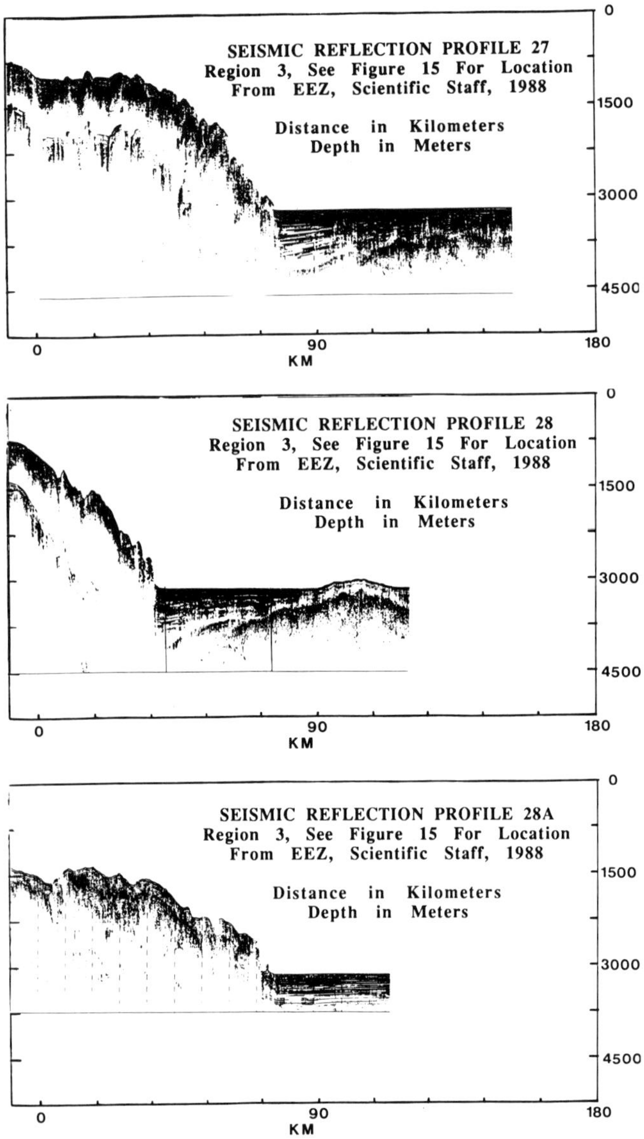

FIG. 18.—Representative seismic-reflection profiles from region 3 (Fig. 15) showing convex-up profiles in region of plate underthrusting. Note trough fill at bases of slopes and irregular thrusted-margin faces. Deformed sediments and sedimentary rocks form mass.

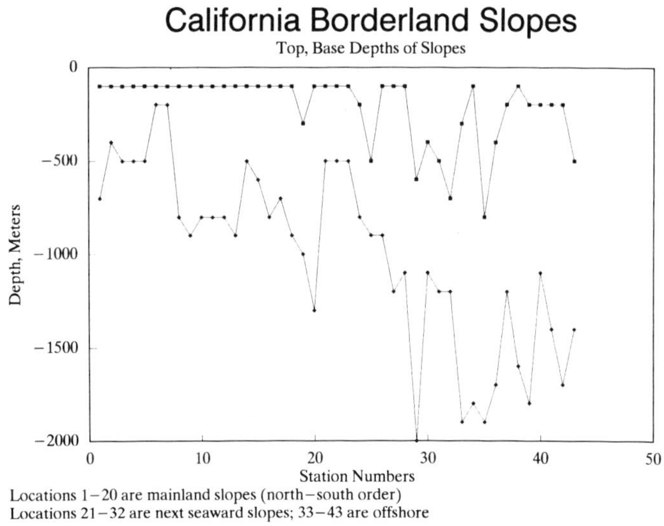

FIG. 19.—Depth to top, base and midslope for all Borderland profiles (locations as in Fig. 2).

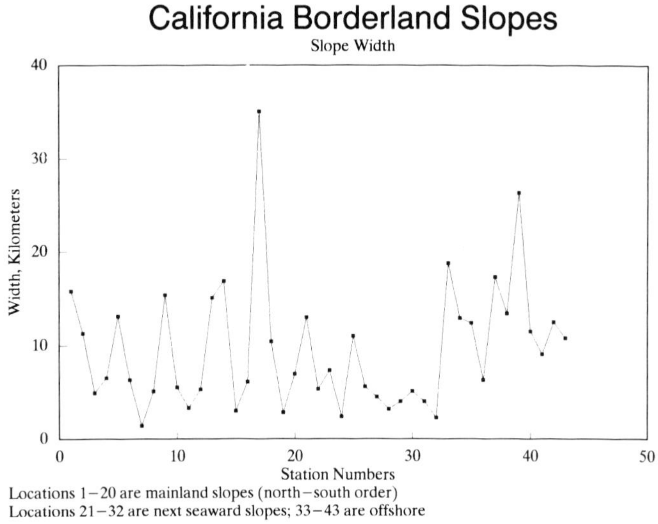

FIG. 20.—Width of slope for Borderland profiles (locations in Fig. 2).

(Santa Monica to Santa Barbara) is under a compressive regime, which may affect the slope widths. The San Diego Trough that parallels the southward-trending coast is filled to its sill, which may be reflected in the wider slope as a broad, gentle, depositional apron at base of slope.

Offshore slope widths range from 2 to 20 km and average 9 km in width. There is a slight correlation with insular-shelf width, but since many of these offshore profiles are of bank slopes and no bank top widths were recorded, the relation is weak. Slope width probably is controlled more by the base-of-slope datum for each slope than by any other factor. Note in Figure 19 that the base of slope ranges from 200 to almost 1,500 m along the mainland slopes alone. As will be seen in a later section, the gradients of the offshore slopes are higher than those of the mainland slopes, but as noted, their base depths are deeper, by as much as 2,000 m. Thus, slope relief is the largest factor in determining width of the slope.

Profile form.—

Profiles within each coastal segment (promontory to promontory) show a progression from concave-up to convex-up profiles (*M-S* values negative to positive, respectively; Fig. 21). Most probably, this indicates a decreasing sediment drape to the south, downcurrent from the main sediment-input locations in the northern sector of the inner Borderland (Schwalbach and Gorsline, 1985). The concave-up profiles are those with depositional wedges and

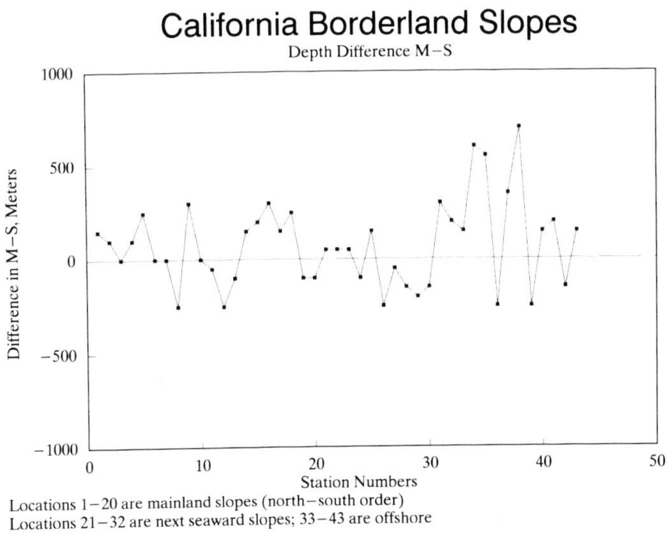

FIG. 21.—Difference in depth of midslope and depth of steepest segment (M-S) for Borderland profiles (locations as in Fig. 2).

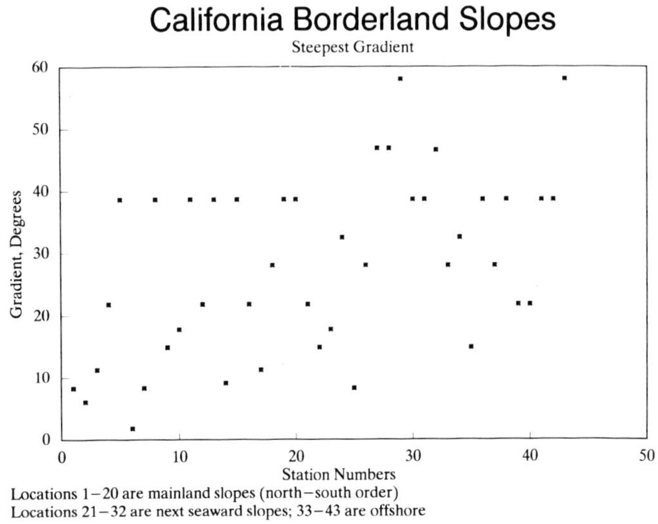

FIG. 23.—Steepest gradients of Borderland slope profiles (locations as in Fig. 2).

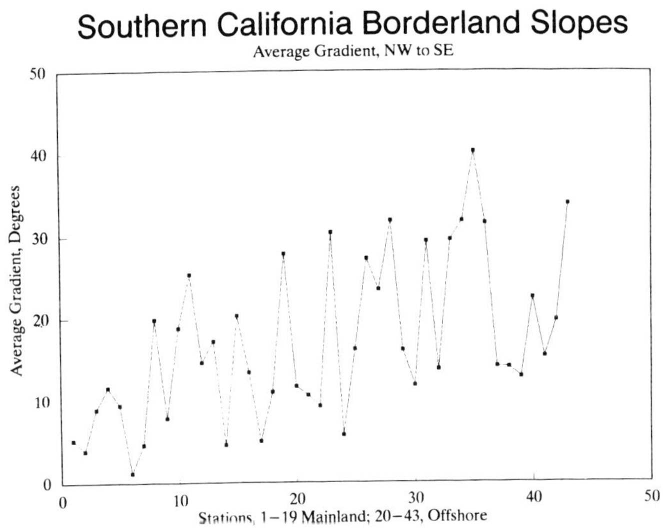

FIG. 22.—Average gradients of slopes arranged in northwest to southeast order (locations as in Fig. 2).

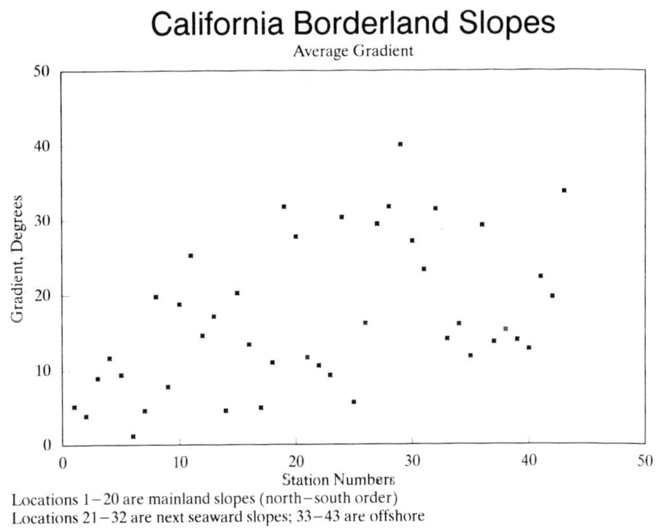

FIG. 24.—Average gradients of Borderland slopes (locations as in Fig. 2).

aprons. Most are primarily fault scarps modified by deposition and mass failures (Haner and Gorsline, 1979).

Offshore profiles tend to be mixed and are either strongly convex or concave up and probably are essentially deposition- or structure-dominated, respectively.

Profile gradient.—

The average gradient of all Borderland profiles from northwest to southeast (Fig. 22) shows increasing steepness to the south. Figure 23 compares the steepest slopes and Figure 24 compares average gradients between mainland and offshore. In both figures the higher gradients are offshore, which is probably primarily indicative of thin-sediment cover on these slopes and the typical absence of a depositional wedge or apron as noted earlier.

Slope provinces.—

The steeper, lower relief slope profiles of the Borderland are, as a result, much narrower than the outer Pacific margin slopes of greater scale and are generally more nearly linear. A few of the offshore profiles become significantly convex up. Most are probably fault scarps, modified to varying degrees by deposition; northern mainland slopes are gentler and draped by sediment or decreased in grade by mass failures.

Two general provinces can be defined on the basis of all characteristics. One is the mainland slope, which receives the bulk of deposition from the mainland sources, particularly in the northern coastal cells of highest stream discharge. The offshore profiles have significantly higher relief (deeper basin-floor datums), are steeper (negligible sediment mantling), and are essentially unmodified structural scarps. Acoustic-reflection profiles are not shown, but generally indicate the paucity of sediment drape on offshore slopes and presence of sediment wedges and slumps on the mainland slopes (Haner and Gorsline, 1979).

SUMMARY AND CONCLUSIONS

The two sets of slope profiles cover a spectrum from active margins of both transform and underthrust types to margin-basin slopes. It is evident that the factor that differentiates these profiles is the tectonic style of the continental margin-ocean floor interaction. Thus, on the active Pacific margin, the underlying control of slope morphology is transform motion between crustal plates versus convergent underthrusting motions.

The general form of slopes can be approximated by the *M-S* parameter, which is the difference in depth between the midslope depth and the depth to the steepest slope segment. For sediment-free or thinly veneered slopes, this factor is near zero (steep straight slope) or weakly positive or negative. Where sedimentation has built a slope-sediment wedge or base-of-slope apron, the values are typically negative and the form is concave up. Where underthrusting occurs along the northern Pacific margin, or where deposition has not altered the lateral fault scarp of the southern slope of regions 1 and 2, the form is negative and convex up.

Slope width is a function of gradient and relief. Along the Pacific Coast, the widest slopes are found north of Point Conception following the trend of decreasing average gradient, whereas base-of-slope depths remain roughly the same. The steep, thinly veneered slopes of the southern portion of the pacific margin with less relief are the narrowest.

It has been interesting to see the responsiveness of slope morphology to regional structural trends and features. It is also evident that the slopes cannot be generalized by a global average gradient or relief. Each region has its own characteristic forms varied by the modifying effects of deposition and associated mass failure on primary, tectonically formed scarps.

ACKNOWLEDGMENTS

I acknowledge the assistance of Janette Murillo in preparing the basic slope measurements. Janet Dodds drafted the various figures not done by computer methods. John Armentrout gave good critical comments on an earlier poster version of the topic, which improved the breadth of the paper. Discretionary funds were used from the generous endowment of the Wilford and Daris Zinsmeyer Chair in Marine Studies.

REFERENCES

ATWATER, TANYA, 1970, Implications of plate tectonics for the Cenozoic tectonic evolution of western North America: Geological Society of America Bulletin, v. 81, p. 3513–3536.

DOYLE, L. J. AND GORSLINE D. S., 1977, Marine geology of Baja California Continental Borderland, Mexico: American Association of Petroleum Geologists Bulletin, v. 61, p. 903–917.

EEZ Scientific Staff, 1988, Atlas of the Exclusive Economic Zone, Western Conterminous United States: U. S. Geological Survey Miscellaneous Investigations Series No. I-1792, 152 p.

EMERY K. O., 1960, The Sea off Southern California: John Wiley & Sons, Inc., New York, 366 p.

EMERY, K. O., AND AUBREY, D. W., 1986, Relative sea-level change from tide gauge records of western North America: Journal of Geophysical Research, v. 91, B14, p. 13941–13953.

EMERY, K. O., AND UCHUPI, ELAZAR, 1984, The Geology of the Atlantic Ocean: Springer-Verlag, New York, 1050 p.

GORSLINE, D. S., AND TENG, L. S., 1989, The California Continental Borderland, *in* Winterer, E. L., Hussong, D. M., and Decker, R. W., eds., The Eastern Pacific Ocean and Hawaii, v. N, The Geology of North America: Geological Society of America, p. 471–487.

HANER, B. E., AND GORSLINE, D. S., 1979, Processes and morphology of continental slope between Santa Monica and Dume Submarine Canyons, southern California: Marine Geology, v. 28, p. 77–87.

HEEZEN, B. C., THARP, MARIE, AND EWING, E. M., 1959, The Floors of the Oceans. Part I, The North Atlantic Ocean: Geological Society of America Special Paper 65, 122 p.

INMAN, D. L., AND FRAUTSCHY, J. D., 1966, Littoral processes and the development of shorelines: Coastal Engineering, Special Conference Proceedings, p. 511–536.

INMAN, D. L., AND NORDSTROM, C. E., 1971, On the tectonics and morphologic classification of coasts: Journal of Geology, v. 79, p. 1–21.

KULM, L. D., PETERSON, C. P., CONNARD, G. G., JOHNSON, S. Y., NIEM, W. A., MCCLAIN, K. J., LOUBERE, P. W. LEWIS, B. T. R., COUCH, R. W., NIEM, A. R., SCHEIDEGGER, K. F., CROSSON, R. S., AND STEWART, R. J., 1984, Western North American Continental Margin and Adjacent Ocean Floor off Oregon and Washington: Ocean Margin Drilling Program, Regional Atlas Series, No. 1, Marine Interscience International, Woods Hole, MA, 32 p.

SCHWALBACH, J. R., AND GORSLINE, D. S., 1985, Holocene sediment budgets for the basins of the California Continental Borderland: Journal of Sedimentary Petrology, v. 55, p. 829–842.

SHEPARD, F. P., 1948, Submarine Geology: Harper and Bros., Inc., New York, 348 p.

SHEPARD, F. P., 1973, Submarine Geology (3rd ed.): Harper and Row, Inc., New York, 551 p.

SHEPARD, F. P., AND DILL, R. F., 1966, Submarine Canyons and other Sea Valleys: Rand-McNally and Co., Chicago, 381 p.

SHEPARD, F. P., AND EMERY, K. O., 1941, Submarine topography off the California coast: Canyons and tectonic interpretations: Geological Society of America Special Paper 31, 171 p.

TENG, L. S., 1985, Seismic stratigraphic study of the California Continental Borderland basins: Structure, stratigraphy and sedimentation: unpublished Ph.D. Dissertation, University of Southern California, Los Angeles, 197 p.

TENG, L. S., AND GORSLINE, D. S., 1989, Late Cenozoic sedimentation in California Continental Borderland basins as revealed by seismic facies analysis: Geological Society of America Bulletin, v. 101, p. 27–41.

UCHUPI, ELAZAR, 1962, Continental margin from San Francisco, California to Cedros Island, Baja California, Mexico: Unpublished Ph.D. Dissertation, University of Southern California, Los Angeles, 232 p.

VEATCH, A. C., AND SMITH, P. A., 1939, Atlantic submarine valleys of the United States and the Congo submarine valley: Geological Society of America Special Paper 7, 101 p. plus maps.

SUBTIDAL STROMATOLITES, OOIDS AND CRUSTED-LIME MUDS AT THE GREAT BAHAMA BANK MARGIN

ROBERT F. DILL

Caribbean Marine Research Center, 610 Tarento Dr., San Diego, California 92106

ABSTRACT: Lithified stromatolites, attaining 2-m heights and associated with cement-encrusted carbonate-mud beds, are found in many inter-island channels of the southern chain of Exuma islands and cays. They occur within migrating flood-tidal, ooid-sand bars and dunes. Reversing tidal and wind-driven currents, with velocities up to 3 kts (150 cm/s), flow across depositional sites for 3 hrs of each 6-h tidal period. Such a high-energy, bank-margin environment is not usually considered to be the site of stromatolite growth or of mud-size particle deposition; however, mud beds occur in water depths of 4 to 8 m. The normal thickness is usually about 10 cm, but one large bed has a total thickness of 1 m. If exposed to current flow for several weeks, the hard crusts become colonized by a diverse microbial community of algae, diatoms and marine plants similar to that growing and trapping ooids on nearby lithified stromatolites. The white lime muds forming the interior of the crusted-mud beds are cohesive, having the consistency of "tooth paste." At first glance, the interior muds appear to be homogeneous; however, they are pelleted. Uncemented ooid sand occurs above and below the laminated, crusted-mud beds, a relation that shows they occupy the same depositional environment. Their age of less than 4 ka and their depth of occurrence argue against the lenticular-bedded muds being relict deposits associated with the Holocene flooding of the Bahamian platform.

Scanning electric microscope (SEM) images reveal that the muds contain pelletoids of aragonite needles. The ooid grains are held in place and cemented by organic remains of the microbial mat. It is clear from the SEM photomicrographs that the surface crusts on the mud beds are actively undergoing diagenetic changes, whereas the encased uncemented mud is not. Strong currents scour and undermine the crusted beds, forming "rip-up clasts" and "mud chips." Microbial mats colonize many of the larger chips, trapping ooid sand. The chips become part of larger stromatolites as they grow and coalesce. Cross sections of lithified stromatolites show that they contain clasts of mud chips incorporated in their laminated internal structure and are thus contemporaneous with stromatolite growth.

The probable source of the lime-mud needles is precipitation from "whitings" and from resuspended biogenic muds. During tropical storms and hurricanes, fine-grained sediment that blankets the Pleistocene surface of the Bahama carbonate platform forms dense clouds of muddy bank water. The ebb tides and offbank, wind-driven currents carry the mud-charged water mass to the bank edge. There, every 6 hours, the muddy water mass comes in contact with flood-tidal oceanic water and intense mixing takes place. It is below the mixing zone of sediment-laden, slightly hypersaline bank and cooler open-oceanic waters that crusted, laminated-mud beds, interbedded with well-developed ooid-sand bodies, lithified stromatolites, hardgrounds and areas of grapestone, are found.

The occurrence of Bahamian crusted muds within ooid-sand dunes, rip-up clasts and lithified stromatolites, all subjected to strong subtidal currents, represents a newly defined association. If seen in ancient deposits, such associations might be interpreted as supratidal or shoreline deposits. Finding these features in high-energy subtidal channels at the eastern margin of the Great Bahama Bank mandates caution when making such interpretations in the geologic record.

INTRODUCTION

The eastern margin of the Great Bahama Bank historically has never received the research interest of its western counterpart, due in part to the difficulty of reaching the area and a lack of pioneering carbonate studies in the region. Therefore, when modern lithified stromatolites were first discovered by Dravis (1983), it came not only as a surprise but also was overlooked by many as a significant find. The importance of Dravis' discovery just west of Eleuthera Island near Schooner Cay became clear when, in 1984, "giant" stromatolites were found 185 km to the south, off Lee Stocking Island (Fig. 1). These "giant" forms are growing in inter-island, subtidal channels, and are associated with large, migrating ooid-sand dunes. The waters in which they grow have normal oceanic salinities and are subject to high-velocity tidal currents. Thus, the marine setting of Dravis' smaller stromatolites was shown not be an isolated occurrence but instead to be widespread and representing an entirely new area for the study of modern stromatolites (Dill and others, 1986; Shinn and Dill, 1986). Subsequent studies revealed that many of the inter-island channels along the eastern margin of the Great Bahama Bank contain stromatolites. Between 1984 and 1991, studies in the stromatolite region also revealed other types of lithified features whose origin and preservation can be attributed to recent marine cementation and a relation with microbial mats (Dill and others, 1986; Dill and Shinn, 1986; Shinn and Dill, 1986; Shinn, 1987; Dill and others, 1989; and Kendall and others, 1990). Stromatolite cementation is associated with a complex microbial community of plants and micro-organisms, which is referred to as a microbial mat. Paradoxically, beds of laminated-aragonite mud also are found interbedded within oolite dunes (Dill and others, 1989). Within this setting, marine cements are forming crusts on laminated aragonitic-mud beds that are not normally considered to be associated with a high-energy environment (Dill and others, 1988; Boardman and others, 1989; Falls and others, 1990). Even more remarkable, it can be demonstrated that these muds are related to lithified 2.5-m-high stromatolites (Dill and others, 1989). Since 1984, research at the Caribbean Marine Research Center (CMRC) has shown that features owing their preservation and structure to marine cements are usually restricted to areas containing migrating sand bodies, are flooded by both open-oceanic and bank waters, and have well-developed microbial mats, a setting for modern stromatolites not studied before Dravis' discovery.

It is important to recognize that the extensive flood-tidal ooid banks and shoals that blanket the eastern margin of the Bahama Bank cover areas that were subaerially exposed less than 5 ka, which suggests a high rate of late Holocene carbonate sediment production and accumulation. Strong reversing tidal currents form natural levees of ooid sand along high-energy tidal channels. Ooid and skeletal sands that spill over the back sides of these levees become stable and are being cemented to form hardgrounds. In the chan-

From Shoreline to Abyss, SEPM Special Publication No. 46

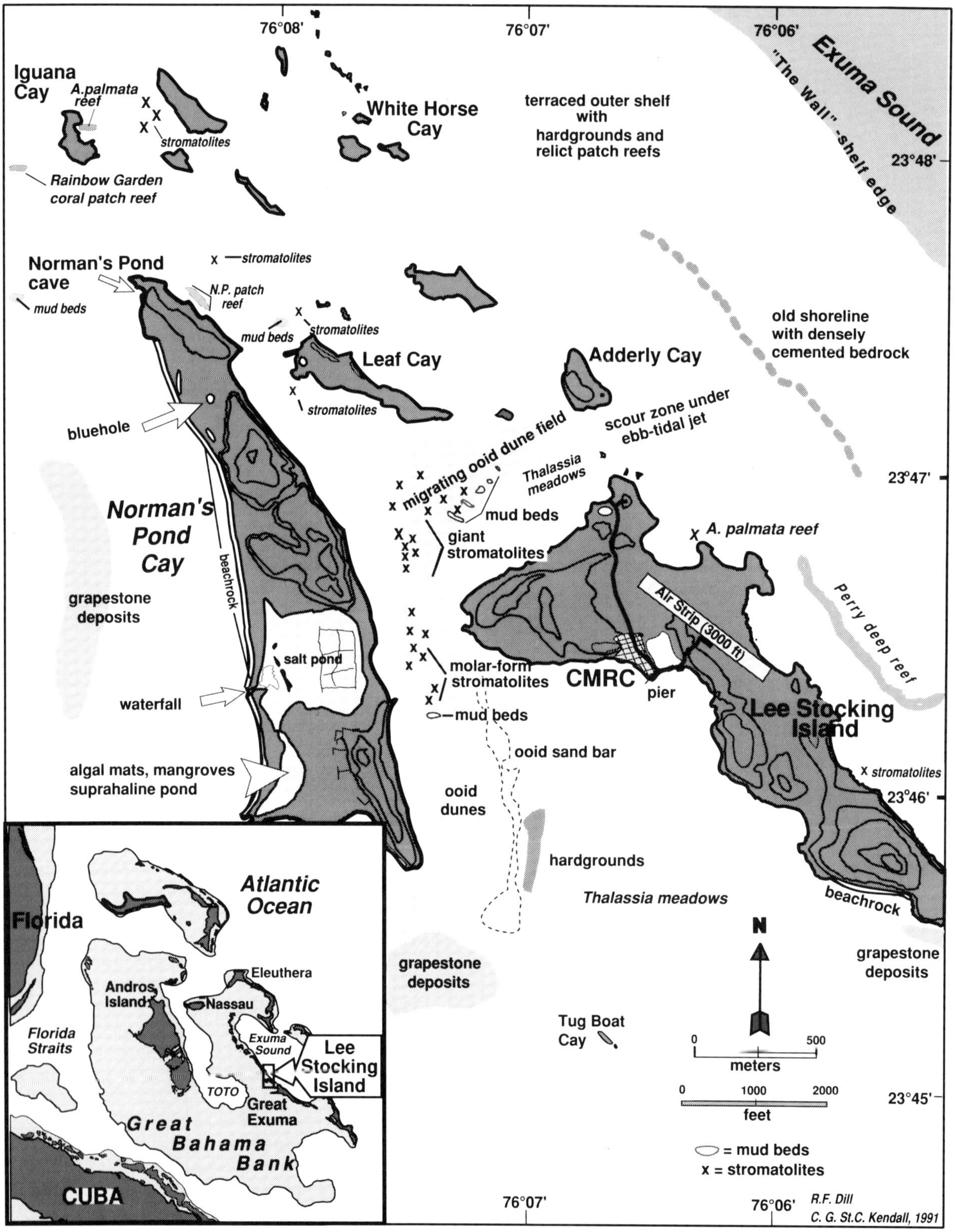

FIG. 1.—Locations of geologic features adjacent to Lee Stocking Island at the south end of the chain of Exuma islands and cays. Inset map of the Bahamas shows relation of Great Bahama Bank, deep platform reentrants of Exuma Sound and Tongue of the Ocean (TOTO) to study area 25 km north of Great Exuma Island.

nel between Lee Stocking Island and Norman's Pond Cay, skeletal and ooid sand forms a large migrating dune field (Fig. 1). Near the eastern edge of the dune area, the seagrass *Thalassia* stabilizes ooid sands bordering the main channel, forming a resistant mat of sediment and roots. Erosion scars ("blow-outs") caused by strong currents reveal that the lower parts of the mat have cement-encrusted roots that in some areas are becoming lithified. In the same area, carbonate-producing plants such as species of *Penicillus, Udotea, Ripocephalus,* and *Halimeda* are much more calcified and crusty than those found outside the area of hardground development (Fig. 1). Grapestones are far more common just beyond the bankward sides of the distributaries of flood-tidal deltas that extend back onto the bank for distances of as much as 25 km. All of these features of enhanced cementation have two common distinctions: 1) restriction to areas where microbial mats colonize the substrate, and 2) location within a zone of turbulent mixing of warm, saline Great Bahama Bank and cooler oceanic Exuma Sound waters. Sedimentary features found in this environment are unique to this region of tidal mixing and characterize this distinctive bank-margin setting (Dill and others, 1989; Aalto and Shapiro, 1990). Distinctive features such as the stromatolites, mud beds, ooids, flood-tidal dunes and hardgrounds are useful in developing a model, based on sedimentary features that define and characterize the eastern Bahamian bank margin. Such a facies model can be used to interpret environmental settings found at ancient carbonate-platform margins (Kendall and others, 1990; Falls and others, 1990).

The Bahamian stromatolites differ in size and shape and provide a model for environmental relations with the chemistry of the sea water and ocean currents. They are in many ways similar, yet distinctly different when compared with several well-studied examples of ancient and modern stromatolites. In particular, those in Shark Bay, Australia (David, 1967; Logan and others, 1974; and Burne and James 1986) and ancient forms that developed during the Paleozoic and earlier rock sequences (Awramik and Margulis, 1976; James 1983; Awramik, 1986; Griffin, 1987; Awramik and Riding, 1988). The extremely interesting hypersaline forms growing in Storrs Lake on San Salvador Island in the Bahamas, described by Neumann and others (1989), provide yet another site for stromatolite development in an area that was a subtidal setting during Holocene lowstands of sea level. The discovery of stromatolites as components of an intertidal fringing reef off Stocking Island in the Exuma Islands (Reid and Browne, 1990, 1991) are another important new Bahamian find which provides another setting for their formation. Stromatolites are now being found in many areas where few field geologists have previously ventured; in Neuman's case, an area of "stinky, dirty-gooey, sticky, slimy" muds of a hypersaline lake. Because of this high diversity of settings and ever increasing range of occurrences the Bahamian studies have become a part of the UNESCO-IGCP Project 261 for Stromatolites. This worldwide effort compares the Bahamian forms with those growing both in highly saline and fresh waters such as Lake Tanganyika (Cohen and Gevirtzman, 1986; Cohen and Thouin, 1987; Casanova, 1987); the Cuatro Cienegas basin, northeast Mexico (Caran and Winsborough, 1986; Winsborough and Golubic, 1987); Nevada (Osborne and others, 1982); Antarctic oasis lakes (Parker and others, 1981); other lakes (Awramik, 1984); in playa lakes in southwestern Australia (Burne and Moore, pers. commun., and Burne and Moore, 1987), and Shark Bay (Logan and others, 1974) and from personal observations (Kendall and others, 1990). These comparative studies demonstrate the differences between the Bahamian stromatolites and those forming in other areas and settings. There are undoubtedly many more discovery sites waiting for those willing to venture into new areas.

DESCRIPTION OF LEE STOCKING AREA

Oceanographic Conditions

The channels in which the features described in this paper occur are 2 to 10 m deep and as much as 500 m wide. Large ooid-dune fields typically occupy channels, forming flood-tidal banks and shoals. Strong currents keep the upper surfaces of the giant ooid-sand dunes and shallow banks in motion throughout most of the tidal cycle. Flood tides bring oceanic water from Exuma Sound onto the eastern margin of the Great Bahama Bank for a distance of about 8 km. The ebb tides and wind-driven circulation bring warm and more saline waters, formed by evaporation and solar heating on the shallow bank, back to the margin. During summer months, the dense water spills over the upper lip of Exuma Sound as strong downwelling currents. In some areas, the currents are strong enough to cause scour depressions in shelf sediments. Undercutting of rock outcrops occurs where the ebb-tidal flow intensifies between inter-island channels. This phenomenon could be an important process in carrying sediments formed at the bank margin into the inter-platform basins. Constrictions by shoals and islands on the bank margin accelerate the flow of tidal currents. Currents around stromatolites and over shallow hardground shoals reach turbulent velocities at maximum peak flow of 150 cm/s. Tides are diurnal, having a maximum sea-level excursion of about 1 m. Currents in the channels between islands at the bank margin usually peak between 60 and 100 cm/s during maximum tidal excursions, and are dependent on wind conditions. At both high and low slack tides, there is a period of no current flow (slack tide) that lasts between 30 and 40 min, during which detailed *in situ* measurements can be made. Following tidal reversal, the currents again gradually increase, reaching peak-flow conditions in about 2 hours. Local current patterns and duration of flow on and off the bank depend on storm intensity and duration of trade winds. As a consequence, the topography of the shallow-bank top is continually changing as the submarine-dune fields migrate. Stromatolites and mud beds are exposed to major changes of environmental conditons as they are covered and exposed by the ooid sands forming dunes.

The waters from Exuma Sound that flood the stromatolites at high tide are exceptionally clear and have normal oceanic salinity of 37 ‰ to 38 ‰. Underwater visibility in the tidal channels during flood tide often exceeds 50 m. In contrast, ebb-tidal waters from the shallow banks have a greenish cast, are more saline (in the summer months as high as 40 ‰), and have a horizontal visibility from 5 to 15 m.

On the bank just south of Lee Stocking (Barraterre Cut), median maximum seawater temperature decreased from 32°C (90°F; range 29° to 33°C) in August of 1987, to 29°C (84°F; range 27° to 31°C) in October. During the same interval, the median minimum seawater temperature was 29°C in August and September (range 28° to 30° and 29° to 30°C, respectively). Minimum seawater temperature begins to decrease in October (median 28°C, range 24° to 30°C). Lowest temperatures were 24°C (75°F) and represent wind-generated, upwelling of deep water from Exuma Sound (Lang and others, 1988; R. I. Wicklund, CMRC, pers. commun.).

A special condition occurs when hurricanes and large tropical storms pass over the region, creating strong storm surges and wave-induced bottom currents. During these periods, the waters of the bank turn a "milky white" from resuspended lime mud previously deposited on the bank during quiet periods (R. I. Wicklund, pers. commun.). Strong ebb-tidal flow created by storm surges carries the sediment-charged waters to the ooid-dune areas at the bank margin where mud beds are found. The present consensus of researchers working in the area is that the bankward-flowing flood tides hold the storm-derived, sediment-charged water at the bank margin long enough for mud deposition to take place in the relatively quiet waters found in troughs in the lee of dune crests. There also must be a mechanism to cause the muds to form aggregates. One hypothesis suggests that the presence of organic material and a change in the temperature cause rapid settling of pellet-size mud aggregates. Another suggested source for the mud-size aggregates of aragonite needles in such high concentrations is that the mixing of carbonate-saturated waters with different physical and chemical concentrations can accelerate the precipitation of a nearly pure aragonite mud during these storm-induced periods of rapid deposition. A different explanation is presented by Boardman and others (1989), who contend that the muds found in the ooid dunes were deposited in a superhaline coastal lagoon formed as sea level crested over the platform's upper lip during the late Holocene transgression.

The relations of organic material within sea water and the solubility and precipitation of carbonate minerals (aragonite) from supersaturated tropical waters have been extensively studied (Chave, 1965; Chave and Suess, 1967, 1970; Chave, 1970; Land, 1967; Land and others, 1967; Macintyre and Playford, 1964; Smith and others, 1968; Land and Moore, 1980). A strong indication exists that organic molecules are able to coat carbonate nuclei and inhibit crystal growth and precipitation, even though the sea water is saturated in carbonate (Mitterer and Cunningham, 1985). Therefore, it is interesting to speculate on whether or not some mechanism is active in the mixing zone that could strip the organic material from carbonate crystals as they are forming. One possibility is the oxidation of organic coatings on developing carbonate nuclei by photosynthetically produced oxygen, which brings about a reaction whereby crystal formation exceeds coating in the supersaturated waters in the mixing zone. If this occurs, it could cause rapid aragonite needle precipitation, and possibly the formation of aragonitic pelletoids, which, once formed, would rapidly settle to create mud beds. Another possibility is that biological activity at the microbial level or even in plankton blooms could reduce available soluble organic material to sufficiently change the chemistry of the supersaturated water and allow the precipitation of carbonate muds. This later mechanism has been suggested as the cause of rapid crystal formation to form "whitings" (Robbins and Blackwelder, 1990). This avenue of reseach certainly deserves more attention with newly developed *in-situ* microprobes and SEM technology.

Chemical Oceanography

A pilot study, to obtain chemical oceanographic data, was conducted in June and July of 1988 near Lee Stocking Island by Falls and others (1990). The primary objective was to compare the carbon- and oxygen-isotopic composition of modern platform carbonates with the isotopic chemistry of platform water. These data permitted the determination of the time of carbonate precipitation and the influence of biological, physical and chemical factors on precipitation. Samples of sediment and water for stable-isotope analysis were collected from areas of subtidal columnar stromatolites, hardgrounds, crusted muds, and in whitings. Analyses for δ^{13} C of total dissolved inorganic carbon (total CO_2) of water samples, δ^{18} O of water samples, and the stable carbon and oxygen isotopes of carbonate samples are being made by Falls at the Stable Isotope Laboratory at the University of South Carolina.

Stable-isotope analyses of modern carbonates from the tidal inlet along the north end of Lee Stocking Island indicate a strong correlation of both carbon and oxygen isotope values with grain size. Carbonate samples used in this study include the soft, recently formed, outer surface of several large columnar stromatolites, bedded muds from within the tidal channel, and marine hardgrounds from the shallow shoals adjacent to the channel. All samples were predominantly aragonite and were divided into mud-, silt- and sand-size particles and bleached of organics prior to isotopic analysis. Isotopic values for all carbonates range between 3.3 to 4.6 ‰ for δ^{13} C and −0.3 to −1.4 ‰ for δ^{18} O. Values of δ^{13} C for the mud-size aragonite from all of the aforementioned samples are depleted relative to the silt- and sand-size ooids, and peloids have isotope values lighter than 4.1 for δ^{13} C and 0.8 ‰ for δ^{18} O. The δ^{13} C values for green calcerous algae from the restricted platform behind Lee Stocking are similar to values for the stromatolite mud. Values of δ^{13} C for the hardground and channel-based muds are intermediate between the δ^{13} C values for the calcareous algae and coarse grained sediment. These data suggest that the green calcareous algae, which flourish on the restricted platform, are a significant source of aragonite mud in and around the channel but are not the only source (Falls and others, 1990).

Carbon-isotopic studies of total dissolved inorganic carbon (total DIC) have primarily been applied to open-ocean water masses (Sackett and Moore, 1966; Deuser and Hunt, 1969; Kroopnick, 1985). Although the oxygen-isotopic composition of platform waters has previously been studied

(Lowenstam and Epstein, 1957), the stable carbon-isotopic analysis of total CO_2 in the ambient sea water has been ignored or only used to normalize ^{14}C data in previous studies of the Bahamian platform and Exuma Sound (Smith, 1940a,b; 1941; Broecker and Takahashi, 1966). Preliminary results from our study of the shallow-water platform around Lee Stocking Island indicate that δ^{13} C values of total CO_2 for the platform water (−2.00 to 0.59-ppm) are significantly depleted relative to the published values for the Atlantic Ocean (Kroopnick, 1985; 1.34 to 2.13-ppm) at roughly the same latitude as our Lee Stocking samples (Fall and others, 1990).

During an extended period of residence on the platform, the carbon-isotopic composition of the water mass is subjected to preferential fixation of ^{12}C by photosynthesis, increased salinity by evaporation and increased temperature by solar heating. The resulting enrichment of the stable-carbon isotope reservoir by photosynthesis is a diel process and is partially compensated by the process of respiration. The abundance of organic material produced on the shallow platform is shown by the greenish cast of the water. During ebb tides, the decaying organics are exported to the bank margin. The combination should result in significant differences between the carbon-isotopic composition of the open-oceanic water occupying the semi-enclosed Exuma Sound and those forming on the Great Bahama Bank.

Water samples were collected using a hypodermic syringe within 2 m of the bottom, in the upper 3 cm of both columnar and molar-form stromatolites, and in the upper few centimeters of the sediment during slack high and slack low tide. The isotopic enrichment observed during slack low tide relative to slack high tide in the channel presumably reflects the exchange between enriched platform water from the bank west of Lee Stocking Island and isotopically lower values for water from the Exuma Sound side of the island.

Giant Lithified Stromatolites

Columns, domes, and biohermal forms.—

The stromatolites near Lee Stocking Island range in height from a few centimeters to 2.5 m. In the central channel, they occur as individual columns growing upward to the height of maximum sediment suspension by strong bottom currents. Vertical accretion is primarily from a buildup of dome-shaped layers of cemented oolitic and skeletal sand. The internal structures viewed in cross sections form hemispherical laminations and digitations. The primary stromatolite form consists of hemispheres and club-shaped columns, which often coalesce to form linear bioherms in areas of abundant and well-developed dunes (Fig. 2). The large submarine-sand dunes and megaripples associated with the giant stromatolites give way to low, starved dunes on the seaward end of Adderly Cut north of Lee Stocking Island (Fig. 1). Here the dunes have troughs that not only contain small developing "juvenile" stromatolites but also a lag deposit of oncolites and shells covered with microbial mats that form a compacted interlocked "hardground" (Fig. 3). The same, outer, seaward parts of the channels also are bottomed by oncolites resting on laminated beds of fine-grained lime mud. In the summer months, most of the stable and exposed areas are covered with microbial mats forming a "hardground." Scour depressions and exposure of bases of many stromatolites in the troughs of dunes reveal they are firmly anchored on a densely cemented, caliche-encrusted paleosol with micro-karst solution pits. This surface is interpreted as the contact between Pleistocene and Holocene deposits. A large concentration of conch shells, some highly eroded, occurs in this cemented basal bed in an area between western most point on Lee Stocking Island and Norman's Pond Cay. The deposit also is related to densely cemented dunes encrusted with crustose coralline algae just west of the main channel. The hard, brownish-red paleosol subsurface is in several places overlain by a cemented skeletal sand and gravel that forms a sheet-like deposit, from which the club and molar-form stromatolites grow. Ballast stones of granite and gneiss have been found incorporated within the cemented layer above the paleosol, showing that they are probably less than 200 years old. Conch shells often form the central nucleus of these stromatolites. In slabbed columns, worm tubes were found, but no internal corals were observed. Large conch (*Strombus gigas*) have been incorporated in some of the columns in the outer part of the channel north of Lee Stocking Island and are often found forming the nuclei of many of the small stromatolites in other areas. It is important to note that some of the conchs have man-made hatchet cuts in their shell that allowed the removal of the animal for food. The cuts prove that stromatolites are forming in the present environment and are not remnants of a prior lower stand of sea level. The importance of this relation will be discussed later.

A striking feature of the Lee Stocking stromatolite columns is their morphology. They are streamlined, with their long axis often tilting toward the direction of the prevailing flood-tidal current. Tidal channels that trend east-west contain similarly oriented stromatolites. The channel north of Lee Stocking Island makes a sharp 90° turn to the south approximately half way along its length, taking on a north-south trend. The tilt and streamlining of stromatolites in this inner part also change to a north-south orientation and are oriented perpendicular to the direction of tidal currents. Apparently erosion, water chemistry, clarity and/or sediment nourishment are the dominant shaping factors. The shapes and lean are clearly hydrotropic, being controlled by water-flow patterns. The processes that cause the morphology of stromatolites are important because ancient forms have been given genetic names and classified by shape and growth forms. In fact, it may well be that the gross shape is not determined by different organisms forming the microbial mat but by physical factors, such as current direction, salinity and temperature. Another possibility is that different growth forms are caused by a preference of some organisms to live in specific physical environments.

Examination of the physical oceanographic processes shows that the different forms found in the Lee Stocking area are not heliotropic, or a morphology dominated by a relation responding to the position of the sun. This is an important difference from other areas and in comparison with ancient forms, where the position of the sun is often

FIG. 2.—Giant stromatolites: (A) Large club-shaped molar-form stromatolite with depressed center. For scale, diver is 6 ft tall. Note scour depression. (B) In general, biohermal walls of coalesced stromatolites have the same trend as migrating dune crests. Note heights of the tallest stromatolites are level and at same depth as crests of the highest dunes. (C) Coalesced stromatolites inclined toward incoming flood-tidal currents; Diver is headed westward away from incoming tidal direction. Scour depressions with ripple marks around stromatolites show refracted current patterns. (D) Exposed stromatolites cover >50 percent of the bottom. Those behind diver are just beginning to emerge (arrow). Upper surfaces of the stromatolites are colonized by a dense growth of green alga (*Batophora*). All photographs were taken in Adderly Cut channel north of Lee Stocking Island at a depth of 5m.

FIG. 3.—Rip-up clasts or mud chips form when mud beds break apart upon being undermined by scour. Mudchips and other, larger debris (small stromatolites, shells, cemented shrimp burrows, rhodoliths, oncolites, and driftwood) are carried to and concentrated in troughs of giant sand waves as lag deposits. (A) Small stromatolites and lag deposits at east end of Adderly Cut channel north of Lee Stocking. This part of the channel is an area of starved migrating dunes that rarely reach crest heights of 1 m. (B) Diver holds 1-m-long measuring rod next to small stromatolite that has become encrusted with epiphites, indicating exposure for a considerable period. (C) Diver hiding in quiet waters in the lee of migrating dune crest (dark arrow) in order to remain stationary in the strong 2-kt current (100 cm/s). Photographs taken at water depth of 8 m in outer part of Adderly Cut channel (D) Crusted-mud chips form a lag deposit along with highly bored cemented oncolites. (E) Tube-shaped cemented burrow resembling those made by *Callianassa* shrimp are abundant in lag deposits in this part of the channel.

quoted as being the dominant controlling factor of morphology (Awramik, 1986).

The surface texture of stromatolites often is an indicator of physical conditions that control morphology. In Adderly Cut, north of Lee Stocking Island, many of the larger stromatolites have overlapping shingle-shaped protrusions facing the incoming flood-tidal waters in east-west trending channels (Fig. 4). The upper surface of these features forms small ledges that have a greater growth rate than areas behind the shingle. With time, a small space forms in the lee of these structures, separating them from the main body of the columnar stromatolite. The differential growth rates cause them to grow up and away from the main body. The pattern of growth is found in cross sections showing the internal structures of large stromatolites. The structures appear as cross-laminations at the edge of one side of the normal dome-like structures. If this different growth pattern, developing on the side of the stromatolite facing the incoming oceanic waters (another hydrotropic manifestation), proves to be common, it could be useful in determining the direction of the bank margin in ancient carbonate platforms with similar stromatolitic growth forms.

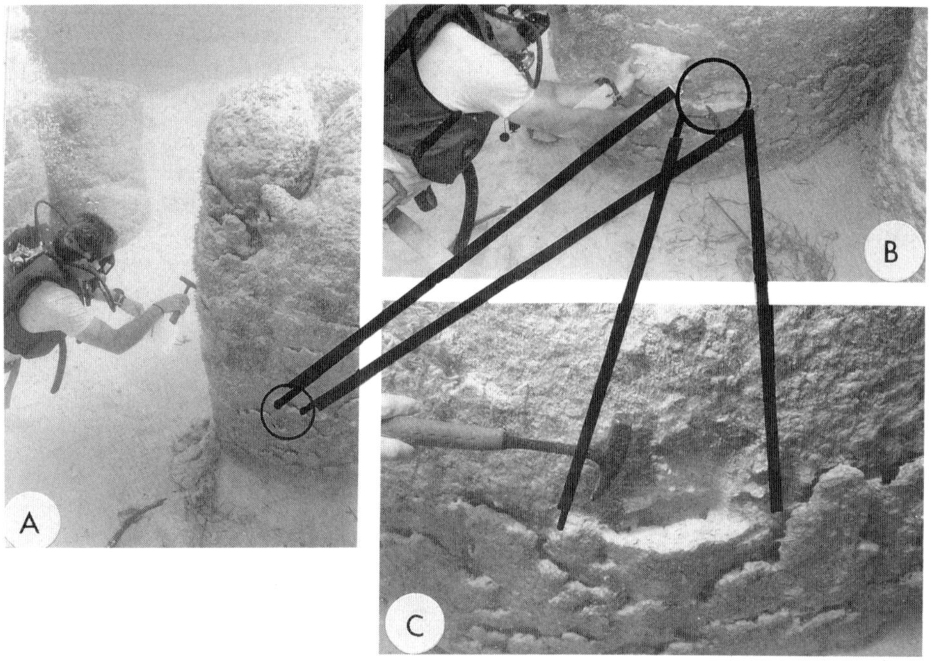

FIG. 4.—(A) Shingled surfaces form on sides of large stromatolites facing incoming Exuma Sound flood-tidal flow. Shingled structure also is reflected in internal fabric of slabbed columns as overlapping shingles on one side of the stromatolite. Lines connect same locations on large column (A) at progressively (B) to (C) closer views. Site located in Adderly Cut north of Lee Stocking Island at depth of 7 m.

The unlithified upper surface of the stromatolites ranges from smooth to pustulate. Observations by the author indicate that the textural variation may be related to the duration of exposure during different seasons and the microbial composition of the recolonizing mat, which is seasonal. The shape and morphology of stromatolites at different locations in the channel appear to be related to the physical processes, whereas their surface textures are affected by a combination of current and dominant organisms forming the microbial mat. In the channels north and east of Lee Stocking, stromatolites form over 50 percent of the bottom area (Fig. 2D). However, they are often covered with migrating sand dunes and are only visible in the scour depressions and troughs between the dunes. Observations beginning in 1984 show that the dunes are migrating, covering, and re-exposing buried stromatolites. Average burial times appear to be about 4 months before a stromatolite is again re-exposed. Vibracoring has shown that much of the area under migrating dune crests is underlain by a hard layer that cannot be cored beyond a depth of a meter or so, probably because of buried stromatolites. Because of the migrating dune-field cover, the total bottom coverage of stromatolites has not been determined, but it probably exceeds the 50-percent figure in the area of giant stromatolites in the channel north of Lee Stocking Island.

The internal structures found in dissected columns are strikingly similar to some early Paleozoic and Precambrian forms. They consist of convex-upward laminations composed of ooid and pelletal sand having structures like those described by Hoffman (1974; 1976) and James (1983) for the Great Slave Lake region of Canada. Laminae are caused by a combination of variations in grain size, cementation and alignment of pores. The laminations occur in two styles: nested, small-scale, convex-upward laminations that form 1- to 4-cm-wide columns, and larger convex-upward laminations that encompass the large portions of columnar stromatolites. Small columns often are joined together by unlaminated cemented sand. Siliceous spicules of probable sponge origin have been observed in some of the sectioned smaller stromatolites. Many laminations thin toward the edges of the stromatolite and in places are truncated by near-vertical wall structures along the sides of the stromatolites with columnar shapes. The sampled larger stromatolites are composed of both coalesced small and large columns as much as 1 m or more high. The maximum synoptic relief, determined by using any one lamination in the stromatolitic structure, is only 16 cm, whereas the total exposed relief of the structure itself may exceed 2 m (Figs 2 and 5).

Molar-form stromatolites.—

Another growth form is exhibited in the north-south-trending channels. Here, stromatolite colonies have sediment-filled depressions and raised rims. These features are now referred to by the more descriptive term of "molar-form" stromatolites (Fig. 5A), as suggested by Griffin (1987). Molar-form stromatolites are usually found in water depths of 8 to 10 m and are usually in less than 1 m high in the north-south-trending channel between Lee Stocking and Norman's Pond Cay. However, they can be higher, as seen in Figure 5A, which is in the area of giant stromatolites north of Lee Stocking Island. The central part of their domed tops, when exposed, has internal voids that are often filled

FIG. 5.—(A) Large molar-form stromatolites coalesce, forming biohermal walls upon reaching a height of about 2 m. Maximum height coincides with maximum crest height of giant sand dunes migrating through the area. These stromatolites have depressed centers and reflect differential growth—faster at the rims and slower in the center. (B) Club-shaped columns covered with *Batophora*. Irregularly shaped dunes reflect turbulent flow of tidal currents. Ripple patterns show direction of last tidal current and are continually changing. End column (at the diver's hand), when sectioned after retrieval, was concrete-hard and structureless due to extensive multiple borings filled with marine cement. Photographs taken in Adderly Cut channel north of Lee Stocking Island at depth of 7.5 m.

with ooid sand, shell debris, and cemented-mud chips (Fig. 6). The mud chips are derived from nearby mud beds and attest to the capability of the strong currents in the channels to suspend and carry cobble-size objects to the top of the stromatolites. The currents certainly control the shape of exposed columns. Suspended bedload causes a higher rate of erosion at the base of exposed columns, giving them a club shape. Most of the molar-form stromatolites are composed of several coalesced columns. The height of the dunes in the area of molar-form stromatolites rarely exceeds 1 m. Once stromatolites reach dune-crest level, they grow laterally, coalescing with their neighbors to form the molar-form shape. The irregular shape also may be related to differential growth rates caused by exposure to warm, high-salinity bottom currents that flow under incoming, cool, oceanic water during the summer ebb-tidal flow off the bank.

Relation of giant stromatolites to dunes and ooid sands.—

Most of the coalesced biohermal walls and columns that exceed 1.5 m parallel the crests of large (1- to 2.5-m-high) ooid-sand dunes. The crests of these dunes migrate with each tidal cycle, covering first and then re-exposing the upper surfaces of stromatolites. The ooid-dune field north of Lee Stocking Island partly covers a field of stromatolites that grows on either a rubble hardground or caliche-encrusted paleosol. The paleosol covers Pleistocene coral heads observed in karst solution depressions forming windows in the paleosol. The coral heads at a depth of about 9 m are partly dissolved, showing that they were exposed to solution during a pre-Holocene exposure and at a time when there was no ooid-sediment cover or stromatolites in the area.

Not all of the stromatolitic heads and columns are every fully exposed at any one time. One rarely sees the area

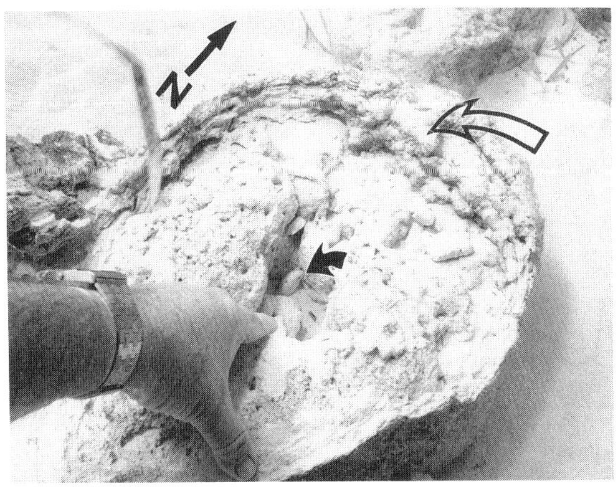

FIG. 6.—Removal of top layers of poorly cemented, smooth, dome-shaped stromatolite (open arrow) shows internal structure. Several small columns have been incorporated within a cap of microbial mat. The central core is a friable, poorly cemented breccia with a matrix of loose ooid sand, shell hash and crusted-mud chips (small black arrow). Larger black arrow points north in north-south trending channel. Note streamlined outer form of the stromatolite. Depth 8 m in the channel between Lee Stocking and Norman's Pond Cay.

where the base joins the underlying Pleistocene bedrock surface. Dunes that migrate back and forth across the surface of the sand body control the production of accumulated microbial-mat deposits. This on-and-off again mechanism is probably the cause of the banding seen in the internal structures of stromatolites (Dravis, 1983; Dill and others, 1986). During maximum tidal flow, suspended and avalanching ooid sand is swept over the crests of dunes to collide with all exposed surfaces of the growing stromatolites. To a diver, these sands look like a submarine blizzard (Dill and Shinn, 1986). During every 6-h tidal cycle, currents transporting ooid sand bombard the stromatolitic surfaces for about 2 hours. Diver observations, underwater video documentation (Shinn, 1986), and our initial staining experiments (Dill and others, 1986; Dill and Shinn, 1986; Shinn and Dill, 1986; Shinn, 1987) show that the stromatolites may receive a dusting of ooid sand during each tidal cycle.

The lithified stromatolites usually are capped by a relatively soft layer of microbial mat containing abundant ooids (Riding and others, 1991). These mats can attain a thickness of as much as 3 cm. The microbial mat is formed by a complex community of plants and microorganisms with mucilaginous appendages that behave like sticky flypaper. This growth strategy, developed in mat organisms that catch ooids efficiently during the tidally induced "blizzards," could have been an evolutionary process. This type of organism would be more successful and more resistant to tidally induced stresses than those without the protection of ooid sand-grain armor. Also, the sand grains would deter some grazers. Distinct layers that are apparent in sections of the stromatolite heads (Fig. 7) develop during successive stages of growth. The open pores between grains provides room for additional growth in a protected area free from bombardment. Thus, when the stromatolites are exposed in the troughs of the migrating sand waves, accretion of one or possibly two layers of ooid sand is possible each day.

Significantly, the tops of the stromatolites that do rise above the sand-dune crests become colonized by the environmentally tough coral *Siderastrea, Sargassum* or gulfweed and a variety of other epiphytes (Fig. 8E). Determining the cause of the maximum growth height and its relation to dune height and sediment-burial depth is important because it provides a modern analog for determining synoptic relief, a value used by paleontologists to describe and classify ancient forms (Walter, 1972; Shinn and Dill, 1986; Shinn, 1987). In general, most of the visible stromatolites are seen in the troughs and scour depressions in the dune field. The observations to date indicate that the maximum growth potential of any stromatolite is controlled by the maximum height of dune crests. Stromatolites that exceed this height suffer surface degradation and a formation hiatus.

FIG. 7.—Sawed cross section of a stromatolite shows internal domed lamination composed of cemented ooids and skeletal sand. Voids within the layers (open arrow), often filled with uncemented ooid sand, form a highly permeable structure. Outer surface is smooth due to sediment-filled, binding microbial mat (dark arrow). Stromatolite located in the channel between Lee Stocking Island and Norman's Pond Cay at a water depth of 7.5 m.

Degraded stromatolites.—

In areas near the sides of the channels, the ooid-sand dunes have been stabilized by dense mats of the marine grass *Thalassia*. This is most commonly seen at the edges of channels or where the amount of ooid sand is insufficient to create dunes (Fig. 9). If sand does not actively strike the stromatolites, they stop growing and are destroyed by bioerosion. Without stress, the cemented tops are colonized by sessile organisms and plants. Crevices and holes in these stromatolites are densely populated by grazing fish and invertebrates (Fig. 10). All exposed surfaces are covered by epibionts, such as species of *Acetabularia, Batophora, Sargassum, Siderastrea,* several varieties of unidentified sponges and crustose coralline algae (Fig. 8E, 9, and 10). Most solid parts of the stromatolites found at the margin of migrating dune fields have been invaded by the boring clam *Pholadea sp.* and the boring sponge *Cliona sp.* In several slabbed sections of degraded stromatolites, the internal dome-like laminations are completely obliterated by multiple borings and cement fill. Some of the "older" (?) larger stromatolites

FIG. 8.—Variations in surface texture of stromatolites depend on time exposure and components of microbial mat-trapping sediment. Texture ranges from smoth (Fig. 7) to pustulant (C). The smooth appearance is deceptive because trapped grains actually cover a rough structure (A) and (B). (D) Recent exposure of a buried stromatolite (open arrow) shows its crinkly, ridge-like surface. Sand grains set in suspension during periods of high currents infill rough hard surfaces (open arrow), forming a smooth, resistant sedimentary surface (dark arrow). (E) Crest of a stromatolitic bioherm colonized by the coral *Siderastrea* (open arrow) and *Sargassum* weed (dark arrow). Other surfaces are coated with an ooid-filled microbial mat and *Batophora*. Water depth is 6 m.

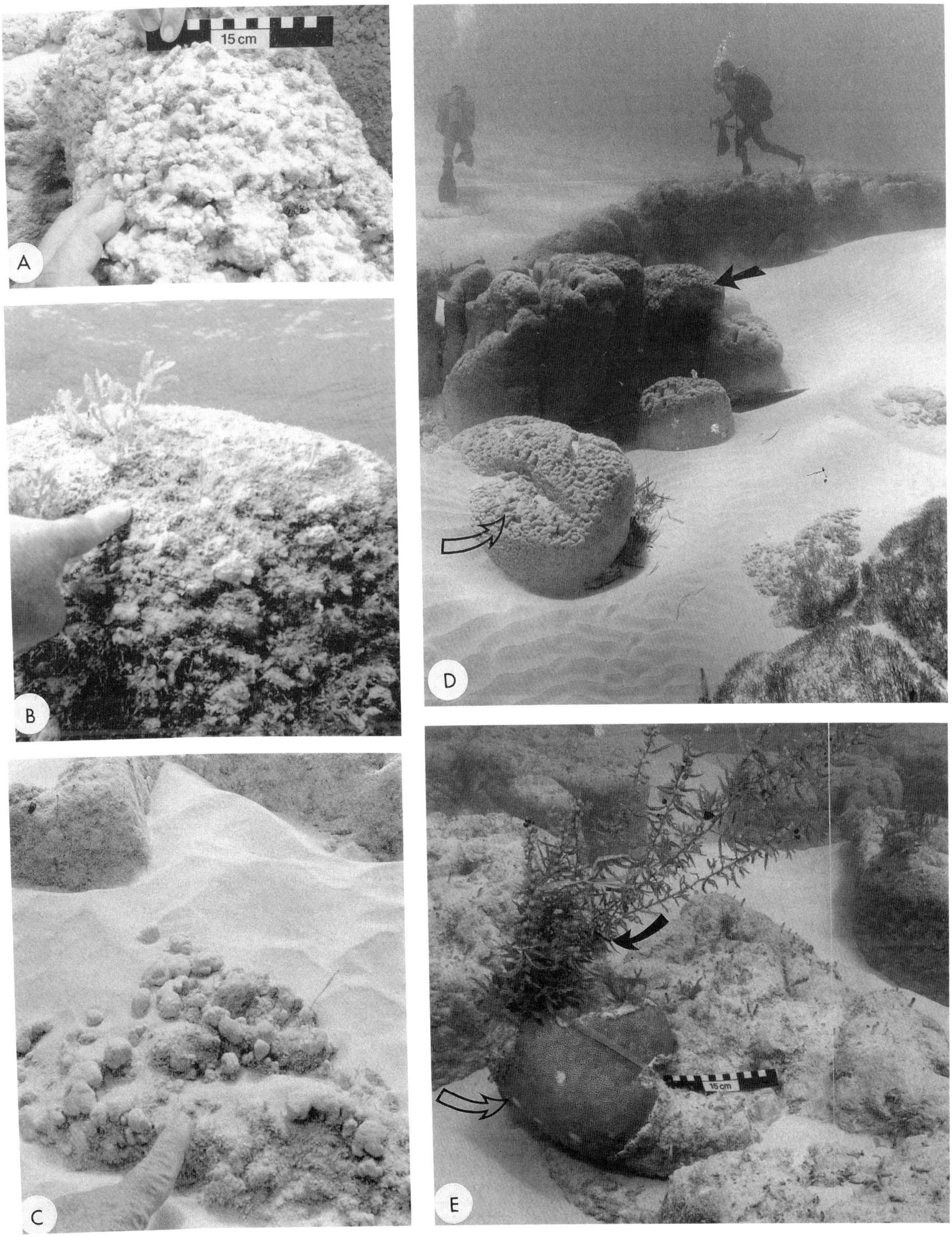
15 cm
A
B
C
D
15 cm
E

FIG. 9.—The south edge of Adderly channel contains lag deposits and low-sand dunes with degraded stromatolites. A sharp transition to a *Thalassia*-stabilized sand-dune area contains relict stromatolites (dark arrow) colonized by a diverse assemblage of sessile organisms. Area below divers is rubble-lag deposit of broken shells, rhodoliths, mud chips, and small stromatolites mixed with ooid and skeletal sand. Water depth is 8 m.

in areas of bottom stabilization have a hard cap of growing coral, which leads one to speculate that some of the patch reefs in areas of lower energy currents might have originally been stromatolite fields that have been colonized by coral after dunes stopped migrating through the area. The lack of healthy growing stromatolites has led the author to conclude that migrating dunes and an abundant supply of sediment are necessary for stromatolite formation in the Lee Stocking region. A similar conclusion was reached by Dravis (1983) for the Schooner Cay area. It could indicate that the conditions forming ooids themselves also control the preservation and development of modern subtidal high-energy stromatolites.

Growth rate.—

The outer surfaces of some stromatolites have been stained with Alizarin Red S to determine short-term, microbial sedimentary-mat growth rates. The staining experiments indicate the potential to accrete a layer a day (Dill and Shinn, 1986; Dill and others, 1986; Shinn, 1987). Another indication of growth rate is from a date of 480 ± 80 yrs on a conch shell forming the base of a 90-cm-high stromatolite. The date indicates that the top of this stromatolite has an average upward growth rate of 0.2 cm per year. The basal core of a 40-cm-high stromatolite collected in Adderly Cut contains an oncolite with a C^{14} age of 3,700 yrs (Aalto and Shapiro, 1990), which would give it an average growth rate of 0.01 cm per year if the oncolite is not a relict feature that formed long before the beginning of stromatolite growth.

Episodic accretion is evidenced by the hiatuses evident in the banding and lamination of the internal structures found in sectioned stromatolites. Periods of active accretion may proceed at 1 cm or more per year (Dill and others, 1986), which is greater than the rate of mat accretion recorded at Shark Bay by Logan and others (1974). Staining experiments have further confirmed that the microbial mats trapping ooids on stromatolites can grow rapidly. Stromatolites stained *in situ* with Alizarin Red S show that a layer of sand approximately one ooid thick (200 μm) had been deposited over the stained surface 24 hours later. Observations and underwater-video documentation show that most stromatolites are dusted with ooid sand during each tidal cycle. Sand thrown against vertical microbial mat-covered surfaces sticks instantly to form a new layer of sediment (A. Jones, pers. commun.).

The potential rates of growth and trapping have been confirmed, but as yet there are no reliable dates on the actual rates of the cementation process or how it proceeds as the stromatolites grow. Nails placed in the sides of the stromatolites in 1985 do not appear in 1990 (6 years later) to

FIG. 10.—Degraded stromatolites form at the south margin of the main channel, where dunes are stabilized by *Thalassia* and are not periodically covered by sand. Stromatolites become colonized by a highly diverse population of epiphites, become burrowed, and are slowly destroyed by bioerosion. (A) Fish (arrow) use stromatolites to hide and for protection from strong currents. (B) A broad meadow of *Thalassia* with low encrusted stromatolites. A cemented hardground crops out through thin skeletal and ooid sand (open arrow). (C) Sponges and algae completely cover stromatolites not periodically buried with sand (arrow). Water depth is 8 m.

have accreted stromatolitic growth around them. Therefore, determination of the actual rate of growth will require much more dating and more detailed long-term measurements.

Microbial Mats

Present studies at CMRC are designed to determine if the microbial mats found on the subtidal stromatolites, mud beds and hardgrounds are in part or even completely responsible for their lithification. Drs. R. Riding from University College in the United Kingdom and S. Awramik from the University of California at Santa Barbara collected mats during an IGCP Project 261 field trip to Lee Stocking in July 1987. Their preliminary identification of the organisms forming the mat community at the top and sides of growing stromatolites reveals a complex assemblage of blue-green algae, epiphytic diatoms, red algae and the dasyclads *Batophora* sp. and *Acetabularia* sp. At the time of their sampling, the dominant organisms forming the mats were microscopic algae and cyanobacteria. Three components of this flora are consistently present, and are frequently mutually associated in creating a complex, layered microbial mat:

1) *Bryopsis*. This unbranched chlorophyte ranges from 50 to 70 μm in diameter and forms a dense felt of recumbent filaments with erect growing tips. The mucilaginous surfaces of these algae readily agglutinates ooids, which commonly cover all but the growing tips of the filaments. Another common, but as yet unidentified chlorophyte with filamentous branches 50 to 70 mm in diameter (but with possible diameters up to 270 mm) forms a dense felt of recumbent and erect filaments. Some of the erect filaments with their growing tips extend up to 1 cm above the mat surface.

2) *Schizothrix*. The dominant filamentous cyanobacterium is a large oscillatoriacian, *Schizothrix* sp., with a well-defined sheath and trichomes up to 7 μm in diameter. Trichomes are blue-green at top surfaces of columns, whereas at depth within the mat (>1 mm), the trichomes are commonly pink. On the sides of stromatolite columns they give a pink color to the surface mats. The cyanobacterium occurs tangled around ooids, is epiphytic on green algal filaments, and is intimately associated with benthic diatoms in gelatinous masses (Fig. 11).

3) *Benthic diatoms*, commonly associated with the filamentous cyanobacterium, also are epiphytic on the *Bryopsis*. Diverse benthic diatoms, commonly associated with *Schizothrix*, appear as patches of golden brown color among the white ooids. Many are epiphytic on green algal filaments and are attached to the ooids and larger algae by means of stalks, tubes, and/or mucilaginous secretions. One species has mucilaginous tubes extending from its body, an adaptation that would facilitate the trapping of ooids. Motile forms glide freely over the surface of stabilized ooids, leaving behind a slime trail that aids in binding the ooids. The average number of diatoms species present in any sample is about 50 (M. Hein and B. Winsborough, pers. commun.). *Mastogloia* is the most well represented genus with 10 species common, followed by *Nitzschia* and *Navicula*, each with about six species. Individual diatoms range in size from small (4 × 8 μm) to long and slender (4 × 200 μm). There are no obvious differences between the diatom populations on the tops and sides of the stromatolites. Population differences do exist, however, in their preferred microhabitat. Cryptic species are concentrated in pockets and interstices of the bound ooid surface and other species project into the current by means of long stalks and tubes.

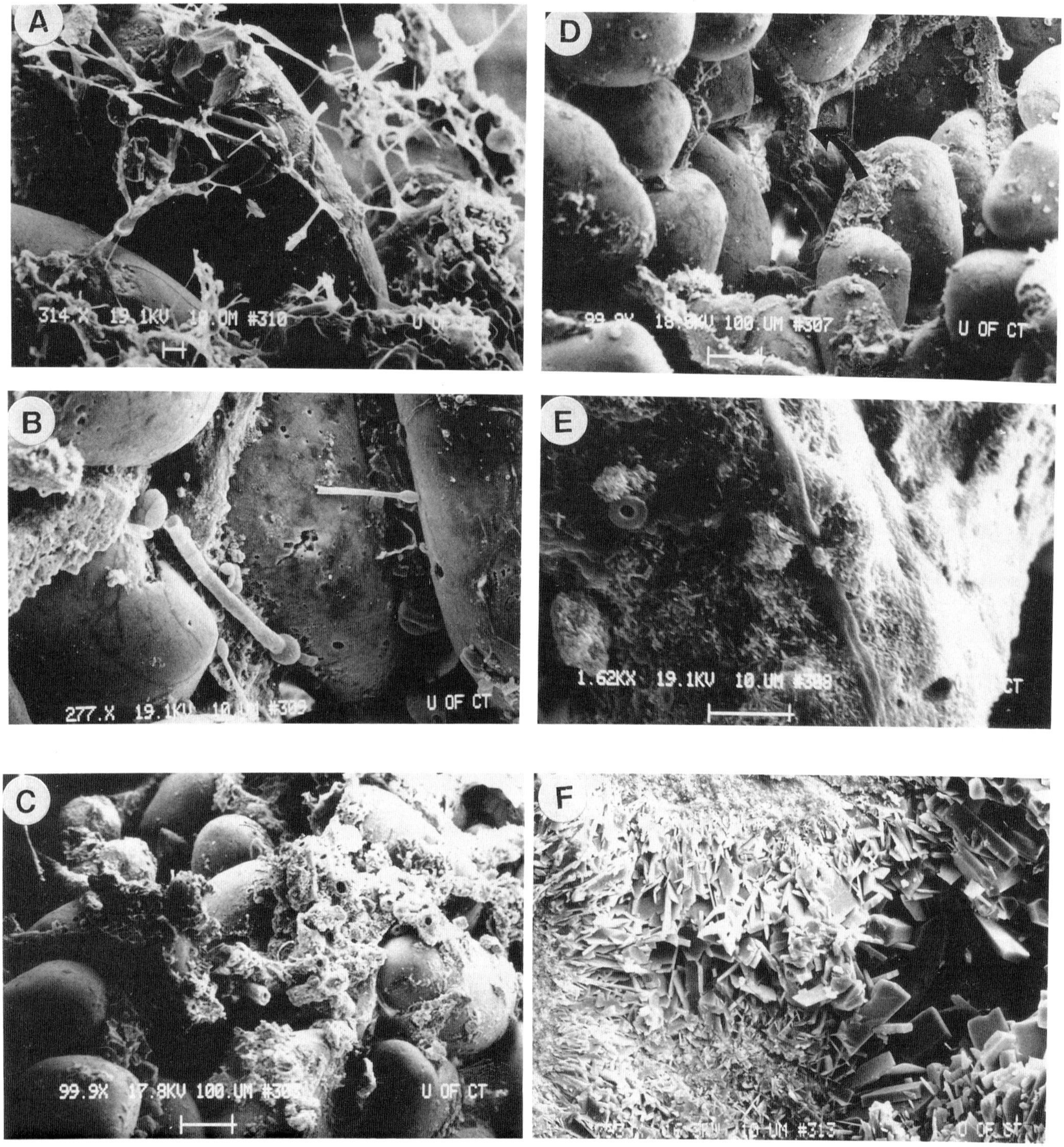

FIG. 11.—SEM photomicrographs of outer part of a stromatolite showing relation of cyanobacteria and ooids. (A) Interwoven filaments of cyanobacteria at outer microbial mat contain ooids (scale bar 10 μm). (B) Tube-like cyanobacteria within pore spaces of cemented stromatolites. Note bored ooids encrusted by tubes composed of aragonite needles. (scale bar 10 μm). (C) and (D) Encrusted ooids tightly held together by microbial-mat cementation and tubes (dark arrow). (Scale bar 100 μm). (E) Enlargement of yoke above black arrow in (D). (Scale bar 10 μm). (F) Bladed-aragonite crystals bind the ooids. Cementation is sufficiently strong to cause breakage of ooids at fractures. (Scale bar 10 μm). Photomicrographs by R. P. Steinen (Dill and others, 1989).

Additional common components are other filamentous cyanobacteria, planktonic diatoms, coccoid cyanobacteria including *Gloeocapsa*, *Gomphosphaeria*, and *Coelosphaerium*, as well as filamentous chlorophytes, rhodophytes, and chrysophytes. Animals associated with the microbial mats commonly include harpacticoid copepods, spinoid polychaetes, foraminifera, cladocerans, nemerteans, ostracodes, polyps of colonial hydroids, encrusting bryozoa, flagellated and testate protozoa and several species of sponges. The taller stromatolites not periodically covered with migrating dunes are often colonized by the coral *Siderastrea* and *Sargassum* weed (Fig. 8E). The green alga *Batophora sp.*, documented by Dravis (1983) as significant to stromatolite growth in the Schooner Cays area, only occurs locally and is seasonal on the stromatolites off Lee Stocking (Figs. 2D and 5B).

Of special interest is a distinctive, branched, filamentous green alga that has fine "hair-like" sticky appendages that appear to have evolved to trap sediment particles (J. West and A. Jones, pers. commun.). Several species of gelatinous colonial chrysophytes and green algae have been collected, which could be important in producing a protective grain-supported structure for growth in what would normally be a hostile environment. Organic cyanobacterial mats are responsible for some trapping of suspended ooids and biogenic sands. However, there are other organisms in the complex consortium of blue-green algae, bacteria, specially adapted diatoms, and as yet unidientified plants and organisms that may be important. Their association probably varies with season and time when buried columns are re-exposed. It is also significant that the micro-ecosystem has adapted to high-energy conditions to protect the mat from further bombardment by suspended grains.

Stromatolite surfaces are colonized by a variety of organisms. The macroalgae *Batophora* (chlorophyte), seaweed *Sargassum* (Phaeophyta), and algae *Laurencia* and *Polysiphonia* (Rhodophyta) are locally conspicuous (Figs. 4, 5, 8, and 9). *Acetabularia* (chlorophyte), which is common in the summer, was not abundant when the stromatolites were sampled in September 1987 or in the winter months of 1988 and 1989. In some samples that have been subject to erosion, the carbonate-encrusted rhizoidal systems of the unidentified siphonaceous chlorophyte are exposed. Colonial hydrozoans, sponges, corals (*Siderastrea*) and calcareous tube worms living on the stromatolites do not constitute a significant part of the structure.

Additional common microbial components are other filamentous cyanobacteria (*Caloxthrix, Phormidium* and *Microcoleus*), coccoid cyanobacteria including *Gomphosphaeria* and *Coeleosphaerium,* coccoid and colonial chrysophytes, and filamentous chlorophytes and rhodophytes. Protists include flagellated and testate protozoans. Microscopic animals in the microbial community include harpacticoid copepods, spionid polychaetes, cladocerans, nemerteans, nematodes, ostracodes, polyps of hydroids, encrusting bryozoans, and at least two species of sponge.

Structurally, the microbial mat can be divided into a several-story system composed of two major components (R. Riding and others, 1991). Both components contain filamentous chlorophytes, diatoms and cyanobacteria. This multi-layer or two-story system is the result of the relatively large size and erect growth habit of the filamentous chlorophyte. A lower story composed of a gelatinous mass of diatoms and cyanobacteria with trapped and bound fine-grained sediment is formed as a result of a relatively non-turbulent micro-environment that exists among the chlorophyte filaments. The filaments act as a baffle and sediment is trapped among the filaments. The surfaces of sand-size grains are colonized by benthic diatoms and cyanobacteria and are tightly bound into the mat (Fig. 11A, B, and C). Although the upper story is composed of the same microbial components, the processes involved in trapping are different. The chlorophyte provides the stable substrate for colonization by benthic diatoms and cyanobacteria. These epiphytes (which excrete mucilage) grow on the chlorophyte and appear to trap and bind sediment directly from the water column. The upper story does not form a real mat, and the amount of sediment trapped in the upper story appears to be less than that collecting at the base of filaments. Size differences between the sediment of the upper and lower stories remain to be determined. Superficial examination indicates that the microbiota from the two levels are similar; however, the lower story may contain more mucilage-rich coccoidal cyanobacteria and diatoms, which may produce more gel. A third story is possible to recognize where colonial hydrozoans, chlorophytes and rhodophytes actively grow, producing higher relief than that of the aforementioned upper story. These erect organisms probably assist trapping by baffling the sediment, which is later bound by the diatoms and cyanobacteria. The third story needs further study because the present work only covers the uppermost layers. The nature of both the endolithic population and organisms observed in SEM photomicrographs of the interiors of stromatolites indicate there is an active endolithic biota that could be affecting marine-cementation processes (Fig. 11).

Cementation Patterns

Preliminary work indicates the stromatolites are cemented by both fibrous and botryoidal aragonite and micritic magnesian calcite (Dill and others, 1986, 1988; K. Aalto and R. Shapiro, pers. commun.). At Lee Stocking the gradient from organically bound sediment to rock extends from the upper surface to a depth of 4 to 5 cm. The cement is mainly acicular aragonite, similar to that observed in other areas of marine cementation. Microscope examination and SEM photomicrographs exhibit binding by aragonitic needles in the form of filamentous growths (Fig. 10) that are probably related to the trapping organisms (Dill and others, 1988).

Although durable enough to resist removal by the strong tidal currents, the upper layers of the microbial mat are easily removed with a knife or crumbled between one's fingers. The laminated mushy layer merges transitionally with cemented laminae of oolite and can be carved with a knife to a depth of approximately 10 cm. The surface can be sawed with a carpenter's saw to a depth of approximately 20 cm (Fig. 12). Below that depth, the stromatolites are sufficiently cemented to resist sawing. Some of the larger ones have central cores that are "concrete hard" (Fig. 13A). The cementation gradient is highly variable. Some stromatolites are well cemented at the surface, whereas others have outer rims of semi-cemented mat (Fig. 8A, B), and still others are found with what appears to be a secondary growth, forming a girdle of soft stromatolite around an extremely hard central core (Fig. 13B).

It is the consensus of scientists studying the Bahamian stromatolites that the cementation is like that in Shark Bay (Burne and James, 1986; S. Awramik and N. James, pers.

FIG. 12.–The sawed stromatolite has a cement-hard central core (black arrow) and soft outer rim (open arrow) that grades in hardness to a point 15 cm toward the center, where it was extremely hard to cut with a carpenter saw. Outer parts could be crumbled in fingers. Location in the channel north of Lee Stocking Island at depth of 5 m.

commun.). Dravis (1983) reports that the stromatolites of Eleuthera transitionally change from a deposit of uncemented ooid sand at the surface to a hard limestone at the core. Marine cementation is also rapid in the ooid-forming environment of the Bahama Bank several kilometers west of the Exuma Islands at Yellow Bank. This region is where marine cementation was first documented to be forming hardgrounds in the Bahamas (Taft and others, 1968). In localized areas, aragonite crystals are tightly bundled to form what has been termed acicular fan druse (Shinn, 1969) or botryoidal aragonite (Ginsburg and James, 1976). Botryoidal aragonite occurs as a partial filling in voids. Micrite-textured Mg calcite is patchy and is more common in finer grained, geopetal internal sediment.

Submarine cementation is believed to be rapid when the rate of carbonate supersaturated water passing through a sediment is high (Land and Goreau, 1970; James and Ginsburg, 1979; Shinn, 1983). The stromatolites in general are highly porous and thus have a high permeability when not covered with a dense microbial mat. This condition takes place when they first emerge from a ooid-dune cover and are thus in the area of maximum current flow near the crest of dunes. At the margin of Exuma Sound, upwelling oceanic water becomes "super" supersaturated with respect to calcium carbonate as it warms by as much as 8°C (17°F) upon reaching the shallow bank. The combination of the solar warming, evaporation and intensive turbulent mixing around stromatolites may be the driving mechanism for the precipitation of marine cement on successive ooid layers and the formation of the marine cements that lithify the stromatolites. This type of marine cementation may be important not only in the growth of ooids, but in grapestone formation and hardground development as well.

Encrusted-Mud Beds

One of the remarkable finds in addition to stromatolites in the Lee Stocking Island region has been their association with laminated mud beds that have accumulated within the high-energy tidal channels (Fig. 1). The mud beds have only been found within the troughs of the migrating ooid dunes and have uncemented ooid sands both below and above their laminated surfaces (Fig. 14). The mud beds are located at depths betwen 4 to 8 m and have been measured with a total thickness of as much as 1 m. However, the thickness of a laminated bed is commonly about 10 cm. Most deposits are comprised of a series of laminated beds 2 to 5 cm thick. Bedding surfaces exposed to sea water become encrusted with millimeter-thick, marine-carbonate cement. If exposed for several weeks, the hard crusts become colonized by the same diverse microbial community of algae, diatoms and marine plants that is growing and trapping ooids on nearby lithified stromatolites, large shells, oncolites and exhumed cemented burrows resembling those of the shrimp *Callianassa* (Fig 3E). The almost pure-white lime muds forming the interior of the crusted beds are soft but cohesive, appearing at first glance to be homogeneous. When rubbed between the fingers, they have the sticky smooth consistency of tooth paste. Once disturbed in this manner, the muds form white clouds of dirty water that drift away with the slightest current. Broken blades of sea grass (*Thalassia*), small pieces of driftwood and shells often are found along bedding planes between crusts that separate individual beds, but very little material other than needle-shaped aragonite mud occurs in the interiors of most beds. This indicates that the individual laminae are deposited at different times, assuming time is needed to develop the crusts and that the muds are deposited in a different hydraulic regime than the plant debris trapped between crusted beds. It is important to emphasize that the mud beds have uncemented ooid sand both above and below their laminated, crusted surfaces. Thus, the muds within the migrating dunes were deposited in the same environment as the ooid sands and are not a relict deposit related to a previous lowered sea level, or a regional environmental setting strikingly different from that presently in the channel.

The thin crusts separating unlithified muds are burrowed and/or bored at both a macro and micro scale as seen in the SEM (Fig. 15). Numerous small burrows are observed, which also indicates a formation of the crust and subse-

FIG. 13.—(A) Many larger stromatolites have a resistant central core that is cement hard from secondary marine cementation (open arrows). Others (B) have a girdle of softer stromatolitic growth (black arrow) growing around the hard central core (open arrow), indicating multiple phases of growth. Quite often the area between the two types is separated by a crevasse as much as 5 cm wide.

quent attack by boring organisms prior to deposition of the next crusted mud layer.

When the fresh mud is gently washed by a stream of water from a pipette, small pelletoids separate from the main body of what appears to be homogeneous mud. The individual pellets have diameters of as much as 50 μm similar to the smaller grains of ooid sands in the area. The size similarity could explain why the fine-mud aggregates are found in similar high-energy hydraulic regimes as the ooid sands and stromatolites. The pelletoids also are seen in SEM photos of the mud and are composed of needle-shaped carbonate crystals (Fig. 15E) (R. Steinen, pers. commun.; Dill, 1988; Dill and others, 1988).

SEM photomicrographs show that the crusts are formed of blunt-ended crystals with their long axis horizontal to the plane of the crust; marine cement binds them together. Crusts have a shiny appearance and have micro-borings, which expose the underlying, randomly oriented, non cemented needle muds. Small rhomboid crystals that look like dolomite occur at the contact between the uncemented and cemented layers. The blunt to rounded ends of the aragonite needles seen in the SEM photomicrographs attest to active resolution of the surface crust and active diagenesis after formation. The encapsulated internal muds still retain their original sharp-ended, needle-shaped crystals even within the pelleted aggregates.

Rip-up clasts form where scour undermines a crusted bed and pieces are broken off by strong currents forming "mud chips" (Figs. 3C and D and 14C). Some of the larger chips are colonized by microbial mats that trap ooid sand, forming small soft stromatolites. Over a 6-month period in 1988, some of these small soft stromatolites were observed to grow and become part of larger structures. Cross sections of lithified stromatolites show that they have internal clasts of mud chips incorporated into their laminated internal structure (Fig. 6), which shows that the stromatolites and the mud chips are contemporaneous and form in the same environmental setting.

A major question immediately arises as to the origin of the aragonitic needles that form the muds because they are not present in any quantity in the waters seen flowing through the interisland channels at the present time. Two possible

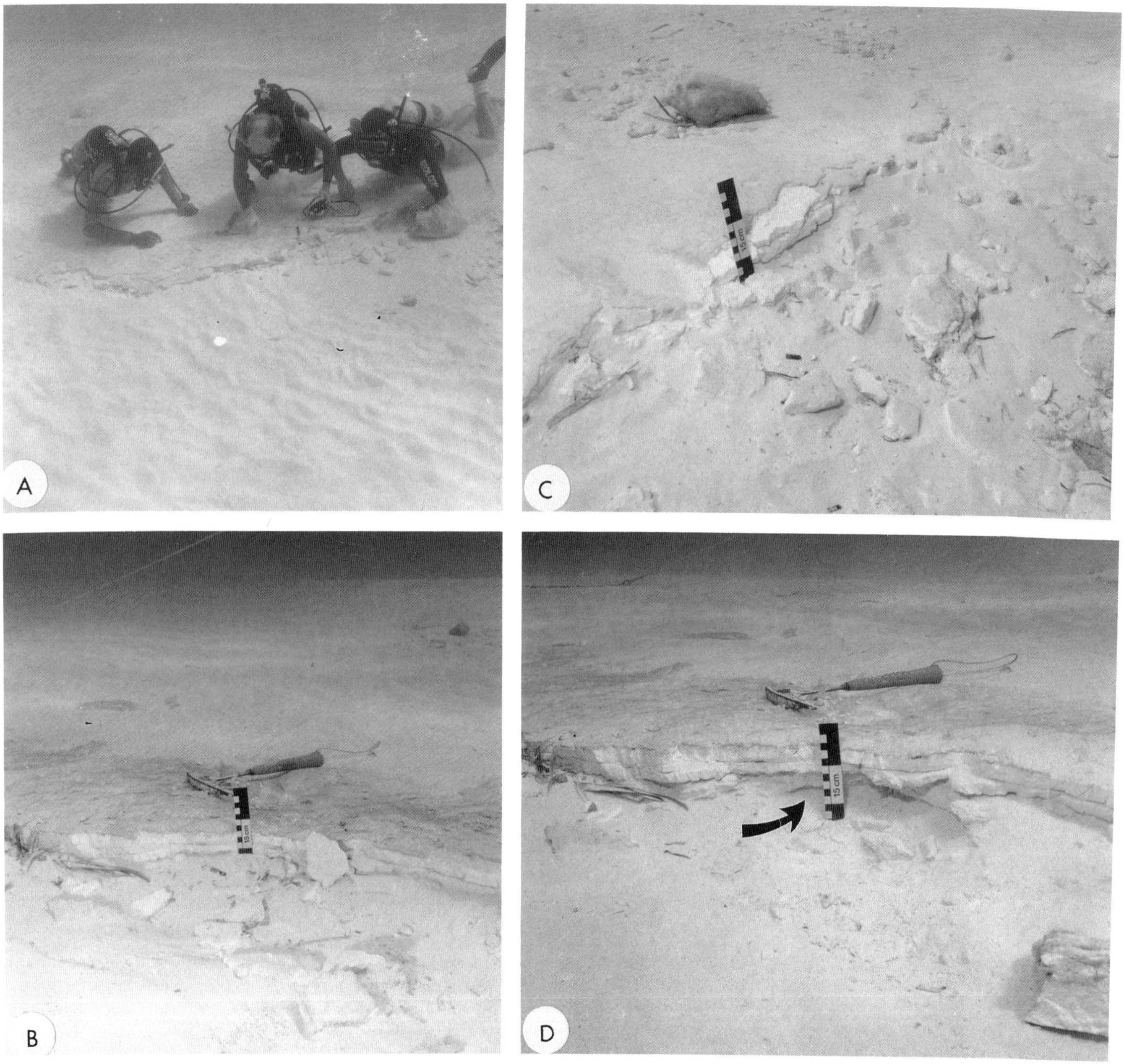

FIG. 14.—Mud beds interbedded within migrating ooid-sand dunes in many of the high-energy tidal channels between bank-margin islands. Mud bed under middle diver in (A) shows exposed edge and internal white, soft, cohesive aragonitic muds. Surface of bed is encrusted with dark grey marine cement. Note dunes in background. Ruler near middle diver's hand in cm. (A) Crusted mud exposed in the trough of a migrating dune north of Lee Stocking Island at depth of 7 m. (B) Closer view of exposed edge of crusted mud showing internal laminations and dark grey cemented crust. Note mud chips. (C) Exposed mud laminations with cemented crust. A small conch shell rests on exposed surface above ruler. (D) Digging below the mud beds near ruler in (B) shows that underlying uncemented ooid sand (dark arrow) and interbedding within the migrating-dune system. Four distinct beds with an average thickness of 2 cm can be seen in this deposit. Bedding surfaces have mm-thick crusts composed of aragonite needles (Fig. 15 A–C). These crusted muds are located in water depths of 8 m in Adderly Cut north of Lee Stocking Island.

FIG. 15.—SEM photomicrographs by R. P. Steinen (Dill and others, 1989) show that muds found interbedded within ooid dunes are composed almost entirely of aragonite needles. (A) Surface crust with burrows acting as windows into uncemented underlying aragonite needles (note blunted ends and preferred orientation of needles; scale bar 10 μm). (B) High power view of mud crust showing arrangements of crystals forming the hard surface (scale bar 1 μm). (C) Closer view of surface showing knobby appearance of crystals. Some have intergrown, indicating they form by precipitation rather than accretion. Most crystals are tangential to the crust surface. The mixture of crystals with both sharp and blunted or rounded ends indicates surface is undergoing both dissolution and precipitation. Large crystals are 1 μm in length and about 0.1 μm in width (scale bar 1 μm). (D) Surface to right of dark arrow in (A) showing surface crust and underlying substrate. Crust is composed of elongated crystals (scale bar 2 μm). (E) Lightly cemented needle mud. Note rhombic crystals that have partly grown around needles of aragonite (scale bar 2 μm). (F) Pelletoid mud from interior of a crusted-mud bed. Pellets can reach diameters of 60 μm (scale bar 10 μm).

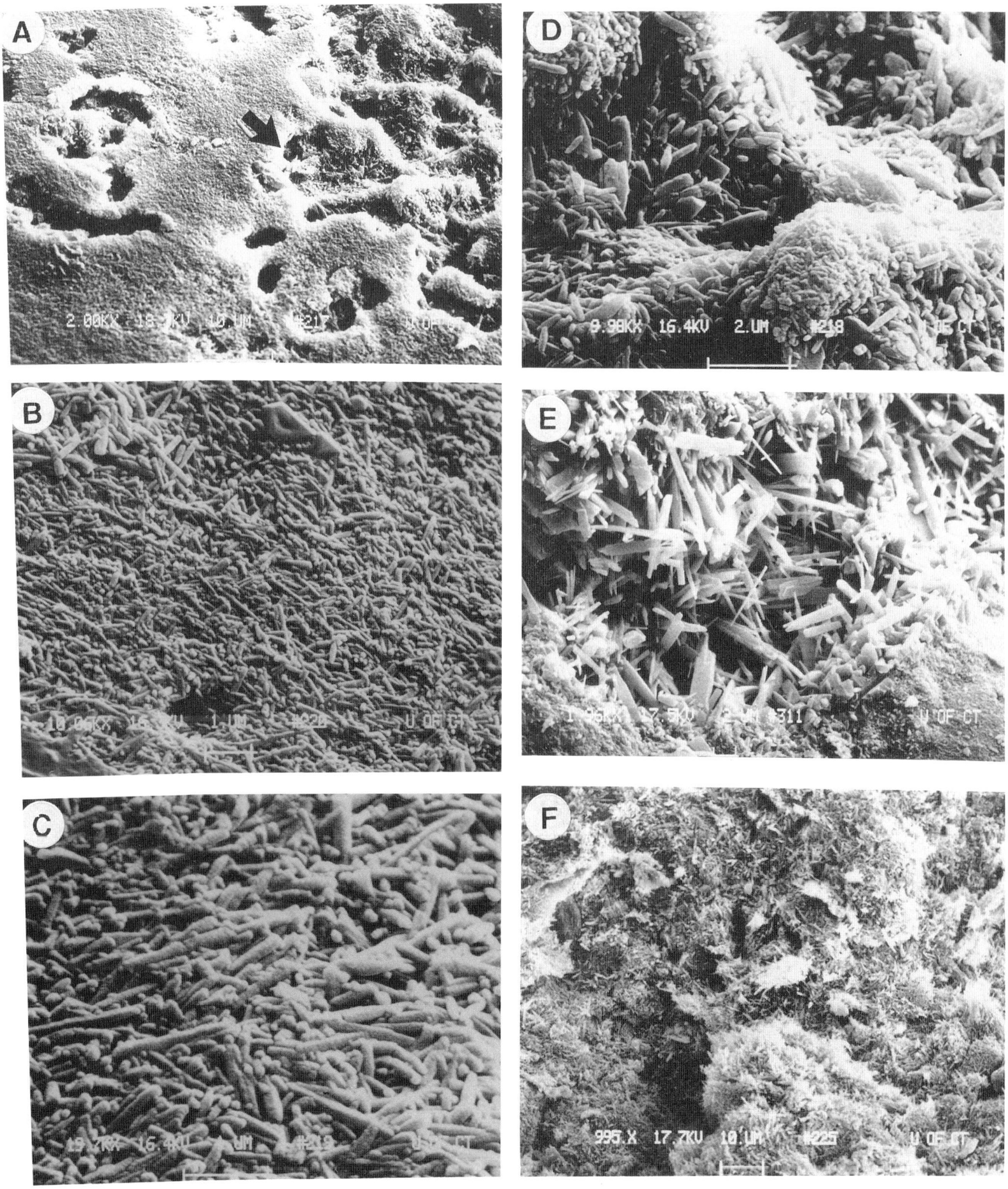
A
B
C
D
E
F
2.00KX
10.UM
#217
10.06KX
1.UM
#220
U OF CT
19.7KX
16.4KV
#219
9.98KX 16.4KV 2.UM #218
#311
U OF CT
995 X 17.7KV 10.UM #225 U OF CT

sources have been suggested, one a precipitation of aragonite needles in the form of whitings and the other the resuspension of muds of biogenic sources, such as remnants of the green algae *Penicillus* and *Rhipocephallus*. During normal periods, muds slowly accumulate in the thin layer of Holocene sediments that blankets the Pleistocene surface of the Bahama carbonate platform (Ginsburg and Lowenstam, 1958; Shinn and others, 1985, 1989; Steinen and others, 1988; Kendall and others, 1990). During tropical storms and hurricanes, this fine-grained sediment is set in suspension as dense clouds of muddy bank water that, in an area such as Florida Bay, are then deposited as a blanketing mud layer over normal sediments and bars after the storms (E. A. Shinn, pers. commun.). A similar resuspension of bank muds has been observed in the Lee Stocking area, where milky-white waters are moved to the bank margins by ebb-tidal currents after large storms (R. I. Wicklund, pers. commun.). At each flood tide, this sediment-laden bank water comes in contact with oceanic water and an intense mixing takes place. The mud beds below this zone of mixing are dominantly formed of aggregates of pelleted aragonite needles, which suggests that it is in this rarely occurring setting that pelletization and deposition of the mud beds take place.

The other possibility is that storms create a condition of precipitation and aggregation when organic material (plankton or suspended bottom material) is mixed. The mixture may cause the formation of flocculated aragonite needles, which then rapidly settle in the troughs of dunes forming the mud beds of nearly pure aragonite. A strong possibility that the aragonite needles are precipitates shown by their δO^{13} values but there is no indication of what causes precipitation to take place or what the processes are that cause aggregation of the needles into pellits.

Chave (1970) showed that carbonate needles are attracted to organic material in the open ocean. It may be that a similar process is responsible for the development of the mud beds. However, it remains to be determined if the muds originated primarily as a precipitate or from resuspended muds of algal origin. Since the muds are primarily formed of needle aragonite that is not commonly found in high concentrations in the water masses that daily flow over the area, then episodic processes must account for their origin. To date, the laminated aragonite-needle mud beds have only been found below areas where there is a mixing of storm-generated, sediment-laden, slightly hypersaline bank waters and the cooler open-oceanic waters. It is in this same setting that one finds not only the crusted, laminated-mud beds, interbedded within well-developed ooid-sand bodies, but also lithified stromatolites and a high production of ooid-sand grains.

Another possibility has been presented by Boardman and others (1989), who described a core taken in the Lee Stocking area mud beds. The core, containing two mud layers, was compared to cores containing similar mud beds and chips found in a tidal channel off Joulters Cay north of Andros Island on the north side of the Great Bahama Bank. Inference was made that the muds formed in a similar environment. The authors assumed that there had been a change in the Lee Stocking area from that of its present status to one like that of Joulters Cay. However, the Joulters Cay mud has been shown by Harris (1979) to be a continuous silt bed. Furthermore, the beds contain both pelleted needles and epibiontic muds, which would make them quite different from those found off Lee Stocking Island (R. Steinen, pers. commun.). These differences therefore make it difficult to justify stating that there was a remarkably different setting at the Lee Stocking site when the muds formed, especially one similar to that found at Joulters Cay. Using the Joulters Cay model, Boardman and his colleagues must speculate that a giant sand bar blocked the entrance of Adderly Cut, forming a large quiet lagoon. Their proposed mechanism for aragonite-mud deposition in this lagoon is precipitation from supersaturated waters trapped behind this protective coastal-dune system. The high salinities were used to justify a lack of burrowing activities by a lagoonal infauna, a condition that prevails in some present coastal lagoons of the Bahamas.

More extensive work in this region by the author and his colleagues make the Boardman and others' hypothesis questionable. There is no evidence of a lagoon in the area of the mud beds off north Lee Stocking within the past 3 ka—a period of time during which sea level reached its present stand and Boardman and others state the lagoon depositing the mud existed. The shallowest mud bed in the core dated by those authors has a C^{14} age of 560 yrs and the lower mud layer an age of 2190 yrs. These dates clearly place the muds as having been deposited in waters nearly as deep as those of today and not at a period of time when sea level was significantly lower than now. Evidence is lacking that would support sand bars formed seaward of the present location of the mud beds east of Lee Stocking during the Holocene rise of sea level. The mud beds in question are not connected; they have elongated lenticular shapes, different thicknesses, and are stratigraphically isolated from each others. The aragonite-mud beds are not found in adjacent grass beds, which extend stratigraphically both above and below the depths where the mud beds would have been deposited if this had been a hypersaline coastal lagoon. The mud beds are limited to an area of mobile ooid dunes and appear to have been deposited in the troughs of the dune system. The mud beds thus are not blanket deposits, as one would expect from a shallow lagoon. The morphology and occurrences indicate that muds were probably deposited in the troughs of migrating dunes and not in a low-energy, bank-bar lagoonal environment, as proposed by Boardman and others (1989).

Grapestones

Extensive deposits of grapestones are found in the distributaries at the bankward edges of flood-tidal deltas. The grapestones are in areas of maximum tidal intrusion of tidal waters from Exuma Sound to the east and along the beaches of the leeward side of most islands and cays just beyond the surf zone. They have not been found in areas where callianassid shrimp are actively reworking sediments or in most *Thalassia* beds. This facies has yet to be adequately studied in the Lee Stocking area. However, we feel that these cemented multigrained sands are important because of their large areal extent and the apparent restriction of these facies to the bank-margin mixing zone. Algal mats commonly are found growing in the areas of grapestones and marked temperature fluctuations are typical in the shallow, centimeter-deep waters that cover them. A distinct temperature gradient from warm at the surface to cool at depth can be felt by inserting one's hand into grapestone deposits.

COMPARISON WITH OTHER HOLOCENE STROMATOLITES

Great Bahama Bank

Following the finding of stromatolites off Lee Stocking, seach for their occurrence elsewhere has continued both on the eastern margin of the Great Bahama Bank and on the western and northern rims as well. Three new sites of stromatolite and mudbed occurrences have been found in channels north of the Lee Stocking Island discovery site. Stromatolites were also found in 1987 in a large area within the Exuma Land and Sea Park near Norman's Cay approximately 93 km north of Lee Stocking Island (E. A. Shinn and R. B. Halley, pers. commun.). However, although the stromatolites are forming there in abundance, the Lee Stocking stromatolites are still the largest and best developed of those seen to date. It is also important to note that lithified stromatolites have not been found in any of the other well-studied areas along the western and northern margins of either the Great or Little Bahama Bank. However, reconnaissance overflights made by Dravis (1983) and research at the CMRC indicate that stromatolites may be forming near the southern rim of the Tongue of the Ocean, approximately 50 km east of Lee Stocking.

The Great Bahama Bank stromatolites now have a demonstrably wide distribution. They are not a unique and uncommon form. Their occurrence in an entirely new setting, one with different sedimentary processes, makes them useful models for comparing modern stromatolites with those that formed in the past, possibly as long ago as 2 billion years. They certainly are not rare, as many early reviewers of our reports stated.

San Salvador Island Saline-Lake Site

Neumann and others (1989) and Paull and others (1989) describe the Mg-calcite stromatolites forming in subtidal hypersaline lagoons on San Salvador, Bahamas. These are forming in yet another setting than those reported in the literature and provide another comparison for use in trying to decipher ancient environments using modern analogs. The San Salvadorian forms are growing in Storrs Lake in water depths of as much as 2 m and attain a height of about 1 m. They have a club-like shape and diameters of as much as 1 m. The stromatolites are draped by a microbial mat with hard cemented parts protruding into the water above the structures by as much as 10 cm. The exterior of the individual stromatolites is densely cemented, whereas the interiors often crumble into loose stacks of millimeter-thick laminae and thicker intervals of thrombolytic, non-layered growth. Sediments enclosing the stromatolites are composed of carbonate, gypsum and organic matter.

Salinities range from 60 to 80 percent and have a seasonal variation that depends on rainfall, wind and solar heating. The Storrs Lake calcified stromatolites are both thrombolytic and stromatolithic and accreted by chemical precipitation at the contact between the microbial mat and the cemented hard surfaces. There is some inclusion of sediments, but unlike in the Lee Stocking stromatolites, the trapping seems to be a secondary process.

The microbial-mat community forming the Storrs Lake stromatolites must have evolved into a carbonate-reducing body within the past 3 ka following the Holocene cresting of sea level over the island shelf. Before that period, the lakes were either fresh or did not exist. It is intriguing to speculate whether or not the cyanobacteria and possibly the bacterial assemblages were in existence in the open sea, or whether they rapidly evolved to live in the harsh environment of a supersaline lake.

Australian Sites

The Bahamian studies have shown that the stromatolites growing and becoming lithified are found in a marine setting that is markedly different than the well-publicized modern marine examples in Shark Bay, western Australia. There, stromatolites are located in a high-salinity, low-current velocity, low-wave energy, and a semi-restricted, intertidal to shallow subtidal lagoon (Logan and others, 1974). However, recent work has indicated that the present setting may not have been the one in which the main stromatolite development took place. The internal cores of the Australian types are old and coated with an outer rind of living microbial mats that may not have been responsible for their formation. There is strong evidence that the region of Shark Bay has undergone tectonic uplift and that the main body of the stromatolites may have formed under conditions similar to those found in the Bahamas (N. James, pers. commun., 1989 and 1990). Until the discovery of the Bahamian sites, Shark Bay was the only known area of open-marine occurrence, resulting in its being the most commonly used and cited modern marine-stromatolite analog for many fossil forms (Logan, 1961; Davis, 1967, 1970; Logan and Cebulski, 1970; Awramik and Vanyo, 1986). However, as Monty (1976, 1977) pointed out, the sedimentary features and the sediments associated with many ancient stromatolites could not have been produced by conditions presently found in Shark Bay. Thus, this classic modern example may not have been respresentative and its use as a comparison to ancient forms may have been misleading.

COMPARISON OF ANCIENT AND MODERN STROMATOLITE DEPOSITS

"Stromatolites" are defined as organosedimentary structures produced by the trapping and binding of sediment. Cyanobacteria (blue-green algae) and their ancestral precursors are generally accepted as being the first undisputed multicellular life on earth capable of building the stromatolitic structures that left a fossil record (Walter, 1976). The modern marine stromatolitic structures off Lee Stocking Island form as the result of the growth and metabolic activities of microbial mats. These, in turn, trap sediment, building a structure that becomes lithified and that will leave a fossil record. This certainly meets the requirements and justifies classifying them as true stromatolites.

Stromatolites have been found occurring in sedimentary rocks on all continents and range in age back to 3.5 Ga (James, 1983; Byerly and others, 1986). Ancient stromatolites dominated the early geologic record for more than 3 Ga and have been an important focus of paleoecologists studying early life on Earth. In most Precambrian rocks, they represent the only fossils to provide evidence of evolution. Sedimentologists have attempted to use these fossil structures as indicators of sedimentary processes and depositional setting (Awramik and Margulis, 1976; Awramik, 1986). They also have been used as stratigraphic marker beds in the simpler ecosystems of the Precambrian and early Paleozoic rocks. Because of their association with mineralization, attempts have been made to use them as index fossils and stratigraphic marker beds with limited success (Hoffman, 1967; Schopf, 1970; Monty, 1973; Walter, 1976; Flugel, 1977; Awramik, 1981). One of the main problems in using stromatolites as index fossils and environmental indicators has been the lack of modern comparative models. Furthermore, as the studies at CMRC indicate, it may well be that evolving species did not create their morphology, but instead the physical characteristics of the setting may be responsible for resultant growth forms. This possibility of deciphering ancient environments has not been extensively applied to the ancient forms and may be important in reevaluation of early deposits containing stromatolites, mud beds and ooid sands.

The physical nature of the sediments and the physical processes forming internal structures indicate that the Bahamian stromatolites have associated sedimentary features and internal structures that may satisfy many of the environmental conditions that Monty (1977) inferred must have existed and been necessary for ancient stromatolite growth, but not exhibited by those at Shark Bay. Thus, as Ginsburg (1986) pointed out in his keynote address before the SEPM Symposium on Stromatolites in 1986, "The Bahamian stromatolites may provide a new avenue of comparative research for interpreting the past environmental processes that led to the development of ancient stromatolites." An existing example of the use of our Bahamian data is the comparison of the Lee Stocking stromatolites with Cambrian examples found in the Death Valley region of California and Nevada, where stromatolites have similar forms and sedimentary associations (Griffin, 1987).

There are also major differences between ancient and recent Bahamian forms (Griffin, 1987; S. Awramik, pers. commun.). One commonly discussed difference is the micritic nature of the central cores of ancient stromatolites (thrombolites), which do not contain internal banding or zonation. The stromatolites we have examined off Lee Stocking are usually formed of banded layers of ooid and skeletal sands in a micritic cement. However, some Lee Stocking stromatolites have central cores that are almost completely micritic in nature with almost no evidence of original ooid sands. Is this difference from other stromatolites found in the same area something that takes place after formation, or is it a diagenetic change, or is it a process related to modern marine boring and re-cementation? At this time these are questions that remain unanswered until many more rock cores or stromatolites in the area are examined and from them it is determined what causes this major difference in what appear to be the same or closely related environments.

If lithified, the cemented crusts found on the micritic mud beds would result in a structure similar to many laminated, micritic, Precambrian limestones that have been classified as stromatolitic. The rounded blocks of undermined, crusted-mud beds contain a pure aragonite-needle mud that appears structureless. If fully lithified, these beds would be micritic in nature as are the ancient forms. If modern forms were lithified and found in ancient limestones, they would be classified as oncolites. Do we then possibly have an unrecognized modern analog that answers some of the questions concerning micritic stromatolites?

SUMMARY

The increased knowledge concerning the mass physical properties of carbonate sediments, internal structures and associated accumulations of lag deposits provides a much broader understanding of the present sedimentary settings that contain stromatolites, ooid sands, grapestones, and mud beds. Bed forms are preserved in many ancient environments and can therefore be used on a comparison basis. Once the processes forming today's bank-margin features are understood, it will be possible to postulate better the conditions that fostered the first world-wide stromatolite development 2 to 3 Ga. It is necessary to emphasize that the Lee Stocking structures are not necessary analogs to all the stromatolites that formed prior to the development of higher life forms in the Paleozoic. However, a knowledge of the Bahamian forms, their depositional setting and their organic associations certainly facilitates the comparison with ancient forms. In many ways the differences may help in understanding how and what conditions prevailed to produce stromatolites, ooids and mud beds in a setting without today's predation, biologically produced coarse-grained sediments, and seas with a chemistry that differs from today's oceans.

Although the types of micro-organisms and the succession of different organisms forming the organic mats may be responsible for the different growth forms found on the surface of the larger stromatolites, present studies suggest that the water-flow patterns may be an equal or even more important factor in controlling the overall shape and growth rate of the giant stromatolitic columns. Also, water clarity, chemistry and nutrient content of both interstitial water and the ambient environment may be as important as currents in determining morphology and species composition of the organic mats that are trapping suspended sediment. Different species names have been given to the different shapes and morphologies of stromatolites found in the fossil record. Our studies indicate that care must be used in applying "morphospecies" concepts to ancient forms.

The high-energy currents carrying a suspended load of ooid sand across a migrating submarine-dune field is not normally conducive to plant growth. The studies at CMRC show that the microbial mats contain organisms with special adaptations and appendages useful for trapping and holding sand grains. It is commonly accepted that these organisms catch and stabilize the ooid sands, using this trapping mechanism as a defense and the resulting protective structure as a way to grow in what would normally be considered a hostile setting. The resulting loosely packed, granular structure acts as a protective shield from the continued bombardment by sand grains. The voids between newly trapped grains also provide additional pore spaces in the building structure and an expanding favorable habitat for more microbial-mat producers to trap more grains. Cementation of the stabilized grains covered by the microbial mat then permits marine cements to form and build an upward-growing structure that is continually washed by larger volumes of nutrient-carrying water. This mechanism could compensate for the adverse setting of sediment "bombardment" and create a more favorable "biological" niche for the mat-forming organisms over ones that colonize flat surfaces. By fostering the development of stromatolitic structures, the organisms would experience a higher rate of nutrient exposure during high currents even in the nutrient-poor tropical oceanic waters. This communal growth strategy would give the stromatolite-building organisms a distinct advantage in obtaining nutrients over the quiet-water organisms. The stromatolitic structure, growing in a high-energy harsh environment, therefore provides a protective habitat from destructive grazing organisms that cannot cope with the high currents, crunchy sand grains and bombardment.

An entirely new mechanism for rapid, fine-grained deposition of mud is proposed to explain the occurrence of pure aragonitic-mud layers within uncemented mobile submarine-dune fields and their relation to stromatolites. An understanding of the causes and occurrences of these mud deposits, interbedded with mobile ooid-sand bodies of the high-energy subtidal channels, could provide geologists, who use bed forms and mudclast associations to interpret ancient paleoenvironments, another setting in which these ancient forms developed. It now appears that fine-grained aragonite needles in dense suspensions, when encountering the bank-margin mixing zone, form aggregates that must rapidly settle and act hydraulically like sand grains. Why this deposition takes place and whether or not it is related to organic material in the water or is solely caused by organisms has yet to be determined. Therefore, intriguing questions remain that can only be answered by further research in the bank-margin mixing zone. This research will require detailed oceanographic understanding of the mixing zone through the application of high-technology instrumentation and *in-situ* micro-geochemistry coupled with SEM images of cements and structures. A much better mapping of the occurrences of stromatolites, hardgrounds, high-energy muds and the cements of the region also will be needed.

ACKNOWLEDGMENTS

My first interest in the Bahamas was kindled by Francis P. Shepard, who on several expeditions brought back spectacular photographs of the deep canyons cutting the Great Bahama Bank carbonate platform and instilled in me a deep desire to know more about how they formed in this carbonate environment. Later on, I became much more involved when Jacques Cousteau provided an opportunity to work in the "blue holes" of Andros Island. I thank these two great men for opening my eyes to this oustanding region for carbonate research.

I gratefully acknowledge the support of Robert Wicklund and Gerri Wenz, CMRC Director and Assistant Director, who have developed the scientific support facilities available through the Caribbean Marine Research Center and have encouraged and supported my scientific efforts at Lee Stocking Island. It is important to stress that this paper could not have been written without the input of my scientific colleagues, in particular my association over the past 6 years with Christopher G. St. C. Kendall, University of South Carolina, Eugene Shinn of the U.S. Geological Survey and Randolph Steinen of the University of Connecticut. The data and thoughts presented herein are ours, but I bear the responsibility for applying them to support the conclusions presented within this paper. In addition I especially thank the students who have worked closely with me in both field and laboratory, in particular those students supported by their own funds and time who have ventured under water to help gather much of the field data.

The generosity of John H. Perry, Jr. (owner of Lee Stocking Island) in making his airplane and island available for our work is greatly appreciated. I thank Shawn Bartlett and all the staff members of CMRC for support, especially for the way in which they have kept the remote island facility functioning as an exceptional research center. Ralph Knight provided aircraft overflights and photographs of the study areas and pointed out many areas of potential study seen on his weekly flights to Lee Stocking Island.

Partial funding for work at CMRC came from grant funds provided by the Office of Undersea Research, NOAA, U.S. Department of Commerce, Emily Stoudt and Susan Longacre, E&P Technology Division of Texaco, and the U.S. Geological Survey. My special thanks are extended to Kenneth Aalto and Russell Shapiro, Humboldt State University; Stanley Awramik, University of California, Santa Barbara; Robert N. Ginsburg, University of Miami; Michael Hein, Water and Air Research, Inc.; Barbara Lidz, U.S. Geological Survey; Robert Riding, University College, Cardiff, U.K.; Anthony Jones and Kevin Kelly, University of Hawaii; Judy Lang and Barbara Winsborough, University of Texas; John Wood, University of California, Berkeley; and Richard Wilkins, all of whom have provided field observations and comments that have led to the conclusions of this paper. Special thanks go to Fred Falls for his geochemical studies. The editorial suggestions of Kenneth Aalto, Randolph Steinen and Robert Osborne have greatly improved this paper and their time spent and offerings are greatly appreciated.

REFERENCES

AALTO, K. R., AND SHAPIRO, R. S., 1990, Petrology of a modern subtidal stromatolite, Lee Stocking Island, Bahamas: Proceedings, 5th Geological Conference of the Bahamas, San Salvador Island, Bahamas, p. 20–21.

AWRAMIK, S. M., 1981, Precambrian columnar stromatolite diversity, reflections of metazoan appearances: Science, v. 174, p. 825–826.

AWRAMIK, S. M., 1984, Ancient stromatolites and microbial mats, *in* Smith,

J. M., and Jones, A. K., eds., Microbial Mats and Stromatolites: Alan R. Liss, Inc., New York, p. 1–22.

Awramik, S. M., 1986, New fossil finds in old rocks: Nature, v. 319, p. 446–447.

Awramik, S. M., and Margulis, L., 1976, *in* Walter, M. R., ed., Stromatolites: Elsevier, Amsterdam, 790 p.

Awramik, S. M., and Riding Robert, 1988, Role of algal eukaryotes in subtidal columnar stromatolite formation: Proceedings, National Academy of Science, Nature, v. 85, Geology, p. 1327–1329.

Awramik, S. M., and Vanyo, J. P., 1986, Heliotropism in modern stromatolites: Science, v. 231, p. 1279–1281.

Boardman, M. R., Lasemi, Zak, and Carney, C. K., 1989, Lime mud in ooid tidal channels: Joulters Cays and Lee Stocking Island, Bahamas: Abstracts with Programs, Geological Society of American Annual Meeting, v. 21, p. A291.

Broecker, W. S., and Takahashi, Taro, 1966, Calcium carbonate precipitation on the Bahama Banks: Journal of Geophysical Research, v. 71, p. 1575–1602.

Burne, R. V., and James, N. P., 1986, Subtidal origin of club-shaped stromatolites, Hamelin Pool: Abstracts with Proceedings, 12th International Sedimentological Congress, Canberra, p. 49.

Burne, R. V., and Moore, L. S., 1987, Microbialites: Organosedimentary deposits of benthic microbial communities: Palaios, v. 2, p. 241–254.

Byerly, G. R., Lowe, D. R., and Walsh, M. M., 1986, Stromatolites from the 3,300–3,500 Myr. Swaziland Supergroup, Barberton, Mountain Land, South Africa: Nature, v. 319, p. 489–491.

Caran, S. C., and Winsborough, B. M., 1986, Depositional environments of the Cuatro Cienegas Basin, northeastern Mexico-inland "sabkhas" and stromatolite lakes: Abstracts with Programs, Geological Society of America Annual Meeting, p. 557.

Casanova, Joel, 1987, Stromatolites et hauts niveaux lacustres Pleistocenes du Bassin Natron-Magadi (Tanzanie-Kenya): Science et Geologie, v. 40, p. 135–153.

Chave, D. E., 1965, Carbonates: Association with organic matter in surface sea water: Science, v. 148, p. 1723–1724.

Chave, D. E., 1970, Carbonate-organic interactions in sea water, *in* Hood, D. W., ed., Organic matter in natural waters: Institute of Marine Science, University of Alaska, Occasional Publication No. 1, p. 373–385.

Chave, D. E., and Suess, Erwin, 1967, Suspended minerals in seawater: Transactions, New York Academy of Science, Series II, v. 29, p. 991–1000.

Chave, D. E., and Suess, Erwin, 1970, Calcium carbonate saturation in seawater: Effects of dissolved organic matter: Limnology and Oceanography, v. 15, p. 633–637.

Cohen, A. S., and Gevirtzman, D. A., 1986, Littoral carbonate sedimentation in modern Lake Tanganyika, Africa: Abstracts with Programs, Geological Society of America Annual Meeting, p. 569.

Cohen, A. S., and Thouin, Catherine, 1987, Nearshore carbonate deposits in Lake Tanganyika: Geology, v. 15, p. 414–418.

Davis, G. R., 1967. Recent and Pleistocene carbonate sedimentation, Eastern Shark Bay, Western Australia: Unpublished Ph.D. Dissertation, University of Western Australia, Perth, Australia, 248 p.

Davis, G. R., 1970, Algal-laminated sediments, Gladstone embayment, Shark Bay, Western Australia, *in* Logan, B. W., Davies, G. R., Read, J. F., and Cebulski, D. E., eds., Carbonate Sedimentation and Environments, Shark Bay, Western Australia: American Association of Petroleum Geologists Memoir 13, p. 85–168.

Deuser, W. G., and Hunt, J. M., 1969, Stable isotope ratios of dissolved inorganic carbon in the Atlantic: Deep-Sea Research, v. 16, p. 221–225.

Dill, R. F., and Shinn, E. A., 1986, Giant stromatolites in subtidal channels of the Bahamas provide new insight to ancient counterparts: Abstracts with Programs, Geological Society America Annual Meeting, p. 585.

Dill, R. F., and Shinn, E. A., Jones, A. T., Kelly, Kevin, and Steinen, R. P., 1986, Giant subtidal stromatolites forming in normal saline waters: Nature, v. 324, p. 55–58.

Dill, R. F., Kendall, C. G. St. C, and Steinen, R. P., 1988, Deposition of carbonate mud beds within high-energy, subtidal sand dunes, Bahamas; Abstracts Programs, Geological Society of America Annual Meeting, v. 20, p. 78–79.

Dill, R. F., Kendall, C. G. St. C., and Steinen, R. P., 1988, Stromatolites, marine sand dunes, and laminated lime-muds deposited in the high-energy, subtidal channels of the Exuma Islands of the Bahamas, a geological enigma: Abstracts and Program, 4th Symposium on the Geology of the Bahamas p. 10–12.

Dill, R. F., Kendall, C. G. St. C., and Shinn, E. A., 1989, Giant Subtidal Stromatolites and Related Sedimentary Features: Field Trip Guidebook T373, 28th International Geological Congress, American Geophysical Union, Washington, D.C., 33 p.

Dravis, J. J., 1983, Hardened subtidal stromatolites, Bahamas: Science, v. 219, p. 385–386.

Falls, W. F., Williams, D. E., Kendall, C. G. St. C, and Dill, R. F., 1990, Stable oxygen and carbon isotope study of recent sediments and cements, Lee Stocking Island, Bahamas: Organic vs. inorganic precipitation: Abstracts with Programs, American Association of Petroleum Geologists Annual Meeting, v. 74, p. 652.

Flugel, E., 1977, Fossil Algae—Recent Results and Developments: Springer-Verlag, Berlin, 375 p.

Ginsburg, R. N., 1986, The endless frontier of comparing living and fossil stromatolites: Abstracts with Programs, Society of Economic Paleontologists and Mineralogists, Annual Mid-Year Meeting, Raleigh, N.C., v. III, p. 43.

Ginsburg, R. N., and James, N. P., 1976, Submarine botryoidal aragonite in Holocene reef limestones, Belize: Geology, v. 4, p. 431–436.

Ginsburg, R. N., and Lowenstam, H. A., 1958, The influence of marine bottom comunities on the depositional environment of sediments: Journal of Geology, v. 66, p. 310–318.

Griffin, K. M., 1987, A comparison of the depositional environments of Upper Cambrian thrombolites and stromatolites of the Nopah Formation, Death Valley Region, California, and modern stromatolites of the Bahamas: Abstracts with Programs, Geological Society of America Annual Meeting, p. 446–447.

Harris, P. M., 1979, Facies Anatomy and Diagenesis of a Bahamian Ooid Shoal: Comparative Sedimentology Laboratory, Division of Marine Geology and Geophysics, Rosenstiel School of Marine and Atmospheric Science, The University of Miami, Miami, Fla., Sedimenta VII, 163 p.

Hoffman, P. E., 1967, Algal stromatolite use in stratigraphic correlation and paleocurrent determination: Science, v. 157, p. 1043–1045.

Hoffman, P. E., 1974, Shallow and deep-water stromatolites in lower Proterozoic platform-to-basin facies changes, Great Slave Lake, Canada: Journal of Earth Science, v. 16, p. 992–1016.

Hoffman, P. E., 1976, Stromatolite morphogenesis in Shark Bay, Western Australia, *in* Walter, M. R., ed., Stromatolites: Elsevier Science Publication, Amsterdam, p. 261–273.

James, N. P., 1983, Reefs, *in* Scholle, P. A., Bebout, D. G., and Moore, C. H., eds., Carbonate Depositional Environments: American Association of Petroleum Geologists Memoir 33, p. 345–454.

James, N. P., and Ginsburg, R. N., 1979, The seaward margin of Belize barrier and atoll reefs: International Assocation of Sedimentology Special Publication No. 3, p. 345–454.

Kendall, C. G. St. C., Dill, R. F., and Shinn, E. A., 1990, Guidebook to the Marine Geology and Tropical Environments of Lee Stocking Island, the Southern Exumas, Bahamas: Carbonate Facies, Geological History, Giant Stromatolites, Oceanography and Biological Associations (5th ed.): KenDill Publishing Co. and the Caribbean Marine Research Center, Riviera Beach, Florida 33404, 106 p.

Kroopnick, P. M., 1985, The distribution of $\delta^{13}C$ of total CO_2 in the world oceans: Deep-Sea Research, v. 32, p. 57–84.

Land, L. S., 1967, Diagenesis of skeletal carbonates: Journal of Sedimentary Petrology, v. 37, p. 841–869

Land, L. S., McKensie, F. T., and Gould, S. J., 1967, Pleistocene history of Bermuda: Geological Society of America, v. 78, p. 993–1006.

Land, L. S., and Goreau, T. F., 1970, Submarine lithification of Jamaican reefs: Journal of Sedimentary Petrology, v. 40, p. 457–462.

Land, L. S., and Moore, C. H., 1980, Lithification micritization and syndepositional diagenesis of biolithites on the Jamaican island slope: Journal of Sedimentary Petrology, v. 50, p. 357–370.

Lang, J. C., Wickerlund, R. I., and Dill, R. F., 1988, Depth- and habitat-related bleaching of *Zooxanthellate* reef organisms near Lee

Stocking Island, Exuma Cays, Bahamas, *in* Proceedings, 6th International Coral Reef Symposium, Australia, v. 3, p. 269–274.

LOGAN, B. W., 1961, Cryptozoon and associated stromatolites from the Recent, Shark Bay, Western Australia: Journal of Geology, v. 69, p. 517–533.

LOGAN, B. W., AND CEBULSKI, D. E., 1970, Sedimentary environments of Shark Bay, Western Australia: American Association of Petroleum Geologists Memoir 13, p. 1–37.

LOGAN, B. W., HOFFMAN, P. M., AND GEBELEIN, C. D., 1974, Algal mats, cryptalgal fabrics and structures, Hamelin Pool, Western Australia: American Association of Petroleum Geologists, Memoir 22, 140 p.

LOWENSTAM, H. A., AND EPSTEIN, SAMUAL, 1957, On the origin of sedimentary aragonite needles of the Great Bahama Bank: Journal of Geology, v. 65, p. 364–375.

MACINTYRE, I G., AND PLAYFORD, R. F., 1964, Dissolved $CaCO_3$ in the Labrador Sea: Journal of Fisheries Research Board, Canada, v. 21, p. 1475–1544.

MITTERER, R. M., AND CUNNINGHAM, ROBERT, JR., 1985, The interactions of natural organic matter with grain surfaces: Implications for calcium carbonate precipitation, *in* Schneidermann, N. and Harris, P. M., eds., Carbonate Cements: Society of Economic Paleontologists and Mineralogists, Special Publication 36, p. 17–31.

MONTY, C. L. V., 1973, Precambrian background and Phanerozoic history of stromatolite communities, an overview: Annual Meeting Society of Geology Belgium, v. 96, p. 585–622.

MONTY, C. L. V., 1976, The origin and development of cryptalagal fabrics, *in* Walter, M. R., ed., Stromatolites: Elsevier, Amsterdam, p. 193–249.

MONTY, C. L. V., 1977, Evolving concepts on the nature and ecological significance of stromatolites, *in* Flugel, E., ed., Fossil Algae: Springer-Verlag, New York, p. 15–36.

NEUMANN, A. C., PAERL, H. A., BEBOUT, B. A., AND MCNEESE, L. R., 1989, Modern stromatolites and associated mats, San Salvador, Bahamas, *in* Mylroie, J. E., ed., Proceedings, 4th Conference of the Geology of the Bahamas: Bahamian Field Station, San Salvador, Bahamas, p. 235–251.

OSBORNE, R. H., LICARI, G. R., AND LINK, M. H., 1982, Modern lacustrine stromatolites, Walker Lake, Nevada: Sedimentary Geology, v. 32, p. 39–61.

PARKER, B. C., SIMMON, S. G. M., JR., LOVE, F. G., SEABURG, K. G., AND WHARTON, R. A., JR., 1981, Modern stromatolites in Antarctic dry valley lakes: Bioscience, v. 31, p. 656–661.

PAULL, C. K., NEUMANN, A. C., BEBOUT, B. M., MCNEESE, L. R., AND SHOWERS, WILLIAM, 1989, Carbon stratigraphy of modern stromatolites, San Salvador, Bahamas: Abstracts with Programs, Geological Society of America Annual Meeting, v. 21, p. 195.

REID, R. P., AND BROWNE, K. M., 1990, Intertidal stromatolites: A new discovery in the Bahamas (abst.): American Geophysical Union Ocean Science Meeting, New Orleans, v. 71, p. 363.

REID, R. P., AND BROWNE, K. M., 1991, Intertidal stromatolites in a fringing Holocene reef complex, Bahamas: Geology, v. 19, p. 15–18.

RIDING, ROBERT, AWRAMIK, S. M., WINSBOROUGH, B. M., GRIFFIN, K. M., AND DILL, R. F., 1991, Bahamian giant stromatolites: Microbial composition of surface mats: Geology Magazine, v. 128, p. 227–234.

ROBBINS, L. L., AND BLACKWELDER, P. L., 1990, Origin of whitings: A biologically induced nonskeletal mechanism: Abstract with Programs, American Association of Petroleum Geologists Annual Meeting, p. 156.

SACKETT, W. M., AND MOORE, W. S., 1966, Isotopic variations of dissolved inorganic carbon: Chemical Geology, v. 1, p. 323–328.

SCHOPF, J. W., 1970, Precambrian micro-organisms and evolutionary events prior to the origin of vascular plants: Biological Reviews, v. 45, p. 319–352.

SCHOPF, J. W., OEHLER, D. Z., HORODYSKI, R. J., AND KUENVOLDEN, K. A., 1971, Biogenicity and significance of oldest known stromatolites: Journal of Paleontology, v. 45, p. 477–485.

SHINN, E. A., 1969, Submarine lithification of Holocene carbonate sediments in the Persian Gulf: Sedimentology, v. 12, p. 109–144.

SHINN, E. A., 1983, Tidal flats, *in* Scholle, P. A., Bebout, D. G. and Moore, C. H., eds., Carbonate Depositional Environments: American Association of Petroleum Geologists Memoir 33, p. 171–210.

SHINN, E. A., AND DILL, R. F., 1986, Giant living lithified columnar stromatolites, Exuma Island, Bahamas: Abstracts with Programs, Society of Economic Paleontologists and Mineralogists, Annual Mid-Year Meeting, Raleigh, N.C., v. III, p. 28.

SHINN, E.A., 1987, Giant Stromatolites, Castles of Sand (12 min., 16-mm film and video: EAS Productions, St. Petersburg, Florida, and Educational Film Department, American Association of Petroleum Geologists, Tulsa, Oklahoma.

SHINN, E. A., STEINEN, R. P., LIDZ, B. H., AND HALLEY, R. B., 1985, Bahamian Whitings; No fish story (abst.): American Association of Petroleum Geologists, v. 69, p. 307.

SHINN, E. A., STEINEN, R. P., LIDZ, B. H., AND SWART, P. K., 1989, Whitings, a sedimentologic dilemma: Journal of Sedimentary Petrology, v. 59, p. 147–161.

SMITH, C. L., 1940a, Calcium carbonate precipitation: Journal of Marine Research, v. 3, p. 171–189.

SMITH, C. L., 1940b, The Great Bahama Bank, 1. General hydrographical and chemical features: Journal of Marine Research, v. 3, p. 147–170.

SMITH, C. L., 1941, The solubility of calcium carbonate in tropical sea water: Journal of Marine Biological Association of the United Kingdom, v. 25, p. 235–242.

SMITH, S. V., DYGAS, J. A., AND CHAVE, K. E., 1968, Distribution of $CaCO_3$ in pelagic sediments: Marine Geology, v. 6, p. 391–400.

STEINEN, R. P., SWART, P. K., SHINN, E. A., AND LIDZ, B. H., 1988, Bahamian lime mud: The algae didn't do it (abst.): Geological Society of America, v. 20, p. A209.

TAFT, W. H., ARRINGTON, D. V., HAIMOVITZ, A., MACDONALD, C. G., AND WOOLHEATER, C., 1968, Lilthification of modern carbonate sediments at Yellow Bank, Bahamas: Bulletin of Marine Science, Gulf and Caribbean, v. 18, p. 762–828.

WALTER, M. R., 1972, Stromatolites and the biostratigraphy of the Australian Precambrian and Cambrian: Paleontological Assocation of London Special Publication 11, 192 p.

WALTER, M. R., 1976, Stromatolites: Elsevier, New York, 790 p.

WINSBOROUGH, B. M., AND GOLUBIC, STJEPKO, 1987, The role of diatoms in stromatolite growth: Two examples from modern freshwater settings: Journal of Phycology, v. 23, p. 194–201.

PROGRESS AND PROBLEMS IN KNOWLEDGE OF CORAL LAGOON TOPOGRAPHY AND ITS ORIGIN IN THE SOUTH PACIFIC, BY WAY OF PINNACLE STUDY

ANDRE GUILCHER

University of Western Brittany, URA-D094, French National Center for Scientific Research, BP 814, 29285, Brest, France

ABSTRACT: Two alternative explanations still compete regarding the formation of atoll lagoons: the classic Darwinian theory and the karstic-saucer theory. Although the deep drillings at Mururoa Atoll, French Polynesia, certainly favor the first alternative, the discussion remains open elsewhere. Progress in knowledge may come from investigations of coral pinnacles or knolls, which dot lagoons in various quantities.

Data on sets of pinnacles in the Tuamotu, Society, and Gambier Islands are discussed. The recent endo-upwelling theory has tried to explain the formation of these features by an ascent of deep oceanic water rich in nutrients through the pervious mass of atolls. Shallow-core borings were made in 1988 into one pinnacle and the adjoining rim of Tikehau Atoll, Tuamotu Islands. Studies of nutrients in the interstitial waters support the endo-upwelling theory. However, preliminary examination of the cores shows the presence of a highly dolomitized Tertiary reef underlying the Holocene coating without intervening Pleistocene rocks. Tikehau thus appears as a rather special structure and the same type of investigations should be continued in other atolls in the Tuamotus.

Finally, the discussion is extended to the Gambier almost-atoll, a structure akin to Truk, Carolines, described in 1970 by Francis Shepard. Here the question of pinnacle origin also exists but is complicated by peculiarities associated with the shape and subsidence of the volcanic basement, differential tilting of the barrier, and absence of a deep passage into it. Thus, the discussion needs to be enlarged to consider diverse parameters of coral reef history. Results expected from multiple borings are a good approach to understanding atoll formation, still in question in spite of various proposals since the middle of the 19th century.

INTRODUCTION

In coral reef geomorphology, there has been, and still is, a debate between those who hold for the well-known Darwinian subsidence theory, leading to the successive fringing reef—barrier reef—atoll stages and shapes (Darwin, 1842), and those who, although they accept a long-continued subsidence to account for the great thickness of oceanic reefs, maintain that the saucer shape of atolls is derived from a karstification of initially horizontal platforms, which occurred during one or several stages of subsidence. This alternative explanation, which has been referred to by Guilcher (1988) as the karstic-saucer theory, has been proposed mainly by Yabe (1942), Asano (1942), Hoffmeister and Ladd (1945) MacNeil (1954), Purdy (1974) and Bourrouilh (1977). It is supported by experiments that show that when limestone blocks with a horizontal flat surface are subjected to an acid treatment, they develop a saucerlike shape with a peripheral rim cut by gullies and a central hollow. Echo soundings in the Belize lagoon in the Caribbean support this theory. Bourrouilh (1977), from observations in many Pacific structures, particularly the Loyalty Islands (New Caledonia), Rennell (Solomons), Makatea (between the Society and Tuamotu Islands), and Nauru, has proposed a sequence of events that is summarized and simplified in Figure 1. Combinations of eustatic shifts of sea level and local tectonic movements explain the different situations.

On the other hand, multiple drillings carried out in different parts (peripheral rim and central lagoon) of Mururoa Atoll, Tuamotu Islands, exploited and synthesized by Buigues (1985) and Aissaoui and others (1986), have demonstrated (Fig. 2) that this particular structure grew only slightly differently from the Darwin model. Its evolution includes a long fringing-reef stage and a short barrier-reef stage, and a central lagoon with a peripheral rim formed long ago, as indicated by the thick sequence of fine sediment deposited within it. Clearly, the central depression is not a recent solution feature. Therefore, the classic Darwinian theory is supported at Mururoa Atoll.

However, one case does not prove that all atolls passed through the same succession of events as Mururoa; moreover, karstic processes occur in many reefs due to Pleistocene sea-level fluctuations (Purdy, 1974; Guilcher, 1988). Francis Shepard (1970) made an important contribution in this field with his handsome description and mapping of Truk almost-atoll, Caroline Islands. It is thus appropriate, in a volume devoted to his memory, to discuss these problems, to examine ways in which current research may shed more light on them, and to suggest places where further investigations would be important. The wide area covered by the Tuamotu, Society and Gambier Islands in French Polynesia is the site of this discussion. There are 76 atolls in the Tuamotus, five in the Societies and two in the Gambiers. Gambier itself is an almost-atoll.

PINNACLES OR KNOLLS IN THE LAGOONS OF THE SOCIETY, TUAMOTU AND GAMBIER STRUCTURES AS POSSIBLE ELEMENTS FOR KNOWLEDGE OF LAGOON ORIGIN

As in the other parts of the coral world (e.g., Diego Garcia Atoll in the Indian Ocean, or Enewetak Atoll in the Marshall Islands), lagoons of the Society and Tuamotu atolls currently contain many coral pinnacles (called carenas in Paumotu language, Tuamotu Islands), which rise steeply from the lagoon floor and often reach the surface. The word "knoll" is used also for these features. Their upper surfaces generally flourish with corals. Manihi, Ahe, Tikehau, Fakarava, Kaukura, Fangataufa Atolls in the Tuamotus may be cited as examples (Figs. 3 and 4). Coarse debris settles as aprons around their bases, e.g., at Mopelia Atoll, Society Islands. Arutua Atoll (Fig. 5) is even more typical. Bellingshausen Atoll, the smallest and northwesternmost in the Societies, includes an incredible number of such knolls of small size (Fig. 6), whereas Mopelia, which is distinctly larger (8×6 km) and has been surveyed in detail (Guilcher and others, 1969), has far fewer (10 to 15) knolls (Fig. 7). Pinnacles are not restricted to atoll lagoons, but also are found in lagoons between a barrier reef and a central vol-

From Shoreline to Abyss, SEPM Special Publication No. 46

ISBN 0-918985-92-7

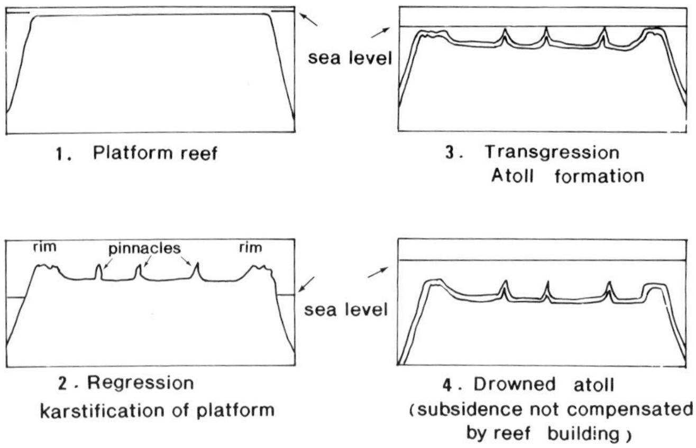

FIG. 1.—Atoll formation in karstic theory (mainly according to Bourrouilh, 1977 simplified).

canic island, as at Maupiti, Society Islands, where they are quite numerous and varied in size and shape (Fig. 8).

A particular feature regarding the pinnacles or knolls, rather common in the Tuamotus, is that their tops frequently rise as much as 1 m above low tide. Examples are shown (Fig. 9) in the Ahe lagoon at the northern part of the Tuamotus. Corals are normally dead on tops of such sites. Hao Atoll, also in the Tuamotus, provides several other good examples. This feature results (Pirazzoli and Montaggioni 1986, 1988; Pirazzoli and others, 1988) from slight Holocene variations of sea level in the Society, Tuamotu and Gambier Islands: sea level, relative to present sea level, reached +0.8/+1.0 m betweeen 5 and 1.25 ka; was never below +0.7 m throughout that interval; and, since 1.5 ka dropped gradually to the present position. Local small variations in these figures could result from geodynamic, isostatic and/or climatic events. No mention will be made here of the tectonically influenced structures on the southwestern side of the old Tuamotu chain, resulting from the "deep and arch" phenomenon due to the formation of the younger Society chain (Pirazzoli and others, 1988, p. 169; Guilcher 1988, p. 86).

In fact, the presence or abundance of pinnacles in lagoons varies according to location. Scilly Atoll, close to Mopelia and somewhat larger (12 × 12 km; Fig. 10), has no pinnacle reaching the surface, and none approaching it. However, echosounding transects and diving (Salvat, 1983; Fig. 11) have revealed that in this lagoon, where nearly 80 percent is between 40 and 60 m deep, pinnacles sometimes more than 30 m high remain at 20 to 30 m below the surface. Thus, observations at the surface and at shallow depths must always be made in deeper parts of the lagoon. More important is the lack of pinnacles in Bora Bora lagoon behind a barrier reef (Guilcher and others, 1969, p. 37 and 39), where 14 transects carried out in 1963 in all parts of the lagoon show that the bottom is extremely flat, except for a very small area in the east (Fig. 12). In the Tuamotu Islands, Taiaro Atoll, investigated in detail by Chevalier and others (1976), has no pinnacles and an extremely poor coral community.

A first working hypothesis to explain these differences could be that they are related by various degrees of connection of lagoons at the surface with the open sea, and the consequencial effects on coral growth. The Polynesian lagoon surfaces are connected to the ocean by two kinds of passages across atoll rims and barrier reefs, called hoas and avas in Paumotu terminology. Hoas (Fig. 9) are very shallow depressions between low islands, or motus; avas (Fig. 9) are deeper passages, allowing ships to enter lagoons, and are thought to be inherited from subaerial gullies cut across the emerged rim during the last Pleistocene glaciation (Guilcher, 1979). As a matter of fact, the depths of the avas have been modified since the postglacial transgression by coral growth: avas (Figs. 7, 13 and 14) are currently very deep (From 30 to 50 m or more) at their outer ends, whereas they commonly become much shallower (5 m or less) and tend to branch out in their inner parts due to recent obstruction by coral patches. The depths of the shallowest places at the inner ends of the avas are 9.5 m at Kauehi, 3.7 m at Tikehau, and 3 to 4 m at Mopelia (Fig. 7). However, the passage has great importance in hydrologic exchanges between ocean and lagoon, since the tidal range (±1 m in the Tuamotus, less, however, in the Societies) controls strong alternating currents (Fig. 14) that permit the exchange of large water masses.

An additional source of flow of oceanic water into the lagoons is the occasional occurrence of heavy southerly or southwesterly swells, generated in the roaring forties of the Southern Ocean and resulting in big washovers along the southern parts of atoll rims. The alternating currents in the avas are then interrupted and replaced by a continuous outflow. Figure 15 shows observations made at Mopelia Atoll in this respect. It remains true, however, that surficial exchanges between ocean and lagoon are more continuous when avas exist.

Avas exist in many Tuamotu atolls, at Mopelia Atoll and in barrier reefs of the Society Islands; they are absent, however, in some of the Tuamotu structures, and from Scilly, Bellingshausen, Tetiaroa and Tupai Atolls in the Societies. The presence of numerous pinnacles coincides with avas at Mopelia Atoll and Maupiti lagoon in the Societies and many large Tuamotu atolls, such as Rangiroa, Ahe, Tikehau, Fakarava, and Arutua (Guilcher, 1988, Fig. 34, p. 54). However, atolls with no avas also occur (Figs. 16 and 17).

Two atolls of the Society Islands, Tetiaroa and Tupai (Figs. 16 and 17), seem to be quite recent, the underlying volcano being probably only a few tens of meters beneath the lagoon and even less at Tupai, where a detrital volcanic block thrown by a hurricane has been collected (Salvat, 1989, p. 83; Diraison and others, pers. commun.). Both lack avas but normally receive oceanic water through hoas and over wide low parts of the rim. Both bear sets of coral pinnacles, but, according to Salvat (1989), they are not flourishing at Tetiaroa. The existence of pinnacles in these atolls of recent origin is not particularly surprising, since pinnacles exist in the nearby lagoon of Maupiti, where the central volcano remains at the surface.

The absence of an ava and a very poor connection through hoas coincide with absence of pinnacles at Taiaro Atoll

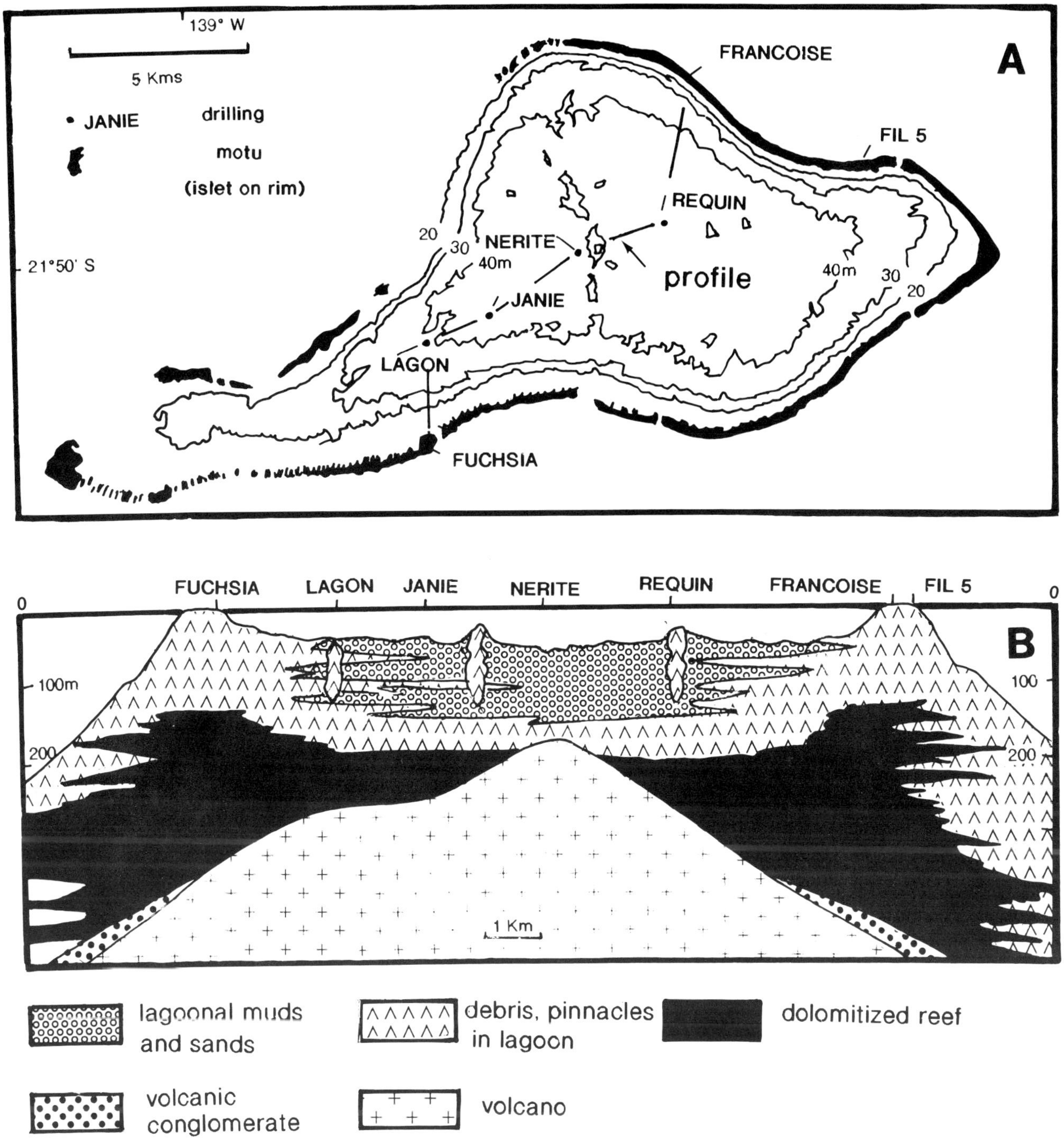

FIG. 2.—Distribution of facies in Mururoa Atoll, according to Buigues (1985) and Aissaoui and others (1986) combined. (A) Map with location of profile and boreholes. (B) Facies. Note lenticular distribution and differentiation into lagoon and peripheral rim at depths from −160 to −170 m.

(Chevalier and others, 1976; map in Guilcher, 1988, p. 151). That structure is a small (4.5 × 3 km) closed atoll with a continuous rim built during a higher Holocene stage of sea level, as discussed by Pirazzoli and others (1988). This rim is breached only by former hoas through which overwash is now restricted to exceptional storms. The salinity is close to 43 per mil, which is 7 per mil more than in the surrounding ocean. Corals of different species (*Pocillopora, Stylocoeniella, Montastraea* and *Platygyra*) have lived before on the shores of the lagoon, when a higher sea-level stand allowed a better connection through hoas. At present, *Porites lobata* is the only species alive and is never found below a depth of 1.5 m. Thus, pinnacle growth is presently impossible, and the bottom of the lagoon is flat, except for slight depressions and low rises from 0.5 to 3 m high: there is no evidence that Taiaro formerly had pinnacles that would

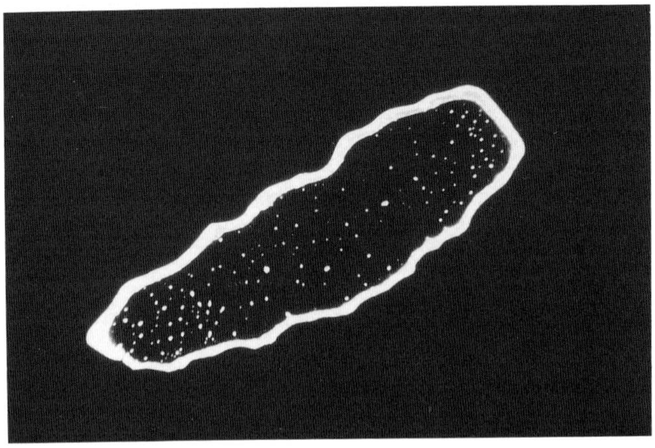

FIG. 3.—Manihi Atoll, Tuamotu Islands, 28 × 8 km, one passage (ava) in the southwest. North at top. Note numerous pinnacles (white dots) in the lagoon. SPOT image processed at SPT Tahiti, CNES/IFREMER/French Polynesian Territory. Reproduced with permission.

FIG. 4.—Fangataufa Atoll, Tuamotu Islands. Detail of surface of large pinnacle (±100 m long) rising in lagoon at sea level. Photograph by B. Salvat.

have subsequently disappeared. It is probable that other "closed" Tuamotu atolls, which still await investigation, resemble Taiaro in this respect.

In contrast, Scilly Atoll has numerous, rather deep pinnacles (Fig. 11) but no avas. It is true that the renewal of water is intensified there during occasional southerly heavy swells, as at Mopelia Atoll (Fig. 15; and Guilcher, 1988, p. 152–153). It is difficult, however, to understand the presence of a very large number of pinnacles at Bellingshausen (Fig. 6), which has no avas, and, conversely, their almost complete absence at Bora Bora (Fig. 12). Bora Bora has a large and exceptionally deep ava (>20 m deep) in addition to hoas. The occurrence of deep avas convinced the U.S. Navy to choose that lagoon as a base during World War II.

More detail is known so far about pinnacles and their environment in two lagoons of the Tuamotus: at Tikehau Atoll (Harmelin-Vivien, 1985), which belongs to the connected type, with one deep passage (ava) on the leeward side and numerous shallow hoas; and Takapoto Atoll (Chevalier and others, 1979; Salvat and Richard, 1985), which has many hoas and no ava. Owing to the good connection at Tikehau between lagoon and open sea, the observed temperatures and salinities are not very different from those in the surrounding ocean (Charpy, 1984). The yearly range of temperature is 4°C (26 to 30°C) in the lagoon as compared to 2.4°C (26.8 to 29.2°C) in the ocean; the range of salinity is 0.6 per mil (36.4 to 35.8 per mil) in the lagoon versus 0.4 per mil (36.3 to 35.9 per mil) in the ocean. The amount of phytoplankton is three times higher in the lagoon than in the ocean, with a maximum of chlorophyll *a* and of particulate carbon content near the bottom.

An example of a pinnacle at Tikehau is given here (Fig. 18) after Harmelin-Vivien, 1985. This survey was made in 1982, before the 1983 hurricane that destroyed many corals and spread rubble on adjacent slopes. It can be seen that the community is luxuriant. The areas slightly emerged at low tide bear many algae such as *Halimeda* and *Caulerpa* on sandy or gravelly substrate. Differences in populations exist between the windward and the leeward sides. The windward side is colonized by shallow *Porites lobata* and *Millepora platyphylla,* followed at depth by *Montipora, Astreopora, Porites, Pavona, Platygyra* and associated algae. Borer mollusks are abundant, representing 40 percent of the biomass. From −10 to −15 m, other coral colonies such as *Montipora, Stylocoeniella, Platygyra,* and bushes of *Acropora formosa* grow on detrital slopes. On the leeward side, branching corals of *Acropora variabilis* are typical of the shallow part, and borers are not as abundant as on the windward side. The fish fauna is very diverse (more than 100 species) but is heterogeneously distributed. As a whole, the living cover is rich on this pinnacle and on the ones that have been surveyed at Tikehau.

Takapoto Atoll (17 × 5.5 km; Chevalier and Denizot, 1979; Chevalier and others, 1979; Salvat and others, 1985), which has an approximate mean depth of 60 m, contains 414 pinnacles in its lagoon in spite of the lack of a deep passage. The pinnacles are especially numerous in the southwest and are between 10 and 100 m in diameter. According to Rougerie (1979, 1985b), who refers to data collected from 1974 to 1978, the water is well oxygenated at all depths (120 to 150 percent of saturation in the upper layers, still nearly 100 percent in the bottom). The contents of dissolved mineral phosphates, nitrates and silicates are 0.10, 0.20 and 0.50 uatg/1, respectively. Chlorophyll *a* concentration is higher than expected and can reach 0.5 mg/m^3 (mean values: 0.3), whereas outside the atoll, the oceanic values are less than 0.1. The salinities are normally much higher than at Tikehau as a result of the confinement: at the surface, up to 39.5 per mil in December 1974 and up to 41.1 per mil in February 1977, whereas the oceanic salinity is about 35.8 to 36 per mil; the differences between surface and bottom are less than 1 per mil. However, still higher salinities would be expected, and it is suggested by Rougerie (1985b), that they are lowered by percolation of oceanic water through the atoll rim and slope.

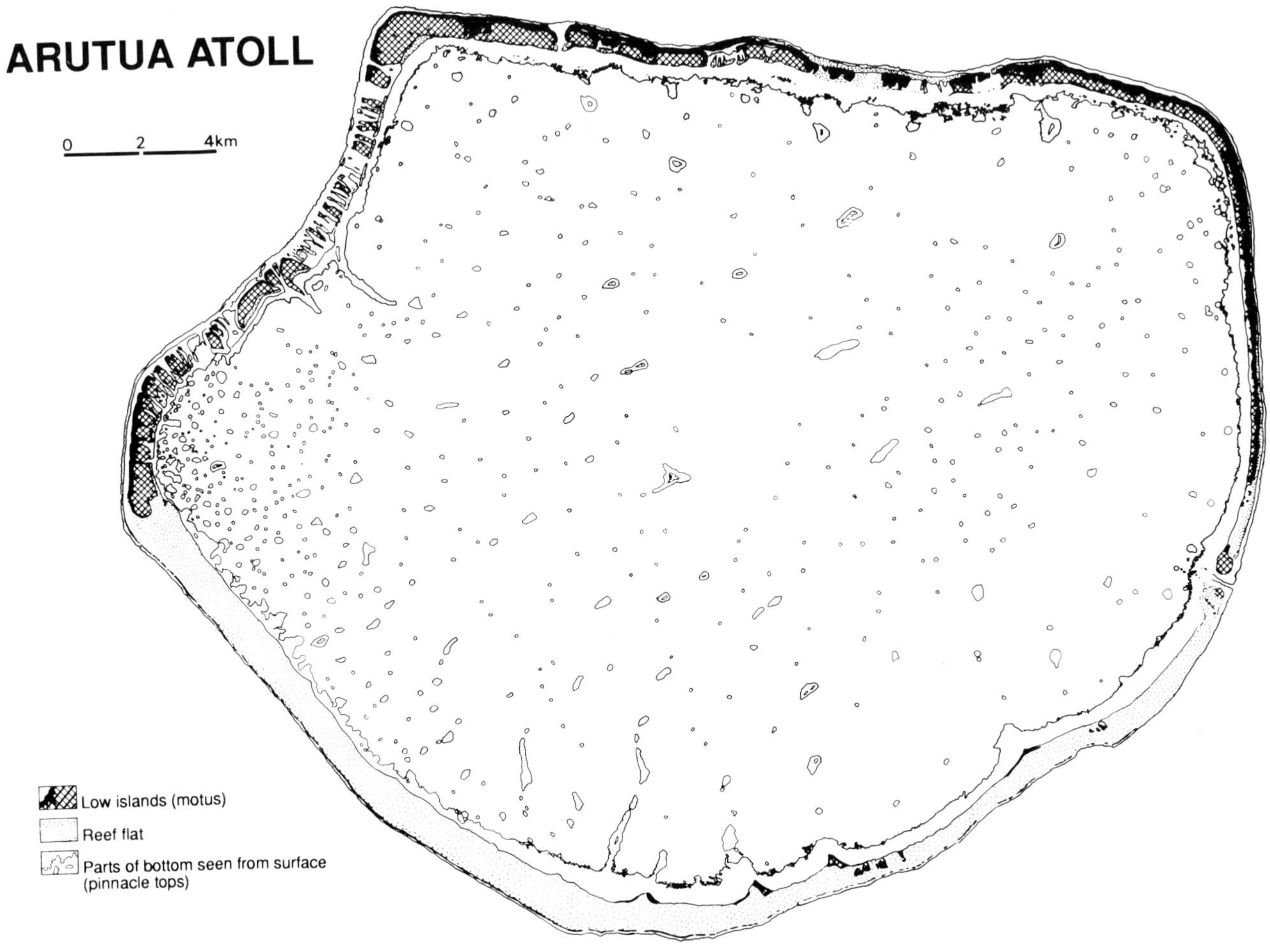

FIG. 5.—Arutua Atoll, Tuamotu Islands. 31 × 25 km. One passage in southeast. Pinnacles in lagoon. North at top. SPOT image processed at SPT Tahiti, CNES/IFREMER/French Polynesian Territory. Reproduced with permission.

FIG. 6.—Bellingshausen Atoll, Society Islands, 3.5 × 3.5 km. Width of view 3 km. No passage, but lowerings on atoll rim. Many pinnacles in lagoon. Photograph by A. Guilcher, 1963.

FIG. 7.—Mopelia Atoll, Society Islands. One passage, 50 m wide and 14 m deep in its narrowest part, 3 to 5 m deep on lagoon side. A few pinnacles visible in lagoon. Photograph by A. Guilcher, 1963.

FIG. 8.—Maupiti Island, Society Islands. Many pinnacles in lagoon between central volcanic mountain and barrier reef. Length of central volcano 3 km. Photograph by A. Guilcher, 1963.

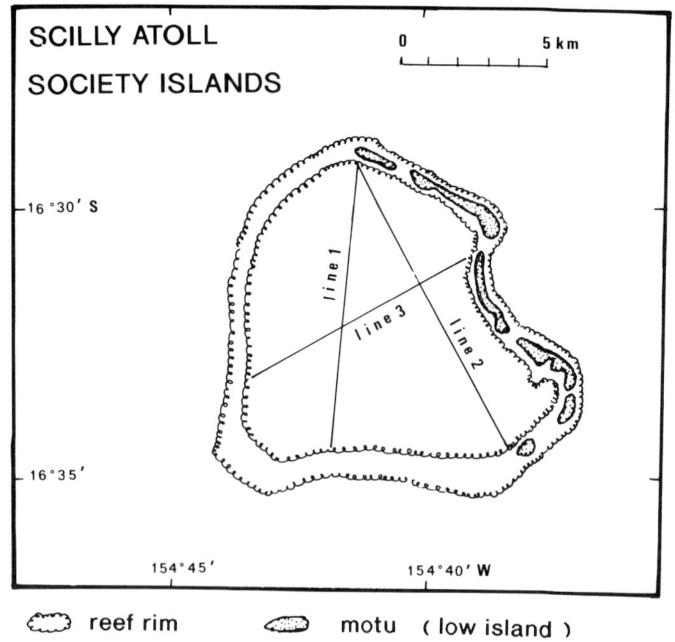

FIG. 10.—Scilly Atoll, Society Islands. Location of sounding lines shown on Figure 11.

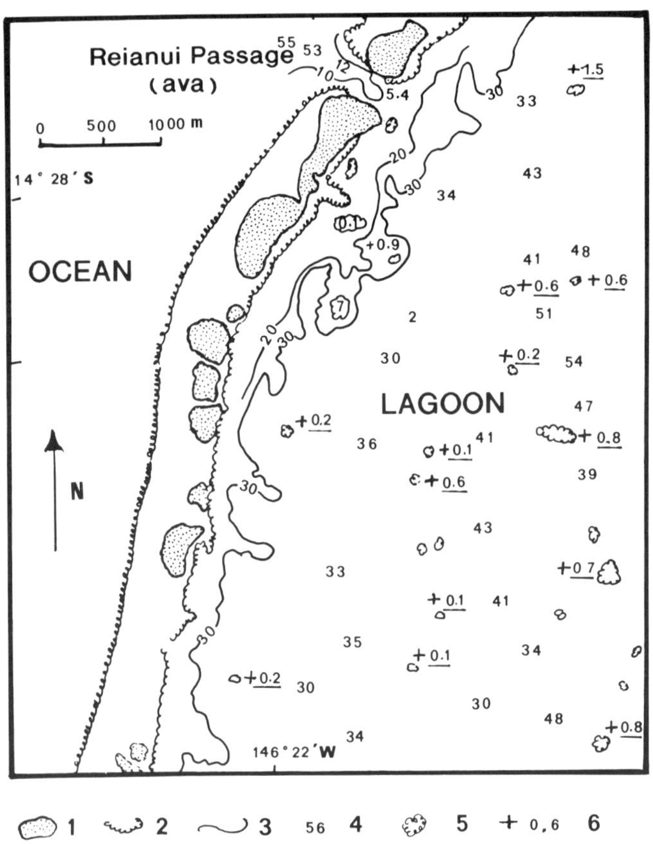

FIG. 9.—Part of Ahe Atoll, Tuamotu Islands, showing pinnacles in lagoon, rising above low spring tide as a result of recent Holocene emergence. Redrawn after French chart 6364 (1963 survey), with permission of Service Hydrographique et Océanographique de la Marine. (1) motus (low isles on atoll rim); (2) reef edge; (3) contours in meters below lowest spring tides; (4) depths in meters below lowest spring tides; (5) pinnacles in lagoon; (6) elevation in meters above lowest spring tides. The deep passage or ava becomes shallower as it enters the lagoon. Hoas, or very shallow gaps, occur between the sandy islands or motus.

Three pinnacles have been described at Takapoto Atoll by Salvat and others (1985). The Orapa pinnacle may be considered as typical. Its upper parts bear coral assemblages including principally *Porites lobata* (40 percent) and *Millepora platyphylla* (20 percent), with many *Tridacnae*. Smaller coral colonies (*Montipora* and *Favia*) live in deeper parts down to 20 m, and other sessile bivalves are found down to the bottom. However, Chevalier and others (1979a) note that the number of species is poorer than in more open atolls such as Tikehau. *Acropora* and *Pavona* are extremely rare, and *Fungia* is absent, these facts contributing to the dominance of *Porites* and *Millepora*, which are better adapted to the high salinity. Thus, there is an impoverishment related to the confinement, although coral growth is still possible on the pinnacle. Another fact is that, as in a number of pinnacles in Takapoto and in other Tuamotu atolls, the Orapa pinnacle bears on its top dead *Porites* in growth position and slightly emerged, a result of the Holocene standstill of sea level above the modern one (Fig. 9). Impoverishment at Takapoto might be ascribed to a decrease of exchange between the lagoon and the ocean, owing to the filling of some hoas, which may or may not be related to the change in relative sea level. In any case, the Orapa pinnacle bears testimony to a rather luxuriant growth of corals in Takapoto lagoon from middle Holocene to present, and the study of other pinnacles in the same atoll leads to a similar conclusion. However, the thickness and total age of the coating of Holocene organisms on these pinnacles are not known, and a Pleistocene (Sangamon?) core remains possible.

Temoe Atoll (Fig. 19), which has been investigated by Pirazzoli (1987) and Pirazzoli and Montaggioni (1987), is

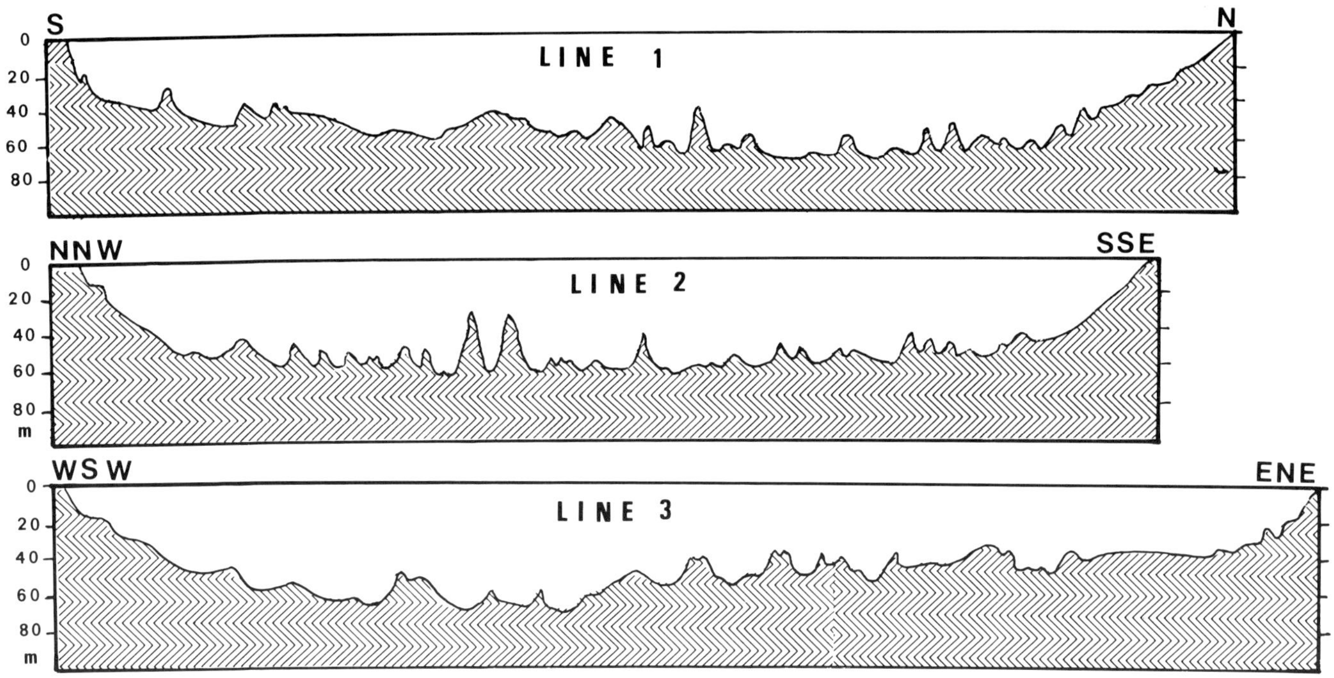

FIG. 11.—Scilly Atoll. Echosounding lines, after Salvat (1983).

an interesting feature with regard to pinnacles. It is the southernmost atoll in French Polynesia (23°20′S) but is north of Elizabeth Atoll in the Tasman Sea, which is the southernmost in the world at 30°S. The Elizabeth Atoll owes its existence to the East Australian current. Temoe is approximately at the same latitude as the Gambier almost-atoll, which will be considered later. As at Takapoto, Temoe has no ava but does contain numerous hoas through which oceanic water can enter only during southerly swell conditions. As in Tuamotu atolls, the sea level stood some 0.8 m higher than now until less than 2 ka. In such conditions, the lagoon (average depth: 13.2 m; about 23 m in a few places) bears a large number of pinnacles (Fig. 19), some of them reaching the surface. No data are available concerning the coral communities on these pinnacles, but their large number is to be taken into account in relation to what occurs in the Gambier almost-atoll. Similarly, Kaukura in the Tuamotus has many pinnacles, although the lagoon is connected to the ocean through hoas only. The lagoon of Elizabeth Atoll at 30°S in the Tasman Sea, a structure of 4 × 8 km, described by Slater and Phipps (1977), is very shallow (1 to 6 m) and is full of pinnacles, as is Temoe, in spite of the latitude.

FIG. 12.—Northeast part of Bora Bora lagoon, Society Islands. Width of view in lagoon in foreground is 8 km. No pinnacle occurs in spite of wide and deep passage located on left in backgkround. Photograph by A. Guilcher, 1963.

Summary.—

Except for the most extreme case of Taiaro Atoll, which may be matched in a few other similar Tuamotu structures not yet investigated in detail, pinnacle growth is possible and pinnacles occur in a very large number of atolls of this archipelago, despite great differences in surface-water flux between lagoon and ocean. On the other hand, pinnacles are not always clearly related to the degree of connection with the open sea. The influence of such connection remains thus to be proven. Also, numerous pinnacles at Bellingshausen (Fig. 6) and several other atolls are surprising. One can wonder whether or not the degree of water exchange by surface oceanic water is in fact the main control of pinnacle growth. The problem of pinnacle age: Holocene growth, or inherited structures with only a thin cover of modern organisms, also requires further investigation.

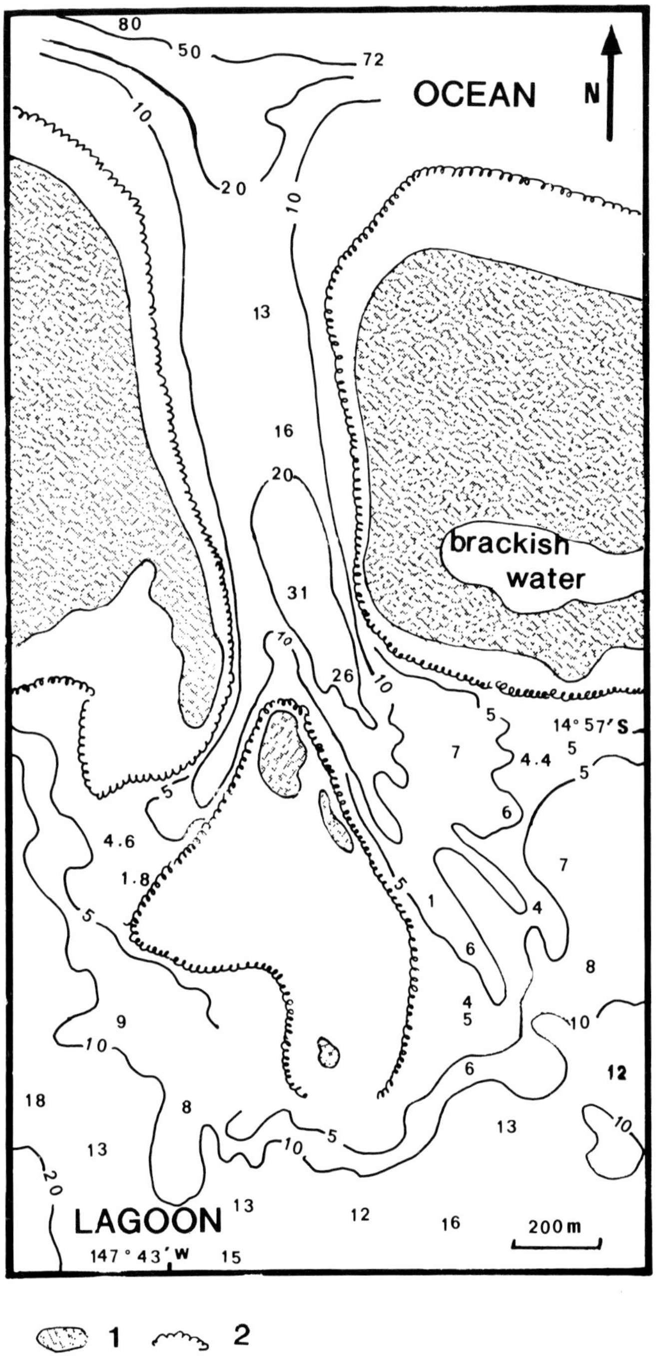

FIG. 13.—Avatoru passage, Rangiroa Atoll, Tuamotu Islands, branching and becoming much shallower as it enters the lagoon. Avatoru means three passages: an outer one and two inner ones. (1) motus; (2) reef edge. Redrawn after French chart 6363 (1966 to 1968 surveys), with permission of Service Hydrographique et Océanographique de la Marine.

FIG. 14.—Arutua Atoll, Tuamotu Islands. Branching passage becoming shallow on lagoon side of the entrance. Width between motus (islands) is 500 m. One pinnacle in lagoon. Strong ebb current when picture was taken. Photograph by A. Guilcher, 1965.

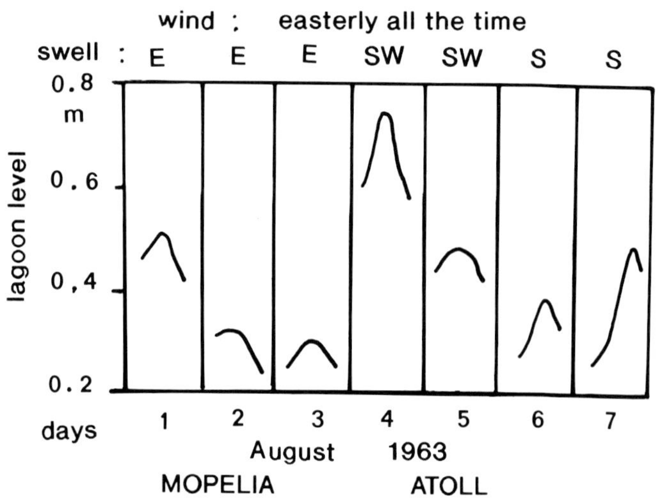

FIG. 15.—Mopelia Atoll, Society Islands. An example of lagoon filling by overwash during heavy southerly swell. Level observations during the day only. Normal low tide +0.2 m; tidal range 0.2 m. After Guilcher and others (1969).

FIG. 16.—Tetiaroa Atoll, Society Islands. No ava present. Note pinnacles and reticulated pattern in western part of lagoon. Width of view 6 km. Photograph by A. Guilcher, 1963.

THE ENDO-UPWELLING THEORY OF PINNACLE GROWTH

A theory has recently been put forward by Rougerie and Wauthy (1986, 1989) that attempts to explain the contemporaneous growth of pinnacles in atoll lagoons. These authors remark that the high productivity observed in Tuamotu lagoons is in strong contrast with the poor content of nutrients and organic matter in the surrounding oceanic sur-

FIG. 17.—Tupai Atoll, Society Islands. No ava present, Note pinnacles in lagoon. Width of view 6 km. Photograph by A. Guilcher, 1963.

face water. Observations at Hao Atoll in 1965 (Michel, 1969) have shown that nitrates and organic matter were abundant in the lagoon and almost absent outside. The same contrast has been observed in 1969 at Mururoa Atoll (Michel, 1969), at Rangiroa Atoll (Michel and others, 1971), and in 1982 at Tikehau Atoll (Charpy, 1984). It seems that life in these four lagoons owes little to the ocean, but that the lagoons export organic matter to the ocean at ebb tide through passages. Formerly, Sargent and Austin (1954) had made similar remarks concerning Rongelap Atoll, Marshall Islands, which was referred to by them as "a flourishing coral atoll in an empty ocean." Also in the Marshalls, a study by the Scripps Institution at Enewetak Atoll (Webb and others, 1975) has shown that the content of oceanic water in mineral nitrogen increases when this water enters the lagoon. Thus, these facts are not restricted to the Tuamotus.

The reason(s) for the observed lagoonal productivity require explanation. The following hypothesis was proposed by Rougerie and Wauthy (1986, 1988). It is well known that the content in nutrients is considerably higher in the ocean at depth than at the surface. It is also known that atoll limestone is very porous and heterogeneous. Recent studies, e.g., by Buddemeier and Oberdorfer (1986), have greatly increased the knowledge of this heterogeneity, especially at Enewetak Atoll, where multiple solution unconformities with large voids have been found in shallow boreholes as much as 80 m below sea level. The traditional

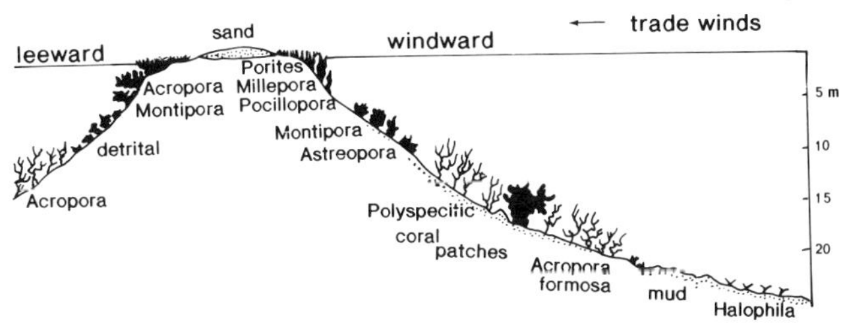

FIG. 18.—A pinnacle in the southwestern part of Tikehau lagoon, Tuamotu Islands. Redrawn from Harmelin-Vivien (1985), with permission of editor of Proceedings of 5th Coral Reef Congress, Tahiti.

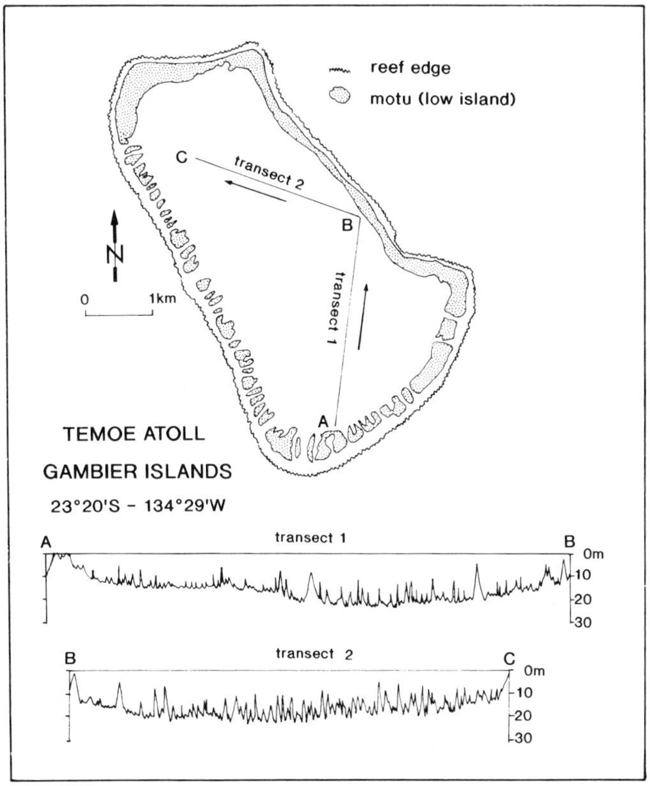

Pinnacles in lagoon of Temoe Atoll, Gambier Islands.
After Pirazzoli, 1987.

FIG. 19.—Pinnacles in Temoe Atoll, Gambier Islands. Reproduced with permission of CERF, Journal of Coastal Research.

Ghyben-Herzberg concept (Wiens, 1962, p. 317–326) has been considerably modified. At Enewetak, "a relatively shallow and variable body of freshwater (typically a few meters or less) is underlain by an extensive (tens of meters) brackish transition which varies in salinity and extent in response to freshwater recharge" (Buddemeier and Oberdorfer, 1986, p. 104). At Mururoa Atoll, studies of the volcanic basement have shown that this basement also is porous, although less so than the superposed limestone. The porosity of the limestone in coral reefs is increased considerably by its well-known cavernous structure. Observations in deep wells at Mururoa have proven that at a depth of 450 m, i.e., near the sides of the volcanic basement, the interstitial water has the same properties as the nearby Antarctic Intermediate Water. Thus, the lower part of the atoll is fed with water rich in nitrates and silicates. Hence Rougerie and Wauthy's idea that this deep water, heated by the geothermal flux originated in the basalt (see Fig. 20) and thus made lighter, is able to rise inside the atoll through the interstices and is the origin of the high productivity in the lagoon. Pinnacles would be located at places where sets of fissures end in the bottom of the lagoon, thus favoring especially the productivity and coral growth. Figures 20 and 21 summarize the process postulated by Rougerie and Wauthy. The pinnacles serve as outlets of feeders by which the atoll ecosystem is maintained. At Enewetak (Buddemeier and Oberdorfer, 1986, p. 109), "an unusually large number

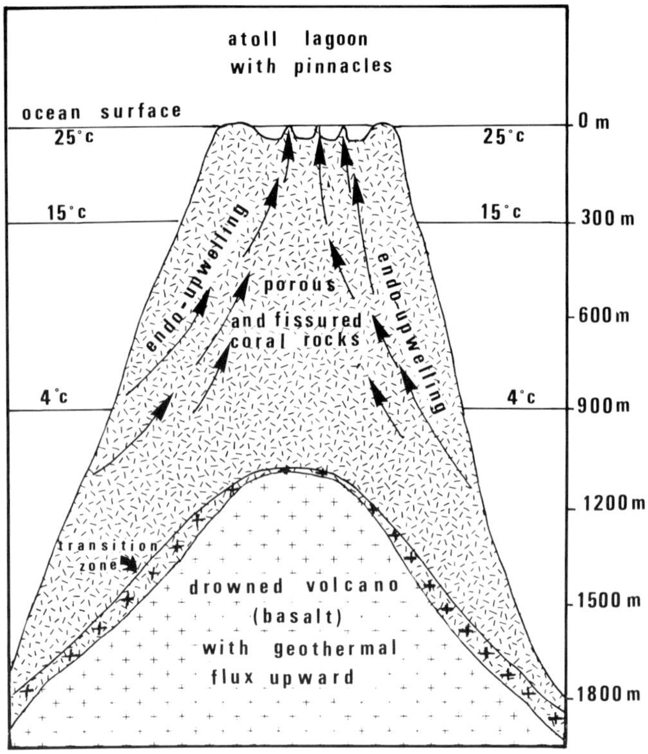

FIG. 20.—The endo-upwelling theory, after Rougerie and Wauthy (1986).

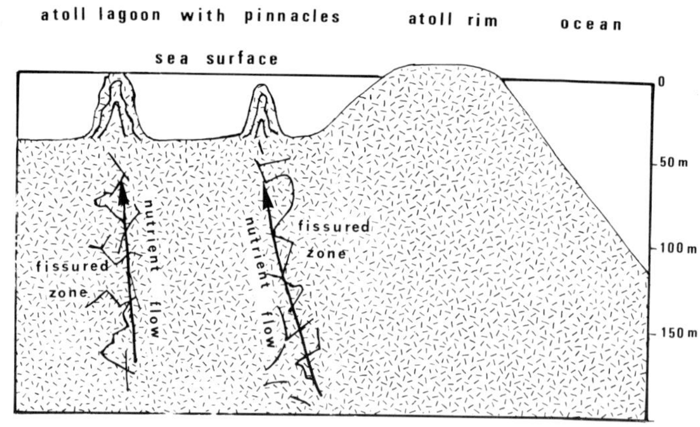

FIG. 21.—Endo-upwelling as a cause of pinnacle formation, after Rougerie and Wauthy (1986).

of lagoon pinnacle reefs developed on Pleistocene outcrops . . . may provide direct hydraulic connections with the karst surface."

This hypothesis is supported by field observations made in the 1950s in the Marshall Islands. Measurements in drill holes in 1952 at Enewetak Atoll (Swartz, 1958) showed that the temperatures in these holes decreased steadily with increasing depth (in contrast to what occurs in continental drill holes) until a minimum of 6.4°C (43.5°F) was reached at depths from 966 to 1,027 m. Observations in oceanic waters close to the atoll showed a close parallel in thermal profiles, "leaving little question that the character of the thermal profile in the atoll is largely a result of cooling by the adjacent ocean" (Swartz, 1958, p. 711). The increase in temperature that was observed lower down in these Enewetak drill holes, as in continental drill holes, should be normally ascribed to the influence of the volcanic basement. Measurements in another drill hole at Bikini Atoll showed the same thermal drop as at Enewetak, before additional heating by the geothermal flux in the more internal parts of the structure.

Such a process also provides an explanation of the transformation of calcite and aragonite into dolomite, which occurs currently at depth in atolls. Fairbridge (1957) suggested the action of water rich in Mg coming from the deep ocean and ascending through the structure after being warmed by the volcanic basement. More recently, Saller (1984) proposed the same mechanism to explain dolomitization in the deep parts of Enewetak, Funafuti and Midway, by precipitation from percolating deep oceanic waters, supersaturated with respect to dolomite. Similarly, at Mururoa Atoll, Aissaoui and others (1986) have shown that dolomitization has been a peripheral, deep phenomenon within the atoll (Fig. 2). According to Aissaoui and others (1986, p. 45), "dolomitization (in Mururoa Atoll) is essentially a process involving dissolution-precipitation. . . . The intimate association of aragonite and calcite dissolution with dolomite growth implies that parental waters be aggressive vis-à-vis calcite and therefore cannot be normal, near surface, seawater." Here again, a lateral movement of deep oceanic water into the older parts of the atolls is required.

A quite different case is the formation of a high pinnacle in the Prony Bay of New Caledonia lagoon, Melanesia (Rougerie, 1985a, p. 125–128). Hot springs (32 to 42°C) and consequent limestone precipitation have built this pinnacle, which rises to the surface from a depth of 33 m. The environment and conditions of growth are different here, but a parallel can be seen.

Thus, it seems that endo-upwelling processes are likely to be in progress in atolls, bringing nutrients to the lagoons, favoring considerably the development of coral life and even allowing it to exist in an otherwise unfavorable oceanic environment. It now remains to be determined whether pinnacles in lagoons currently are built under the influence of endo-upwelling, or whether, according to the karstic-saucer theory (Fig. 1), they represent remnants of former, wider, flat-topped coral edifices, built during high Quaternary sea levels, etched and sculpted during the Wisconsinian regression, and now only coated by a rather thin cover of living corals.

To address this problem, a shallow boring in a large pinnacle of the southwestern part of Tikehau lagoon, Tuamotu Islands, was completed in December 1988, and four other borings were made at the same time into the outer edge of the nearby atoll rim (Fig. 22). Preliminary results indicate that in all borings, the interstitial waters are very rich in dissolved inorganic nutrients: the silicates and nitrates are 10 times more abundant than in lagoonal and oceanic surface waters, and the nitrates are six times more abundant. These values are close to those observed in the Antarctic Intermediate Water at the depth of 500 m (Rougerie and

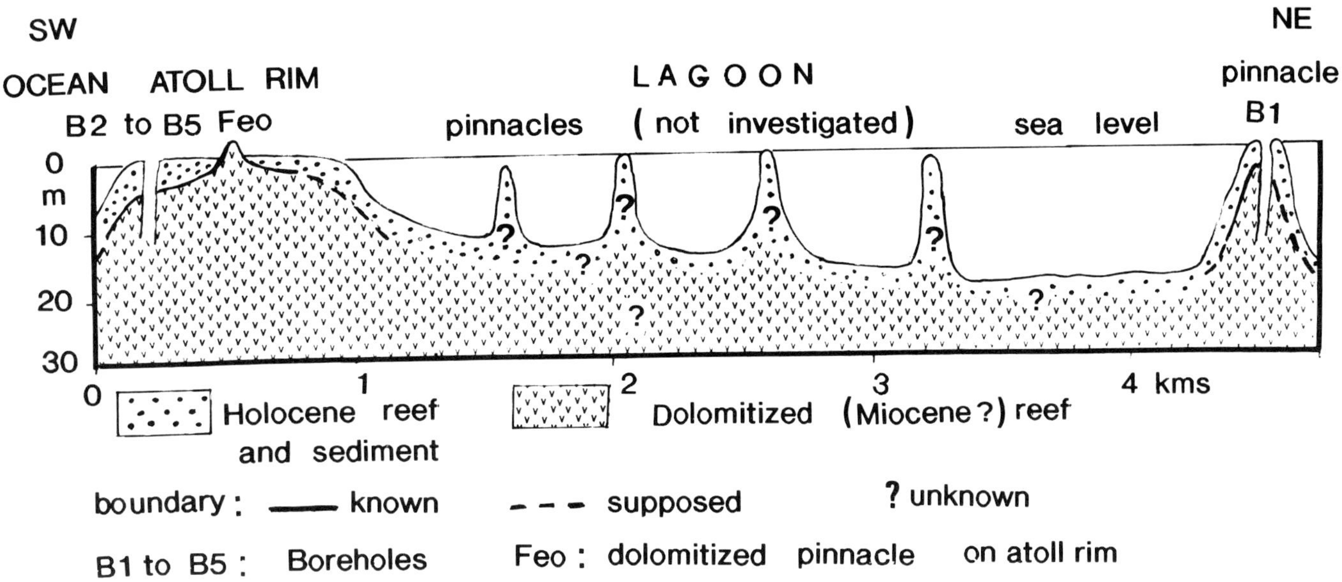

FIG. 22.—Provisional results of borings at Tikehau Atoll, Tuamotu Islands (Montaggioni, pers. commun., 1990).

Wauthy, pers. commun.). Such results strongly support the endo-upwelling theory.

On the other hand, the cores extracted from the pinnacle and the atoll rim, examined by L. Montaggioni, have recovered massive colonies of *Porites* of Holocene age in the upper 2 m. Immediately below the *Porites* are (Fig. 22) deeply dolomitized, calcitized coral reefs, quite similar to the "feos" standing on the rim in the vicinity and appearing to be older than Pleistocene, perhaps as old as Miocene (Montaggioni, pers. commun.). These provisional results need to be checked by further study of the cores. Presently, it appears that the Pleistocene is absent, perhaps because Tikehau Atoll was emergent at that time due to uplift, and was later submerged. Such a sequence may seem strange at first, but is rather likely when compared with what is known at the phosphated Mataiva Atoll, close to Tikehau in the Tuamotus (Montaggioni, 1989).

Thus, interstitial waters found in the Tikehau borings support the endo-upwelling theory; pinnacle formation may be related to this process. However, the sections of Tertiary (?), highly dolomitized rocks just below the Holocene do not provide clear conclusions on the succession of events (although feeding fissures could have persisted in the same places inside the atoll since the Tertiary). Tikehau, as at Mataiva and other nearby atolls, may have had a special history because of the vicinity of the more recent chain of volcanoes of the Society Islands, which could have caused tectonic instability. Borehole data from atolls of the Tuamotus located outside that area (e.g., Manihi, Ahe or Takapoto) would be most valuable.

THE GAMBIER ALMOST-ATOLL AS ANOTHER POSSIBLE SITE FOR PINNACLE INVESTIGATIONS

The Gambier almost-atoll (Fig. 23) is the finest and most typical after the Truk almost-atoll that was studied by Shepard (1970). Although its submarine topography has not been surveyed with the degree of accuracy available for the structure at Truk, it is reasonably well known in most areas from soundings and large-scale charts of the French Hydrographic Office. Successive descriptions and interpretations have been given by Dana (1853), Agassiz (1906), Davis (1928), Brousse (1974), and Brousse and others (1974).

As at Truk, the almost-atoll consists of scattered volcanic islands, among which the main ones are Mangareva (+441 m elevation), Taravai (+256 m), Aukena (+198 m), Akamaru (+246 m) and Kamaka (+166 m). Several other volcanic islets are more than 50 m high. The lagoon in which these islands lie is surrounded by a barrier reef 90 km long without any true, well-defined passage or ava, but only wide depressions, especially in the northwest where ships of 7-m draught can enter. An essential difference from Truk is that the part of the barrier reef rising to sea level, or from 1 to 3 m above it as in the similar other structures, includes only 42 percent of its total length. These high segments are located in the northwest, north, northeast and east, and in the western promontory at Tokorua. Elsewhere, the barrier reef, although always prominent above the lagoon bottom, is drowned under 3 to 15 m of water. At Truk, in contrast, the barrier is drowned for only 5 km in the southwestern corner out of a total length of 200 km, with, however, about 50 true passages or avas available for ships. As at Truk, the Gambier motus, or low isles on the barrier, are located on the windward side and are almost absent on the western leeward side, even where the barrier would have provided an adequate support at sea level.

The lagoon has been accurately sounded during recent surveys, except, unfortunately, for three gaps (Fig. 23) in the south of Kamaka, the southwest of Taravai, and principally the northernmost part, which is known only, according to the charts, as being "full of coral knolls." Where it is surveyed, it is sometimes fairly deep, exceeding 50 m over a rather large area between Mangareva and Aukena Islands, down to 63 m in two places; a smaller deep (70 m) area exists at the southeast end of Mangareva. The

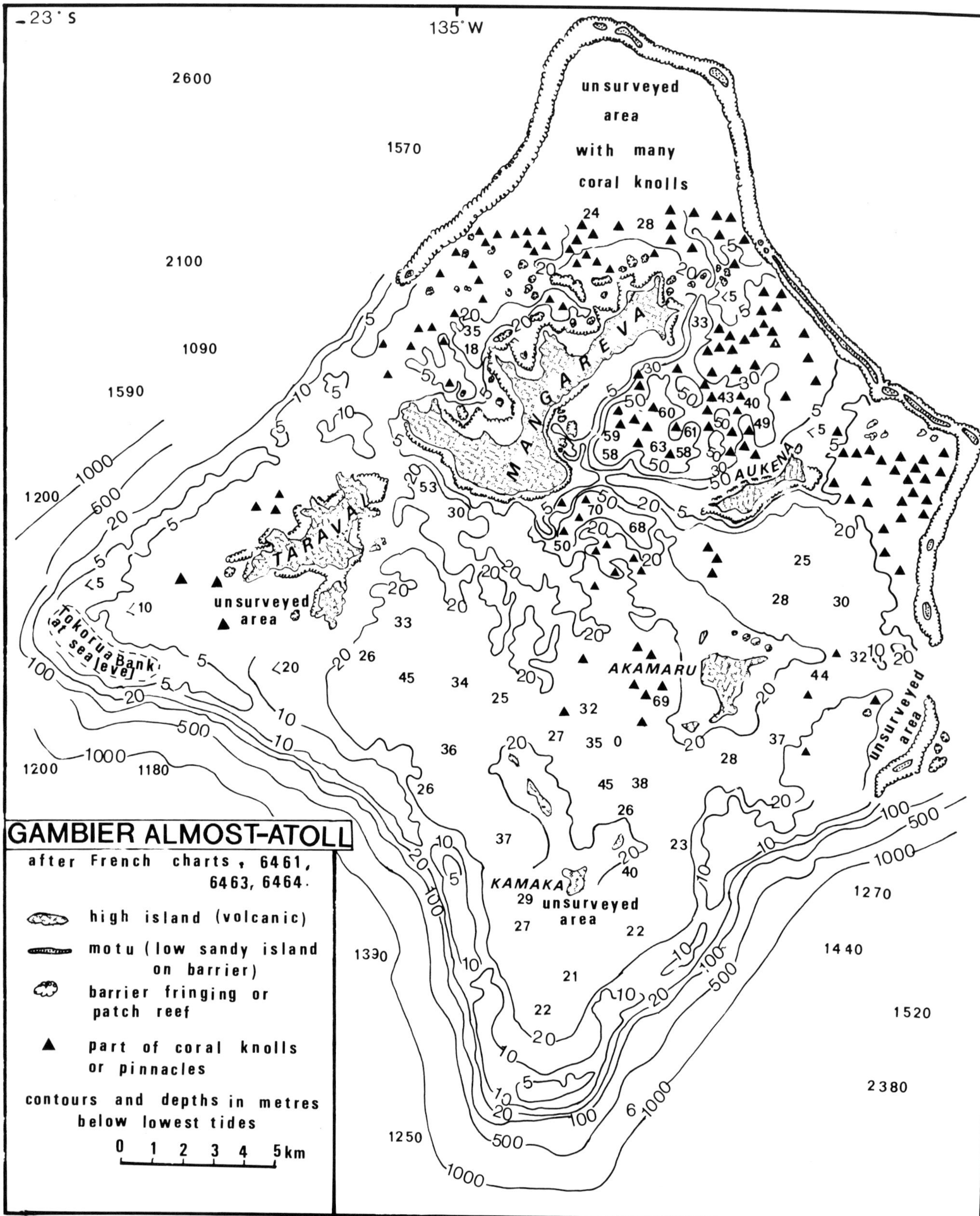

FIG. 23.—Generalized map of Gambier almost-atoll. Reproduced with permission of Service Hydrographique et Océanographique de la Marine.

southern half, although somewhat shallower, generally exceeds 20 m and sometimes 40 m in depth. The depth of the lagoon in its different parts has no positive correlation with the surrounding barrier, drowned or not, as the shallower part of the lagoon corresponds with the area where the barrier is drowned, and its deeper part is located where the barrier is at sea level or above.

A host of pinnacles has been found in the lagoon. Only some of them are shown on Figures 23 and 24. They rise from and to very different depths, as they also do in the Society Islands, e.g., at Scilly (Fig. 11) and Maupiti (Fig. 8), and at Diego Garcia Atoll, Indian Ocean (Stoddart and Taylor, 1971, Fig. 14). Whereas in the northwest area of Mangareva and the east of Aukena the surrounding depths are moderate or even shallow, between Mangareva and Aukena and in the southeast part of Mangareva the pinnacles can be 40 or 50 m high. Their steepness is not well known, however, since they have not been investigated accurately by diving, as at Tokehau (Fig. 18).

The volcanic islands of Gambier consist mostly (Brousse, 1974) of tholeiitic basalts in association with very subordinate (<0.2 percent) alkaline basalts. The K/Ar age of the Gambier volcano is 5 ma (Brousse, 1985). This volcano is included in a volcanic chain in which the youngest element is Pitcairn (0.6 ma), and older ones are Mururoa Atoll (10 to 11 ma), the Duke of Gloucester Atolls, and Hereheretue Atoll (±18 ma assumed). The presence of the two small atolls of Temoe (Fig. 19) and Portland slightly submerged to the southeast of Gambier, both certainly older than Gambier, is perhaps to be understood by the superposition of two chains of structures in that area (Pirazzoli and others, 1988).

The Gambier almost-atoll was explained by subsidence as early as 1853 (Dana, 1853, p. 94), and this interpretation also was accepted by Davis (1928, p. 375–376). Agassiz (1906, p. 64) thought that landforms of these volcanic islands result only from erosion. Although recent ideas and research point to subsidence as the main phenomenon in

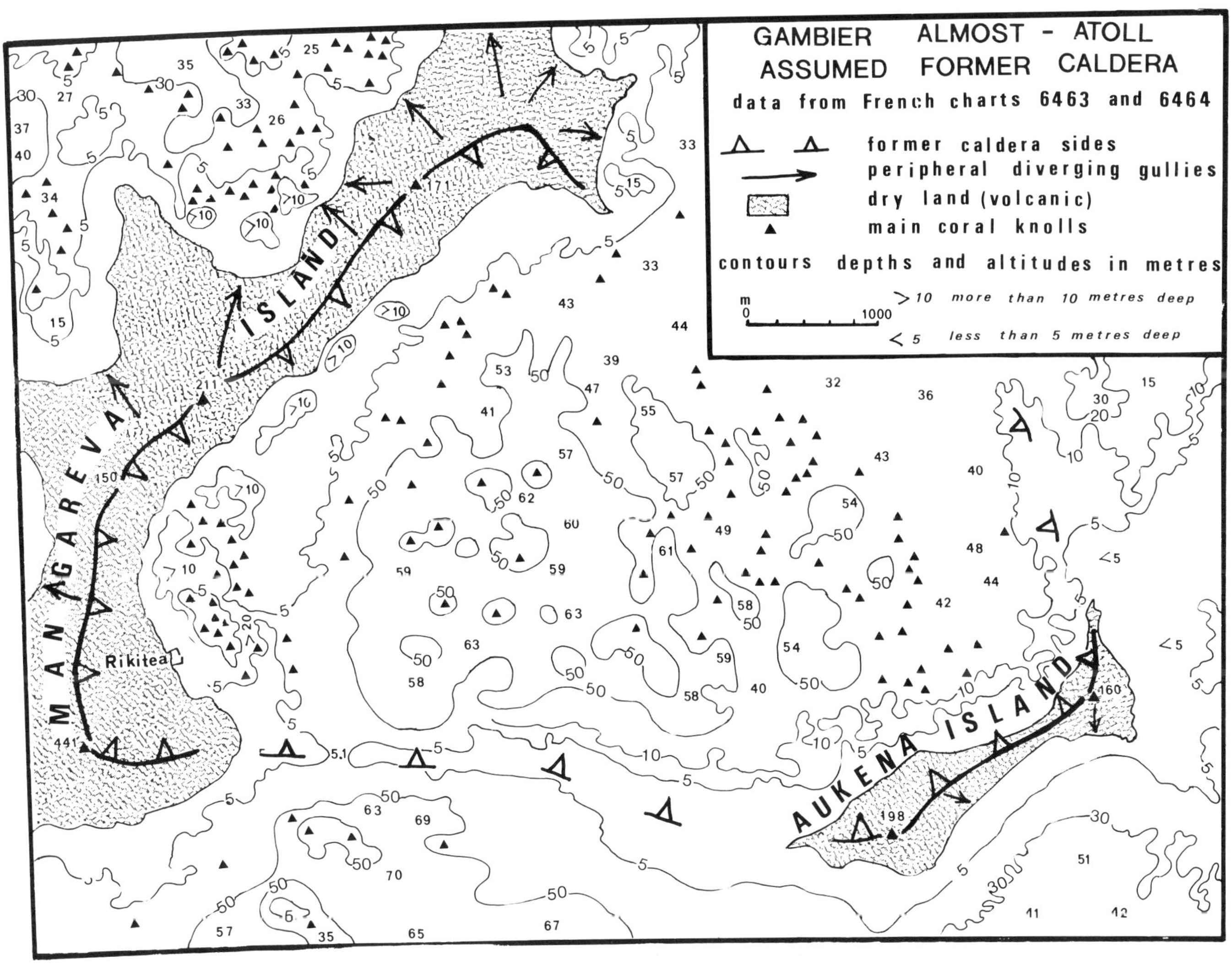

FIG. 24.—Gambier almost-atoll. Assumed to be a former caldera between Mangareva and Aukena Islands. Reproduced with permission of Service Hydrographique et Océanographique de la Marine.

Gambier, Brousse and others (1974, p. 88) remarked that some features in the outline of the archipelago should be remnants of calderas, as Agassiz suggested. The critical landforms (Fig. 24) are Mangareva Island, Aukena Island and the depression between them, which greatly resembles a drowned caldera. The shape of the asymmetric escarpments on the islands, the peripheral diverging gullies, the shallow connection between the southern parts of the two islands, and the deeply submerged central area, are all in accord with the presence of a caldera. More calderas may exist off the southeastern coast of Mangareva, where depths of as much as 70 m are recorded, and off the west coast of Akamaru (cf. island outline and adjoining depths of as much as 69 m). Collapses of calderas are known elsewhere in the region, as in the Society Islands (Diraison and others, pers. commun.).

Still, the problem in terms of coral reef geomorphology is to know how reef features were added in response to the volcano and its caldera(s?) during its subsidence and shifting Quaternary sea levels. A more accurate knowledge of the pinnacles can add significantly to our understanding of Gambier's history of evolution toward an atoll, hence in atoll formation in general.

A first point is the partial drowning of the barrier. It shows that tectonic movements have continued here until quite recently, probably until the Holocene. Otherwise, the coral growth should have been much more uniform around the lagoon. However, this partial drowning cannot be due to a general tilting to the south because of its complicated pattern. One must imagine a local drowning off Taravai Island, its interruption at Tokorua Bank, and a more extensive drowning in the southern half, with perhaps local differences as suggested by the bathymetry.

A second problem is the absence of any true passage or ava, which would be narrow, deep and well localized within the barrier. Although such passages are sometimes lacking in atoll rims (e.g., at Scilly and Bellingshausen, discussed earlier), they are almost always present in barrier reefs. In the present stage of coral reef research (Guilcher, 1988, p. 55), passages appear as heirs of subaerial valleys cut across atoll rims and barriers during the low Quaternary stages of sea level. As supposed by Darwin as early as 1842 (p. 100), more recently by Guilcher (1979), and by Ruzie and Gachon (1985, p. 388) from drillings at Mururoa Atoll, such passages can have been filled by Holocene sediment on the windward side of the structures and may have had more chance to persist on the leeward side, where sediment supply is poorer. At Gambier, however, there is apparently no trace of a passage, even on the western leeward side.

An explanation of the anomaly may be that, during low glacial sea levels, the fresh water of the emerged lagoon escaped completely to the ocean through permeable karst. Although there are a number of passages now at Truk, Shepard (1970, p. 1910) thought that many features in Truk lagoon result from Pleistocene karstification, particularly because depressions in the lagoon are currently much deeper than the nearby passages. The same conclusion was reached from investigations in the Mayotte lagoon, Western Indian Ocean (Guilcher, 1965) and from work by Purdy (1974) in Belize.

The third problem to be solved at Gambier is to what extent the lagoonal features and particularly the pinnacles might result from a Pleistocene solution of limestone. The deepest parts of the lagoon seem to be remnants of calderas; if so, they must have drained through subterranean joints during low sea-level stands. To what extent did these depressions receive calcareous sedimentation during high Pleistocene sea levels, when coral growth was certainly possible? Answers to these questions could come, in part, from cores in some of the high pinnacles. Such cores could show whether the pinnacles are Pleistocene residual features—either former pinnacles or remnants of bottom sedimentation, the facies giving an answer to the enigma—or of Holocene construction. The answer can of course be different at different sites; pinnacles located elsewhere than in former calderas, e.g., between Mangareva and the northwestern part of the barrier, should also be investigated. Brousse and others (1974, p. 70–76) have shown that the shapes, widths and heights of the pinnacles are quite varied: some are elongated and adapted to the local topography, whereas others are quite sharp and narrow, with a diameter of only 2 or 3 m at their tops, particularly in the northernmost area of the lagoon, which has not been surveyed by the Hydrographic Office. The shape at the surface may be recent and not necessarily informative of internal structures.

Pinnacle investigations in the Gambier lagoon, where little sedimentological information is available, could provide interesting material relating to lagoonal evolution in atolls and behind barrier reefs. The Gambier coral structures thus remain one of the places in the Pacific where further studies would be important in resolving such questions.

CONCLUSIONS

Further research is required to solve the problem of atoll origin. Besides deep and multiple drillings such as those carried out in Mururoa lagoon, which are valuable but very expensive, much shorter and less costly boreholes can be made in great numbers in lagoon pinnacles. They can show if these pinnacles were recently built by endo-upwelling of oceanic water rich in nutrients in lagoons that previously existed, or if they are remnants of a Pleistocene antecedent platform. The first boreholes of this kind have now been made in Tikehau lagoon, Tuamotu Islands. Preliminary results are quite conclusive with respect to the endo-upwelling process, since the interstitial waters have much higher concentrations of inorganic nutrients than the lagoonal and oceanic surface waters, pointing to their deep origin. This is most encouraging. However, the cores extracted from the pinnacle and the adjoining rim gave unexpected (provisional) results, with Tertiary dolomites close to the surface below a thin Holocene cover. Tikehau is perhaps an atypical structure; atolls located in other areas inside the Tuamotu archipelago might be more suitable. Such research is in its early stages and should be continued in different lagoons. The Gambier almost-atoll deserves to be included in these investigations, because it belongs to another kind of structure (caldera), in which the problems of origin, although parallel, involve particular parameters. Lastly, it would be beneficial if this research were to be an international effort, with an extension to other atoll groups in the Pacific such as the Federated States of Micronesia and Kiribati. The investigations already in progress in the southeast Pacific would then serve as a beginning for a more comprehensive research effort.

REFERENCES

AGASSIZ, A., 1906, Reports of the scientific results of the expedition to the Eastern Tropical Pacific. V. General report of the expedition: Memoirs of the Museum of Comparative Zoology, Harvard University, v. 33, p. 1–75.

AISSAOUI, D. M., BUIGUES, D., AND PURSER, B. H., 1986, Model of reef diagenesis: Mururoa Atoll, French Polynesia, *in* Schroeder, J. H., and Purser, B. H., eds., Reef Diagenesis: Springer-Verlag, Berlin, p. 27–52.

ASANO, D., 1942, Coral reefs of the South Sea Islands: Tohoku Imperial University, Geological and Paleontological Institute, Report No. 39, p. 7–19 (in Japanese).

BOURROUILH, F., 1977, Géomorphologie de quelques atolls dits "soulevés" du Pacifique Ouest et Sud-Ouest, origine et évolution des formers récifales actuelles: Mémoires du Bureau de Recherches Géologiques et Minières, Orléans, v. 89, p. 419–439.

BROUSSE, R., 1974, Géologie et pétrologie des Iles Gambier: Cahiers du Pacifique, v. 18, p. 9–19.

BROUSSE, R., 1985, The age of the islands in the Pacific Ocean: Volcanism and coral build up: Proceedings, 5th International Coral Reef Congress, Tahiti, v. 6, p. 389–400.

BROUSSE R., CHEVALIER, J. P., DENIZOT, M., AND SALVAT, B., 1974, Etude géomorphologique des Iles Gambier: Cahiers du Pacifique, v. 18, p. 20–119.

BUDDEMEIER, R. W., AND OBERDORFER, J. A., 1986, Internal hydrology and geochemistry of coral reefs and atoll islands: Key to diagenetic variations, *in* Schroeder, J. H., and Purser, B. H., eds. Reef Diagenesis: Springer-Verlag, Berlin, p. 91–111.

BUIGUES, D., 1985, Principal facies and their distribution at Mururoa Atoll, French Polynesia: Proceedings, 5th International Coral Reef Congress, Tahiti, v. 3, p. 249–255.

CHARPY, L., 1984, Quelques caractéristiques de la matière organique particulaire du lagon, *in* L'atoll de Tikehau, premiers résultats, Centre ORSTOM de Tahiti, Notes et Documents Océanographiques, v. 22, p. 13–34.

CHEVALIER, J. P., BAGNIS, R., POLI, G., RICHARD, B., AND SALVAT, B., 1976, Etude géomorphologique et biologique de l'atoll fermé de Taiaro (Tuamotu, Polynésie Française): Cahiers du Pacifique, v. 19, p. 169–288.

CHEVALIER, J. P., AND DENIZOT, M., 1979, Les organismes constructeurs de l'atoll de Takapoto: Journal de la Société des Océanistes, v. 35, p. 31–34.

CHEVALIER, J. P., DENIZOT, M., RICARD, M., SALVAT, B., SOURNIA, A., AND VASSEUR, P., 1979, Géomorphologie de l'atoll de Takapoto: Journal de la Société des Océanistes, v. 35, p. 9–18.

DANA, J. D., 1853, On coral reefs and islands: Dodd and Mead, New York, 143 p.

DARWIN, C., 1842, The structure and distribution of coral reefs: Smith Elder & Co., London, 214 p.

DAVIS, W. M., 1928, The coral reef problem: American Geographical Society, Special Publication 9, 596 p.

FAIRBRIDGE, R. W., 1957, The dolomite question: Society of Economic Paleontologists and Mineralogists Special Publication 5, p. 125–178.

GUILCHER, A., 1965, Coral reefs and lagoons of Mayotte Island, Indian Ocean, and of New Caledonia, Pacific Ocean: Colston Research Society, Proceedings, 17th Symposium at Bristol, Butterworths, London, p. 21–45.

GUILCHER, A., 1979, Origin of passes across rims and barrier reefs: Proceedings, XIVth Pacific Science Congress, Khabarovsk, Commission CD, p. 113.

GUILCHER, A., 1988, Coral reef geomorphology: John Wiley and Sons, New York, 228 p.

GUILCHER, A., BERTHOIS, L., DOUMENGE, F., MICHEL, A., SAINT REQUIER, A., AND ARNOLD, R., 1969, Les récifs coralliens de Mopelia et de Bora-Bora (Iles de la Société) et quelques autres récifs et lagons de comparaison (Tahiti, Scilly, Tuamotu occidentales). Morphologie, sédimentologie, fonctionnement hydrologique: Mémoires ORSTOM, Paris, v. 38, 106 p.

HARMELIN-VIVIEN, M., 1985, Tikehau Atoll, Tuamotu archipelago: Proceedings, 5th International Coral Reef Congress, Tahiti, v. 1, p. 211–268.

HOFFMEISTER, J. E., AND LADD, H. S., 1945, Solution effects on elevated limestone terraces: Geological Society of America Bulletin, v. 56, p. 809–818.

MACNEIL, F. S., 1954, The shape of atolls: An inheritance from erosion forms: American Journal of Science, v. 252, p. 402–427.

MICHEL, A., 1969, Plancton du lagon et des abords extérieurs de l'atoll de Mururoa: Cahiers du Pacifique, v. 13, p. 81–132.

MICHEL, A., COLIN, C., DESROVIERS, R., AND OUDOT, C., 1971, Observations sur l'hydrologie et le plancton des abords des passes de l'atoll de Rangiroa: Cahiers de l'ORSTOM, série Océanographie, v. 11, p. 375–402.

MONTAGGIONI, L. F., 1989, Le soulèvement polyphasé d'origine volcano-isostatique: clef de l'évolution post-Oligocène des atolls du nord-ouest des Tuamotus (Pacifique central): Comptes-Rendus de l'Académie des Sciences de Paris, v. 309, sér. II, p. 1591–1598.

PIRAZZOLI, P. A., 1987, A reconnaissance and geomorphological survey of Temoe Atoll, Gambier Islands, South Pacific: Journal of Coastal Research, v. 3, p. 307–323.

PIRAZZOLI, P. A., AND MONTAGGIONI, L. F., 1986, Late Holocene sea level changes in the Northwest Tuamotu Islands, French Polynesia: Quaternary Research, v. 25, p. 350–368.

PIRAZZOLI, P. A., AND MONTAGGIONI, L. F., 1987, Les Iles Gambier et l'atoll de Temoe (Polynésie Française): anciennes lignes de rivage et comportement géodynamique: Géodynamique, v. 2, p. 13–25.

PIRAZZOLI, P. A., AND MONTAGGIONI, L. F., 1988, Holocene sea level changes in French Polynesia: Palaeogeography, Palaeoclimatology, Palaeoecology, v. 68, p. 153–175.

PIRAZZOLI, P. A., MONTAGGIONI, L. F., SALVAT, B., AND FAURE, G., 1988, Late Holocene sea level indicators from twelve atolls in the central and eastern Tuamotus, Pacific Ocean: Coral Reefs, v. 7, p. 57–68.

PURDY, E. G., 1974, Reef configuration: Cause and effect, *in* Laporte, L. F., ed., Reefs in Time and Space: Society of Economic Paleontologists and Mineralogists Special Publication 18, p. 9–76.

ROUGERIE, F., 1979, Caractéristiques générales du milieu liquide lagonaire de l'atoll de Takapoto: Journal de la Société des Océanistes, v. 35, p. 35–45.

ROUGERIE, F., 1985a, Le lagon sud-ouest de Nouvelle-Calédonie: spécificité hydrologique, dynamique et productivité: Office de la Recherche Scientifique et Technique Outre-Mer, Paris, Etudes et Synthèses, 234 p.

ROUGERIE, F., 1985b, Main characteristics of the Takapoto lagoon and internal functioning: Proceedings, 5th International Coral Reef Congress, Tahiti, v. 1, p. 363–367

ROUGERIE, F., AND WAUTHY, B., 1986, Le concept d'endo-upwelling dans le fonctionnement des atolls-oasis: Oceanologica Acta, v. 9, p. 133–148.

ROUGERIE, F., AND WAUTHY, B., 1988, The endo-upwelling concept: a new paradigm for solving an old paradox: Proceedings, 6th International Coral Reef Symposium, Townsville, Australia, v. 3, p. 21–26.

ROUGERIE, F., AND WAUTHY, B., 1989, Une nouvelle hypothèse sur la genèse des phosphates d'atolls: le rôle du processus d'endo-upwelling: Comptes-Rendus de l'Académie des Sciences de Paris, v. 308, sér. II, p. 1043–1047.

RUZIE, G., AND GACHON, A., 1985, Contribution of geophysical techniques in carbonate studies of the atolls. Application to the investigations of the structures of Mururoa Atoll, French Polynesia: Proceedings, 5th International Coral Reef Congress, Tahiti, v. 6, p. 381–388.

SALLER, A. H., 1984, Petrologic and geochemical constraints on the origin of subsurface dolomite, Enewetak Atoll: An example of dolomitization by normal seawater: Geology, v. 12, p. 217–220.

SALVAT, B., 1983, La faune benthique du lagon de l'atoll de Scilly, Archipel de la Société: Journal de la Société des Océanistes, v. 39, p. 515.

SALVAT, B., 1989, Ecologie et environnement corallien en Polynésie Française, *in* Ofai, no. 8, Tahiti and Moorea, p. 76–89.

SALVAT, B., AND RICHARD, G., 1985, Takapoto Atoll, Tuamotu archipelago: Proceedings, 5th International Coral Reef Congress, Tahiti, v. 1, p. 323–362.

SARGENT, M. C., AND AUSTIN, T. S., 1954, Biologic economy of coral reefs: U.S. Geological Survey Professional Paper 260E, p. 293–300.

SLATER, R. A., AND PHIPPS, C. V. G., 1977, A preliminary report on the coral reefs of Lord Howe Island and Elizabeth Reef, Australia: Proceedings, 3rd International Coral Reef Symposium, Miami, v. 2, p. 313–318.

SHEPARD, F. P., 1970, Lagoonal topography of Caroline and Marshall Islands: Geological Society of America Bulletin, v. 81, p. 1905–1914.

STODDART, D. R., AND TAYLOR, J. D., eds., 1971, Geography and Ecology of Diego Garcia Atoll, Chagos Archipelago: Atoll Research Bulletin, v. 149, 237 p.

SWARTZ, J. H., 1958, Geothermal measurements on Eniwetok and Bikini Atolls: U.S. Geological Survey Professional Paper 260U, p. 711–739.

WEBB, K. L., DUPAUL, W. D., WIEBE, W., SOTTILO, W., AND JOHANNES, R. E., 1975, Enewetak Atoll: Aspects of the nitrogen cycle on a coral reef: Limnology and Oceanography, v. 20, p. 198–210.

WIENS, H. J., 1962, Atoll Environment and Ecology: Yale University Press, New Haven, 532 p.

YABE, H., 1942, Problems of the coral reefs: Tohoku Imperial University, Geological and Paleontological Institute, Report No. 39, p. 1–6 (in Japanese).

GEOLOGIC DEVELOPMENT AND DEPOSITIONAL HISTORY OF THE FLORIDA MIDDLE GROUND: A MID-SHELF, TEMPERATE-ZONE REEF SYSTEM IN THE NORTHEASTERN GULF OF MEXICO

GREGG R. BROOKS AND LARRY J. DOYLE

Department of Marine Science, University of South Florida, 140 Seventh Avenue South, St. Petersburg, FL 33701

ABSTRACT: The Florida Middle Ground (FMG), the northernmost living coral reef in the Gulf of Mexico, was investigated by high-resolution seismic-reflection methodology, surface-sediment analyses, and direct observation by submersible to identify controls and processes responsible for recent geologic development, and to determine how it relates to development of the surrounding west Florida continental margin, a non-reef-rimmed carbonate platform. Modern reef growth is developed on a foundation consisting of a dead reef that probably was formed during one or more sea-level highstands of the Quaternary. Initial reef growth probably was controlled by bathymetry and regional circulation patterns that served to recruit Caribbean fauna to the FMG.

Surface sediments are distributed in indistinct zones exhibiting a patchy distribution controlled by physical processes and the rugged bathymetry. Sediments throughout the FMG consist dominantly of molluscan shell hash derived from both the reef ridges and surrounding shelf environment. Barnacle fragments are the only major constituent found associated with reef ridges and not the surrounding shelf sediments. Barnacles probably are prevalent on the FMG because of the suitable sites for attachment provided by the dead-reef foundation. Thus, barnacles provide the only major sedimentological indicator of a transition from an open-shelf to reefal environment. The lack of a clear distinction between sediment types of the two environments may have implications for those investigating ancient carbonate environments.

Recent geologic development of the FMG parallels that of the west Florida margin. Driven by high-frequency sea-level fluctuations, similar carbonate sediments are produced in both settings during periods when the shelf surface is flooded. Thick surface sediments on the FMG, relative to the surrounding shelf, indicate reef ridges act to trap sediments much like shelf-edge reefs trap sediments on a reef-rimmed carbonate platform surface. Reef growth and sediment accumulation cannot keep pace with sea-level rise, which results in the drowning of FMG reefs along with the remainder of the west Florida carbonate margin.

INTRODUCTION

Modern reefs are commonly viewed in terms of lush tropical coral reefs typical of the Caribbean and equatorial Pacific regions. Research has shown that modern reefs, as well as those preserved in the ancient rock record, are not limited to shallow tropical waters, but occur in a variety of settings (James, 1983). Consequently, the inhabiting organisms responsible for maintaining reef growth are highly variable and, once incorporated into surrounding sediments, form deposits characteristic of each particular reefal environment.

The Florida Middle Ground (FMG), located on the outer west Florida continental margin (Fig. 1) at approximately 28°30′N latitude, has the distinction of being the northernmost living coral reef in the Gulf of Mexico (Jordan, 1952; Back, 1972; Cheney and Dyer, 1974). It covers over 700 km^2 of the midshelf and ranges from 25 to 40 m in depth (Fig. 2). The FMG is characterized by two north-northwest-trending parallel ridges over 30 km in length, separated by a broad, flat central valley as much as 7 km wide (Fig. 2). At the southern end of the central valley, several isolated knolls, pinnacles and small ridges rise (Fig. 2). The surrounding region is a drowned karst plain exhibiting a rolling topography of limestone with sinkholes and solution basins (Brooks, 1974). The FMG occurs at a break in regional gradient, which increases from a seaward slope of 0.01° from the east to 0.06° to the west (Hilde and others, 1981).

The FMG is situated in a complex area of mixing between the Caribbean-derived Loop Current, the West Florida Estuarine Gyre and Florida Bay Waters (Fig. 1). The water is generally clear, of constant salinity (36±0.5°/oo), and classified as north tropical in nature (Austin and Jones, 1974; Cheney and Dyer, 1974). Average annual bottom temperatures fluctuate between 16° and 20°C with the lowest values occurring in January and February and the highest in September and October.

The objectives of this study were to define the controls and processes responsible for the development of the FMG, to determine its relation to the development of the surrounding west Florida margin, and to compare it with other modern reef systems. Results will be useful for identifying similar "atypical" reef environments in both the modern and ancient sedimentary records.

METHODS

The data base for this project consisted of a combination of high-resolution seismic-reflection data, surface-sediment sample analyses and video data recorded from a manned submersible. Seismic data were collected with a 500 Joule EG&G Uniboom by Texas A&M University. Transects were oriented in an east-west direction normal to the strike of reef ridges. Surface-sediment samples were collected seasonally during each of four cruises from October 1978 to July 1979. The samples, obtained by divers, grab sampler and dredge hauls, were taken from four benchmark stations previously determined to represent a typical biological assemblage of the region. During November 1978, closely spaced samples were recovered along transects in the vicinity of three of these stations. A total of 178 samples was analyzed. Twenty-eight samples from transects across the Middle Ground, taken in 1973, were analyzed for sediment constituents. Textural and calcium-carbonate data from these samples were determined previously by Birdsall (1978) and are incorporated into the interpretation of results. The sample locations are shown in Figure 2. Direct observations were made and video data recorded on FMG reefs via submersible.

Seismic data were analyzed using sequence and facies analysis as described in Mitchum and others (1977).

From Shoreline to Abyss, SEPM Special Publication No. 46

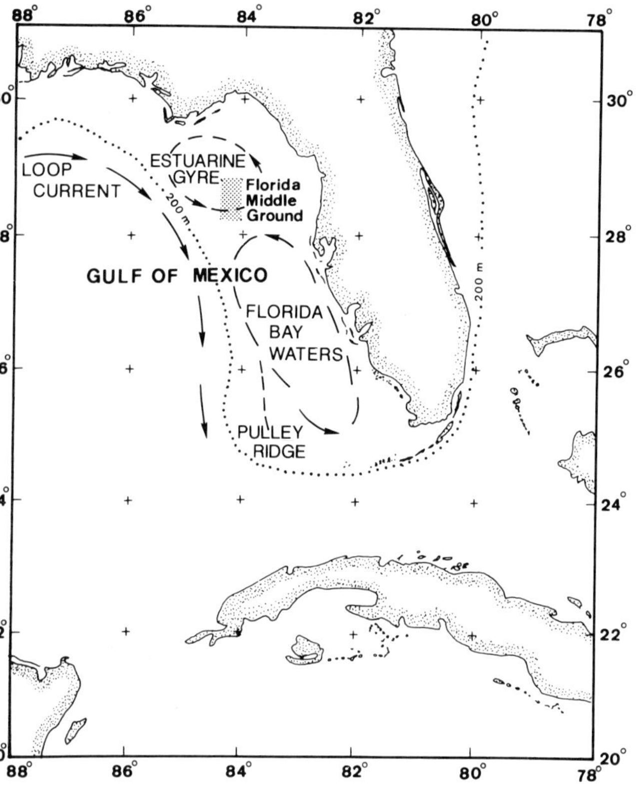

FIG. 1.—Location map showing the Florida Middle Ground on the west Florida shelf. Also shown are the current systems influencing the study area, and the Pulley Ridge biohermal complex on the southern shelf.

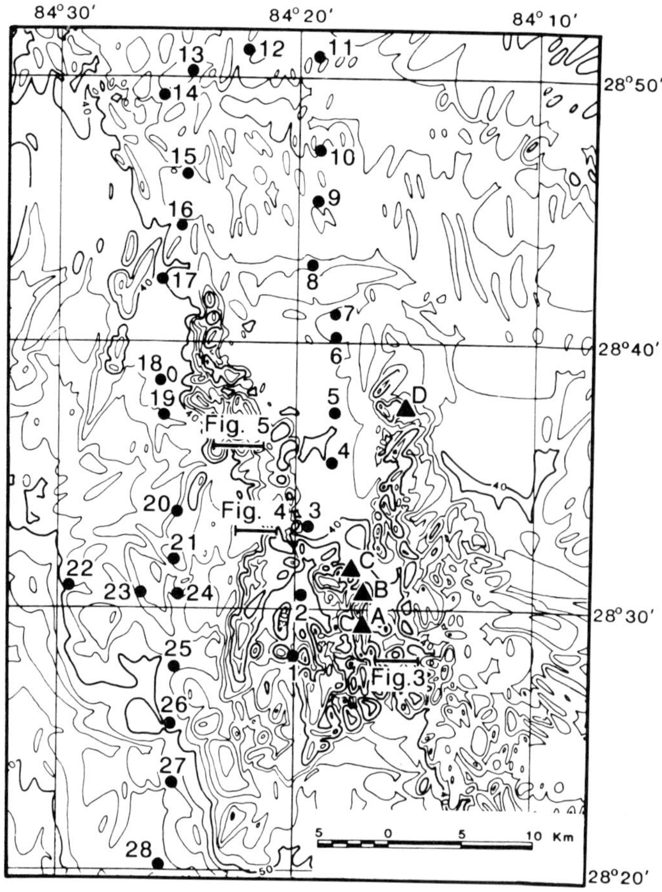

FIG. 2.—Bathymetric map of the Florida Middle Ground showing sampling sites and locations of seismic profiles shown in Figures 3–5. Contour interval 2 m. ● denotes surface-sample transect sites occupied by Birdsall (1978). ▲ denotes Benchmark sites, A = 491, B = 481, C = 151, D = 247.

Surface-sediment samples were analyzed for texture, calcium-carbonate content, constituent composition and mineralogical composition. Grain size was determined for all samples by the seive and pipette method (Folk, 1965). Calcium-carbonate content was determined by acid leaching as described in Milliman (1974). Sediment constituent composition was determined by the whole-grain point-count method (Carver, 1971). For each sample, at least 100 particles were identified from each whole phi (ϕ) interval of the sand-size fraction. This method gives a 95 percent chance of finding any constituent type comprising at least 3 percent of the total sample (Dennison and Hay, 1967), which obviates bias that may occur due to size differences. Sediment constituent composition was normalized to 100 percent for each sample. An acetate impression of the internal structure was made on 10 selected samples according to the technique described by Milliman (1974) to identify constituents too degraded for identification under reflected light. Each sample was analyzed and constituents identified by transmitted light. Surface texture of individual sediment grains was identified for 10 representative samples. At least 100 grains from each sample were selected by the point-count method (Carver, 1971) and described as "fresh," "dull," or "worn," according to criteria described in Wanless (1977). "Fresh" refers to grains showing no weathering at all, but exhibiting detailed surface features. "Worn" refers to grains that have essentially lost all of their surface features, but may be identifiable by shape. "Worn" grains commonly exhibit a rough, pitted and/or bored surface. The term "dull" refers to grains that have lost little material to the weathering process, but have lost their gloss, are discolored, or otherwise altered from their "fresh" appearance.

The mineralogical composition of 10 samples selected from different environments was determined by X-ray diffraction techniques (Grim, 1968; Carver, 1971). Amino-acid racemization dates were determined for selected samples. Both a complete assemblage and a monospecific molluscan assemblage were dated from each of nine samples. The complete assemblage was used to gain a general idea of the age of the sediment constituents collectively. Monospecific molluscan material was used to obtain a more specific age for the mollusc fraction, which is by far the dominant fraction in FMG surface sediments.

RESULTS AND DISCUSSION

Seismic Data

As a result of the high reflectivity of FMG carbonate-rich sediments, penetration of the signal beneath the sea floor was greatly inhibited. On the average, approximately 20 ms (two-way time) or less of penetration yielded useful data. Three distinct seismic sequences are present, each bounded by an unconformable surface (Figs. 3–5).

The bottom sequence, termed sequence I, is characterized by a very irregular, highly reflective surface (Figs. 3–5). This unit represents the acoustic basement in the study area. Surface irregularities are shown by discontinuous, truncated and occasional hyperbolic reflectors, though often their identification is inhibited by signal noise. This surface is interpreted to represent a karst surface, the irregular nature resulting from dissolution of the underlying carbonate rock during one or more periods of lowered sea level. Sequence I may correspond to the Miocene (?) karst surface underlying the south Florida outer shelf and cropping out on the upper slope along the southwest corner of the platform. Holmes (1985) reports that this surface is covered by a thin veneer of sediment on the shelf and, like sequence I underlying the FMG, abruptly increases in slope from the relatively flat (0.07° seaward dip) inner shelf to a gently sloping (0.2° seaward gradient) outer shelf.

The middle sequence, sequence II (Figs. 3–5), is bounded on top by a flat-lying surface and on the bottom by the irregular surface of sequence I. Sequence II is less than 5 to 10 ms thick. Seismic facies are dominantly reflection free. Where internal reflectors are present, they are generally parallel to subparallel and discontinuous. Once again, signal noise inhibits the identification of seismic facies. The extremely flat surface and lack of internal reflectors suggest that this sequence formed by uniform deposition independent of underlying bottom relief, possibly representing the Holocene, or earlier, transgressive sand sheet. Based upon the broader data base, Hilde and others (1981) report that this sequence downlaps and eventually pinches out against the surface of sequence I along a north-northwest-south-southeast line through the study area. Hence, sequence II does not appear to exist landward of the study area.

Sequence III (Figs. 3–5) is bounded by the sea floor at the top and the relatively flat surface of sequence II at the base. It is from less than 5 to 10 ms thick with thickest deposits occurring along the flanks of ridges (Figs. 3–5) and as lenses in the plain between ridges (Fig. 3). As with sequence II, seismic facies are dominantly reflection free, and where present, internal reflectors are parallel to sub-

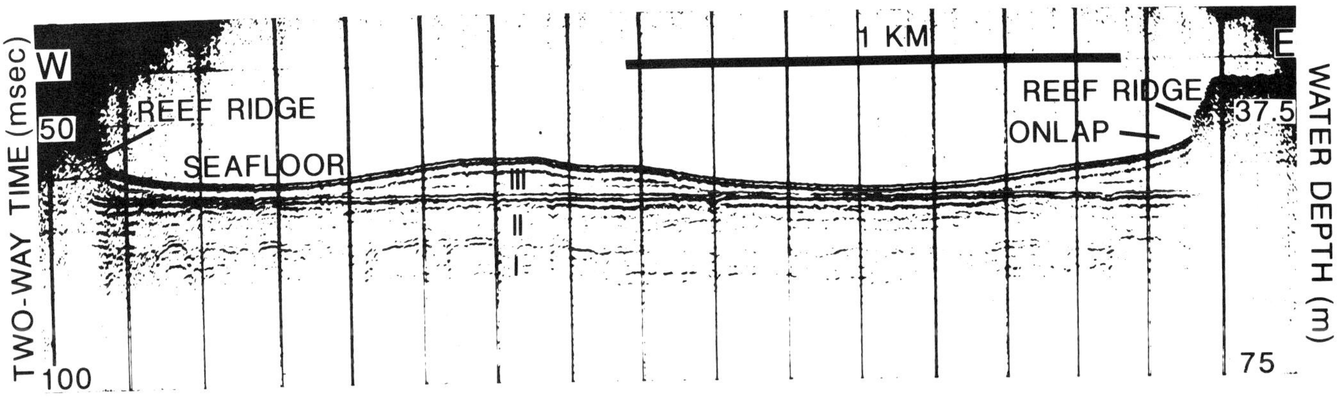

FIG. 3.—Seismic profile showing two reef ridges separated by a broad plain (see Fig. 2 for location). Note the presence of three sequences (I, II and III) bounded by unconformities, and onlapping of the surface unit against ridges as well as the thickening of the unit in the center.

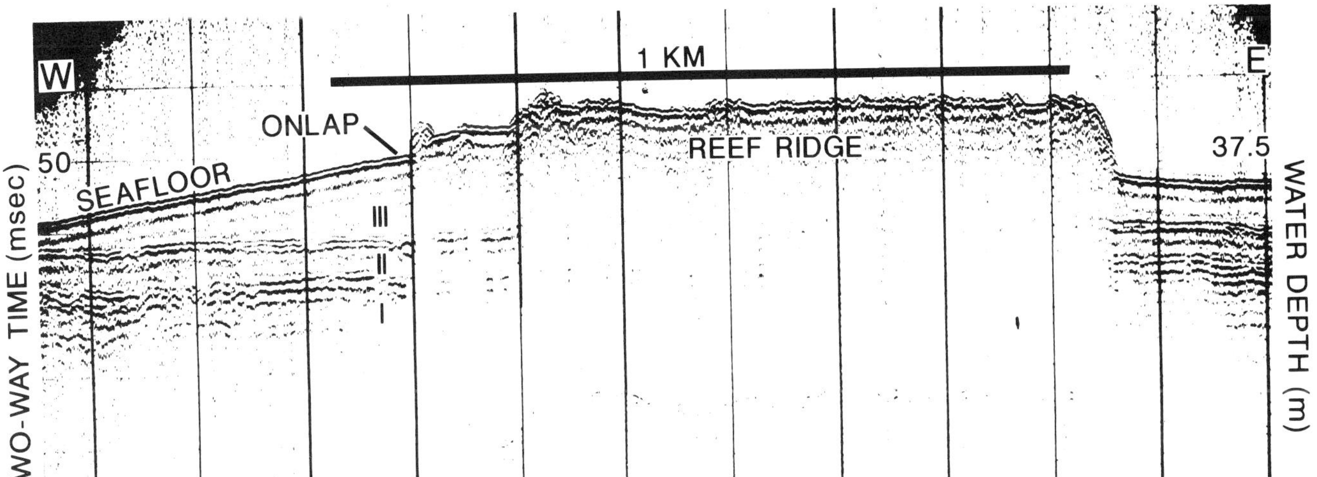

FIG. 4.—Seismic profile showing sequences I, II and III, and the reef ridge with onlapping sediments (see Fig. 2 for location).

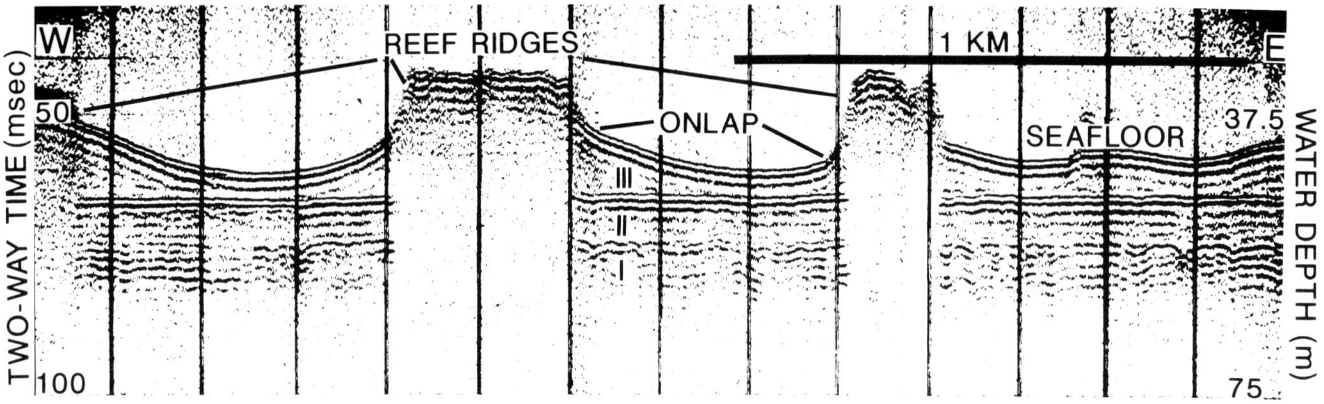

FIG. 5.—Seismic profile showing sequences I, II and III, and reef ridges with onlapping sediments (see Fig. 2 for location).

parallel and discontinuous. Reflectors appear to lap out against FMG reef ridges, forming sediment wedges more than 7 ms thick (Fig. 3). Sediments of sequence I are interpreted to have been deposited as sheet drape. The thick wedges of sediments along ridge flanks suggest that sediments are either derived from the ridges and shed off along the flanks, or that physical processes are transporting sediments and depositing them against the ridges. The presence of thick sediment lenses, some in excess of 5 ms, in the broad, flat plains (Fig. 3) indicates that sediments are being "ponded" between reef ridges.

Reef ridges (Figs. 3–5) are acoustically transparent features with highly reflective surfaces. Ridges are tens to hundreds of meters wide and rise from 5 to 15 ms above the surrounding sea floor. Ridges do not appear to be accumulating an appreciable sediment cover.

Sediments

Results of sedimentological analyses are discussed in terms of both areal distribution patterns and temporal changes, on a seasonal basis, of parameters including texture, major sediment constituents, calcium-carbonate content and age dating of selected samples.

Textural distribution patterns are shown in Figures 6 and 7. Sediment distribution is patchy on both a regional scale throughout the FMG as shown by Figure 6, and locally as shown by the closely spaced samples along transects at benchmark stations (Fig. 7). Mean grain sizes of transect samples range from −0.9 ϕ (very coarse sand) on a ridge crest in the vicinity of BM-481 (Figs. 2 and 7), to 3.4 ϕ (very fine sand) on the open shelf immediately northwest of the westernmost ridge (Fig. 6).

Coarsest grained sediments are found on ridge crests with mean sizes consistently in the very coarse (−1 to 0 ϕ) to coarse (0–1 ϕ) sand range (Fig. 7). Immediately adjacent to ridge crests and in bathymetric depressions within ridges, mean sizes decrease to dominantly fine sand (2–3 ϕ; Fig. 7). With few exceptions, sediments in the central plain and on the shelf adjacent to the ridges have mean sizes from fine (2–3 ϕ) to very fine (3–4 ϕ) sand (Fig. 6). The fining of sediments away from the ridge crests suggests either that the ridge crests are the origin for coarse-grained sediments or that fine-grained sediments are being winnowed from the ridge crests, leaving only the coarser grained sediments behind. The occurrence of finer grained sediments in bathymetric depressions within ridges (Fig. 7) supports the contention that fines are being winnowed from exposed portions of ridge crests and suggests that the rugged bathymetry associated with the FMG may be a major control on sediment distribution. The few instances of relatively coarse mean

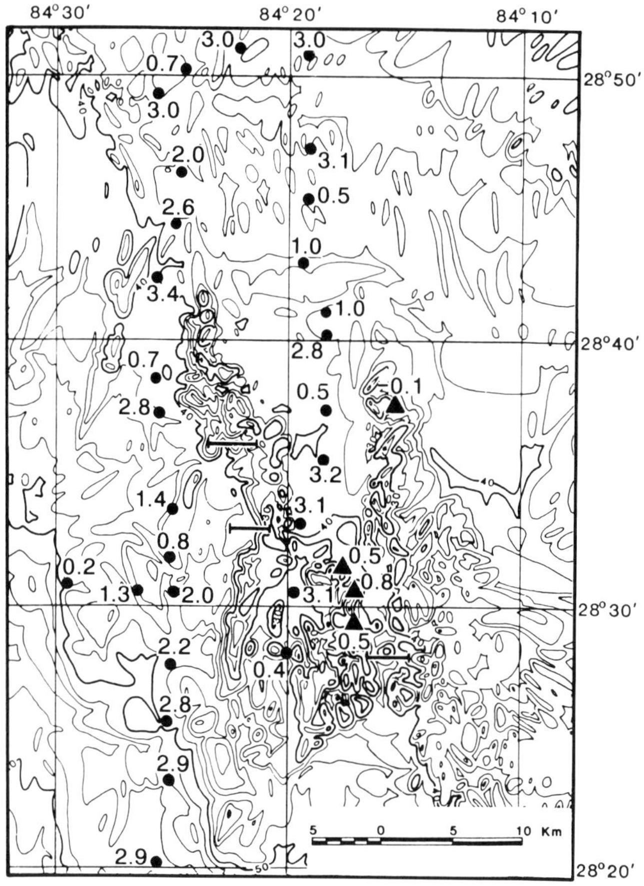

FIG. 6.—Bathymetric map of the FMG showing a general decrease in mean grain size away from the reef ridges. Values expressed as mean phi (ϕ).

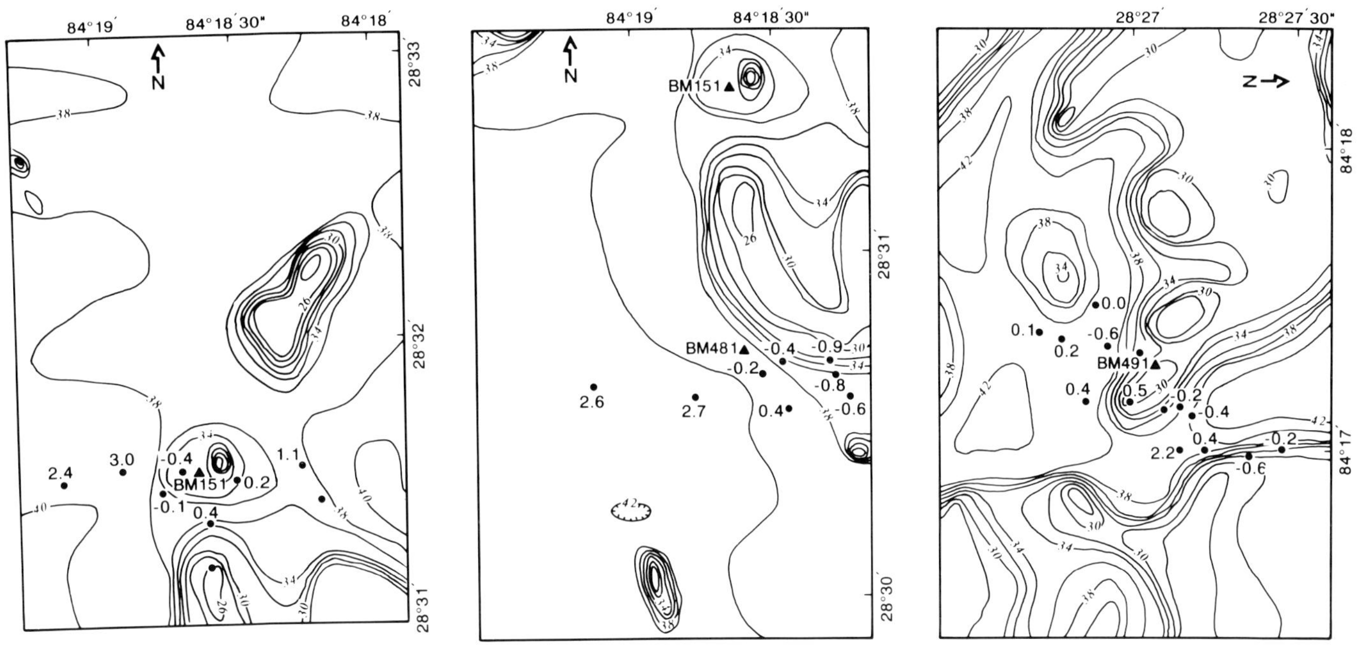

FIG. 7.—Bathymetric maps of the vicinity of Benchmark stations 151, 481 and 491, on reef ridges showing a general decrease in mean grain size abruptly away from ridge crests and in bathymetric depressions within ridges. Values expressed as mean phi (ϕ).

grain sizes on the central plain and shelf sediments adjacent to ridges may reflect localized *in-situ* production of relatively coarse-grained biogenic calcium carbonate.

The mean grain size of sediments collected from benchmark stations on ridge crests shows a tendency toward coarsening from October 1978 to June 1979 (Fig. 8). Mean sizes for the October 1978 sampling period range from 0.3 ϕ (coarse sand) to 1.4 ϕ (medium sand). Mean sizes for the two stations (151 and 247) sampled during January 1979 were both −0.6 ϕ (very coarse sand). Mean sizes for the June 1979 sampling period ranged from −1.0 ϕ (granule/very coarse sand) to 0.3 ϕ (coarse sand). This tendancy toward coarsening may be a result of progressively greater winnowing by bottom currents between October 1978 and June 1979, or alternatively, a greater input of production of coarser grained sediments within this time period. The latter possibility is doubtful because coarse-grained sediments consist dominantly of biogenic calcium carbonate, and an increase in the biogenic production of calcium carbonate of sufficient magnitude to produce the observed trends is unlikely. We suggest that this progressive tendancy toward coarser mean grain sizes is a result of greater winnowing. It is probably more a result of a gradual increase in bottom-current activity, or a series of high-energy events (such as winter-cold fronts), rather than one single event because of the incremental increases from October 1978 to January 1979 and then again from January 1979 to June 1979 at stations 151 and 247.

The distribution of calcium carbonate in sediments is similar to that of mean grain size in that it is quite patchy (Figs. 9 and 10). Values for the closely spaced samples collected along transects in the vicinity of each bench mark range from 78 to 99 percent calcium carbonate (Fig. 10). Highest values, most greater than 90 percent, are found on the shallow-ridge crests, whereas lower values (70–90 percent) are found in deeper areas adjacent to ridge crests and bathymetric depressions within ridges (Fig. 10). Lowest percentages are found farthest from the reef ridges with values as low as 18 percent $CaCO_3$ reported for the northernmost sample in the study area (Fig. 9). The non-calcium-carbonate component consists dominantly of fine-grained quartz sand. The apparent decrease in calcium-carbonate content away from the reef ridges suggests that biogenic calcium-carbonate production is greatest on reef-ridge crests. Alternatively, this tendancy may be explained by the lack of input of quartz sand on ridge crests. The occurrence of fine quartz sand in shelf samples, especially north of the reef ridges, may simply mask the calcium-carbonate content. The lack of quartz sand on ridge crests could be an artifact of winnowing of finer grained sediments, as mentioned previously.

Mean percentages of calcium-carbonate content for samples collected at different periods throughout the year show no appreciable change (Fig. 8). Values are all greater than 90 percent $CaCO_3$, ranging from 91 to 99 percent. This suggests that, although sediments on reef crests may be winnowed at a greater rate during certain times of the year, as indicated by the sediment textural distribution, the relative input of calcium-carbonate material to the sediment remains consistent and is independent of the winnowing process.

Constituent composition of FMG sediments (Fig. 11) is dominated by molluscs with an average of 36 percent; followed by barnacles at 8 percent; benthic foraminifera, annelid tubes and bryozoa at 3 to 4 percent; and the coralline alga *Goniolithon* spp. at 2 percent. Planktonic foraminifera,

Station 151

10/78
Grain Size
mean = 0.3ϕ

Percent Carbonate
mean = 98.2

```
75%-

                   x
50%-               x
                   x
                   x
25%-x              x
    x        x     x
    x  x     x     x
    1  2  3  4  5  6
Sediment Constituents
```

01/79
Grain Size
mean = -0.6ϕ

Percent Carbonate
mean = 91.3

```
                   x
    x              x
    x              x
    x              x
    x  x     x     x
    x  x  x  x  x  x
    1  2  3  4  5  6
Sediment Constituents
```

06/79
Grain Size
mean = -1.0ϕ

Percent Carbonate
mean = 94.1

```
    x
    x              x
    x              x
    x  x     x  x  x
    x  x  x  x  x  x
    1  2  3  4  5  6
Sediment Constituents
```

Station 247

10/78
Grain Size
mean = 1.2ϕ

Percent Carbonate
mean = 98.7

```
75%-

                   x
50%-               x
    x              x
    x              x
25%-x              x
    x  x           x
    x  x  x  x  x  x
    1  2  3  4  5  6
Sediment Constituents
```

01/79
Grain Size
mean = -0.6ϕ

Percent Carbonate
mean = 91.3

```
                   x
    x              x
    x              x
    x  x           x
    x  x  x  x  x  x
    1  2  3  4  5  6
Sediment Constituents
```

06/79
Grain Size
mean = -1.0ϕ

Percent Carbonate
mean = 95.3

```
    x              x
    x              x
    x              x
    x  x           x
    x  x     x     x
    x  x  x  x  x  x
    1  2  3  4  5  6
Sediment Constituents
```

Station 481

10/78
Grain Size
mean = 1.4ϕ

Percent Carbonate
mean = 98.7

```
75%-

                   x
50%-               x
    x              x
    x              x
25%-x              x
    x  x           x
    x  x  x  x  x  x
    1  2  3  4  5  6
Sediment Constituents
```

06/79
Grain Size
mean = -0.3ϕ

Percent Carbonate
mean = 93.0

```
75%-


50%-
    x
    x              x
25%-x  x           x
    x  x     x  x  x
    x  x  x  x  x  x
    1  2  3  4  5  6
Sediment Constituents
```

Station 491

10/78
Grain Size
mean = 0.9ϕ

Percent Carbonate
mean = 98.8

```
75%-

                   x
50%-               x
                   x
    x              x
25%-x              x
    x  x           x
    x  x  x  x  x  x
    1  2  3  4  5  6
Sediment Constituents
```

06/79
Grain Size
mean = 0.1ϕ

Percent Carbonate
mean = 96.2

```
75%-


50%-
    x
    x              x
25%-x              x
    x  x     x  x  x
    x  x  x  x  x  x
    1  2  3  4  5  6
Sediment Constituents
```

1 = mollusc, 2 = barnacle, 3 = benthic foraminifera, 4 = annelid, 5 = bryozoa, 6 = other

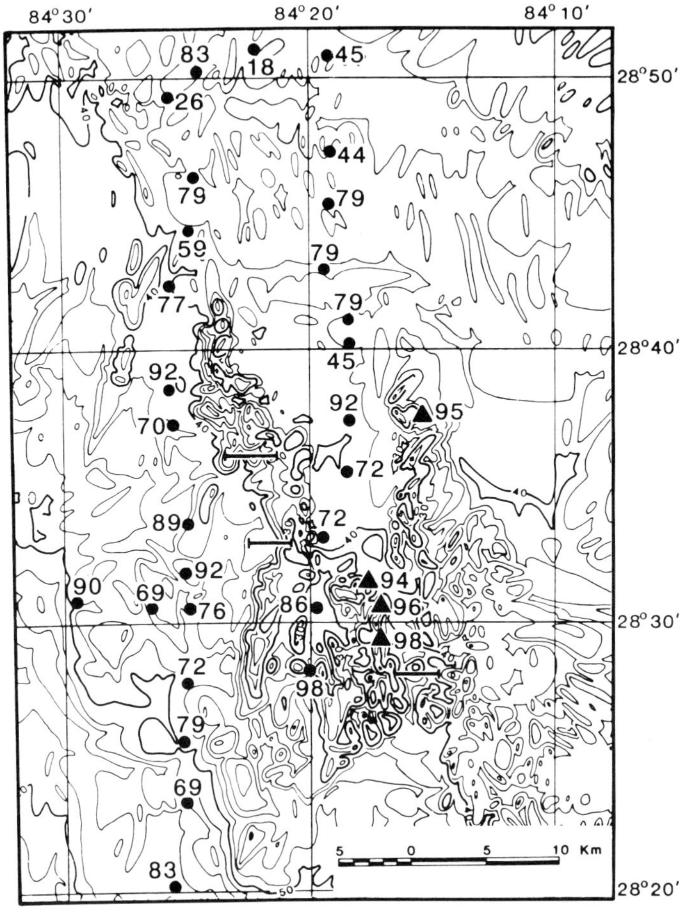

FIG. 9.—Bathymetric map of the FMG showing a general decrease in calcium-carbonate content away from the reef ridges. Values are expressed as the percentage of sediments consisting of calcium carbonate.

echinoids, octacorals, schleractinians, ostracods, arthropods, sponge spicules, complex skeletal particles (molluscs encrusted with bryozoa or annelid tubes), rock fragments, pelletoids, aggregates, and quartz were identified in trace amounts. Typically, quartz is present only in trace quantities; however, some samples, especially those collected north of the reef ridges, are dominated by quartz. This is supported by the percent $CaCO_3$ data (Fig. 9), as the insoluable residue consists almost entirely of fine quartz sand.

The abnormally high percentage of grains unidentifiable under reflected light (38 percent as shown in Fig. 11) can be attributed to the method of identification and/or the degraded condition of the grains. Because grains were analyzed from each whole phi (ϕ) interval, the grains in the finer fractions (2–3 ϕ and 3–4 ϕ) often consisted of fragments too small to identify. When these fine fractions constituted a large portion of the sediment sample, the entire sample would have a large unidentifiable fraction. Additionally, the quality of individual grains often was found to be quite poor. Such poor condition could be a reflection of age, or rapid degradation by physical, chemical or biological processes. Results of acetate-peel analyses indicate that the composition of grains unidentifiable under reflected light is essentially the same as for the identifiable fraction, with the exception of a slight increase in *Goniolithon* spp. grains. An increase in *Goniolithon* spp. probably reflects the error associated with identifying *Goniolithon* spp. because of the lack of distinguishing external morphological characteristics, especially when in such a degraded condition as observed here.

Areal distributions of the three most abundant constituents, namely molluscs, barnacles and benthic foraminifera, are shown in Figures 12 and 13. Mollusc fragments comprise from 7 to 55 percent of FMG sediments. They are by far the dominate carbonate constituent present. With few exceptions, molluscs comprise from 30 and 50 percent of the sediment collected along transects in the vicinity of benchmark stations (Fig. 13). Lower values are generally confined to samples collected in deeper areas away from or between ridge crests, or from protected areas within ridges created by the rugged bathymetry. Percentages of sediments consisting of mollusc fragments in the central plain and on the shelf immediately surrounding the FMG ridges is not appreciably different from those of samples from on the ridges (Fig. 12). In fact, the two highest percentages (53 and 55 percent) were reported for sediments collected on the shelf immediately west of the westernmost ridge. Lowest values, ranging from 7 to 39 percent, were found for sediments north of the ridges (Fig. 12), and are interpreted to be a reflection of the high percentage of fine-grained quartz sands (Fig. 9).

Molluscan content for benchmark stations varies little between October 1978 and June 1979 (Fig. 8). Values range from approximately 30 to 50 percent, which is consistent with transect samples collected from ridge crests. No evidence of seasonal variability exists.

Barnacle fragments average approximately 8 percent of FMG ridge-crest sediments, second only to molluscs (Fig. 11), but are highly variable from station to station (Figs. 14 and 15). Barnacles range from 0 to 13 percent of closely spaced samples collected along benchmark transects (Fig. 15). Highest values, commonly greater than 5 percent, often are found on shallow-ridge crests, whereas lowest values, generally no more than a few percent, are found on deeper areas away from ridge crests (Fig. 15), or in bathymetric depressions within ridges.

Lowest concentrations of barnacles occur in the central plain, as well as north and southwest of the ridges, where only trace amounts were found (Fig. 14). The distribution is quite patchy however, as values from 5 to 10 percent are reported for samples collected at considerable distances away from ridge crests (Fig. 14).

Barnacle fragments show no appreciable change in percentage from October 1978 to June 1979 (Fig. 8). Values

FIG. 8.—Mean grain size, calcium-carbonate content and relative percentages of sediment constituents for Benchmark stations 151, 247, 481 and 491 during different sampling periods.

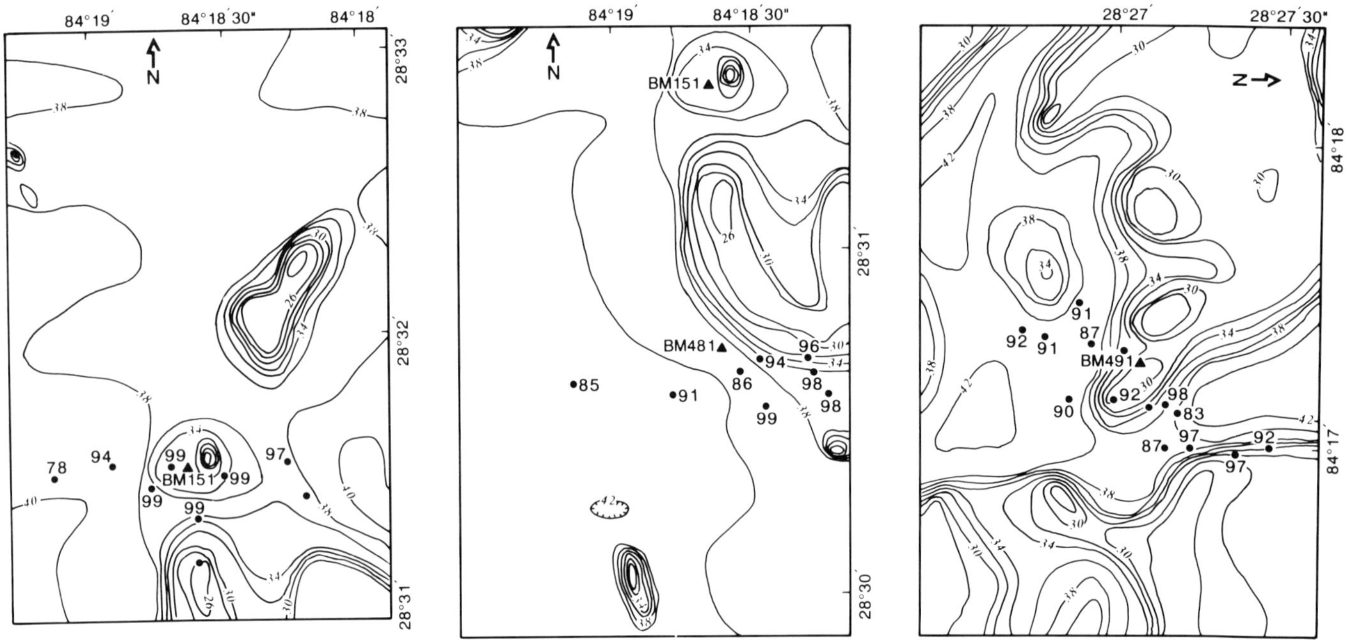

FIG. 10.—Bathymetric maps of the vicinity of Benchmark stations 151, 481 and 491, on reef ridges showing a general decrease in calcium-carbonate content abruptly away from ridge crests and in bathymetric depressions within ridges. Values are expressed as the percentage of sediment consisting of calcium carbonate.

range between 5 and 15 percent, which is consistent with transect sample values reported for ridge crests.

Benthic foraminifera average approximately 4 percent of FMG sediments (Fig. 11) but range from 0 to 15 percent of the sample (Figs. 16 and 17). Values for samples along benchmark transects are quite variable, with a weak tendancy to increase in deeper samples away from ridge crests (Fig. 17). Both highest (15 percent) and lowest (0 percent) values were recorded for these closely spaced benchmark transect samples (Fig. 17), illustrating the patchy areal distribution.

Samples collected on and near reef ridges including the central plain contain consistently less than 10 percent benthic foraminifera, with most less than 7 percent. Samples collected away from reef ridges, with the exception of the northernmost quartz-rich samples, have somewhat higher values, some of which exceed 10 percent (Fig. 16).

Benthic foraminifera percentages show no evidence of variability from October 1978 to June 1979 (Fig. 8). Values are consistently in the 5 percent or less range for all sampling periods.

In general, distribution patterns of the three most abundant carbonate constituents reflect their origin, physical dispersion and bathymetry. The dominance of mollusc fragments throughout the study area indicates that molluscs are a common organism to both reefal and shelf environments. The lower percentages reported for sediments collected in bathymetric depressions within ridge crests probably reflects dilution by finer grained carbonate-producing organisms. Mollusc fragments are more common in the coarse-grained fraction. Consequently, in exposed areas such as ridge crests, winnowing of finer grained material results in an increase in the percentage of molluscan debris. The low percentages of molluscan material in the northernmost samples probably reflects dilution by the large quantities of fine-grained quartz sands.

The distribution of barnacle fragments is similar to that of molluscs in that percentages are highest on ridge crests. It differs from that of molluscs, however, in that lowest percentages are reported for samples collected farthest from reef ridges, suggesting an association of barnacles with reef ridges. Like molluscs, barnacles are more common in the coarser size fractions, which may explain why percentages are much greater on shallow, exposed portions of ridge crests than in deeper, more protected areas within ridges. The distribution of benthic foraminifera is patchy, but generally opposite that of barnacles. Because benthic foraminifera occupy the finer size fractions, their distribution is more heavily influenced by low to moderate-energy processes relative to what is required to distribute coarser grained material, hence the patchy distribution, especially when considering the rugged bathymetry. Their apparent increase in shelf samples may reflect winnowing from ridge tops, or alternatively, may be an indication that benthic foraminifera are not associated exclusively with reef ridges.

Surface-texture analysis indicates that, on average 76 percent of the grains are "worn," 20 percent are "dull" and 4 percent are "fresh." Figure 18 illustrates examples of each category as well as a typical constituent assemblage. The general deteriorated condition probably is responsible for the large number of grains unidentifiable under reflected light (Fig. 11), and may be indicative of intense weathering or old age. Amino-acid racemization ages range from 3,700±1,500 yrs to 7,500±2,500 yrs. Youngest sediments

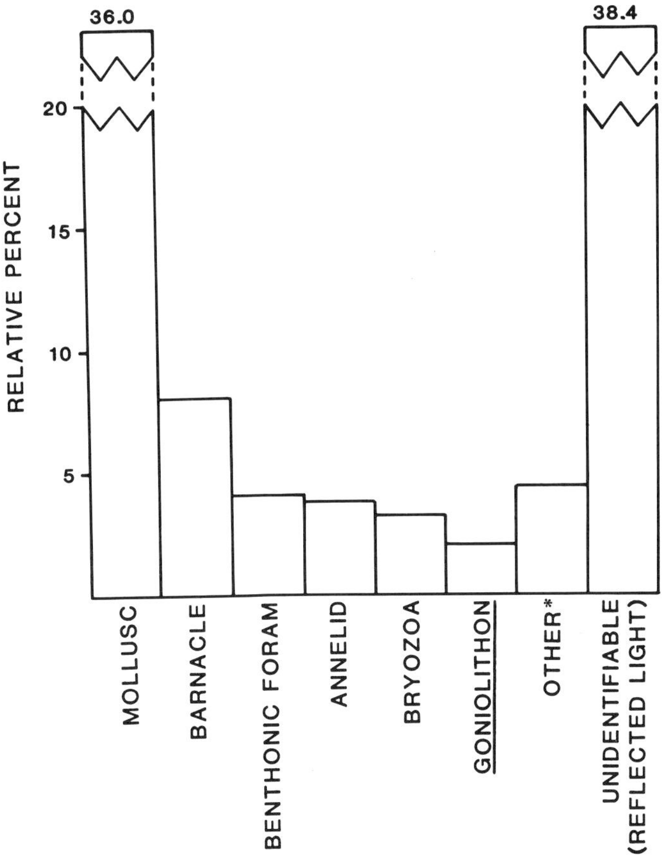

* INCLUDES: PLANKTONIC FORAMINIFERA, ECHINOIDS, OCTOCORALS, SCLERACTINIANS, OSTRACODS, CRABS, SPONGE SPICULES, COMPLEX SKELETAL (MOLLUSCS ENCRUSTED WITH BRYOZOA OR ANNELIDS), ROCK FRAGMENTS, PELLETOIDS, AGGREGATES, & QUARTZ.

FIG. 11.—Constituent composition of FMG sediments.

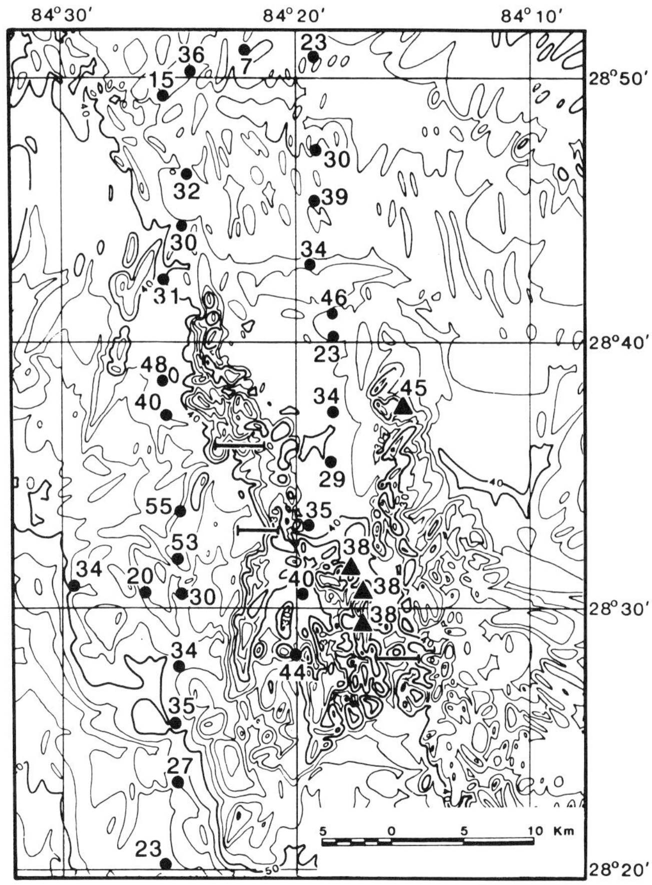

FIG. 12.—Bathymetric map of the FMG showing the distribution of mollusc shells in sediments. Values expressed as the relative percentage of sediment consisting of mollusc shells or fragments.

are found on the ridge crests, whereas oldest sediments are found on the central-plain and open-shelf environments.

Clay minerals constitute less than 5 percent of FMG sediments and, therefore, are relatively unimportant from a volumetric standpoint. The only clay mineral detected was hydrobiotite, a member of the illite group, and most common in temperate to subtropic regions (Grim, 1968). Blackened grains were found to be common in samples collected away from reef ridges, especially in the northernmost portion of the study area. Mineralogically, the blackened grains consist almost entirely of calcium carbonate. Their blackened appearance is the result of postdepositional processes occurring in the open-shelf but not reefal environment.

Video Data

Direct observations from submersible and video tape reviews reveal a variety of characteristics associated with FMG reef ridges, ridge flanks and the central plain/open shelf. The reef top exhibits rugged bathymetry covered extensively by *Millepora* sp. Less than 50 percent is covered by sediments, the remainder consisting of exposed reefal material. Sediment covering on ridges is confined principally to bathymetric depressions and sheltered, low-energy areas. Sediments consist dominantly of coarse sand, gravel and reef rubble.

Reef flanks consist, for the most part, of large blocks of reefal material appearing to be calving from reef margins. Blocks average approximately 1 m on a side, but reach more than 3 m in some cases. Reef flanks are occupied by living assemblages dominated by coralline algae, *Millepora* sp., and sponges. Surface sediments drape over calved reef blocks and biologic assemblages. Sediments consist of coarse sand and rubble rich in *Madracis* sp. fragments and appear to be transported from the ridge tops, where they are deposited as a drape over associated ridge flanks.

The central plain separating the two reef ridges and the surrounding open shelf exhibits similar characteristics. Sand waves and ripples exist in a variety of orientations. Some are characterized by dark, finer grained material in their troughs, whereas others were reported to be covered by various types of algae. Sediments consist of a fine sand to silty sand and appear to be heavily bioturbated.

In all environments of the FMG, sand waves and ripples

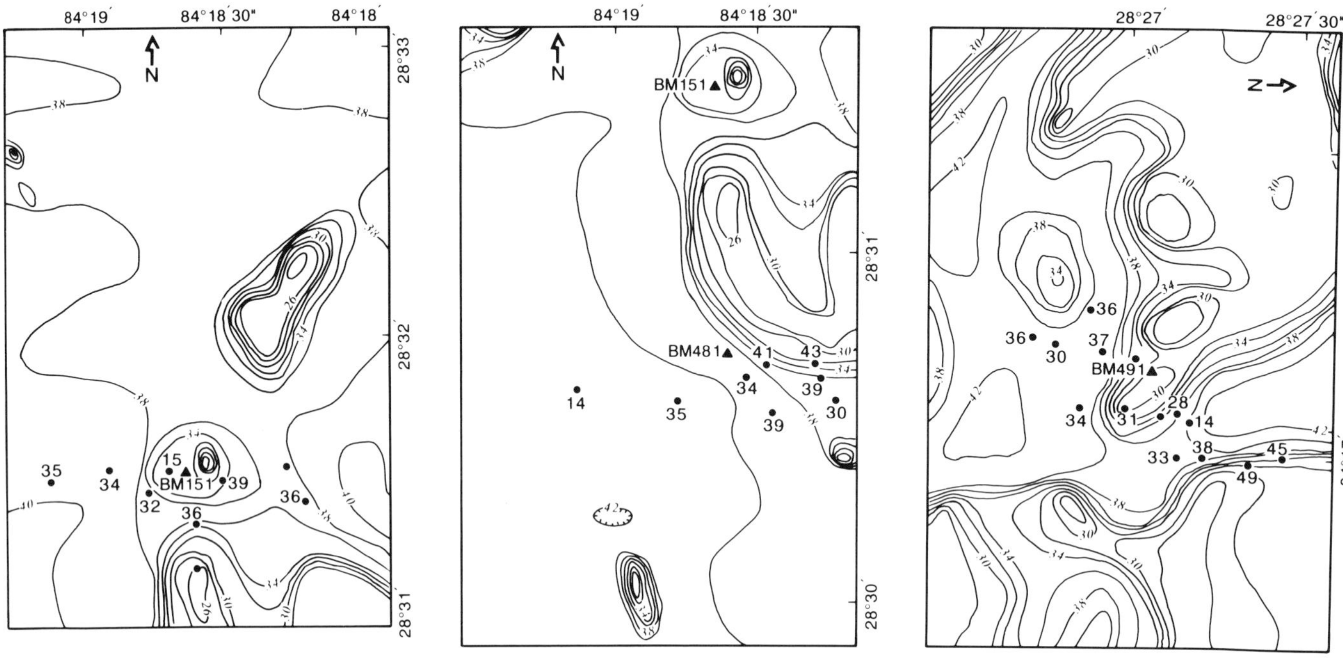

FIG. 13.—Bathymetric maps of the vicinity of Benchmark stations 151, 481 and 491, on reef ridges showing the distribution of mollusc shells in sediments. Values expressed as the relative percentage of sediment consisting of mollusc shells or fragments.

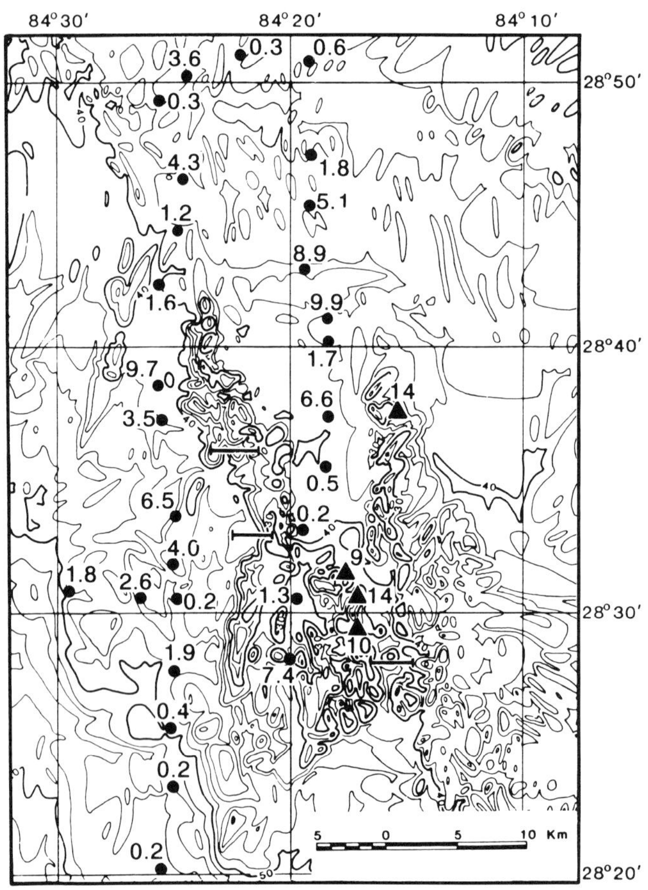

FIG. 14.—Bathymetric map of the FMG showing a general decrease in barnacle fragments away from the reef ridges. Values expressed as the relative percentage of sediment consisting of barnacle fragments.

were noted having north, northeast and northwest orientations. Bottom currents were measured directly flowing from north to south. Current velocity was not measured. Observations of a variety of bedforms suggest a complex physical environment consisting of omni-directional flow of variable velocities.

Geologic Development

Modern reef growth is limited to the tops of the highest ridge crests, built upon a dead-reef foundation. This dead-reef foundation may have formed during one or more of the sea-level highstands of the Pleistocene, or to some extent, during the early Holocene transgression. Bloom (1974) suggests that modern reef complexes commonly form as thin veneers on Pleistocene reefal buildups. These reefal buildups are not planed off during subaerial exposure, because their high permeability permits water to be easily transmitted through the framework. Controlling factors responsible for initiating original reef growth are not clearly understood. Growth may have been initiated by the break in bottom gradient between the relatively flat-lying shelf to the east and the gently seaward-dipping surface to the west. This explanation would be consistent with that postulated for the Pulley Ridge central-shelf biohermal complex discovered on the south Florida shelf. Pulley Ridge (Fig. 1), which is from 70 to 90 m deep, corresponds to a slight but rather abrupt break in slope of the underlying Miocene(?) surface interpreted to represent post-Miocene, westward tilting of the west Florida platform (Holmes, 1981).

Original growth also may have developed preferentially on an underlying karst surface, as suggested by Price (1954). The irregular surface of sequence I is interpreted to represent a karst surface that developed by dissolution of the

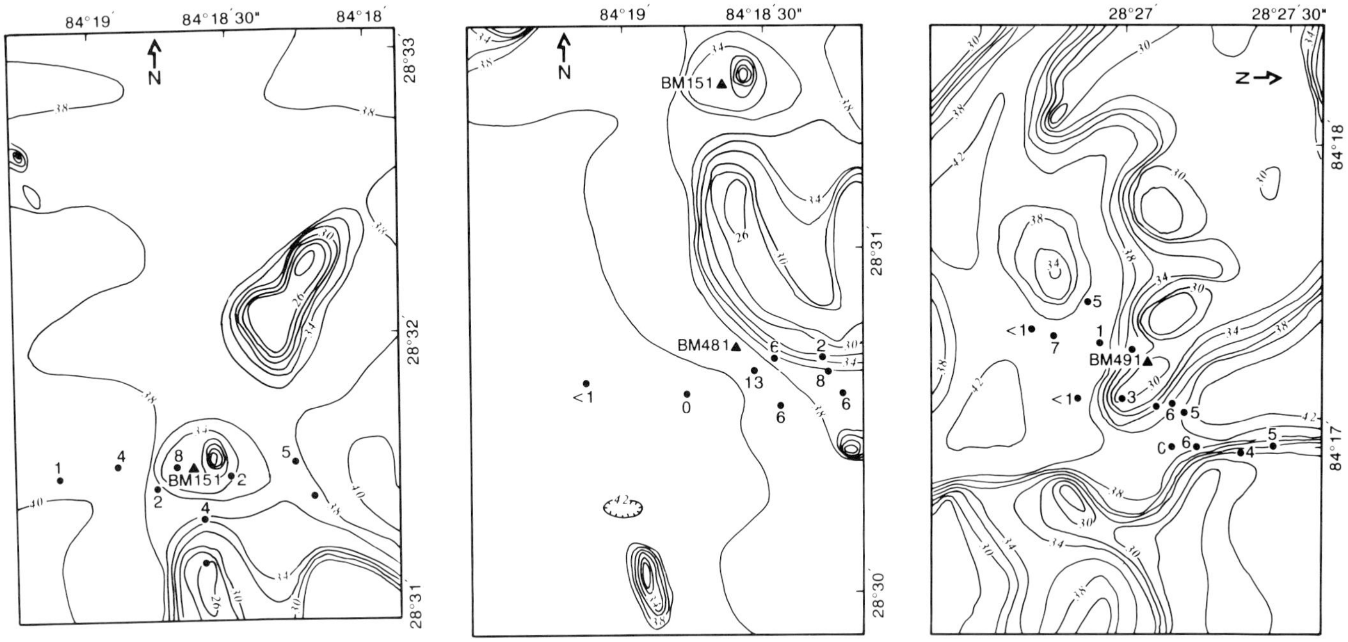

FIG. 15.—Bathymetric maps of the vicinity of Benchmark stations 151, 481 and 491, on reef ridges showing a general decrease in barnacle fragments abruptly away from ridge crests and in bathymetric depressions within ridges. Values expressed as the relative percentage of sediment consisting of barnacle fragments.

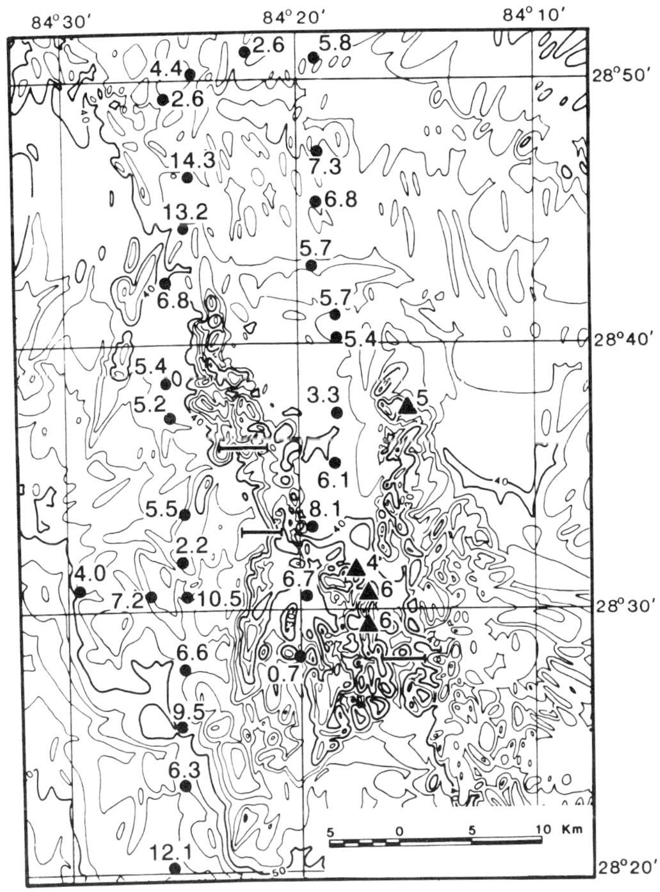

FIG. 16.—Bathymetric map of the FMG showing a general increase in benthic foraminifera away from the reef ridges. Values expressed as the relative percentage of sediment consisting of benthic foraminifera.

underlying limestone during one or more sea-level lowstands when the area was subaerially exposed. Reef growth may have developed during an ensuing sea-level rise, on a bathymetric high or slight break in gradient created by karstification. Modern reefal development on karst surfaces is thought to be common, with different stages of karstification often resulting in a specific reef type (Purdy, 1974). It also has been suggested that FMG ridges occupy what were once the banks of a river that was active during a period of lower sea level (Jordan, 1952). During the subsequent sea-level rise, these river banks may have created a bathymetric high optimum for reefal development. Hence, paleodrainage patterns also may have controlled the parallel-ridge configuration of the FMG.

Initial development as well as continued maintenance of FMG reefs also may have been controlled to a large extent by regional circulation patterns. The FMG occupies a junction between three major water masses; the Gulf Loop Current, the west Florida Estuarine Gyre and Florida Bay Waters (Fig. 1). It has been suggested (MAFLA, 1975) that Caribbean reefal organisms become entrained in the Loop Current, which transports them to the junction of these water masses where they settle out to form part of the biologic community. Consequently, this biologic recruitment may regularly provide species that would not normally be able to reproduce, given the relatively low water temperatures (MAFLA, 1975), thereby continuously "stocking" the FMG with Caribbean reefal fauna.

It is likely that Loop Current recruitment occurs to some extent at present, and perhaps, in conjunction with bathymetry, may have controlled initial FMG reef development. Considering the midshelf setting and the volume of reef growth, we feel the most reasonable interpretation is that

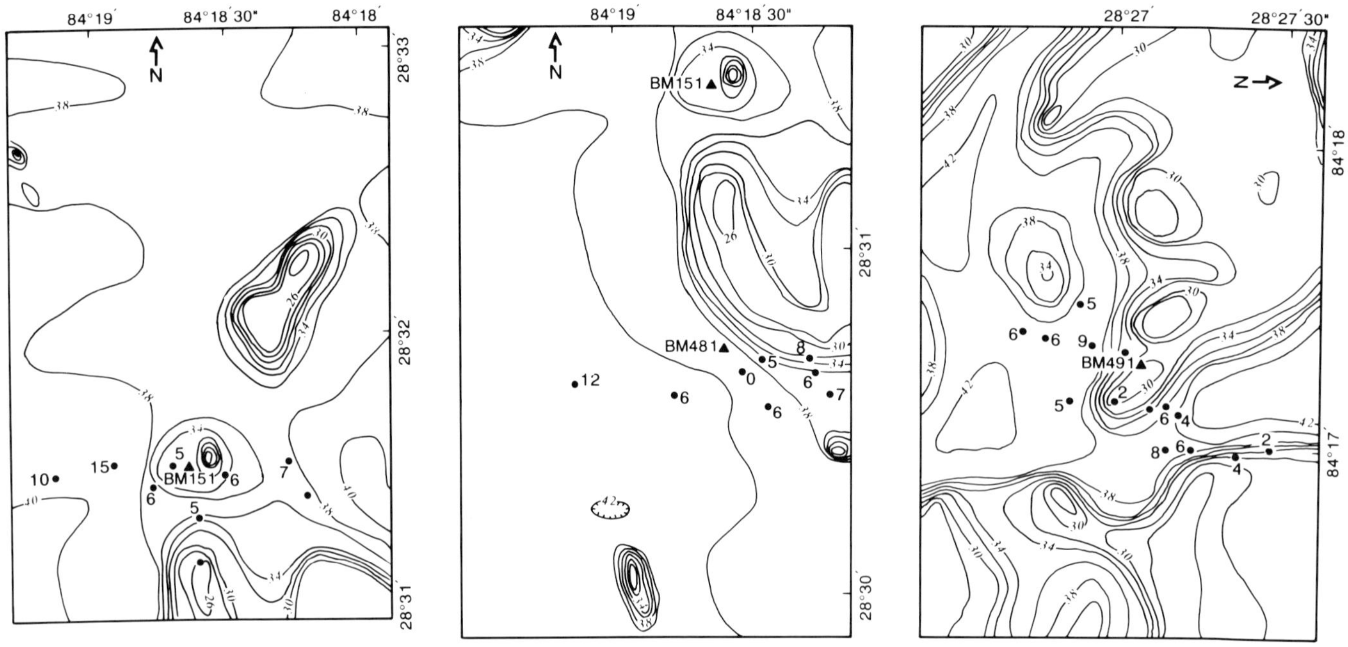

FIG. 17.—Bathymetric maps of the vicinity of Benchmark stations 151, 481 and 491 on reef ridges showing a general increase in benthic foraminifera abruptly away from ridge crests and in bathymetric depressions within ridges. Values expressed as the relative percentage of sediment consisting of benthic foraminifera.

original reef development occurred during one of the Pleistocene sea-level highstands and not during the early Holocene transgression, and that primary bathymetric controls were the karst surface of sequence I and the slight but abrupt seaward increase in slope of the same surface. The break in slope and karst surface of sequence I could have offered a suitable site for reef development, even in a midshelf setting. If initial development occurred during the early Holocene, up to 12 m of vertical growth would have had to develop on the smooth, flat surface of sequence II since being flooded, which is unlikely. Initial reef development on the banks of an ancient river is also unlikely, because there is no evidence of extensive channeling in the subsurface as would be expected if a river of this scale were to have occupied the region.

Modern sediments on the FMG have originated from two sources, the reef ridges and surrounding shelf. Shelf sediments consist of biogenic carbonates, principally a molluscan shell hash, and relatively fine-grained quartz sands. Reef-derived sediments consist of biogenic carbonates similar to those produced on the adjacent shelf. They are produced by biogenic processes on reef ridges and by destruction of the reef framework. "Calving" of large blocks and reef rubble commonly observed on reef flanks may be a consequence of erosion by bottom currents, undermining by biologic organisms, or a combination of the two. Despite the reef destruction, very few lithoclasts were identified as surface-sediment components, suggesting that either these blocks are not eventually broken down to sand-size grains, or that they are broken down to their individual components, making their lithoclastic origin unrecognizable. Another possibility is that sediment contribution directly by biogenic-carbonate production occurs at much faster rate than reef breakdown. We believe this unlikely due to the large quantity of blocks and rubble observed on reef flanks.

Components of reef-derived sediments are similar to the carbonate fraction of those surrounding the FMG, with the exception of barnacles. Barnacles, a major component of FMG reef-derived sediments, are rare in associated shelf sediments. We suggest that barnacles are common on the FMG because of the optimum sites for attachment provided by exposed reefal material. Barnacles probably attach to the lower, "dead" portions of the reef, the same structure that provides the foundation for modern reef growth. This attachment may actually be a form of bioerosion that aids in the destruction of reef ridges. Hence, the barnacle component of reef-derived sediments provides a sedimentological indicator of the transition from an open-shelf to reefal environment.

Sediments on and surrounding FMG reef ridges are not zoned with respect to reef environments, but are well distributed, controlled by physical processes, and bathymetry. Patchy sediment-distribution patterns suggest that biogenic-carbonate sediments produced on ridge crests are winnowed, where exposed, leaving relatively coarse carbonate sands. The relatively thin sediment cover observed on ridges indicates this winnowing process to be quite active. Currents exceeding 100 cm/s have been observed along ridge crests (Back, 1972). Sediments entrained in the current probably are transported to ridge flanks and the nearby shelf/central valley where they mix with shelf-derived sediments. Finer grained ridge-crest sediments in bathymetric depressions, or otherwise sheltered from currents, are not winnowed but continue to accumulate. Ridge-crest sediments

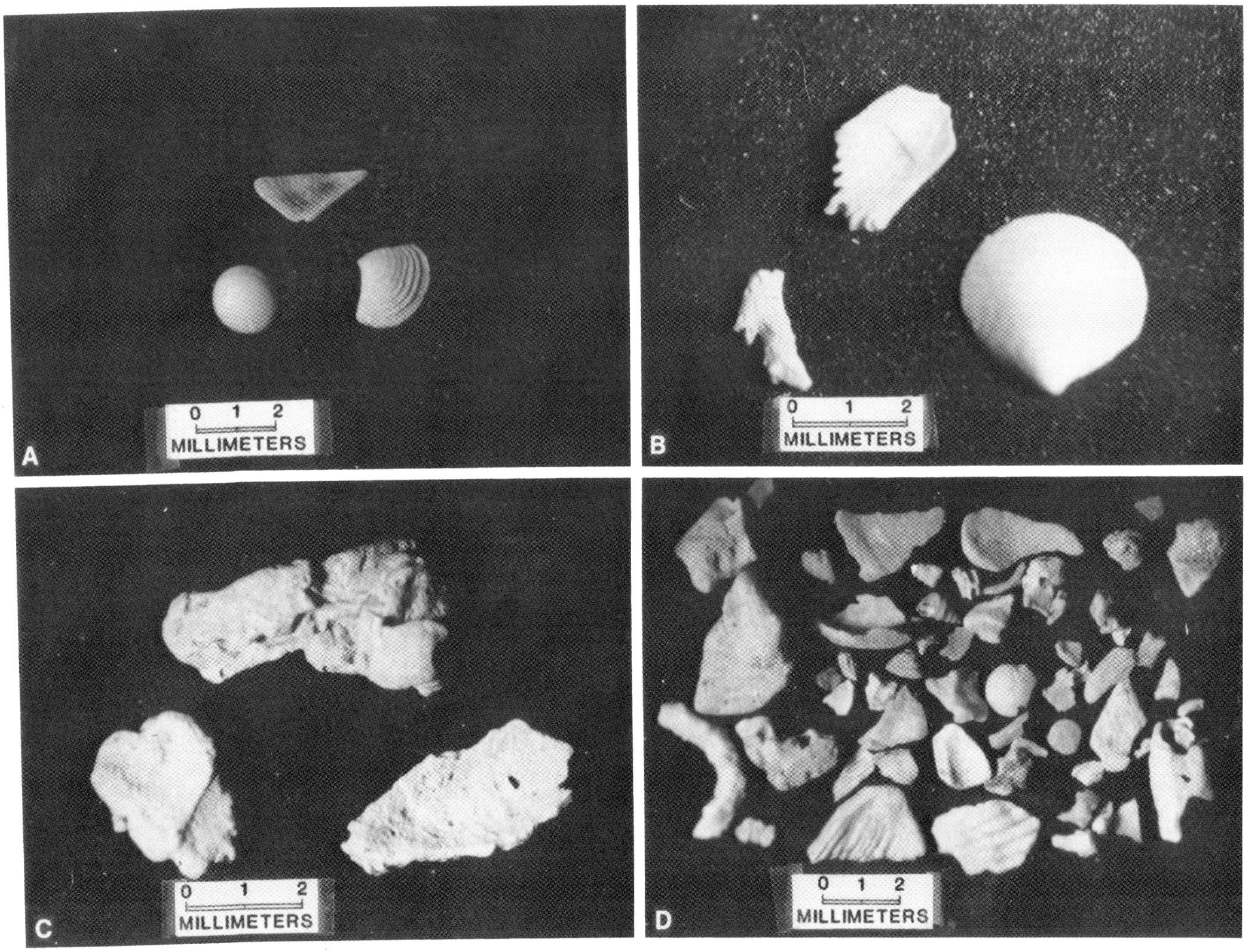

FIG. 18.—Photographs of FMG sediments showing fresh (A), dull (B) and worn (C) constituents, as well as a typical sediment assemblage (D).

deposited on ridge flanks are probably those too coarse to be transported in suspension by currents. They are transported off the ridge margin, where they form a sediment drape covering eroded reef-flank blocks and rubble. Shelf-derived sediments "onlapping" onto reef flanks by bottom currents may augment reef-derived input. This has been suggested by Hilde and others (1981), based upon an apparent asymmetry of reef-flanking sediments indicating unidirectional flow. These asymmetric patterns were not apparent here; furthermore, input from both ridge crests and the adjacent shelf, two opposing directions, would require periods of bidirectional flow. Multiple orientations of bedforms on ridge crests and the shelf/central valley show that this is quite possible.

The central valley between reef ridges functions as a sediment sink (Brooks, and others, 1980). Fine-grained sediments winnowed from reef ridges remain in suspension and are transported to these areas where they mix with shelf-derived sediments. Reef ridges act to shelter these areas from physical processes that are responsible for continuously redistributing sediments on open portions of the shelf (Neurauter, 1980; Logue, 1990). The result is the continued accumulation of a mixture of sediments derived from reef ridges and the adjacent shelf, producing deposits in excess of 5 m thick compared to the thin veneer (<1–3 m) for the surrounding shelf (Doyle and Sparkes, 1980).

Development of the FMG is consistent with that of the west Florida margin as a whole. Numerous investigations have shown that recent development of the west Florida margin is driven by high-frequency sea-level fluctuations, biogenic-carbonate production and physical processes (Doyle and Sparkes, 1980; Neurauter, 1980; Holmes, 1981; Doyle and Holmes, 1985; Holmes, 1985; Mullins, and others, 1987; Mullins and others, 1988; Brooks and Holmes, 1989, 1990; Logue, 1990). During sea-level transgressions and high-stands when the shelf is flooded, shallow-water carbonate sediments are produced and deposited on the shelf surface. Sediments are redistributed periodically by a variety of physical processes including storm-generated currents, tidal currents and landward intrusion of the Loop Current (Neu-

TABLE 1.—MAJOR SEDIMENT CONTRIBUTORS FROM VARIOUS AREAS

Skeletal Fragments (%)	Florida Middle Ground[1]	Florida Reef Tract[2]	Alacran Reef[2]	Andros, Bahamas[2]	Abaco, Bahamas[2]	Ragged Islands[2]	Hogsty Reef[2]	Courtown Cays[2]	Albuquerque Cays[2]	Roncader Bank[2]	Serrana Bank[2]	West Florida Shelf Surrounding Florida Middle Ground[3]
Coral	2	20	26	24	27	12	27	35	30	25	33	Tr
Molluscs	30	12	7	6	11	18	22	10	9	15	5	47
Foraminifera	3	6	8	12	11	13	5	3	2	3	3	10
Coralline Algae	1	10	11	33	10	—	19	21	21	24	24	Tr
Halimeda	Tr	30	40	17	16	39	2	28	32	17	17	Tr
Misc. Skeletal	—	7	1	6	5	—	1	—	1	1	1	15
Barnacle	8	?	?	?	?	?	?	?	?	?	?	Tr

[1]This study
[2]Milliman (1974)
[3]Brooks, G. R. (unpublished data)

rauter, 1980). Storm-generated currents probably have the most profound effect on sediment distribution. These currents can be generated by hurricanes, tropical storms and winter-cold fronts that average every 5 to 10 days during the winter (Niiler, 1976). In addition to being redistributed throughout the shelf, sediments may be transported offshelf where they settle onto the slope and eventually are transported to the slope base by gravity flows (Doyle and Holmes, 1985; Mullins and others, 1987; Brooks and Holmes, 1989, 1990). Bathymetry plays a role in this process as offshelf transport is facilitated by the lack of a continuous shelf-edge reef rim. Offshelf transport may be one factor responsible for the occurrence of such a thin sediment veneer (less than 3 m) overlying the Miocene (?) limestone surface on the shelf. Another factor may be the relatively slow production of mollusc-rich carbonate sediments in this setting compared to the more prolific production of coral- and algal-rich sediments common to lower latitude carbonate platforms.

As a consequence of the relatively slow accumulation of carbonate sediments on the shelf surface during sea-level highstands, erosion of these sediments is more complete during lowstands. Hence, with the exception of modern sediments, no appreciable accumulations (from a volumetric standpoint) are present overlying the Miocene (?) surface.

The FMG is likewise controlled by sea level, with carbonate-sediment production occurring during sea-level highstands when the surface is flooded. Physical processes are important for distributing sediment, but bathymetry is important on the FMG more for trapping sediments, rather than aiding in their export, as described for the non-reef-rimmed, shelf-slope break. Hence, FMG reefs act to impound sediments much like a shelf-edge reef functions to impound sediments on the platform surface. In this instance, however, accumulation is insufficient to keep pace with rising sea level. The FMG therefore, like the remainder of the west Florida shelf, is slowly being drowned.

In comparison to what are considered "typical" modern reef systems, the FMG is unique. Its midshelf, high-latitude setting and preponderance of non-reefal sediments, which lack zonation, show that in terms of carbonate-sediment production and accumulation, the FMG is more like an open-shelf than coral reef environment. Table 1 compares the dominant FMG sediment components with those of other reef systems and the surrounding west Florida shelf. FMG sediments are unlike those of most reef systems and with the exception of barnacle fragments, virtually indistinguishable from surrounding shelf carbonate sediments. This subtle difference in sediment type has implications for the interpretation of ancient open-shelf versus reefal environments. Based solely upon sedimentology, the two environments may be virtually indistinguishable.

SUMMARY AND CONCLUSIONS

Modern reef growth on the Florida Middle Ground is controlled by bathymetry, in that the foundation is a dead reef that originally developed during one or more of the Pleistocene sea-level highstands. Controls on original reef growth are not fully understood but are probably a combination of bathymetry and circulation patterns that supplied the FMG with Caribbean reefal fauna. This recruitment may be occurring at present, thereby acting to continually "stock" the FMG with Caribbean fauna. Sediments on the FMG are derived from two sources, the reef ridges and surrounding shelf environment. Sediments, dominantly a molluscan shell hash, are virtually indistinguishable between the two sources with the exception of barnacle fragments in reef-ridge sediments. Barnacles are common reefal constituents probably because of the suitable sites for attachment provided by the dead-reef foundation. Barnacles, therefore, provide the principle sedimentological indicator of a transition from shelf to reefal environment. Sediments are not zoned, as in many reefal settings, but have a patchy distribution controlled by physical processes and the rugged bathymetry. A variety of types and orientations of bedforms indicates numerous, different, flow directions and velocities, probably driven by tides and storms. Geologic development of the FMG parallels that of the surrounding west Florida margin, driven principally by high-frequency sea-level fluctuations. FMG reef ridges act to trap sediments much in the same manner as shelf-edge reef systems, but sediment accumulation and reef growth cannot keep pace with sea-level rise. The result is that the FMG is being drowned along with the remaining west Florida continental margin.

ACKNOWLEDGMENTS

We gratefully acknowledge the help of all of the participants in the project. We thank T. Hilde for supplying the seismic data. We also thank R. H. Osborne for many helpful comments.

REFERENCES

AUSTIN, H. M., AND JONES, J. I., 1974, Seasonal variations of physical oceanographic parameters on the Florida Middle Ground and their relation to zooplankton biomass on the west Florida shelf: Quarterly Journal of the Florida Academy of Science, v. 37, p. 16–32.

BACK, R. M., 1972, Recent depositional environment of the Florida Middle Ground: Unpublished M.S. Thesis, Florida State University, Tallahassee, Florida, 237 p.

BIRDSALL, B. C., 1978, Eastern Gulf of Mexico, continental shelf phosphorite deposits: Unpublished M.S. Thesis, University of South Florida, St. Petersburg, Florida, 87 p.

BLOOM, A. L., 1974, Geomorphology of reef complexes, *in* Laporte, L. F., ed., Reefs in Time and Space: Society of Economic Paleontologists and Mineralogists Special Publication 18, p. 1–8.

BROOKS, G. R., AND HOLMES, C. W., 1989, Recent carbonate slope sediments and sedimentary processes bordering a non-rimmed platform: Southwest Florida continental margin, *in* Crevello, P. D., Wilson, J. L., Sarg J. F., And Read, J. F., eds. Controls on Carbonate Platform and Basin Development: Society of Economic Paleontologists and Mineralogists Special Publication 44, p. 259–272.

BROOKS, G. R., AND HOLMES, C. W., 1990, Modern configuration of the southwest Florida carbonate slope: Development by shelf margin progradation: Marine Geology, v. 94, p. 301–315.

BROOKS, G. R., PARKER, D. M., STEINMETZ, J. C., AND DOYLE, L. J., 1980, Unusual ponding of sediments on a deep-water reef (abst.): American Association of Petrolum Geologists, v. 64, p. 682.

BROOKS, H. K., 1974, General and geotechnical characteristics of the surficial deposits on the west Florida shelf, *in* Smith, R. E., ed., Procedures in Marine Environment Implications of Offshore Drilling, Eastern Gulf of Mexico: State University System of Florida Institute of Oceanography, St. Petersburg, Florida, p. 301–307.

CARVER, R. E., 1971, Processes in Sedimentary Petrology: Wiley-Interscience, New York, 653 p.

CHENEY, D. P., AND DYER, J. P., III, 1974, Deep water benthic algae of the Florida Middle Ground: Marine Biology, v. 27, p. 185–190.

DENNISON, J. M., AND HAY, W. W., 1967, Estimating the needed sampling area for subaquatic ecologic studies: Journal of Sedimentary Petrology, v. 41, p. 706–708.

DOYLE, L. J., AND HOLMES, C. W., 1985, Shallow structure, stratigraphy and carbonate sedimentary processes of the west Florida upper continental slope: American Association of Petroleum Geologists Bulletin, v. 69, p. 1133–1144.

DOYLE, L. J., AND SPARKS, T. N., 1980, Sediments of the Mississippi, Alabama and Florida (MAFLA) continental shelf: Journal of Sedimentary Petrology, v. 50, p. 905–916.

FOLK, R. L., 1965, Petrology of Sedimentary Rocks: Hemphills, Austin, Texas, 159 p.

GRIM, R. E., 1968, Clay Mineralogy: McGraw-Hill, New York, 596 p.

HILDE, W. C., SHARMAN, GEORGE, WARSI, WARIS, LEE, C. S., FEELEY, MARY AND MEYER, MIKE, 1981, Section 1: Mapping and sub-bottom profiling, *in* Northern Gulf of Mexico Topographic Features Study: Texas A&M University, College Station, Texas, v. 5. Final Report submitted to the U.S. Bureau of Land Management, p. 6–20.

HOLMES, C. W., 1981, Late Neogene and Quaternary geology of the southwestern Florida shelf and slope: U.S. Geological Survey Open-File Report 81-79, 29 p.

HOLMES, C. W., 1985, Accretion of the south Florida platform, late Quarternary development: American Association of Petroleum Geologists Bulletin, v. 69, p. 149–160.

JAMES, N. P., 1983, Reef environment, *in* Scholle, P. A., Bebout, D. G., and Moore, C. H., eds., Carbonate Depositional Environments: American Association of Petroleum Geologists Memoir 33, p. 346–462.

JORDAN, G. F., 1952, Reef formation in the Gulf of Mexico off Apalachicola Bay, Florida: Geological Society of America Bulletin, v. 63, p. 741–744.

LOGUE, K. T., 1990, Bedforms and surface sediments of the northwest Florida inner continental shelf: Unpublished M.S. Thesis, University of South Florida, St. Petersburg, Florida, 87 p.

MAFLA, 1975–76, Baseline monitoring studies, Mississippi, Alabama, Florida Outer Continental Shelf: U.S. Bureau of Land Management Contract No. 08550-CT5-30, v. IV, 217 p.

MILLIMAN, J. D., 1974, Marine carbonates: Springer-Verlag, New York, 375 p.

MITCHUM, R. M., VAIL, P. R., AND SANGREE, J. B., 1977, Stratigraphic interpretation of seismic reflection patterns in depositional sequences: American Association of Petroleum Geologists Memoir 26, p. 117–133.

MULLINS, H. T., GARDULSKI, A. F., WISE, S. W., AND APPLEGATE, J., 1987, Middle Miocene oceanographic event in the eastern Gulf of Mexico: Implications for seismic stratigraphic succession and Loop Current/Gulf Stream circulation. Geological Society of America Bulletin, v. 98, p. 702–713.

MULLINS, H. T., GARDULSKI, A. F., HINCHEY, E. J., AND HINE, A. C., 1988, The modern carbonate ramp-slope of central west Florida: Journal of Sedimentary Petrology, v. 58, p. 273–290.

NEURAUTER, T. W., 1980, Bedforms on the west Florida shelf as detected with side scan sonor: Unpublished M.S. Thesis, University of South Florida, St. Petersburg, Florida, 120 p.

NIILER, P. P., 1976, Observations of low frequency currents on the west Florida shelf: Memoires Societe Royale des Sciences de Liege, p. 331–358.

PRICE, W. A., 1954, Shorelines and coasts on the Gulf of Mexico, *in* Galtsoff, P. S., ed., Gulf of Mexico, Its Origin, Waters and Marine Life: U.S. Fish and Wildlife Bulletin, v. 87, p. 36–54.

PURDY, E. G., 1974, Reef configuration: Cause and effect, *in* Laporte, L. F., ed., Reefs in Time and Space: Society of Economic Paleontologists and Mineralogists Special Publication 18, p. 9–16.

WANLESS, H. R., 1977, Baseline monitoring studies, Mississippi, Alabama, Florida Outer Continental Shelf, 1975–1976: U.S. Bureau of Land Management, Contract No. 08550-CT5-30, 14 p.

PART IV
SUBMARINE-CANYON AND -FAN STUDIES

INITIATION PROCESSES AND FLOW EVOLUTION OF TURBIDITY CURRENTS: IMPLICATIONS FOR THE DEPOSITIONAL RECORD

WILLIAM R. NORMARK

U.S. Geological Survey, 345 Middlefield Road, Menio Park, California 94025

AND

DAVID J. W. PIPER

Atlantic Geoscience Centre, Geological Survey of Canada, Bedford Institute of Oceanography, P.O. Box 1006, Dartmouth, Nova Scotia, Canada B2Y 4A2

ABSTRACT: Interpretations of sea-level change and source-area tectonism from the character of turbidite deposits require knowledge of the sediment source, the flow-initiation process, how the turbidity current evolved during flow, and what flow phases were associated with deposition. The flow responds to shape and size characteristics of both the erosional pathway and the morphology of previous deposits; these, in turn, reflect factors such as basin size, shape, and tectonic activity. The initiation of turbidity currents generally involves either transformation from mass failures or one of a variety of fluid-flow events involving ignitive flow. The variability of these initiating processes with geologic setting is illustrated by analysis of three case studies involving 1) initiation by storm surge on a carbonate-reef slope, 2) a seismically triggered failure of a line source within mid-slope sediment, and 3) a bedload delta showing both hyperpycnal flow of river bedload and failure of unstable prodelta sediment. General criteria for the initiation of turbidity currents are given.

Depositional and erosional features of the turbidite deposits themselves allow recognition of changes in flow processes as the turbidity currents traverse the basin. High-concentration basal flow beneath the main turbulent turbidity current will leave a distinctive depositional record from more energetic flow, particularly in proximal and channel settings. Turbidity currents of different sizes and sediment distribution interact quite differently with local fan morphology. Many large, muddy turbidity currents are much thicker than levee-channel relief; the upper parts of such flows are "stripped off" across levees and can become rechannelized farther downslope. In restricted basins, the basin margin acts like a channel wall for larger flows, thus maintaining competence over long distances. Thinner, sandier flows flush out channel systems, and a large proportion of sediment in some turbidite deposits may be derived from bed erosion. A broad range of local bedforms, especially sediment waves and scours, reflects changes in the character of turbulence and competence within quasi-steady flows. This paper emphasizes the broad range of size, sediment distribution, and depositional processes in turbidity currents that largely reflects the variety of initiation processes. This variability means that caution is needed when interpreting depositional conditions of ancient turbidite sediment with ill-defined morphologic setting.

INTRODUCTION

Historical Perspective

The first commemorative volume for Francis Shepard a quarter century ago (Miller, 1964) marks a logical starting point for reviewing our current concepts on the initiation and flow evolution of turbidity currents. That volume presents what were then some of the latest ideas on the processes operating within, and origins of, submarine canyons by colleagues and coworkers of Shepard. Shepard himself was beginning to spend more effort on studying the deposits of the turbidity currents that move through the canyons. His enthusiasm for sea-going research and willingness to share his results stimulated our own work on deep-sea fans when we were graduate students together at Scripps. He was especially interested in the character of the fan valleys that extend from the mouths of submarine canyons (Shepard and Buffington, 1968; Shepard and others, 1969), and he had some appreciation of the erosive and transport capability of turbidity currents (Fig. 1). By 1965, he already was the first to document the existence of a horseshoe-shaped meander in a submarine-fan channel; in fact, his map of the meander on the Monterey Fan shows few differences from current maps based on "state-of-the-art" data from SEABEAM sounding and GLORIA sidescan images (Shepard, 1966; Eittreim and others, 1983; EEZSCAN 84 Scientific Staff, 1986; Normark and Gutmacher, 1988).

In the mid-1960s, ideas concerning the nature of turbidity currents were still strongly influenced by the original experiments of Kuenen (1950, 1955), in which sediment transport was maintained by the high density of the current. Bagnold (1962) demonstrated that at high concentrations grain-to-grain collisions prevented turbulent transport of sediment; this damping effect started at concentations of about 10 percent (linear concentration $>10^{-1}$), putting an upper bound on turbidity-current density (Fig. 2). Turbidite sequences were perceived as poorly sorted deposits; well-sorted sand beds on the ocean floor were interpreted by Hubert (1964) as contourites. Nevertheless, evidence was accumulating for tractive flow in turbidity currents similar to that in rivers (Sanders, 1965). The work of Heezen and Ewing (1952), Menard (1964), and Morgenstern (1967) emphasized catastrophic sediment failure as the origin of turbidity currents. In the first commemorative volume, Chamberlain (1964) proposed sediment failure for the initiation of turbidity currents in the head of Scripps Canyon. A decade later, Inman and others (1976) showed that storm surges and rip-current-like flows were capable of initiating some flows in Scripps Canyon.

Our knowledge of the structure, dimensions, and processes of deep-water turbidite systems underwent a substantial improvement beginning in the mid-1960s with the technological advances provided by continuous seismic-reflection profiling, 3.5-kHz high-resolution profiling, and side-looking-sonar imaging systems. Until these systems were available, our understanding of fan processes was based on gross (fan-wide) morphologic resolution and limited data on the surface sediment of submarine fans.

Echo sounding and sediment sampling were relatively effective tools in mapping submarine canyons and their uppermost (recent) sediment fill, especially in upper-slope and shelf depths. Little was known, however, about the mor-

From Shoreline to Abyss, SEPM Special Publication No. 46

FIG. 1.—(A) Francis P. Shepard and large transported shale clast in sandy fan-valley fill, Miocene Doheny Channel, southern California; (B) Large shale block in San Jose Canyon, Baja California, Mexico. Photographs courtesy of J. R. Curray.

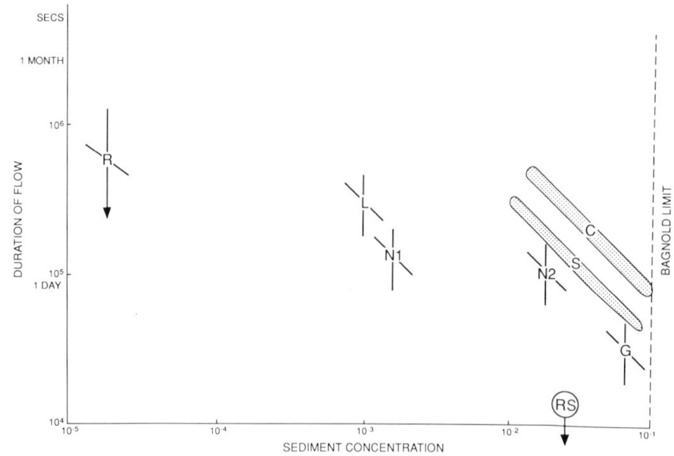

FIG. 2.—Plot of log duration against flow concentration for well-known turbidity currents. Crosses indicate range of estimates where flow density has been constrained: stippled lines indicate events for which flow density is not constrained: C (turbidite volume >30 km^3) = Contessa turbidite, Miocene of northern Appenines; G (>160 km^3) = 1929 Grand Banks turbidity current (Hughes Clarke and others, 1990); L = Wisconsinan muds, Laurentian Fan (Stow and Bowen, 1980); N1 (0.02 km^3), N2 (0.03 km^3): turbidites I and II, Navy Fan (Bowen and others, 1984); R = Reserve Fan (lacustrine) (Normark, 1989); RS (0.00002 km^3) = surge-type flows in Rupert Inlet inferred to form from slope failure (Hay, 1987a,b); S (20 km^3) = turbidite on Silver Abyssal Plain described by van Tassel (1981).

phologic features of the deeper water turbidite deposits at the mouths of the canyons. H. W. Menard had studied a few submarine fans, and his 1955 paper introduced such concepts as bankful flow in leveed submarine channels, the effect of Coriolis deflection in developing asymmetric levees, and the implication that flow expansion occurs through hydraulic jumps when turbidity currents exit from submarine canyons. Menard's (1964) book on submarine geology was an excellent summary of the existing concepts about turbidity-current flow and initiation processes, and Komar (1969) developed some of these ideas to derive flow properties for turbidity currents using channel-shape and size characteristics.

The understanding of flow processes has always been hampered by the extreme difficulty of monitoring turbidity currents in the sea. Our knowledge of processes is based on inferences from marine deposits and morphology, augmented with observations in lakes, fjords, and flumes, and the application of numerical modeling. This approach has meant that the variety of turbidity-current flow processes may well have been underestimated. Geologists working on ancient rock successions have had an even narrower window on flow processes because of the lack of information on seabed morphology and large-scale geometry in most ancient deposits. Walker's (1973) review provides a very useful perspective of how the work on submarine fans in the 1960s was instrumental in forming our current concepts on turbidity-current phenomena.

The concept of turbidity-current flow has been defined in subtly different ways. As a flow process, it implies a gravity flow in which grain dispersion is maintained by fluid turbulence. As a natural phenomenon, it implies a discrete flow of sediment and fluid in which much of the transport is in turbulent suspension, but where there might be significant, high-concentration sediment transport by debris-flow or grain-flow processes (Lowe, 1982; Norem and others, 1989). We use the term in the latter sense.

Scope of Paper

An increasing body of evidence shows that many, if not most, turbidity currents are quasi-steady flows of duration of as much as many tens of hours (Fig. 2) and that much of the sediment transport takes place in flows with sediment concentrations that are either well below the Bagnold limit (Bagnold, 1962) or approach it only near the base of the flow. This concept is radically different from the catastrophic, high-density surges (sediment concentration 6×10^{-1}) envisaged by Kuenen (1950) and Heezen and Ewing (1952). Evidence for prolonged flow, which has generally resulted from the study of modern fan systems, includes 1) the common occurrence of regular bedforms in both channels and on levees (Malinverno and others, 1988; Normark and others, 1980); 2) the predominance of well-sorted sediments with "tractional" bedding structures (Sanders, 1965; Lowe, 1982); 3) the regular sinuous courses of some turbidity-current channels (Kastens and Shor, 1985; Manley and Flood, 1988); and 4) the volume of individual turbidites relative to conduit cross-sectional areas (Bowen and others, 1984; Hughes Clarke and others, 1990). The concept of prolonged flow should be equally applicable to depositional processes in ancient turbidite systems as well. The lack of recognizable large-scale bedforms in most ancient systems is a consequence of their size and the lack of distinctive internal features on coarse bedforms (Piper and Kontopoulos, 1990). Small- and medium-scale tractional bedforms are abundant (Lowe, 1982, 1988). The largest correlatable ancient turbidites are on a scale that requires long flow duration for their emplacement.

There is considerable interest at present in interpreting the relative effects of sea-level changes and tectonic processes on the sequence stratigraphy of turbidite deposits (Weimer and Link, 1991). To understand adequately the linkage between such external controls and the resulting deposits requires an understanding of processes. The nature of source sediment for turbidite events and how sediment is stored in the source area influences the initiation process and the nature of the flow. Turbidity currents are initiated by various triggers (including earthquakes, storms, and floods). Following initiation, the evolution of a turbidity current during flow reflects the influence of previous currents in creating the pathway for the current. The morphology of the resulting turbidite deposit is a function of the interaction of turbidity-current flow with local and basinwide relief features that, in turn, reflect factors such as basin size, shape, and active tectonism. In this paper, we examine these factors and emphasize the broad range in size, sediment distribution, and depositional process of turbidity currents.

Our review of the broad range in the character of turbidity currents is in three sections that reflect both the nature of the data and the degree of change in our understanding of the processes and products of these currents. First, we consider the range of initiation processes, including fluid-flow and sediment-failure mechanisms; recent experimental results and detailed case studies have provided key new insights in this area. We then consider the flow of turbidity currents in channels. Much of our initial understanding of turbidity currents was based on derivation of flow properties from consideration of channel characteristics; work since the first Shepard commemorative volume has emphasized that the channel properties reflect a rather limited range of flow conditions that might exist within any given turbidite system. Last, we consider the deposits left by turbidity currents. Depositional bedforms and the shape of individual turbidite beds yield additional insights to turbidity-current processes away from channels; this relatively new area of investigation should become a major focus of future turbidite studies. Following this discussion, the paper then examines the implications for facies distribution in the resulting turbidite deposits.

INITIATION AND FLOW EVOLUTION

Turbidity currents can result from a variety of oceanic (water-column) and sediment-failure events. Direct observation of the initiation of turbidity currents remains an elusive goal, but the last 25 years have resulted in reasonable documentation of several initiation mechanisms. In this section, we consider turbidity currents generated by the injection of concentrated-sediment suspension resulting from river discharge, storm surges, and those that are the result of mass failures on deltaic and basin slopes (Table 1).

Concentrated Sediment-Suspension Injection

River Discharge.—The direct or hyperpycnal flow of turbid rivers and tailings discharge into lakes has long been known

TABLE 1.—LIST OF TURBIDITY-CURRENT INITIATION PROCESSES WITH LITERATURE EXAMPLES.

Process and Trigger	Dominant Sediment Type	Flow	Deposits	References
Fluid entrainment into debris flows	Commonly mud or mixed	High-concentration base, thick muddy top	Poorly sorted, muddy turbidite with sorted mud top	Hampton, 1972
Ignitive flows from sediment failure	Commonly silty mud	Accelerating fluid flow, dependant on bed erosion	Well-sorted deposits showing evidence of steady fluid flow[1]	Hughes Clarke, 1988
Hyperpycnal-bedload flow from rivers and glacial meltwater	Mainly sand and coarse silt	Accelerating fluid flow, may derive mud from bed erosion	As above	Prior and Bornhold, 1989
Ignitive flow of sand in canyon heads, triggered by storms	Sorted sand	Accelerating fluid flow, may derive mud from bed erosion	As above	Fukushima and others, 1985
Ignitive flow of carbonate detritus through reef gaps, triggered by storms	Calcareous detritus	Accelerating fluid flow; may erode reef-front carbonate and biogenic mud from slopes below reef	As above	Dengler and others, 1984a,b
Ignitive flow derived from suspended sediment (shelf turbid layer or river discharge)	Mud	Low-density flow. Needs topography to concentrate and maintain	?Thin, muddy turbidites	Concept reviewed by McCave, 1972
Ignitive flow during eruption of airfall tuffs	Fine tephra	Low-density flow. Needs topography to concentrate and maintain	?	See discussion by Lowe, 1988

[1]Character depends on relative proportions of source material and material eroded from the bed and on the grain-size distribution of these two components.

to result in turbidity currents (Gould, 1951; Lambert and others, 1976; Weirich, 1984). In the marine environment, the greater density of salt water means that substantial sediment concentrations must be achieved to generate a turbidity current. Tailings discharge has been observed to form turbidity currents in the marine environment (Hay and others, 1982; Hay, 1987a,b). Prior and others (1987) observed low-salinity current events in a fjord prodelta channel that may represent hyperpycnal flow of river bedload material down a steep prodelta slope in a turbidity current. Shepard and others (1977) and Gennesseaux and others (1971) also report episodic current flow and sand transport down steep channels off flooding rivers. Wherever fluid-flow-generated currents have been measured near their source, they are of considerable duration and fluctuating velocity. In some cases, velocity has been observed to generally increase with time to a peak value.

For a turbidity current to develop from an initial suspension flow, the suspension flow must maintain a sufficient velocity (through inertial effects or external forcing) until there is a sufficient gradient for a turbidity current to be shelf sustaining. The maintenance of forcing resulting from a hydraulic head along a channel, which can be considered using the physics of jet flow, does not provide an adequate explanation for long-distance flow of turbidity currents. Bagnold's (1962, 1966) concept of autosuspension proposed that if the bottom fell away faster than the sediment could settle out, then the turbulent energy in the flow would increase, resulting in an accelerating turbidity current down steep slopes. Such a mechanism applies only to that part of a turbidity current in which the sediment is transported in suspension, although there may be transfer of shear stress from this component to a basal higher concentration part of the flow. This concept has been refined by Parker (1982) to define conditions under which such acceleration permits new sediment to be entrained, a condition referred to as ignition, because this new sediment provides additional power to the current. Fukujima and others (1985) and Parker and others (1986) show that as long as the initial conditions of flow were above the critical ignitive threshold, the spatial evolution of a flow is not sensitive to variability in the initial conditions, since the behavior of the flow in velocity-discharge space converges on a single path (the convergence line). Thus, variations in currents measured near source areas may not be maintained downslope in turbidity currents that have experienced ignition.

The general range of conditions required for ignition are illustrated in Figure 3, based on the numerical modeling of Parker and others (1986). The high-sediment concentrations required for ignition are most likely in relatively thin flows and a steep slope (e.g., S>0.05 [3°]) is essential for ignition. The critical flow velocity required for ignition is relatively insensitive to variables other than grain size, ranging from 1 m/s for fine sand to 0.2 m/s for medium silt. Critical sediment concentrations for ignition are of the order of 10^{-2} (volume concentration), in the same range as sediment concentration required for hyperpycnal flow of rivers (Gilbert, 1983), but are very sensitive to flow thickness and gradient. On steep slopes (S>0.2), relatively thin, silty flows of concentration around 10^{-4} may ignite.

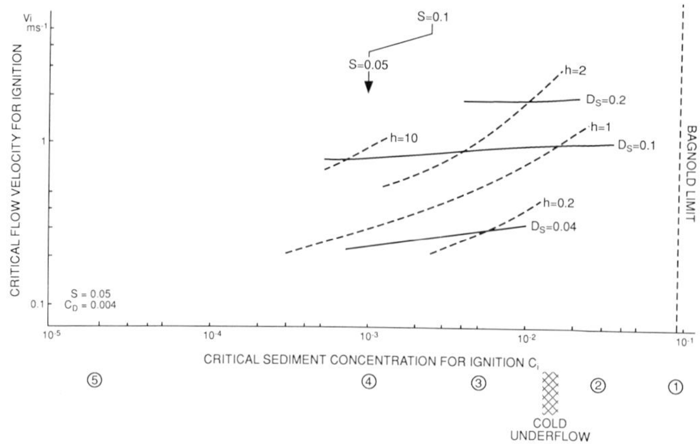

FIG. 3.—Schematic plot illustrating the conditions under which ignitive flow may occur. Plot shows critical velocity (V_i) and critical concentration (C_i) for ignition to occur for various flow thicknesses (h, meters) and representative grain sizes (D_s, mm); on a slope of $S = 0.05$, assuming a drag coefficient of $C_D = 0.004$ (graph based on four-equation model results of Parker and others, 1986). Also shown is the approximate offset in the concentration scale resulting from an increase in slope to $S = 0.1$. Shown for comparison are some representative suspended-sediment concentrations: 1 = Huanghe River, China, in flood (Wright and others, 1988); 2 = Sustina River, Alaska, in flood (Hoskin and Burrell, 1972); 3 = maximum mesured discharge into Glacier Bay, Alaska (Hoskin and Burrell, 1972); 4 = ignitive condition for sand in the head of La Jolla Canyon (Fukushima and others, 1985); 5 = nearshore-sediment concentration after flood of Santa Clara River (Drake and others 1972). The "cold underflow" is the approximate sediment concentration in cold river water required to produce a marine underflow (Gilbert, 1983).

Measured suspended-sediment concentrations in some rivers in flood (cases 1 and 2 in Fig. 3) are sufficient for both hyperpyncal flow into sea water and ignition. Prior and Bornhold (1989) argue that high-concentration bedload discharge in some steep-gradient rivers will continue flowing down submarine slopes as an inertial flow. Observations of high-concentration discharge at ice margins, however, suggest that any hyperpycnal flow is soon terminated by the buoyant upwelling of the main discharge as sediment settles out (Syvitski and others, 1987). At present, there is no consensus among those who work on river-mouth processes as to the importance of hyperpycnal flow. The observation that cable breaks off rivers such as the Magdalena, Congo, Markham, and the fan deltas of the Gulf of Corinth (Heezen and Hollister, 1971, ch. 6) correlate with times of peak discharge suggests that turbidity currents occur at these times. Such data do not distinguish between hyperpycnal flow and rapid progradation leading to prodelta failure of the type recognized by Syvitski and Hein (1991). Observations of ancient deltas (e.g., McCabe, 1978) provide evidence for proximal turbidity-current flow on delta-front channels, but again it is unclear whether this results from hyperpycnal flow or failure of mouth-bar sediment. Ignition can result from the introduction of suspended sediment in rivers at the top of steep prodelta slopes, or as the result of resuspension of marine sediment at the top of the continental slope.

Storm Surges.—Ignitive flow of coastal sand down submarine canyons has been documented by Inman and others (1976) and Shepard and others (1977) for La Jolla Canyon, where rip currents associated with storm surges catastrophically remove sand from canyon heads. Sand loss occurs approximately annually: typically 10^5 m^3 of sediment is lost, substantially less than the average size of turbidite deposits on La Jolla Fan (Piper, 1970). Fukushima and others (1985) have shown that numerical models of ignitive flow can predict observations in the canyon (case 4, D_s = 0.15 in Fig. 3). Similar processes have been observed in other canyons (Reimnitz, 1971) and can potentially occur wherever there is a narrow shelf and canyons intersect the littoral-drift system. Such environments would have been much more common during Pleistocene lowstands of sea level.

Turbidity currents have also been inferred to have formed from a storm surge across the reef front off Kahe Point, Oahu, Hawaii (Dengler and others, 1984a,b). Hurricane Iwa passed through the Hawaiian Islands on 23 November 1982, causing damage to a series of telecommunication cables in water depths greater than 1,100 m and disruption of a set of current-meter moorings that had been deployed on the slope above 800 m (Dengler and others, 1984b). Figure 4 shows the position of the shallowest damaged telecommunications cable and the four moorings recovered after the storm passed. The displacement of the current meters to deeper water occurred in a series of flow episodes with peak current speeds of as much as 220 cm/s. In general, downslope flow during these episodes was in the range of 70 to 135 cm/s (figs. 2 and 3 in Dengler and others, 1984b).

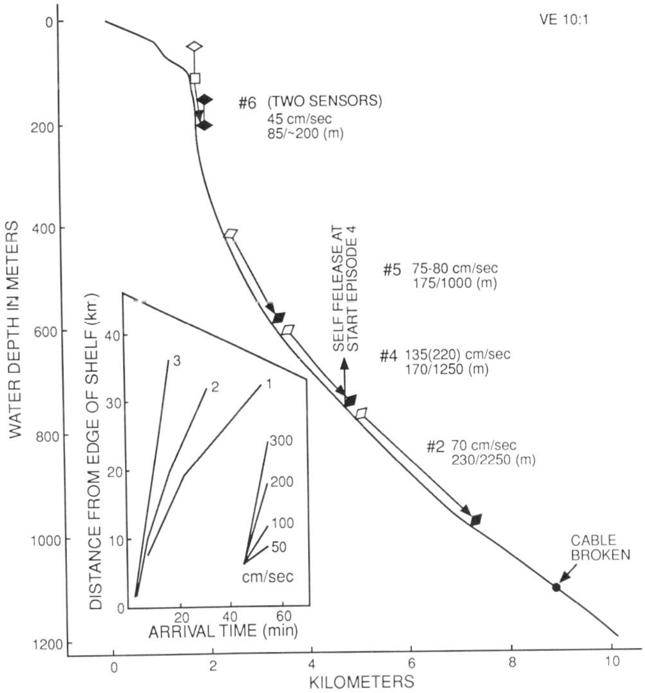

FIG. 4.—Initiation case history 1: turbidity current produced by Hurricane Iwa. Downslope profile offshore Kahe Point, Oahu, showing positions of four current-meter moorings (numbered 2, 4, 5, and 6) before and after a series of flow episodes. Peak recorded current speeds and ratio of total downslope/horizontal displacement given for each mooring. Inset shows frontal speed of three main turbidity-current episodes correlated between the lower three moorings. For further explanation, see text. Based on Dengler and others (1984a,b).

The velocity of the front of three of the main turbidity-current episodes was between 50 and 300 cm/s with frontal speeds decreasing downslope. The initiation of each displacement event of the current meters corresponded with the occurrence of high-velocity flow, and the water temperatures recorded at the mooring sites during the flow episodes were generally 2° to 4°C above the ambient temperature before displacement (fig. 4 in Dengler and others, 1984b).

Dengler and others (1984a) noted additional facts that support turbidity-current flow as the cause for displacement of the moorings, including 1) later detailed surveys indicate no significant change in the bathymetry of the slope below the reef front; 2) the shallowest mooring had two current meters (Fig. 4), and the upper meter showed onshore flow whereas the lower meter showed offshore (downslope) flow, suggesting fluid-flow initiation and not sliding of the sediment at the anchor site; 3) the drag characteristics of the mooring configuration and anchor weights are consistent with displacement by currents having the observed speeds; and 4) both the float housings and flat surfaces on the current-meter housings had accumulated significant amounts of sediment, something not observed in any similar deployments before and after those affected by Hurricane Iwa. Mass failure of reef-front talus and carbonate sediment cannot be ruled out as a trigger for some of the turbidity-current events, but storm-surge flow appears to be the likely initiation process for the three main turbidity-current episodes that can be correlated between the moorings.

Carbonate turbidites and debris flows are common on carbonate margins (McIlreath and James, 1979; Cook and others, 1983). The effects reported by Dengler and others (1984a,b) could well be common in other areas. In the better known Bahamas system, for example, Hubbard and others (1976) observed fair-weather transport of platform-derived sand through forereef chutes, and Hine and others (1981) presented arguments for storm transport of carbonate sand off the banks. Hooke and Schlager (1980) argue that the gullied slopes off the margin are eroded by turbidity currents. Deep-water carbonate-turbidite units are frequently quite distinct in their distribution from debris flows (Bornhold and Pilkey, 1971). Debris flows probably result from reef collapse and may generate turbidity currents; this collapse may be seismically triggered or induced by overloading (Mullins, 1983). Turbidite frequency on the basin floor is commonly one per several thousands of years (Mullins, 1983); so, if basinal turbidites are related to storms, probably only exceptional storms result in large-volume turbidite deposits. Many basinal turbidites show multiple entry points (Mullins, 1983), consistent with either a storm or seismic trigger.

Sediment-Failure Initiation Processes

Although most turbidity currents are commonly thought to be seismically triggered, there are remarkably few studies that document the relation between sediment failure and turbidity-current flow. There is an extensive literature, both observational and modeling, on seismic triggering of slumps, slides, and debris flows on land. Such failures in lakes or the sea can locally result in surge-type grain flows or debris flows (Johnson, 1970).

Flow Models.—Flume experiments by Hampton (1972) provided a conceptual model for the evolution of a turbidity current from a debris flow: turbulence at the head of the debris flow leads to entrainment of water and the development of a separate turbidity current above the debris flow. This current might then move out ahead of the debris flow. Other entrainment models are reviewed by Weirich (1988), who demonstrated that hydraulic jumps in debris flows may be the most important conversion mechanism. Observations of debris flows on land (Costa and Williams, 1984) indicate that local turbulence associated with flow irregularities also may result in significant reduction in sediment concentration. Modeling of snow avalanches (e.g., Gubler, 1989) has provided some insights into sediment slides and turbidity currents (Dengler and Wilde, 1987; Norem and others, 1989); the avalanche proper corresponds to a debris flow or grain flow and the resulting snow cloud to a turbidity current (Middleton and Hampton, 1976).

Lowe (1982) argued that high-density turbidity currents transport a wide range of grain sizes, but that only medium sand and finer can be transported in full suspension by flow turbulence. Coarse sand–fine gravel may be suspended in high-concentration currents, whereas coarser gravel is moved by grain-flow or debris-flow processes. These three grain-size populations may become segregated during deposition. The two coarser populations require transport at or above the Bagnold limit and are thus restricted to proximal environments.

Some field observations support the concept that debris flows may deposit sediment only in a proximal environment before generating an accelerating turbidity current that may erode early debris-flow deposits. Piper and others (1985) showed that a debris-flow deposit associated with seismically triggered sliding on the Scotian Slope has a downslope erosional edge and is overlain by a discontinuous sandy turbidite. Aarseth and others (1989) recognized deep scours downslope from slide scars on steep slopes in Norwegian fjords. Lowe (1982) interpreted bedforms at the top of proximal coarse sandy deposits, interpreted as deposits from a high-concentration flow, as resulting from reworking by the associated low-density turbidity current that transported finer sand and mud.

There have been many attempts to model the behavior of surge-type flows, both debris flows and high-density turbidity currents, using both flume experiments (Luthi, 1980, 1981) and numerical modeling with various rheological assumptions (Karlsrud and Edgers, 1982; Kirwan and others, 1986; Dengler and Wilde, 1987; Norem and others, 1989). The common feature of such models is that they show a progressive decrease in flow-velocity downslope.

1929 Grand Banks Event.—Some of the issues involved in the seismic initiation of turbidity currents can be examined from the case study of the 1929 Grand Banks turbidity current. The 1929 event on the Laurentian Fan (Fig. 5), as interpreted by Heezen and Ewing (1952) and Heezen and others (1954), has long been regarded as a type example of a catastrophic, seismically triggered turbidity current. Side-scan imagery and the location of cable breaks show that large areas of the continental slope and rise in water depths of more than about 600 m failed as a result of the magnitude 7.2 earthquake. Failure did not take place in a single large slide; rather, there were numerous shallow slides, separated by less disturbed areas of sea bed. Debris-flow deposits are widespread on the continental slope (Piper and Normark, 1982; Piper and others, 1988) and overlie turbidite deposits in the fan valleys. Failed sediments included gassy proglacial and Holocene muds, older Pleistocene sediments exposed by erosion on the continental slope, and perhaps some coarse proglacial sediments. The location of cable breaks and of turbidite deposits shows that essentially independent turbidity currents flowed down at least five discrete valley systems, separated by levees that were not

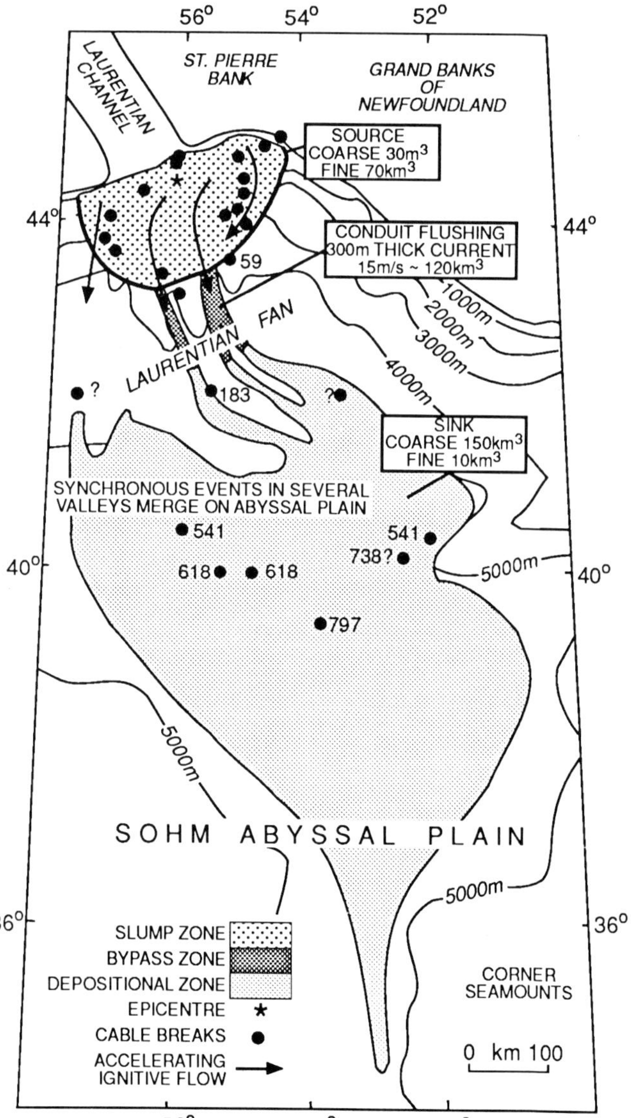

FIG. 5.—Initiation case history 2: the 1929 event on the Laurentian Fan, showing the location of cable breaks (for delayed breaks, numbers are in minutes after earthquake), and the source, bypass, and depositional zones with approximate sediment budgets. For further explanation, see text. Based on Piper and others (1988).

inundated by the flow. The main turbidite deposit on the Sohm Abyssal Plain was thus fed through several different pathways.

The volume of the turbidite (at least 160 km^3, principally sand) compares poorly with the estimated volume of failed sediment (maximum of 100 km^3, principally fine-grained) (Piper and Aksu, 1987). Hughes Clarke (1988), working in the main conduit of Eastern Valley on Laurentian Fan, showed that 1) maximum flow velocity can be estimated from the times of cable breaks, 2) the increase in wave length of gravel bedforms downvalley (Hughes Clarke and others, 1990, fig. 9) requires that the flow was accelerating for at least 100 km from the source area, and 3) flow thickness of a few hundred meters is estimated from erosional trimlines. Sediment concentration is constrained by the Bagnold limit (related to energy lost from grain collisions within the flow) and by cross-flow slopes, implying that flow duration was at least 4 and more probably more than 11 hours. The evidence of flow acceleration and the presence of erosional channels and scours on the valley floor suggest that the "missing" sand in the sediment-budget calculations was eroded from older deposits on the valley floor (Hughes Clarke and others, 1990). Numerous authors (the earlier ones reviewed by Menard, 1964; the more recent ones by Kirwan and others, 1986) have attempted to model the flow as a surge, but this is inconsistent with the evidence for prolonged and quasi-steady flow provided by the regular gravel bedforms and scours on the valley floor and by the sediment-flux requirements.

Hughes Clarke (1988) argues that to maintain flow over a period of several hours, there must have been a continuous process of transformation of debris flows to turbidity currents. He presents arguments that silty sediment put into suspension following widespread slope failure could have flowed ignitively, with the convergent-valley pattern on the slope above the Laurentian Fan allowing the concentration, maintenance, and acceleation of such ignitive flow. The failure may have involved sliding away from the feather edge of a cap of sediment rich in gas hydrates (G. Neave, pers. commun., 1990); removal of the lithostatic load of this cap could have allowed degassing of dissolved gas in the underlying sediment, thus providing a resuspension mechanism.

In summary, the 1929 earthquake triggered sediment failure along about 100 km of continental margin. Flows of many hours duration evolved through ignitive processes from muddy debris flows and eroded coarse sediment while accelerating through the fan-valley conduits. There were at least five discrete flows down different valleys that formed an amalgamated deposit on the Sohm Abyssal Plain.

Other Seismically Triggered Turbidity Currents.—Turbidity currents triggered by large earthquakes may be recognized from their synchronous development in several different drainage basins. Kastens (1984) recognized debris flows and turbidites that could be correlated between separated basins of the Calabrian ridge that can be associated with large earthquakes with a 1,500-yr recurrence interval. Adams (1989) has shown that Holocene turbidite units described by Griggs and Kulm (1970) from Cascadia Channel can be correlated with turbidites in other dispersal systems off the Washington coast and related to very large, subduction-related earthquakes with a 300-yr recurrence interval. As in the 1929 Grand Banks event, turbidity currents that travel down different drainage systems eventually may merge on the abyssal plain. Similar line-source seismoturbidites can be inferred from the ancient record. For example, the Contessa megabed in the Marnoso-arenacea Formation of northern Italy is a single turbidite layer whose minimum volume is about 30 km^3 (Ricci Lucchi and Valmori, 1980). This volume of material of calcareous mud and sand of presumed upper-slope origin is unlikely to have been supplied through a submarine canyon for this basin-wide (>25 km by >120 km) deposit. Mutti and others (1984) document other very large-volume turbidite beds from Italy and Spain that represent large-scale failure of slope areas. Amalgamated beds in ancient basin-plain sequences may be evidence of such events.

Some general predictions can be made for the frequency of seismic triggering of slides (Fig. 6), although not every seismically triggered slide produces a turbidity current. The magnitude-frequency relation for earthquakes in different tectonic regimes is reasonably well known, as are epicentral depths. There are engineering guidelines for the magnitude of peak horizontal accelerations (on bedrock) with variation in distance from the earthquake epicenter and earthquake magnitude (Fig. 6A; Idriss, 1985); these guidelines underestimate accelerations on unconsolidated sediment. The relation of earthquake size to landslide distribution on land was documented by Keefer (1984). Horizontal acceleration is the principal seismic-triggering mechanism for sediment failure on steep slopes (Fig. 6B), although liquefaction through cyclic loading may also be important. The "static factor of safety" (Booth and others, 1985) is a measure of the susceptibility of sediment to failure in the absence of an earthquake: it is low for rapidly accumulating sediment or sequences with gas in the shallower levels and is high for slowly accumulating sediment or sequences with diagenetic hardening. Some studies of slumps have estimated the horizontal accelerations required to initiate slumping in different areas. Figure 6C shows the relation between an estimated static factor of safety, slope angle, and magnitude of nearby earthquakes for several areas of slumping described in the literature.

The magnitude-frequency relation for earthquakes in a particular area is important in determining the size of sediment failure that may be generated. Infrequent large earthquakes (with few small earthquakes), such as occur in some subduction systems, allow thick sediment accumulations to build up between failures (Adams, 1989). In contrast, frequent but smaller earthquakes (such as occur in some extensional and strike-slip margins) may result in many smaller failures (Anastasakis and Piper, in prep.).

Earthquakes are not the only triggers for failure of upper-slope sediments. Failure also may be induced by storm waves (Lee and Edwards, 1986), as may have been the case for one or more of the turbidity currents generated by Hurricane Iwa as described earlier (Dengler and others, 1984a,b). Such a process may have been relatively more important during eustatic lowstands of sea level, when many deltas developed near the top of continental slopes; in the Holo-

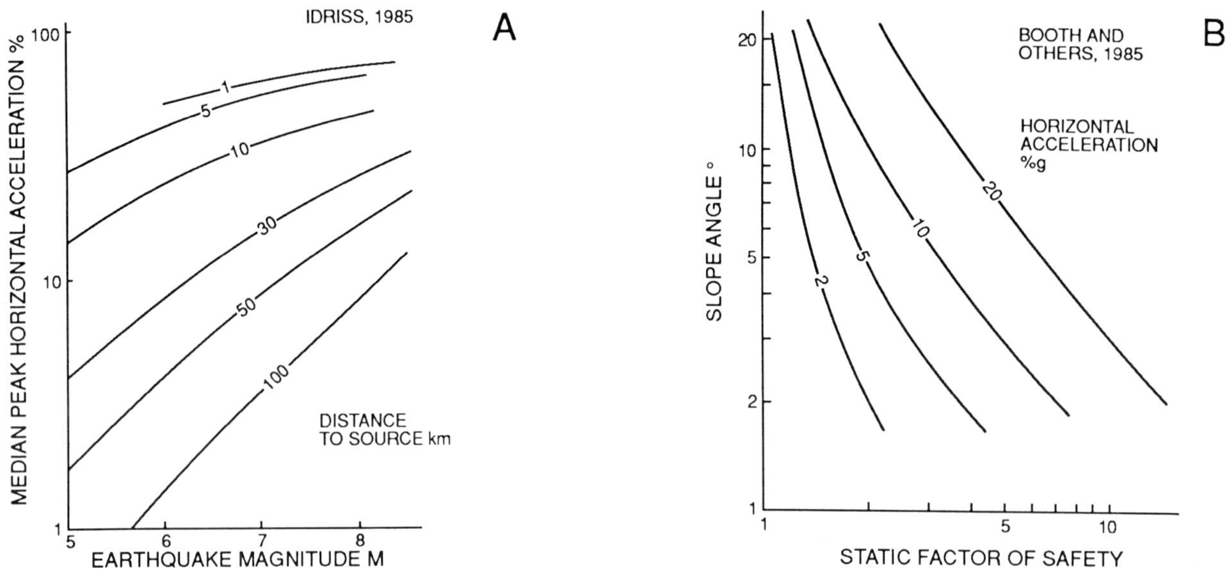

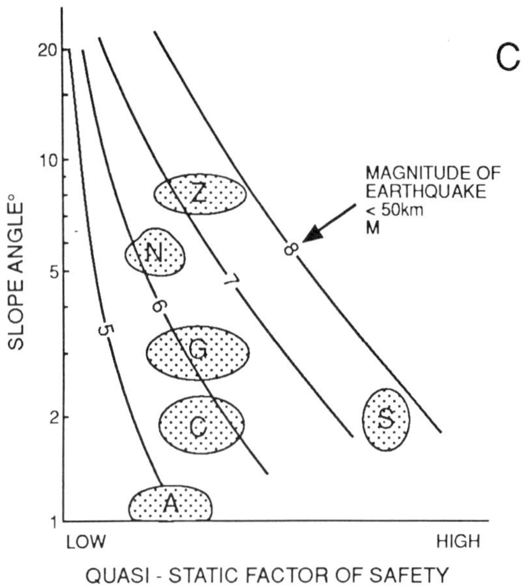

FIG. 6.—Schematic plots showing conditions leading to earthquake-triggered slumps. Note that these plots are designed to illustrate concepts and that caveats in the original sources must be noted. (A) From Idriss (1985), the median peak horizontal accelerations at various distances from the focus of earthquakes of different magnitudes. Note that these estimates are for accelerations developed on bedrock. (B) From Booth and others (1985), the horizontal accelerations required to produce a high probability of failure on a particular slope. Assumes bulk density of 1.5 and no cyclic effects on pore pressure. Factor of safety is essentially a measure of the intrinsic stability of the sediment: it can be estimated using the nomogram of Booth and others (1985). It is least for sediment with high-pore pressures resulting from underconsolidation or elevated gas content. (C) Data from plots (A) and (B) combined to show effect of horizontal acceleration resulting from an earthquake with an epicenter at 50-km distance on sediments of varying stability and slope. Plotted on this graph are a number of known failure events: A = Copper River delta, Alaska (Hampton and others, 1978); C = head of Cascadia Channel (Adams, 1989); G = Grand Banks 1929 failure (Piper and others, 1988); N = Northern Israel slope and S = Southern Israel slope (Almagor and Wiseman, 1977); Z = Zakynthos Basin (Anastasakis and Piper, pers. commun.). Note that effects such as variability in bulk density and of cyclic loading on pore pressure are not incorporated into this plot: for this reason, only a relative scale is used for the factor of safety.

cene, however, this process is probably less likely to be a major source of turbidity currents.

Turbidity-Current Initiation on Deltas

Because sediment strength and seabed slope are important in determining whether sediments fail, rapidly accumulating deltaic sediment, which tends to be underconsolidated and has relatively low strength, is particularly susceptible to failure, especially where the deposit is prograding across steep slopes. Such deltaic environments also may be the sites of inflow of sediment-laden river discharge that can form a turbidity current directly. The high-sediment concentrations associated with high-bedload-discharge rivers suggest that they may be the most likely to initiate turbidity currents.

During eustatic lowstands of sea level, many rivers discharged at the top of steep continental slopes. Far fewer river deltas extend to the upper slope in the Holocene. Thus, the best documentation for the initiation of turbidity currents through sediment failure comes from the study of active fjord deltas. The smaller size, relatively protected fjord environment, and shallower water depths of these deposits allow more detailed mapping and an ability to monitor some of the processes involved. These fjord-delta studies provide a useful analog to processes operating on larger passive-margin deltas during periods of lowstands of sea level.

Fjord Deltas of British Columbia.—The potential complexity of turbidity-current initiation in deltaic settings is well illustrated by observations made on fjord-head deltas in British Columbia, Canada (Fig. 7). The rivers are fed by melting snow and have a high-bedload discharge. Most deltas have a braided distributary pattern, such as the Klinaklini in Knight Inlet (Syvitski and Farrow, 1983). The prodelta slopes of these braid deltas, on which silty muds

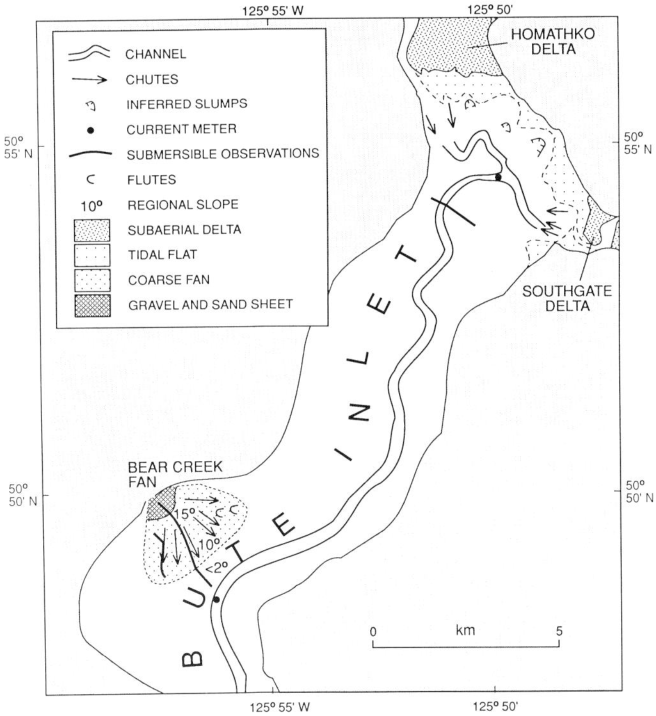

FIG. 7.—Initiation case history 3: initiation processes on British Columbia fjord-head deltas. For further explanation, see text. Based on Syvitski and Farrow (1983), with additions from Prior and others, (1986, 1987), Syvitski and others (1988), and Prior and Bornhold (1989).

settling from the discharge plume accumulate, have closely spaced chutes floored with coarse sediment, similar to those described by Prior and others (1981). Syvitski and others (1987) ascribe such chutes to small slides or local liquefaction generated through a combination of wave-induced cyclic loading and oversteepening following progradation. They note that the presence of downslope levees indicates the development of turbidity currents. The prodelta silty mud units commonly show evidence of sliding; the best example is the Kitimat delta, where sliding in the 1970s has been described by Prior and others (1982).

Prodelta chutes and larger delta-front channels, such as off the Homathko River delta at the head of Bute Inlet (Fig. 7), lead to channels on the fjord basin floors (Prior and others, 1986). The channels decrease in depth and width downslope and are commonly sinuous. Their upper reaches are erosional and their lower reaches are flanked by levees. The channels pass downslope into the ponded basin floor. Sediment in channels consists of sandy turbidite beds. Side-scan sonar shows crenulated ridged features (probably sediment waves) on the channel floors (Prior and others, 1986, Fig. 5). Overbank sediments are silt and mud. Debris-flow deposits are found in proximal settings. It is not clear to what extent the turbidites are derived from failure of delta-front sand in chutes, from the slope failure of silty mud deposits, or the hyperpycnal flow of river sediment into the fjord during flooding. Monitoring of the main channel in the Homathko prodelta slope by Prior and others (1987) provided tentative evidence for hyperpycnal flow in that some powerful current events in the channel were associated with reduced salinities. Current-meter observations of Prior and others (1987) suggest several sand-transporting events during each annual flood period. On some fan deltas, such as the Bear Creek Fan (Fig. 7), flow features extend across the fan up to the river mouth, and there is no evidence for sediment instability that might have triggered turbidity currents (Prior and Bornhold, 1989).

Itirbilung Fjord, Baffin Island.—Current-meter observations also have been made by Syvitski and Hein (1991) off a bedload delta in Itirbilung Fjord, Baffin Island. An instrument array was moored in a submarine channel on the prodelta slope. Over a 39-day period, nine strong, near-bottom flow events were detected that lasted between 1 and 6 hours with current velocities as much as 0.36 m/s and that were associated with a slight increase in bottom-water temperature, suggesting entrainment of surface water. Fluvial-discharge events were observed as peak down-fjord velocities recorded at a moored meter 11 m below the sea surface and were inferred from climatic modeling. In detail, however, there is no correspondence between the fluvial-discharge events and the observed near-bottom flows. Syvitski and Hein (1991) inferred, therefore, that the strong, discrete near-bottom flows were initiated by prodelta sliding of coarse-grained sediment in delta-front chute failures and that weaker multiple-flow events are indicative of slides in mud.

Hyperpycnal Flow and Slumping: Two Mechanisms of Turbidity-Current Initiation.—Fjord-head deltas are the site of transfer of fluvial sediment to turbidite systems. The regularity of prodelta and basin-floor channels (c.f. Hay, 1987b), the extension of flow features all the way up to the river mouth (Prior and Bornhold, 1989), the evidence of frequent delta-front currents transporting sand, and the low-salinity events identified by Prior and others (1987) all suggest that at times of flood there may be hyperpycnal flow of bedload and a portion of the suspended load of rivers down prodelta chutes and channels. Suspended loads in high-discharge rivers (examples 1-3 in Fig. 3) are sufficient to flow hyperpycnally into sea water; the sediment concentrations in bedload at times of flood must be even higher. The relatively low frequency of turbidity currents in many settings may indicate that they are generated by unusual flow conditions.

Other observations on prodelta morphology and sediment indicate frequent sliding and sediment liquefaction, a function of the rapid rates of progradation on steep slopes. Failure may be triggered by wave loading, low tides, seismicity, or overloading by progradation. Slide movement on delta slopes related to storm-wave reactivation can be relatively frequent, as recently observed on the large Huanghe delta, China (Prior and others, 1989). Although debris flows result from such failure (Prior and others, 1982), direct evidence of their evolution into turbidity currents is lacking.

Deltas thus provide a point source for turbidity-current flow because of the concentration of the associated sediment supply and the lower shear strength of rapidly deposited deltaic sediments. More observations and theoretical work are needed to establish the relative importance of hyperpycnal flow in deltaic settings.

FLOW CHARACTERISTICS IN CHANNELS

Introduction

Natural deep-sea turbidite systems receive turbidity currents of a wide range of sizes (Piper and Normark, 1983). These turbidity currents vary in the total amount of sediment transported, duration, grain-size distribution of transported sediment, velocity, and thickness. Smaller currents will tend to deposit their load in more proximal environments; larger currents may transport sediment to distal environments. The morphology of deep-sea fan systems reflects the cumulative erosional and depositional effects of a large number of turbidity currents, each of which interacts with channel conditions in different ways.

Conduit Flushing

The erosional form of many submarine canyons and upper-fan valleys, both modern and ancient, suggests that some currents that flow through them are erosional and therefore probably ignitive. Hughes Clarke and others (1990) demonstrate from chord lengths of gravel waves that the 1929 Grand Banks current accelerated over a distance of 100 to 200 km, and they postulate that much of the sediment in the 1929 deposit was eroded from the floor of the fan valley. Piper (1970) demonstrated that large Holocene turbidite beds on La Jolla Fan contain coarse sand eroded from Pleistocene deposits. Furthermore, individual turbidite units on La Jolla Fan are substantially larger than the amount of sand lost during a single event in the canyon heads. Thus, the volume of large turbidite beds cannot be equated directly to the volume of sediment released in a single initiating event; a large current may resuspend and transport substantial amounts of deep-water sediment deposited by previous, probably smaller, currents. The "parking areas" concept proposed by Mutti and others (1984) for seismoturbidites is probably a specialized case of conduit flushing.

Channelized Flow

Submarine canyons lead to fan valleys and channels where the evidence for flow processes is recorded in a wider range of depositional products than is observed in canyons. Channel overflow, erosion of channel walls, lateral migration of channels, and large-scale depositional bedforms are just some of the features that provide insight to flow characters that can be used to deduce flow conditions.

Limited field evidence shows that turbidity currents that transport a high proportion of sand are thinner than those that are predominantly of mud. At a sharp bend on the fan valley of Navy Fan, Piper and Normark (1983) showed that turbidity current I, with 3 percent sand, was 25 m thick, whereas turbidity current II, which deposited 42 percent sand and overall transported about 50 percent more sediment, was only 10 m thick. The 1929 Grand Banks turbidity current, transporting predominantly sand, was only a few hundred meters thick, but had a velocity in excess of 15 m/s. Pleistocene muddy turbidity currents on Laurentian Fan spilled over levees up to 900 m above the valley floor, but are inferred to have flowed at less than 1 m/s (Stow and Bowen, 1980).

Field evidence also shows that there is a vertical gradient in the grain size of sediment transported by a turbidity current. This has been determined from the decrease in sediment grain size passing from valley floors and walls to adjacent levees in both deep-water fans and tailings deposits in lakes (Normark and Dickson, 1976; Hess and Normark, 1976).

These two general trends are a consequence of the dynamics of turbidity-current flow together with the effects of entrainment of water at the top of the flow and ignitive conditions for sediment resuspension at the base. They are confirmed by flume experiments (Parker and others, 1987) and can be predicted by physical modeling (Stacey and Bowen, 1988a,b). Not only do flows show an upward decrease in sediment size, but there is also a corresponding upward decrease in sediment concentration. Flow velocity generally decreases upward through a flow, although basal frictional drag produces an additional velocity minimum at the bed. Because of the gradients in properties of natural flows, any modeling that considers a turbidity current as a layer-averaged flow is a substantial simplification of reality, although such a simplification is commonly necessary to make numerical modeling feasible (Komar, 1971, 1975; Parker and others, 1986).

Rapid entrainment of sea-water, accompanied by reduction in velocity, an increase in concentration of sediment at the base of the flow, and deposition may take place in three settings: 1) on steep slopes at high velocities as the Froude number approaches unity, entrainment will result in an increase in flow thickness, thereby inhibiting the development of supercritical flow; 2) where constraints to flow expansion are removed (e.g., by increase in width of a valley); and 3) where there is an abrupt decrease in gradient.

Different flows will react in different ways to seafloor relief. A small flow may be confined within a fan-valley system; a larger flow may be essentially channelized, but spill over the levees (Hay, 1987b). Thick flows may experience flow stripping of the upper part of the flow at abrupt bends in the channel (Piper and Normark, 1983), and a very thick flow may move downslope with the fan valley acting merely as a local roughness element (in cases where the flow thickness greatly exceeds the channel depth).

The role of Coriolis force in creating a significant cross-flow gradient is much greater for slow, muddy turbidity currents than for fast, sandy flows. Features that result from Coriolis effects, such as levee asymmetry and the location of spillover points, therefore will be influenced by variations in both flow thickness and sediment load.

In rivers, bankful stage is exceeded by the approximately 1-yr flood, but because of the accommodation provided by flood plains and the lack of entrainment of ambient fluid (air) into the river flow, maximum flow depths are rarely more than double bankful depths. There is, at present, no quantitative information for marine deposits regarding what factors control the size of valley-levee systems, or what proportion of turbidity currents reach bankful stage.

On mature passive-margin fans such as the Amazon and Mississippi, there are highly sinuous fan channels with continuous levees. Such channels appear in equilibrium with bankful or smaller flows: there is no evidence for slightly larger flows that would be expected to breach levees on sharp bends, although the possibility of very thick flows oblivious to channel relief cannot be excluded. This suggests 1) that there is a process such as overspill on bends (Hay, 1987a,b) on the upper-fan valley that regulates the thickness of turbidity currents flowing down the channel and 2) that flows are relatively slow, hence of relatively low density and long duration. The rivers associated with these sinuous-channel systems that feed these large fans transport only very fine sand and mud to their mouths; periodic hyperpycnal flow of such material could yield turbidity currents of rather uniform characteristics.

Turbidity currents initiated by seismic failure (e.g., Holocene deposits of Navy Fan and Cascadia Channel), or by bedload-delta processes such as hyperpyncal flow or mouth-bar failure (e.g., Var Fan, Pleistocene of Navy Fan) appear to be much more variable in size, and variably erode or deposit on different parts of the fan system, thus yielding a more complex morphology (Piper and Normark, 1983; Savoye and Piper, 1990). This variability will result in a much wider levee system, with lower back-levee gradients, than one formed by a succession of just bankful flows.

The depth of channel-levee systems is controlled by the elevation of valley-spillover points and the amount of entrainment and consequent flow thickening that will take place downflow from the spillover points. Spillover points develop at abrupt changes in channel direction, and will tend to occur where channel gradients decrease, so that there is an increase in flow thickness. Spillover points may develop as a result of sediment failure or the erosion of flute-like scours. Spillover points will tend to erode deeper if there are flows sufficiently large that significant suspended sand can spill over to yield a new ignitive flow with erosional capability. Spillover of sediment that includes sand will, in many cases, allow the spillover sheet to behave ignitively, e.g., to accelerate and erode a new channel (as on the Var Fan). This tendency of sheet-like spillover flow to reorganize results in the formation of headless channels, particularly along basin margins where the topography tends to concentrate flow and interaction of topography and Coriolis tilting yields maximum flow depths (Normark, 1985). In contrast, spillover of muddy sediment will lead to sheet-like deposition of mud from a flow that continues to decelerate (as seen in turbidite II on Navy Fan; Piper and Normark, 1983).

The behavior of the flow remaining in the channel that has lost its upper part through flow stripping is difficult to predict. For very thick flows that do not "see" channel morphology, flow continuity means that spillover has little dynamic effect. At the other extreme, flows that are largely confined within the channel will overspill without major impact on the channelized flow (Hay, 1987b). For flows so thin that only a very small proportion of their sediment load is lost by spillover will be almost unaffected. For flows of intermediate size (such as turbidite I of Piper and Normark, 1983), the loss of the upper half of the flow by spillover yields a flow that decelerates rapidly, since there is a disproportionately high-sediment concentration.

Levee deposition downslope from spillover points at elevations above the channel floor greater than at the spillover point represents either flow thickening resulting from gradient reduction, or deposition from flows that are substantially thicker than the depth of channel relief. Such levees are thus likely to be broad and to show little cross-levee variability in sediment properties.

The deposition of sediment waves on levees (Normark and others, 1980) requires flow thicknesses substantially greater than channel depth, in order that channel proceses do not interfere with the uniformity of the sediment waves. Mature and persistent wave forms will eventually control the elevation of spillover points (through troughs in the waves) for smaller flows.

The differences between large sandy, mixed, and muddy turbidity currents are summarized in Figure 8. A large muddy flow (Fig. 8A) will move slowly and thicken rapidly where there is gradient reduction. Once leaving the confines of a feeding canyon or slope valley, it will thicken rapidly at the gradient reduction at the base of slope and flow in a

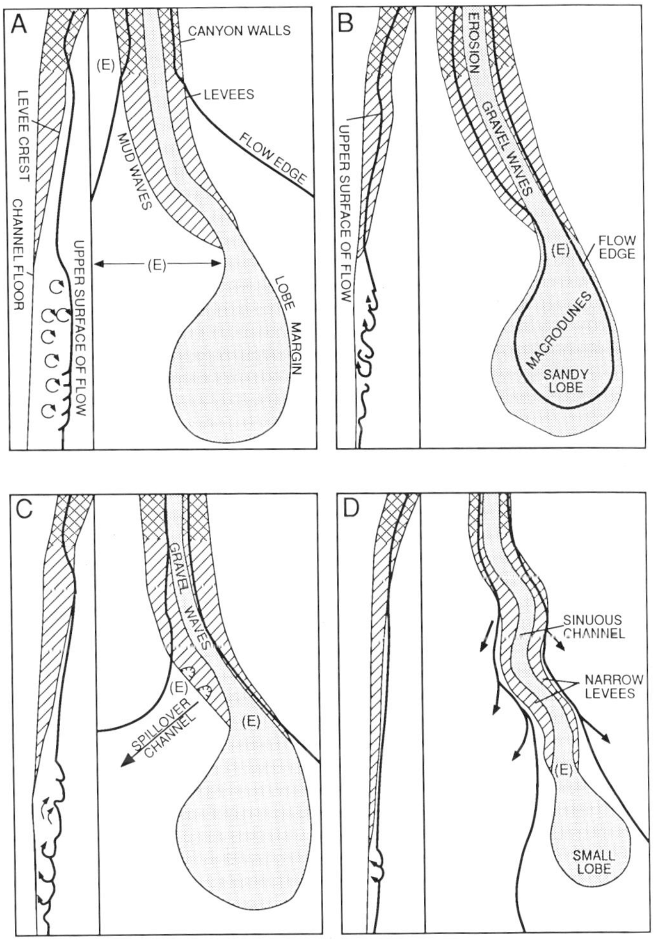

FIG. 8.—Diagrams illustrating flow of different types of turbidity current through a simple canyon/leveed channel/lobe fan system. Maps illustrate flow distribution and areas of bedform erosion and deposition. Profiles (along left side) show schematic longitudinal gradient, flow thickness, bed erosion and zones of increased entrainment of ambient water along axis of flow. On maps, (E) = zones of major entrainment; arrows indicate spillover of the muddy top of a flow. (A) Large muddy flow. (B) Large sand-rich flow. (C) Large mixed-sediment flow. (D) Series of repeated small muddy flows that just fill the fan valley.

sheet-like manner hardly "feeling" the underlying morphology. Such flows may produce muddy sediment waves on large levees. A large sandy flow (Fig. 8B) will erode in the slope valley and upper-fan valley, will be thin enough to remain confined within the levees of the upper-fan valley, and will deposit through flow expansion only in the levee termination zone. Large mixed flows (Fig. 8C) will show intermediate characteristics. Velocity may be sufficient to result in erosion or bypassing through the upper-fan valley with sand deposition in the valley-termination zone. At the same time, spillover of sandy material will lead to erosion on the back side of the levee and reconcentration of spillover into headless channels.

Evolution of Channel Systems Through Time

The evolution of many turbidite depositional systems through time can be analyzed by consideration of the controls on fluid entrainment, flow distribution, and deposition previously described. Early in the evolution of basins, active faulting results in steep slopes that commonly lead to relatively flat-floored basins. Such steep slopes promote both sediment failure and ignitive flow, resulting in the erosion of slope valleys. Where such valleys meet the basin floor, gradient reduction and flow expansion resulting from loss of valley confinement result in rapid deposition from all types of flows. This initiates the process of levee progradation that is seen in many fans (e.g., Navy Fan: Normark and others, 1979; Laurentian Fan: Normark and others, 1983).

The subsequent evolution of such systems depends on the dominant type of turbidity current. In sandy systems, such as for the latest Pleistocene deposits on Navy Fan, fed from a prograding braid delta, Coriolis effects are minor but sand lobes at the channel terminations reduce gradient and divert flow along lobe margins in an apparently random manner (Normark and others, 1979, their fig. 12). The presence of some mixed flows leads to the buildup of low levees; significant levee erosion occurs at spillover points resulting from the stepwise progradation along lobe margins, and the muddier portions of mixed flows deposit thick muds on the distal fan.

Where the dominant turbidity currents are muddy, as in the case of many passive-margin fans, e.g., the Amazon or Mississippi fans, thick flows (perhaps resulting from slide initiation) will lead to deposition of sheet-like mud units over large areas of the fan, the deposition of muddy sediment waves, and a steady aggradation of the fan surface. If turbidity flows result predominantly from hyperpycnal flow from rivers, the smoothing effect of the ignitive mechanism on velocity and discharge may result in relatively uniform flows that build up levees to a bankful condition, so that overflow results in rapid deceleration and thus steep levee backsides.

Where many flows are of mixed character, there is generally the opportunity for both pure muddy and pure sandy flows as well. These may occur in settings where deltas (or glacial outwash) provide sorted coarse sediment at the river mouth (potentially yielding sandy flows) but finer grained plume deposits accumulate on steep prodelta slopes that are susceptible to failure. The resulting muddy debris flows may evolve into ignitive flows that can develop a mixed character by conduit erosion. Well-developed fans of this type include the Laurentian and Var Fans. Such Northern-Hemisphere fans commonly show a strong left hook (Bellaiche and others, 1986) at the end of the slope valley or erosive upper-fan valley, and the presence of multiple channels leading to the valley termination zone. These have evolved through time from a system in which the slope valley built a simple depositional lobe with one or more distributary channels (Normark and others, 1983; Savoye and Piper, 1990). Fast, thin, coarse-grained flows will remain confined within a channel, and thus will not tend to erode levees; but they may be fast enough to cause significant channel-floor erosion. This effect will increase as the proportion of coarse sand and fine gravel increases. Aksu and Piper (1987) have proposed that the abundance of long leveed valleys in latitudes with a Pleistocene temperate glacial source results from the ability of turbidity currents derived from a coarse glaciogenic source to flush out their conduit.

Slower, mixed flows will show more Coriolis tilt effects, and flows of appropriate size and velocity may result in rapid asymmetric-levee progradation at the end of the slope valley. This asymmetric progradation will tend to divert thinner sandy flows, which flush out the channel, to the left. The maintenance and development of such an embryonic left hook will depend on the relative abundance of different-size flows; those that tend to deposit as a result of flow expansion at the end of the slope valley (these will stabilize the levees) and those that are sufficiently powerful to erode a spillover channel once they are no longer confined within the channel. The development of multiple channels within the main leveed-fan valley is also the result of flows of varying sizes causing either channel erosion, or building up low levees beside the eroded channels. Large muddy flows on such complex mixed systems will, as on other fans, tend to produce regional aggradation of muddy sediment.

Turbidity currents on any given volume can produce distinctly different bedding geometries within the resulting deposit, depending on the size and shape of the basin (Figs. 8 and 9). Where the fan is building on oceanic crust with few or no outcrops of the volcanic basement, the turbidity currents will spread laterally from the channels until the thinned flow is unable to maintain sediment in suspension. The lower part of the turbidy current will continue to follow the channel(s) to their termination, effectively carrying the coarsest sediment in the flow to the lobe areas of the system. In this case, the morphology of the fan is the result of unrestricted growth processes; aggradation in one sector of the fan will eventually lead to deflection of the flows, either through levee breeching or compensation, to a new area (Bouma, 1962; Mutti and Sonnino, 1981). For basins formed on continental crust, especially narrow, elongate troughs in foreland-basin environments, the larger turbidity currents will, in spreading laterally, reach the basin margins before thinning or entraining sufficient water to decrease flow competence drastically. In such cases, the maintained flow integrity can result in substantial coarse sediment being carried to the most distal part of the basin.

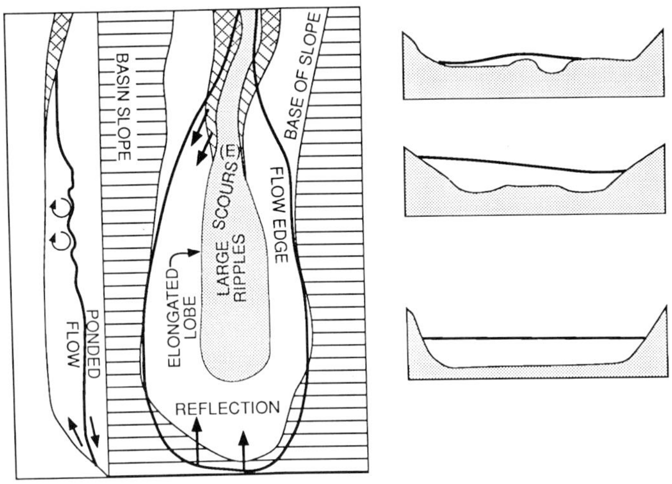

FIG. 9.—Diagram illustrating a large, mixed-sediment turbidity current flowing in a confined basin in northern hemisphere, thus with preferential overflfow to righthand side (looking downflow). Contrast with an open continental-margin flow (as illustrated in case C, Fig. 8). Symbols as in Figure 8.

Relatively thick lobe sequences can develop. In such basins, flow reflection (Pickering and Hiscott, 1985) and ponding of mud suspensions (McCave and Jones, 1988) may be important processes.

TURBIDITY-CURRENT BEDFORMS: IMPLICATIONS FOR FLOW PROCESSES

Bedforms in turbidite systems provide evidence for the importance of prolonged turbulent-fluid flow in the deposition of turbidite beds. They also have the potential for providing quantitative information on the characteristics of the flow. In studies of fluvial flow, there is a good quantitative understanding of the relation between bedforms, sediment grain size, and flow conditions (Middleton and Southard, 1984, ch. 7). Furthermore, large-scale, bed-roughness elements such as riffles and pools appear to have a predictable relation with flow parameters (Keller and Melhorn, 1978). Although data are insufficient at present, features resulting from fluid flow in turbidity currents may in the future provide a means of inferring the scalar properties of individual turbidity currents. This will require an understanding of the relation of bedform parameters to grain size, rate of sediment fallout, and flow conditions (Lowe, 1988).

Scours

Large-scale scours within turbidite deposits are, perhaps, the prime example of features that went unobserved because of their scale, until the use of deeply towed imaging systems provided morphometric details from modern-fan surfaces. These large-scale erosional features, which commonly have horizontal dimensions of tens to hundreds of meters, are generally too small to be clearly resolved using conventional, surface-ship echosounding or reflection-profiling techniques. Even with deep-tow data, the first observed turbidity-current scours were interpreted as isolated remnants of a braided-channel pattern (Normark, 1970). In outcrop studies, erosional features of this size are generally misidentified as channels (Mutti and Normark, 1987).

The initial interpretation of large, isolated depressions on modern-fan surfaces as scours (giant flutes) was based on a detailed survey of the upper-fan-channel/lower-fan-transition area on Navy Fan off southern California (Normark and others, 1979). More than 20 individual flute-shaped depressions were mapped using side-scan sonar images; these ranged in size from 20 m to as much as 400 m across (Table 2). The maximum depths of these scours is generally proportional to their width; the 400-m scour is about 20 m deep at its upslope edge. The scours of Navy Fan were limited to the area around and immediately downslope from the termination of the main leveed-fan valley on the upper fan but upslope of the small depositional lobes found at the end of the unleveed distributary channels (Fig. 10). The floors

TABLE 2. DIMENSIONS AND LOCATIONS OF SCOURS ON MODERN TURBIDITE SYSTEMS

Location	Gradient	Number Observed	Largest[1] Depth/length (m)	References	Notes
1. Eastern Valley, Laurentian Fan	0.025–0.017	1 +	100/1500	Shor and others, 1990	Many smaller scours
2. Levee-overflow zone, Var Fan	~0.012	4+	90/2000 40/1000 20/800	Malinverno and others, 1988	Irregular sediment-wave area
3. Channel-lobe transition zone, San Lucas Fan	0.028	8 large many small	60/600 50/500	Normark, 1970	Scours misidentified as remnants of braided channels
4. Channel-lobe transition zone, Navy Fan	0.017	>20	20/300 12/200	Normark and others, 1979	Smaller scours are commonly nested in groups of 2–5
5. Monterey East Valley, Monterey Fan	~0[2]	1	0–15/400 Headwall scarp is quite variable	Hess and Normark, 1976	Only one deep-tow track across this part of fan
6. Reynidsjup Channel	0.035	2+	30/2000	Lonsdale and Hollister, 1979	May be a contour-current feature
7. Channel-termination zone, Mississippi Fan	0.0025	many	±2/>2000	O'Connell and others, 1990	Relief too small to determine
8. Line Islands apron	~0[3]	3	70/3100	Normark and Spiess, 1976	Not turbidity-current scours; result of bottom-water flow

[1]Measured from top of lip to deepest point and to end of closed depression in downstream direction; depth (m)/length (m).
[2]Scour lies in saddle in upper reach of spillover channel.
[3]Scours occur in saddle between volcanic ridges; this example is not from a turbidite system but is used to illustrate that relatively slow, but prolonged, bottom-water flow can produce very large scours.

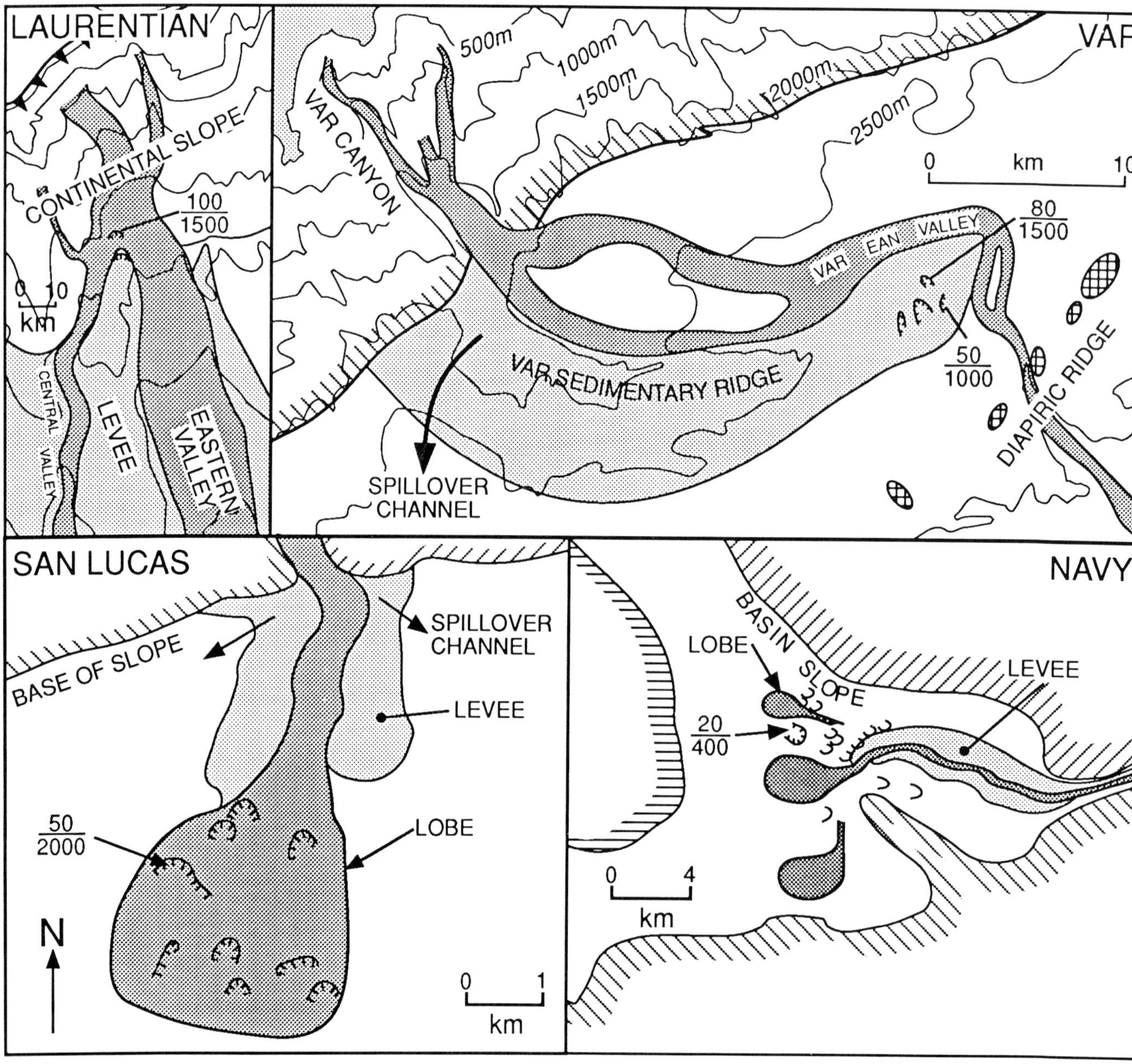

FIG. 10.—Sketch maps illustrating morphology, depth and setting of scours on fans 1–4 of Table 2.1: Laurentian Fan; 2: Var Fan; 3: San Lucas Fan; 4: Navy Fan. Numbers show depth and length of largest scour (in meters).

of several of the scours were photographed during the deep-tow studies, but a thin mud cover obscured the substrate. Deep-towed and surface-ship 3.5-kHz profiles suggest that the scours occur in a relatively sandy part of the fan.

On the San Lucas Fan south of the tip of Baja California, even deeper scour features occur in a similar setting immediately downslope from the termination of the main leveed-fan valley on the upper fan. These examples do not appear to exhibit regular flute shapes (Fig. 10) but this, in part, reflects the significantly higher altitude of the deep-tow vehicle during the survey of San Lucas Fan. The higher altitude (and consequent higher grazing angle of the acoustic signal) for the side-scan sensors greatly reduced the definition of the upper edges of the features, resulting in especially poor resolution of the more gently sloping sides. Both the Navy and San Lucas Fan scours occur immediately downstream from the area where turbidity currents undergo rapid flow expansion with the loss of the levee/channel conduit that had confined the flow farther upslope (Fig. 11). Mutti and Normark (1987) suggest scour features are indicative of this channel-lobe transition area on submarine fans.

The largest scours (with maximum horizontal dimensions of 1 km or greater) observed to date have been found on the floors of broad turbidite channels, including those on Laurentian and Monterey Fans (Table 2; Fig. 10; Shor and

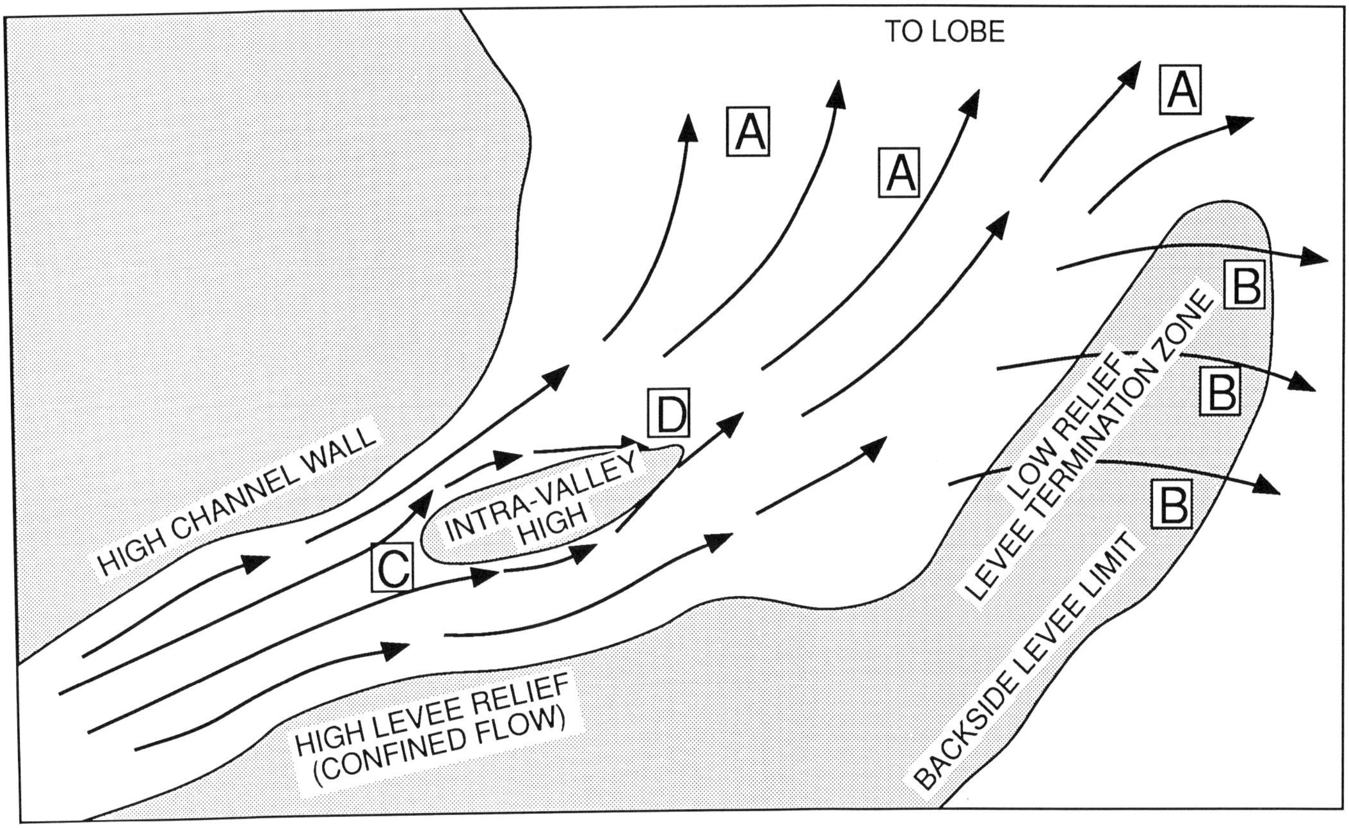

FIG. 11.—Diagram showing four causes of turbulence resulting in scour formation (c.f. Table 2). A: Rapid flow expansion, generally associated with widening of "channels" or decrease in gradient. Specific sites for different types of flow are shown as (E) in Figure 8. B: Thinning of flow over a levee; C: Turbulence upstream of obstacle to flow. D: Turbulence at reattachment point downstream of obstacle to flow.

others, 1990; Hess and Normark, 1976). Little except the size and shape is known about the Monterey Fan example, but Shor and others (1990) document several types of scours on the floor of the Eastern Valley of Laurentian Fan. The largest scour they describe is nearly 100 m deep and over a kilometer across. This flute-shaped feature is the best known of all large scours because it has been described using SEABEAM sounding and observations from the ALVIN submersible in addition to deep-towed side-looking sonar imaging. This giant flute has cut through a conglomerate layer into interbedded mudstone and sand in a reach of the valley just downstream from a region of significant flow expansion (an approximate doubling of valley width). Shor and others (1990) suggest that this flute formed by flow detachment, resulting in enlarging a smaller initial scour. They ascribe the formation to the passage of the turbidity current generated by the 1929 Grand Banks earthquake and deduce that the time scale of formation is on the order of minutes, not hours.

In addition to the giant flute, Shor and others (1990) observed other types of erosional depressions in Eastern Valley (Fig. 11). A series of depressions, as much as 30 m deep and 500 m across, is aligned along axis and spaced 1.5 to 3 km apart. They suggest that such features might be analogous to the pools formed in bedrock rivers. In the vicinity of these aligned erosional depressions, they also observe a scour of similar dimensions that lies about 1 km upvalley from the nose of a downvalley-trending elongate ridge near the western margin of Eastern Valley. The erosional depression could have developed from turbulence (vortices) formed during separation of the turbidity-current flow to either side of the ridge. This mode of formation is also favored by Normark and Spiess (1976) to explain the 60- to 80-m-deep flute-shaped depressions that lie 1 to 3 km upflow from small volcanic knobs that protrude above the surface of the Line Islands archipelagic apron.

All of the scour examples discussed here occur in, or immediately downslope from, areas of the turbidite deposits, where the fan or channel gradients indicate that flows could be supercritical using layer-averaged flow models. In addition, all occur in areas within or downslope from regions where rapid flow expansion is implied, associated with an increased turbulence within the flow (Fig. 11). The scours forming upstream from flow obstructions, however, indicate that scours might occur in any area of the fan if the local gradient is steep enough to maintain supercritical flow.

The scour examples referred to by Mutti and Normark (1987) from ancient turbidite sequences range from 0.5 m (large "normal" flutes) to more that 10 m across and are found downslope from or close to the termination of major channels. These scours are recognized by a mud drape that clearly demonstrates that the current forming the erosional depression did not then deposit sediment and fill it. Comparison with modern fans suggests that there is an abrupt flattening of the gradient in this area related to deposition from the turbidity currents as they spread laterally after leaving the leveed channel. In the cases where a rapid change in slope occurs, turbulence might be related to an hydraulic jump in the turbidity currents that also results in flow expansion.

Sediment Waves

The imaging of many deep-sea fans by side-scan sonar systems in the last decade has shown that coarse fan-valley and lobe sediments have a variety of large-scale depositional bedforms (Table 3, Fig. 12). Gravel waves, with chord lengths of several tens of meters and heights of a few meters, have been imaged in several modern fan valleys (Fig. 12F, G). The somewhat sinuous, and in places, bifurcating pattern of gravel-wave crests in plan view (Fig. 12G) is similar to that of smaller scale bedforms formed beneath turbulent flows. Where seen in cross section from submersible (Shor and others, 1990), the gravel appears crudely graded and includes contorted rip-up clasts of underlying mudstone. In distal areas, gravel waves are overlain by sand ribbons similar to those described from continental shelves. In the proximal parts of sandy lobes, sandy or fine gravelly bedforms (macrodunes; Hughes Clarke and others, 1990) with mean wavelength of several hundreds of meters and relief of less than 5 m occur in lanes (Fig. 12H). Higher resolution side-scan sonar in fjord channels (Prior and others, 1986) shows less regularly developed, ridged features, probably sediment waves built of coarse sand, with a wavelength of 10 m (Fig. 12E).

The internal structure of these coarse-grained bedforms is generally not known. Such bedforms have been recognized only very rarely in the ancient record (Winn and Dott, 1979), probably because the gravel waves have only a massive or poorly graded structure without internal stratification (Shor and others, 1990). Massive, lenticular, conglomerate beds in the ancient record with lateral dimensions of tens to hundreds of meters (such as the pods described by Johnson and Walker, 1979) may be the equivalent of the modern gravel waves (Piper and Kontopoulos, 1990). Cross-stratified sets of coarse sand and fine gravel at the top of proximal turbidites (Fig. 12B, C), that Lowe (1982) interpreted to result from reworking by the low-density part of a composite turbidity current, are smaller scale features than the large bedforms recognized on modern fans using side-scan sonar.

The quantitative significance of variations in coarse bedforms is at present poorly understood. Malinverno and others (1988) show that the wavelength of bedforms decreases from the axis to the margins of the flat floor of the Var canyon and fan valley, suggesting that the wavelength may be a measure of bed shear stress (velocity; Hughes Clarke, 1988); this is supported by observations of analogous bedforms in glacial-lake outbursts (Baker, 1973). The overall greater size of bedforms formed in the 1929 event compared with the slower 1979 Var turbidity current (Malinverno and others, 1988) is further evidence of this relation.

Regular sediment waves also are developed in predominantly fine-grained sediments, principally on levees. These resemble contourite sediment waves in that they propagate upcurrent. Wavelengths are typically more than 1 km and heights tens of meters (Table 3). For example, large-scale sediment waves on the main western levee of the Monterey Fan (Fig. 12D) commonly reach a 1.5-km wavelength. A transect of piston and long gravity cores (to 7 m) shows that the sediment from the upper part of the sediment-wave field is dominantly silt and mud with thin (1 to 10 cm generally) sand and sandy silt beds (Normark and others, 1985; Mutti and Normark, 1987). These coarser beds are quite variable in thickness even within the width of the core sample (10 cm); many beds pinch out within the core. The variability in bed thickness appears to increase with distance from the levee crest, indicating an increase in flow velocity and turbulence as the turbidity currents flow across the levee slope.

The significance of size variation in these large fine-grained sediment waves is not well understood. On the Var Fan

TABLE 3.—DIMENSIONS OF SEDIMENT WAVES IN SUBMARINE-FAN SYSTEMS. EXAMPLES FROM ANCIENT ROCK OUTCROPS SHOWN IN ITALICS.

Location	Material	Internal Structure	Wavelength (m)	Height (m)	Reference
Coarse Sediment					
Eastern Valley, Laurentian Fan	Conglomerates	Massive, poorly graded	30–80	1–5	Piper and others, 1988; Hughes Clarke, 1988
Valley-termination zone, Laurentian Fan	? Coarse sand	?	300	~4	Piper and others, 1988
Var Fan	Gravel	?	35–100	1.5–5	Malinverno and others, 1988
Corinth Graben, Greece	Gravel	Massive, poor grading	50–100	2–6	Piper and Kontopoulos, 1990
Lago Sofia, Chile	Gravel	Bedded and cross-bedded	?	4	Winn and Dott, 1979
Fine Sediment					
Upper-fan levee, Monterey Fan	Mud with silt and fine-sand interbeds	Internal reflectors onlap upcurrent bedforms#	300–2100	5–40	Normark and others, 1980
Var Fan					
Upper levee			500–3000	10–30	Savoye and others (pers. commun.)
Lower levee			3000–5000	30–50	
Labrador margin					
Lower slope			500–2000	5–20	Praeg and Schafer, 1989
Upper slope			500–3000	10–30	
Reynidsjup Channel*			500	20	Lonsdale and Hollister, 1979

*May result from contourite flow.

#Internal reflectors are subparallel to sediment-wave surface; onlapping of reflectors implies upslope migration of the bedforms.

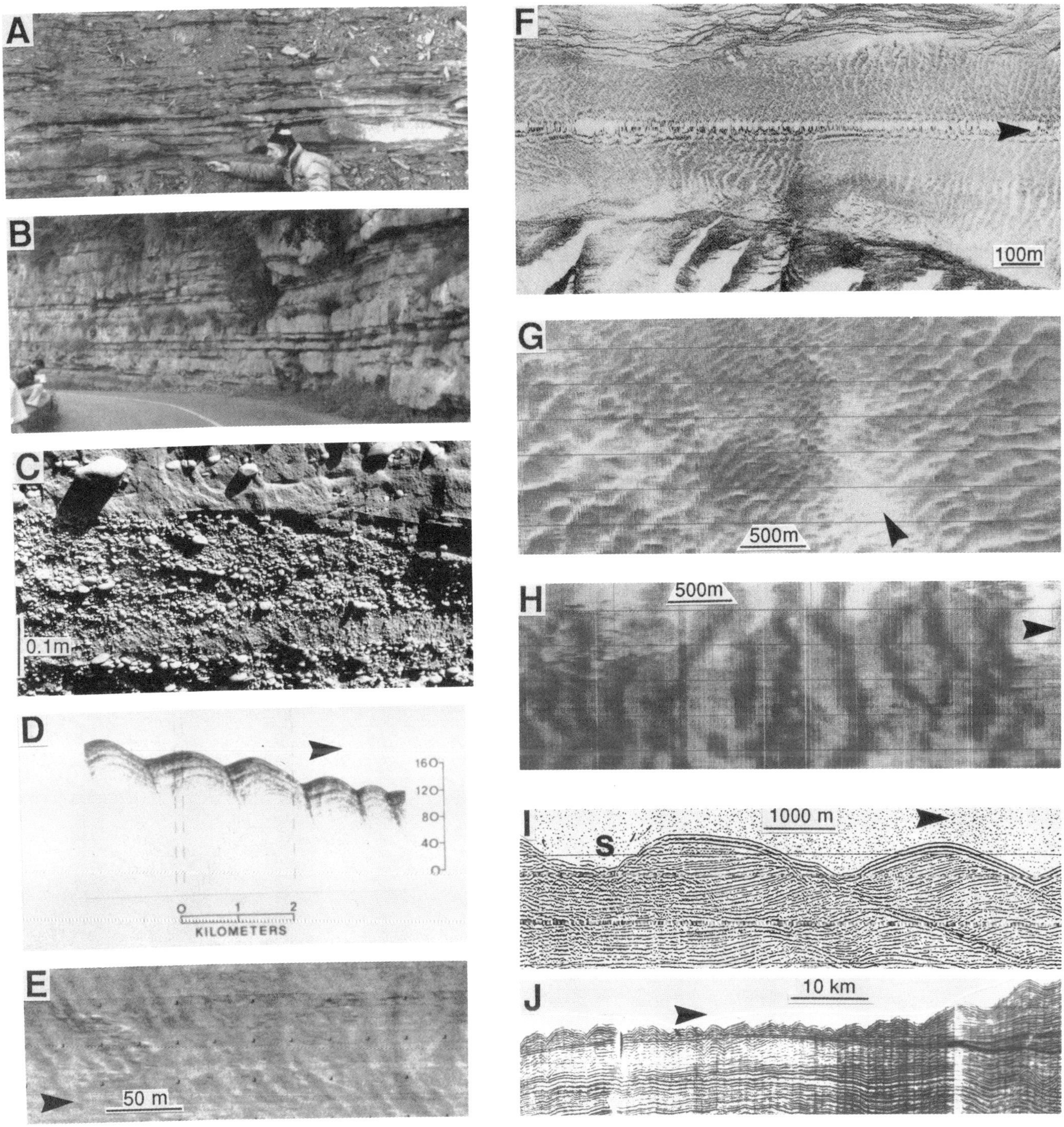

FIG. 12.—Photographs illustrating the variety of wave-like bedforms in turbidites. Arrows indicate flow directions. Scales in meters except where indicated. (A) Ripples, Hecho Group, Spain (Mutti and others, 1988); (B) Megaripple-size wavy bedding, Piedmont Basin, Italy (Cazzola and others, 1981); (C) Small gravel waves, Quaternary, Corinth graben, Greece (Doutsos and Piper, 1990); (D) Mudwaves, Monterey Fan, built of mud with thin sand beds, in 3.5-kHz profile (Normark and others, 1980); (E) Coarse-sand waves, turbidite channel, floor of Bute Inlet, B.C. (Prior and others, 1986; (side-scan image courtesy of D. B. Prior); (F) Small gravel waves, Var Fan valley (photo courtesy of Michel Voisset and Bruno Savoye, 1990); (G) Large gravel waves, 1929 Grand Banks turbidite, Laurentian Fan (Hughes Clarke and others, 1990); (H) "Macrodunes" 5 m high, with wavelength 300 m, 1929 Grand Banks turbidite, Laurentian Fan (Hughes Clarke and others, 1990); (I) Mud waves, Var Fan (profile courtesy of Bruno Savoye and Laurence Droz). s = scour (c.f. Fig. 10). (J) Contour-current sediment waves, Hamilton Spur, Labrador margin (Myers and Piper, 1988).

(Fig. 12I) and the fans off the Labrador Shelf (Table 3), waves become larger downcurrent. On the levee and adjacent midfan of Monterey Fan, sediment waves have wavelengths that are inversely proportional to the slope angle. Few core samples are available, but the 3.5-kHz reflection profiles suggest that there are no marked differences in grain size and bedding characteristics in the areas with well-developed sediment waves.

IMPLICATIONS FOR FACIES DISTRIBUTION IN TURBIDITE SYSTEMS

The variation in initiating processes and flow characteristics of turbidity currents is mirrored in the variability of depositional facies (Figs. 13 and 14). In this case, we use the term facies in a generic sense and do not attempt to explain any specific facies scheme, e.g., Mutti and Ricci Lucchi (1972). Rather, we point out some factors that control the composition and distribution of sediment in a turbidite system. Flow processes are related to the local depositional environment, and for purposes of this discussion we follow the concept of fan elements (Mutti and Normark, 1987).

Facies Variation with Elements

Channel Element.—The least continuous record of deposition within a turbidite system tends to be in the channels, which are conduits for sediment reaching the fan. Under certain flow conditions, erosion and/or non-deposition will occur. In addition, the preservation of channel facies in turbidite systems is a consequence of base-level fluctuations and fan aggradation. Thus, flow variability alone does not explain much of the observed variability in channel facies. Sandy and mixed flows have the potential to erode channels (including cutting scours) or leave a coarse lag molded into bedforms. In complex channel-levee systems, there may be various possible levels of spillover (e.g., from incised channel to valley floor or from valley floor over levees where the height varies along channel), all of which present the opportunity for flow stripping to remove the top of the sandy flow and the resulting potential for rapid deposition. Thick muddy flows that hardly "feel" channel morphology will have higher velocities within large channels and may transport and deposit small amounts of finer sand.

Levee and Overbank Element.—Levee morphology on modern deep-sea fans varies from irregularly eroded levees (Navy Fan), narrow levees with steep backsides (Amazon, Rhone), broad smooth levees (parts of the Laurentian Fan), to broad smooth levees with sediment waves (Var, Monterey). There is a corresponding variation in the observed facies. Discontinuous sands with wavy bedding and climbing ripples are found on levees on which overflow of mixed flows takes place, with rapid flow expansion of a turbulent sand-transporting current causing the irregular deposition. The narrow levees of many passive-margin fans have abundant thin (<5 cm) very fine sand-silt laminae (e.g., Damuth and Kumar, 1975). Broad smooth levees generally consist of silty muds and reflect steady deposition from thick muddy currents that do not experience significant flow expansion across the levee (Stow and Bowen, 1980).

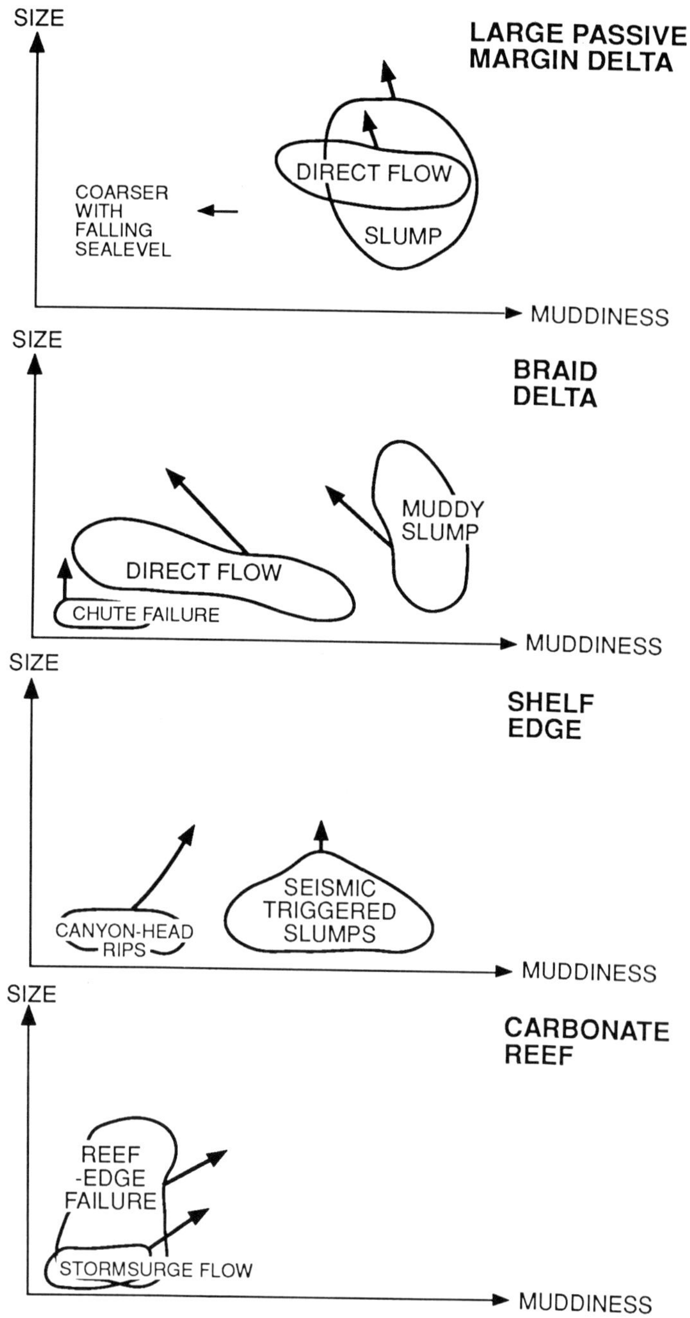

FIG. 13.—Conceptual graphs showing sediment type and size of turbidity currents resulting from different types of initiation process in a variety of sedimentary settings. Arrows indicate modifications resulting from conduit flushing. Further modifications take place downfan as a result of deposition of coarse sediment and flow stripping.

Lobe Element.—Steady muddy flows inferred to develop on passive-margin fans will experience flow expansion and deposit sediment at the end of a channel-levee system. Such lobes will aggrade steadily and be prograded over by the channel-levee system (Manley and Flood, 1988). Mixed or sandy flows of variable size also will experience flow expansion on the lobe, but large sandy flows may initially erode rather than deposit (Savoye and Piper, 1990). Characteristic sequence development is unlikely in lobes built

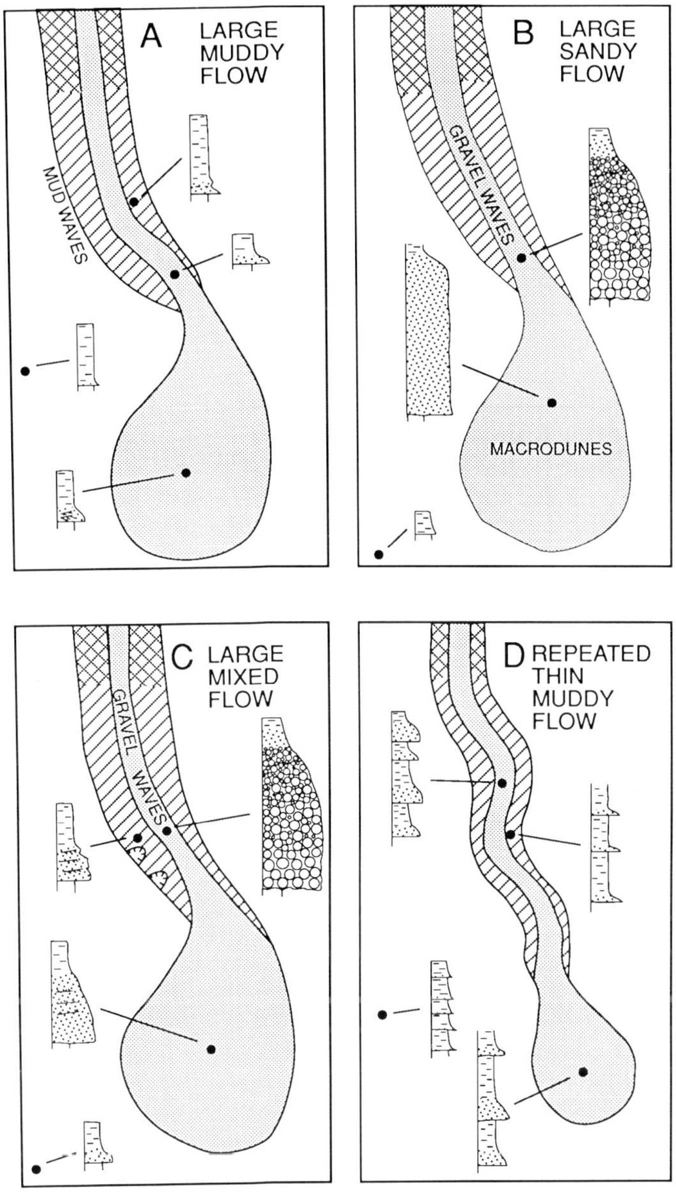

FIG. 14.—Diagrams illustrating relation of type of turbidite deposit to flow of different types of turbidity current through a simple canyon/leveed-channel/lobe-fan system. Compare with Figure 8, which shows flow processes in the same systems.

by such currents. Thick muddy flows will generally experience little modification in response to passing through the channel-lobe transition zone, unless the transition zone corresponds to a regional gradient change; mud aggradation may occur anywhere in the basin.

Distal-Fan Basin-Floor Element.—The deposition of sediments on distal parts of the fan is highly dependant on the upstream flow behavior of turbidity currents; the majority of turbidity currents in a system never reaches the distal parts of a basin (Piper and Normark, 1983, their figs. 8 and 9). Thus, it is only the very large flows that control the accumulation of distal-fan sediment. In smaller, elongate basins, such flows may spread laterally until they fill the basin. The entire basin then serves as a channel in the distal areas (Fig. 9), and vertical-flow expansion resulting from gradient changes probably provides the major control on deposition.

Fan Facies Reflecting Changes in Initiation Processes or Source Materials

Allocyclic changes such as sea-level eustasy or climatically modified sediment delivery cause variations in fan facies that can be analyzed using the concepts that we develop in this paper.

Effect of Sea-Level Change.—The role of eustatic sea-level change is to change the types of flow-initiation process and hence the types of turbidity currents flowing in the basin. Falling sea level, if accompanied by fluvial incision, may promote hyperpycnal flow of turbidity currents into the basin; if canyon incision occurs, the resulting increase in local slopes may trigger slumps (Weimer, 1989), aided by release of gas from hydrates if present (Kayen and Lee, 1989). Depending on the nature of the delta, this may result in either mixed or muddy flows of variable size. Eustatic lowstands provide the time of maximum direct input of fluvial sediment, which in sandy fans will lead to a flushing out of channel systems. During rising sea level, seismic triggering of flows will predominate, leading to occasional thick muddy flows.

Sea-level changes also may influence coastal morphology, so that the locus of sediment supply is changed. This is illustrated by the Monterey Fan off central California. The fan is fed by two major submarine-canyon systems in the Monterey Bay area and several smaller canyons farther south (Shepard and Dill, 1966; Hess and Normark, 1976). Sea-level fluctuation has had a pronouncned effect on the growth of the fan because much of the sediment for the fan has come from the north. During periods of low sea level (to depths approaching the current shelf edge), the more northern Ascension Canyon intercepts sediment moving south by littoral transport. The head of this canyon, however, consists of numerous branching gullies that only incise the outermost shelf. Therefore, during high sea-level stands, this sediment source feeds the Monterey Canyon farther south because the heads of this canyon extend across the shelf to near the beach itself. Thus, the major sediment input to the fan has periodically switched between the two canyon systems during the late Pleistocene, resulting in a relatively complicated history of channel/valley development on the fan.

Effect of Changes in Sediment-Input.—The role of changes in sediment delivery is to influence the type of turbidity current that is formed. Changes in sediment delivery result from changes in climate in the source area, changes in coastal morphology resulting from sea-level fluctuations, and changes in the type of initiating process. We illustrate these concepts with Navy Fan in the California Continental Borderland. In the late Pleistocene, Navy Fan was fed by the inferred braided delta of the Tijuana River prograding into the head of Coronado Canyon. This would have yielded either sandy or mixed flows, resulting either from bedload deltaic processes or seismic triggering. Thicker flows would have been modified by spillover at the base of slope in San Diego Trough, so that thin sandy flows dominated the growth of the fan. These produced the progradation pattern dominated by sandy lobes, and distal basins also accumulated sandy turbidites (Normark and others, 1979). In the Holocene, sea-level rise separated the Tijuana delta from the continental slope and the only turbidity currents to reach Navy Fan probably resulted from seismic triggering of Pleistocene prodeltaic sediment or of basin-slope sediment farther north (upslope) in San Diego Trough. The relative proportions of sand derived from the source area and from conduit flushing are unclear, but by the time the two flows studied in detail by Piper and Normark (1983) reached Navy Fan, one was muddy and the other mixed. Neither flow was in equilibrium with the fan morphology; sand was deposited mainly in the fan valley and on one abandoned lobe; mud accumulated in the distal basin.

Fan Facies and Basin Characters

We have shown that a consideration of initiation processes and turbidity-current flow provides insight into the development of morphological features of fans and the facies developed on fans. Changes in initiation process and source material result in changes in facies. Finally, we have illustrated how the same considerations also provide insight into how fans vary with tectonic setting and sea-level change.

It is not our purpose to endorse any particular classification of fans, nor to provide an exhaustive explanation of fan variability. We have demonstrated, however, that an understanding of sedimentary processes is relevant to the study of basin architecture. The following tectonic classification of fans proposed by Shanmugam and Moiola (1988) provides a useful base for this discussion:

Type 1.—Immature passive-margin fans with a narrow shelf and steep gradients have a dominance of sand supply, principally from fan deltas. Rapid progradation of delta-mouth sediment leading to failure through oversteepening, possible hyperpycnal flow of sandy-river bedload, and seismic triggering of prodeltaic slides are likely, yielding turbidity currents of variable size and sediment composition (Syvitski and Farrow, 1989). Because most deltas are small and earthquakes tend to be frequent but not of very high magnitude, very large turbidity currents are unlikely.

Type 2.—Mature passive-margin fans have low sand-to-mud ratios in their sediment supply and fan morphology suggests that hyperpycnal flow from rivers is important in fan evolution. Because of the generally aseismic nature of such margins, sea-level changes are predominant in triggering sediment failure; very large muddy flows may develop in this way, leading to basin aggradation. Except at such times, flows tend to be uniform and of low velocity, permitting steady growth of narrow levees, progradation of channel-levee systems, and uniform aggradation of lobes.

Type 3.—Active-margin fans may resemble those of immature passive margins. In other cases, however, there is a more mature fluvial supply of sediment and a higher proportion of mud. Steep slopes and large-magnitude (M>8) earthquakes result in seismic triggering being dominant at least during eustatic highstands, so that turbidity currents tend to be of variable size and sediment composition. Very large mixed flows may develop, with complex erosional and depositional behavior.

Type 4.—Mixed-setting fans were distinguished by Shanmugam and Moiola (1988) as fans that occur in basins bounded by both active and passive margins. They point out that the sedimentary development of such fans is generally controlled by one or another of the margin types; as such, we do not consider them further.

CONCLUSIONS

The basis for our improved understanding of turbidity-current phenomena since the original Shepard memorial volume stems primarily from 1) the ability to monitor some flow processes (but generally only in shallow-water settings) and 2) to observe finer scale features on the surface of turbidite deposits than was possible 25 years ago. Both of these areas reflect the tremendous improvement in the technology available to marine geologic studies. Nevertheless, the interpretations presented in this paper, whether about initiation events, flow thickness, how flow conditions control sediment distribution in a turbidite system, or nearly all phases of flow in turbidity currents, are based on very few direct observations of deep-water deposits. No significant deep-water event has ever been directly monitored. This paper has suggested that there are several areas in which we need to focus research to better understand the evolution of turbidite facies. The most important of these are 1) the frequency and nature of hyperpycnal flow of rivers into deep-water turbidite basins along continental margins; 2) the range in characteristics, the depositional setting (channel floor, overbank, etc.), and the hydrodynamic significance of large-scale bedforms in turbidite systems; and 3) the investigation of individual turbidites to understand better the dynamics of turbidity currents. The key conclusion is that the evolution of deep-sea turbidite systems is probably a consequence of prolonged flows of turbidity currents behaving in a relatively predictable, albeit extremely varied, manner. The catastrophic 'dump-truck' model of a turbidity current is definitely inappropriate for most settings.

ACKNOWLEDGMENTS

Our understanding of turbidity currents has evolved over the past 20 years since we first worked with Fran Shepard and has been influenced by a wide range of studies with many co-workers and through several key forums, particularly COMFAN I (1982) and COMFAN II (1988) hosted by Arnold Bouma and Emiliano Mutti, respectively. We especially thank Tony Bowen and John Hughes Clarke for discussions that lead to the development of some of the concepts in this paper and Dave Prior and Jim Syvitski for provision of unpublished manuscripts. We also benefitted from discussions of ongoing work with Bruno Savoye, George Anastasakis, Emiliano Mutti, Gian Zuffa, Kate Moran, Franco Ricci Lucchi, and Fabio Trincardi. The manuscript was improved through reviews by Rick Hiscott, Homa Lee, and Jim Syvitski.

REFERENCES

AARSETH, I., LONNE, O., AND GISKEODEGAARD, O., 1989, Submarine slides in glaciomarine sediments in some western Norwegian fjords: Marine Geology, v. 88, p. 1–21.

ADAMS, J., 1989, Turbidites off the Oregon-Washington margin record paleoearthquakes on the Cascadia subduction zone: Geological Survey of Canada Paper 89-1F, p. 37–43.

AKSU, A. E., AND PIPER, D. J. W., 1987, Late Quarternary sedimentation in Baffin Bay: Canadian Journal of Earth Sciences, v. 24, p. 1833–1846.

ALMAGOR, G., AND WISEMAN, G., 1977, Analyses of submarine slumping on the continental slope of the southern coast of Israel: Marine Geotechnology, v. 2, p. 349–388.

BAGNOLD, R. A., 1962, Autosuspension of suspended sediment: Proceedings, Royal Society of London, Ser. A, v. 265, p. 315–319.

BAGNOLD, R. A., 1966, An approach to the sediment transport problem from general physics: U.S. Geological Survey Professional Paper 422-1, 37 p.

BAKER, V. R., 1973, Paleohydrology and sedimentology of Lake Missoula flooding in eastern Washington: Geological Society of America Special Paper 144, 73 p.

BELLAICHE, G., COUTELLIER, V., DROZ, L., AND MASSON, P., 1986, Deep sea and Martian channels: Deep Sea Research, v. 33, p. 973–980.

BOOTH, J. S., SANGREY, D. A., AND FUGATE, J. K., 1985, A nomogram for interpreting slope stability of fine-grained deposits in modern and ancient marine environments: Journal of Sedimentary Petrology, v. 55, p. 29–36.

BORNHOLD, B. D., AND PILKEY, O. H., 1971, Bioclastic turbidite sedimentation in Columbus Basin, Bahamas: Geological Society of America Bulletin, v. 82, p. 1341–1354.

BOUMA, A. H., 1962, Sedimentology of Some Flysch Deposits: Elsevier, New York, 168 p.

BOWEN, A. J., NORMARK, W. R., AND PIPER, D. J. W., 1984, Modelling of turbidity currents on Navy submarine fan, California Continental Borderland: Sedimentology, v. 31, p. 169–185.

CAZZOLA, C., FONNESU, F., MUTTI, E., RAMPONE, G., SONNINO, M., AND VIGNA, B., 1981, Geometry and facies of small, fault controlled deep-sea fan systems in a transgressive depositional setting (Tertiary Piedmont Basin, northwestern Italy), *in* F. Ricci Lucchi, ed., Excursion Guidebook; 2nd European Regional Meeting of International Association of Sedimentologists, Bologna, p. 5–56.

CHAMBERLAIN, T. K., 1964, Mass transport of sediment in the heads of Scripps submarine canyon, California, *in* Miller, R. L., ed., Papers in Marine Geology. MacMillan, New York, p. 42–64.

COOK, H. E., HINE, R. C., AND MULLINS, H. T., 1983, Platform margin and deep-water carbonates: Society of Economic Paleontologists and Mineralogists Short Course Notes No. 12, 573 p.

COSTA, J. E., AND WILLIAMS, G. P., 1984, Debris flow dynamics: U.S. Geological Survey Open-File Report 84-606 (Videotape only, no text).

DAMUTH, J. E., AND KUMAR, M., 1975, Amazon Cone: Morphology, sediments, age, and growth pattern: Geological Society of America Bulletin, v. 86, p. 863–878.

DENGLER, A. T., AND WILDE, P., 1987, Turbidity currents on steep slopes: Application of an avalanche-type numeric model for ocean thermal energy conversion design: Ocean Engineering, v. 14, p. 409–433.

DENGLER, A. T., NODA, E. K., WILDE, P., AND NORMARK, W. R., 1984a, Slumping and related turbidity currents along proposed OTEC cold-water-pipe route resulting from Hurricane Iwa: Proceedings, 1984 Offshore Technology Conference, OTC 4702, p. 475–480.

DENGLER, A. T., WILDE, P., NODA, E. K., AND NORMARK, W. R., 1984b, Turbidity currents generated by Hurricane Iwa: Geo-Marine Letters, v. 4, p. 5–11.

DOUTSOS, THEODOR AND PIPER, D. J. W., 1990, Listric faulting, sedimentation, and morphological evolution of the Quaternary eastern Corinth rift, Greece: First stages of continental rifting: Geological Society of America Bulletin, v. 102, p. 812–829.

DRAKE, D. E., KOLPACK, R. L., AND FISCHER, P. J., 1972, Sediment transport on the Santa Barbara-Oxnard shelf, Santa Barbara Channel, California, *in* Swift, D. J. P., Duane, D. B., and Pilkey, O. H., eds., Shelf Sediment Transport. Dowden Hutchinson and Ross, Stroudsburg, Pennsylvania, p. 307–331.

EEZSCAN 84 SCIENTIFIC STAFF, 1986, Atlas of the Exclusive Economic Zone, Western Coterminous United States, scale 1:500,000: U.S. Geological Survey Miscellaneous Investigations Series, 1-1792, 152 p.

EITTREIM, S. L., EMBLEY, R. W., AND NORMARK, W. R., 1983, Monterey sea-valley meander revisited: EOS, Transactions, American Geophysical Union, v. 64/65, p. 730.

FUKUSHIMA, Y., PARKER, GARY, AND PANTIN, H. M., 1985, Prediction of ignitive turbidity currents in Scripps submarine canyon: Marine Geology, v. 67, p. 55–81.

GENNESSEAUX, M., GUIBOUT, P., AND LACOMBE, H., 1971, Enregistrement de courants de turbidite dans la Vallee Sous-Marine du Var (Alpes-Maritimes): Comptes Rendus Academie Sciences, Paris, Tome 273, Ser. D, p. 2456–2459.

GILBERT, R., 1983, Sediment processes of Canadian Arctic fjords: Sedimentary Geology, v. 36, p. 147–175.

GOULD, H. R., 1951, Some quantitative aspects of Lake Mead turbidity currents: Society of Economic Paleontologists and Mineralogists Special Publication 2, p. 34–52.

GRIGGS, G. B., AND KULM, L. D., 1970, Sedimentation in Cascadia deep-sea channel: Geological Society of America Bulletin, v. 81, p. 1361–1384.

GUBLER, H., 1989, Comparison of three models of avalanche dynamics: Annals of Glaciology, v. 13, p. 82–89.

HAMPTON, M. A., 1972, The role of subaqueous debris flow in generating turbidity currents: Journal of Sedimentary Petrology, v. 42, p. 775–793.

HAMPTON, M. A., BOUMA, A. H., CARLSON, P. R., MOLNIA, B. F., CLUKEY, E. C., AND SANGREY, D. A., 1978, Quantitative study of slope instability in the Gulf of Alaska: Proceedings, 10th Offshore Technology Conference, OTC 3314, 12 p.

HAY, A. E., 1987a, Turbidity currents and submarine channel formation in Rupert Inlet, British Columbia. I. Surge observations: Journal of Geophysical Research, v. 92, p. 2875–2882.

HAY, A. E., 1987b, Turbidity currents and submarine channel formation in Rupert Inlet, British Columbia. II. The roles of continuous and surge type flow: Journal of Geophysical Research, v. 92, p. 2883–2900.

HAY, A. E., BURLING, R. W., AND MURRAY, J. W., 1982, Remote acoustic detection of a turbidity current surge: Science, v. 217, p. 833–835.

HEEZEN, B. C., AND EWING, MAURICE, 1952, Turbidity currents and submarine slumps and the 1929 Grand Banks earthquake: American Journal of Science, v. 250, p. 849–873.

HEEZEN, B. C., AND HOLLISTER, C. D., 1971, The Face of the Deep: Oxford University Press, New York, 659 p.

HEEZEN, B. C., ERICSON, D. B., AND EWING, MAURICE, 1954, Further evidence for a turbidity current following the 1929 Grand Banks earthquake: Deep Sea Research, v. 1, p. 193–202.

HESS, G. R., AND NORMARK, W. R., 1976, Holocene sedimentation history of the major fan valleys of Monterey Fan: Marine Geology, v. 22, p. 233–251.

HINE, R. C., WILBER, R. J., DANE, J. M., NEUMANN, A. C., AND LORENSON, K. R., 1981, Offbank transport of carbonate sands along open, leeward bank margins: Northern Bahamas: Marine Geology, v. 42, p. 327–348.

HOOKE, R. L., AND SCHLAGER, W., 1980, Geomorphic evolution of the Tongue of the Ocean and the Providence Channels, Bahamas: Marine Geology, v. 35, p. 343–366.

HOSKIN, C. M., AND BURRELL, D. C., 1972, Sediment transport and accumulation in a fjord basin, Glacier Bay, Alaska: Journal of Geology, v. 80, p. 539–551.

HUBBARD, D. K., WARD, L. G., FITZGERALD, D. M., AND HINE, A. C., 1976, Bank margin morphology and sedimentation, Lucaya, Grand Bahama Island: Coastal Research Division Technical Report 7-CRD, University of South Carolina, 36 p.

HUBERT, J. F., 1964, Textural evidence for deposition of many western North Atlantic deep sea sands by ocean-bottom currents rather than by turbidity currents: Journal of Geology, v. 72, p. 757–785.

HUGHES CLARKE, J. E., 1988, The geological record of the 1929 "Grand Banks" earthquake and its relevance to deep-sea clastic sedimentation: Unpublished Ph.D. Dissertation, Dalhousie University, Halifax, Nova Scotia, 171 p.

HUGHES CLARKE, J. E., SHOR, A. N., PIPER, D. J. W., AND MAYER, L. A., 1990, Large-scale current-induced erosion and deposition in the path of the 1929 Grand Banks turbidity current: Sedimentology, v. 37, p. 613–629.

IDRISS, I. M., 1985, Evaluating seismic risk in engineering practice: Proceedings, 11th International Conference on Soil Mechanics and Foundation Engineering, Balkema, Rotterdam, v. 1, p. 255–320.

INMAN, D. L., NORDSTROM, C. E., AND FLICK, R. E., 1976, Currents in submarine canyons: An air-sea-land interaction: Annual Review of Fluid Mechanics, p. 275–310.

JOHNSON, A. M., 1970, Physical Processes in Geology: Freeman Cooper, San Francisco, 577 p.

JOHNSON, B. A., AND WALKER, R. G., 1979, Paleocurrents and depositional environments of deep water conglomerates in the Cambro-Ordovician Cap Enrange Formation, Quebec Appalachians: Canadian Journal of Earth Sciences, v. 16, p. 1375–1387.

KARLSRUD, K., AND EDGERS, E., 1982, Some aspects of submarine slope instability, *in* Saxov, S., and Nieuwenhuis, J. K., eds., Marine Slides and Other Mass Movements, Plenum Press, New York, p. 61–81.

KASTENS, K. A., 1984, Earthquakes as triggering mechanism for debris flows and turbidites on the Calabrian Ridge: Marine Geology, v. 55, p. 13–33.

KASTENS, K. A., AND SHOR, A. N., 1985, Depositional processes of a meandering channel on Mississippi Fan: American Association of Petroleum Geologists Bulletin, v. 69, p. 190–202.

KAYEN, R. E., AND LEE, H. J., 1989, Effect of sea-level fall-induced gas-hydrate decomposition on stability of continental slope, Beaufort Sea: EOS, Transactions, American Geophysical Union, v. 70, p. 1152.

KEEFER, D. A., 1984, Landslides caused by earthquakes: Geological Society of America Bulletin, v. 95, p. 406–421.

KELLER, E. A., AND MELHORN, W. N., 1978, Rhythmic spacing and origin of pools and riffles: Geological Society of America Bulletin, v. 89, p. 723–730.

KIRWAN, A. D., Jr., DOYLE, L. J., BOWLES, W. D., AND BROOKS, G. R., 1986, Time-dependent hydrodynamic models of turbidity currents analyzed with data from the Grand Banks and Orleansville events: Journal of Sedimentary Petrology, v. 56, p. 379–386.

KOMAR, P. D., 1969, The channelized flow of turbidity currents with application to Monterey deep-sea fan channel: Journal of Geophysical Research, v. 78, p. 4544–4558.

KOMAR, P. D., 1971, Hydraulic jumps in turbidity currents: Geological Society of America Bulletin, v. 82, p. 1477–1488.

KOMAR, P. D., 1975, Supercritical flow in turbidity currents: A discussion: Journal of Sedimentary Petrology, v. 45, p. 747–749.

KUENEN, P. H., 1950, Turbidity currents of high density: 18th International Geological Congress, London, 1948, Report, Part 8, p. 44–52.

KUENEN, P. H., 1955, Experiments in connection with turbidity currents and clay suspensions, *in* Whittard, W. F., and Bradshaw, R., eds., Submarine Geology and Geophysics: Butterworths, London, p. 47–74.

LAMBERT, A. M., KELTS, K. R., AND MARSHALL, N. F., 1976, Measurement of density underflows from Walensee, Switzerland: Sedimentology, v. 23, p. 87–105.

LEE, H. J., AND EDWARDS, B. E., 1986, Regional method to assess offshore slope stability: Journal of Geotechnical Engineering, ASCE, v. 112, p. 489–509.

LONSDALE, P., AND HOLLISTER, C. D., 1979, Cut-offs at an abyssal meander south of Iceland: Geology, v. 7, p. 597–601.

LOWE, D. R., 1982, Sediment gravity flows. II. Depositional models with special reference to the deposits of high-density turbidity currents: Journal of Sedimentary Petrology, v. 52, p. 279–297.

LOWE, D. R., 1988, Suspended-load fallout rate as an independent variable in the analysis of current structures: Sedimentology, v. 35, p. 765–776.

LUTHI, S., 1980, Some new aspects of two-dimensional turbidity currents: Sedimentology, v. 28, p. 97–105.

LUTHI, S., 1981, Experiments in non-channelized turbidity currents and their deposits: Marine Geology, v. 40, p. M59-M68.

MALINVERNO, A., RYAN, W. B. F., AUFFRET, G., AND PAUTOT, G., 1988, Sonar images of the path of recent failure events on the continental margin off Nice, France: Geological Society of America Special Paper 229, p. 59–76.

MANLEY, R. L., AND FLOOD, R. D., 1988, Cyclic sediment deposition within Amazon deep-sea fan: American Association of Petroleum Geologists Bulletin, v. 72, p. 912–925.

McCABE, P. J., 1978, The Kinderscoutian delta (Carboniferous) of northern England: A slope influenced by density currents, *in* Stanley, D. J., and Kelling, G., eds., Sedimentation in Submarine Canyons, Fans and Trenches: Dowden, Hutchinson and Ross, Stroudsburg, Pennsylvania, p. 116–126.

McCAVE, I. N., 1972, Transport and escape of fine-grained sediment from shelf areas, *in* Swift, D. J. P., Duane, D. B., and Pilkey, O. H., eds., Shelf Sediment Transport, Process and Pattern: Dowden Hutchinson and Ross, Stroudsburg, Pennsylvania, p. 225–248.

McCAVE, I. N., AND JONES, K. P. N., 1988, Deposition of ungraded muds from high-density non-turbulent turbidity currents: Nature, v. 333, p. 250–252.

McILREATH, I. A., AND JAMES, N. P., 1979, Facies models. 12. Carbonante slopes *in* Walker, R. G., ed., Facies Models: Geological Association of Canada, p. 133–143.

MENARD, H. W., 1955, Deep-sea channels, topography, and sedimentation: American Association of Petroleum Geologists Bulletin, v. 39, p. 236–255.

MENARD, H. W., 1964, Marine Geology of the Pacific: McGraw-Hill, New York, 271 p.

MIDDLETON, G. V., AND HAMPTON, M. A., 1976, Subaqueous sediment transport and deposition by sediment gravity flows, *in* Stanley, D. J., and Swift, D. J. P., eds. Marine Sediment Transport and Environmental Management: Wiley, New York, p. 197–218.

MIDDLETON, G. V., AND SOUTHARD, J. B., 1984, Mechanics of Sediment Movement: Society of Economic Paleontologists and Mineralogists Short Course Notes No. 3, 401 p.

MILLER, R. L., 1964, Papers in Marine Geology: Macmillan, New York, 531 p.

MORGENSTERN, N. R., 1967, Submarine slumping and the initiation of turbidity currents, *in* Richards, A. F., ed., Marine Geotechnique: University of Illinois Press, Urbana, p. 189–220.

MULLINS, H. T., 1983, Modern carbonate slopes and basins of the Bahamas; *in* Cook, H. E., Hine, A. C., and Mullins, H. T., eds., Platform Margin and Deep-Water Carbonates: Society of Economic Paleontologists and Mineralogists Short Course Notes No. 12, p. 4-1 to 4-138.

MUTTI, E., AND NORMARK, W. R., 1987, Comparing examples of modern and ancient turbidite systems: Problems and concepts, in Leggett, J. K., and Zuffa, G. G., eds., Marine Clastic Sedimentology: Concepts and Case Studies: Graham and Trotman, London, p. 1–38.

MUTTI, E., AND RICCI LUCCHI, F., 1972, Le torbiditi dell' Appenino settentrionale: Introduzione all' analisi di facies: Memorie della Societa Geologica Italiana, v. 11, p. 161–199.

MUTTI, E., SEGURET, M., AND SGAVETTI, M., 1988, Sedimentation and deformation in the Tertiary sequences of the southern Pyrenees; Am. Assoc. Petrol. Geol. Mediterranean Basins Conf. Guidebook Field Trip No. 7, Nice, Special Publication of the Institute of Geology, University of Parma, 157 p.

MUTTI, E., AND SONNINO, M., 1981, Compensation cycles: A diagnostic feature of sandstone lobes (abst.): International Association of Sedimentologists, 2nd European Meeting, Bologna, p. 120–123.

MUTTI, E., RICCI LUCCHI, F., SEGURET, M., AND ZANZUCCHI, G., 1984, Seismoturbidites: A new group of resedimented deposits: Marine Geology, v. 55, p. 103–116.

MYERS, R. A., AND PIPER, D. J. W., 1988, Seismic stratigraphy of Late Cenozoic sediments in the northern Labrador Sea: A history of bottom circulation and glaciation: Canadian Journal of Earth Sciences, v. 25, p. 2059–2074.

NOREM, H., LOCAT, J., AND SCHIELDROP, B., 1989, An approach to the physics and the modelling of submarine flow slides: Report of the Norges Geotekniske Institut 522090-2, 28 p.

NORMARK, W. R., 1970, Growth patterns of deep sea fans: Unpublished Ph.D. Dissertation, University of California, San Diego, 165 p.

NORMARK, W. R., 1985, Local morphologic controls and effects of basin geometry on flow processes in deep marine basins, *in* Zuffa, G. G., ed., Provenance of Arenites: NATO-ASI Series, D. Riedel, Dordrecht, p. 47–63.

NORMARK, W. R., AND DICKSON, F. H., 1976, Sublacustrine fan sedimentation in Lake Superior: American Association of Petroleum Geologists Bulletin, v. 60, p. 1021–1036.

NORMARK, W. R., AND GUTMACHER, C. E., 1988, Sur submarine slide, Monterey Fan, central California: Sedimentology, v. 35, p. 629–647.

NORMARK, W. R., AND SPIESS, F. N., 1976, Erosion on the Line Islands archipelagic apron: Effects of small scale topographic relief: Geological Society of America Bulletin, v. 87, p. 286–296.

NORMARK, W. R., GUTMACHER, C. E., CHASE, T. E., AND WILDE, P., 1985, Monterey Fan, Pacific Ocean, *in* Bouma, A. H., Normark, W. R., and Barnes, N. E., eds. Submarine Fans and Related Turbidite Systems: Springer-Verlag, New York, p. 79–86.

NORMARK, W. R., HESS, G. R., STOW, D. A. V., AND BOWEN, A. J., 1980, Sediment waves on the Monterey Fan levee: A preliminary physical interpretation: Marine Geology, v. 37, p. 1–18.

NORMARK, W. R., PIPER, D. J. W., AND HESS, G. R., 1979, Distributary channels, sand lobes and mesotopography of Navy submarine fan, California Borderland: Sedimentology, v. 24, p. 749–774.

NORMARK, W. R., PIPER, D. J. W., AND STOW, D. A. V., 1983, Quarternary development of channels, levees and lobes on middle Laurentian Fan: American Association of Petroleum Geologists Bulletin, v. 67, p. 1400–1409.

O'CONNELL, SUZANNE, RYAN, W. B. F., AND NORMARK, W. R., 1991, Evolution of a fan channel on the outer Mississippi Fan: evidence from side-looking sonar: *in* Weimer, P., and Link, M. H. (eds.), Seismic Facies and Sedimentary Proceses of Submarine Fans and Turbidite Systems: Springer-Verlag, New York, p. 365–382.

PARKER, GARY, 1982, Conditions for the ignition of catastrophically erosive turbidity currents: Marine Geology, v. 46, p. 307–327.

PARKER, GARY, FUKUSHIMA, Y., AND PANTIN, H. M., 1986, Self-accelerating turbidity currents: Journal of Fluid Mechanics, v. 171, p. 145–181.

PARKER, GARY, GARCIA, M., FUKUSHIMA, Y., AND YU, W., 1987, Experiments on turbidity currents over an erodible bed: Journal of Hydraulic Research, v. 25, p. 123–147.

PICKERING, K. T., AND HISCOTT, R. N., 1985, Contained (reflected) turbidity currents from the Ordovician Cloridorme Formation, Quebec, Canada: An alternative to the antidune hypothesis: Sedimentology, v. 32, p. 373–394.

PIPER, D. J. W., 1970, Transport and deposition of Holocene sediment on La Jolla deep sea fan, California: Marine Geology, v. 8, p. 211–227.

PIPER, D. J. W., AND AKSU, A. E., 1987, The source and origin of the 1929 Grand Banks turbidity current inferred from sediment budgets: Geo-Marine Letters, v. 7, p. 177–182.

PIPER, D. J. W., AND KONTOPOULOS, N., 1990, Large gravel waves of turbidite origin: Ancient analogues of features seen by sidescan sonar (abst.): Geological Association of Canada—Mineralogical Association of Canada Joint Annual Meeting, Vancouver, B.C., p. A107.

PIPER, D. J. W., AND NORMARK, W. R., 1982, Acoustic interpretation of Quaternary sedimentation and erosion on the channelled upper Laurentian Fan, Atlantic margin of Canada: Canadian Journal of Earth Sciences, v. 19, p. 1974–1984.

PIPER, D. J. W., AND NORMARK, W. R., 1983, Turbidite depositional patterns and flow characteristics, Navy submarine fan, California Borderland: Sedimentology, v. 30, p. 681–694.

PIPER, D. J. W., FARRE, J. A., AND SHOR, A. N., 1985, Late Quaternary slumps and debris flows on the Scotian slope: Geological Society of America Bulletin, v. 96, p. 1508–1517.

PIPER, D. J. W., SHOR, A. N., AND HUGHES CLARKE, J. E., 1988, The 1929 Grand Banks earthquake, slump and turbidity current: Geological Society of America Special Paper 229, p. 77–92.

PIPER, D. J. W., SHOR, A. N., FARRE, J. A., O'CONNELL, SUZANNE, AND JACOBI, ROBERT, 1985, Sediment slides and turbidity currents on the Laurentian Fan: Side scan sonar investigations near the epicenter of the 1929 Grand Banks earthquake: Geology, v. 13, p. 538–541.

PRAEG, D., AND SCHAFER, C. T., 1989, Seabed features of the Labrador slope and rise near 55°N revealed by SEAMARC I sidescan sonar imagery: Geological Survey of Canada Open-File Report 2028, 28 p.

PRIOR, D. B., AND BORNHOLD, B. D., 1989, Submarine sedimentation on a developing Holocene fan delta: Sedimentology, v. 36, p. 1053–1076.

PRIOR, D. B., AND BORNHOLD, B. D., AND JOHNS, M. W., 1986, Active sand transport along a fjord-bottom channel, Bute Inlet, British Columbia: Geology, v. 14, p. 581–584.

PRIOR, D. B., WISEMAN, W. J., AND BRYANT, W. R., 1981, Submarine chutes of the slopes of fjord deltas: Nature, v. 209, p. 326–328.

PRIOR, D. B., BORNHOLD, B. D., COLEMAN, J. M., AND BRYANT, W. R., 1982, Morphology of a submarine slide, Kitimat Arm, British Columbia: Geology, v. 10, p. 588–592.

PRIOR, D. B., BORNHOLD, B. D., WISEMAN, W. R., Jr., AND LOWE, D. R., 1987, Turbidity current activity in a British Columbia fjord: Science, v. 237, p. 1330–1333.

PRIOR, D. B., SUHAYDA, J. N., LU, N.-Z. BORNHOLD, B. D., KELLER, G. H., WISEMAN, W. J., WRIGHT, L. D., AND YANG, Z.-S. 1989, Storm wave reactivation of a submarine landslide: Nature, v. 341, p. 47–50.

REIMNITZ, ERK, 1971, Surf-beat origin for pulsating bottom currents in the Rio Balsas submarine canyon, Mexico: Geological Society America Bulletin, v. 83, p. 81–90.

RICCI, LUCCHI, FRANCO, AND VALMORI, E., 1980, Basin-wide turbidites in a Miocene, "over-supplied" deep-sea plain: A geometrical analysis: Sedimentology, v. 27, p. 241–270.

SANDERS, J. E., 1965, Primary sedimentary structures formed by turbidity currents and related resedimentation mechanisms, *in* Middleton, G. V., ed., Primary Sedimentary Structures and their Hydrodynamic Interpretation: Society of Economic Paleontologists and Mineralogists Special Publication 12, p. 192–219.

SAVOYE, BRUNO, AND PIPER, D. J. W., 1990, What controls on depositional architecture of the Var Canyon-Fan system? Evidence from sidescan sonar images and high resolution and very high resolution seismic profiles: Abstracts, 13th International Sedimentological Congress, Nottingham, p. 478.

SHANMUGAM, G., AND MOIOLA, R. J., 1988, Submarine fans: Characteristics, models, classification and reservoir potential: Earth Science Reviews, v. 24, p. 383–428.

SHEPARD, F. P., 1966, Meander in valley crossing a deep-ocean fan: Science, v. 154, p. 385–386.

SHEPARD, F. P., AND BUFFINGTON, E. C., 1968, La Jolla submarine fanvalley: Marine Geology, v. 6, p. 107–143.

SHEPARD, F. P., AND DILL, R. F., 1966, Submarine Canyons and other Sea Valleys. Rand McNally, Chicago, 381 p.

SHEPARD, F. P., DILL, R. F., AND VON RAD, U., 1969, Physiography and sedimentary processes of La Jolla submarine fan and fan-valley, California: American Association of Petroleum Geologists Bulletin, v. 53, p. 390–420.

SHEPARD, F. P., MCLOUGHLIN, P. A., MARSHALL, N. F., AND SULLIVAN, G. G., 1977, Current-meter recordings of low speed turbidity currents: Geology, v. 5, p. 297–301.

SHOR, A. N., PIPER, D. J. W., HUGHES CLARKE, J., AND MAYER, L. A., 1990, Giant flute-like scour and other erosional features formed by the 1929 Grand Banks turbidity current: Sedimentology, v. 37, p. 631–645.

STACEY, M. W., AND BOWEN, A. J., 1988a, The vertical structure of density and turbidity currents: Theory and observation: Journal of Geophysical Research, v. 93, p. 3528–3542.

STACEY, M. W., AND BOWEN, A. J., 1988b, The vertical structure of turbidity currents and a necessary condition for self-maintenance: Journal of Geophysical Research, v. 93, p. 3543–3553.

STOW, D. A. V., AND BOWEN, A. J., 1980, A physical model for the transport and sorting of fine-grained sediment by turbidity currents: Sedimentology, v. 27, p. 31–46.

SYVITSKI, J. P. M., BURRELL, D. C., AND SKEI, J. M., 1987, Fjords, Pro cesses and Products: Springer-Verlag, New York, 379 p.

SYVITSKI, J. P. M., AND FARROW, G. E. 1983, Structures and processes in bayhead deltas, Knight and Bute Inlets, British Columbia: Sedimentary Geology, v. 35, p. 217–244.

SYVITSKI, J. P. M., AND FARROW, G. E. 1989, Fjord sedimentation as an analogue for small hydrocarbon-bearing fan deltas, *in* Whateley, M. K. G., and Pickering, K. T., eds. Deltas: Sites and Traps for Fossil Fuels: Geological Society of London Special Publication 41, p. 21–43.

SYVITSKI, J. P. M., AND HEIN, F. J., 1991, Sedimentology of an Arctic Basin: Itirbilung Fjord, Baffin Island, Northwest Territories: Geological Survey of Canada Paper 91-11, 61 p.

SYVITSKI, J. P. M., SMITH, J. N., CALABRESE, E. A., AND BOUDREAU, B. P., 1988, Basin sedimentation and the growth of prograding deltas: Journal of Geophysical Research, v. 93, p. 6895–6908.

VAN TASSEL, J., 1981, Silver Abyssal Plain carbonate turbidite: Flow characteristics: Journal of Geology, v. 89, p. 317–333.

WALKER, R. G., 1973, Mopping up the turbidite mess, *in* Ginsburg, R. N., ed., Evolving Concepts in Sedimentology. Johns Hopkins University Press, Baltimore, p. 1–37.

WEIMER, P., 1989, Sequence stratigraphy of the Mississippi Fan (Plio-Pleistocene), Gulf of Mexico: Geo-Marine Letters, v. 9, p. 185–272.

WEIMER, PAUL, AND LINK, M. H., 1991, Seismic Facies and Sedimentary Processes of Submarine Fans and Turbidite Systems: Springer-Verlag, New York, 460 p.

WEIRICH, F. H., 1984, Turbidity currents: Monitoring their occurrence and movement with a three-dimensional sensor network: Science, v. 224, p. 384–387.

WEIRICH, H., 1988, Field evidence for hydraulic jumps in subaqueous sediment gravity flows: Nature, v. 332, p. 626–629.

WINN, R. D., Jr., AND DOTT, R. H., Jr., 1979, Deep-water fan-channel conglomerates of Late Cretaceous age, southern Chile: Sedimentology, v. 26, p. 203–228.

WRIGHT, L. D., WISEMAN, W. J., BORNHOLD, B. D., PRIOR, D. B., SUHAYDA, J. N., KELLER, G. H., YANG, Z.-S., AND FAN, Y. B., 1988, Marine dispersal and deposition of Yellow River silts by gravity-driven underflows: Nature, v. 332, p. 629–632.

TECTONIC EVOLUTION OF SUBMARINE CANYONS ALONG THE CALIFORNIA CONTINENTAL MARGIN

H. GARY GREENE,[1] SAMUEL H. CLARKE, JR.[1] AND MICHAEL P. KENNEDY,[2]

[1]*U.S. Geological Survey, 345 Middlefield Road, Menlo Park, California 94025; and* [2]*California Divison of Mines and Geology, Los Angeles, California*

Abstract: The development of submarine canyons along active-plate margins commonly is influenced by tectonic processes. Recent studies of submarine canyons along the transform margin of western North America show that the origin and subsequent evolution of many canyons are correlatable with plate motion and plate-margin deformation. Elements of canyon morphology such as bends and meanders commonly are controlled by faults and folds that are relatable to the structural fabric of the continental shelf and slope. Some canyon heads that appear to be displaced from their lower reaches are explainable as the result of movement along strike-slip faults associated with the plate margin.

Many submarine canyons along the California margin are not associated with large rivers and thus may owe their origins either to pre-Holocene fluvial or structural processes. Some modern canyons appear to be associated with pre-Pleistocene ancestral canyons. Because of both vertical and horizontal tectonic movements during the past 20 Ma, some California submarine canyons have been repeatedly filled and exhumed; the most recent exhumation began during the latest lowstand of sea level and continues today.

Canyons that today have their upper reaches on the continental slope or outermost shelf, distant from large rivers or other sources of sediment supply, commonly appear to have been laterally displaced along offshore faults. Palinspastic reconstructions along these faults commonly reveal a genetic relation between such canyons and canyons heading nearshore, from which they were offset. For example, detailed studies of the Ascension-Monterey Submarine Canyon system in Monterey Bay suggest that several smaller canyons on the outer shelf and upper slope have been displaced northwestward from the headward part of Monterey Canyon by right slip along offshore faults of the Palo Colorado-San Gregorio, Ascension, and Monterey Bay fault zones. Many other canyons on the California margin have developed along, or had their courses abruptly altered by, structural zones, owing either to canyon cutting along a zone of weakness or to fault displacement. Mass wasting associated with zones of faulting and slumping, which may have been seismically induced, also may affect canyon form. Clearly, submarine canyons along the California margin commonly owe their origin and morphologic development to influences other than fluvial erosion during sea-level lowstands. A chief influence has been the San Andreas fault system.

INTRODUCTION

Many different hypotheses for the origin of submarine canyons evolved during the 60 years before the popularization of plate-tectonic theory. For example, Spencer (1898, 1903) emphasized the role of glaciation, Smith (1902) stressed ocean currents, and Davis (1934) suggested that canyons are scoured by outflowing bottom currents or wave buildup. Daly (1936) and Kuenen (1937) proposed that canyons are cut by turbidity currents, Johnson (1939) that artesian sapping has a role, and Bucher (1940) that tsunamis are agents of submarine-canyon cutting. Most of these authors sought a single process to explain the origin of submarine canyons. Some more mobilistic concepts, such as continental uplift, as proposed by Hull (1912), and diastrophism, as proposed by Wegener (1924) and de Andrade (1937), were attributed possible roles in the origin of submarine canyons. The role of geologic structure in the development of submarine canyons was discussed by Shepard (1937, 1938b, c, 1957), Shepard and Emery (1941, 1946), Emery (1960), and Shepard and Dill (1966), and Starke and Howard (1968) subsequently suggested that submarine canyons are polygenetic in origin and illustrated this concept in their discussion of Monterey Canyon. However, before the advent of global tectonics and sea-floor-spreading theory, submarine canyons were thought to be formed primarily by fluvial processes, and most canyons to have been cut by subaerial erosion during times of Pleistocene lowstands of sea level (Shepard and Emery, 1941).

Nearly 60 submarine canyons cut the continental margin of California. Many (28) of these are known to be associated with structures that suggest that their origin and/or development was influenced by geologic structure (Shepard, 1937, 1938c, 1939; Shepard and Emery, 1941, 1946). Other canyons are suspected of being structurally controlled, on the basis of their morphology; but, pending further investigation to determine their origin, the role of tectonics in their evolution remains unclear.

The continental margin of California is mostly situated along the active, obliquely convergent transform boundary between the Pacific and North American crustal plates (Fig. 1). This boundary is defined by the San Andreas fault system, a zone of strike- and oblique-slip faults that ranges in width from 50 to 350 km and extends for more than 1,100 km along most of the length of California (Fig. 1). Since its inception during middle Miocene time (20 Ma), more than 300 km of right-lateral displacement have occurred along the San Andreas fault system (Turner and others, 1970; Matthews, 1973; Huffman and others, 1973).

Submarine canyons were formed, filled and exhumed along the tectonically unstable margin of California throughout the Neogene. Many canyons owe their origin and subsequent modification to Neogene tectonism, and some appear to have been displaced from their point of origin by right slip along the San Andreas fault system. Uplift and subsidence due to wrenching within this fault system can add yet another element of complexity to canyon history. Thus, active canyons along the California margin commonly have been modified considerably by sedimentary and deformational processes since their formation.

Profound differences exist between canyons formed on active-plate margins and those formed on passive margins. Submarine canyons on the passive margin of the Eastern United States commonly are straight, except for meandering lower-slope channels. By contrast, submarine canyons of the Western United States, especially those along the

From Shoreline to Abyss, SEPM Special Publication No. 46

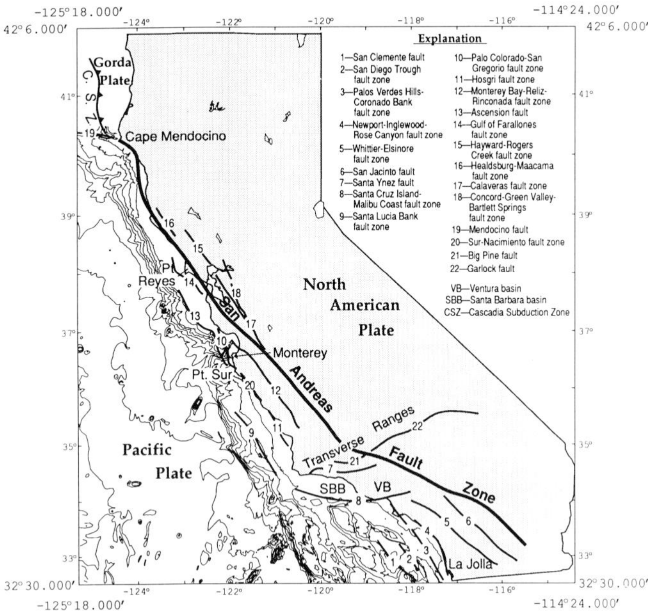

FIG. 1.—Major tectonic features along the California continental margin. Circled numbers refer to faults and fault zones listed on the figure. Contour interval is 500 m.

transform margin of California, exhibit widely varying channel morphology. We believe that these differences in morphology reflect the differences in tectonic environment between active and passive margins. In this report, we describe the histories of selected California submarine canyons to illustrate the characteristics of canyons formed along tectonically active margins.

TECTONIC SETTING

The northwest-trenching San Andreas fault system extends for more than 1,100 km from the northern Gulf of California to Cape Mendocino in northern California (Fig. 1). It is one of the most thoroughly mapped, longest, and most seismically active transform-fault systems in the world (Jennings and others, 1975; Wallace, 1990). The fault system can be divided into southern, central, and northern sections that differ from one another in breadth and orientation.

The southern section trends northwest and comprises a 350-km-wide zone of anastomosing, chiefly right-lateral, strike-slip faults (Fig. 1). Faults in this broad zone include the offshore San Clemente, San Diego Trough, Palos Verdes Hills-Coronado Bank, and Newport-Inglewood-Rose Canyon fault zones, as well as the onshore Whittier-Elsinore fault zone, the San Jacinto fault, and the San Andreas fault proper.

Farther north, the 70-km-wide central section is located in the area of the Transverse Ranges, where the San Andreas fault curves westward (Fig. 1). A series of west-trending faults occurs west of this bend that departs in orientation from the principle structural grain of the region. The most important of these faults in relation to canyon development are the onland Santa Ynez fault, which bounds the Santa Barbara and Ventura Basins along their northern margins, and the offshore Santa Cruz Island-Malibu Coast fault zone, which bounds these basins along their southern margin (Fig. 1).

The northern section trends northwest and is 50 to 100 km wide; it is widest south of the San Francisco Bay region (Fig. 1). Faults and fault zones composing the San Andreas fault system on the continental margin south of San Francisco include the offshore Santa Lucia Bank fault, the Palo Colorado-San Gregorio and Hosgri fault zones, the Monterey Bay-Reliz-Rinconada fault zone, and the San Andreas fault proper (Fig. 1). Except for parts of the Palo Colorado-San Gregorio, Hosgri, and Monterey Bay fault zones that exhibit thrust faulting, the San Andreas fault system in this region consists of subparallel, right-lateral, strike-slip faults (Jennings and others, 1975; Jennings, 1977; McCulloch, 1989a; Greene, 1977; Greene and others, 1989; McCulloch and Greene, 1990). In the Monterey Bay region, this system includes the offshore Ascension fault. North of San Francisco Bay, the San Andreas fault system comprises the offshore Gulf of Farallones fault zone, the onshore-offshore San Andreas fault, and the onshore Hayward-Rodgers Creek-Healdsburg-Maacama and Calaveras-Concord-Green Valley-Bartlett Springs fault zones (Fig. 1). These components are principally northwest-trending, right-lateral, strike-slip faults that are subparallel to the San Andreas fault zone (Jennings and others, 1975; Jennings, 1977; McCulloch, 1989b). The nature of the north terminus of the San Andreas fault system in the vicinity of Cape Mendocino is unclear. North of Cape Mendocino, the California continental margin is dominated by north-northwest- to north-trending reverse and thrust faults reflecting east-northeast- to east-directed compression associated with Gorda-North American plate convergence (Fig. 1; Clarke, 1990).

The San Andreas fault system is seismically active. Moderate earthquakes occur regularly along faults of the system and probably initiate mass movement on unstable slopes throughout the margin. Many submarine canyons off California appear to have been modified by this process.

The California continental margin has been extensively modified by faulting, and fault movement in this region has profoundly influenced canyon evolution. Some submarine canyons are formed by fault displacement, whereas others have been modified during their development by faulting. In some places, ocean currents and waves, as well as entrained sediment, may be funneled by fault-controlled points, such as those near Monterey and La Jolla (Fig. 1), to form powerfully erosive, sand-laden flows that contribute to canyon formation. Earthquake-generated turbidity currents also may be highly erosive.

Submarine canyons commonly are formed along faults because the shearing associated with fault displacement tends to fracture rock, so that it is less resistant to erosion. For example, nearly straight canyon axes, as exemplified by San Clemente Canyon, suggest fault control by high-angle strike-slip faulting (Shepard, 1938b). In addition, the apparent displacement of canyon heads from their original sources of sediment supply, as in the cases of Monterey-Ascension Canyons (Greene, 1977) and Scripps and La Jolla Canyons (Clarke and others, 1983), suggests control by strike-slip faulting, especially where active strike-slip faults are known to cut across canyon axes. Thus, canyon morphology commonly suggests the type of structural control of canyon development.

DISCUSSION

In the following discussion, we use the Ascension-Monterey Canyon system as a model for comparison with other canyons, primarily because it is one of the most studied canyons along the California margin (Davidson, 1897; Shepard, 1938a, d, 1941, 1948b; Shepard and Emery, 1941, 1946; Woodford, 1951; Dill and others, 1954; Martin, 1964; Shepard and Dill, 1966; Martin and Emery, 1967; Starke and Howard, 1968; Normark, 1970; Jensen, 1976; Greene, 1977, 1990; Greene and Clark, 1979; Normark and others 1980; Howell and others, 1980; Greene and others, 1989; Eittreim and others, 1989; Greene and Hicks, 1990). In addition, Monterey Canyon is the largest and, probably, the most tectonically complex of the California submarine canyons (Greene, 1990), and it was one of the first canyons extensively studied by Francis Shepard.

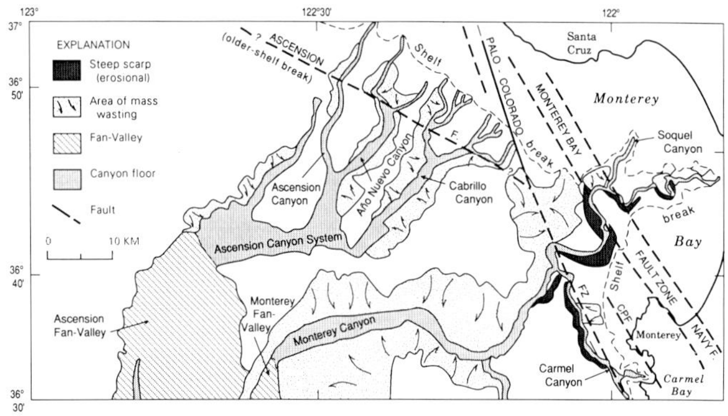

FIG. 2.—Geomorphic sketch map of Ascension-Monterey Canyon based on recently collected National Oceanic and Atmospheric Administration/ National Ocean Survey SeaBeam data, showing major faults of the Monterey Bay region. Faults are shown as heavy, solid lines where well located, dashed where inferred and queried where questionable. The light dashed line is modern shelf break. CPF indicates Cypress Point fault. Modified from Greene and others (1989) and Greene and Hicks (1990).

Submarine Canyons Associated with the Northern Section of the San Andreas Fault System

Ascension-Monterey Canyon System: A Model.—The undersea topography of the Monterey Bay region is as diverse as, and even more spectacular than, that seen on shore. In the words of Davidson (1897), who first described Monterey Canyon.

> ". . . at Monterey Bay there is a complete breaking down of the Coast Ranges from 25 miles [40 km] with the mountains receding well inland. Into this broad and deep bight heads the finest of submerged valleys."

Subsequent studies (Greene, 1977; Greene and Hicks, 1990) have established that the Ascension-Monterey Canyon system that comprises Monterey Submarine Canyon, Monterey Fan valley, and the Monterey Fan dominates the submarine topography and exhibits much greater relief than canyon systems on shore.

Monterey Canyon is generally sinuous; its axis deeply incises the floor of Monterey Bay and the adjacent upper continental slope (Fig. 2). Three distinct meanders have been identified in the upper reaches of the canyon, each apparently the result of a somewhat different process (Greene, 1990).

The axis of Monterey Canyon has a distinct "gooseneck" meander about halfway between the canyon head and the interesection of Monterey and Soquel Canyons (Fig. 2). Recently collected SeaBeam data (Greene and others, 1989; Greene and Hicks, 1990) reveal that an ancient landslide has dammed the axis of Monterey Canyon at this point and that sediment has ponded immediately upstream from the slide. The meander apparently resulted from breaching of this dam and subsequent erosion around the slump mass.

A second well-defined meander also incises Neogene sedimentary and Mesozoic granitic basement rocks in the vicinity of the intersection of Soquel and Monterey Canyons. Upstream and downstream reaches of this meander trend northwest, and the orientation of these reaches has been interpreted as fault controlled (Greene, 1977; Greene and Hicks, 1990). Several faults of the Monterey Bay fault zone that cross the canyon in this area have moved resistant basement rocks into the canyon axis during the past several million years (Greene, 1977). Turbidity currents have flowed along the bounding faults and eroded around the margin of this resistant structural block, forming the meander. A large slump mass along the north wall of Monterey Canyon at the apex of this meander, between faults of the Monterey Bay fault zone (Fig. 2), may have resulted from fault movement.

The downstream reach of this structurally controlled meander curves into a third meander, which is gently bowed and has prominent walls along its south side (Fig. 2). This meander begins upstream, where the axis of the canyon has been offset westward by faulting, and in it the trend of the canyon axis curves from northwest to nearly westward. The meander is abruptly terminated just upcanyon from the confluence of Monterey and Carmel Canyons, where the axis of Monterey Canyon makes a right-angle turn to trend southward toward the mouth of Carmel Canyon. This abrupt change in trend of the axis of Monterey Canyon is at the approximate point where the canyon axis is crossed by the Palo Colorado-San Gregorio fault zone (Greene and others, 1973; Greene, 1977), and we propose that the orientation of the canyon here is the result of faulting.

Carmel Canyon is relatively straight in comparison with Monterey Canyon. It has three heads, two of which are located in Carmel Bay (Fig. 2). Both of these canyon heads cut Cretaceous granitic rock (biotite granodiorite porphyry of the Monterey mass of Ross, 1976). The first two heads merge just outside Carmel Bay and continue for 2 to 3 km westward from shore to join the principal canyon axis in a right-angle turn to the northwest. From this intersection the main canyon axis continues northwestward. The third canyon head lies 3 km south of this intersection, on trend with the principal axis of Carmel Canyon (Fig. 3). The change in axial trend from westward to northwestward, as well as the orientation of the third canyon head, is believed to be

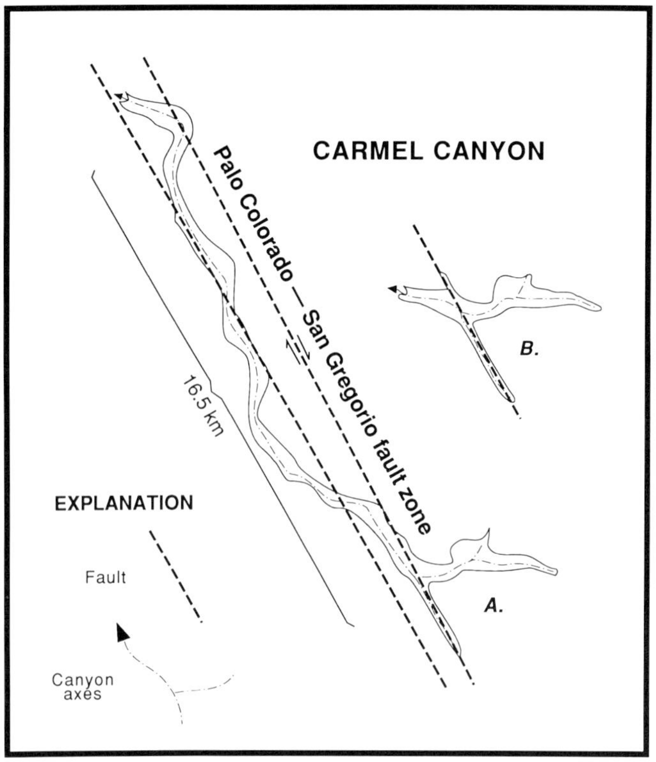

FIG. 3.—Schematic illustration of proposed Carmel Canyon head offset along the Palo Colorado-San Gregorio fault zone. (A) Present configuration of Carmel Canyon. (B) Proposed past alignment of the mouth and heads of Carmel Canyon when 16.5 km of right-lateral slip on the Palo Colorado-San Gregorio fault zone is removed. From Greene and Hicks (1990).

fault controlled because they occur along faults of the Palo Colorado-San Gregorio fault zone (Fig. 3; Greene and others, 1973; Greene, 1977; McCulloch and Greene, 1990). The northwest-trending segment of the canyon axis is aligned with this fault zone, and the canyon head in Carmel Bay appears to have been offset right laterally along the fault zone. Near the intersection of Monterey and Carmel Canyons, the axis of Carmel Canyon is oriented generally east-west, similar to its orientation in Carmel Bay. Greene and others (1989) proposed that these west-trending segments were once aligned but were subsequently offset from each other a distance of 16.5 km by right slip along the Palo Colorado-San Gregorio fault zone (Fig. 3).

At its intersection with Monterey Canyon, the trend of Carmel Canyon changes from northwesterly to southerly as the canyon broadens into a rounded, gently dipping trough. Shepard and Dill (1966) and Greene (1977) proposed that Carmel Canyon joins Monterey Canyon in a distinct hanging valley. Although this hanging valley is not evident from bathymetric data, a steep, rounded chute has been defined by recently collected SeaBeam data (Fig. 2). Moreover, recent *Alvin* dives suggest that steep, stairstep-like cliffs or scarps are present, reminiscent of a modified hanging-valley geomorphology (C. Harrold, pers. commun., 1988).

The northeast wall of Carmel Canyon is dissected by a few relatively straight, west-dipping drainage channels and a few slumps. In contrast, the opposite wall of the canyon is characterized by steep cliffs that extend to the crest of the bedrock ridge. This difference in canyon-wall morphology may be related to differing bedrock lithologies, with the northeast wall composed of resistant Cretaceous granitic rocks and the southwestern wall and adjacent bedrock ridge composed of Jurassic metamorphic rocks and Cretaceous sandstone assigned to the Franciscan Complex (Greene, 1977).

The Ascension Canyon system, which generally trends southwest, is composed of three major distributary channels (Fig. 2). Recently collected SeaBeam data disclose that the system has a total of 11 heads, in contrast to the nine heads reported previously by Greene (1977) and the eight heads reported by Nagel and others (1986). The southeastern, main channel (also known as the Cabrillo Canyon system) bifurcates upslope to form two arms, each of which, in turn, divides into four heads that extend almost to the distal edge of the continental shelf.

The southeastern set of four canyon heads is approximately equally spaced, separated from one another by a distance of about 1.8 km. The southernmost head in this set is unusual in that, downcanyon, it initially trends southeast, then southwest, and then west-northwest. This boxlike configuration appears to result from faulting, because the Palo Colorado-San Gregorio fault zone is aligned with the most shelfward, southeast-trending head, and the change in trend from southwestward to northwestward occurs along the trace of the Ascension fault (Fig. 2). The other canyon heads in the southern set are relatively straight. Ascension fault control is also expressed as "kinks" in the distributary channels associated with the heads of the Ascension submarine-canyon system.

Monterey Canyon is situated mostly on the Salinian block of Page (1970). The canyon appears to have been formed subaerially and subsequently offset northward a considerable distance by right-lateral movement along the San Andreas and associated strike-slip faults (Greene, 1977; Greene and others, 1989; Greene and Hicks, 1990). Monterey Canyon is an old feature, because the deep gorge that extends just onshore from the canyon head is filled with 10-million-year old (Miocene) sedimentary rocks (Starke and Howard, 1968).

Monterey Canyon is being exhumed, and this exhumation reflects Holocene tectonic uplift. Earthquakes in this region probably initiate slumping and turbidity flows within the canyon. In addition, continued strike-slip faulting probably is displacing a segment of the deeper part of the canyon northwestward (Greene and Hicks, 1990).

Tectonic Evolution: The Role of Faulting in the Formation of the Ascension-Monterey Canyon System.—The origin of submarine canyons that are limited to the continental slope or that just notch the continental shelf, far removed from a major sediment source, is problematic and has led to much speculation (Shepard and Dill, 1966). In areas where clear evidence of a fluvial origin for such "headless" submarine canyons is absent and strike-slip faulting is known to have been active, palinspastic reconstructions can, in many places, demonstrate alignments of canyon axes with features that formerly could have supplied sediment and/or focused the erosional processes needed for canyon formation. Such reconstructions have been proposed to explain the present day canyon configurations and fault patterns in the Monterey Bay region (Greene and others, 1973; Greene, 1977), and they may be useful in other continental margins known to have been subjected to strike-slip faulting. In the following discussion, evidence is presented to show that the headless canyons on the continental slope north of Monterey Bay, two anomalous meanders in Monterey Canyon, and the abrupt change in orientation of lower Monterey and Carmel Canyons have resulted from right-lateral faulting (Greene, 1977). Presently inactive(?) submarine canyons on the continental slope northwest of Monterey Bay appear to have been displaced from their original positions near the lower part of Monterey Canyon. Meanders in Monterey Canyon and the change in course of lower Monterey and Carmel Canyons probably have resulted from erosion along fault zones where associated shear has weakened the rocks.

Right-lateral strike slip began in southern California Continental Borderland approximately 20 Ma (Atwater, 1970). In the Monterey Bay region, motion along the Ascension fault likely preceded movement on the San Andreas fault (Greene, 1977). The lower part of Monterey Canyon was eroded subaerially during Zemorrian uplift in the Monterey Bay region and then was displaced to the north by right slip on the Ascension fault (Fig. 4a; Clark and Rietman, 1973; Greene, 1977). The displaced canyon, inactive and buried at the time of its displacement, was later rejuvenated through excavation as Pioneer Canyon, which is located 110 km northwest of the present Monterey Canyon, during the Pleistocene (Greene, 1977).

The unnamed, headless submarine canyon that is today on the continental slope between Pioneer and Ascension Canyons formed between middle and late Miocene time in the same way as Pioneer Canyon (Fig. 4b; Greene, 1977). The presence of unfaulted Pliocene(?) and older sediment overlying the Ascension fault in seismic-reflection profiles suggests that activity along this fault ceased between 14 and 10 Ma. At that time, stress was apparently transferred eastward, and strike slip was initiated along the Palo Colorado-San Gregorio fault zone (Fig. 4c, d).

Faults within the Monterey Bay fault zone are shown in seismic-reflection profiles to displace Quaternary and older rocks (Greene, 1977). The close age and geometric relation of the Monterey Bay fault zone to the Palo Colorado-San Gregorio fault zone (Fig. 1) imply that activity began nearly contemporaneously on both fault zones (Fig. 4e).

Physiographic development of the modern Monterey Bay region began with initiation of activity along faults within the Palo Colorado-San Gregorio and Monterey Bay fault zones. Monterey Peninsula, a sliver between two active faults (Navy and Cypress Point faults, Fig. 2), may have begun to migrate northwestward at that time (Fig. 4e, f). Right-lateral displacement along the Palo Colorado-San Gregorio fault zone moved the older channels of lower Monterey Canyon and the unnamed buried canyon northwestward, away from the upper parts of these channels. Ascension, Año Nuevo, and Cabrillo Canyons (Figs. 2, 4e), all northwest of the present Monterey Canyon, may have formed in this way. They all developed seaward of the upper reach of Monterey Canyon and were displaced from it by progressive offset along the Palo Colorado-San Gregorio fault zone. Canyons and fans west of the Palo Colorado-San Gregorio fault zone were displaced northwestward throughout late Tertiary and Quaternary time, and offset along faults of the Palo Colorado-San Gregorio and Monterey Bay fault zones continues today (Fig. 4g, h).

Total offset on this fault system of approximately 110 km is suggested by the present separation of Monterey and Pioneer Canyons (Fig. 4h), although work by Clark and others (1984) suggests a total displacement of about 180 km. This offset has taken place since early Miocene time, since the cessation of Farallon plate subduction and the initiation of northward migration of the Mendocino triple junction (McKenzie and Parker, 1967; Morgan, 1968; Atwater, 1970). If Pioneer, unnamed, Ascension, Año Nuevo, and Carbillo Canyons have been displaced from Monterey Canyon as indicated, right-lateral motion appears to have accelerated since about 7 Ma. Such an accelerating offset would explain the relatively close spacing (1 to 4 km) of the heads of Ascension, Año Nuevo, and Cabrillo Canyons relative to that (about 30 km) between those of Pioneer and Ascension Canyons. The greater distance between the heads of Pioneer and Ascension Canyons resulted from offset along Ascension fault before about 7 Ma. Other evidence also suggests a change in offset rate during late Miocene and Pliocene time. An increase in offset rate along coastal and near-coastal faults beginning about 12 to 10 Ma, as proposed by Hein (1973) and Fox and others (1985), may coincide with the transfer of stress from the Ascension fault eastward to the Palo Colorado-San Gregorio fault zone. A change in the direction of motion between the North American and Pacific plates at about that time also was postulated by Silver (1974).

As previously mentioned, the meanders in Monterey Canyon were probably initiated by right-lateral slip along faults associated with the Palo Colorado-San Gregorio fault zone. The meander within the Monterey Bay fault zone (Fig. 2) appears to be a consequence of erosion in areas weakened by faulting, and it is enhanced by the presence of resistant basement rocks at shallow depths within a fault-bounded sliver that is skirted by the meander.

Buried, headless canyons that had been displaced from their upper reaches along offshore faults were subsequently exhumed during the lowstand of sea level in Wisconsin(?) time. Differential compaction of the older fill in these canyons probably formed depressed axial troughs that would be preferentially followed by streams transporting sediment to the Wisconsin shoreline (near the edge of the present continental shelf). Excavation of these canyons continued throughout the Pleistocene and may be continuing today in Pioneer, Ascension, Año Nuevo and Cabrillo Canyons.

Sur Canyon.—Sur Canyon is a major submarine canyon 60 km south of Monterey Canyon. The several heads of Sur Canyon cut the southern part of the Sur platform approximately 5 km south of Point Sur and 5 km offshore (Figs. 1 and 5). The Sur platform is a flat-topped bedrock surface composed of metamorphic rock of the Franciscan Complex and the Sur Series (Greene, 1977). This platform projects westward from the coastline at Point Sur, and the upper

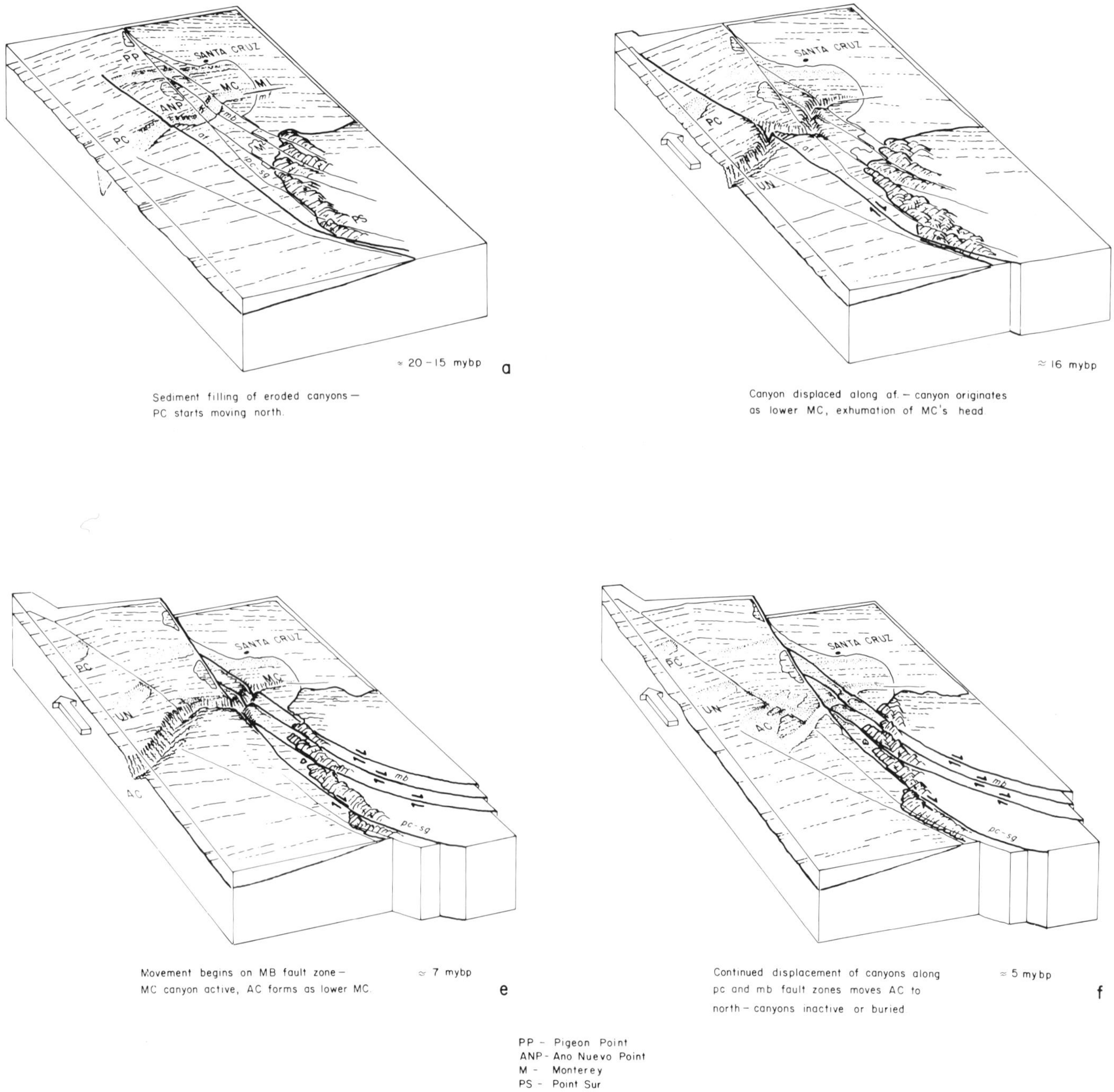

FIG. 4.—Palinspastic reconstructions, showing the tectonic evolution of the Ascension-Monterey Submarine Canyon system. Monterey Canyon was progressively offset along faults of the Palo Colorado-San Gregorio, Ascension, and Monterey Bay fault zones during the period between 20 and 0.2 Ma. Modified from Greene (1977) and Howell and others (1980). Physiography by Tau Rho Alpha.

flanks of the platform radiate outward conically. Except where it is cut by submarine canyons, the upper part of this cone is smooth and relatively undissected (Fig. 5).

Sur Canyon is sinuous but trends generally west for more than 50 km from its head to the point where the main channel broadens and bends abruptly southward (Fig. 5). This channel has three major meanders, two near the head of the main channel and one 14.5 km farther downchannel. The canyon floor averages about 0.25 km in width between the upper meanders and the abrupt southward bend, where the channel widens to 0.75 km.

The origin of the meanders in the upper part of the channel is unclear. However, an anticlinal ridge among several northwest-trending folds in this area (Fig. 5; McCulloch and Greene, 1990) may have influenced development of the second meander. The upper and lower reaches of this meander trend northwest and are aligned along the synclinal axes flanking this anticline. Farther downcanyon, near

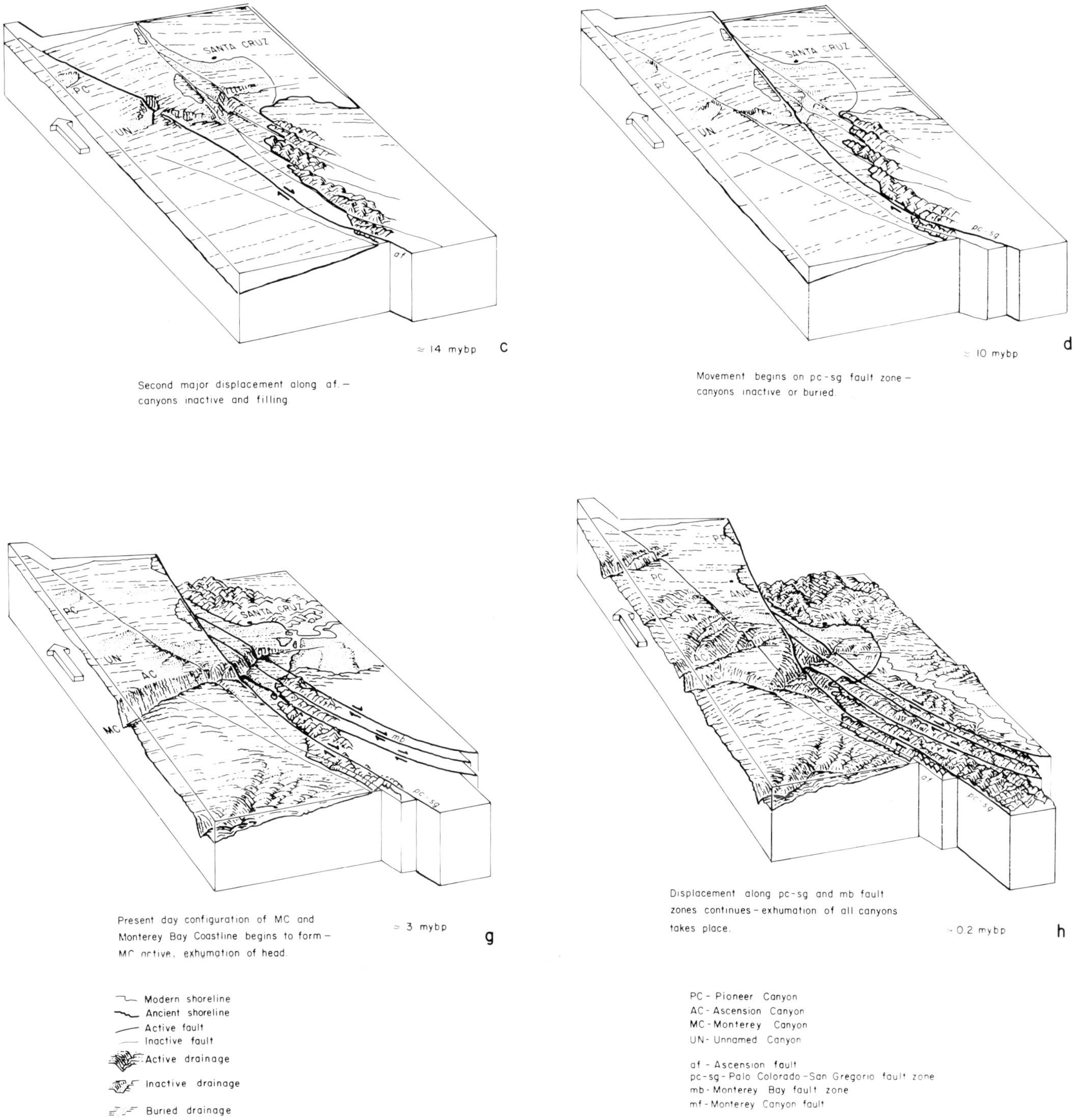

the middle of the west-trending reach of the canyon, is a spectacular example of a meander produced by slumping (Fig. 5). Here, a large slump (more than 2.5 km^2 in area) has dropped to the canyon floor from the north wall of the canyon. The tabular block apparently obstructed the canyon axis, so that turbidity currents that sweep the canyon were forced to erode around the slump mass. The distinct square shape and steep cutbanks of this meander reflect the form of the slump and strong erosive forces.

Previous geologic mapping in the region (Greene, 1977; McCulloch and Greene, 1990) shows that several north-trending faults or fault zones cross the canyon near the abrupt southward bend in the canyon axis (Fig. 5). North-trending cliffs that may be associated with faulting are present along the east side of the canyon below this bend (Fig. 5). However, these cliffs are distorted by landslides, and thus their relation to structure cannot be clearly determined from bathymetry alone. Terracelike areas also are present at several

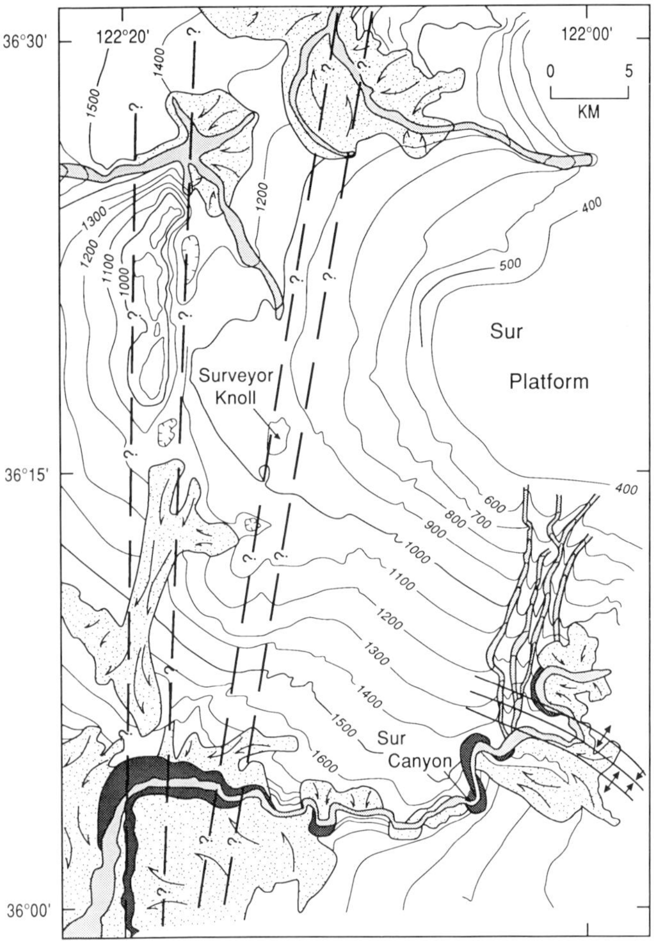

FIG. 5.—Geomorphic sketch map of Sur Canyon and the adjacent continental slope based on recently collected National Oceanic and Atmospheric Administration/National Ocean Survey SeaBeam data showing structural relationships of faults, folds and slumps to the submarine canyon morphology. Explanation of symbols is shown in Figure 2. Contour interval is 100 m. Faults are shown as heavy, dashed lines where inferred, queried where questionable. Folds are shown as thin solid lines with arrows. Modified from Greene and others (1989).

levels along the line of cliffs, suggesting back-and-forth migration of the active channel, as well as net westward migration of the channel axis.

Using SeaBeam bathymetry, Greene and others (1989) have inferred the presence of two zones of faulting: one crossing the Sur channel at its abrupt southward bend and the second crossing the channel about 5 km farther east (Fig. 5). The eastern fault of the set nearest the channel bend extends northward through a north-trending linear swale into the head of an unnamed canyon or gully. This fault, as mapped from bathymetry, is more than 50 km long. It may be expressed to the south by the upper cliff along the east wall of the south-trending main channel of Sur Canyon (Fig. 5). Near its north end, the fault is aligned with a north-trending distributary channel and with a depression resembling a sag pond.

The western fault in this zone is questionably identified but was interpreted by Greene and others (1989) to extend more than 20 km northward from the major bend in Sur Canyon to the west flank of a prominent bedrock ridge. It, too, may be expressed bathymetrically in the cliff immediately east of the south-trending main channel of Sur Canyon (Fig. 5). The abrupt southward bend in this channel may have resulted from uplift along the east side of the fault. The total length of this fault exceeds 40 km.

The eastern fault zone also is composed of two faults but is less well defined than the fault zone to the west (Fig. 5). The western fault in this zone is bathymetrically expressed by a steep, west-facing scarp along the west flank of Surveyor Knoll. The fault trends north-northeast and is aligned with and extends into an unnamed canyon or gully that debouches into Monterey Canyon. It is extended southward from Surveyor Knoll on the basis of aligned bathymetric features (e.g., bathymetric highs, notches in contours, heads of slumps) to cross Sur Canyon 5.5 km east of the southward bend of the main channel.

The eastern fault of this zone is bathymetrically defined by a topographic saddle between Surveyor Knoll and the upper continental slope. The fault extends southward to cross Sur Canyon about 7 km east of the bend (Fig. 5). Throw on the fault is interpreted to be up to the west, and the sea knoll is interpreted as a horst. The fault may extend northward into an area of mass wasting associated with an unnamed tributary to Monterey Canyon, where it appears to truncate the head of a small channel. A series of topographic highs, possibly composed of bedrock, is linearly distributed between the eastern and western faults of this zone, between Surveyor Knoll and Sur Canyon. Farther south, Sur Canyon may be offset slightly by the fault zone (Fig. 5).

Bodega Canyon.—Bodega Canyon is a small slope canyon that heads near the shelf edge in about 110 m of water northwest of Point Reyes (Fig. 1). The upper part of this canyon trends northwest, parallel to the San Andreas fault zone (Fig. 6). At a depth of about 500 m, however, the canyon makes a sharp, nearly right-angle turn to the west and continues due west from there, cutting the continental rise at a depth of about 3,500 m.

The axis of the northwest-oriented, headward part of Bodega Canyon is aligned with a syncline and fault (here called the Bodega Canyon fault) mapped along the northeast edge of Cordell Bank (Fig. 6; Wagner and Bortugno, 1982; McCulloch, 1987, 1989a, b; McCulloch and Greene, 1990). The Bodega Canyon fault is part of a larger, northwest-trending fault-and-fold belt (here called the Gulf of Farallones fault zone; Fig. 1) that extends along the central axis of the Gulf of Farallones and that merges with, or is truncated by, the San Gregorio fault zone farther to the south (McCulloch, 1987, 1989a, b; McCulloch and Greene, 1990). The Bodega Canyon fault and Bodega Canyon define the north and northeast edges of the granitic Cordell Bank.

Bodega Canyon appears to be fault controlled, carved along a synclinal depression and a fault zone. Erosion followed a zone of incompetent faulted rock, skirting the adjacent, resistant granitic rocks of the north tip of Cordell Bank. The age of Bodega Canyon is unknown; however, large-scale displacement is thought to have occurred along the San Gregorio fault zone since early Miocene time, and

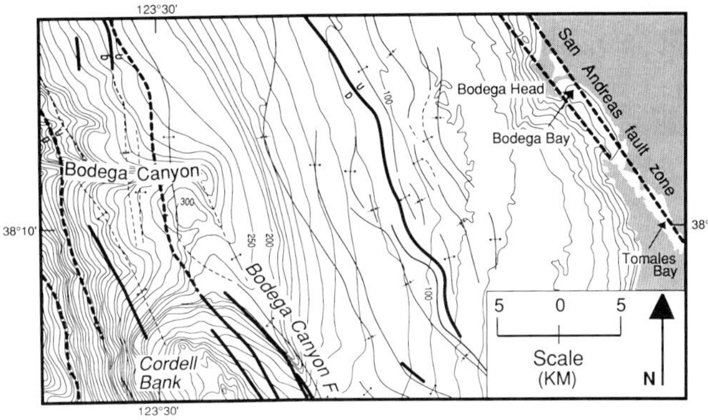

FIG. 6.—Bathymetry and structure of the Bodega-Tomales Bay area, showing structural control of Bodega Canyon. Faults are shown as heavy, solid lines where well located, dashed where inferred, and dotted where concealed. Folds are shown as solid, thin lines where well located, and dashed where inferred. Contour interval is 10 m at depth from 0 to 200 m, and 50 m at depths greater than 200 m. Structure symbols same as Figure 5. Modified from McCulloch (1989b) and McCulloch and Greene (1990).

this canyon may have originally been cut farther to the southeast.

Navarro and Arena Canyons.—Offshore from Point Arena are two small canyons that head at the upper slope, just beneath a well-defined bathymetric break marking the shelf edge (Fig. 7). These canyons are oriented generally east-west and extend downslope to a water depth of about 3,300 m. The head of Navarro Canyon lies in 500 m of water, and the head of Arena Canyon notches the shelf edge at a water depth of about 250 m.

The headward part of Navarro Canyon may be structurally controlled because it is aligned with the Navarro structural discontinuity of McCulloch (1987, 1989b), which delineates a break in the magnetic-anomaly pattern and Neogene structural grain. Arena Canyon heads near, and swings to a trend subparallel to, the discontinuity (Fig. 7). The influence, if any, of this discontinuity on canyon development is unclear; the Navarro structural discontinuity may mark the north tip of the granitic Salinian terrane (Silberling and others, 1984; Howell and others, 1985; McCulloch, 1987, 1989b). If so, early development of Navarro Canyon may have been directed northwestward, then westward, around resistant granitic rock that underlie the outer shelf and slope to the southeast.

Noyo Canyon.—Noyo Canyon notches the seaward edge of the continental shelf in about 150 m of water and cuts the adjacent continental slope northwest of Fort Bragg (Fig. 8). This canyon has a short (5 km) headward reach that trends northwest before making an abrupt right-angle bend to the west and continuing westward to the base of the continental slope to a depth of about 3,000 m.

The pronounced hook in the head of Noyo Canyon probably results from fault movement, because the San Andreas fault trends along this part of the canyon axis (McCulloch, 1989b). Like Bodega Canyon to the south, the headward part of Noyo Canyon incises rocks that have been sheared by movement along the San Andreas fault and thus are relatively easily eroded. The age of this canyon is unknown, although right slip along the San Andreas fault may have displaced its lower reaches from a point farther southeast.

Mendocino-Mattole Canyon System.—The Mendocino-Mattole Canyon area, which is underlain by the Mendocino triple junction—the intersection of the Gorda, Pacific, and North American crustal plates (Fig. 1)—is tectonically complex (Clarke, 1988; McLaughlin and others, 1988). To the north, generally northeast-southwest convergence between the Gorda and North American plates is accommodated by subduction of the Gorda plate along the Cascadia subduction zone, accompanied by deformation of the leading edge of the overriding North American plate. To the south, Pacific-North American plate relative motion is accommodated by right-lateral strike slip along faults of the San Andreas fault system, much of it along faults of the Mattole shear zone (Fig. 9) along the east flank of the King

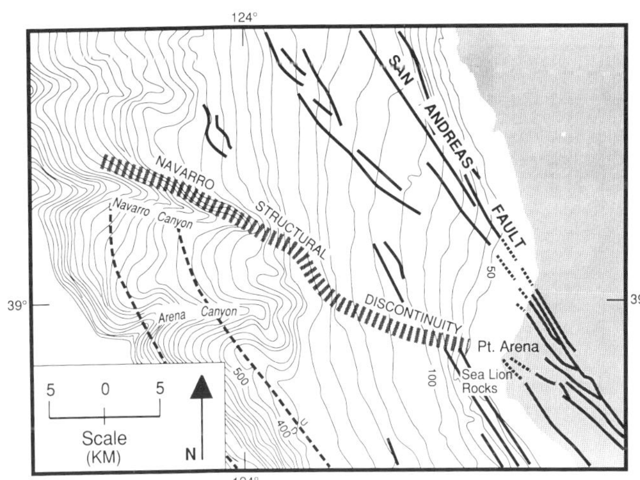

FIG. 7.—Bathymetry and structure of the Navarro and Arena Canyon area, showing relation of canyons to major structural features. Contour interval is 10 m at depth of 50 m to 200 m, and 50 m at greater depths. Faults are shown as heavy, solid lines where well located, dashed where inferred, and dotted where concealed. Modified from McCulloch (1989b).

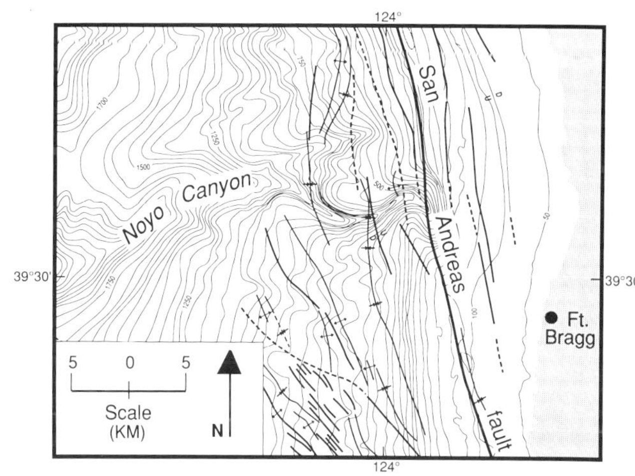

FIG. 8.—Bathymetry and structure in the Noyo Canyon area, showing relation between the San Andreas fault system and canyon morphology. Contour interval is 50 m. Structure symbols same as Figure 6. Modified from McCulloch (1989b).

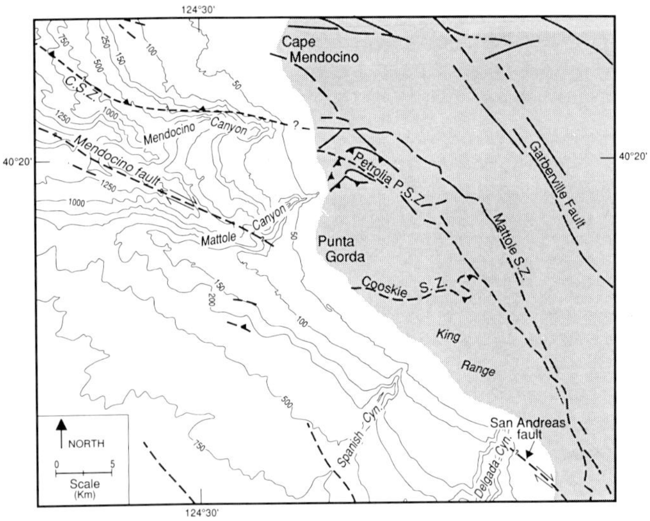

FIG. 9.—Bathymetry and structure of the Cape Mendocino area, showing structural control of the Mendocino-Mattole Canyon system. CSZ indicates seaward edge of the Cascadia subduction zone. Contour interval is 50 m. Faults are shown as heavy, solid lines where well located, dashed where inferred, and queried where questionable; arrows indicate sense of movement and thrust faults are indicated by barbs on upper plate. Modified from Clarke (1990).

Range, and farther east along the Garberville and Lake Mountain (not shown on Fig. 9) fault zones (Lisowski and Prescott, 1989). Relative motion between the Pacific and Gorda plates is accommodated by right slip on the Mendocino fault.

Mendocino and Mattole Canyons cut deeply into the continental shelf and head at or near the shoreline between Cape Mendocino and Punta Gorda. These canyons converge at a depth of about 1,500 m and continue westward as a single channel along the base of the Gorda Escarpment. Potential relations between structure and canyons in this area, as well as their tectonic implications for the submarine topography of California, were first discussed by Shepard (1937, 1983c, 1939, 1957).

Mendocino Canyon has a generally arcuate form that is convex to the north (Fig. 9). The canyon head lies in 50 m of water approximately 3 km from shore. Mendocino Canyon trends southwest until it reaches a depth of about 1,250 m, where the canyon bends sharply northwest for 6 km before turning westward and converging with Mattole Canyon.

The head of Mattole Canyon lies 0.5 km offshore in a water depth of only 20 m. It extends southwestward for about 6.5 km before abruptly bending to the northwest for 3 km, then turning southwestward again. At 750 m of water depth, Mattole Canyon again bends northwestward and continues downslope for 16 km, where it makes a sharp turn to the north to converge with Mendocino Canyon (Fig. 9).

The headward part of Mendocino Canyon is cut into extensively sheared rocks where north-dipping thrust faults of the Cascadia subduction zone extend across the upper continental slope and shelf; its position is thus structurally controlled (Clarke and Field, 1989; Clarke, 1990). Elements of this fault zone extend landward as the Petrolia shear zone (Clarke, 1988; McLaughlin and others, 1988). Likewise, linear west- to northwest-trending parts of Mattole Canyon are developed along the landward extension of the Mendocino fault, which is represented on land along the north flank of the King Range by the Cooskie shear zone (Clarke and Field, 1989; Clarke, 1988, 1990; McLaughlin and others, 1988).

Eel Canyon.—Eel Canyon cuts the northern California margin off the Eel River, 20 km southwest of Humboldt Bay (Fig. 10). The landward end of this canyon is bowl-shaped, more than 15 km wide, and has at least five heads that cut the continental shelf at a water depth of 60 m, approximately 7 km from the coast. A distinct, right-angle meander occurs on the continental slope about 15 km downslope from the canyon head. Here, the west-trending canyon axis bends southward for 7 km and westward for 4 km before bending sharply northward for 7 km and then turning westward to extend onto the lower continental slope.

Major structural zones appear to have controlled the development of submarine canyons that cut the continental shelf in the southern Eel River basin area. Eel Canyon was developed in sheared rocks, largely argillite and graywacke, of the Coastal belt of the Franciscan Complex where these rocks are cut by the Russ fault and other faults of the False Cape shear zone (not shown) mapped onshore (Clarke, 1990). The canyon is bounded on both the north and south by north-dipping thrust faults (Clarke and Field, 1989; Clarke, 1990). The bowl-shaped head of the canyon probably is sheared throughout, and the shape of the canyon head may be the result of mass wasting within these rocks.

The pronounced meander in Eel Canyon may result from erosion around a more resistant mass of rock that has been structurally emplaced. The upslope leg of this meander is obliquely aligned with a zone of thrust faulting (Clarke and Field, 1989; Clarke, 1990). The canyon may have encountered a resistant ridge and followed more easily eroded, faulted rocks around this resistant mass. The downslope leg of the meander was influenced by northwest-trending growth folds that caused the canyon to bend northward before resuming its westward trend to the base of the continental slope.

Submarine Canyons Associated with the Transverse Segment of the San Andreas Fault

Arguello Canyon.—Arguello Canyon is a linear, northeast-trending submarine canyon. It has five well-defined heads that notch the distal edge of the continental shelf just west of Point Arguello at a depth of about 110 m (Fig. 11). These heads trend slightly more westerly than the main channel and converge with it at depths between 1,500 and 1,750 m. The main channel bends southwestward at a depth of about 1,750 m, forming a fairly distinct crook. Arguello Canyon is unusual among California submarine canyons in that levees are well developed in its headward parts. These levees occur on the southwest-dipping continental slope at depths ranging from about 250 to 1,300 m (Clarke and others, 1983; McCulloch, 1989c).

Arguello Canyon is situated in a region where the tec-

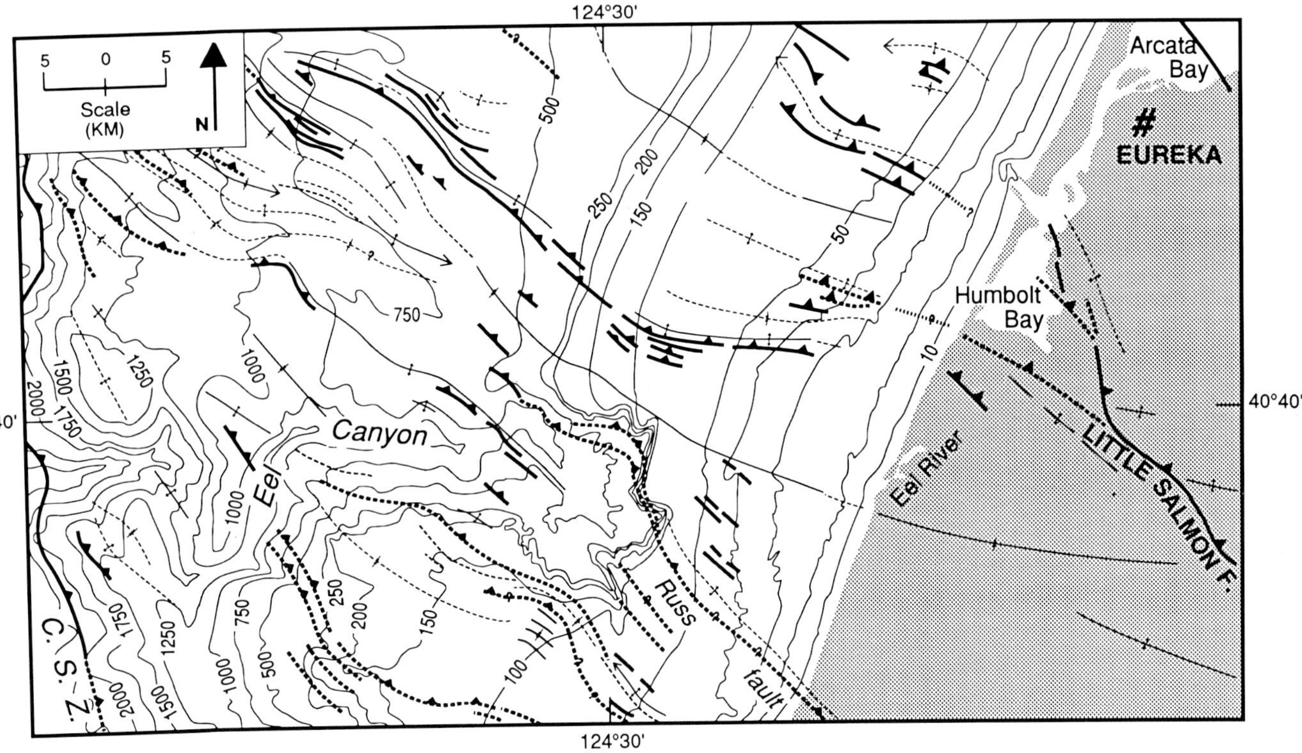

FIG. 10.—Bathymetry and structure offshore from the Eel River, showing structural control of Eel Canyon. CSZ indicates seaward edge of Cascadia subduction zone. Bowl shape of Eel Canyon head suggests active mass wasting. Contour interval is 10 m to 50 m depth, and 50 m at greater depths. Structural symbols same as Figure 6, except that thrust faults are indicated by barbs on upper plate. Modified from Clarke (1990).

tonics associated with the San Andreas fault system change from predominantly compressional south of the Santa Ynez fault to predominantly strike slip to the north. Many low-angle reverse and thrust faults have been mapped in the Point Arguello-Point Conception area (Clarke and others, 1983; McCulloch, 1989c). Clarke and others (1983) have defined three major imbricate thrust sheets or nappes that vary considerably in complexity; some are severely folded

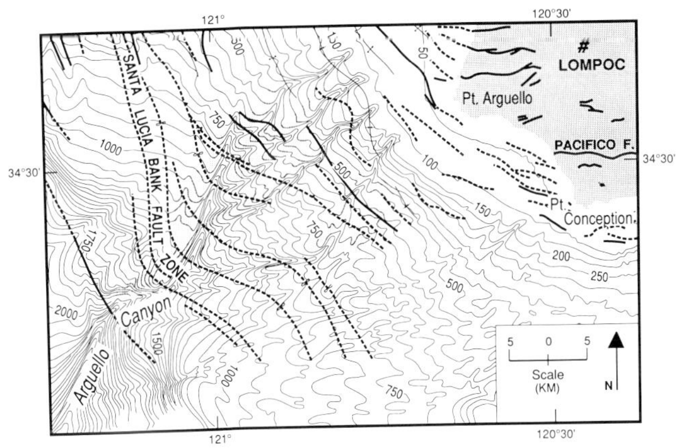

FIG. 11.—Bathymetry and structure offshore from Point Arguello, showing relation between geologic structures and morphology of Arguello Canyon. Contour interval 50 m. Structure symbols same as Figure 6, except that thrust faults are indicated by barbs on upper plate. Modified from McCulloch (1989c).

and faulted, whereas others are relatively undeformed. Clarke and others (1983) postulated that these thrust sheets were pushed westward during late Pleistocene time. Decoupling and westward thrusting of late Tertiary bedrock over older basement rocks resulted in uplift of the sea floor. Levees mapped in the headward part of the Arguello Canyon are restricted to one thrust sheet; Clarke and others (1983) have speculated that these levees formed at greater depths and were then elevated on the thrust sheet during its westward transport. Levee growth subsequently may have been enhanced by deposition from sediment-laden rivers during the Pleistocene, when sea level was much lower than at present (Clarke and others, 1983). The crook in the lower part of Arguello Canyon lies between two active faults (McCulloch, 1989c) and may mark the boundary between two of the nappes.

Hueneme-Mugu Canyon System.—The Hueneme-Mugu Canyon system cuts into late Tertiary and Pleistocene sedimentary rocks (Vedder and others, 1986) along the southern margin of the transverse segment of the San Andreas fault system, just north of the Santa Cruz Island-Malibu Coast fault zone (Fig. 12). Like Arguello Canyon, this system has formed largely in a compressional tectonic regime. The canyon system has five heads, the westernmost of which is Hueneme Canyon and the easternmost Mugu Canyon, and between them are three unnamed canyons (Fig. 12). The heads of Hueneme and Mugu Canyons lie near the shoreline in water depths of 10 m, whereas the intermediate canyons have their heads in water depths of between 40 and

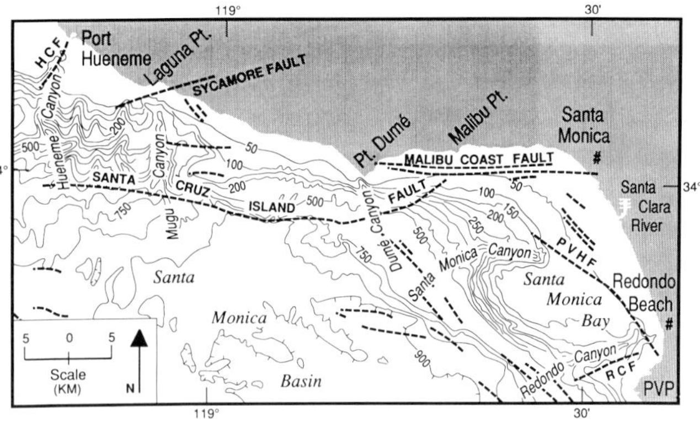

FIG. 12.—Bathymetry and geologic structure to the Port Hueneme to Santa Monica Bay area, showing relation of geologic structures to the Hueneme-Mugu and Redondo Canyon systems. Contour interval 50 m. Heavy dashed lines indicate faults. HCF = Hueneme Canyon fault; PVHF = Palos Verdes Hills fault; RCF = Redondo Canyon fault; PVP = Palos Verdes Peninsula. Modified from Vedder and others (1986).

60 m on the outer continental shelf (Fig. 12). All of these canyons are narrow, less than 15 km long, and relatively straight; they trend south and debouch in fairly shallow water (about 720-m depth) on the flat-floored Santa Monica Basin (Nardin, 1983). Mugu Canyon alone has a distinct meander.

The Hueneme-Mugu Canyon system appears to have been eroded by turbidity currents in sediment transported onto the continental shelf and upper slope by outflow from the Santa Clara River (not shown). Most of the offshore structural and geomorphic features apparent in the region today began to form during north-south compression associated with major tectonism in the middle Pleistocene (Greene and others, 1978). The nearly southward trend of the Hueneme Canyon fault in the basement rocks beneath Hueneme Canyon suggests that the canyon began its growth along a fault trace (Greene and others, 1978; Vedder and others, 1986). Similarly, the meander in the head of Mugu Canyon was postulated by Greene and others (1978) to be fault controlled (Fig. 12). These workers noted that the Sycamore fault extends southward offshore to form the west-trending headwall of Mugu Canyon, and speculated that the canyon head may have been offset left laterally along this fault.

Submarine Canyons Associated with the Southern Section of the San Andreas Fault System

Redondo, San Pedro, and San Gabriel Canyons.—Redondo Canyon is a short (17 km), relatively straight, west-southwest-trending canyon that cuts the continental shelf in the southern part of Santa Monica Bay, just north of the Palos Verdes Peninsula (Fig. 12). Its head lies near the shoreline at a depth of 10 m, and it loses physiographic expression at a depth of about 750 m, where it debouches onto the east flank of the Santa Monica Basin (Fig. 12; Haner, 1971). The headward part of the canyon cuts unconsolidated marine shelf and slope deposits of late Pleistocene age, whereas a short distance to the west the canyon incises rock of Miocene age. The shoreward margin of the head of Redondo Canyon lies along a steep, west-facing scarp of the Palos Verdes Hills fault, a northwest-trending, right-slip fault that cuts Quaternary and older strata (Vedder and others, 1986). The Palos Verdes Hills fault is thought to be part of a lengthy (440 km long) zone of active right-lateral faulting that extends from northern Santa Monica Bay southward across to Palos Verdes Peninsula to join the Coronado Bank fault offshore (Fig. 1; Clarke and others, 1985, 1987); the Coronado Bank fault, in turn, has been postulated (Legg and Kennedy, 1979) to continue southward to join the Agua Blanca fault zone in Baja California. If so, this zone of wrench faulting constitutes a major strand of the San Andreas fault system in southern California.

Redondo Canyon is also fault controlled. A generally west-trending, south-over-north reverse fault, the Redondo Canyon fault, that extends along most of the south wall of the canyon (Yerkes and others, 1967) appears to dictate the trend of the canyon axis.

San Pedro Canyon is just south of Point Fermin on the Palos Verdes Peninsula (Fig. 13). The head of this canyon notches the distal shelf edge of San Pedro Bay at a water depth of 50 m. The canyon is short (about 11 km long), trends west, and loses its physiographic expression at a depth of about 750 m, where it debouches into Santa Monica Basin. The canyon has two heads, one that extends northeast from the canyon axis, and one that extends southeast parallel to the shelf edge and to the regional structural grain (Fig. 13).

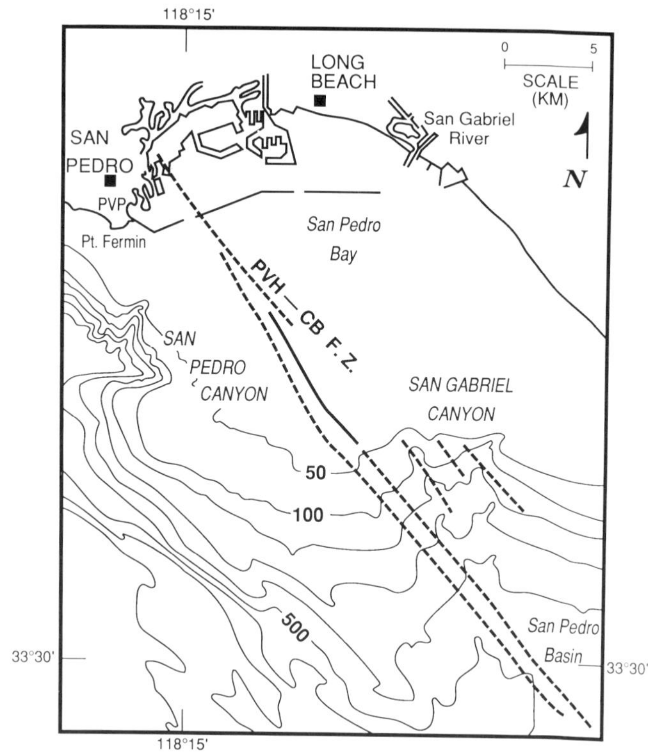

FIG. 13.—Bathymetry and geologic structure in the San Pedro Bay area, showing relation of the Palos Verdes Hills-Coronado Bank fault zone (PVH-CB FZ) and associated faults to San Pedro and San Gabriel Canyons. Contour interval 50 m to a depth of 100 m, and 100 m at greater depths. Structure symbols same as in Figure 12. PVP = Palos Verdes Peninsula. Modified from Clarke and others (1987).

No faults have been mapped along or within San Pedro Canyon, and its origin is undetermined. However, south-trending faults extending seaward from Point Fermin (Vedder and other, 1986) may trancate the head of San Pedro Canyon. Like Redondo Canyon, headward parts of this canyon incise rocks of Miocene age, whereas the lower part cuts rocks of Pliocene and younger age (Vedder and others, 1986).

San Gabriel Canyon is generally aligned north-south and heads in about 40 m of water at the southern edge of the San Pedro Bay shelf, from which it extends south-southwestward for 20 km and debouches into San Pedro Basin (Fig. 13). San Gabriel Canyon has two heads; the western head extends northwestward from the main channel and cuts Miocene marine sedimentary rocks. The eastern head hooks northeastward from the main channel and cuts unconsolidated marine deposits of Quaternary age. Several active faults (Vedder and others, 1986) lie in the vicinity of these canyon heads; thus, faults appear to impart control on the orientation of the upper reaches of the canyon. Because the heads of Redondo and San Pedro Canyons appear to be associated with faults of the Palos Verdes Hill-Coronado Bank fault zone and because no major river or sediment source is present in the vicinity of the heads of these canyons, we speculate that right-lateral offset along this fault zone may have moved the canyons northward, away from their point of origin. San Gabriel Canyon is associated with the San Gabriel River and could have been the original conduit through which sediment was supplied for the erosion of Redondo and San Pedro Canyons. In this case, as much as 45 km of right slip may have occurred along the Palos Verdes Hills-Coronado Bank fault zone since the formation of Redondo Canyon.

San Clemente Canyon.—San Clemente Canyon is a nearly linear, northwest-trending canyon that heads in 1,000 m of water 7 km off the southeast tip of San Clemente Island (Fig. 14). The canyon is 22 km long and debouches into the deepest part of the San Clemente Basin at a depth of 2,050 m. It cuts undifferentiated volcanic and sedimentary rocks of Miocene age throughout most of its length, although the canyon head is in Miocene volcanic rocks that crop out on San Clemente Island (Vedder and others, 1986).

The San Clemente fault (Shepard, 1938b) extends along the east flank of San Clemente Island and along the axis of the canyon (Fig. 14). The canyon itself cuts sheared and severely disturbed rocks within the San Clemente fault zone

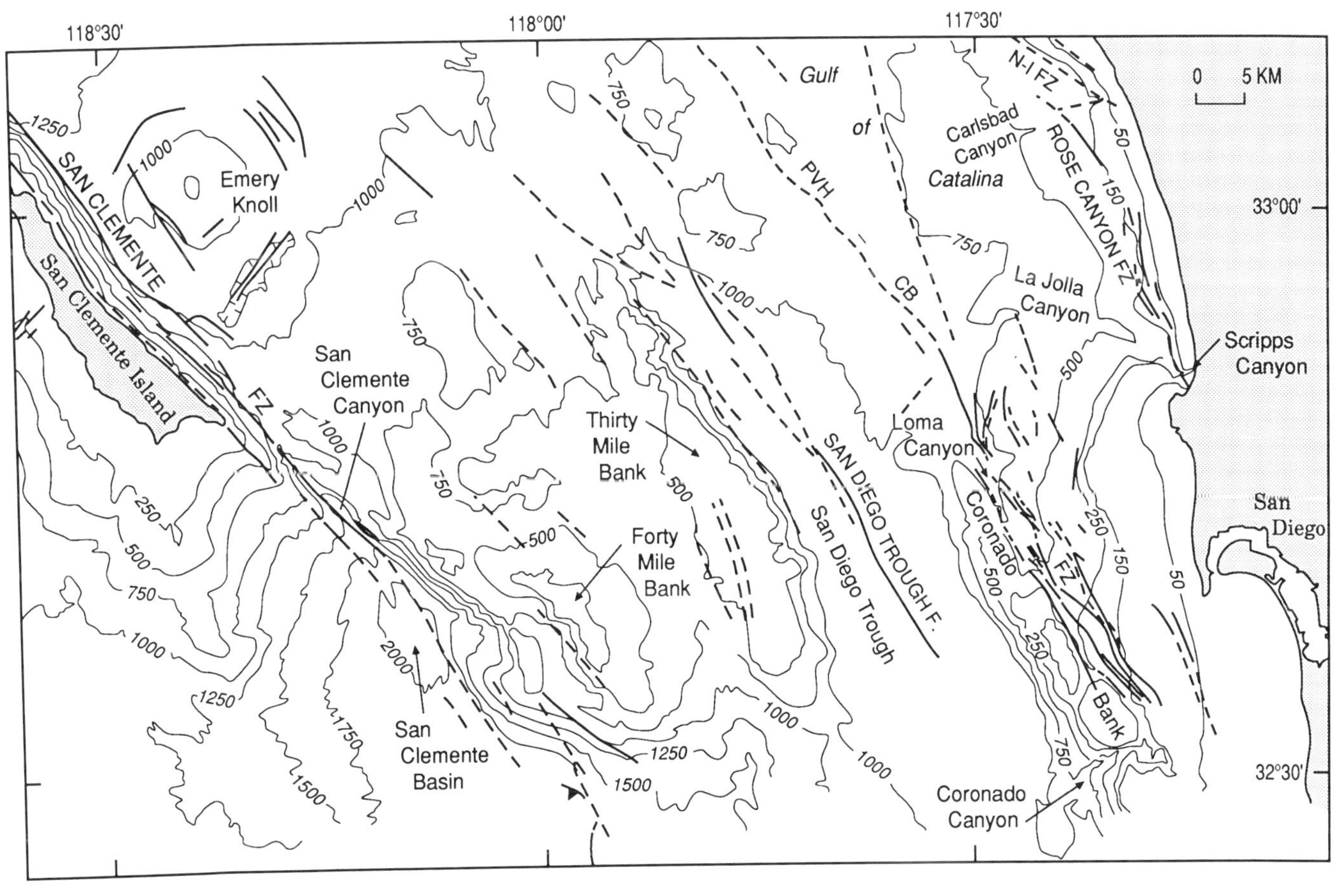

FIG. 14.—Bathymetric and geologic structure in the area between San Clemente Island and San Diego. Note the relation between structure and the morphology of San Clemente Canyon, Carlsbad Canyon, Coronado Canyon and La Jolla-Scripps-Loma Submarine Canyon system. Contour interval is 50 m to a depth of 250 m and is 250 m at greater depths. Faults are shown as heavy, solid lines where well located, dashed where inferred. PVH-CB FZ = Palos Verdes Hills-Coronado Bank fault zone; N-I FZ = Newport-Inglewood fault zone. Modified from Clarke and others (1987) and Vedder and others (1986).

(Shepard, 1938b). It is probably one of the clearest examples of a fault-controlled canyon to be found anywhere along the California continental margin.

Carlsbad Canyon.—Carlsbad Canyon is 45 km north of San Diego on the eastern margin of the Gulf of Santa Catalina (Fig. 14). The canyon is relatively linear, west-southwest trending, and extends from its head at the 30-m isobath for at least 12 km into the eastern part of the gulf. The axis of the upper part of the canyon is aligned with a fault for 4 km and is cut transversely by active elements of the Rose Canyon fault zone at a depth of approximately 420 m. Folded Miocene strata have been uplifted to the southwest of the Rose Canyon fault and juxtaposed with Pliocene and Quaternary strata to the northeast (Clarke and others, 1987). A linear tributary to Carlsbad Canyon extends for at least 5 km north-northwestward along this structural contact; the orientation of this tributary clearly is fault controlled (see Clarke and others, 1987).

La Jolla-Scripps-Loma Canyon System.—The three submarine canyons composing the La Jolla-Scripps-Loma Canyon system (Shepard, 1961) originate within 20 km of the coast off San Diego (Fig. 14). The head of La Jolla Canyon is at the shoreline in about 10 m of water; the canyon trends northwestward for 6 km, at which point it bends westward and extends another 15 km before curving arcuately into the San Diego Trough. It loses physiographic expression at a depth of 1,100 m. Scripps Canyon, a tributary to the headward part of La Jolla Canyon, is a short (2 km long), southwest-trending canyon that intersects La Jolla Canyon about 2 km seaward of its head. Scripps Canyon also heads at the shore in about 10 m of water, just offshore from the Scripps Institution of Oceanography. Shepard (1948a) first described the morphology of Scripps Canyon, and Shepard and Buffington (1968) described the submarine-fan valley associated with Scripps and La Jolla Canyons.

Loma Canyon enters La Jolla Canyon about 37 km downcanyon from the intersection of La Jolla and Scripps Canyons. Loma Canyon is 41 km long and trends northwest. The head of Loma Canyon lies at a depth of about 170 m off the southeast end of Coronado Bank, very near that of Coronado Canyon, which marks the southeast limit of Coronado Bank. Loma Canyon forms the east physiographic boundary of Coronado Bank.

La Jolla Canyon is carved into undifferentiated sedimentary rocks of Quaternary, early Tertiary, and Late Cretaceous age (Clarke and others, 1987). The head of this canyon is cut by a youthful normal fault (north side down), and the upthrown block of the fault forms the southwest canyon wall; elements of the active right-lateral, strike-slip Rose Canyon fault zone define at least part of the north wall (Fig. 14; Kennedy and others, 1980a, b). Upper Cretaceous sedimentary rocks are exposed in canyon walls to the southeast and on the shelf offshore of Point La Jolla. Eocene sedimentary rocks crop out in the northeast wall of the canyon and on the shelf northward to Carlsbad Canyon (Clarke and others, 1987). Development of the upper part of La Jolla Canyon thus has been controlled by faulting.

Scripps Canyon is eroded primarily into Eocene sedimentary rocks (Clarke and others, 1987). The junction of Scripps and La Jolla Canyons is immediately downslope of the intersection of the canyon axis with the Newport-Inglewood-Rose Canyon fault zone (Fig. 14). Clarke and others (1983, p. 11) stated, "Scripps Canyon appears to be the right-laterally offset head of La Jolla submarine canyon." We believe that Scripps Canyon originally was the head of the La Jolla Canyon, so that La Jolla Canyon had a single head that was linear and oriented west-southwest, and that Neogene right-lateral displacement along the Newport-Inglewood-Rose Canyon fault zone (Fig. 1) has offset Scripps Canyon at least 4 km to the south.

La Jolla Canyon is structurally complex below its intersection with Scripps Canyon, at depths between 450 and 750 km, and is bounded by many normal faults, erosional scarps, and slump scarps (Kennedy and others, 1980a, b; Clarke and others, 1987). Clarke and others (1983) suggested that La Jolla Canyon formed in a graben that resulted from tension associated with wrench faulting along the Rose Canyon and Coronado Bank fault zones that bound this area (Fig. 14).

Loma Canyon is aligned with faults in the southern part of the Palos Verdes Hills-Coronado Bank fault zone (Clarke and others, 1987). In the area of Coronado Bank, the eastern edge of this fault zone is composed of echelon normal faults downthrown to the southwest; these faults, which define the northeast flank of Coronado Bank and Loma Canyon, juxtapose undifferentiated sedimentary rocks of Tertiary and Quaternary age to the northeast with rocks of Miocene age on Coronado Bank (Clarke and others, 1987). Loma Canyon, therefore, has formed along sheared and fractured rocks within the Palos Verdes Hills-Coronado Bank fault zone, eroding its course along the contact between rocks of differing age and lithology. Right-lateral strike slip along the Coronado Bank fault zone also may have offset the headward part of Coronado Canyon, southeast of Coronado Bank.

CONCLUSIONS

We attempt to relate canyon development to the recent tectonic history of a continental margin. Because Monterey Canyon is well studied and its origin has been clearly tied to the regional tectonic history (Greene, 1977; Greene and Clark, 1979; Howell and others, 1980; Greene and others, 1989; Greene, 1990; Greene and Hicks, 1990), we use this canyon as a model for comparison with other California submarine canyons.

Monterey Canyon is a Miocene canyon with a headward part that is well incised into Salinian granitic basement rocks. Although the canyon is similar in scale to the Grand Canyon of the Colorado River, today only the small, low-discharge Salinas River enters Monterey Bay near the head of the canyon. Because the head of Monterey Canyon is old and primarily cuts Salinian basement rocks, which have been displaced northwestward along the San Andreas fault, canyon cutting probably was initiated some distance to the southeast of its present position. The lower part of Monterey Canyon crosses the Palo Colorado-San Gregorio fault zone, which marks the west edge of Salinia in the offshore Monterey Bay area (Fig. 2). Here, the canyon morphology changes abruptly from a V-shaped valley with steep relief east of the boundary to a broad, U-shaped, more subdued valley to the west. This change results from the juxtaposition of differing lithologic units that have differing resistance to erosion—resistant granitic rocks to the east, and melange of the Franciscan Complex, which is relatively easily eroded and subject to mass wasting, to the west.

Right-lateral strike slip along the Palo Colorado-San Gregorio and other faults in the Monterey Bay region has strongly influenced canyon development in the Monterey Bay region in still other ways. Most apparently, reaches of both Monterey Canyon and Sur Canyon to the southeast have followed linear zones of weakness in rocks that have been sheared and fractured, some along the Palo Colorado-San Gregorio fault zone. In addition, right slip along the Palo Colorado-San Gregorio and other offshore faults apparently has successfully offset the lower parts of Monterey Canyon throughout Neogene time. This process is thought to be responsible for the apparently "headless" canyons now present on the continental slope north of Monterey Canyon (Fig. 2). In addition, two heads of Carmel Canyon, nearby in Carmel Bay, are thought to have been right laterally offset along this fault zone.

Meander formation also may reflect a structural influence on canyon development. Monterey Canyon has a meander that deeply incises the shelf of Monterey Bay (Fig. 2). This meander is formed along strike-slip faults of the Monterey Bay fault zone that have slivered resistant granitic rocks into the canyon axis. Canyon erosion was constrained to skirt this resistant structural wedge before rejoining the lower channel.

Monterey Canyon thus offers clear examples of many of the ways in which geologic structure and regional tectonics can influence the development of submarine-canyon morphology. Conversely, canyon morphology has been used to infer geologic structures and processses that have influenced canyon development in other parts of the California continental margin, and most of these inferences have been confirmed subsequently from precision bathymetric, side-scan-sonar, and seismic-reflection data.

Canyons that develop along major structural or tectonic boundaries may be clearly oriented, wholly or partly, along these boundaries, with their linearity suggesting structural control of the canyon trend. Notable examples in California include, in addition to Monterey and Carmel Canyons, axial parts of Mendocino, Mattole, and Eel Canyons, and the headward parts of Noyo and Bodega Canyons in northern California; the axial part of Hueneme Canyon near Santa Barbara; axial parts of Redondo, Loma, and San Clemente Canyons, and a conspicuous tributary to Carlsbad Canyon, in the California Continental Borderland. Canyons that cross such boundaries may show evidence of progressive displacement of their lower reaches as in Monterey Canyon, or abrupt changes in the orientation of the canyon axis, as in Carmel and Sur Canyons near Monterey Bay and in Mugu Canyon and the Scripps-La Jolla Canyon system in southern California. Canyons with heads that form abruptly at or near faults may have their origin in slope processes that were localized in response to fault movement, although such a causal relation is difficult to establish in the absence of detailed knowledge of both fault and canyon history. Such canyons include Noyo Canyon in northern California and elements of the Hueneme-Mugu Canyon system and Redondo Canyon in southern California.

Meanders may have a clear structural origin, as in Monterey Canyon, or their origin may be more subtle, as in Eel Canyon, where meander development appears to result from fold growth across the canyon axis. Meander development also may be forced by mass failure of part of the canyon wall into the canyon axis. This phenomenon is particularly apparent in canyons that are not at grade and thus are actively downcutting, such as the upper and middle parts of Monterey and Sur Canyons. Slumps in those canyons have effectively dammed the canyon axes, forcing erosion around the slump mass and creating a meander.

Except for Eel and Mendocino Canyons, which exist today in an area that is dominated by convergence and northeast-southwest compression, all the canyons described here have formed in a region characterized principally by translational (right slip) faulting associated with the San Andreas fault system. Arguello Canyon, in contrast, is thought to have been elevated on a series of thrust sheets after its early development. Prominent flanking levees in the upper part of this canyon system are considered unlikely to have formed in the present tectonic environment, which is characterized by ongoing uplift of the headward part of the canyon, but, instead, to be relict features formed largely before uplift and westward-thrust transport.

Many submarine canyons along the active California continental margin have had their origin and evolution shaped strongly by structural and tectonic processes. Canyons that cross conspicuous structural boundaries in a tectonic environment characterized by large translational movement, such as in California, may have been displaced great distances from their original sources, or segmented, so that one part of the canyon system is displaced from an originally adjacent part. Seemingly anomalous canyon segments, such as apparently "headless" canyons, on the California margin may have undergone such displacement. Moreover, much of the canyon physiography directly reflects both the underlying geologic structure and tectonic processes. Therefore, careful examination of canyon physiography can provide vital information about margin structure and tectonics.

ACKNOWLEDGMENTS

We thank Robert Osborne, Donn Gorsline, Herman Karl and David Drake, whose comments substantially improved the manuscript. The assistance of Karen Hicks in manuscript and illustration preparation, and that of Phyllis Swensen in the drafting of illustrations are also much appreciated.

We also acknowledge Fran Shepard's classic studies of submarine canyons. Fran strongly emphasized the role of geologic structures in the formation of submarine canyons and published on this topic as early as 1938. Although many of Fran's papers were written prior to the advent of plate tectonics, his ideas on the structural formation of submarine canyons are readily compatible with recent plate-tectonic models.

Fran was enthusiastic that we held in common concepts

regarding the tectonic evolution of submarine canyons, and we have attempted to write this paper as Fran would have written it were he with us today. Many of his early concepts have been incorporated in this paper, but we have gone farther in associating these concepts with recent ideas of fault movement along the San Andreas transform fault, the major tectonic boundary between the North American and Pacific plates. We, therefore, dedicate this paper to our very dear friend, Francis P. Shepard.

REFERENCES

ATWATER, TANYA, 1970, Implications of plate tectonics for the Cenozoic tectonic evolution of western North America: Geological Society of America Bulletin, v. 81, p. 3513–3536.

BUCHER, W. H., 1940, Submarine valleys and related geologic problems of the north Atlantic: Geological Society of America Bulletin, v. 51, p. 489–512.

CLARK, J. C., AND RIETMAN, J. D., 1973, Oligocene Stratigraphy, Tectonics, and Paleogeography Southwest of the San Andreas Fault, Santa Cruz Mountains and Gabilan Range, California Coast Range: U.S. Geological Survey Professional Paper 783, 18 p.

CLARK, J, C., BRABB, E. E., GREENE, H. G., AND ROSS, D. C., 1984, Geology of Point Reyes Peninsula and implication for San Gregorio fault history, *in* Crouch, J. K., and Bachman, S. G., eds., Tectonics and Sedimentation along the California Margin: Pacific Section, Society of Economic Paleontologists and Mineralogists, v. 38, p. 67–86.

CLARKE, S. H., Jr., 1988, Late Cenozoic deformation of the California continental margin north of Cape Mendocino: Implications for Mendocino triple junction locations: Geological Society of America, Abstracts and Programs, v. 20, p. A382.

CLARKE, S. H., Jr., 1990, Map showing geologic structures of the northern California continental margin: U.S. Geological Survey Miscellaneous Field Studies Map MF-2130, scale 1:250,000.

CLARKE, S. H., AND FIELD, M. E., 1989, Geologic map of the northern California continental margin—Map 7A (Geology), *in* Greene, H. G., and Kennedy, M. P., eds., California Continental Margin Geologic Map Series, California Division of Mines and Geology, scale 1:250,000.

CLARKE, S. H., Jr., GREENE, H. G., FIELD, M. E., AND LEE, W. H., 1983, Reconnaissance Geology and Geologic Hazards of Selected Areas of the Southern California Borderland: U.S. Geological Survey Open-File Report 83-62, 78 p.

CLARKE, S. H., Jr., GREENE, H. C., AND KENNEDY, M. P., 1985, Identifying potentially active faults and unstable slopes offshore, *in* Ziony, J. I., ed., Evaluating Earthquake Hazards in the Los Angeles Region—An Earth-Science Perspective: U.S. Geological Survey Professional Paper 1360, p. 347–373.

CLARKE, S. H. Jr., GREENE, H. G., KENNEDY, M. P., AND VEDDER, J. G., 1987, Geologic map of the inner-southern California continental margin—Map 1A (Geology), *in* Greene, H. G., and Kennedy, M. P., eds., California Continental Margin Geologic Map Series, California Division of Mines and Geology, scale 1:250,000.

DALY, R. A., 1936, Origin of submarine canyons: American Journal of Science, v. 31, p. 401–420.

DAVIDSON, G., 1897, The submerged valleys of the coast of California, U.S.A., and lower California, Mexico: California Academy of Sciences Proceedings, Series 3, v. 1, p. 73–103.

DAVIS, W. M., 1934, Submarine mock valleys: Geographical Review, v. 24, p. 297–308.

de ANDRADE, C. F., 1937, Os vales submarinos Portugueses e o diastrofismo das Berlenges e de Estremadura (with English summary): Casa Portuguesa, p. 237–249.

DILL, R. F., DIETZ, R. S., AND STEWART, H. B., Jr., 1954, Deep-sea channels and delta of the Monterey submarine canyon: Geological Society of America Bulletin, v. 65, p. 191–194.

EITTREIM, S. L., EMBLEY, R. W., NORMARK, W. R., GREENE, H. G., MCHUGH, C. M., AND RYAN, W. B., 1989, Observations in Monterey Canyon and Fan Valley using the Submersible *Alvin* and a Photographic Sled: U.S. Geological Survey Open-File Report 89-291, 28 p.

EMERY, K. O., 1960, The Sea off Southern California: John Wiley & Sons, New York, 366 p.

FOX, K. F., Jr., FLECK, R. J., CURTIS, G. H., AND MEYER, C. E., 1985, Implications of the northwesterly younger age of the volcanic rocks of west-central California: Geological Society of America Bulletin, v. 96, p. 647–654.

GREENE, H. G., 1977, Geology of the Monterey Bay Region: U.S. Geological Survey Open-File Report 77-718, 347 p., 9 plates.

GREENE, H. G., 1990, Regional tectonics and structural evolution of the Monterey Bay region, central California, *in* Garrison, R. E., Greene, H. G., Hicks, K. R., Weber, G. E., and Wright, T. L., eds., Geology and Tectonics of the Central California Coastal Region, San Francisco to Monterey: Pacific Section, American Association of Petroleum Geologists, Volume and Guidebook, GB67, p. 31–56.

GREENE, H. G., AND CLARK, J. C., 1979, Neogene paleogeography of the Monterey Bay area, California, *in* Cole, M. R., TerBest, H., and Armentrout, J., eds., Cenozoic Paleogeography of the Western United States: Pacific Section, Society of Economics Paleontologists and Mineralogists, Pacific Coast Paleogeography Symposium 3, p. 277–296.

GREENE, H. G., LEE, W. H., MCCULLOCH, D. S., AND BRABB, E. E., 1973, Faults and earthquakes in the Monterey Bay region, California: U.S. Geological Survey Miscellaneous Field Investigations Map MF-518, Scale 1:200,000.

GREENE, H. G., STUBBLEFIELD, W. L., AND THEBERGE, A. E., Jr., 1989, Sea Floor Geology of the Monterey Submarine Canyon System and Adjacent Areas, Offshore Central California, Results of NOAA SeaBeam Survey (Descriptive report for *Surveyor* cruise): U.S. Geological Survey Open-File Report 89-221, 33 p., 4 map sheets.

GREENE, H. G., AND HICKS, K. R., 1990, Ascension-Monterey Canyon system: History and development *in* Garrison, R. E., Greene, H. G., Hicks, K. R., Weber, G. E., and Wright, T. L., eds. Geology and Tectonics of the Central California Coastal Region, San Francisco to Monterey: Pacific Section, American Association of Petroleum Geologists, Volume and Guidebook, GB67, p. 229–249.

GREENE, H. G., WOLF, S. C., AND BLOM, K. G., 1978, The Marine Geology of the Eastern Santa Barbara Channel with Particular Emphasis on the Ground-Water Basins Offshore from the Oxnard Plain, Southern California: U.S. Geological Survey Open-File Report 78-305, 104 p., 13 plates.

HANER, B. E., 1971, Morphology and sediments of Redondo submarine fan, southern California: American Association of Petroleum Geologists Bulletin, v. 82, p. 2413–2432.

HEIN, J. R., 1973, Deep sea sediment source areas: Implications of variable rate of movement between California and the Pacific Plate: Nature, v. 241, p. 40–41.

HOWELL, D. G., CROUCH, J. K., GREENE, H. G., MCCULLOCH, D. S., AND VEDDER, J. G., 1980, Basin development along the late Mesozoic and Cainozoic California margin: A plate tectonic margin of subduction, oblique subduction, and transform tectonics, *in* Ballance, P. F., and Reading, H. G., eds., Sedimentation in Oblique-Slip Mobile Zones: International Association of Sedimentologists Special Publication Number 4, Blackwell Scientific Publications, p. 43–62.

HOWELL, D. G., JONES, D. L., AND SCHERMER, E. R., 1985, Tectonostratigraphic terranes of the circum-Pacific region, *in* Howell, D. G., eds., Tectonostratigrraphic Terranes of the Circum-Pacific Region: Circum-Pacific Council for Energy and Mineral Resources Earth Sciences Series, No. 1, p. 3–30.

HUFFMAN, O. F., TURNER, O. L., AND JACK, R. N., 1973, Offset of late Oligocene-early Miocene volcanic rocks along the San Andreas fault in central California, *in* Kovach, R. L., and Nur, A., eds., Proceedings, Conference on Tectonic Problems of the San Andreas Fault System: Stanford University Publications in the Geological Sciences, v. 13, p. 368–373.

HULL, EDWARD, 1912, The Sub-Oceanic Physiography of the North Atlantic Ocean: Edward Stanford, London, 41 p.

JENNINGS, C. W., 1977, Geologic Map of California, Geologic Data Map No. 2: California Division of Mines and Geology, scale 1:750,000.

JENNINGS, C. W., STRAND, R. G., ROGERS, T. H., STINSON, M. C., BURNETT, J. L., KAHLE, J. E., STREITZ, ROBERT, SWITZER, R. A., AND MOAR, R. R., 1975, Fault map of California with locations of volcanoes, thermal springs, and thermal wells: California Division of Mines and Geology, Geologic Data Map 1, scale 1:750,000.

JENSEN, J. A., 1976, A photographic reconnaissance of the microrelief in the Monterey and Carmel submarine canyons, California: Unpublished M.S. Thesis; Naval Postgraduate School, Monterey, California, 142 p.

JOHNSON, D. W., 1939, The Origin of Submarine Canyons: Columbia University Press, New York, 126 p.

KENNEDY, M. P., CLARKE, S. H., GREENE, H. G., AND LEGG, M. R., 1980a, Recency and character of faulting offshore metropolitan San Diego, California (area 3 of 3, Point La Jolla to Mexico-U.S. border): California Division of Mines and Geology Map Sheet 42, scale 1:50,000.

KENNEDY, M. P., GREENE, H. G., CLARKE, S. H., AND BAILEY, K. A., 1980b, Recency and character of faulting offshore metropolitan San Diego, California (area 2 of 3, Cardiff-by-the-Sea to Point La Jolla): California Division of Mines and Geology Map Sheet 41, scale 1:50,000.

KUENEN, P. H., 1937, Experiments in connection with Daly's hypothesis on the formation of submarine canyons: Leidsche Geologische Mededelingen, v. VIII, p. 316–351.

LEGG, M. R., AND KENNEDY, M. P., 1979, Faulting offshore San Diego and northern Baja California, *in* Abbott, P. L., and Elliott, W. J., eds., Earthquakes and Other Perils, San Diego Region: San Diego Association of Geologists Field Trip Guidebook, p. 29–46.

LISOWSKI, MICHAEL, AND PRESCOTT, W. H., 1989, Strain accumulation near the Mendocino triple junction, California: EOS, Transactions, American Geophysical Union, v. 70, p. 1332.

MARTIN, B. D., 1964, Monterey submarine canyon, California: Genesis and relationship to continental geology: Unpublished Ph.D. Dissertation, University of Southern California, Los Angeles, California, 249 p.

MARTIN, AND EMERY, K. O., 1967, Geology of Monterey Canyon, California: American Association of Petroleum Geologists Bulletin, v. 51, p. 2281–2304.

MATTHEWS, V., 1973, Pinnacles-Neenach correlation: A restriction for models of the Transverse Ranges and the big bend in the San Andreas fault: Geological Society of America Bulletin, v. 84, p. 683–688.

MCCULLOCH, D. S., 1987, Regional geology and hydrocarbon potential of offshore central California, *in* Scholl, D. W., Grantz, Arthur, and Vedder, J. G., eds., Geology and Resource Potential of the Continental Margin of Western North American and Adjacent Ocean Basins—Beaufort Sea to Baja California: Circum-Pacific Council for Energy and Mineral Resources Earth Science Series, Houston, Texas, v. 6, p. 353–401.

MCCULLOCH, D. S., 1989a, Evolution of the offshore central California margin, *in* Winterer, E. L., Hussong, D. M., and Decker, R. W., eds., The Geology of North America, v. N: The Eastern Pacific Ocean and Hawaii: Geological Society of America, Boulder, Colorado, p. 439–470.

MCCULLOCH, D. S., 1989b, Geologic map of the north-central California continental margin—Map No. 6A (Geology), *in* Greene, H. G., and Kennedy, M. P., eds., California Continental Margin Geologic Map Series: California Division of Mines and Geology, scale 1:250,000.

MCCULLOCH, D. S., 1989c, Geologic map of the south-central California continental margin—Map No. 4A (Geology), *in* Greene, H. G., and Kennedy, M. P., eds., California Continental Margin Geologic Map Series: California Division of Mines and Geology, scale 1:250,000.

MCCULLOCH, D. S., AND GREENE, H. G., 1990, Geologic map of the central California continental margin—Map No. 5A (Geology), *in* Greene, H. G., and Kennedy, M. P., eds., California Continental Margin Geo logic Map Series: California Division of Mines and Geology, scale 1:250,000.

MCKENZIE, D. P., AND PARKER, R. L., 1967, The North Pacific: An example of tectonics on a sphere: Nature, v. 216, p. 1276–1280.

MCLAUGHLIN, R. J., SLITER, W. V., CLARKE, S. H., Jr., MCCULLOCH, D. S., FREDERICKSEN, N. O., AND INGEBRETSON, D. C., 1988, Implications of onshore and offshore structure for location and Cenozoic evolution of the Mendocino triple junction: Geologic Society of America, Abstracts with Programs, v. 20, p. A382.

MORGAN, W. J., 1968, Rises, trenches, great faults and crustal blocks: Journal of Geophysical Research, v. 73, p. 1959–1982.

NAGEL, D. K., MULLINS, H. T., AND GREENE, H. G., 1986, Ascension submarine canyon, California—Evolution of a multi-headed canyon system along a strike-slip continental margin: Marine Geology, v. 73, p. 285–310.

NARDIN, T. R., 1983, Late Quaternary depositional systems and sea level change—Santa Monica and San Pedro basins, California Continental Borderland: American Association of Petroleum Geologists Bulletin, v. 67, p. 1104–1124.

NORMARK, W. R., 1970, Channel piracy on Monterey deep-sea fan: Deep Sea Research, v. 17, p. 837–846.

NORMARK, W. R., HESS, G. R., STOW, D. A., AND BOWEN, A. J., 1980, Sediment waves on the Monterey Fan levee: A preliminary physical interpretation: Marine Geology, v. 37, p. 1–18.

PAGE, B. M., 1970, Sur-Nacimiento fault zone of California: Continental margin tectonics: Geological Society of America Bulletin, v. 81, p. 667–690.

ROSS, D. C., 1976, Reconnaissance geologic map of pre-Cenozoic basement rocks, northern Santa Lucia Range, Monterey County, California: U.S. Geological Survey Miscellaneous Field Studies Map MF-750, scale 1:25,000.

SHEPARD, F. P., 1937, Horizontal shift faulting off Punta Gorda, California (abst.): Geological Society of America Proceedings, 1936, p. 101.

SHEPARD, F. P., 1938a, California submarine canyons (abst.): Geological Society of America Bulletin, v. 49, p. 1900–1901.

SHEPARD, F. P., 1938b, San Clemente submarine fault (abst.): Geological Society of America Proceedings, 1937, p. 252.

SHEPARD, F. P., 1938c, Structural trends off the California coast (abst.): American Association of Petroleum Geologists Bulletin, v. 22, p. 1716.

SHEPARD, F. P., 1938d, Submarine canyons off the California coast: California Journal of Mines and Geology, v. 34, p. 298–310.

SHEPARD, F. P., 1939, Tectonic development off the California coast: Oil Weekly, v. 93, p. 69.

SHEPARD, F. P., 1941, Non-depositional physiographic environments off the California coast: Geological Society of America Bulletin, v. 52, p. 1869–1886.

SHEPARD, F. P., 1948a, Detailed studies of submarine canyons at La Jolla, California (abst.): Geological Society of America Bulletin, v. 59, p. 1351.

SHEPARD, F. P., 1948b, Investigation of the head of Monterey submarine canyon: Scripps Institution of Oceanography, Submarine Geology Report 1, 15 p.

SHEPARD, F. P., 1957, Northward continuation of the San Andreas fault: Seismological Society of America Bulletin, v. 37, p. 263–266.

SHEPARD, F. P., 1961, Submarine Canyons of the San Diego-La Jolla Area: Geological Society of America Special Paper 68, 54 p.

SHEPARD, F. P., AND BUFFINGTON, E. C., 1968, La Jolla submarine fan valley: Marine Geology, v. 6, p. 107–143.

SHEPARD, F. P., AND DILL, R. F., 1966, Submarine Canyons and Other Sea Valleys: Rand McNally, Chicago, Illinois, 381 p.

SHEPARD, F. P., AND EMERY, K. O., 1941, Submarine topography off the California coast: Geological Society of America Special Paper 31, p. 103–112.

SHEPARD, F. P., AND EMERY, K. O., 1946, Submarine topography off the California coast: Journal of Geology, v. 54, p. 306–321.

SILBERLING, N. J., JONES, D. L., BLAKE, M. C., Jr., AND HOWELL, D. G., 1984, Lithotectonic terrane map of the western conterminous United States, Part C, *in* Silberling, N. J., and Jones, D. L., eds., Lithostratigraphic Terrane Maps of the North American Cordillera: U.S. Geological Survey Open-File Report 84-523, p. C-1 to C-33.

SILVER, E. A., 1974, Basin development along translational continental margins, *in* Dickinson, W. R., ed., Geologic Interpretations from Global Tectonics and Applications for California Geology and Petroleum Explortion: Bakersfield, California, San Joaquin Geological Society Short Course, p. 6.1–6.5.

SMITH, W. S. R., 1902, The submarine valleys of the California coast: Science, v. 15, p. 670–672.

SPENCER, J. W., 1898, On the continental elevation of the glacial epoch: Geological Magazine, New Series, Decade IV, v. 5, p. 32–38.

SPENCER, J. W., 1903, Submarine valleys off the American coast and in the north Atlantic: Geological Society of America Bulletin, v. 14, p. 207–226.

STARKE, G. W., AND HOWARD, A. D., 1968, Polygenetic origin of Monterey submarine canyon: Geological Society of America Bulletin, v. 79, p. 813–826.

TURNER, D. L., CURTIS, G. H., BERRY, F. A., AND JACK, R. N., 1970, Age relationship between the Pinnacles and Parkfield felsites and felsite clasts in the southern Temblor Range, California: Implications for San Andreas fault displacement: Geological Society of America, Abstracts with Program, Cordilleran Section, v. 2, p. 154–155.

VEDDER, J. G., GREENE, H. G., CLARKE, S. H., AND KENNEDY, M. P., 1986, Geologic map of the mid-southern California continental margin—Map No. 2A (Geology), *in* Greene, H. G., and Kennedy, M.

P., eds., California Continental Margin Geologic Map Series: California Division of Mines and Geology, scale 1:250,000.

WAGNER, D. L., AND BORTUGNO, E. J., compilers, 1982, Geologic map of the Santa Rosa Quadrangle, Regional Geologic Map Series, Map No. 2A: California Division of Mines and Geology, scale 1:250,000.

WALLACE, E. W., ed., 1990, The San Andreas Fault System: U.S. Geological Survey Professional Paper 1515, 283 p.

WEGENER, ALFRED, 1924, The Origin of Continents and Oceans: English translation from 3rd German edition: E. P. Dutton, and Company, New York, 212 p.

WOODFORD, A. O., 1951, Stream gradients and Monterey sea valley: Geological Society of America Bulletin, v. 62, p. 799–852.

YERKES, R. F., GORSLINE, D. S., AND RUSNAK, G. A., 1967, Origin of Redondo submarine canyon, southern California, *in* Geological Survey Research, 1967; U.S. Geological Survey Professional Paper 575-C, p. C97-C105.

A PROPOSED CLASSIFICATION OF BASIN PLAINS

ORRIN H. PILKEY AND CLAUDIA HOKANSON
Duke University, Department of Geology, Durham, North Carolina 27708

ABSTRACT: A classification of basin plains is proposed using three basin parameters: 1) the extent to which sand layers cover the basin-plain floor; 2) the regional-shape trends of sandbody distribution on the plain; and 3) the distribution of sediment from canyon-mouth entry points (sand sources) around the plain margin.

INTRODUCTION

Basin plain is a general term for ponds of sediment found at the outer margin of cross-continental-margin sediment-transport systems. Abyssal plain is the specific term used when such oceanic plains are found on the true ocean-basin floor. Basin plains are similar to lakes in the sense that sediment in them, like water in lakes, is ponded by sills or other obstructions that prevent further flow out of the basin.

Abyssal plains and the mechanisms responsible for their formation were first recognized by Heezen (1959) and Ericson and others (1961). Reviews of basin-plain sedimentation include Horn and others (1971, 1972), Pilkey and Cleary (1986), and Pilkey (1987).

Basin plains are traditionally held to be the outer fans or distal portions of fans (Rupke, 1980; Nelson and Nilsen, 1984). Implicit in this distal-fan concept is that basin-plain sediment is fine grained and must contain few sand layers. Both assumptions are untrue (Pilkey and others, 1980; Pilkey, 1987). In most basin plains the boundary between the lower fan or lower rise and the basin plain marks a very striking change in sediment type. Beginning with any basin-plain boundary, sand content usually increases dramatically over the adjacent non-plain areas. In fact, most basin plains are sandier than any other adjacent offshelf sedimentary environment. A major practical ramification of the fact that basin plains are sandy is that their ancient counterparts should be emphasized rather than avoided in hydrocarbon exploration programs.

Over a time span of 15 years, we have studied 13 basin plains (mostly abyssal plains) in the Western North Atlantic, Caribbean and Mediterranean Basins (Table 1, Fig. 1) More than 1,200 relatively closely spaced piston cores were used for this investigation. The areas studied include the Columbus Basin (Bornhold and Pilkey, 1971), the Navidad Basin (Sieglie and others, 1976), the Hispaniola-Caicos Abyssal Plain (Bennetts and Pilkey, 1976; Ditty and others, 1977), the St. Croix Basin, the Hatteras Abyssal Plain (Cleary and Connolly, 1974; Cleary and Pilkey, 1977; Elmore and others, 1979; Sparks 1979), the Silver Abyssal Plain (van Tassel, 1981), the Tongue of the Ocean (Kier and Pilkey, 1971), the Puerto Rico Trench (Doull, 1983), the Blake Bahama Basin, (Schaeffer, 1984), the Algero-Provencal and South Balearic basins of the Balearic Abyssal Plain (Schorsch, 1980) and the Grenada Basin (Herrick, 1983). Sand-layer geometry of the Sohm Abyssal Plain is based on Horn and others (1971), plus unpublished data from the Lamont-Doherty Geological Observatory.

A summary of these studies, by Pilkey (1987), concludes that basin-plain sediment columns are dominated by turbidites. Hemipelagic muds (not to be confused with turbiditic muds) usually comprise less than 10 to 20 percent of the sediment column; debris flows are a distant third in importance. Accumulation of sediment in a basin plain is a function of 1) shoreline position (sea level) on the adjacent continental margin, 2) tectonic framework, 3) ratio of the drainage-basin area to the area of the basin-plain floor (DA/BA), and 4) type of fan development adjacent to the plain. The geometry of sand layers in basin plains is a function of 1) arrangement of entry points around the edge of the basin plain and 2) the ratio DA/BA (Pilkey, 1987). Arrangement and geometry of sand layers varies widely between and among plains.

The purpose of this paper is to propose a classification for basin plains. To demonstrate the basis for the classification, two individual basins and their sand-layer characteristics are discussed. The sand-layer geometry in 13 basin plains is then examined and a classification derived.

BASIN PLAINS CONSIDERED

The size of the basin plains used in this study varies from 200 km^2 for Navidad Basin to areas exceeding 100,000 km^2 for the Sohm Abyssal Plain, the Hatteras Abyssal Plain and the Algero-Provencal Basin Plain (Table 1). Also shown in Table 1 is the ratio of the drainage area to the basin-plain area (DA/BA). Drainage area refers to the subaerial portion of a continent, or of an island, that furnishes sediment to a basin plain. Submarine-drainage patterns, i.e., submarine-canyon paths, are often imperfectly known; thus, it is sometimes difficult to determine what portion of a continent furnishes sediment to a basin plain.

The ratio DA/BA can be an important factor in understanding basin-plain sedimentation. A large drainage area presumably can furnish more sediment to the basin plain and also may have the potential to produce relatively large single-flow events. Such large-flow events are required to form flat plains. That is, a flat plain can be formed only if flow events are occasionally large enough to cover the entire basin-plain floor and fill the topographic lows produced by uneven rates of sedimentation associated with smaller events (Pilkey, 1988).

Values for DA/BA are highly variable (Table 1), ranging from 18.5 for the tiny Navidad Basin Plain to 0.7 for the Columbus Basin Plain. The Navidad Basin Plain has a high DA/BA ratio because most of the island of Hispaniola drains into it. Only the submerged south rim of the Bahama Banks furnishes sediment to the Columbus Basin; hence, its small DA/BA.

From Shoreline to Abyss, SEPM Special Publication No. 46

TABLE 1.—WATER-DEPTH AREA and DA/BA of 13 ABYSSAL PLAINS USED IN THIS CLASSIFICATION (FROM PILKEY, 1987).

Basin Plain	Depth (m)	Area (km^2)	Ratio of Drainage Area to Basin-Plain Area (DA/BA)
Sohm	5,200	660,400	3.8
Hatteras	5,400	181,200	4.3
Balearic			
Algero-Provencal	2,600	168,700	1.5
South Balearic	2,600	33,750	9.3
Grenada	2,980	20,800	2.0
Silver	5,500	19,000	Small (3) ??
Blake-Bahama	5,400	14,750	?
Columbus	2,200	9,320	0.7
Tongue of the Ocean	1,500	5,660	Large (?)
Hispaniola-Caicos	4,100	5,000	4.5
Puerto Rico Trench	7,800	4,750	4.7
St. Croix	4,500	750	4.0?
Navidad	4,350	200	18.5

INDIVIDUAL BASINS

The data base for the classification of basin plains is illustrated by two examples of sand-layer distribution: that in the Hatteras Abyssal Plain, a longitudinal, type 4 basin, and that in the Columbus Basin Plain, a radial, type 3 basin (Tables 2 and 3). The Hatteras Abyssal Plain (Fig. 2) is a giant plain encompassing 180,000 km^2 (Table 1) and extending as an elongated basin for more than 1,000 km in a north-south direction along the southeastern United States continental margin. The Columbus Basin is much smaller (9,000 km^2) and is tucked into a horseshoe shaped reentrant of the Bahama Banks (Fig. 3).

The Hatteras Abyssal Plain

The Hatteras Abyssal Plain (Fig. 2) has a DA/BA ratio of 4.3. Lying adjacent to a passive continental margin, and the plain receives dominantly, but not exclusively, non-calcareous turbidites. Turbidity currents enter the basin through the Hatteras Canyon (Cleary and Connolly, 1974), the Wilmington Canyon System and the Sohm Gap (Pilkey and Cleary, 1986). All three of these major sediment sources are located at the north end of the plain. A minor sediment source is Cat Gap, located at the south end of the plain, where overflows from the Blake Bahama Abyssal Plain occur (Fig. 1). The Sohm Gap source contributes flows from the Sohm Abyssal Plain.

Previous sedimentologic studies on the Hatteras Abyssal Plain indicate a general north-to-south movement of tur-

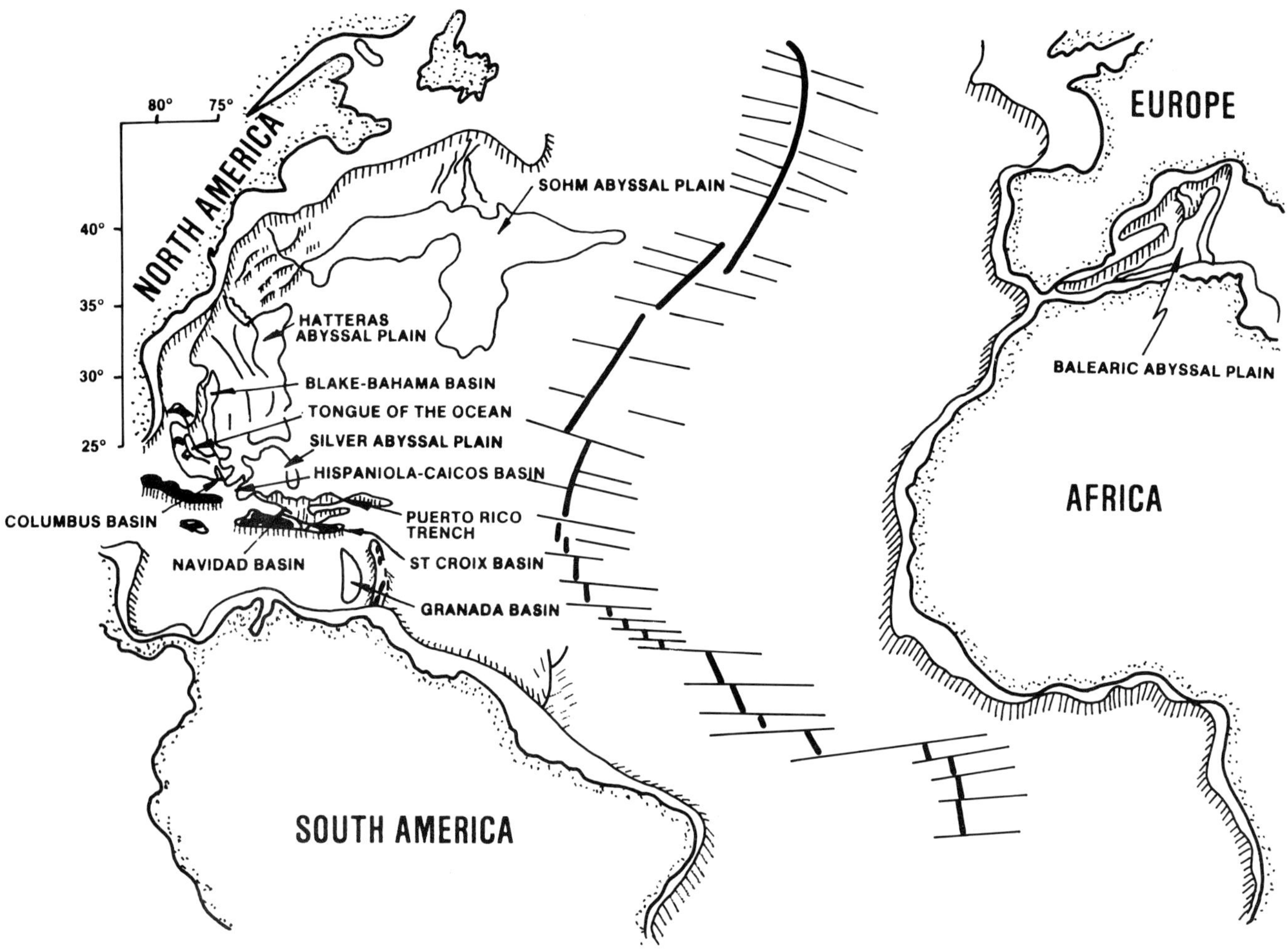

FIG. 1.—Diagrammatic illustration of the North Atlantic Ocean basin, showing mid-Atlantic Ridge and location of the 13 basin plains studied. The Balearic Abyssal Plain is composed of the Algero-Provencal and South Balearic basins.

TABLE 2.—ELEMENTS OF THE BASIN-PLAIN CLASSIFICATION PROPOSED IN THIS STUDY.

Basin Type and Subclassification	Sand Coverage	Sandbody Shape	Examples Studied
Longitudinal			
Type 1	Overall	Wedge	
Type 2	Overall	Sheet	Navidad
Type 3	Partial	Wedge	Silver
Type 4	Partial	Axial	Hatteras, Sohm, Blake-Bahama
Lateral			
Type 1	Overall	Wedge	Puerto Rico Trench
Type 2	Overall	Sheet	
Type 3	Partial	Wedge	
Radial			
Type 1	Overall	Wedge	Hispaniola-Caicos, and South Balearic
Type 2	Overall	Sheet	Tongue of the Ocean
Type 3	Partial	Wedge	Columbus, and Algero-Provencal
Semi-Radial			
Type 1	Overall	Wedge	St. Croix, Grenada
Type 2	Overall	Sheet	
Type 3	Partial	Wedge	

TABLE 3.—RELATIVE SIZE OF DRAINAGE AREA TO BASIN-PLAIN FLOOR AREA FOR EACH OF THE CATEGORIES OF THE BASIN-PLAIN CLASSIFICATION.

Entry-Point Distribution	Subclassification	Drainage Area/Basin Area
Longitudinal	Type 1	Large (>3)
	Type 2	Very Large
	Type 3	Small (<3)
	Type 4	"Giant" Basin
Lateral	Type 1	Large (>3)
	Type 2	Very Large
	Type 3	Small (<3)
Radial	Type 1	Large (>3)
	Type 2	Very Large
	Type 3	Small (<3)
Semi-Radial	Type 1	Large (>3)
	Type 2	Very Large
	Type 3	Small (<3)

bidity currents with a northern proximal area and a southern distal area (Horn and others, 1971; Elmore and others, 1979). In general, sand layers are found in the north half of the plain and the south half is sand free. Isopachs of total centimeters of sand per meter of core, average sand-layer thickness in each core, and sand/shale ratio (lutite thickness, both hemipelagic and turbiditic, was halved as an approximation of compaction to shale) indicate that sand accumulation is greatest adjacent to the major source areas and along the central north-south axis of the plain (Fig. 2A–C, respectively). The sand concentration results from combined flows from the Wilmington Canyon and Sohm Gap. The isopach of the range in thickness of sand layers in each core shows the sand accumulation to be concentrated in the region of the basin axis (Fig. 2D). It seems likely that sand-layer thickness would be most variable where turbidity currents from different sources interfinger.

The number of sand layers per meter of core is greatest west of the basin centerline (Fig. 2E) and the average sand-layer thickness (Fig. 1B) in this area is comparatively thin (16–18 cm), which may indicate that small turbidity currents are deflected southward before reaching the center of the basin plain. The variability also may reflect lateral thinning of sand layers from flows along the main axis, or possible bypassing of the nearby source and basin centerline areas by some flows. Most likely, it is a combination of these.

Clearly, the principal area of sand deposition is down the central axis of the basin, and turbidites along the axis may be considered proximal. Turbidites with a distal character are found east of the main-flow axis, between the two canyon sources and at the south end of the plain. The tongue-shaped, basin-center area of deposition is termed axial in the basin-plain classification described later.

Columbus Basin

Columbus Basin (Fig. 3) has a DA/BA of 0.7 and is located in a passive tectonic setting that receives exclusively carbonate sediment on three sides from the Great Bahama Bank. Long Island lies northeast of the basin and numerous smaller islands encircle it to the north and west. Turbidity currents enter the basin plain from all directions except east (Bornhold and Pilkey, 1971). The typical flow is quite small due to the restricted source area. Sand is, in general, restricted to the edges of the plain, as evidenced by centimeter-thick sand layers per meter of core, sand-shale ratio, and average sand-layer thickness (Fig. 3A, B and C). Two cores from the center of the basin plain, the more northern of which was recovered from a small topographic rise, contain no sand layers. Sand layers in the central-plain area were generally less than 5 cm thick.

Sand layers are thickest around the edges of the plain (Fig. 3D). This probably reflects the proximity of the source area and possibly some ponding and interfingering of flows restricted to the peripheral area. Sand-layer frequency is greatest in the northwest and south areas of the basin (Fig. 3E). Both maxima are probably the result of interfingering of flows from different source entry points, since they seem to occur between the areas of greatest average sand-layer thickness, which presumably represent main-flow axes.

BASIN-PLAIN CLASSIFICATION

The proposed classification, shown in Table 2, is based on three basin-plain parameters. One of the elements involves basin external characteristics; the others describe characteristics of the basin-plain fill. The basic elements are 1) the distribution of the canyon-mouth entry points, 2) the extent to which sand layers cover the basin-plain floor, and 3) the intra-basin or regional-shape trends of sandbody distribution on the basin plain.

Basin plains are first classified according to the configuration of their basin-entry points, i.e., the location of the mouths of submarine canyons around the plain margins (Fig. 4). The plains are further divided into "types" according to various possible combinations of sand coverage and the basin's composite sandbody shape (Figs. 5–8).

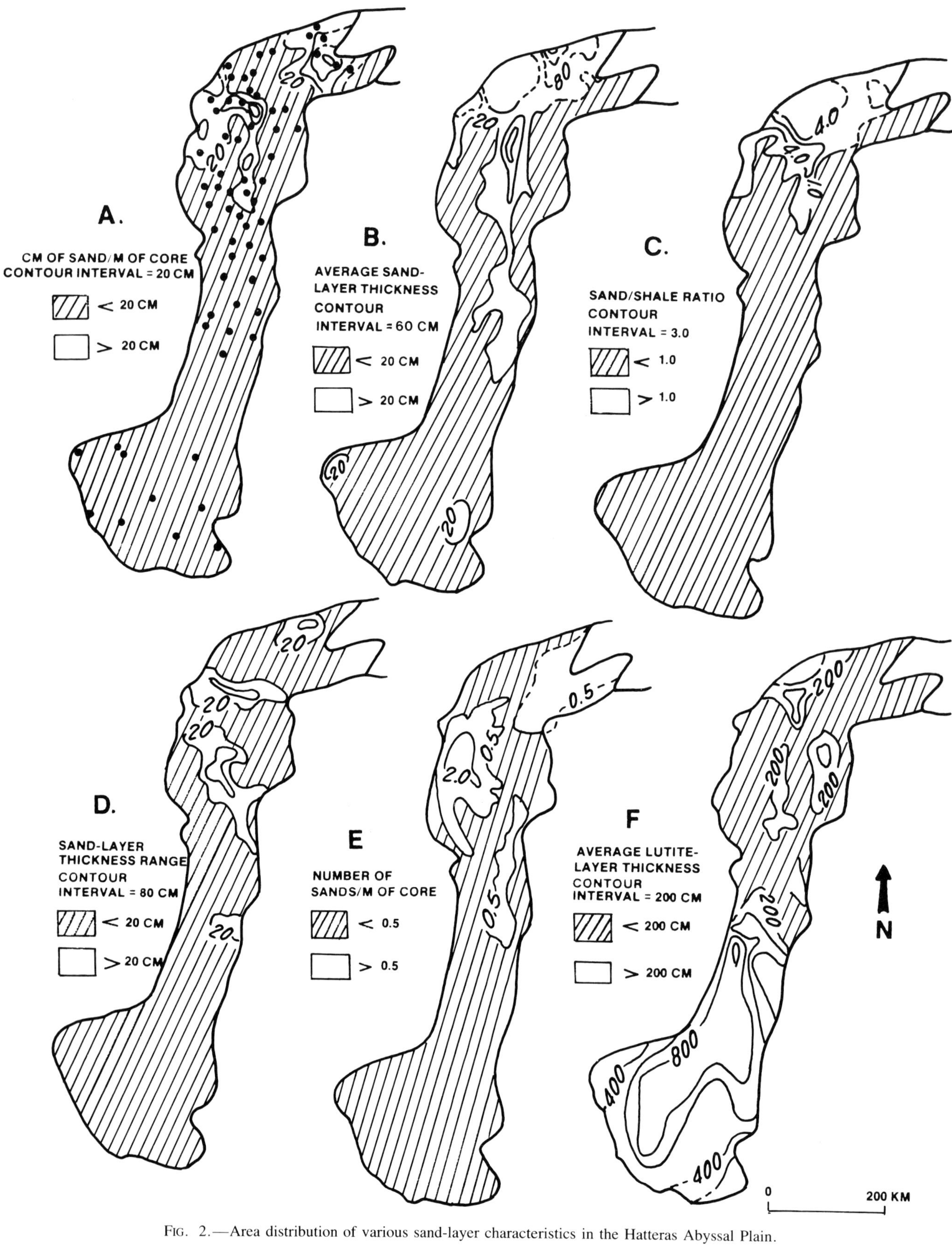

FIG. 2.—Area distribution of various sand-layer characteristics in the Hatteras Abyssal Plain.

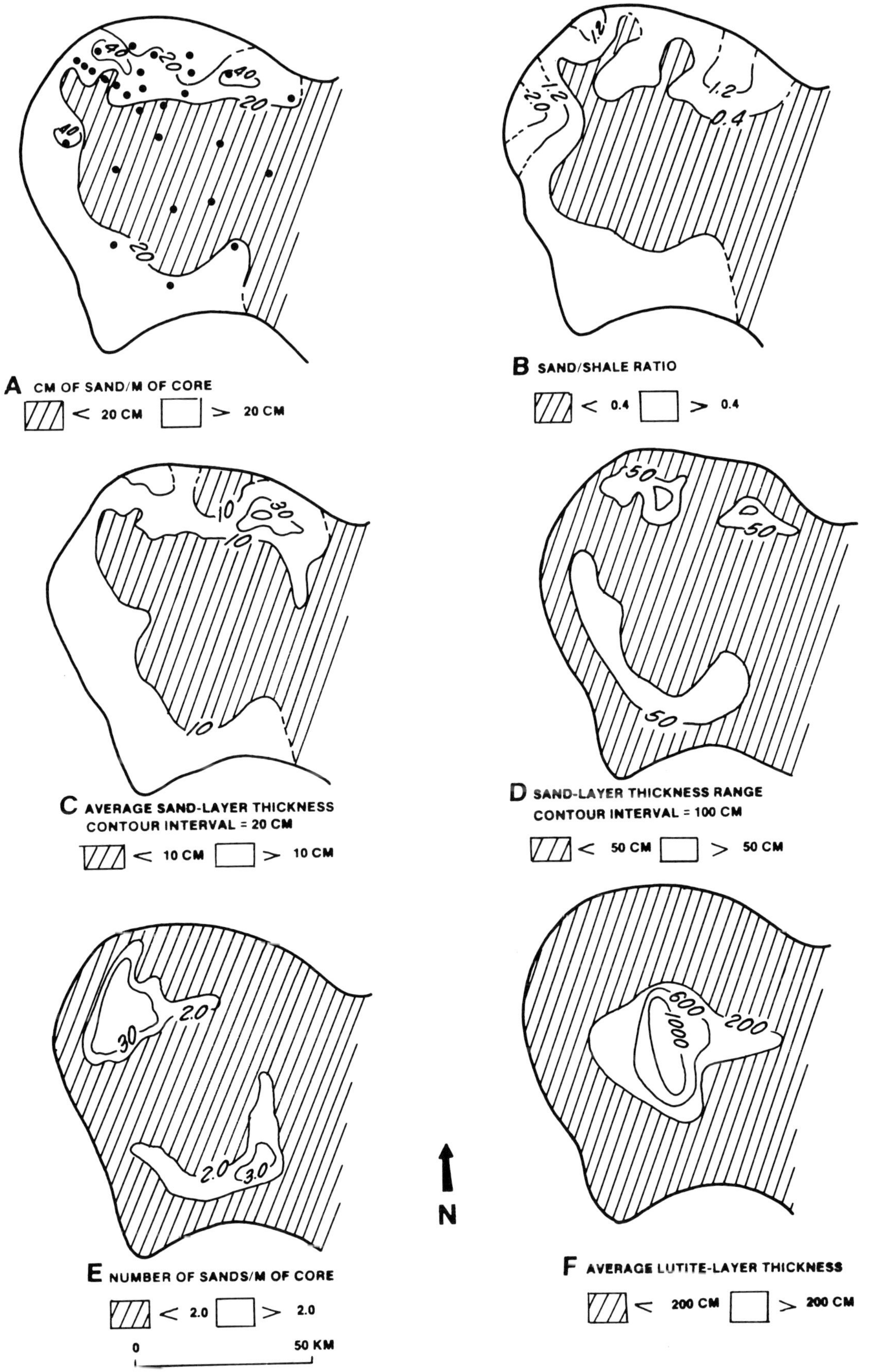

FIG. 3.—Areal distribution of various sand-layer characteristics in the Columbus Basin (After Bornhold and Pilkey, 1971).

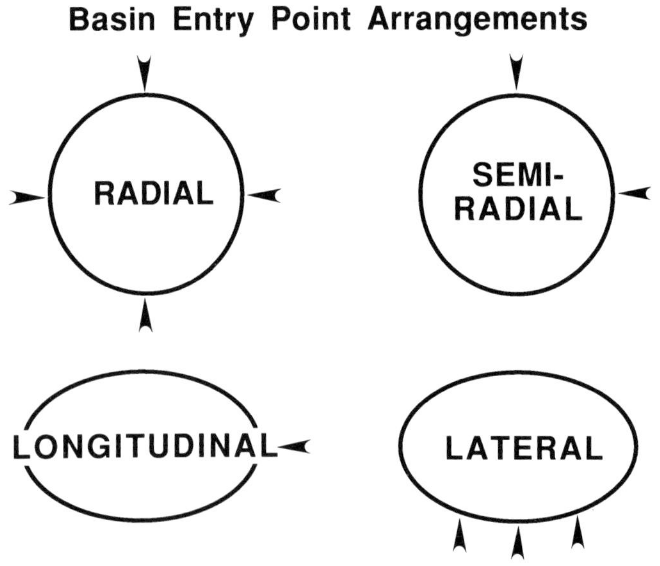

FIG. 4.—Diagrammatic illustration of the four major arrangements of canyon mouths entering the floors of basin plains.

Basin plain-entry point distribution.—

Basin entry-point distribution describes the distribution of sand-bearing sources around the margins of the basin plains. It is important to emphasize that the entry-point distribution does not refer to location of subaerial-source areas, to continental shelves bordering source areas, or to the dis-

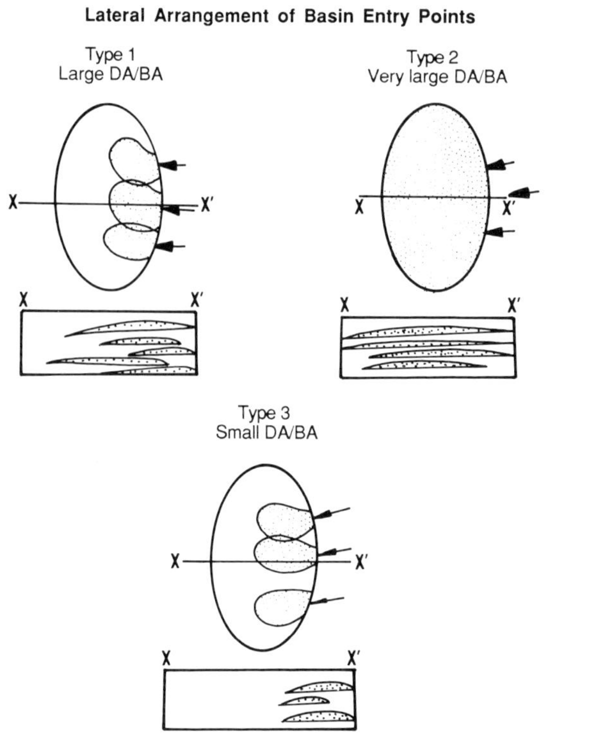

FIG. 6.—Diagrammatic characterization of the types of basin plains with lateral arrangement of basin-entry points.

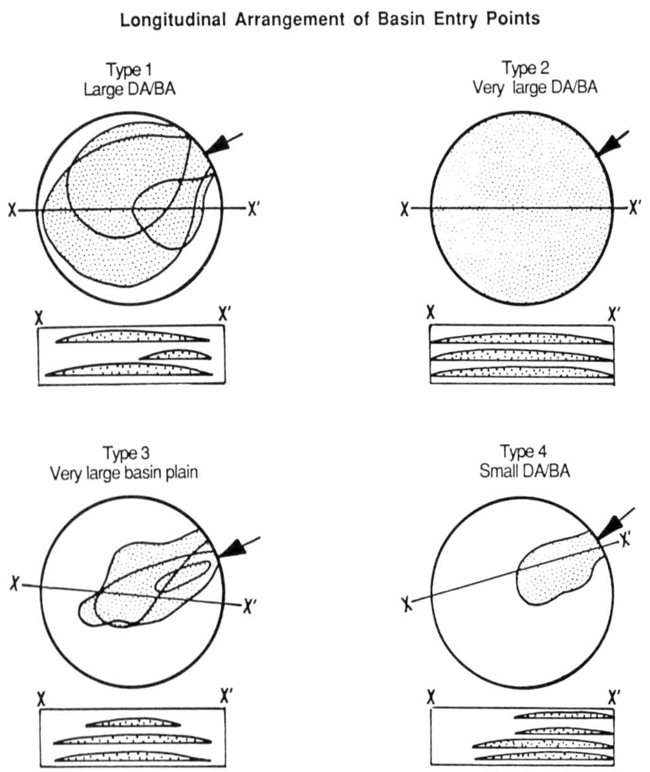

FIG. 5.—Diagrammatic characterization of the types of basin plains with longitudinal arrangement of basin-entry points.

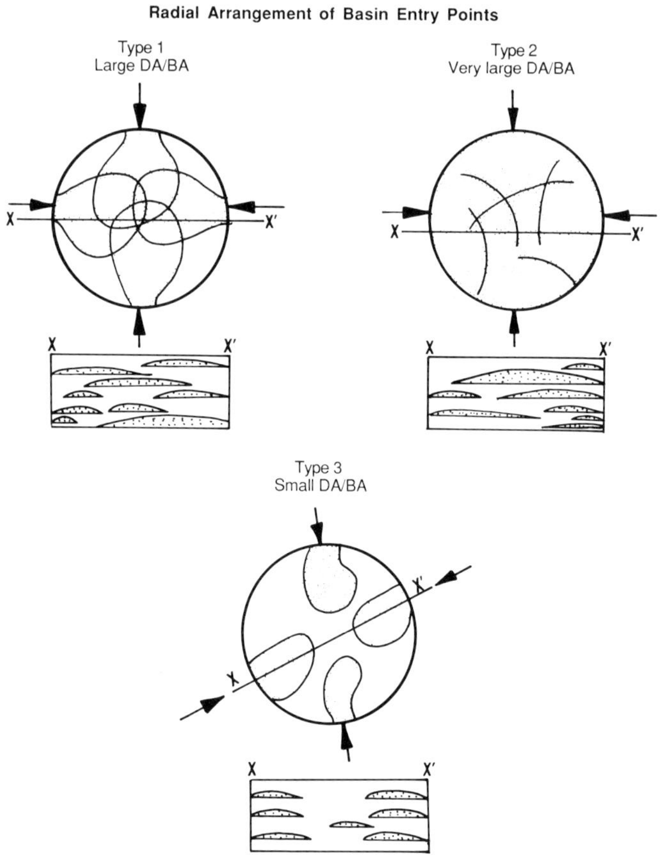

FIG. 7.—Diagrammatic characterization of the types of basin plains with radial arrangement of basin-entry points.

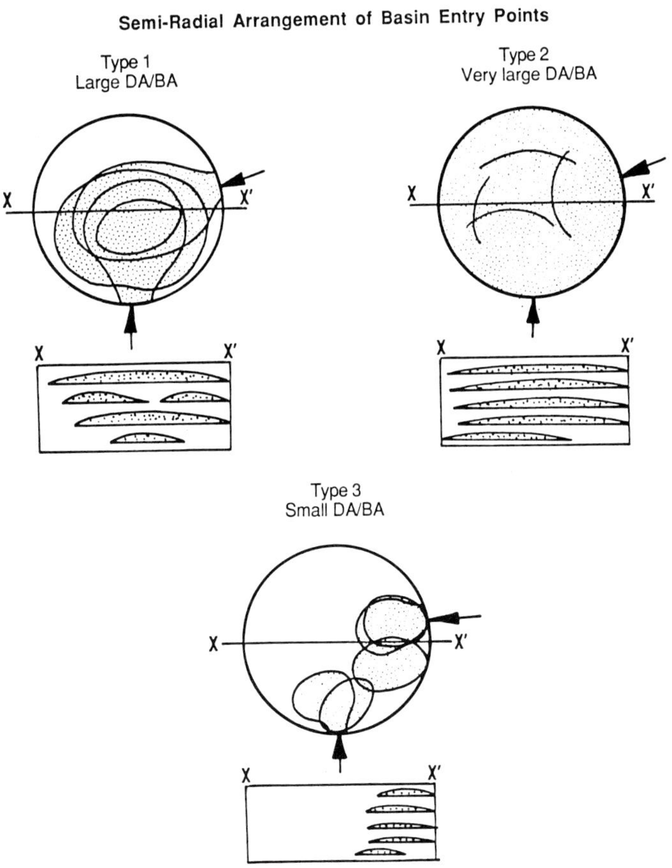

FIG. 8.—Diagrammatic characterizations of the types of basin plains with semi-radial arrangement of basin-entry points.

tribution of submarine-canyon heads incised in adjacent continental slopes. Basin entry-point distribution refers only to the distribution of sediment-flow sources (in most cases, submarine-canyon mouths) on the basin plain proper at the same depth as the basin plain.

Location of subaerial-drainage basins may bear little apparent relation to the basin entry-point distribution. In our study, the best example of this is the Hatteras Abyssal Plain, where little material is contributed directly from the adjacent southeastern continental United States. The combination of the broad Blake Plateau and the Gulf Stream have long prevented sediment escape to the Hatteras plain. Instead, all of the important canyon streams supplying sand to this plain debouch from canyon mouths entering the northern one-third of the plain, and much of the sediment is derived from the northeastern United States.

Four general types of basin entry-point distributions may be identified (after Pilkey, 1987) for the purpose of basin-plain classification: radial, semi-radial, longitudinal and lateral (Fig. 4). Examples of each category are listed in Table 2.

The radial entry-point distribution refers to the situation where there are sources on at least three sides of the margin of the basin plain. A plain with entry points on two sides is called semi-radial. If all the material comes from one side or one end of the basin plain, the plain is assigned to the longitudinal category. Laterally filled plains also receive their sand from one side only, but the basin is distinctly elongated and the entry points are along one of the longer dimensions of the plain (as in the Puerto Rico Trench).

Basins with radial, semi-radial and longitudinal entry-point distributions can be any shape. Basin shape and entry-point distribution are independent of one another, except for lateral-source distribution basins, which are strongly elongated basins, mainly leading to deep-sea trench floor-basin plains. All except one of the observed basin plains (Table 1) that fall into the radial and semi-radial category are more or less equi-dimensional in shape, whereas three out of four with longitudinal entry-point distributions are elongate. The Tongue of the Ocean is a highly elongate basin plain that falls into the radial category, because it is essentially surrounded by the Bahama Bank source. The Sohm Abyssal Plain is considered to be a longitudinally filled basin, although it does not have a simple elongate shape.

Sand coverage.—

Sand coverage refers simply to the areal extent of sand-layer distribution relative to the areal extent of the basin. If sand layers are present over all of the basin plain, sand coverage is termed "overall." If part of the basin plain has *no* sand, sand coverage is termed partial. If sand coverage is partial, the possibility exists for wedge- and axial-shaped sand distributions in the basin plain. If sand coverage is overall, the possibility exists for wedge- and sheet-shaped sand distributions in the basin plain. Thus, four basin "types" are theoretically possible for each of the four categories related to basin-entry points.

Sand-coverage is largely a function of the size of the subaerial-source area contributing sediment to the basin plain relative to the areal extent of the depositional area (DA/BA; Tables 1 and 2). Table 3 lists the general DA/BA characteristics for each of the basin types recognized in the classification. Virtually all of the plains with small (<3) DA/BA ratios exhibit partial sand-layer coverage.

Sandbody Shape

Sandbody shape refers to the difference in the percentage of sand in layers in the sediment column from one side of a basin plain to another. There are three possible composite sandbody shape categories: sheet, wedge, and axial.

Sheet composite sandbody shape refers to evenly distributed sand-layer frequency and thickness over the entire plain floor, which always falls in the overall sand-coverage category. Two observed examples of such distribution are in Tongue of the Ocean and Navidad Basin, which appear to have large DA/BA area ratios.

Wedge-shaped sand-layer distribution refers to the common situation where sand layers are consistently thicker (and lutite sequences consistently thinner) on one portion of a basin plain relative to other portions. Wedge-shaped sand distribution may involve partial sand coverage of the basin plain (Columbus Basin, Algero-Provencal Basin and the Silver Abyssal Plain) or overall sand coverage (Puerto Rico, Grenada Basin and the South Balearic Basin), and is most often a characteristic of a basin with a small DA/BA ratio.

Axial distribution of sand layers is a special category for a more-or-less tongue-shaped, basin-wide distribution of sand with the region of thickest sand layers extending along the central axis of the plain. The sand layers thin toward the sides of the basin floor. This distribution is unique to basin plains with longitudinal orientation of basin-entry points (Table 1) and to very large basin plains.

It is important to emphasize that the composite sand-body-shape classification under discussion is related to turbidite-sand layers, not complete turbidites. The distinction between sand layers and turbiditic lutite sequences in our study was purely visual.

In the Hatteras Abyssal Plain, an example of a basin with an axial distribution of sand layers, mud portions of individual turbidites apparently thicken in all directions (laterally and distally) away from the axis of maximum sand thickness (Elmore and others, 1979). Perhaps the axial distribution of sand-layer thickness is a "requirement" stemming from the ability of turbidity currents to spread sand-size materials greater distances downstream or in a down-flow direction than in a lateral direction away from the axis of the flow. Such a "requirement" would apply only to large basin plains.

PROPOSED CLASSIFICATION

Not all of the combinations possible under the proposed three-tier system of classification of basin plains have been observed in this investigation. Table 2, therefore, includes both real and "imagined" basin plains. Our view of modern basins is somewhat restricted because of the small number of basins observed in detail. Other basin-plain literature is generally not useful because dimensions of sand layers are not reported in a fashion useful for this classification.

Idealized diagrams of the various types of basins in the classification proposed in this paper are shown in Figures 5 through 8. Shaded areas represent the plan view of sand-layer distribution on the plain. Below each circular plan view is a diagrammatic cross-sectional view of sand-layer distribution. The various basin types are further described in Table 2.

Four basin types are noted for longitudinally arranged basin-entry points and only three types for the other entry-point categories. The so-called axial sand-layer distribution was exclusively found in giant elongated basins with longitudinal entry points.

In this classification, basins are first divided into four categories depending on the basin entry-point configuration. Basin plain entry-point configuration and the ratio of drainage area to basin-plain area are the two major factors responsible for the distribution, thickness and frequency of sand layers. The combination of the second and third tiers of the classification allows further division into types 1, 2, 3 and 4. The second and third tiers are basin-sand coverages and sandbody shape, respectively.

All type 1 basins have overall sand coverage and a wedge sandbody distribution. That is, the basin-plain sediment column has sand layers essentially everywhere, but there is a marked difference in the average thickness of sand layers in one portion of the basin relative to that in the remainder of the plain. All type 2 basins are characterized by overall sand coverage and sheet sandbody shape. Type 3 basins exhibit partial sand-layer coverage and a wedge-shaped sand-layer distribution.

The Silver Abyssal Plain does not fit the proposed classification, and is more accurately described as a southern arm of the giant Hatteras Abyssal Plain. The Silver Plain has two sources of sediment: the adjacent carbonate bank to the west, and to the north, where a large amount (90%), of sediment spills over an unnamed abyssal gap from the Hatteras Abyssal Plain. The northern source contributes no sand, since it is more than 700 km from the major sand sources for the Hatteras Abyssal Plain.

In the other 12 basin plains studied, the major sediment sources are the adjacent continental or insular margins, whose sediments are unfiltered by other abyssal plains.

THE PROPOSED CLASSIFICATION AS A PREDICTIVE TOOL—A SUMMARY

Knowledge of sand-layer distribution can provide clues to paleography and be a powerful predictive tool to aid in exploration for petroleum-bearing sand reservoirs. If basin-plain sand-layer geometry is known, basin external char-

TABLE 4.—GENERALIZATIONS ABOUT THE EXPECTED DISTRIBUTION OF SAND LAYERS IN VARIOUS TYPES OF BASIN PLAINS

Basin Type and Subclassification		Proximal/Distal Differences	Location of Thickest Sands	Location of Thickest Lutites	Location of Highest Sand-Layer Frequency
Longitudinal	Type 1	Moderate	Near Entry Points	Distal Regions	Near Entry Points
	Type 2	None	Uniform	Uniform	Uniform
	Type 3	Extreme	Near Entry Points and Along Axis of Turbidity-Current Flow	Distal Regions and Laterally from Axis of Flow	Margins of Axial Sand Tongue
	Type 4	Extreme	Near Entry Points	Distal Regions	Near Entry Points
Lateral	Type 1	Moderate	Near Entry Points	Distal Regions and Margins Between Entry Points	Basin Center
	Type 2	None	Uniform	Uniform	Uniform
	Type 3	Extreme	Near Entry Points	Distal Regions and Between Entry Points	Near Entry Points
Radial	Type 1	Moderate	Near Entry Points	Basin Margins	Basin Center
	Type 2	None	Uniform	Uniform	Uniform
	Type 3	Extreme	Near Entry Points	Basin Center	Near Entry Points
Semi-Radial	Type 1	Moderate	Near Entry Points and Basin Center	Basin Margin Between Entry Points	Distal Regions
	Type 2	None	Uniform	Uniform	Uniform
	Type 3	Extreme	Near Entry Points	Distal Regions	Near Entry Points

acteristics such as size of drainage area and arrangement of basin-entry points can be estimated. If basin external characteristics are known from other lines of evidence, sand-layer distribution within the basin plain can be approximated.

Table 4 lists generalizations concerning the geometry of sand layers in the various basin types. The table shows expected proximal/distal differences in average sand-layer thickness, the expected location of the thickest sand layers, the location of the thickest muds, and the basin location where sand-layer frequency (measured by the number of sand layers per meter of sediment column) would be expected to be highest.

Tables 2, 3 and 4 provide the basis for speculation regarding the paleogeographic setting of a basin and/or speculation as to sand-layer distribution. If sand-layer distribution data listed in Tables 2 or 4 are available, DA/BA ratios and entry-point distributions may be estimated. If the paleogeographic data in Table 3 are available, sand-layer characteristics listed in Tables 2 and 4 can be approximated.

Paleogeographic information is available for many basins; therefore, the DA/BA ratio may be approximated, which would provide the basis for first approximations of the sand-layer distribution using the type of information from Tables 2, 3 and 4. Paleogeographic reconstruction can provide the basis for large-scale basin-wide sand-layer characterization. A large DA/BA suggests an approximately uniform sand-layer distribution. A small DA/BA suggests strong proximal-distal differences in sand-layer character. To refine basin characteristics beyond these parameters, large-scale prediction requires some knowledge of, or estimate of, the location of basin-entry points.

Widespread use of this classification as a predictive tool for either basin external or internal character probably requires the solution of another problem. Basin-plain sediments probably are much more common in the geologic record than is suggested in the geologic literature. The aforementioned widespread belief that basin plains are not sandy because they are located at the distal ends of fans is a geological "urban legend" that is completely unsupported by our studies. In reality, much sand bypasses entire fan systems and is deposited on the plain, which certainly has been shown to be the case in our studies of the Western North Atlantic basin plains. Probably many turbiditic sequences in ancient rocks, now assumed to be fans simply because they are sandy, are actually basin-plain deposits.

REFERENCES

BENNETTS, K., AND PILKEY, O. H. 1976, Characteristics of three turbidites, Hispaniola-Caicos Basin: Geological Society of American Bulletin, v. 87, p. 1291–1300.

BORNHOLD, B. D., AND PILKEY, O. H., 1971, Bioclastic turbidite sedimentation in Columbus Basin, Bahamas: Geological Society of America Bulletin, v. 82, p. 1341–1354.

CLEARY, W. J., AND CONNOLLY, J. R., 1974, Petrology and origin of deep-sea sands: Hatteras Abyssal Plain: Journal of Sedimentary Petrology, v. 37, p. 44–59.

CLEARY, W. J., AND PILKEY, O. H., 1977, Morphology and sediments of three ocean basin entry points, Hatteras Abyssal Plain: Journal of Sedimentary Petrology, v. 47, p. 1157–1170.

DITTY, P. S., HARMON, C. J., PILKEY, O. H., BALL, M. M., AND RICHARDSON, E. S., 1977, Mixed terrigenous-carbonate sedimentation in the Hispaniola-Caicos turbidite basin: Marine Geology, v. 24, p. 1–20.

DOULL, M. E., 1983, Turbidite sedimentation in the Puerto Rico Trench Abyssal Plain: Unpublished M.S. Thesis, Duke University, Durham, North Carolina, 93 p.

ELMORE, R. D., PILKEY, O. H., CLEARY, W. J., AND CURRAN, H. A., 1979, Black shell turbidite, Hatteras Abyssal Plain, Western Atlantic Ocean: Geological Society of America Bulletin, v. 90, p. 1165–76.

ERICSON, D. B., EWING, M., WOLLIN, G., AND HEEZEN, B. D., 1961, Atlantic deep sea sediment cores: Geological Society of America Bulletin, v. 72, p. 193–286.

HEEZEN, B., THARP, M., AND EWING, M., 1959, The Floors of the Ocean, Part I, The North Atlantic: Geological Society of America Special Paper 65, 122 p.

HERRICK, D., 1983, Turbidite sedimentation in the Grenada Basin Plain: Unpublished M.S. Thesis, Duke University, Durham, North Carolina, 72 p.

HORN, D. R., EWING, M., AND DELACH, M. N., 1971, Turbidites of the Hatteras and Sohm Abyssal Plains, Western North Atlantic: Marine Geology, v. 11, p. 287–323.

HORN, D. R., EWING, J. I., AND EWING, M., 1972, Graded-bed sequences emplaced by turbidity currents north of 20° North in the Pacific, Atlantic and Mediterranean: Sedimentology, v. 18, p. 247–275.

KIER, J. S., AND PILKEY, O. H., 1971, The influence of sea level changes on sediment carbonate mineralogy, Tongue of the Ocean, Bahamas: Marine Geology, v. 11, p. 189–200.

NELSON, C. H., AND NILSEN, T. H., 1984, Modern and Ancient Deep Sea Fan Sedimentation: Society of Economic Paleontologists and Mineralogists Short Course Notes No. 14, p. 37–96.

PILKEY, O. H., 1987, Sedimentology of basin plains, *in* Weaver, P. E., and Thomson, J., eds., Geology and Geochemistry of Abyssal Plains: Geological Society Special Publication No. 31, p. 1–12.

PILKEY, O. H., 1988, Basin plains; Giant sedimentation events: Geological Society of America Special Paper 229, p. 93–99.

PILKEY, O. H., LOCKER, S. D., AND CLEARY, W. J., 1980, Comparison of sand layer geometry on flat floors of ten modern depositional basins: American Association of Petroleum Geologists Bulletin, v. 64, p. 841–856.

PILKEY, O. H., AND CLEARY, W. J., 1986, Turbidite sedimentation in the northwestern Atlantic Ocean, *in* Vogt, P. R., and Tvoholke, B. E., eds., The Western North Atlantic Region: The Decade of North American Geology, Geological Society of America, vol. M, p. 437–450.

RUPKE, N. A., 1980, Deep clastic seas, *in* Reading, H. G., ed., Sedimentary Environments and Facies: Elsevier, New York, p. 372–415.

SCHAEFFER, C. L., 1984, Sedimentation in the Blake-Bahamas Abyssal Plain: Unpublished M.S. Thesis, Duke University, Durham, North Carolina, 86 p.

SCHORSCH, L., 1980, The Orleansville turbidite, southern Balearic Basin, western Mediterranean Sea: Unpublished M.S. Thesis, Duke University, Durham, North Carolina, 93 p.

SIEGLIE, G. A., FROELICH, P. N., AND PILKEY, O. H., 1976, Deep sea sediments of Navidad Basin: Correlation of sand layers: Deep Sea Research, v. 23, p. 89–101.

SPARKS, T. N., 1979, Sedimentary structures in Hatteras Abyssal Plain sediments: Unpublished M.S. Thesis, Duke University, Durham, North Carolina, 65 p.

VAN TASSEL, T., 1981, Silver Abyssal Plain carbonate turbidite: Flow characteristics: Journal of Geology, v. 89, p. 317–333.

AN ENTRENCHED THALWEG CHANNEL ON THE RHONE FAN: INTERPRETATION FROM A SEABEAM AND SEAMARC I SURVEY

SUZANNE O'CONNELL
Department of Earth and Environmental Sciences, Wesleyan University, Middletown, Connecticut 06457
WILLIAM R. NORMARK
U.S. Geological Survey, 345 Middlefield Road, Menlo Park, California 94025
WILLIAM B. F. RYAN
Lamont-Doherty Geological Observatory, Palisades, New York 10964
AND
NEIL H. KENYON
Institute of Oceanographic Sciences, Brook Road, Wormley, Godalming, Surrey, GU8 5UB, United Kingdom

ABSTRACT: A recent breach in the western (right-hand, looking downcurrent) levee of the main valley on the Rhone Fan resulted in a rejunevated thalweg within the valley-floor sediment fill and formation of an incipient, sandy, depositional lobe high on the fan. The new thalweg exhibits an unusual cuspate or scalloped form in contrast with the abandoned thalweg downstream of the avulsion that exhibits the more typical meandering or sinusoidal form. The cusps have a radius of curvature that is usually about 0.25 km. Long-range (GLORIA) and high-resolution, deep-tow (SeaMARC I) side-looking sonar images provide details concerning the surface texture (through acoustic-backscatter patterns) and morphology of the relatively young entrenched channel and its surroundings. These data suggest that mass wasting, seen as small gullies on channel walls and deposits on the thalweg floor, accompanied the downcutting of the thalweg and played a major role in forming the cusp-shape morphology. The side-looking sonar images also confirm the widespread occurrence of sediment failure on the main levee areas, as suggested by previous work based on seismic-reflection profiles.

INTRODUCTION

Improvements in swath-mapping techniques such as multibeam echo sounding and side-looking sonar systems provide new insights into understanding sediment-transport processes on deep-sea fan systems. For example, these new tools have shown: extensive networks of meandering channels on the Amazon (Damuth and others, 1983) and Indus (Kenyon and others, 1990) Fans; major bedforms and erosional features of valley floors and levees of the Laurentian (Hughes Clarke and others, 1990) and Var (Savoye and Piper, 1990) Fans; effects of topographic obstructions on the course of channels of the Ebro-Valencia Fan system (Maldonado and others, 1985; O'Connell and others, 1985); and both submarine benches and an elaborate distributary-channel geometry on the Mississippi Fan (O'Connell, 1986; Twichell and others, 1991). It is important to understand how these morphologic features are related to mass wasting and other sediment gravity-flow processes, including debris flows and turbidity currents, that supply fans along continental margins. The plan view of channel and bedform features, especially when accompanied by sediment samples and/or direct observations, permits some quantitative assessment of possible flow processes for turbidity currents (Flood and Damuth, 1987; Normark and Piper, this volume).

The Rhone Fan provides evidence of a submarine-channel morphology not previously described in detail: extremely arcuate (short radius of curvature), cuspate- or scallop-shaped bends with a morphology reminiscent of Martian channels rather than Earth's river systems (Bellaiche and others 1986b; O'Connell, 1986). To date, this configuration has not been observed in spite of extensive seafloor coverage with side-looking sonar and swath-mapping systems, and is particularly well developed in the most recently active, entrenched thalweg within the main depositional-valley fill on the upper Rhone Fan. Our purpose here is to present new GLORIA, SeaMARC I, and SeaBeam acoustic images along a segment of the valley system between water depths of 1,600 m and 2,200 m (Figs. 1 and 2).

The initial interpretations of the depositional processes important on the Rhone Fan are based on SeaBeam bathymetry and seismic-reflection profiles (Bellaiche and others, 1983; Droz and Bellaiche, 1985; Bellaiche and others, 1986a). The available GLORIA and SeaMARC I data provide detailed information concerning both the plan view and surface textures (acoustic-backscatter contrasts resulting from local morphologic relief and sediment grain size) of the fan-valley area. These data are consistent with earlier interpretations of large-scale processes important to the development of the Rhone Fan system and provide additional insights regarding fan sedimentation.

PREVIOUS STUDIES AND REGIONAL GEOLOGY

The Rhone Fan is the largest turbidite system in the western Mediterranean Sea. It is elongate to the south, reaching a maximum width of 200 km and thickness of 1,500 m in the mid-fan area (Fig. 1). It is difficult to interpret the distal boundaries of the fan because of topographic complexities resulting from diapirism and poorly defined lateral contacts with adjacent turbidite systems (such as the Ebro-Valencia Fan complex) in the Mediterranean basin. Sparse reflection-profiling coverage over the southern half of the fan resulted in a wide range of estimates of the fan length that vary from approximately 300 km (Droz and Bellaiche, 1985) to 440 km (Normark and others, 1985). Using the larger estimate, the Rhone Fan covers a surface area of about 7 $\times$ 10^4 km^3 (Barnes and Normark, 1985).

The Rhone Fan was initiated during the Pliocene and continued to develop through the Pleistocene. It received sediment through the Grand-Rhone, Arles, and Petite-Rhone Canyons (Droz and Bellaiche, 1985). The Petite-Rhone Canyon is the most recently active conduit; but during the present highstand of sea level, most of the fluvially sup-

From Shoreline to Abyss, SEPM Special Publication No. 46

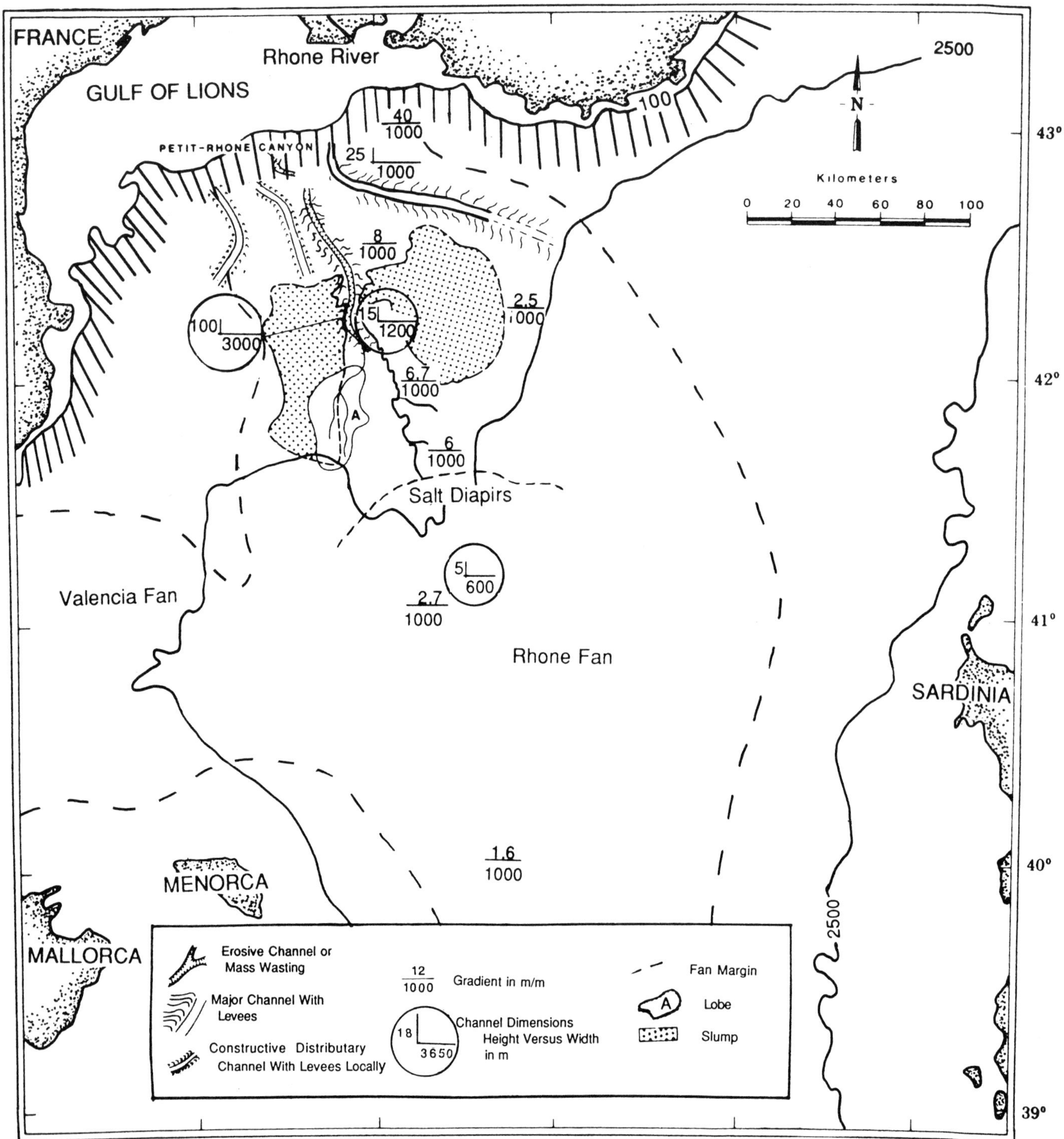

FIG. 1.—Morphometric map of the Rhone Fan (modified from Normark and others, 1985). Hachured pattern below 100-m isobath indicates major relief of continental slope.

plied terrigenous sediment is trapped in the Rhone Delta and does not reach the fan (Droz and Bellaiche, 1985).

In the study area, the Rhone Fan valley developed as a classic example of a depositional leveed-valley system (Normark, 1978; Droz and Bellaiche, 1985; Normark and others, 1985). This leveed-valley system dominates the morphology of the modern fan, but exhibits an avulsion event that resulted in a new channel and abandonment of the leveed-valley system below the avulsion point (42°11′N; 5°W; Figs. 2 and 3). The leveed-valley system provides an elevated pathway for turbidity currents whose outward-facing levee slopes include some of the steepest known de-

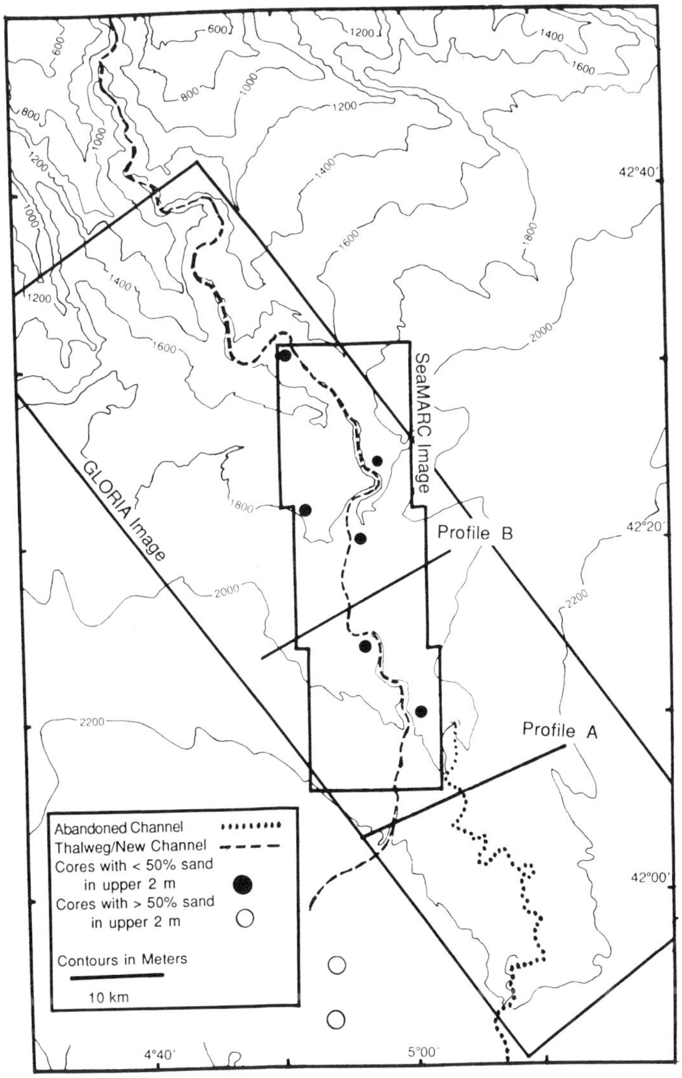

FIG. 2.—Generalized bathymetric map of part of the upper Rhone Fan showing the location of the entrenched thalweg with its extension through the levee breach and the meandering thalweg of the abandoned main valley (modified from Bellaiche and others, 1983). Large northwest-trending rectangle is area of GLORIA image (Fig. 3). North-trending offset rectangle shows area of SeaBeam bathymetric and SeaMARC I data (Figs. 4 and 5). Profiles A and B are presented in Figure 6. Piston core locations are from Mear (1984).

positional slopes on submarine fans. Coutellier and others (1984) and Droz and Bellaiche (1985) showed that large areas of the upper fan on these outward-facing levee slopes have undergone mass wasting.

The internal architecture and morphology of the Rhone Fan result from the stacking of numerous elongate, lenticular units, each consisting of distinctive channel-levee complexes on the upper fan, as observed on seismic-reflection profiles. The axes (i.e., the channels) of these channel-levee complexes have shifted laterally through time, with shifts being more pronounced downslope from the innermost fan (Droz and Bellaiche, 1985).

Several types of turbidity-current channel forms are recognized in this study. We use the terms fan-valley complex or channel-levee complex in a seismic stratigraphic sense to describe that part of the fan characterized by levee wedges flanking a channel conduit. Seismic-reflection profiles are required to determine the extent and thickness of the levee-wedge sequences. We use the term (leveed or depositional) fan valley to refer to the morphologic expression of a levee-wedge sequence of a channel-levee complex. We prefer the term "fan valley" because it was introduced to encourage workers to distinguish fan-channel features from submarine canyons (Shepard, 1965). Leveed valleys characterize the upper reaches of many submarine fans and are depositional, generally elevated pathways for turbidity currents moving from the erosional canyons into the basin area (Normark, 1978). The term "channel" is used to identify smaller scale (generally non-leveed or with small levees) features, including thalwegs (the line of maximum depth) and distributaries, that can be erosional or depositional in character.

METHODS

Ship Navigation

The SeaBeam and SeaMARC I data were collected in 1984 on the *R V Conrad*, and the associated survey lines were positioned using transit-satellite fixes. The GLORIA images, collected in 1983 on the French vessel *N. O. Suroit*, were also positioned using transit-satellite fixes, and the images were digitally processed to improve the appearance and to fit with the French SeaBeam bathymetric map (Bellaiche and others, 1983). Figure 2 shows the respective areas of the two side-looking sonar surveys used in this study.

SeaBeam Bathymetry

SeaBeam is a hull-mounted multi-narrow beam (2.25° beam widths both along and across track directions for each of 16 beams) echo sounder that is computer controlled to obtain sounding information simultaneously from a strip of ocean floor where the width is about two-thirds of the water depth (Renard and Allenou, 1979). During this study, 10-m contours were generated in real time and displayed at an initial map scale of 1:100,000 along the track of the ship. All position and sounding data were recorded on a 9-track magnetic tape for later contouring with edited and improved position data. Post-cruise activities included merging the SeaBeam data from the *R. V. Conrad* with the more extensive SeaBeam coverage of Bellaiche and others (1983). The SeaMARC I images and SeaBeam bathymetry then were adjusted to match; i.e., there was no direct position control for locating the deeply towed SeaMARC I system. Figure 4 shows the merged SeaBeam bathymetry for the SeaMARC I survey area (Figs. 2 and 5).

Side-Looking Sonar

Morphology, surface texture, and structural features are interpreted from the side-looking sonar images that record acoustic-backscatter properties of the sea floor. Two different side-looking systems were employed in this study, GLORIA and SeaMARC I.

GLORIA, a long-range system, operates at a frequency of about 6.5 kHz with a swath width of 30 km and is towed near the sea surface. GLORIA's acoustic parameters and mode of operation are described by Somers and others (1978). Areas of high acoustic backscatter appear as lighter gray parts of the acoustic images (Fig. 3A).

SeaMARC I is a medium-range system, towed near the sea floor. Three main acoustic sensors are on the SeaMARC I vehicle: port and starboard side-looking sonar transducers operating at 27 and 30 kHz, respectively, and a 4.5-kHz downward-looking subbottom profiler. Data were collected both at 5- and 2-km-width swaths. The specifications and mode of operation of SeaMARC I have been described by Kosalos and Chayes (1983), Chayes (1983), and Farre (1984). High acoustic backscatter on SeaMARC I images appear as darker gray to black acoustic returns.

OBSERVATIONS

Leveed Fan-Valley System

The modern Rhone Fan valley-levee system forms the spine of the fan from the continental slope to the point of avulsion (Figs. 2 and 3). The channel-levee deposits rise as much as 500 m above the surrounding sea floor (Fig. 6, and Droz and Bellaiche, 1985). SeaBeam bathymetric mapping (Bellaiche and others, 1983, and this study) shows a large, leveed-fan valley with highly arcuate cuspate bends in the upper-fan area down to a water depth of approximately 1,950 m (Figs. 2 and 3). This is the main valley of the modern Rhone Fan and is 2 to 5 km wide with an average downfan gradient of 7.3 m/1,000 m (0.4°). It is flanked by asymmetric levees. A short, nearly straight channel cuts through the right-hand (looking downstream) levee of the fan valley (Fig. 2). Below water depths of approximately 1,950 m, the main leveed fan-valley complex continues southeast as an abandoned meandering segment (Figs. 2 and 3). Figure 7 shows the levee and valley-fill longitudinal gradients relative to the thalweg and new channel depths.

Piston cores from the depositional lobe immediately downfan from the channel leading from this avulsion point (Fig. 2) contain more coarse-grained sediment in the upper 2 m (Mear, 1984) than cores from the levees and overbank areas of the main fan valley that are predominantly mud and laminated silty clay. These cores have C^{14} ages ranging from 130±50 yrs to 5,250±100 yrs (Mear, 1984), suggesting that the avulsion event is relatively recent.

The levee crests of the main fan valley are well defined on the SeaBeam bathymetry, but are only partially distinguishable on the side-looking sonar images. Well-developed, arcuate, fault-scar morphology is exhibited along the levee crests with the concave scars directed both toward and away from the valley axis (Figs. 3, 5, and 8). The locations of outward-facing submarine slides were previously identified on the basis of SeaBeam bathymetry (Bellaiche and others, 1983) and a transparent acoustic-facies character on 3.5-kHz profiles (Coutellier and others, 1984). These slides are only locally identifiable on the GLORIA image (Fig. 3A and B), possibly because a relatively thin veneer of unconsolidated sediment results in weak acoustic-backscatter properties.

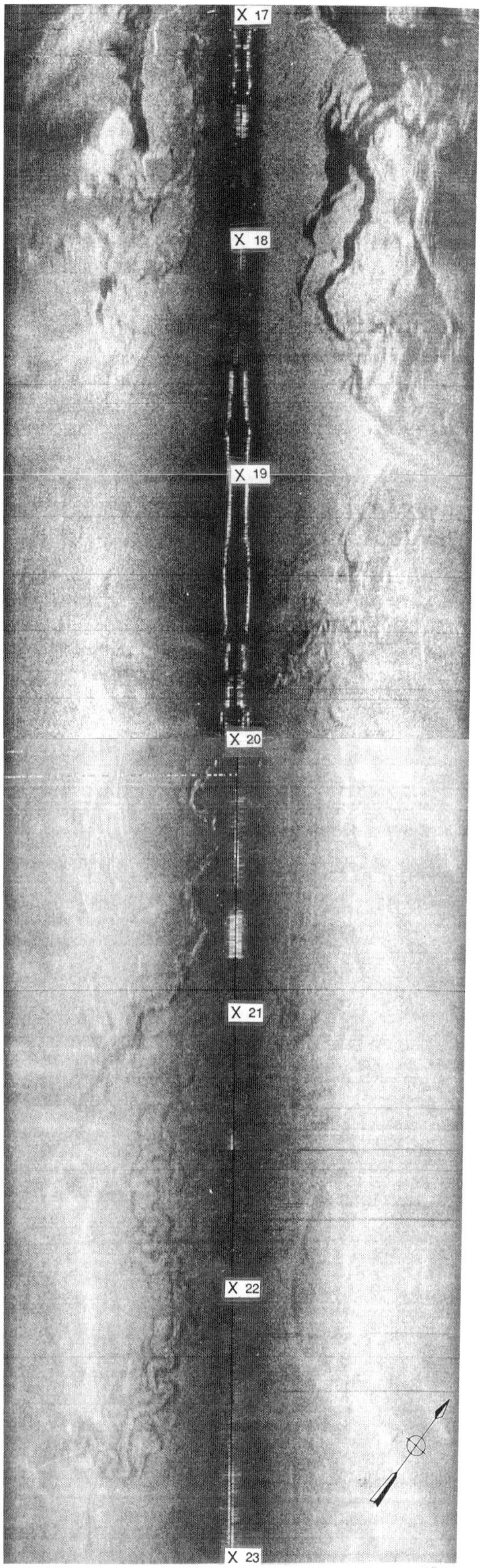

B)

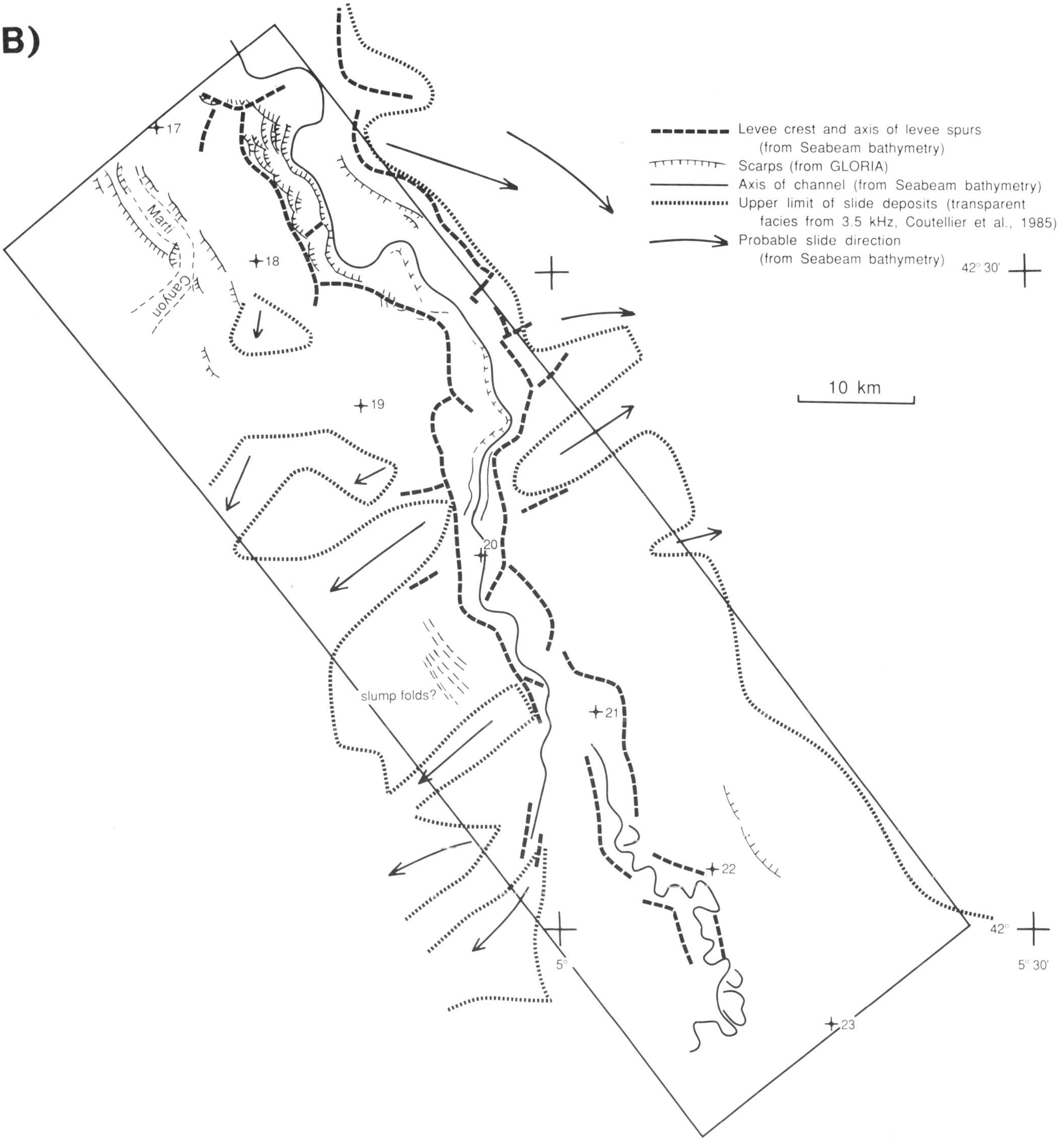

FIG. 3.—(A) GLORIA image of Rhone Fan; location shown in Figure 2. Numbers in the nadir indicate time marks during survey and are used to provide a tie to interpretation in (B). Arrow indicates north. (B) Interpretation of GLORIA image using additional information from SeaBeam bathymetry.

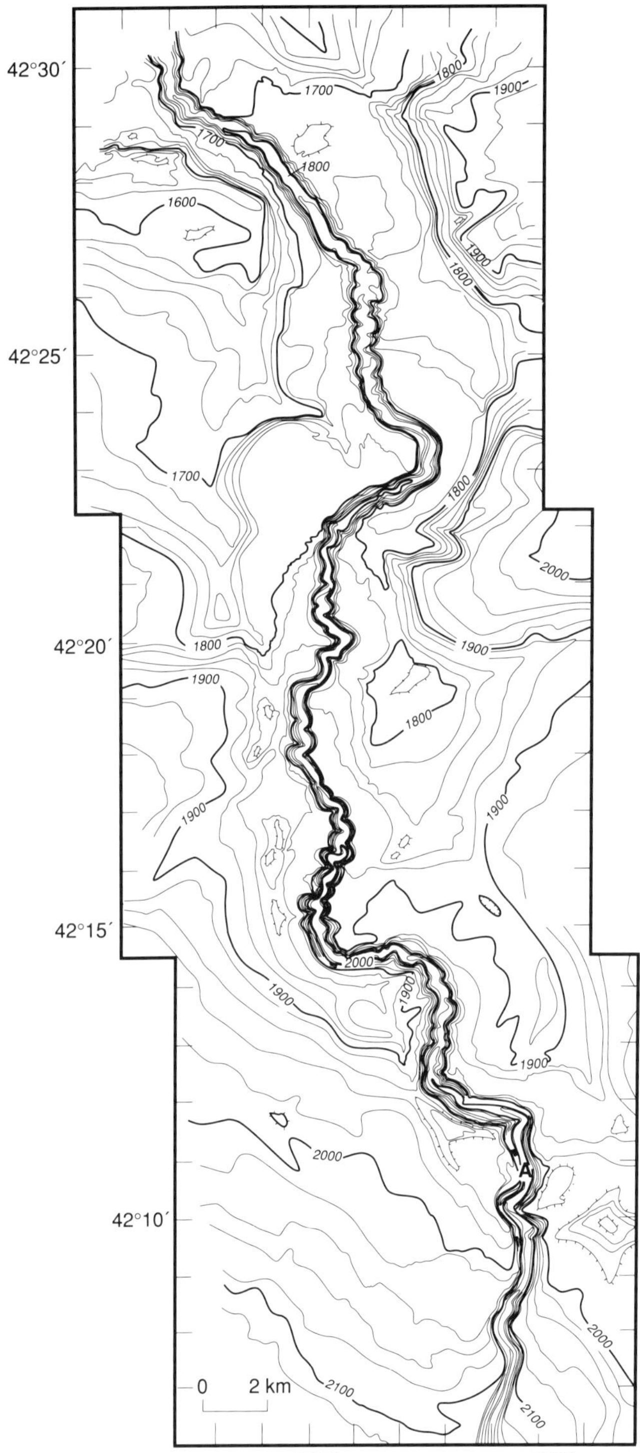

FIG. 4.—Bathymetric map of fan valley extending from Petite-Rhone Canyon; modified from SeaBeam bathymetric map of Bellaiche and others (1983) by incorporation of SeaBeam swath taken during *R V Conrad* survey. Location shown in Figure 2.

The outer levees exhibit acoustic-backscatter patterns that suggest sediment waves or slump folds (Figs. 3, 5, and 8). These features are best documented (more clearly imaged) on the right-hand side of the fan valley, below a water depth of about 1,900 m. The channel formed by avulsion is straighter and incised into the sediment-waves/folds field (Fig. 3). In contrast, the older remnant channel preserves a highly sinuous, meandering path (Bellaiche and others, 1983; Coutellier, 1985; Droz and Bellaiche, 1985). This channel is approximately 30 m deep and less than 1 km wide, with a sinuosity of 2.3. Multiple cutoffs and oxbows are present (Fig. 3), resembling channels observed in GLORIA images of the Amazon (Damuth and others, 1983) and Indus Fans (Kenyon and others, 1990).

Modern Thalweg Channel

Bathymetry.—A readily mapped thalweg channel is entrenched into the fill of the main fan valley (Figs. 3 to 8). The incised channel is narrow, with steep walls and has tight, commonly gullied and terraced, cusp-shaped rims (Figs. 3, 4, 5, 8, and 9). Widths of the thalweg channel vary from 650 to 1,500 m, and there is no consistent downslope change in width. Relief of the thalweg channel (floor to rim) ranges from 85 to 170 m, gradually increasing downstream until the channel floor attains a water depth of 2,120 m (Figs. 3, 4, and 5), where the avulsion occurred. At this location, the thalweg course abruptly changes to the south after cutting through the steep flank of the leveed-fan valley. Beyond the confines of the fan valley (e.g., downstream from the avulsion point), the channel that is continuous with the entrenched thalweg assumes a straighter path. Relief on this extension of the thalweg, which has become the new main channel, decreases abruptly.

The walls of the thalweg within the main valley are very steep, commonly between 20° and 32°. Segments of the thalweg have an extremely tight curvature (Figs. 4, 5, and 9). As with the borders of the main fan valley, the walls are cuspate in form but differ in having radii of curvatures that vary from 0.15 to 0.85 km, with most being close to 0.25 km. Sharply pointed cusps are common (Figs. 4, 5, and 9). The cusps point toward and protrude into the center of the cusp-bounded arcuate forms of the opposite wall. Few cusps are observed near and below the point of avulsion.

Both the main fan-valley floor and the floor of the thalweg upstream from the avulsion have similar gradients of approximately 7.3 m/1,000 m or 0.4°. The gradient of the new channel at and below the avulsion point is more variable, however, ranging from 3.5 m/1,000 m (0.2°) to 87 m /1,000 m (5.0°; Fig. 7). Outside the main fan valley and in a projection perpendicular to the general fan-valley, the seafloor gradient of the levee slopes ranges from 10 m/ 1,000 m (0.6°) to 59 m/1,000 m (3.4°).

Surface textures.—SeaMARC I side-looking sonar images are interpreted in Figure 5. Backscatter patterns generally

FIG. 5.—Interpretation of the SeaMARC I sonographs and selected 4.5-kHz profiles; location shown in Figure 2. See Figures 8 and 9 for examples of original side-looking sonar data

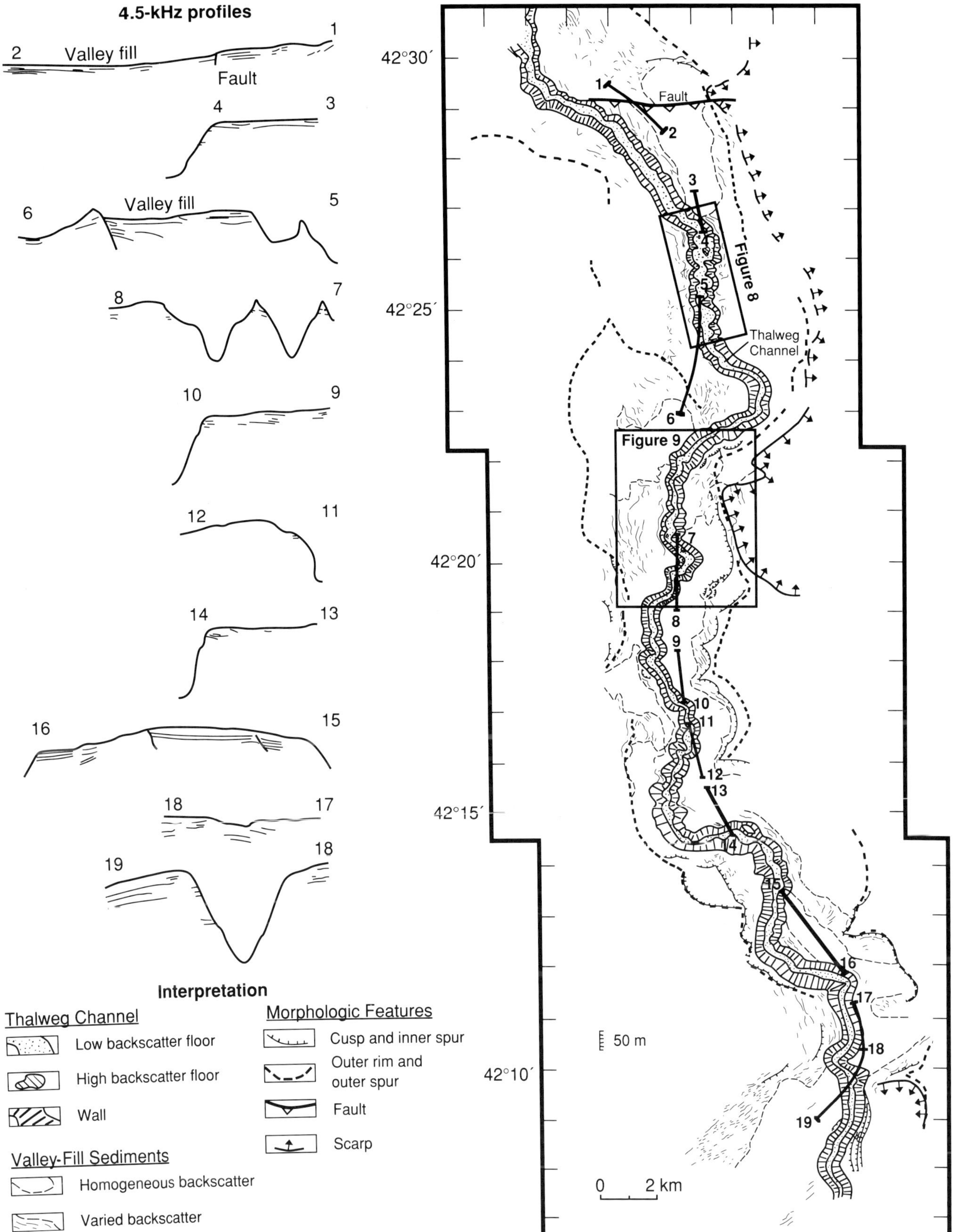
4.5-kHz profiles
1
2 Valley fill
Fault
4 3
6 Valley fill 5
8 7
10 9
12 11
14 13
16 15
18 17
19 18
Interpretation
Thalweg Channel
Low backscatter floor
High backscatter floor
Wall
Valley-Fill Sediments
Homogeneous backscatter
Varied backscatter
Morphologic Features
Cusp and inner spur
Outer rim and outer spur
Fault
Scarp
42°30′
42°25′
42°20′
42°15′
42°10′
4°55′
5°00′
Fault
Figure 8
Thalweg Channel
Figure 9
50 m
0 2 km

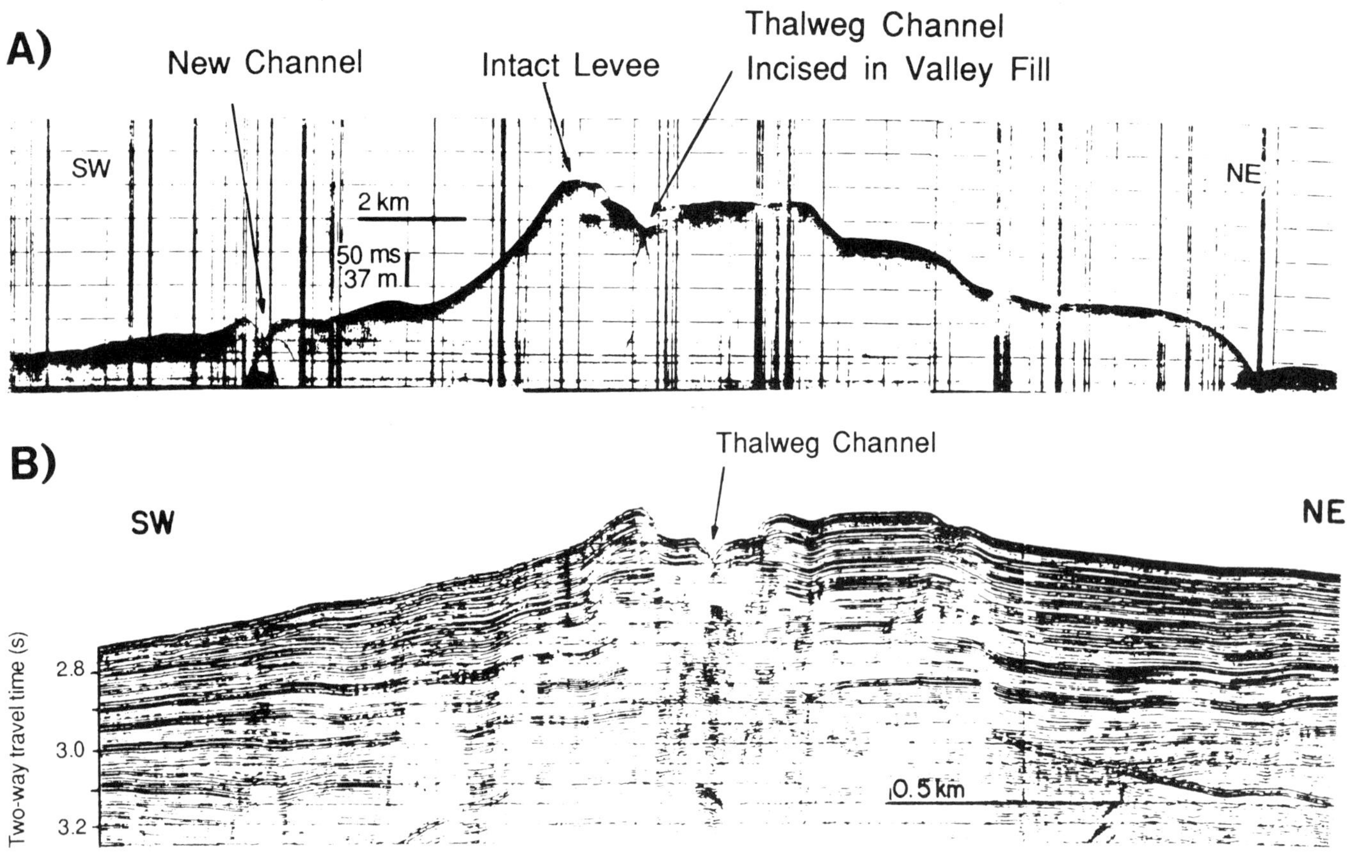

FIG. 6.—Seismic-reflection and echosounding profiles across Petite-Rhone Fan valley. (A) 12-kHz echosounding profile south (downstream) of the avulsion showing 1) abandoned fan-valley fill (flat area) bounded by levees that are highest on the right (looking downcurrent; to left in this profile), 2) abandoned thalweg incised into the valley fill, and 3) new channel incised into fan sediment west of the fan valley. (B) Airgun profile north (upslope) of avulsion point showing levees, valley-floor fill, and entrenched thalweg of the Petite-Rhone channel-levee complex (from Droz and Bellaiche, 1985). Locations shown in Figure 2.

can be assigned to three categories: the thalweg, the floor of the main fan valley, and fan-valley walls and levees (Fig. 5). The thalweg displays a highly variable character, especially on the 2-km-swath-width images (Fig. 9). The thalweg floor is predominantly featureless and flat with low acoustic backscatter (Fig. 9A), although there are some small, higher backscatter patches (Fig. 9B). Small gullies, many of which join together just below the rim, indent the channel walls (Fig. 9C). Between the gullies and the channel floor, a strong acoustic backscatter band is observed (Fig. 9D) that is probably a result of a steepening of the base of the channel walls. The gullies terminate on the face of the wall as hanging gullies, or their lower reaches are buried by mass-wasted debris, whose rough surface may account for the strong backscatter returns.

Terraces (Fig. 9E) are relatively flat surfaces that parallel the cusps on the channel margin including the elongate spurs (Fig. 9F) and are continuous from cusp to cusp. The terraces widen at small reentrants into and beyond the rim (Fig. 9G).

Low-backscatter surface sediments (Fig. 9H) form about 30 percent of the imaged area. These areas are more prevalent east of the thalweg channel, where they occur in patches on the fan-valley floor covering 1 to 4 km^2. High-resolution (4.5 kHz) seismic-reflection profiles show both parallel subbottom reflectors and prolonged acoustic returns, as well as small changes in surface relief within these otherwise featureless areas.

Surficial sediment with distinct backscatter contrasts characterizes nearly half of the survey area. This class of surface texture can be subdivided into two kinds. The first includes straight to curvilinear morphologic features with a large length-to-width ratio that are parallel or subparallel to the fan-valley axis. These cover large areas and resemble the "wrinkle fields" seen on SeaMARC I images in the Valencia Valley (O'Connell and others, 1985). This texture also could result from flat-lying sediment layers intersecting a somewhat irregular surface of the main valley fill. The second kind of surface texture includes low-relief features that form within the valley fill at high angles, commonly perpendicular, to the rim of the thalweg. Many of these features, which might be short ridges, have a wavy or kinked surface trend.

Some scarps (Figs. 3, 5, and 6) face away from the axis of the fan valley (Bellaiche and others, 1986a) and are common within the eastern side of the fan-valley fill. Relief of over 100 m with gradients as steep as 10° is observed. The relief may be along a single cliff or in a series of steps.

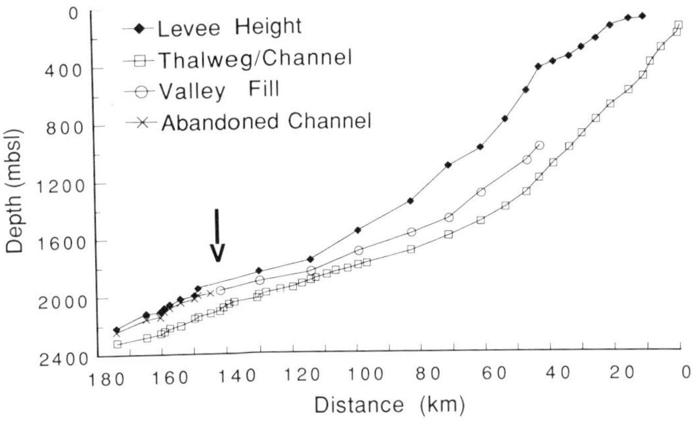

FIG. 7.—Downfan gradients of fan-valley fill, right-hand (western) levee crest, and thalweg of Petite-Rhone valley. Gradients are approximate because only the general valley trend and distance from upper slope (150-m contour) are used. Depth of valley fill is taken from right side of thalweg. Below avulsion point, the depth of new channel is compared to gradients of abandoned-fan valley.

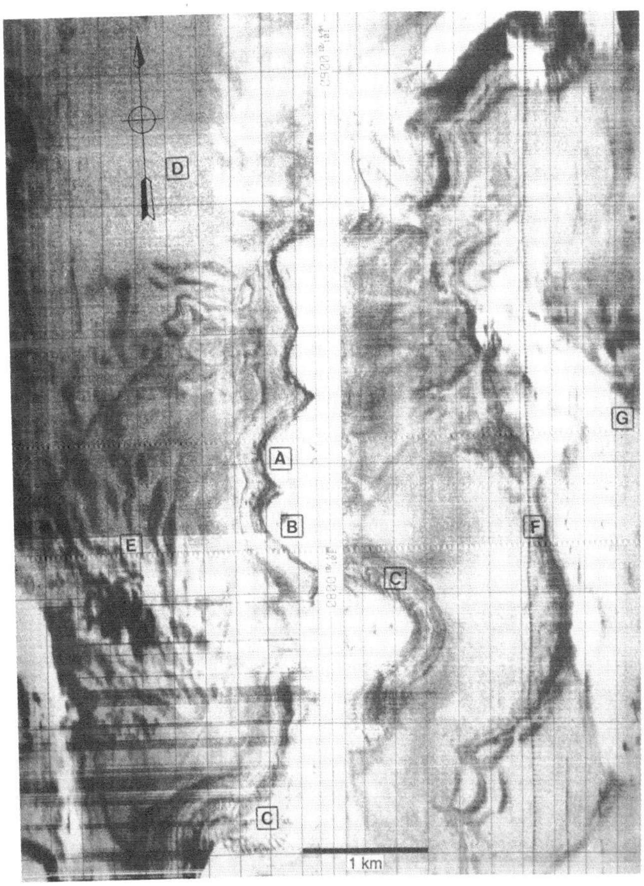

FIG. 8.—Detailed image of fan-valley fill and incised thalweg. Location shown in Figure 5. Letters denote the following: A, flat thalweg floor; B, relief attributed to slump deposits; C, gullies on thalweg wall; D, uniform acoustic backscatter; E, varied backscatter; F, curvilinear inner walls of the main fan valley; and G, slump scarps facing away from main fan valley (from SeaBeam and GLORIA data given in O'Connell, 1986).

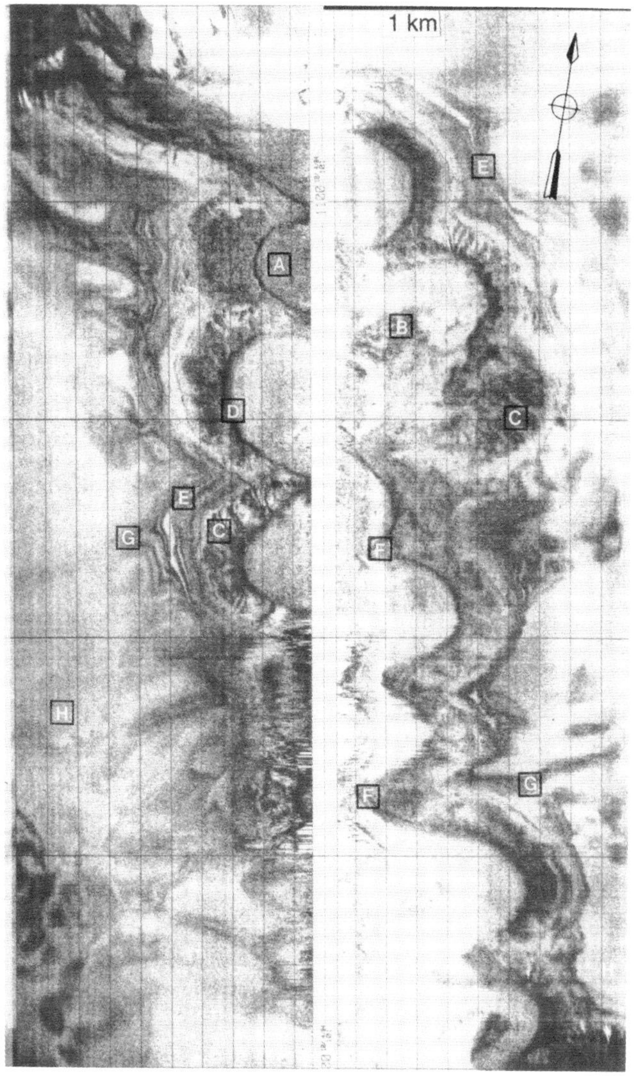

FIG. 9.—Detailed image of thalweg. Location shown in Figure 5. Letters denote the following: A, flat thalweg floor; B, relief attributed to slump deposits on thalweg floor; C, gullies on channel wall; D, high backscatter at base of thalweg wall; E, erosional terraces; F, spurs; G, rim reentrant; and H, main fan-valley fill.

Subsurface structure.—The 4.5-kHz profiles, with penetration between 10 and 30 m, show that most of the seafloor area within the fan-valley fill is underlain by parallel to subparallel subbottom reflectors (Fig. 5). Reflectors commonly outcrop on the thalweg walls. In some profiles, where part of the main fan-valley fill has been removed (in the thalweg), there is a bench-like appearance to the channel wall, suggesting terraces (Fig. 9).

In deeper penetration airgun seismic-reflection profiles, the topographic highs confining the fan-valley fill are predominantly transparent, with the valley floor between them showing a few subparallel reflectors overlying a reverberant channel-fill return (Fig. 6; Droz and Bellaiche, 1985; Bellaiche and others, 1986a). This association of transparent levees and reverberant channel fill, when mapped in plan view, forms lenticular acoustic units that are the major seismic stratigraphic divisions of the fan (Droz and Bellaiche, 1985) and may represent discrete depositional intervals or stages of growth (Mutti and Normark, 1991).

DISCUSSION

Submarine channels and fan valleys generally fall into two major classes, depositional (perched above the general level of the surrounding fan surface) and erosional (entrenched into the underlying fan sediments). Depositional fan-valley and channel systems commonly are associated with submarine fans, on oceanic crust, that are supplied by large rivers (summaries in Normark, 1978, and the wall-chart in Barnes and Normark, 1985). Perched-valley floors are confined between levees that have built above the surface of the adjacent fan. A typical thalweg within depositional valley floors is commonly at a level coincident with or above the level of the surrounding fan. The thalwegs of entrenched (erosional) fan valleys and channels are incised into and lie below the surface of the surrounding fan. Because of a new base level for the Rhone Fan-valley system provided by the avulsion, entrenchment of the thalweg upstream from the point of avulsion is actively modifying the previous depositional system. The perched nature of the main valley floor provided much of the relief relative to the new base level and enhanced the potential for erosional downcutting of the thalweg once avulsion occurred. This example from the Rhone Fan represents a relatively young stage of entrenchment compared to the entrenchments on the La Jolla and Monterey Fans (Normark and Piper, 1969; Normark, 1970a,b). In these systems, the entrenchment has affected all of the original fan-valley-floor areas and cut into the levees on both sides where they are preserved, although the La Jolla Fan presently lacks distinctive levee relief.

The walls of the entrenched thalweg on the Rhone Fan are exceedingly steep (as much as 30°), especially for submarine slopes of partially lithified material. The walls are steepest near their bases, least steep at the rims with the surface of the pre-existing (pre-erosion) valley-fill level, and in places, display relatively flat terraces. The level of the terraces remains nearly constant through several bends, suggesting that this level is not a local phenomenon, but instead marks either a resistant bed(s) that has significant lateral continuity and a gentle basinward dip or a stillstand in the downcutting episode. The difference in steepness with elevation of the wall may be attributed to the state of compaction or lithification of the incised-valley fill. The gentler sloping upper walls may correlate with less consolidated sediment or sandy and silty sediment more easily subject to gravitational collapse. These continuous, flat, terrace surfaces do not appear to be constructional terraces or bars observed on the insides of bends within some depositional channels. In general, depositional bars are not continuous from bend to bend, do not maintain a uniform level between channel bends, and form on alternate sides of the channel on adjacent bends.

Where deeply entrenched into the main valley fill, the thalweg has walls that are cuspate in plan view. The radius between cusps is more or less constant for a particular major bend, with a slight tendency for the radius to enlarge near the downchannel completion of a bend. The walls between cusps are concave toward the thalweg. Where adjacent thalweg-wall reentrants join on the same side of the channel, a cusp projects into the concave reentrant on the opposite side. The cusps narrow where they project into the opposite cusp. Terraces along the rim of a reentrant continue out onto the cusp, wrap around its tip, and continue around the rim to the adjacent cusp.

The reentrants between cusps appear to grow larger as the thalweg entrenches deeper into the valley fill. The enlargement increases the distance between opposite rims and decreases the size of the cusps. A logical sequence would be for the cusp to become so narrow that it collapses, leaving an irregular mound of sediment with high acoustic backscatter on the channel floor (e.g., possibly at letter B in Fig. 9).

The cuspate shape is not common in other submarine channels. Morphologically, the cuspate form of the thalweg most resembles the channels observed on Mars (Sharp and Malin; 1975; MARS CHANNEL WORKING GROUP, 1983; Bellaiche and others, 1986b) and fluvial-channel morphology attributed to groundwater sapping (Laity and Malin, 1985). The Martian channels are as much as 1,100 m deep, which is much greater than the thalweg on the Rhone Fan. However, like the thalweg, the inner-wall slopes become steeper near their bases, exceeding 50° locally. The steepness of the Martian channel walls is interpreted to mean that they are incised into high-strength material, although not bedrock (Tanaka and MacKinnon, 1987).

Shaping of the thalweg walls by a process similar to groundwater-sapping effects on land is possible. Recent observations from the manned submersible *Alvin* have documented fluid discharge along the base of an entrenched thalweg in the Ascension Fan valley and in the erosionally deepened Monterey Fan valley, both on Monterey Fan (Embley and others, 1990). The fluid-discharge sites are marked by chemosynthetic biota that are supported by H_2S in the discharged fluid. The source of the H_2S is thought to be reduced organic matter in the fan-valley and levee sediment, and the gas produced helps to drive fluid circulation. Excavation of the valley fill by erosion provides direct outlets to the sea floor. The production of gas from the reduction of organic matter in fan-valley sediment, as observed on the Monterey Fan (Embley and others, 1990), should be even more likely on the Rhone Fan. The rapid deposition of fluvial sediment (as opposed to the littoral-drift sources for much of the sediment on Monterey Fan) with a relatively high organic content would enhance the process.

In the Rhone Fan valley, the erosion of the walls is accomplished, at least in part, by numerous small gullies that incise the walls. These gullies commonly coalesce below the channel rim. On land, gullies are a common agent of erosion, a mechanism leading to wall retreat, particularly on steep walls. In addition, the evidence of sediment deposits on the thalweg floor suggests periodic slumping, or other erosive processes, causing loss of material from the walls.

In subaerial fluvial systems, incised channels form in response to either decreased erosional resistance or increased erosional force (Schumm and others, 1984). The most common causes of decreased erosional resistance are reduction of vegetation or a disruption of the cohesive forces associated with the sediment. Common causes of increased erosional forces include: constriction of flow, decrease in sediment concentration of flow, steepening of gradient, increased discharge, and decreased sediment load (Schumm and others, 1984). With the exception of vegetative changes, all of the factors that contribute to subaerial-channel entrenchment also might apply to submarine systems. The common element in all of these causes is that more erosion is taking place. In the Rhone Fan, the effects of this erosion on the thalweg channel are fairly dramatic, as seen by the steepness and height of the confining walls, the relatively straight path of the newly incised channel as compared to the highly sinuous path of the abandoned channel, and the unusual shape of the thalweg walls. The incision is related to the avulsion of the western levee, allowing the turbidity currents to flow down the steep former levee slope to a base level more than 100 m below the surface of the fan-valley fill. The change in gradient at the levee crossing perturbed the gradient for the entire channel system, causing headward erosion in the thalweg in an attempt to readjust the gradient, as has been suggested for several other submarine fans (Normark, 1970b,; Normark and Piper, 1969). This readjustment led to the removal of large amounts of sediment by rapid erosion of the valley fill.

The thalweg of the Rhone Fan is apparently a good example of a juvenile entrenched system, one that has formed by breaching the main fan-valley levee, and, as demonstrated by the flatness of the floor and almost identical gradient for the thalweg and unexcavated fan-valley fill, one that may be at or near equilibrium again.

There are at least two possibilities for initiating avulsion of the Rhone Fan valley channel: 1) collapse of the confining levee through mass wasting, as widely documented for much of the upper Rhone Fan valley, or disruption resulting from faulting (i.e., perhaps on a structure similar to the east-trending fault located farther north in Fig. 5); or 2) cutting through the fan-valley levee by an overflowing, erosive turbidity current, in a manner similar to a fluvial crevasse.

CONCLUSIONS

Francis Shepard is well remembered for his work on submarine canyons, but his pioneering efforts to characterize the fan-valley and channel systems that extend from the canyons is not as frequently cited, even though, for example, he was the first to document unequivocally a horseshoe-shaped meander in a fan valley (Shepard, 1966). His detailed map of the meander through careful echosounding work is difficult to appreciate in light of the elaborate, sinuous, channel patterns so easily documented by side-looking sonar systems available today. Shepard and Buffington (1968) provided the first detailed bathymetric map of an entire fan-valley system. The presentation of his bathymetric data in many cases does not provide as striking an image as those provided by side-looking sonar, which in addition to seeing the large-scale relief provides information on subtle relief and sedimentary properties. Nevertheless, his efforts to provide reliable maps of the valley and channel systems associated with submarine turbidite systems become even more impressive as we continue to survey submarine fans with ever more advanced imaging systems.

This study of the elevated, depositional fan valley on the Rhone Fan describes the abandonment of a meandering thalweg channel and the incision of a straighter, more deeply entrenched channel with cuspate walls. The creation of the new channel through a breach in the main valley levee leads to the deposition of coarse-grained sediment, some perhaps redistributed from the main valley fill, on a new depositional lobe high on the fan. This process is somewhat unusual in that it occurred during the present highstand of sea level.

The confining walls of the entrenched thalweg are steep and probably near the angle of repose of partly to moderately compacted sediment. Along the rim, multiple gullies that incise the wall coalesce before reaching the thalweg floor, suggesting that the wall is being actively eroded. The relatively rare occurrence of sediment talus piles along the base of the slope suggests that mass wasting of large areas on the steep slope is infrequent or that the flow within the thalweg is able to flush the system effectively.

ACKNOWLEDGMENTS

The shipboard assistance of B. Alonso, D. Chayes, I. Dinz, J. Freitag, B. Gallagher, J. Lorenzo, K. Kastens, A. Malinverno, C. H. Nelson, A. Palanques, M. Somers, J. Revie, M. Pelras, and officers and crew of the *R V Conrad* and *N. O. Suroit* are gratefully acknowledged. Funding for the *R V Conrad* study was provided by the Joint Committee of Science and Technology of the U.S. Spain Treaty of Friendship, the Comision Asesora do Investigacion Cientifica y Techica (Spain), and the U.S. National Science Foundation (grant OCE 81-18069). Funding for the *N. O. Suroit* study was provided by Société Nationale Elf-Aquitaine and the National Environment Research Council (U.K.). This is Lamont-Doherty Geological Observatory contribution number 4743.

REFERENCES

BARNES, N. E., AND NORMARK, W. R., 1985, Diagnostic parameters for comparing modern submarine fans and ancient turbidite systems, *in* Bouma, A. H., Normark, W. R., and Barnes, N.E., eds., Submarine Fans and Related Turbidite Systems: Springer-Verlag, New York, p. 13–14.

BELLAICHE, G., COUTELLIER, V., AND DROZ, LAURENCE, 1986a, Seismic evidence of widespread mass transport deposits in the Rhone deep-sea fan: Their role in the fan construction: Marine Geology, v. 71, p. 327–340.

BELLAICHE, G., COUTELLIER, V., DROZ, LAURENCE, AND MASSON, P., 1986b, Deep-sea and Martian channels: Deep-Sea Research, v. 33, p. 973–980.

BELLAICHE, G., ORSOLINI, P., PETIT-PERRIN, B., BERTHON, J. L., RAVENNE, C., COUTELLIER, V., DROZ, L., ALOISI, J. C., GOT, H., MEAR, Y., MONACO, A., AUZENDE, J. M., BEUZART, P., AND MONTI, S., 1983, Morphologie au sea-beam del'eventail sous-marine profond de Rhone (Rhone deep sea fan) et de son canyon afferent: Comptes Rendus de l'Academie des Sciences Paris, v. 296, p. 579–583.

CHAYES, D. N., 1983, Evolution of SeaMARC I, *in* Proceedings, Third

Working Symposium on Oceanographic Data systems: Maryland, IEEE Computer Society Press, p. 103–108.

COUTELLIER, V., 1985 Mise en evidence et role des mouvements gravitaires dans l'evolution de la marge continentale: Unpublished Ph.D. Dissertation, Universite Pierre et Marie Curie-Paris VI, Paris, France, 204 p.

COUTELLIER, V., BELLAICHE, G., DROZ, LAURENCE, AND MEAR, Y., 1984 La pente continentale du golfe du Lion et l'eventail sous-marin profond du Rhone: Cartographie des facies acoustiqes obtenus par sondage 3.5 kHz: Comptes Rendus de l'Academie des Sciences, Paris, v. 299, p. 717–720.

DAMUTH, J. E., KOWSMANN, R. O., FLOOD, R. D., BELDERSON, R. H., AND GORINI, M. A., 1983, Age relationships of distributary channels on Amazon deep-sea fan: Implications for fan growth pattern: Geology, v. 11, p. 470–473.

DROZ, LAURENCE, AND BELLAICHE, G., 1985, The Rhone deep-sea fan: Morphostructure and main growth pattern: American Association of Petroleum Geologists Bulletin, v. 69, p. 460–479.

EMBLEY, R. W., EITTREIM, S. L., MCHUGH, C. H., NORMARK, W. R., RAU, G. H., HECKER, BARBARA, DEBEVOISE, A. E., GREENE, H. G., RYAN, W. B. F., HARROLD, C., AND BAXTER, C., 1990, Geological setting of chemosynthetic communities in the Monterey fan-valley system: Deep-Sea Research, v. 37, p. 1651–1667.

FARRE, J. A., 1984, The importance of mass-wasting processes on the continental slope: Unpublished Ph.D. Dissertation, Columbia University, New York, 227 p.

FLOOD, R. D., AND DAMUTH, J. E., 1987, Quantitative characteristics of sinuous distributary channels in the Amazon deep-sea fan: Geological Society of America Bulletin, v. 98, p. 728–738.

HUGHES CLARKE, J. E., SHOR, A. N., PIPER, D. J. W., AND MAYER, L. A., 1990, Large-scale current-induced erosion and deposition in the path of the 1929 Grand Banks turbidity current: Sedimentology, v. 37, p. 613–629.

KENYON, N. H., HUNTER, P. M., WEAVER, P. P. G., AYUB, A., INAM, A., RASUL, N., CRAMP, A., CARLSON, P. R., COOPER, A. K., AND STEVENSON, A. J., 1990, High input and low input turbidite systems: Examples from the Indus fan and the Bering Sea basin (abst.): 13th International Sedimentological Congress, Nottingham, p. 265.

KOSALOS, J. G., AND CHAYES, D. N., 1983, A portable system of ocean bottom imaging and charting, *in* Proceedings, Third Working Symposium on Oceanographic Data Systems: Maryland, IEEE Computer Society Press, p. 629–656.

LAITY, J. E., AND MALIN, M. C., 1985, Sapping processes and the development of theater-headed valley networks on the Colorado Plateau: Geological Society of America Bulletin, v. 96, p. 203–217.

MALDONADO, A., GOT, H., MONACO, A., O'CONNELL, S., AND MIRABILE, L., 1985, Valencia Fan (northwestern Mediterranean): Distal depositional fan variant: Marine Geology, v. 62, p. 295–319.

MARS CHANNEL WORKING GROUP, 1983, Channels and valleys on Mars: Geological Society of America Bulletin, v. 94, p. 1035–1054.

MEAR, J., 1984, Sequences et unites sedimentaires du glacis rhodanien (Mediterranee Occidentale): Unpublished Ph.D. Dissertation, University of Perpignan, Perpignan, France, 218 p.

MUTTI, EMILIANO, AND NORMARK, W. R., 1991, An integrated approach to the study of turbidite systems, *in* Weimer, Paul, and Link, M. H., eds., Seismic Facies and Sedimentary Processes of Submarine Fans and Turbidite Systems: Springer-Verlag, New York, p. 75–106.

NORMARK, W. R., 1970a, Channel piracy on Monterey deep-sea fan: Deep-Sea Research, v. 17, p. 837–846.

NORMARK, W. R., 1970b, Growth patterns of deep-sea fans: American Association of Petroleum Geologists Bulletin, v. 54, p. 2170–2195.

NORMARK, W. R., 1978, Fan valleys, channels, and depositional lobes on modern submarine fans: Characters for recognition of sandy turbidite environments: American Association of Petroleum Geologists Bulletin, v. 62, p. 912–931.

NORMARK, W. R., AND PIPER, D. J. W., 1969, Deep-sea fan valleys, past and present: Geological Society of America Bulletin, v. 80, p. 1859–1866.

NORMARK, W. R., BARNES, N. E., AND COUMES, F., 1985, Rhone Fan, Mediterranean, *in* Bouma, A. H., Normark, W. R., and Barnes, N. E., eds., Submarine Fans and Related Turbidite Systems: Springer-Verlag, New York, p. 151–156.

O'CONNELL, SUZANNE, 1986, Anatomy of modern submarine depositional and distributary systems: Unpublished Ph.D. Dissertation, Columbia University, New York, 302 p.

O'CONNELL, SUZANNE, ALONSO, B., KASTENS, K. A., MALDONADO, ANDRES, MALINVERNO, A., NELSON, C. H., PALANQUE, A., AND RYAN, W. B. F., 1985, Morphology and evolution of a mid-ocean channel, the Valencia Valley (NW Mediterranean): GeoMarine Letters, v. 5, p. 149–157.

RENARD, V., AND ALLENOU, J. P., 1979, SeaBeam, multi-beam echo-sounding in "Jean Charcot:" Description, evaluation, and first results: International Hydrographic Review, v. 56, p. 35–67.

SAVOYE, BRUNO, AND PIPER, D. J. W., 1990, What controls on depositional architecture of the Var Canyon-Fan system? Evidence from side-scan sonar images and high resolution and very high resolution seismic profiles (abst.): 13th International Sedimentological Congress, Nottingham, p. 478.

SCHUMM, S. A., HARVEY, M. D., AND WATSON, C. C., 1984, Incised channels: Morphology, dynamics and control: Water Resources Publication, Littleton, Colorado, 200 p.

SHARP, R. P., AND MALIN, M. C., 1975, Channels on Mars: Geological Society of America Bulletin, v. 86, p. 593–609.

SHEPARD, F. P., 1965, Types of submarine valleys: American Association of Petroleum Geologists Bulletin, v. 49, p. 304–310.

SHEPARD, F. P., 1966, Meander in valley crossing a deep-ocean fan: Science, v. 154, p. 385–386.

SHEPARD, F. P., AND BUFFINGTON, E. C., 1968, La Jolla submarine fan-valley: Marine Geology, v. 6, p. 107–143.

SOMERS, M. L., CARSON, R. M., REVIE, J. A., EDGE, R. H., BARROW, B. J., AND ANDREWS, A. G., 1978, GLORIA II—An improved long-range side-scan sonar, *in* Proceedings, Institute of Electrical Engineering on Offshore Instrumentation and Communications: Oceanology International Technical Session J, BPS Publications Ltd., London, p. 16–24.

TANAKA, K. L., AND MACKINNON, D. J., 1987, Development of the Chryse hydrologic system, Mars (abst.); Lunar and Planetary Science Institute, v. XVIII, p. 996–997.

TWICHELL, DAVE, KENYON, N. H., PARSON, L. M., AND MCGREGOR, B. A., 1991, Depositional patterns of the Mississippi Fan surface, Gulf of Mexico: Evidence from GLORIA II and high resolution seismic profiles, *in* Weimer, Paul, and Link, M. H., eds., Seismic Facies and Sedimentary Processes of Submarine Fans and Turbidite Systems: Springer-Verlag, New York, p. 349–363.

PART V
GEOMORPHOLOGIC AND TECTONIC STUDIES

GENETIC GLOBAL GEOMORPHOLOGY: A PROSPECTUS

ELAZAR UCHUPI AND K. O. EMERY

Woods Hole Oceanographic Institution, Woods Hole, Massachusetts 02543

ABSTRACT: Earth's surface morphology is primarily the result of interaction between plates moved by seafloor spreading and/or intraplate tectonic/magmatic processes. Once the youthful endogenetic (tectonic/magmatic) terranes are isolated from inter/intraplate influences as a result of long-continued lateral migration or changes in geometry, exogenetic processes (erosion and deposition) subdue and reduce the original relief. The rate of this modification and the nature of the geologic processes involved in it are controlled by climate, which may change with time or with migration of the plates across climatic zones, and by oscillations in sea level. Whether a terrane can reach an old-age stage in the geomorphic cycle depends upon its isolation from plate activity long enough for non-tectonic processes to complete its degradation. The low relief of ancient terranes in Precambrian shields is a clear indication that the geomorphic cycle can come to completion.

INTRODUCTION

The first-order relief features of the Earth's surface are the continents and ocean basins, including marginal seas. Areas above sea level with a mean elevation of +840 m occupy 29.2 percent of the surface, and the oceanic areas with an average depth of −3,729 m (Menard and Smith, 1966) constitute the remaining 70.8 percent. The crust in the land areas, containing more light granitic and gneissic lithotypes and more silicic material than that in oceanic basins, is zero to a few kilometers thick at the continental/oceanic crustal boundary near the continental slope, less than 20 km thick beneath the outer continental shelves and more than 70 km thick in the Himalayan suture region, the Tibetan Plateau, and the Altiplano in Peru (Emery and Uchupi, 1984, p. 201; Meissner, 1986, p. 1–7). According to Meissner, continental crust is more variable and inhomogeneous than oceanic crust and displays complex interactions with the mantle, having magma generation and differentiation of its own. Oceanic crust is much more mafic than continental crust and is only about 5 km thick. At distances from oceanic-subduction complexes, oceanic crust remains fairly constant in thickness with time from its creation along the mid-ocean ridge and other spreading axes to the continental margins, where it may be as old as 185 ma. The boundary between continental and oceanic crusts varies considerably. It may be marked by an outer high or be imperceptible with one crustal type grading into the other along divergent (passive) margins; be a series of linear highs and lows of continental and oceanic affinities along translation margins; or be an underthrust contact along convergent margins. Oceanic and continental provinces also differ in that oceanic basins are transitory, so that they and their geologic record (except for strands of ophiolites obducted onto the continents) ultimately are destroyed by the same subduction that leads to continental collison. In contrast, continents and their complex geologic records span nearly 4 billion years, although their outlines change by removal of pieces by rifting and addition of these pieces elsewhere by subduction of oceanic crust and by continental collision.

Geologic processes that modify tectonic relief also differ within the first-order relief features of the Earth. Erosional processes dominate on land, although low areas can become important depocenters of continental sediments. In oceanic basins deposition dominates, although erosion by gravitational processes that are active below sea level (slumping, debris flow, and turbidity currents) and oceanic currents can provide much sediment for deposition along continental margins. Except locally, where igneous rocks are broken into talus by faulting and mass movements, most degradational processes (mechanical, chemical, and biological) have little effect on igneous terranes below sea level. Wave action is significant only near sea level and biochemical-carbonate aggradation occurs only in the photic zone of carbonate platforms and reefs. This reversal from dominant erosion on continents to dominant deposition in oceanic basins is illustrated by the transfer via streams, glaciers and wind to the ocean floor of much of the sediment produced by erosion on land. In addition to these detrital components, the dissolved loads of streams brought to the ocean are then withdrawn from solution by plants, whose tissue and skeletal parts are used as food by small and large animals, and the remains of both plants and animals are added to the detrital minerals as ocean-floor sediment. For example, Emery and Milliman (1978) estimated the total contributions of land-derived detrital sediment to the oceans via streams to be 8,010 × 10^6 tons per year and the dissolved components as 3,830 × 10^6 tons per year. A subsequent compilation by Milliman and Meade (1983) increased the estimates of detrital component to 13,500 × 10^6 tons per year in suspension plus 1,000 to 2,000 × 10^6 tons per year for dissolved material. To this must be added perhaps as much or more detrital sediment brought by glaciers and wind and a much lesser contribution of dissolved material by ground water from land.

The ocean itself provides detrital and dissolved sediment by wave erosion and solution along shore (especially from sea cliffs and calcareous reefs) for subsequent deposition. These detrital components approximate 30 × 10^6 tons per year for the Atlantic (Emery and Uchupi, 1984, p. 688) and perhaps 150 × 10^6 tons per year for the World Ocean. Additional contributions of marine sediment are from abrasion and solution of shallow-water platforms, leading to the building of some continental shelves or parts of them, carbonate-platform accretion by biochemical processes in shallow water, and chemical precipitation in shallow water (evaporites) and in deeper water (manganese oxide, phosphorite, etc.). Regardless of their source from land or ocean floor, sediment tends to concentrate in the deeper areas of the ocean.

From Shoreline to Abyss, SEPM Special Publication No. 46

GEOMORPHOLOGY

Qualitative description of topography of land areas (second-order relief features) and its causative agents was followed by compilation first into local and eventually into regional maps. The next step was a search for a reasonable sequence of topographic forms of the land surface—a genetic classification of land forms. Much of this was synthesized by William Morris Davis (1909). The impact of this geomorphologist/geographer on the study of the evolution of land forms was greater than that of any other person (Thornbury, 1957, p. 10). Davis' concept of the geomorphic cycle stated that the evolution of a landscape follows a systematic sequence, which he designated initial, youth, maturity, and old age. This idea that differences in land forms can be explained in terms of differences in geologic structure, geomorphic processes, and stages of development has deeply influenced most geomorphologists. The Davisian school of geomorphology is based on the premise that a relatively rapid initial tectonic uplift was followed by a period of stability that permitted the cycle to come to completion and allowed the initial uplift to be reduced to a nearly featureless surface. Support was provided by Chamberlin (1913–1921) in a series of short articles spanning nearly a decade, by Umbgrove (1947) and his epochs of folding, and by legions of physical-geology textbook writers who used their sequential diagrams. An extension to shoreline erosion and deposition following the Davisian concept also was produced by Johnson (1919). Examples of topography produced by different agents and illustrating the different erosional and depositional stages on rocks of different internal structure soon were published (Salisbury and Atwood, 1908; Dake and Brown, 1925).

This concept of geomorphology has not met with universal approval. It was recognized early that complications caused by renewed tectonism, change of erosional or depositional agents, or other factors introduced uncertainties in use of the Davis and Johnson geomorphic cycles. For example, Walther Penck (1924) and his students believed that rapid uplift followed by a long period of stability was not the norm. They stated that the initial uplift of the land was extremely slow and was followed by an accelerated uplift. Such a sequence could prevent a landscape from passing through the stages of development described by Davis and Johnson. Other geologists questioned this concept of long periods of stability that allow the cycle to proceed to completion (Shepard, 1923). Scheidegger (1987, p. 1) also stated that it is not possible to give a 'linear' (initial, youth, maturity, and old age) classification of geomorphic features. It is because of these and other factors that Shepard (1937) proposed his classification of shorelines of *Primary Origin*—generally non-marine origin of the shore by land agents, and *Secondary Origin*—shorelines modified by marine erosion and deposition. Thus arose a general consensus among geologists that such processes as tectonics, extent of weathering (mechanical and chemical), drainage, and bioerosion are so varied that the geomorphic history of any given region cannot be reconstructed so simply.

Herein, we reconsider the present configuration of the Earth (its geomorphology) within an all-encompassing plate-tectonic concept. Under this unifying concept, initial relief is created primarily as a result of interaction of plate movements (divergence, convergence, and translation) and secondarily as a result of intraplate magmatic/tectonic processes, including passage of a crustal plate over a fixed mantle hotspot (Crough, 1979, 1981; Morgan, 1983), changes in plate geometry (Cloetingh, 1988), and push forces originating in mid-ocean ridges (Zobach and others, 1986). Once a region moves away from these influences, the young high-relief topography is subdued by erosion and deposition on land and by deposition in associated oceanic basins. Isostatic sinking of positive tectonic features in oceanic basins because of thermal decay (Sclater and others, 1971) also diminishes tectonic relief. Oscillation of sea level (mainly tectonically induced) through time (Haq and others, 1987) modifies this trend, as does the migration of plates across different climatic zones. Additional complications result from epeirogenic motions of continents that indicate that they may be decoupled from the oceans and behave independently from them. For example, in Africa, five erosional surfaces have risen to elevations above 3 km (Holmes, 1965, p. 605–609), much higher than the estimated maximum rise in sea level of 600 m during the Cretaceous (Hancock and Kauffman, 1979). Comparison of the deviations of sea level between different continents also indicates major episodes of differential subsidence and uplift for Europe, the Americas, Africa, Australia, and Asia (Harrison and others, 1983). Consequently, at any given time, the topographic expression of an area reflects conditions existing at that time, so that the final morphology is a summation of all these events. The configuration of a given region may not change in the systematic manner described by Davis, but a geomorphic cycle does progress in a general direction leading to a low-relief region.

GEOMORPHIC CYCLE

General

Toward the end of the Paleozoic Era, two-thirds of the Earth was occupied by a wide oceanic basin and the other third by the megacontinent Pangaea (Coney and others, 1980). Consolidation of Pangaea resulted in a discontinuous and sinusoidal band of mountain chains wrapped around older cratonic blocks and stable platforms (Hurley, 1974). This consolidation resulted from the northward drift of Gondwana and detached fragments and their collision with Laurasia (Lefort, 1989, p. 106). Detached fragments are represented by a complex of allochthonous terranes welded onto the deformed Paleozoic Appalachian miogeocline. They differ from one another in their stratigraphy, structure, metamorphic histories, faunas, mineral deposits, and paleomagnetic character (Williams and Hatcher, 1982). These unique characteristics may be responsible in part for minor topographic differences along the North American Atlantic seaboard, or they may indicate former glacial processes at the north and fluvial ones at the south.

The present morphology of the Earth's surface is a consequence of the Mesozoic fragmentation of Pangaea, dispersion of the fragments, opening of the Tethys and Atlantic Oceans, terrane accretion including subduction of mid-ocean ridges around the periphery of the Pacific Ocean, and secondary back-arc spreading in the Balearic-Liguro-Provençal Basin of the western Mediterranean and basins in the western Pacific. Associated with this dispersion and consolidation was a 600-m rise in sea level during the Cretaceous that inundated much of the continental areas. Consolidation since the Cretaceous has led to nearly complete closure of the Tethys and to high-standing continents, possibly displaying their greatest relief of all geologic time. Opening of the Greenland-Norwegian Sea, the Atlantic, and western Tethys occurred in three distinct stages: rifting, when the site of continental separation displayed a complex mosaic of half-grabens and horsts and synsedimentary faulting; a thermally induced uplift phase and subsequent erosion of the rift system during continental decoupling; and a seafloor-spreading phase, when the continental margins became stable and domains of thermal subsidence and exogenetic processes. Fragments rifted away from Gondwana in southern Pangaea and microcontinents present in the Tethys during the Mesozoic drifted northward, colliding and accreting onto eastern Asia and western North America. Toward the end of the Mesozoic, the counterclockwise-rotating Africa collided with Eurasia, closing the western Tethys. It is this complex motion of plate movements that has produced the present plate configuration, and distribution of continents and endogenetic and exogenetic realms. This Wilsonian cycle of continental fragmentation, dispersion, convergence, and collision has been repeated numerous times throughout Earth's history. In such a regime of never-ending Wilsonian cycles, a region can be reduced to low relief only by being isolated from the cycle; the longer the isolation, the greater the degradation.

If Earth in common with other terrestrial planets developed a global tectonic grid early in its history, as postulated by Bryan (1986) and others, this orthogonal or near-orthogonal fracture pattern may have controlled the orientation of rifting trends, orientation of spreading axes, and position of transform faults in the incipient Atlantic and other marginal seas. That pattern may continue to guide spreading directions in the present Atlantic Ocean. Only after oceanic crust becomes detached from the continents can spreading geometry reflect true mantle-flow directions. However, deviations of sea level from mean positions on different continents indicate that continents may act partly independently of ocean basins. Thus, the close linkage between continents and oceanic basins postulated by Bryan for a "mature" basin such as the Atlantic must terminate early in its seafloor-spreading history. Transcurrent motion along intra-continental shears (the Amazon shear system and the Pisco-Juruá fault that had transported the Guyana Precambrian shield westward relative to the Brazilian Precambrian shield) was reactivated during the separation of North America from Africa-South America during the Mesozoic. This reactivation included intrusion of more than 100,000 km^3 of volcanic rocks into Paleozoic evaporites of the Amazon trough, a 100-km southwestward translation of northern South America, which led to shearing of the subducted Nazca Plate and assistance in forming the Nazca Ridge (Szatmari, 1983). The 130- to 150-km-long Benue Trough, encroaching on the African Precambrian shield and deformed during the Santonian, is along a left-lateral shear that, together with those in Dahomey, Togo, Ghana, the Ivory Coast, and Liberia, may extend northeastward across Africa to the eastern Mediterranean (Neev and others, 1982). Deformation along the Benue Trough shear resulted from convergence of the Northwest African and Southeast African plates to form the present continent (Burke and Dewey, 1974).

Two antagonistic processes were involved in the land/seascape evolution since the breakup of Pangaea: endogenetic (tectonism/magmatism) that formed the primary morphology, and exogenic (erosion and deposition) that modified it. Although these processes occur concurrently as long as a particular region is in the realm of plate activity, tectonism still dominates (see Scheidegger, 1987, p. 1–6, for a discussion concerning the taxonomy of a landscape). Plate activity probably is not continuous, but the degree of its periodicity has yet to be resolved. Tectonism along plate boundaries probably is a process of accumulating and releasing stress. The rates at which these stresses are built and released are yet to be resolved. Along belts of divergence in the mid-ocean ridge, where crustal thicknesses are small, stresses probably are released from hundreds to at most a thousand years (ARGO—RISE GROUP, 1988). In contrast, along subduction zones, stresses on overlying thick continental crusts may build for millions of years before release, perhaps explaining why deformation in the Rocky Mountains did not begin until the latest Cretaceous and peaked in the Paleocene, even though North America began moving westward during the Middle Jurassic. Such a lag also may be responsible for the quiescence that followed, allowing the uplifted region to be eroded to low relief before being again rejuvenated by normal faulting during the Neogene. Stresses where oceanic crust is being subducted beneath oceanic crust probably may be relieved at time intervals intermediate between those of margins and mid-ocean ridges. It is within this framework of dominant processes that we later describe the Earth's morphology. Where endogenetic processes dominate or have dominated until recent geologic time, a region can be considered to be in the initial to early mature stage of the cycle; and where exogenetic processes are extensive, a region can be considered to be in the mature to old stages of the geomorphic cycle (Figs. 1 and 2). Later, we describe the geohistory and distribution of the areas where these two provinces predominate. This geohistory is described in some detail so as to illustrate that endogenesis on the continents is the consequence of seafloor spreading, a process rarely considered in previous geomorphic studies.

Endogenetic Provinces (Initial to Early Maturity)

Interplate Processes—Plate Boundaries

General evolution.—Endogenic processes take three forms: divergence (or extension), convergence (or compression), and translation. Most movements along plate boundaries are a mixture of more than one form of tectonic transport. Geomorphologic features associated with extension regimes vary considerably. The East African Rift system is a complex of non-lapping, partially overlapping, and facing half-grabens defined by a series of master faults with cross-trending nor-

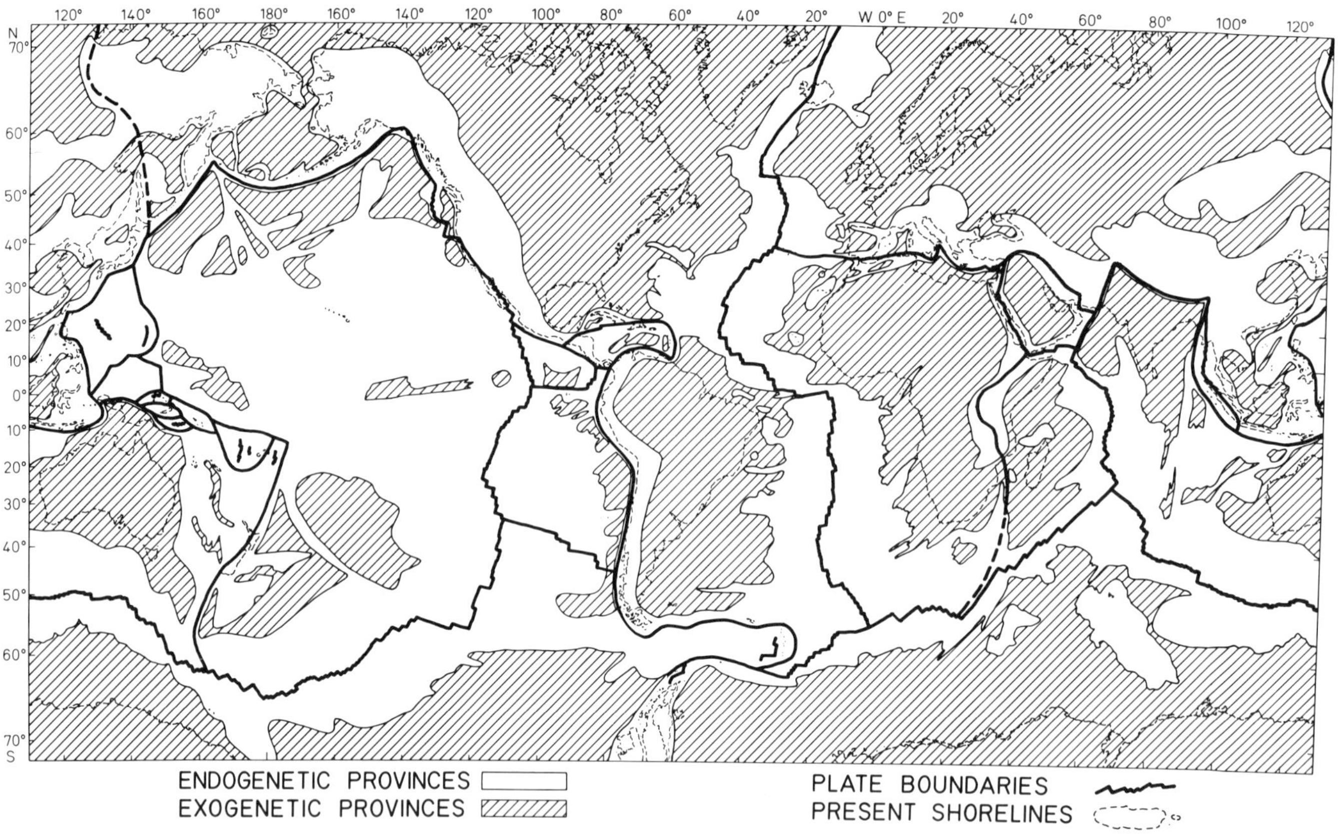

FIG. 1.—Areas containing second-order relief features of endogenetic and exogenetic origin superimposed on first-order relief features (continents and ocean basins). Compiled from numerous publications discussed in the text and UNESCO (1976). Base map from Espinosa and others (1981), with plate boundaries from Morgan (1972), Burke and Wilson (1976), and Howell and others (1985).

mal faults connecting the partially overlapping master faults that form accommodation zones (Rosendahl, 1987). At the Afar Triangle near the northern end of the East African Rift system, extensional deformation occurred in phases of short duration between the Oligocene and Pleistocene (Pilger and Rösler, 1976). The series of northwest-trending en-echelon volcanic highs rising above the sediment floor is considered to be emerged seafloor-spreading axes (Stieltjes, 1973). The northwest-trending basement high, Danakil horst, that separates the Afar Triangle from the Red Sea may be a microplate (Le Pichon and Francheteau, 1978) formed during reorganization of plate boundaries at the triple junction of the African, Arabian, and Somalian plates. Extensional features of the geologically young Red Sea range from a low-relief spreading axis between Lat. 15°30'N and 20°00'N, to a series of large deeps spaced at intervals of 50 km floored by oceanic crust from Lat. 28°00'N to 23°30'N, to rifted terrane at the southern and northern ends of the Red Sea rift (Cochran, 1983). Along the edges of the rift is a 1,000 to 3,000-m-high, seaward-facing hinge along which are exposed Pan-African strata and volcanics and accreted exotic terranes, including ophiolites (Stern and others, 1984; Stoeser and Camp, 1985). Toward the northwest, the rift terminates in a series of northeast-trending, asymmetric, rhomboid, en-echelon half-grabens formed by lateral motion along the 'leaky' Dead Sea transform (Ben-Avraham, 1985). Diverging from the north end of the Red Sea rift are the Gulf of Suez, another rift structure, and the Gulf of Aqaba (Elat), part of the Dead Sea left-lateral shear (Ben-Avraham, 1985). The Gulf of California is a Pacific analog of the Red Sea. It differs from the Red Sea in that the gulf developed in a transform setting, whereas the Red Sea evolved in an extensional regime. As in the Equatorial Atlantic, the Gulf of California spreading axis is dominated by short spreading segments separated by transform faults. The somewhat wider Gulf of Aden (Cochran, 1981) represents the next cycle in oceanic-basin formation followed by the Norwegian-Greenland Sea north of Iceland.

In the much wider and more mature Pacific, Indian and Atlantic Oceans, where their distal parts are isolated from continental influences, the mid-ocean ridge displays its most pronounced relief, a morphologic style controlled mostly by rates of spreading. Along-strike continuity of the mid-ocean ridge is broken by transform faults that juxtapose old lithosphere against the ridge axis (see Macdonald and others, 1988, for discussion of ridge-axis discontinuities). In the slowly spreading Mid-Atlantic Ridge, where thermal processes are quite variable and release of stresses takes longer (magma accretion is focused at discrete centers, Lin and others, 1990), the topography is quite rough, the cross-sectional area is smaller than in the Pacific, and the median valley is deep and well developed (see Ballard and van Andel, 1977, and Kong and others, 1988, for detailed descriptions of rift-valley morphology). In contrast, the much

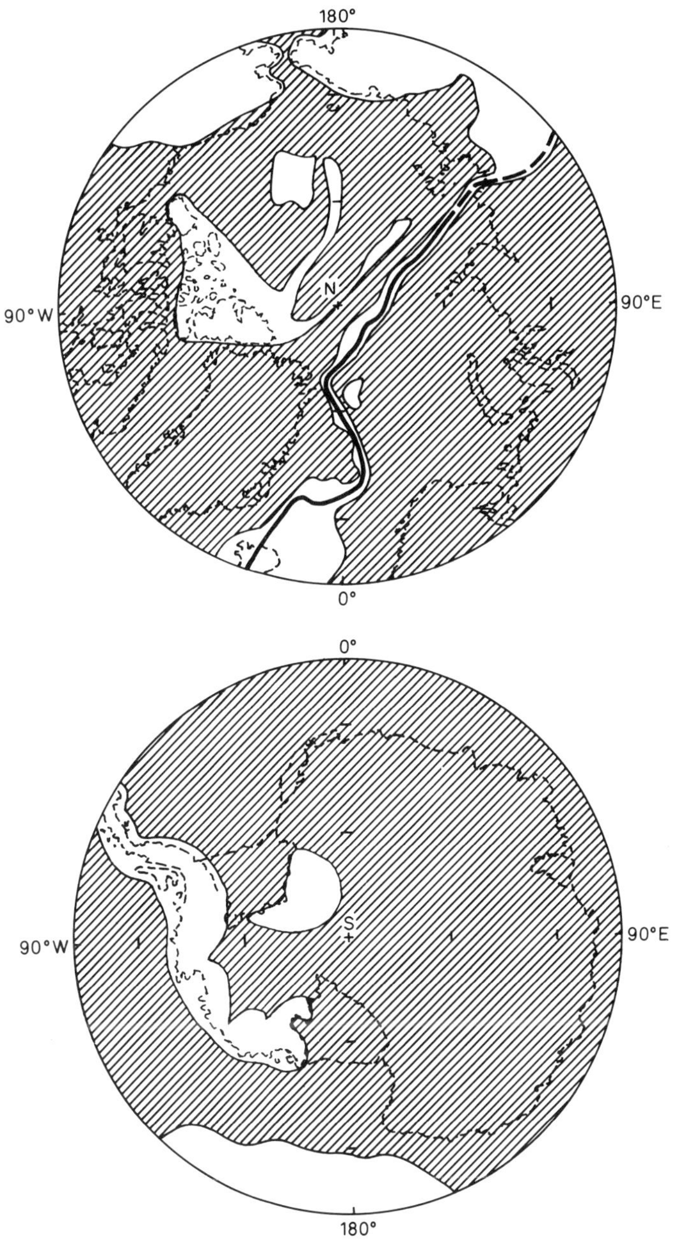

FIG. 2.—Areas containing second-order relief features of endogenetic and exogenetic origin in the polar regions. Compiled from publications discussed in the text. Base map from Anonymous (1984). Symbols as on Figure 1.

faster spreading East Pacific Rise (where thermal processes are more constant and stress is released at a much faster rate) has a broader and greater cross section than the Atlantic spreading belt, the median valley is poorly developed or missing, and its position may be occupied by a median high. This part of the mid-ocean ridge system also is characterized by overlapping and propagating axes and possibly more extensive fields of hydrothermal activity than the Mid-Atlantic Ridge. Motion of a propagating rift/failing rift/migrating transform-zone system leaves in its wake a complex pattern of morphologic trends oblique to both relative and absolute plate motion (Kleinrock and Hey, 1989). Lin and Morgan (pers. commun.) report that mantle upwelling in slowly spreading ridges has a plume-like structure, whereas sheet-like structures dominate beneath faster spreading ridges. Differences in rate of spreading and morphology between fast- and slowly spreading ridges may be due to these different injection geometries. Points of magmatic injection in time may be controlled by periodically spaced gravitational instability of the mantle (Whitehead and others, 1984; Crane, 1985). Other morphologic variations may result from the interaction of mid-ocean ridges and continental edges. For example, when the northern Norwegian-Greenland Sea opened as the Knipovich Ridge propagated northward, it collided with the continental Spitsbergen Shear Zone, the line of continental breakup (Crane and others, 1986). This led to thermal asymmetries on either side of the Knipovich Ridge that yielded slower spreading rates toward the east, and the emplacement of oceanic plateaus at the intersection (Yermath Plateau).

Why do faster spreading rates occur in the Pacific than the Atlantic? Is it a reflection of the subduction regime that rims the Pacific with the subducting oceanic crustal slab driving the spreading, whereas in the Atlantic more energy is needed to overcome the inertia to drive the continents westward over the subducted Pacific oceanic crust? Or, are the subduction belts merely a reflection of thermal convection within the mantle, as suggested by computer modeling (Bercovici and others, 1989)? Finally, the spreading belts show a linear age progression; namely, laterally sinking isostatically, underthrusting the continents and oceanic plates (including the Caribbean and Scotian Seas in the Atlantic) in the Pacific, and being buried by sedimentary detritus from the continents throughout much of the Atlantic.

Pacific Ocean.—The Pacific Ocean is rimmed by subduction complexes that have produced a heterogeneous mixture of endogenetic forms. The margins consist of continental borderlands, or marginal seas with secondary spreading axes and trenches. Only off northwestern North America are there Atlantic-type continental rises. On the Pacific margin of Antarctica are at least four microcontinents (Antarctic Peninsula, Ellsworth Mountains-Whitmore Mountains, Thurston Island-Eights Coast, and Marie Byrd Land) that accreted to the continent during the late Paleozoic(?), Mesozoic and possibly Cenozoic (Dalziel and Grunow, 1985). The tip of South America south of Lat. 37°S is an allochthonous terrane welded onto the continent as a result of subduction of the ancestral Pacific from Middle Devonian time to the Triassic (Forsythe, 1982). The formation of the Cape Fold Belt, an intraplate orogen in South Africa, by vertical movements may be related to this subduction episode. This belt, together with the Sierra de la Ventana in the Sierras de Buenos Aires, the Sierras Subandinas, the folds in the Falkland Islands, the Antarctic Peninsula, the South Shetland Islands, South Orkneys (see later discussion), and the Hunter-Bowen orogen of Australia, are segments of an extensive zone of deformation, the Late Paleozoic-Triassic Gondwanide orogen. When South America began to rift away from Africa during the Jurassic, subduction was reinitiated along the west coast of Gondwana, leading to the emplacement of silicic volcanics as far east as South Africa (shallow-dipping lithosphere) 196 to 191 ma; silicic volcanics along the west coast of South America (steeply-dipping lithosphere) 166 ma; emplacement of a volcanic chain along the west coast of South America 150 ma; formaton of a back-arc basin (Rocas Verdes Basin) 141 ma; and deformation of the subduction complex involving basin closure and crustal shortening in the late Albian. Sometime in post-Cretaceous time, this north-trending Andean-West Antarctic Cordillera was disrupted to form the North and South Scotia Ridges, Drake Passage, and the Scotia Sea Basin (Dalziel, 1974). Volcanism along the South Sandwich volcanic arc at the eastern end of the Scotia Sea probably began 7 ma (Hill and Barker, 1980). Seismic data indicate that the Benioff Zone along the west coast of South America varies along the strike from flat to inclined segments, separated by flexures in the zone (Bovis and Isacks, 1984). Subduction of aseismic ridges (Juan Fernández, Nazca, and Carnegie) added buoyancy to the subducted plate, uplifting parts of the overriding South American Plate and creating gaps in the Quaternary volcanic chain (Barazangi and Isacks, 1979; Hasagawa and Sacks, 1981; Nur and Ben-Avraham 1983a, Bovis and Isacks, 1984). The collision of the Marcus Necker Ridge, Magellan Seamount, and Louisville Ridge with the Izu and Tonga Trenches in the western Pacific created similar volcanic gaps near their points of collision (Nur and Ben-Avraham 1983a). Interaction along belts of plate convergence has produced a belt of tectonism approximately 700 km wide east of the trench in the South American margin. Since the middle Miocene, the central Andes Mountains have experienced intense magmatic activity and deformation (Burchfiel, 1980). Folds, reverse faults, and thrust faults with a relative motion to the east have developed along the eastern margin of the Andes and the sub-Andean zone. Present relief of the mountain chain is due to late Cenozoic orogenesis. Farther north, Columbia and Ecuador west of the Central Cordillera consist of oceanic terranes accreted since the Late Cretaceous (Case and others, 1971).

The morphology of Baja California and mainland Mexico is the product of Paleozoic to Cenozoic accretion and subsequent modificaton by exogenetic processes. The continental borderland off the Baja California peninsula is a Triassic to Cretaceous terrane composed of ophiolite, and oceanic-arc and forearc facies (Gastil, 1985). The northwestern coast of the peninsula is a Jurassic to Cretaceous arc terrane sutured to the eastern peninsula during the mid-Cretaceous and to the borderland to the west during the Turonian. The flysch terrane along the axis of the peninsula is a late Paleozoic to Cretaceous apron deposited along the western edge of the North American craton, and the Ballenas Terrane along the east-central edge of Baja California is a deep-water clastic and siliceous wedge that became attached to the craton since the mid-Paleozoic. The San Felipe Terrane to the northeast may be part of the craton. The broad Pacific margin of Mexico consists of a complex mosaic of Mesozoic sedimentary and island-arc submarine volcanic rocks and small terranes of pre-Mesozoic strata covered inland by continental Cenozoic volcanics (Campa U., 1985). These terranes, which may be a southward continuation of the Cordilleran suspect terranes, range in age from late Paleozoic at the northern end to Early Cretaceous at the southern end. The Mexican Thrust Belt west of the Pacific margin consists of several tectonostratographic terranes deformed during the Late Cretaceous to Early Tertiary Laramide Orogeny and block faulted (southern extension of the Basin and Range province) at the northwest end during the Neogene. The Guatemala Transverse Zone, the boundary between the North American and Caribbean plates, forms the southern terminus of the thrust belt and the Pacific margin.

Western North America north of Mexico and Baja California is another area of diffused plate tectonics where magmatic activity, metamorphism, and deformation have extended a considerable distance inland of the plate boundary. Here, deformation extended more than 1,000 km into the North American Plate (Burchfiel, 1980). According to Coney and others (1980), the vast mosaic of terranes that accreted onto the North American craton during Mesozoic and Early Cenozoic times have undergone large-scale concurrent and post-accretionary horizontal translations and rotations around vertical axes, motions that continue to the present. Following the Late Cretaceous to Early Tertiary deformation and magmatism (Laramide orogeny) and post-tectonic erosion, the Rockies (including the Colorado Plateau) were block faulted and uplifted during the Neogene, creating most of the relief seen today (Madole and others, 1987). During the early Cenozoic, a large region of the Pacific Northwest as far south as northwestern Montana underwent arc volcanism and regional north-south block faulting. By Miocene time, the region from southeastern Washington to southeastern California (including the Rio Grande Rift that separates the Great Plains from the Colorado Plateau and the Basin and Range to the west) underwent extensional deformation accompanied by magmatic activity (Davis, 1980). The voluminous Columbia River basalts were extruded then within a brief 3-million-year period between 16.5 and 13.4 ma. This plate extension and magmatism may be a back-arc phenomenon (Davis, 1980), or related to subduction of the Farallon Plate beneath the North American Plate (Atwater, 1970).

The Aleutian Island Arc along the boundary between the Pacific Plate and the Bering Sea, a marginal sea north of the arc, may be the result of an interaction with the rafted Umnak Plateau that caused the subduction zone to jump southward from the present continental margin (Ben-Avraham and Cooper, 1982). The Bering Sea itself is believed to be a part of the Kula plate being subducted along the margin that was trapped when the subduction zone jumped southward. The Shirshov and Bowers ridges were rafted on this oceanic plate. The much larger Arctic Basin is believed to have originated through the isolation of part of the Kula Plate by the collision and suturing of Alaska-Chukotka with Eurasia (Churkin and Trexler, 1980). The major topographic features of the Arctic Ocean include the Eurasian and Amerasian Basins, the Nansen Ridge, a spreading ridge initiated 63 to 40 ma, the Lomonosov Ridge, a piece of the Barents shelf carried away by the Nansen Ridge, the Alpha Cordillera due to Mesozoic deformation and subduction related to the North Atlantic development, and the Chukchi Complex attached to the shelf as a continental block superimposed on the edge of the oceanic basin (Clark, 1981). Clark suggested that the Amerasian Basin may be a Paleozoic low modified during the Mesozoic by tectonic activity along the Alpha Ridge. The Jurassic or Early Cretaceous Eurasian Basin is underlain by oceanic basement and was formed by seafloor spreading via the Alpha Ridge, or rotation of northern Alaska away from the Canadian Arctic Islands, rifting of the Kolyma Massif away from the Canadian Arctic Islands, or asymmetric spreading along an axis beneath the continental rise in the eastern Beaufort Sea (Grantz and others, 1981). The Tertiary Eruekan orogen of northernmost North America (Sverdrup Basin) is the result of Greenland's pivotal movement relative to North America, which formed a braided Cenozoic plate boundary in the Ellesmere and Axel Heiberg Islands of the Canadian Archipelago. Plastic flow of carboniferous evaporites has added to the structural complexities of the region (de Paor and others, 1989).

The Mesozoic tectonic history of northeast Asia (western end of the North American plate) involved the convergence and collision of the Okhotsk, Omolonsk, Bureinsk, and Khankaisk microcontinental sialic terranes with the Siberian platform from the mid-Jurassic to Early Cretaceous (Nur and Ben-Avraham, 1983b; Parfenov and Natal'in, 1985). After their accretion, the terranes were folded and faulted during the late Mesozoic opening of the North Atlantic and Canadian Basin in the Arctic, indicating the presence of a zone of compression between the Eurasian and North American plates. Farther south in Sakhalin and Hokkaido Island of northern Japan, endogenetic construction is due to latest Cretaceous or Paleogene subduction and collision tectonics along the boundary between the Eurasian and North American plates (Kimura and others, 1983). The Hidaka orogeny in Hokkaido is described by Okada (1983) as due to multi-phase collision during the Oligocene to latest Miocene with extensive deep-water gravity-flow deposits emplaced in front of the collision suture. In central Japan paleomagnetic studies suggest that the Miono area drifted a long distance northward after deposition in the equatorial or subequatorial regions and was attached to the autochthonous Hida terrane (Mizutani and Hattori, 1983). During the Late Cretaceous, acidic volcanics were extruded, overlapping both terranes. In southwestern Japan, plate convergence juxtaposed an accretionary southern terrane of a Pacific-type of tectonic environment against a northern terrane of a marginal-sea environment. The two terranes are separated by the Kurosegawa Tectonic Zone (Hada and Suzuki, 1983).

The Yamato Ridge near the center of the Sea of Japan divides the area into the Japan Basin at the west and Yamato Basin at the east. The high extends southwestward to connect with the Honshu shelf via Oki Bank. Both basins were formed by seafloor spreading, with the Japan Basin opening 45 to 35 ma and the Yamato Basin 15 to 10 ma (Kobayashi, 1985). At the northwestern margin of the Philippine Basin are the Ryukyu Islands and Ryukyu Trench extending from Kyushu to Taiwan. Ages of igneous rocks on the arc indicate that it has experienced at least five magmatic episodes from the earliest Paleozoic to early Pleistocene. The earliest may have been a mid-ocean magmatic episode that formed an ophiolite complex that later accreted to the Eurasian continental margin. The pre-Miocene strata are folded and commonly intruded, but the youngest sedimentary strata are relatively undisturbed (Kobayashi, 1985). West of the Ryukyu Arc is the more than 2,000-m-deep Okinawa Trough, whose sediment-deformation pattern indicates that extensional tectonics are dominant. Taiwan, at the junction of the Eurasian and Philippine plates, is the product of episodic subduction and collision that began during the Cretaceous about 87 ma (Chingchang and others, 1985).

China is a composite of tectonostratigraphic terranes that can be grouped into a Pacific domain assembled between the late Paleozoic and early Mesozoic and a Tethys domain, whose accretionary ages range from late Paleozoic in the north to Tertiary in the south. Interaction of the Pacific Indian and Eurasian plates has led to continued interplate deformation of the terranes with different rates of intraplate movement causing local extension and development of fault basins (Xiong and Coney, 1985). Southeastern Asia consists of several blocks having different geologic histories and separated by major faults and sutures that contain dismembered ophiolites (Hutchinson, 1982). Southeast Asia, Sumatra, western Borneo, and adjacent shallow seas consist of terranes of Central Tibet, Southern Tibet, Indo-China, Western South Asia, and Western Burma joined to Southern Asia along the Son Ma-Song Da suture belt (Stauffer, 1985). The Philippine Plate (enclosed by the Nankai Trough and Ryukyu Trench at the northwest, the Philippine Trench at the west, and the Bonin and Mariana Trenches at the east and southeast) consists of three terranes: from east to west the West Philippine Basin of the late Mesozoic to early Tertiary, the Shikoku and Parece Basins of about 30 to 17 ma, and the Mariana Trough of about 6 ma (Uyeda and McCabe, 1983). All three basins may have been formed by an extensional back-arc-spreading process (Karig, 1971), or as a combination of entrapment of a pre-existing ocean consisting of the Kula and Pacific Plates (the West Philippine Basin), a spreading axis (the Central Basin Fault), and the change of a long transform fault along the eastern margin to a subduction belt when the Pacific Plate changed its direction of motion from north-northwest to west-northwest about 40 ma, trapping the West Philippine Basin and making the abrupt change in the trend of the Emperor-Hawaiian Seamount Chain. Younger basins were developed by episodic back-arc spreading (Uyeda and Ben-Avraham, 1982). The Philippine Archipelago itself is a composite of volcanic arcs, ophiolite fragments, and microcontinents that were welded prior to the late Miocene (McCabe and others, 1985). The South China Basin west of the Philippines, an 'Atlantic-type' marginal basin bounded by passive continental margins to the north and south, is the result of: 1) mid-Jurassic through mid-Cretaceous Andean-type subduction with northern Palawan-Reed Bank-Luconia Shoals representing the forearc areas of this margin (subduction, however, continued beneath western Borneo through the Late Cretaceous and early Tertiary); 2) rifting of a Proto-China margin from latest Cretaceous or Paleocene to early Oligocene; 3) seafloor spreading in the eastern half of the basin from late Oligocene to early Miocene, a time during which northern Palawan-Reed Bank-Luconia Shoals were separated from Asia; and 4) collision of the above blocks with the Palawan convergent margin from the early Miocene to late Miocene (Taylor and Hayes, 1983). Subduction of South China Basin crust beneath the Manila Trench led to collision of the Philippines with Taiwan and northern Palawan during the late Miocene. The Molucca Sea Plate south of the Philippines, extending from Mindanao to the Sula thrust, is being subducted along its western margin beneath the Sangihe Arc and along its eastern margin beneath the Halmahera Arc. It thus represents the collision zone between these two converging arcs (Moore and Silver, 1983). Indonesian region north of Australia, a complex of deep-sea trenches, troughs and volcanic islands, is located in the zone of convergence of three megaplates: Eurasia (disrupted internally), Pacific moving west-northwest relative to Eurasia, and India-Australia. Within the broad zone of interaction of these megaplates are microplates with complex movements.

In New Guinea and western Melanesian Arcs, Hamilton (1979) proposed that parts of southeastern Asia are fragments from Australia transported thousands of kilometers westward by seafloor spreading in the Pacific Ocean during the Triassic and Jurassic. Australia broke away from Antarctica during the Late Cretaceous, the Caroline Basin dates from the Oligocene, the Bismark and Solomon Seas are younger, and a brief extension in the Woodlark Basin occurred during the Miocene. Opening of the Coral Sea began during the Paleocene or earliest Eocene, concurrently with opening of the Tasman Sea and separation of New Caledonia and New Zealand from Australia (Milsom, 1985). Toward the west, Australia and New Guinea are in near contact with the Banda Arc collision zone. The only Mesozoic oceanic crust in the area may be represented by ophiolite masses thrust over New Guinea from the north and east. The Solomon Islands that form part of the Melanesian Island Arc and marginal-basin complex (which extends across the southwestern Pacific from New Ireland to Roga and Kermadec Islands) are due to interaction between the Australian and Pacific Plates with three and possibly four microplates sandwiched within the collision zone (Coulson, 1985). Caroline Basin is a plate formed as a back-arc basin to the northward-dipping subduction of the Indo-Australian Plate beneath the Central Highlands of New Guinea (Hegarty and Weissel, 1988), and the Norfolk Ridge system and its margins are a result of breakup of the eastern margin of Gondwana during the Late Cretaceous to early Eocene, oblique collision along a plate boundary during the early Eocene to early Oligocene, and sporadic intraplate volcanic activity after the early Oligocene (Eade, 1988). In New Caledonia (a continental fragment broken away from Australia), the Eocene collision belt behaved like a transform with overthrusting to the northeast and obduction of the Pacific Plate at the southern end of the island (Brothers and Little, 1988). The Loyalty Islands 100 km east of New Caledonia and separated from it by a more than 2,000-m-deep trench consist largely of uplifted coral reef atop Tertiary basalt and dolerite.

Seismicity in the Tonga and Kermadec ridges northeast of New Zealand indicates that subduction beneath these ridges is being displaced gradually toward the east, with displacement being much greater at the northern end where subduction is 4 to 3 ma (Dupont, 1988). The geologic history of New Zealand from the Carboniferous to mid-Cretaceous is one of subduction of the Pacific coast of Gondwana, extensive mid-Cretaceous to present lateral movements along the Alpine Fault Zone, and development of the microplates that comprise the region (Korsch and Wellman, 1988).

Indian Ocean.—Tectonism from the northeastern Indian Ocean to the western Mediterranean Sea is the result of the breakup of Gondwana, opening of Tethys and Bay of Biscay, accretion of fragments onto Asia and Europe, and continental collision (India at the eastern end, and Betic/Rif systems [Alboran Plate] and inverted Atlas Rift at the western end) as Tethys closed (Biju-Duval and others, 1977; Nur and Ben-Avraham, 1983b, Curray and others, 1982; Ross and others, 1986). With these accretions, collisions, massive gravitational sliding, sediment transfer from one plate onto another, and late tectonic normal faulting, a complex mountain chain extending from Asia to Iberia was formed. Associated with the Himalayas collision zone are the Baikal and Shansi graben complexes linked to the zone by a system of strike-slip faults and folds (Burke, 1980). Probably, these extensional structures resulted from the eastward migration of China away from the arc zone between India and China (Molnar and Tapponnier, 1975). Collision of the Arabian platform and the Eurasian Plate along the Zagros Crush Zone led to uplift of the northeastern edge of the platform, pronounced subsidence south of the crush zone, and emplacement of the Zagros Mountains by gravitational sliding atop the Precambrian Hormuz Salt. Uplift along the northerly-dipping subducted plate also has uplifted the Makram margin. As a result of this uplift, the margin sedimentary strata have slipped seaward along a décollement to form a series of ridges parallel to the coast (Ross and others, 1986, and references therein). The Oman Mountains at the northeast end of the Arabian Peninsula trend northward along the Oman Line into the Zagros Mountains. They are the result of obduction of oceanic crust and oceanic facies from the southeast over a Mesozoic carbonate platform during the Turonian and early Maestrichtian and compression during the mid-Tertiary that formed the folds along the western side of the Musandam Peninsula. Folds east of the Musandam Peninsula may be due to diapirism of Upper Cretaceous(?) mud resting on the Mesozoic carbonate platform or oceanic crust obducted over the platform (Ricateau and Riché, 1980). After the Miocene, vertical tectonics dominated, with the Oman Mountains rising and the Musandam Peninsula subsiding to form the Strait of Hormuz connecting the Persian Gulf and the Gulf of Oman (Searle and others, 1983). In the western Tethys, deformation as a result of compression was accompanied and followed by extension to create the rift system in western Europe, the southern limit of which evolved during the early Miocene into a seafloor-spreading phase that produced the Balearic Basin. The Pyrenean orogenic system diverting northwest to west from the Alpine system formed during the latest Cretaceous to Oligocene by southward subduction in the Bay of Biscay (Western Pyrenees) and northward subduction in the Eastern Pyrenees. At the western end of the lengthy Tethys suture are the Miocene Azores Islands along the crest of the Mid-Atlantic Ridge. Lateral and convergent motion along the Azores-Gibraltar Fracture Zone uplifted oceanic basement to form linear highs of seamount dimensions. Some of these highs may have risen to shallow water to serve as foundations for carbonate accretions that later were drowned when the highs sank.

Atlantic Ocean.—In the Atlantic, the loop-shaped Scotia Arc that rims the Scotia Sea was produced by a combination of subduction and easterly translation. Easterly displacement of southernmost South America and the Antarctic Peninsula took place during the Oligocene to late Miocene (Dalziel and Elliot, 1973). Deformation on the accretionary wedge is most intense where fragments of the volcanic arc were transported approximately 1,500 km eastward and incorporated into the wedge. A similar type of deformation by a combination of translation and subduction is responsible for the Gulf of Mexico-Caribbean region. Allochthonous terranes transported into these marginal seas include northern Mexico south of the Texas Lineament that rotated counterclockwise relative to North America (Urrutia-Fucugauchi, 1981). The Yucatan block migrated into its present position from either the Gulf of Mexico or the Pacific, as did the Chortis block (southern Guatemala, Honduras, El Salvador, Nicaragua, and the Nicaragua Rise). The megashear separating the Mexican and Yucatan blocks, along which the Trans-Mexican Volcanoes were constructed during the Quaternary, is an eastward prolongation of the Tamayo Transform developed during the opening of the Gulf of California in the Pliocene (Aubouin and others, 1982). Volcanism along the shear is related in some way to the subduction of the Cocos Plate. Southern Central America (Costa Rica and Panama) is an oceanic crustal block of Pliocene to Quaternary volcanoes and oceanic terranes of Pacific origin thrust over the Caribbean Plate (Bowin, 1975). Final closure of the Caribbean Sea from the Pacific was as late as the Pliocene (Keigwin, 1978). As the Caribbean Plate migrated into the Atlantic, its northern and southern edges were deformed by slight convergence of the Americas and its eastern edge by convergence with the North American Plate. Translation and oblique extension along the Guatemala Transverse Zone (that cuts across northern Central America and abuts the Caribbean nappes blanketed by Holocene carbonates and displaced by the eastern Yucatan scarp during the Miocene) has created a deep chasm (Cayman Trough) in a carbonate/volcanic high to separate the Nicaragua Rise and Cayman Ridge. Interaction between the Caribbean and North American plates also fragmented the Bahama Platform during Cretaceous to Paleogene time, creating a mosaic of highs and lows. Subsequent progradation of sediment from the bank tops has healed some of this fragmentation (Eberli and Ginsburg, 1987).

Intraplate Processes

Mid-Plate.—Scattered throughout many plates are topographic highs of various origins. Some volcanoes display linear trends and apparently were formed sequentially above stationary hotspots as oceanic crust moved across them (Burke and Wilson, 1976). For example, west-northwesterly trends of volcanoes in the Pacific were formed during the past 42 ma, when the plate rotated clockwise around a pole near Lat. 69°N, Long. 68°W (Duncan and Clague, 1985). Edifices trending northward also were formed in the same manner between 80 and 42 ma, when the Pacific rotated around a different pole at Lat. 17°N, Long. 107°W. The classic example of such a linear island chain is the Emperor Seamount Chain-Hawaiian Ridge, with the seamount chain formed 70 to 42 ma as the Pacific Plate moved northward, and the Hawaiian Ridge formed during the past 42 ma as the Pacific moved west-northwestward. Other features associated with hotspots are the Galapagos Platform and Cocos and Carnegie Ridges in the eastern Pacific (Hey and others, 1977) and the Louisville Ridge, a 4,300-km-long Late Cretaceous and Cenozoic chain in the southwest Pacific (Lonsdale, 1988). Construction chronology of the northwest-trending 4,200-km-long Line Island Chain of seamounts, atolls and elongated ridges, whether built synchronously or by age-progressive volcanism, is yet to be resolved. Other linear features possibly formed by plate migration over hotspots include the Gulf of Alaska volcanic chains, Caroline Islands, Islas Revillagigedos, island chains of French Polynesia, New Hebrides-Samoa Lineament, and the Musician Seamounts.

Some highs scattered throughout the Indian Ocean range from continental origin (Seychelles Bank and Exmouth Plateau) to highs showing some continental affinity (northern and southern extension of the Chagos-Laccadive Ridge, the southern extension of the Seychelles and Fortune banks, the Scott Plateau, Mozambique Ridge, Wallaby Plateau, and possibly the extreme southern end of the Kerguelen-Heard Plateau). Other highs that are volcanic in origin include Ninetyeast Ridge, Mascarene Plateau, Madagascar Ridge, the Ob, Lena, and Marion Dufresne Seamount Chain, and Broken Ridge (Schlich, 1982). Many ridges that formed near sea level were capped by shallow-water carbonates, and then subsided at rates similar to those of oceanic crust. The junction between the Southeast and Southwest Indian Ridges is characterized by a series of en echelon scarps facing the slowly spreading (8 mm/yr) Southwest Indian Ridge axis. The scarps intersect at nearly right angles the spreading fabric formed on the Antarctic Plate at the Southeast Indian Ridge, producing north-pointing 'spurs' of ocean floor. Discontinuity of this topographic grain may be due to intermittent developments of magmatic convection cells beneath the eastern end of the Southwest Indian Ocean adjacent to the more continuous process beneath the faster spreading Central Indian ridge (25 mm/yr) and Southeast Indian Ocean (30 mm/yr) (Parson and others, 1986).

In the Atlantic Ocean, the Agulhas and Falkland Plateaus, the marginal plateaus along eastern Brazil, Flemish Cap, Orphan Knoll, Galicia Bank, and Rockall Plateau, display continental affinities. Other highs were formed by mid-plate volcanism (Bermuda Rise, Canary Island Ridge, and Cape Verde Rise) or by igneous activity along a leaky fracture zone (Beata Ridge in the Caribbean Sea, New England Seamount Chain-White Mountain Magma Series in North America, Cameroon Volcanic Chain and Walvis Ridge), or by a short period of intense localized volcanic activity along the Mid-Atlantic Ridge (Meteor Swell-Islas Orcadas Rise, the J-Anomaly Ridge-Madeira Tore Ridge, Ceara Rise-Sierra Leone Rise, and Rio Grande Rise), or by hotspots such as the active one at Iceland, the Azores, and the Tristan de Cunha, Fernando da Noronha and Trindade Ridges off eastern Brazil.

Duncan (1984) ascribed the New England Seamount Chain to migration over a hotspot, an origin postulated by Morgan (1983) for most topographic highs in the Atlantic. Morgan believed that the White Mountain Magma Series, the New England Seamount Chain, the Corner Rise, and the Cruiser-Hyeres-Great Meteor Seamounts east of the Mid-Atlantic Ridge follow closely his predicted track of the Great Meteor hotspot. According to Morgan, the great Meteor track, together with the Deccan-Laccadive-Chagos-southern Mascarene-Mauritius-Reunion track in the Indian Ocean, are the best examples of a hotspot track migrating from one plate to another by crossing a mid-ocean ridge. Recent uplift of the western Rio Grande Rise is due to massive Eocene volcanism (Gamboa and Rabinowitz, 1988). Volcanic peaks that rose above sea level were eroded and served as foundations for carbonate deposition. Ridges such as the J-Anomaly Ridge-Madeira Tore Ridge and Walvis Ridge also rose above sea level, were eroded, and supported reefs as they subsided below sea level (Pastouret and Goslin, 1974; Supko and Perch-Nielsen, 1977; Tucholke and Ludwig, 1982). Other positive elements, such as the highs along the Agulhas Fracture Zone or the highs along the equatorial megashear between northern Brazil and the Gulf of Guinea, formed by uplift along zones of translation (St. Peter and St. Pauls Rocks along this shear are in part continental) and abandonment of spreading ridges due to changes in geometry (Burdwood Ridge in the Scotia Sea, Aves Swell? in the Caribbean, the Bay of Biscay seamounts rejuvenated by the Paleogene Pyrenean orogeny, a ridge atop the Iceland Plateau and Aegir Ridge off Norway). Remnants of the triple junction between the Bay of Biscay spreading axis and the Mid-Atlantic Ridge are the northwest-trending King's Trough, Peake and Freen Deeps and the northeast-trending Palmer and Azores-Biscay ridges. The Blake Spur magnetic-anomaly ridge off the eastern United States formed by excessive volcanism when the Mid-Atlantic Ridge spreading belt jumped eastward during the Jurassic, and the Ran Ridge in the Labrador Sea and the ridges in Rockall Trough and the Balearic Basin became buried and left no topographic expression. Highs formed by secondary back-arc spreading also occur in the western Pacific and Scotian Sea in the South Atlantic. Reykjanes Ridge is believed to be the result of channelized flow down the axis of the ridge from the Iceland hotspot to the north. Morgan (1978) proposed a similar channelized flow from a hotspot to a nearby mid-ocean ridge crest at Bouvet and Ascension Islands in the South Atlantic; Scotland-Greenland Ridge in the North Atlantic; Disco Island basalts in Baffin Bay; Balleny Islands off Antarctica; Rodriguez, Amsterdam and St. Paul Islands in the Indian Ocean; and Darwin and Wolf Islands 200 km north of the main group of the Galapagos Islands.

The numerous isolated seamounts scattered throughout the oceans appear to have been formed at or near mid-ocean ridges. Analysis of wide-beam, SeaBeam, and map-count data from the eastern and southern Pacific yielded an average of 5,400 ± 650 seamounts per million square kilometers of which 170 ± 17 have reliefs of more than 1 km (Smith and Jordan, 1987). The height of a seamount probably is proportional to the hydraulic head with a 285-m-high volcanic edifice having a magma source at a depth of 6 km below the sea floor. Some of them once rose above sea level to be eroded to platforms for calcium-carbonate accretion, displaying a linear trend from fringing and barrier reefs to atolls. Some subsided faster than carbonate was deposited, to form guyots. At least 100 guyots have been reported from the western Pacific, 14 of which have Cretaceous reefs that drowned about 100 ma (Hamilton, 1956; Heezen and others, 1973). According to Menard (1984), the interval between the appearance of a volcanic island and its disappearance as an atoll or guyot may range up to 30 Ma, depending on the size of the island, the occurrence of reefs, and the development of the reefs with time as the plate migrates latitudinally. There is an additional interval before anomalous atolls transform to guyots. Thus, 100 Ma ago, a group of atolls and coral banks atop a broad elevated area extended from the Magellan Seamounts to the Mid-Pacific Mountains to the Geisha Seamounts (Menard, 1984). This island-dotted Cretaceous terrane (Darwin Rise) that was subjected to volcanism and sill intrusion is comparable with the Polynesian terrane (also known as the South Pacific Superswell; McNutt and Fischer, 1986), except that no abyssal flood basalts have been reported to date from the Polynesian plume area (Menard, 1984). The South Pacific Superswell, a broad area in French Polynesia east of the Darwin Rise, is 250 to 750 m shallower than lithosphere of the same age in the North Pacific. These anomalous depths south of the Marquesas Fracture zone are consistent with a thermal source having a change in depth along the fracture zone that delineates a lithospheric discontinuity (McNutt and Fischer 1986). In the Atlantic Ocean, Great Meteor Seamount and Mytilus Seamount have yielded Miocene shallow-water carbonates, and the Newfoundland Seamounts have Cretaceous fossils (Sullivan and Keen, 1977). Although not sampled, the Fogo Seamounts off western Grand Banks are believed to be capped by Lower Cretaceous strata like the nearby J-Anomaly Ridge (Emery and Uchupi, 1984, p. 334–335). In addition to serving as platforms for reefs, mid-plate seamounts with their potential for intense open hydrothermal circulation (now or in the past) make ideal sites for hydrothermal precipitation (Bonatti and Joensuu, 1966; Ohmoto, 1978; Hekinian and others, 1981; Lonsdale and others, 1982; Malahoff and others, 1982; Malahoff, 1988; Uchupi and Ballard, 1989).

Exogenetic Provinces (Maturity to Old Age)

In areas of cratons and stable platforms not involved in the Mesozoic to Recent plate interaction, exogenetic processes have reduced the topography nearly to peneplains, exposing the roots of the tectonic (endogenetic) features that were active in plate margins. Although erosion and deposition are active in regions where endogenetic activity is at its maximum, continuous to episodic tectonism gives these regions a youthful appearance. It is along "trailing edges" of continents (Atlantic-type continental margins) that extensive terrigenous aprons prograde from sites of tectonism, extending their maximum distance and developing a low-relief terrane; a complex of estuaries, lagoons and barrier islands along the coast; wide shelves; continental slopes; continental rises; deep-sea fans; and, finally, abyssal plains that extend to the distal ends of the ocean basins, where they slowly encroach onto the isostatically subsiding mid-ocean ridge. At distal ends of oceanic basins away from continental influences, low-productivity areas and subsequently low pelagic rates of sedimentation (such as in the Sargasso Sea) allow oceanic basement to crop out over large areas. Where terrigenous input is large enough and the oceanic basin is narrow, as in the Norwegian-Greenland seas, the margin sediment may completely bury the flanks of the ridge, with the sediment extending onto the median valley.

So-called passive margins generally display three morphologic types: a wide and shallow shelf, a low-relief canyon-entrenched continental slope, and a well-developed rise/abyssal plain where a siliciclastic regime predominates. Shallow platforms bordered by nearly vertical, relatively canyon-free slopes and poorly developed continental rises/abyssal plains occur in carbonate regimes. At high latitudes and formerly glaciated areas, shelves are irregular and characterized by a coastal trough, deep channels oriented at high angles to the coast, a chain of banks on the outer shelf, ınd/or an outer shelf hundreds of meters deep. Unique fea-res of the Atlantic Ocean are bottom-current-controlled limentary forms such as outer ridges, drifts and lower inental-rise hills.

e present morphology of the margins is a summation of ee clearly recognizable factors: climate, tectonic re- ınd geometry of the ocean basin. During its early rior to seafloor spreading, the basin is controlled t a c nental infra-rift phase, where non-faulted sags de- ve g or near the incipient continental/oceanic crus- tal v followed by a rifting phase dominated by syn- sedı nt faulting (Falvey, 1974). The East African Rift syste Triangle, Red Sea, Gulf of Aden, and Nor- wegia ınd Sea are analogs of the young Atlantic from in ier rifting to a well-established seafloor-spread- ing regi tially, rifting is restricted to the inner mar- gin, but rogresses, rifting becomes concentrated on the outer n in, creating a pronounced low separated from the rest of ıargin by a high-relief hinge in the basement rock. If c conditions are appropriate, this structural low cou ec me the site of evaporite deposition, as hap- pened rn North America, northwestern Africa, the Gulf o, eastern Brazil, and equatorial western Af- rica. ci ation becomes less restricted, more open-water sha onates predominate, as also happened in these are n the continents migrated northward away from a c ıucive to carbonate deposition, extensive carbon- me limited to southern latitudes, and mid-latitudes sites of siliciclastic deposition. In contrast, off out n Africa and southern South America, and in the ian Archipelago and the Labrador/Baffin Bay, the and late rift sequence and drift sequence are essen- t y siliciclastics. The relatively deep Blake Plateau off southeastern United States is the product of Jurassic to arly Cretaceous shallow-water carbonate deposition, until ecoming drowned near the end of the Early Cretaceous. Easterly surface outflow via the east-trending Swannee Channel in southern Georgia led to emplacement of the east-trending clastic wedge (Charleston Bump) at the northern end of the plateau during the Late Cretaceous to Paleogene and to northerly flow via the Straits of Florida and emplacement of the north-trending wedge off the present coast during the Cenozoic. Lateral migrations of the Gulf Stream in response to changes in sea level during the Cenozoic influenced the emplacement of the coastal wedge.

Deposition of sediment also is controlled by tectonic activity, leading to deposition of fans (Bengal and Indus) in the Indian Ocean during the rise of the Himalayas. Another factor that influenced deposition on the margin was change in sea level (tectonically and eustatically induced). During high sea-level stands, sediment deposition was restricted to the inner margin, and during lowstands, when the shore migrated seaward to the shelf's edge, the continents underwent fluvial erosion, the continental slope was cut back by submarine processes (turbidity currents, slumping and gravitational sliding) and sediment was deposited on the continental rise and adjacent abyssal plains. During the Paleogene, this sedimentary regime was affected by another factor: bottom currents. With the latitudinal growth of the oceanic basins and initiation of thermohaline circulation, the margins underwent extensive erosion that caused steep slopes fronting the carbonate platforms to retreat landward as much as 15 km (Dillon and Popenoe, 1988). During latest Eocene to earliest Oligocene, the frontal scarp (Blake Escarpment) of the platform became deeply eroded by a southerly-flowing deep-water current that removed much of the high-energy reefal facies and exposed the lagoonal facies landward of the reef (Dillon and Popenoe, 1988). The Bahama Escarpment fronting the Bahamas and extending from the southern Blake Plateau nearly to the Puerto Rico Trench also underwent similar erosion by bottom currents, exposing the low-energy lagoonal facies (Freeman-Lynde and others, 1981). This erosion was a combination of mechanical degradation of the lower scarp and collapse of the scarp above. Recent investigations on the Florida Escarpment suggest that not all of this erosion was mechanical, but part of it could have been chemical, caused by escaping saline waters along submarine springs at the base of the scarp (Paull and others, 1990). As bottom-current activity slowed, the outer margins became sites of deposition by bottom currents, leading to the development of various topographic forms. Other factors that controlled deposition on the continental margin include the Cretaceous euxinic event, when most of the oceanic basins became sites of black-shale deposition; siliceous deposition during the Eocene as a result of changes in circulation that enhanced upwelling and increased productivity in the surface waters; and isolation of the Mediterranean Sea during the Miocene, leading to its dessication and the deposition of Messinian evaporites, deep erosion of the basin margins, and its subsequent flooding during Pliocene. A more recent event was the refrigeration during latest Cenozoic time, when mid-latitude shelves became glaciated. Shelves farther south were under the influence of a fluvial regime, with carbonate platforms undergoing a phase of karstification. Glacial activity still persists at extreme high northern and southern latitudes. The present geomorphology of passive exogenetic margins is related to these events when sea level was about

130 m lower than today and to the Holocene transgression that began at about 11 ka ago and caused sea level to reach its present level at about 5 ka.

Exogenetic provinces are not tectonically inert and are sites of endogenetic processes, but erosion and deposition are so persistent that these endogenetic processes have limited effect on later geomorphology. In the mid-continent region of the United States, this contemporaneous tectonism, probably of plate origin, is controlled by early geologic features, such as the mid-continent rift that extends from Kansas to the Great Lakes; the St. Lawrence River rift; failed arms of the Precambrian to Eocambrian triple junctions (aulocogens) in the Mississippi Embayment and southern Oklahoma; and zones of weakness and crustal boundaries and local basement inhomogeneities (Hinze and others, 1980). Rejuvenation of the Appalachian Paleozoic suture was caused by seaward tilting of basement beneath the continental margin and uplift along the landward edge of the inclined plane. Also affecting the margin are tectonic events like the "Abaco" slump event described from the carbonate margin off the southeastern United States (Leg 101 Scientific Party 1988) and the 1929 earthquake event off Canada that caused extensive slumping, gravity sliding and subsequent turbidity currents in the Laurentian Fan off Canada. Not all turbidity currents, however, are triggered by seismic events, but some result from rapid deposition of unstable sediments, as, for example, events on the deltas of the Magdalena River in the Caribbean Sea and the Congo River in equatorial Africa. Another tectonic deformation on continental margins is plastic flow of evaporites by overthrusting and diapirism (Permian of northwestern Europe, North Sea, Barents Sea, and off eastern Greenland(?), Mesozoic in the Canadian Maritime Provinces, northwestern Morocco, the Carolinas, Gulf of Mexico, eastern Brazil, and Angola) and of Cenozoic mud (north slope of Alaska, Niger and Mississippi deltas and Amazon Fan). Vertical flow of evaporites may produce gravitational displacement of surface sediment, as reported from the Carolina Trough (Dillon and Popenoe, 1988). Breaching of the evaporites produced the observed brine falls and pools in the Gulf of Mexico (Rezak and Bright 1981), with brine layers in intradiapir locations creating small euxinic basins (McKee and others, 1978). Uplift along the margin off eastern Mexico and slippage along a décollement led to formation of the Mexican Ridges in the western Gulf of Mexico.

As reported by Working Group 3 (1988), rainfall that raises the water table above sea level on continents can produce significant underflow into the oceans. Along passive margins containing evaporites, this flow may be aided by negative buoyancy as meteoritic waters become so saturated with salt that they become denser than sea water and sink to the lowest aquifers. They escape into the deep ocean where the aquifers crop out on the continental slope. This flow, at a rate of about 10 km^3/yr (comparable to the flow of the Potomac River), is believed to be eroding the Yucatan carbonate platform edge at a rate of hundreds of meters per thousand years (Hanshaw and Back, 1980). Such flow also may be responsible for eroding submarine canyons off New Jersey and western Florida (Robb, 1984; Paull and others, 1990). These hydrologic processes also act along active margins on endogenetic provinces. Their active tectonic setting creates additional features in the process, such as dewatering by progressive compaction in response to accretionary buildup and pronounced deformation at the toe of the wedge. Pore water also may escape via mud volcanoes (Westbrook and Smith, 1983; Breen and others, 1986; ODP Leg 110 Scientific Party, 1987) in front of the complex, through various types of faults, and along permeable units within the accreted wedge. Subduction of the wedge also can increase the pore-fluid pressure, tending to drive out pore water by various passageways, flows that may lead to hydrofracturing of the wedge.

Other factors that have affected not only the exogenetic but also the endogenetic realm include encounters with extraterrestrial objects (comets and/or meteorites), volcanic explosions (Mount St. Helens, Krakatoa, and Thera about 1450 BC), and tsunamis and catastrophic floods such as the Missoula flooding of the Columbia Plateau at 22 and 14 ka (Baker, 1973, p. 11–12) that have produced geological as well as biological effects.

DISCUSSION AND CONCLUSIONS

As illustrated by Figures 1 and 2, the surface morphology of the earth can be separated into two provinces, one where endogenetic (tectonic/magmatic) processes dominate, and a second where exogenetic (erosion and deposition) processes dominate. Both provinces display secondary regions of various reliefs, providing evidence that both processes operate to some degree in both provinces. Endogenetic provinces occur along or near plate boundaries and recently abandoned boundaries, or are associated with intraplate processes. These provinces may be narrow or may extend a considerable distance onto adjacent plates as diffuse tectonism. Exogenetic provinces are best developed along divergent margins of continents that are distant from plate boundaries. Exogenetic processes are faster above sea level and quite slow below sea level, so that morphologic features as old as the Cretaceous can survive in oceanic basins. Oceanic basins rimmed by endogenetic provinces (perimeter of Pacific Ocean) receive relatively little detritus from adjacent continents so that endogenetic provinces can be traced from their mid-ocean ridge points of creation to the continental margins. In the Pacific, continental rises are only well developed on the northeast corner of the basin. In oceanic basins surrounded by exogenetic provinces (e.g., Atlantic Ocean, parts of the Indian Ocean, Southern Ocean off Antarctica, and Arctic Ocean), detritus from land can reach the distal ends of the basins, onlapping the isostatically sinking flanks of mid-ocean ridges. Endogenetic processes appear to occur episodically rather than as a continuous mode. Even on the mid-ocean ridge, periods of quiescence may be so long that volcanic terrane emplaced in the neovolcanic zone can be covered to some degree by pelagic sediments. On continents with their thick and relatively inert crusts, tectonism distant from the adjacent plate boundary may be short lived and preceded by long periods of stability when stresses build or accumulate before being released. These long periods of stability may permit exogenetic processes to degrade the endogenetic provinces to low relief (e.g., the Laramide erosional surface in the Rocky Mountains after the Paleocene orogenesis that was rejuvenated by block faulting during the Miocene). Because plates may migrate across several climatic zones during the geomorphic cycle and because sea level oscillates, tectonically and eustatically, the linear geomorphic trend described by Davis (1909) and others rarely occurs. Geomorphologic evolution is one of tectonic relief followed by erosional/depositional destruction of that relief. If a region remains within the endogenetic realm, tectonism ultimately cancels the effects of erosion and deposition. A region removed from the realm of tectonism can be reduced to low relief. In Precambrian shields, exogenetic processes have been so extensive that the roots of plate sutures have been exposed, and the geomorphology is the end result of many events. At any given time, the geomorphology displayed by a region reflects processes active at that particular time or during the near past. In both oceanic and continental realms, much of the present geomorphology is the result of Pleistocene events modified to some degree by Holocene ones.

The endogenetic and exogenetic provinces of Figures 1 and 2 were compiled into a summary diagram (Fig. 3) that shows true areas at all latitudes (avoiding the distortions inherent in flat maps of the globe). The diagram also combines all areas of both kinds of province for both land and ocean so as to provide a general impression of the global relation of the provinces and the distribution of land and ocean. For example, 36 percent of the total area of the Earth's endogenetic provinces is on land, in contrast with the 29 percent of the Earth's surface that is land. Correspondingly, 64 percent of the area of endogenetic provinces is within the 71 percent of the Earth's surface that is ocean. This relation suggests that tectonic topography is less rapidly obliterated by erosion on land than it is buried by sediment on the ocean floor. Similarly, 22 percent of the Earth's exogentic provinces are on land, versus 78 percent on the ocean floor, in contrast with the 29 percent of the Earth's surface that is land and 71 percent that is ocean. Note that most of the area of exogenetic provinces on land is in the Northern Hemisphere and most of that in the ocean are in the Southern Hemisphere, in accord with the distribution of global land and ocean areas. Note also that the endogenetic provinces on land are nearly double (1.92 x) those of the exogenetic provinces; in contrast, they are nearly equal (0.97 x) on the ocean floor. This relation accords with other evidence from river discharge of sediment that most products of erosion on land are transferred to and deposited on the ocean floor.

Genesis of Earth's geomorphology is controlled by two events: first, ones due to internal factors, where a surface is created at the mid-ocean ridge and consumed at subduction zones, and second, a symmetrical pattern of external forces, where land forms created by internal activities are changed to various forms in equilibrium with them. This harmony of creation and re-creation, what Umbgrove (1947) called the "pulse of the earth," is the fruit of the balance of the dynamic relation between endogenetic and exogenetic processes. Globally, the endogenetic and exogenetic provinces may appear to develop uniformly, but at a regional level they reveal pronounced inhomogeneities with

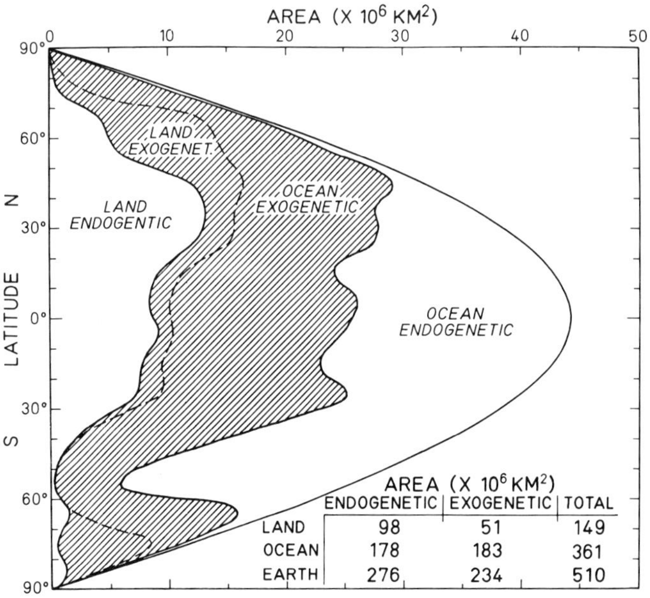

FIG. 3.—Summary diagram of endogenetic and exogenetic provinces in Figures 1 and 2 measured in 10° latitudinal belts for which the areas of land and ocean were compiled by E. Kossinna in 1921 (see Sverdrup and others, 1942, p. 13).

endogenetic regions being temporarily buried under their erosional products, followed by deformation again in the near future, and the continuity of broad sediment aprons is disrupted by remnants of older regions temporarily resurrected by adjustments to plate stresses. It is within such a framework that future studies in global-scale geomorphology should concentrate. Such studies should be done within the domain of the three-dimensional spherical models of convection in the Earth's mantle, as described by Bercovici and others (1989). The cylindrical plumes and downwelling planes described by these models probably are episodic and migratory (relative to the moving lithosphere), explaining the vertical oscillations of the Earth's surface with time. Recent studies indicate that large intracrustal deformation structures hundreds to more than 1,000 km in length may reflect mantle motions (Burchfiel and others, 1989). For example, Africa's unusual morphology of basins and swells, where volcanic activity has persisted throughout the Neogene, can best be explained as a result of mantle convection. Africa apparently came to rest with reference to mantle-convective circulation in the Early Neogene, and as it did, its surface mimicked the mantle convection impinging on the undersurface of the lithosphere to form its basin-and-swell morphology with the swells developed over upwelling mantle (Burke and Wells, 1989, and references therein). Many of the smaller structures within the large zones on intracrustal deformation over horizontal distances of several hundred kilometers, however, appear to be separated from the upper mantle by zones of detachment within the continental crust and are not coupled to motion of the mantle (Burchfiel and others, 1989). What is the minimum structural size reflecting mantle motion, and what is the relation of the detached structures to this motion have yet to be resolved. In a way, the scheme that we propose is a combination of the Davis 'forward looking' system where topography evolves linearly, and Walther Penck's 'backward looking' system where topographic forms provide clues to the endogenetic processes that produced the original topography before it became modified by exogenetic processes (see von Engeln, 1942, p. 69–73 and p. 256–268, for a discussion of these American and German schools of geomorphology). It is our contention that, although plate motions (divergence, convergence and translation) produce different topographic forms, each terrane evolves linearly.

Whether a geomorphic cycle comes to conclusion depends upon changes in plate geometry that in turn affect climate, base level and tectonic rejuvenation. We postulate, like Davis, that uplift in plate interiors away from zones of convergence is followed by relatively long periods of stability. These orogenic events probably occur more frequently near plate boundaries. Deformation can thus be visualized as a wave front diminishing gradually away from the plate boundary. Mid-ocean ridges where tectonic events are much more frequent (in the order of tens of hundreds of years) are zones of continuous deformation. Tectonism along margins of translation probably is comparable to those of mid-ocean ridges, and along convergent oceanic plates is intermediate between convergent margins and mid-ocean ridges. Like Penck, we believe that the morphology of a given site is the expression of the styles of deformation and degradation. If plate geometries change, the style of deformation also changes, which, in turn, affects the processes of degradation. Complexities in the geomorphic cycle also arise by migration of plates across climatic zones, leading to changes in the style of degradation and producing different landforms with time even though the style of deformation remains constant. There is a need to look at morphology at a global scale. We tend to focus our attention so much on what is in front of us that we fail to see the Earth's morphology in larger perspective. Our focus in the present study has been the distribution and geohistory of the endogenetic and exogenetic global provinces. In future studies special emphasis will be placed on their geomorphic evolution. We plan to do this within a qualitative/intuitive realm. Lest we forget, the concept of continental drift was the creation of imagination refuted temporarily by those who divested themselves of their intuition.

ACKNOWLEDGMENTS

Discussions over the years with Dave Aubrey, Marty Kleinrock, John Milliman, Hans Schouten, Debbie Smith and Brian Tucholke of the Geology and Geophysics Department at Woods Hole Oceanographic Institution and Bill Dillon, Bob Oldale, Peter Popenoe and Bill Schwab of the U.S. Geological Survey Marine Geology Branch at Woods Hole were extremely helpful in the writing of this report. Much of the material described was compiled by Uchupi in the course of teaching marine geology in the MIT-WHOI Education Joint Program. Thanks also are extended to Bob Ballard, Director of the Center for Marine Exploration at Woods Hole, for making it possible for Uchupi to participate in cruises and/or post-cruise analyses of video data recorded by ARGO in the East Pacific Rise north of the Clipperton Fracture Zone, Mohns Ridge, Reykjanes Ridge (NR-1 cruise), Tyrrhenian Basin in the western Mediterranean, Tongue of the Ocean in the Bahamas, and the *RMS Titanic* and battleship *Bismark* sites in the western and eastern North Atlantic. Appreciation is due the Woods Hole Oceanographic Institution's Ocean Industry Program for defraying some of the costs of preparing this report. Useful comments were provided by and, who read preliminary drafts of the manuscript. This paper is contribution number 7186 of the Woods Hole Oceanographic Institution.

REFERENCES

ANONYMOUS, 1984, General Bathymetric Chart of the Oceans (GEBCO), scale 1:35,000,000 at the equator: Ottawa, Canadian Hydrographic Service, 1 sheet.

ARGO-RISE GROUP, 1988, Geological mapping of the East Pacific Rise axis (10°19'-11°53'N) using the ARGO and ANGUS imaging systems: Canadian Mineralogist, v. 26, p. 467–488.

ATWATER, TANYA, 1970, Implications of plate tectonics for the Cenozoic evolution of western North America: Geological Society of America Bulletin, v. 81, p. 3513–3536.

AUBOUIN, J., AZEMA, J., CARFANTAN, J.-C., RANGAN, C., TANDY, M., AND TOURMON, J., 1982, The Middle America Trench in the geological framework of Central America, *in* Aubouin, J., and von Huene, R., and others, Initial Reports of the Deep Sea Drilling Project, v. 67: U.S. Government Printing Office, Washington, D.C., p. 747–755.

BAKER, V. R., 1973 Paleohydrology and sedimentology of Lake Missoula flooding in eastern Washington: Geological Society of America Special Paper 144, 79 p.

BALLARD, R. D., AND VAN ANDEL, T. H., 1977, Morphology and tectonics of the inner rift valley at latitude 36°50'N on the Mid-Atlantic Ridge: Geological Society of America Bulletin, v. 93, p. 3195–3210.

BARAZANGI, MUAWIA, AND ISACKS, B. L., 1979, Subduction of the Nazca plate beneath Peru: Evidence from spatial distribution of earthquakes: Royal Astronomical Society Geophysics Journal, v. 57, p. 537–555.

BEN-AVRAHAM, ZVI, 1985, Structural framework of the Gulf of Elat (Aqaba): Journal of Geophysical Research, v. 90, p. 703–726.

BEN-AVRAHAM, ZVI, AND COOPER, A. K., 1982, The early evolution of the Bering Sea by collision of ocean rises and north Pacific subduction zones: Geological Society of America Bulletin, v. 92, p. 484–495.

BERCOVICI, DAVE, SCHUBERT, GERARD, AND GLATZMAIER, G. A., 1989, Three-dimensional spherical models of convection in the Earth's mantle: Science, v. 244, p. 950–955.

BIJU-DUVAL, B., DERCOURT, J., AND LE PICHON, X., 1977, From the Tethys Ocean to the Mediterranean seas: A plate tectonic model of the evolution of the Western Alpine System, *in* Biju-Duval, J., and Montadert, L., eds., Structural History of the Mediterranean Basins: 25th Congress, Commission Internationale pour l'Exploration Science Méditerranee: Editions Technip, Paris, p. 143–164.

BONATTI, ENRICO, AND JOENSUU, OIVA, 1966, Deep-sea iron deposits from the South Pacific: Science, v. 154, p. 643–645.

BOVIS, MICHAEL, AND ISACKS, B. L., 1984, Hypocentral trend surface: Probing the geometry of Benioff zones: Journal of Geophysical Research, v. 89, p. 6153–6170.

BOWIN, C. O., 1975, Caribbean gravity field and plate tectonics: Geological Society of America Special Paper 169, 79 p.

BREEN, N. A., SILVER, E. A., AND HUSSONG, D. M., 1986, Structural styles of an accretionary wedge south of the island of Sumba, Indonesia revealed by SeaMARC II side-scan sonar: Geological Society of America Bulletin, v. 97, p. 1250–1261.

BROTHERS, R. N., AND LITTLE, A. R., 1988, Regional geology of Caledonia, *in* Nairn, A. E. M., Stehli, F. G., and Uyeda, Seiya, eds., The Ocean Basins and Margins, v. 7B, The Pacific Ocean: Plenum Press, New York, p. 325–374.

BRYAN, W. B., 1986, Tectonic controls on initial continental rifting and the evolution of young ocean basins—a planetary perspective: Tectonophysics, v. 132, p. 103–115.

BURCHFIEL, B. C., 1980, Tectonics of noncollisional regimes—The modern Andes and the Mesozoic Cordilleran Orogen of the Western United States, *in* Continental Tectonics: National Academy of Sciences, Washington, D.C., p. 65–95.

BURCHFIEL, B. C., QUIDONG, DENG, MOLNAR, PETER, ROYDEN, LEIGH, YIPENG, WANG, PEIZHEN, ZHANG, AND WEIQI, ZHANG, 1989, Intracrustal detachment within zones of continental deformation: Geology, v. 17, p. 748–752.

BURKE, KEVIN, 1980, Intracontinental rifts and aulocogens, *in* Continental Tectonics: National Academy of Sciences, Washington, D.C., p. 42–49.

BURKE, KEVIN, AND DEWEY, J. F., 1974, Two plates in Africa during the Cretaceous: Nature, v. 239, p. 387–390.

BURKE, KEVIN, AND WELLS, G. L., 1989, Trans-African drainage system of the Sahara: Was it the Nile?: Geology, v. 17, p. 743–747.

BURKE, KEVIN, AND WILSON, J. T., 1976, Hotspots on the Earth's surface: Scientific American, v. 235, no. 2, p. 46–57.

CAMPA U. M. F., 1985, The Mexican Thrust Belt, *in* Howell, D. G., ed., Tectonostratigraphic Terranes of the Circum-Pacific Region: Circum-Pacific Council for Energy and Mineral Resources Earth Science Series Number 1, Houston, p. 299–313.

CASE, J. E., DURAN, S. L. G., LOPEZ, R. A., AND MOORE, W. R., 1971, Tectonic investigations in western Colombia and eastern Panama: Geological Society of America Bulletin, v. 82, p. 2685–2712.

CHAMBERLIN, T. C., 1913–1921, Diastrophism and the formative processes: Journal of Geology, v. 21, p. 517–522, 523–533, 577–587, 673–682; v. 22, p. 131–144, 266–274, 516–528; v. 26, p. 193–197; v. 28, p. 1–17, 126–157, 473–504, 655–701; v. 29, p. 391–415, 679–700.

CHINGCHANG, BIQ, SHYU, C. T., CHEN, J. C., AND BOGGS, SAM, JR., 1985, Taiwan: Geology, geophysics, and marine sediments, *in* Nairn, A. E. M., Stehli, F. G., and Uyeda, Seiya, eds., The Ocean Basins and Margins, v. 7A, The Pacific Ocean: Plenum Press, New York, p. 503–550.

CHURKIN, MICHAEL, JR., AND TREXLER, J. H., 1980, Circum-Arctic plate acretion—Isolating part of a Pacific plate to form the nucleus of the Arctic basin: Earth and Planetary Science Letters, v. 48, p. 356–361.

CLARK, D. L., 1981, Geology and geophysics of the Amerasian Basin, *in* Nairn, A. E. M., Churkin, Michael, Jr., and Stehli, F. G., eds., The Ocean Basins and Margins, v. 5, The Arctic Ocean: Plenum Press, New York, p. 599–634.

CLOETINGH, SIERD, 1988, Intraplate stresses: A tectonic cause for third-order cycles in apparent sea level?, *in* Wilgus, C. K., Hastings, B. S., Posamentier, H. W., Kendall, G. St. C., Ross, C. A., and Van Wagoner, J. C., eds., Sea-Level Changes: An Integrated Approach: Society of Economic Paleontologists and Mineralogists Special Publication 42, p. 19–30.

COCHRAN, J. R., 1981, The Gulf of Aden: Structure and evolution of a young ocean basin and continental margin: Journal of Geophysical Research, v. 86, p. 263–288.

COCHRAN, J. R., 1983, A model for the development of the Red Sea: American Association of Petroleum Geologists Bulletin, v. 67, p. 41–69.

CONEY, P. J., JONES, D. L., AND MONGER, J. W. H., 1980, Cordilleran suspect terranes: Nature, v. 288, p. 329–333.

COULSON, F. I., 1985, Solomon Islands, *in* Nairn, A. E. M., Stehli, G. H., and Uyeda, Seiya, eds.,The Ocean Basins and Margins, v. 7A, The Pacific Ocean: Plenum Press, New York, p. 607–682.

CRANE, KATHLEEN, 1985, The spacing of rift axis highs: Dependence upon diapiric processes in the underlying asthenosphere?: Earth and Planetary Science Letters, v. 72, p. 405–414.

CRANE, KATHLEEN, SUNDVOR, E., FOUCHER, J. P., AND BUCK, R., 1986, The role of fracture zones on the thermal evolution of the northern Norwegian-Greenland Sea: EOS, Transactions, American Geophysical Union, v. 67, p. 1221.

CROUGH, S. T., 1979, Hotspot epeirogeny: Tectonophysics, v. 61, p. 321–333.

CROUGH, S. T., 1981, Mesozoic hotspot epeirogeny in eastern North America: Geology, v. 9, p. 2–6.

CURRAY, J. R., EMMEL, F. J., MOORE, D. G., AND RAITT, R. S., 1982, Structure, tectonics, and geological history of the northeastern Indian Ocean, *in* Nairn, A. E. M., Stehli, F. G., and Uyeda, Seiya, eds., The Ocean Basins and Margins, v. 7A, The Pacific Ocean: Plenum Press, New York, p. 399–512.

DAKE, C. L., AND BROWN, J. S., 1925, Interpretation of Topographic and Geologic Maps with Special Reference to Determination of Structure: McGraw-Hill Book Co. Inc., New York, 355 p.

DALZIEL, I. W. D., 1974, Evolution of margins of the Scotia Sea, *in* Burk, C. A., and Drake, C. L., eds., The Geology of Continental Margins: Springer-Verlag, New York, p. 567–579.

DALZIEL, I. W. D., AND ELLIOT, D. H., 1973, The Scotia Arc and Antarctic margin, *in* Nairn, A. E. M., Stehli, F. G., and Uyeda, Seiya, eds., The Ocean Basins and Margins, v. 7A, The Pacific Ocean: Plenum Press, New York, p. 171–246.

DALZIEL, I. W. D., AND GRUNOW, A. M., 1985, The Pacific margin of Antarctica: Terranes within terranes within terranes, *in* Howell, D. G., ed., Tectonostratigraphic Terranes of the Circum-Pacific Region: Circum-Pacific Council for Energy and Mineral Resources Earth Science Series Number 1, Houston, p. 555–564.

DAVIS, G. A., 1980, Problems of interplate extensional tectonics, western United States, *in* Continental Tectonics: National Academy of Sciences, Washington, D.C., p. 84–95.

DAVIS, W. M., 1909, The geographical cycle, *in* Geographical Essays: Ginn & Co., Boston, p. 249–276.

DILLON, W. P., AND POPENOE, PETER, 1988, The Blake Plateau basin and Carolina trough, *in* Sheridan, R. E., and Grow, J. A., eds., The Geology of North America, v. I-2, The Atlantic Continental Margin, U.S.: Geological Society of America, p. 291–328.

DUNCAN, R. A., 1984, Age of progressive volcanism in the New England seamounts and the opening of the North Atlantic: Journal of Geophysical Research, v. 89, p. 9980–9990.

DUNCAN, R. A., AND CLAGUE, D. A., 1985, Pacific plate motions recorded by linear volcanic chain, *in* Nairn, A. E. M., Stehli, F. G., and Uyeda, Seiya, eds., The Ocean Basins and Margins, v. 7A, The Pacific Ocean: Plenum Press, New York, p. 89–122.

DUPONT, J., 1988, The Tonga and Kermadec ridges, *in* Nairn, A. E. M., Stehli, F. G., and Uyeda, Seiya, eds., The Ocean Basins and Margins, v. 7B, The Pacific Ocean: Plenum Press, New York, p. 375–410.

EADE, J. V., 1988, The Norfolk Ridge system and its margins, *in* Nairn, A. E. G., Stehli, F. G., and Uyeda, Seiya, eds., The Ocean Basins and Margins, v. 7B, The Pacific Ocean: Plenum Press, New York, p. 303–324.

EBERLI, G. P., AND GINSBURG, R. N., 1987, Segmentation and coalescence of the Cenozoic carbonate platforms, northwest Great Bahama Bank: Geology, v. 15, p. 75–79.

EMERY, K. O., AND MILLIMAN, J. D., 1978, Suspended matter in surface waters: Influence of river discharge and of upwelling: Sedimentology, v. 25, p. 125–140.

EMERY, K. O., AND UCHUPI, ELAZAR, 1984, The Geology of the Atlantic Ocean: Springer Verlag, New York, 1050 p.

ESPINOSA, A. F., RINEHART, WILBUR, AND THARP, MARIE, 1981, Seismicity of the Earth 1960–1980, scale 1:46,460,600 (Heezen, B. C. and Tharp, Marie, 1977, World Ocean Panorama), Office of Naval Research, 1 sheet.

FALVEY, D. A., 1974, The development of continental margins in plate tectonic theory: Journal of the Australia Petroleum Exploration Association, v. 14, p. 95–106.

FORSYTHE, ROBERT, 1982, The Late Paleozoic to Early Mesozoic evolution of southern South America: A plate tectonic interpretation: Journal of the Geological Society of London, v. 139, p. 671–682.

FREEMAN-LYNDE, R. P., CITA, M. B., JADOUL, FLAVIO, MILLER, E. L., AND RYAN, W. B. F., 1981, Marine geology of the Bahama Escarpment: Marine Geology, v. 44, p. 119–156.

GAMBOA, L. A. P., AND RABINOWITZ, P. D., 1988, The evolution of the Rio Grande Rise in the southwest Atlantic Ocean: Marine Geology, v. 58, p. 35–58.

GASTIL, GORDON, 1985, Terranes of peninsular California and adjacent Sonora, *in* Howell, D. G., ed., Tectonostratigraphic Terranes of the Circum-Pacific Region: Circum-Pacific Council for Energy and Mineral Resources Earth Science Series Number 1, Houston, p. 273–283.

GRANTZ, ARTHUR, EITTREIM, STEPHEN, AND WHITNEY, O. T., 1981, Geology and physiography of the continental margin north of Alaska and implications for the origin of thc Canada Basin, *in* Nairn, A. E. M., Churkin, M., Jr., and Stehli, F. G., eds., The Ocean Basins and Margins, v. 5, The Arctic Ocean: Plenum Press, New York, p. 439–492.

HADA, S. AND SUZUKI, T. S., 1983, Tectonic environments and crustal section of the outer zone of southwest Japan, *in* Hashimoto, M., and Uyeda, S., eds., Accretion Tectonics in the Circum-Pacific Regions: Terra Publishing Co., Tokyo, p. 207–218.

HAMILTON, E. L., 1956, Sunken Islands of the Mid-Pacific Mountains: Geological Society of America Memoir 64, 97 p.

HAMILTON, WARREN, 1979, Tectonics of the Indonesian region: U.S. Geological Survey Professional Paper 1078, 346 p.

HANCOCK, J. M., AND KAUFFMAN, E. G., 1979, The great transgression of the Late Cretaceous: Geological Society of London Quarterly Journal, v. 136, p. 175–186.

HANSHAW, B. B., AND BACK, WILLIAM, 1980, Chemical mass-wasting on the northern Yucatan peninsula by groundwater dissolution: Geology, v. 8, p. 222–224.

HAQ, B. U., HARDENBOL, JAN, AND VAIL, P. R., 1987, Chronology of fluctuating sea levels since the Triassic: Science, v. 235, p. 1156–1167.

HARRISON, C. G. A., MISKELL, K. J., BRASS, G. W., SALZMAN, E. S., AND SLOAN, J. L., II, 1983, Continental hypsography: Tectonics, v. 2, p. 357–377.

HASAGAWA, AKIRA, AND SACKS, I. S., 1981, Subduction of the Nazca plate beneath Peru from seismic observations: Journal of Geophysical Research, v. 86, p. 4971–4980.

HEEZEN, B. C., MATTHEWS, J. L., CATALANO, R., NATLAND, J., COOGAN, A., THARP, M., AND RAWSON, M., 1973, Western Pacific guyots, *in* Heezen, B. C., MacGregor, I. D., and others, Initial Reports of the Deep Sea Drilling Project, v. 20: U.S. Government Printing Office, Washington, D.C., p. 653–702.

HEGARTY, K. A., AND WEISSEL, J. K., 1988, Complexities in the development of the Carolina Plate region, western equatorial Pacific, *in* Nairn, A. E. M., Stehli, F. G., and Uyeda, Seiya, eds., The Ocean Basins and Margins, v. 7B, The Pacific Ocean: Plenum Press, New York, p. 277–302.

HEKINIAN, R., FEVRIER, M., BISCHOFF, J. L., PICOT, P., AND SHANKS, W. C., 1981, Sulfide deposits from the East Pacific Rise near 21°N: Science, v. 207, p. 1433–1444.

HEY, RICHARD, JOHNSON, G. L., AND LOWRIE, ALLEN, 1977, Recent plate motions in the Galapagos Area: Geological Society of America Bulletin, v. 88, p. 1385–1403.

HILL, I. A., AND BARKER, P. F., 1980, Evidence for Miocene back-arc spreading in the central Scotia Sea: Royal Astronomical Society Geophysical Journal, v. 63, p. 427–440.

HINZE, W. J., BRAILE, L. W., KELLER, G. R., AND LIDIAK, E. G., 1980, Models for midcontinent tectonism, *in* Continental Tectonics: National Academy of Sciences, Washington, D.C., p. 73–83.

HOLMES, ARTHUR, 1965, Principles of Physical Geology (rev. ed.): George Ronald Publishing Co., New York, 1288 p.

HOWELL, D. G., SCHERMER, E. R., JONES, D. L., BEN-AVRAHAM, ZVI, AND SCHEIBNER, ERWIN, 1985, Preliminary tectonostratigraphic terrane map of the circum-Pacific Region: Circum-Pacific Council for Energy and Mineral Resources, Houston, scale 1:17,000,000.

HURLEY, P. M., 1974, Pangaeaic orogenic system: Geology, v. 2, p. 373–376.

HUTCHINSON, C. S., 1982, Southeast Asia, *in* Nairn, A. E. M., and Stehli, F. G., eds., The Ocean Basins and Margins, v. 6, The Indian Ocean: Plenum Press, New York, p. 451–512.

JOHNSON, D. W., 1919, Shore Processes and Shoreline Development: John Wiley & Sons Inc., New York, 584 p.

KARIG, D. E., 1971, Structural history of the Mariana Island arc system: Geological Society of America Bulletin, v. 82, p. 323–344.

KEIGWIN, L. D., JR., 1978, Pliocene closing of the Isthmus of Panama, based on biostratigraphic evidence from nearby Pacific and Caribbean Sea cores: Geology, v. 6, p. 630–634.

KIMURA, G., MIYASHITA, S., AND MIYASAKA, S., 1983, Collision tectonics in Hokkaido and Sakhalin, *in* Hashimoto, M., and Uyeda, S., eds., Accretion Tectonics in the Circum-Pacific Regions: Terra Publishing Company, Tokyo, p. 123–134.

KLEINROCK, M. C., AND HEY, R. N., 1989, Detailed tectonics near the tip of the Galapagos 95.5°W propagator: How the lithosphere tears and a spreading axis develops: Journal of Geophysical Research, v. 94, p. 13,801–13,838.

KOBAYSHI, KAZUO, 1985, Sea of Japan and Ryukyui Trench-arc back-arc system, *in* Nairn, A. E. M., Stehli, F. G., and Uyeda, Seiya, eds., The Ocean Basins and Margins, v. 7A, The Pacific Ocean: Plenum Press, New York, p. 419–458.

KONG, L. S. L., DETRICK, R. S., FOX, P. J., MAYER, L. A., AND RYAN, W. B. F., 1988, The morphology and tectonics of the Mark area from SeaBeam and SeaMARC I observations (Mid-Atlantic Ridge 23°N): Marine Geophysical Research, v. 10, p. 59–90.

KORSCH, R. J., AND WELLMAN, H. W., 1988, The geological evolution of New Zealand and the New Zealand region, *in* Nairn, A. E. M., Stehli, F. G., and Uyeda, Seiya, eds., The Ocean Basins and Margins, v. 7B, The Pacific Ocean: Plenum Press, New York, p. 411–484.

LEFORT, J.-P., 1989, Basement correlation across the North Atlantic: Springer-Verlag, New York, 148 p.

LEG 101 SCIENTIFIC PARTY, 1988, Leg 101—An overview, *in* Austin, J. A., Jr., Schlager, W., Palmer, A. A., and others, Proceedings, Initial Reports (Part A), Ocean Drilling Program, 101: Ocean Drilling Program, College Station, Texas, p. 455–472.

LE PICHON, XAVIER, AND FRANCHETEAU, JEAN, 1978, A plate tectonic analysis of the Red Sea—Gulf of Aden area: Tectonophysics, v. 46, p. 369–406.

LIN, J., PURDY, G. M., SCHOUTEN, H., SEMPERE, J.-C., AND ZERVAS, C., 1990, Evidence from gravity data for focused magmatic accretion along the Mid-Atlantic Ridge: Nature, v. 344, p. 627–632.

LONSDALE, PETER, 1988, Geography and history of the Louisville hotspot chain in the southwest Pacific: Journal of Geophysical Research, v. 93, p. 3078–3104.

LONSDALE, PETER, BATIZA, RODEY, AND SIMKIN, TOM, 1982, Metallogenesis at seamounts on the East Pacific Rise: Marine Technology Society Journal, v. 16, p. 54–60.

MACDONALD, K. C., FOX, P. J., PERRAM, L. J., EISEN, M. F., HAYMON, R. M., MILLER, S. P., CARBOTTE, S. M., CORMIER, M.-H., AND SHORE, A. N., 1988, A new view of the mid-ocean ridge from the behavior of ridge-axis discontinuities: Nature, v. 335, p. 217–225.

MADOLE, R. F., BRADLEY, W. C., LOEWENHERZ, D. S., RITTER, D. F., RUTTER, N. W., AND THORN, C. E., 1987, Rocky Mountains, *in* Graf, W. L., ed., Geomorphic Systems of North America: Geological Society of America Centennial Special, v. 2, p. 211–258.

MALAHOFF, ALEX, 1988, Polymetallic sulfide formation and gold enrichment of hydrothermal vents of Pacific seamounts: Program, Underwater Mining Institute meeting, Woods Hole Oceanographic Institution, Oct. 1988, 3 p.

MALAHOFF, ALEX, MCMURTRY, G. M., WILTSHIRE, J. C., AND YEH, H.-W., 1982, Geology and chemistry of hydrothermal deposits from active submarine volcano Loihi: Nature, v. 298, p. 234–239.

MCCABE, ROBERT, ALMASCO, J. N., AND YUMUL, GRACIANO, 1985, Terranes of the central Philippines, *in* Howell, D. G., ed., Tectonostratigraphic Terranes of the Circum-Pacific Region: Circum-Pacific Council for Energy and Mineral Resources Earth Science Series Number 1, Houston, p. 421–435.

MCKEE, T. R., JEFFREY, M., PRESLEY, B. J., AND WHITEHOUSE, U. G., II, 1978, Holocene sediment geochemistry of continental slope and intraslope basin areas, northwest Gulf of Mexico, *in* Bouma, A. H., Moore, G. H., and Coleman, J. M., eds., Framework, Facies, and Oil-Trapping Characteristics of the Upper Continental Margin: American Association of Petroleum Geologists Studies in Geology No. 7, p. 313–326.

MCNUTT, MARCIA, AND FISCHER, KAREN, 1986, The South Pacific Superswell: EOS, Transactions, American Geophysical Union, v. 67, p. 1220.

MEISSNER, ROLF, 1986, The Continental Crust. A Geophysical Approach: Academic Press, New York, 426 p.

MENARD, H. W., 1984, Darwin reprise: Journal of Geophysical Research, v. 89, p. 9960–9968.

MENARD, H. W., AND SMITH, S. M., 1966, Hypsometry of ocean basin provinces: Journal of Geophysical Research, v. 71, p. 4305–4325.

MILLIMAN, J. D., AND MEADE, R. H., 1983, World-wide delivery of river sediment to the oceans: Journal of Geology, v. 91, p. 1–21.

MILSOM, JOHN, 1985, New Guinea and the western Melanesian arcs, *in* Nairn, A. E. M., Stehli, F. G., and Uyeda, Seiya, eds., The Ocean Basins and Margins, v. 7A, The Pacific Ocean: Plenum Press, New York, p. 551–605.

MIZUTANI, SHINJIRO, AND HATTORI, ISAMU, 1983, Hida and Mino: tectonostratigraphic terranes in central Japan, *in* Hashimoto, M., and Uyeda, S., eds., Accretion Tectonics in the Circum-Pacific Regions: Terra Publishing Co., Tokyo, p. 169–178.

MOLNAR, PETER, AND TAPPONNIER, PAUL, 1975, Cenozoic tectonics of Asia: Effects of a continental collision: Science, v. 189, p. 419–426.

MOORE, G. F., AND SILVER, E. A., 1983, Collision Process in the northern Molucca Sea, *in* Hayes, D. E., ed., The Tectonic and Geologic History of Southeast Asian Seas and Islands, Part 2: American Geophysical Union, Geophysical Monograph 27, p. 360–372.

MORGAN, W. J., 1972, Plate motions and deep mantle convection, *in* Shagam, R., Hargrove, R. B., Morgan, W. J., Van Houten F. B., Burk, C. A., Halland, H. D., and Hollister, L. C., eds., Studies in Earth and Space Sciences: Geological Society of America Memoir 132, p. 7–22.

MORGAN, W. J., 1978, Rodriguez, Darwin, Amsterdam . . . , A second type of hotspot island: Journal of Geophysical Research, v. 83, p. 5355–5360.

MORGAN, W. J., 1983, Hotspot tracks and the early rifting of the Atlantic: Tectonophysics, v. 94, p. 123–139.

NEEV, DAVID, HALL, J. K., AND SAULL, J. M., 1982, The Pelusium magashear across Africa and associated lineament swarms: Journal of Geophysical Research, v. 87, p. 1015–1030.

NUR, AMOS, AND BEN-AVRAHAM, ZVI, 1983a, Displaced terranes and mountain building, *in* Hsu, K. J., ed., Mountain Building Processes: Academic Press, New York, p. 73–84.

NUR, AMOS, AND BEN-AVRAHAM, ZVI, 1983b, Break-up and accretion tectonics, *in* Hashimoto, M., and Uyeda, S., eds., Accretion Tectonics in the Circum-Pacific Regions: Terra Publishing Company, Tokyo, p. 3–18.

ODP LEG 110 SCIENTIFIC PARTY, 1987, Expulsion of fluids from depth along a subduction-zone décollement horizon: Nature, v. 326, p. 785–788.

OHMOTO, H., 1978, Submarine calderas: A key to the formation of volcanogenic massive sulfide deposits: Mining Geology, v. 28, p. 219–231.

OKADA, HAKUYU, 1983, Collision orogenesis and sedimentation in Hokkaido, Japan, *in* Hashimoto, M., and Uyeda, S., eds., Accretion Tectonics in the Circum-Pacific Regions: Terra Publishing Company, Tokyo, p. 91–106.

PAOR, DE, D. G., BRADLEY, D. W., INSENSTADT, GLORIA, AND PHILLIPS, S. M., 1989, The Arctic Eurokean orogen: A most unusual fold-and-thrust belt: Geological Society of America Bulletin, v. 101, p. 952–967.

PARFENOV, L. M., AND NATAL'IN, B. A. 1985, Mesozoic accretion and collision tectonics of northeastern Asia, *in* Howell, D. G. ed., Tectonostratigraphic Terranes of the Circum-Pacific Region: Circum-Pacific Council for Energy and Mineral Resources Earth Science Series Number 1, Houston, p. 363–373.

PARSON, L. M., PATRIAT, P., AND SEARLE, R. C., 1986, Evolution of the Indian Ocean Triple Junction—Evidence from the SE-Indian/SW-Indian-Ridge trace: EOS, Transactions, American Geophysical Union, v. 67, p. 1221.

PASTOURET, L., AND GOSLIN, J., 1974, Middle Cretaceous sediments from the eastern part of Walvis Ridge: Nature, v. 248, p. 495–496.

PAULL, C. K., SPIESS, F. N., CURRAY, J. R., AND TWICHELL, D. C., 1990, Florida Canyon: One of the Florida Escarpment's enigmatic submarine box canyons: Geological Society of America Bulletin, v. 102, p. 502–515.

PENCK, WALTHER, 1924, Die Morphologische Analyse: J. Engelhorn's Nachforshung, Stuttgart, 283 p.

PILGER, A., AND RÖSLER, A., 1976, General aspects with special reference to Afar. Temporal relationship in the tectonic evolution of the Afar depression (Ethiopia) and the adjacent Afro-Arabian rift system, *in* Pilger, A., and Rösler, A., eds., Afar Between Continental and Oceanic Rifting: E. Schweizerbart'sche verlagsbuchhandlung, Stuttgart, p. 1–25.

REZAK, RICHARD, AND BRIGHT, T. J., 1981, Seafloor instability at East Flower Garden Bank, northwest Gulf of Mexico: Geo-Marine Letters, v. 1, p. 97–103.

RICATEAU, R., AND RICHÉ, P. H., 1980, Geology of the Musandam Peninsula (Sultanate of Oman) and its surroundings: Journal of Petroleum Geology, v. 3, p. 139–152.

ROBB, J. M., 1984, Spring sapping on the lower continental slope, offshore New Jersey: Geology, v. 12, p. 278–282.

ROSENDAHL, B. R., 1987, Architecture of continental rifts with special reference to East Africa: Annual Review of Earth and Planetary Sciences, v. 15, p. 445–503.

ROSS, D. A., UCHUPI, ELAZAR, AND WHITE, R. S., 1986, The geology of the Persian Gulf—Gulf of Oman: A synthesis: Review of Geophysics, v. 24, p. 537–556.

SALISBURY, R. D., AND ATWOOD, W. W., 1908, The Interpretation of Topographic Maps: U.S. Geological Survey Professional Paper 60, 84 p. and 170 plates.

SCHEIDEGGER, A. E., 1987, Systematic Geomorphology: Springer-Verlag, New York, 285 p.

SCHLICH, R., 1982, The Indian Ocean: Aseismic ridges, spreading centers, and oceanic basins, *in* Nairn, A. E. M., and Stehli, G. F., eds., The Ocean Basins and Margins, v. 6, The Indian Ocean: Plenum Press, New York, p. 51–148.

SCLATER, J. G., ANDERSON, R. N., AND BELL, M. L., 1971, Elevation of ridges and evolution of the central eastern Pacific: Journal of Geophysical Research, v. 76, p. 7888–7915.

SEARLE, M. P., JAMES, N. P., CALON, T. J., AND SMEWING, J. D., 1983, Sedimentological and structural evolution of the Arabian continental margin in the Musandam Mountains and Dibb zone, United Arab Emirates: Geological Society of America Bulletin, v. 94, p. 1381–1400.

SHEPARD, F. P., 1923, To question the theory of periodic diastrophism: Journal of Geology, v. 31, p. 599–613.

SHEPARD, F. P., 1937, Revised classification of marine shorelines: Journal of Geology, v. 45, p. 602–624.

SMITH, D. K., AND JORDAN, T. H., 1987, The size distribution of Pacific Seamounts: Geophysical Research Letters, v. 14, p. 1119–1122.

STAUFFER, P. H., 1985, Continental terranes in Southeast Asia: Pieces of which puzzle?, *in* Howell, D. G., ed., Tectonostratigraphic Terranes of the Circum-Pacific Region: Circum-Pacific Council for Energy and Mineral Resources Earth Science Series Number 1, Houston, p. 529–539.

STERN, R. J., GOTTFRIED, DAVID, AND HEDGE, C. E., 1984, Late Precambrian rifting and crustal evolution in the northeastern desert of Egypt: Geology, v. 12, p., 168–172.

STIELTJES, LAURENT, 1973, Evolution tectonique récente du rift d'Asal, T.F.A.I.: Revue de Géographie Physique et de Géologie Dynamique, v. 15, series 2, p. 425–436.

STOESER, D. B., AND CAMP, V. E. 1985, Pan-African microplate accretion of the Arabian Shield: Geological Society of America Bulletin, v. 96, p. 817–826.

SULLIVAN, K. D., AND KEEN, C. E., 1977, Newfoundland Seamounts, *in* Baragar, W. R. A., Coleman, C. L., and Hall, J. M., eds., Volcanic Regimes of Canada: Geological Association of Canada Special Paper 16, p. 461–476.

SUPKO, P. R., AND PERCH-NIELSEN, KATHARINA, 1977, General synthesis of central and South Atlantic drilling results, leg 39, Deep Sea Drilling Project, *in* Supko, P. R., Perch-Nielsen, Katharina, and others, Initial Reports of the Deep Sea Drilling Project, v. 39: U.S. Government Printing Office, Washington, D. C., p. 1099–1131.

SVERDRUP, H. U., JOHNSON, M. W., AND FLEMING, R. H., 1942, The Oceans, Their Physics, Chemistry, and General Biology: Prentice-Hall, Inc., Englewood Cliffs, New Jersey, 1087 p.

SZATMARI, PETER, 1983, Amazon rift and Pisco-Jurua fault; Their relation to the separation of North America from Gondwana: Geology, v. 11, p. 300–304.

TAYLOR, BRIAN, AND HAYES, D. E., 1983, Origin and history of the South China Sea Basin, *in* Hayes, D. E., eds., The Tectonic and Geologic History of Southeast Asian Seas and Islands, Part 2: American Geophysical Union Geophysical Monograph 27, p. 23–56.

THORNBURY, W. D., 1957, Principles of Geomorphology: John Wiley & Sons, Inc., New York, 618 p.

TUCHOLKE, B. E., AND LUDWIG, W. J., 1982, Structure and origin of the J-Anomaly Ridge, western North Atlantic: Journal of Geophysical Research, v. 87, p. 9389–9407.

UCHUPI, ELAZAR, AND BALLARD, R. D., 1989, Evidence of hydrothermal activity on Marsili Seamount, Tyrrhenian Basin: Deep-Sea Research, v. 36, p. 1443–1448.

UMBGROVE, J. H. F., 1947, The Pulse of the Earth: Martinus Nijhoff, The Hague, 358 p.

UNESCO, 1976, Geological World Atlas, scale 1:10,000,000: UNESCO, Commmission for the Geological Map of the World, 22 sheets.

URRUTIA-FUCUGAUCHI, JUAN, 1981, Paleomagnetic evidence for tectonic rotation of northern Mexico and the continuity of the Cordilleran orogenic belt between Nevada and Chihuahua: Geology, v. 9, p. 178–183.

UYEDA, SEIYA, AND BEN-AVRAHAM, ZVI, 1982, Origin and development of the Philippine Sea: Nature, v. 240, p. 176–178.

UYEDA, SEIYA, AND MCCABE, R., 1983, A possible mechanism of episodic spreading of the Philippine Sea, *in* Hashimoto, M., and Uyeda, S., eds., Accretion Tectonics in the Circum-Pacific Regions: Terra Publishing Company, Tokyo, p. 291–306.

VON ENGELN, O. D., 1942, Geomorphology. Systematic and Regional: The Macmillan Co., New York, 655 p.

WESTBROOK, G. K., AND SMITH, M. J., 1983, Long décollement and mud volcanoes: Evidence from Barbados Ridge complex for the role of high prewater pressures in the development of an accretionary wedge: Geology, v. 11, p. 279–283.

WHITEHEAD, J. A., JR., DICK, H. J. B., AND SCHOUTEN, H., 1984, A mechanism for magmatic accretion under spreading centres: Nature, v. 312, p. 146–148.

WILLIAMS, HAROLD, AND HATCHER, R. D., JR., 1982, Suspect terranes and accretionary history of the Appalachian orogen: Geology, v. 10, p. 530–536.

WORKING GROUP 3, 1988, Fluid circulation in the crust and the global geochemical budget, *in* Report of the Second Conference on Scientific Drilling "Cosod II": European Science Foundation and Joint Oceanographic Institutions for Deep Earth Sampling, p. 67–86.

XIONG, JI, AND CONEY, P. J., 1985, Accreted terranes of China, *in* Howell, D. G., ed., Tectonostratigraphic Terranes of the Circum-Pacific Region: Circum-Pacific Council for Energy and Mineral Resources Earth Science Series Number 1, Houston, p. 349–361.

ZOBACK, M. L., NISHENKO, S. P., RICHARDSON, R. M., HASENGAWA, H. S., AND ZOBACK, M. D., 1986, Mid-plate stress deformation and seismicity, *in* Vogt, P. R., and Tucholke, B. E., eds., The Geology of North America, v. M, the Western North Atlantic Region: Geological Society of America, p. 297–312.

DEVELOPMENTS IN UNDERSTANDING THE TECTONIC EVOLUTION OF THE CALIFORNIA CONTINENTAL BORDERLAND

MARK R. LEGG
ACTA Inc., Torrance, California 90505

ABSTRACT: Understanding of the tectonic evolution of the California Continental Borderland has developed relatively slowly because of the region's complex geology and its submarine character. For more than a century, scientists have recognized the dynamic character of borderland geology and tectonics—noting its basin and range physiography, identifying faults along steep linear escarpments bounding the offshore islands and submarine banks, and measuring earthquakes within the region. Exploration for offshore petroleum and other mineral resources with attendant technological developments in earth sciences generated maps of the subbottom geologic structure and confirmed interpretations of late Cenozoic deformation. The plate-tectonics paradigm provided an overall framework that constrained models of the regional tectonic evolution, while explaining the diverse geologic observations within this complex submarine province; borderland normal faulting, of basin and range character, was replaced by dextral strike-slip faulting associated with the modern San Andreas fault system.

Yet, more than half of the entire region remains unexplored in detail, particularly that part west of Baja California, Mexico. Increased understanding of borderland tectonics requires a multidisciplinary and international effort. Recent technological advancements, including seafloor swath mapping and three-dimensional multichannel seismic-reflection profiling, allow major advancements in this tectonic understanding. The Miocene development of the Pacific-North America plate boundary involved the passage of the East Pacific Rise spreading center along the western edge of the borderland and the consequent lengthening of the dextral San Andreas transform-fault system. Broad-scale features of basin development and widespread volcanism during this epoch may be explained by regional oblique extension. Partly due to complexity associated with clockwise rotation of elongate crustal blocks and undefined deep crustal structure, palinspastic reconstructions are poorly constrained, and mechanisms of the plate-boundary reorganization remain to be described in detail. Recent detailed investigations of parts of the inner borderland reveal a systematic change from oblique extension in the southern region, west of northern Baja California, to oblique covergence in the north, west of southern California and the Transverse Ranges. This character is evident in both the late Cenozoic geologic structure, as mapped by seismic-reflection profiles, and in the present seismicity, as shown by earthquake focal mechanisms. Because its tectonic evolution includes each of the three major types of plate-boundary interaction—subduction, rifting, and transform faulting—the California Continental Borderland provides an exceptional opportunity for studies of these kinds of processes along an active continental margin.

INTRODUCTION

The California Continental Borderland is a mostly submarine geologic province that extends from Point Arguello on the north to Bahía Sebastian Vizcaíno and Cedros Island on the south (Fig. 1; Moore, 1969). Based upon broad reconnaissance study, Moore (1969) divided the borderland into five major structural zones: 1) the northern zone of west-trending Transverse Ranges structure (Santa Barbara Channel and northern Channel Islands platform); 2) southern zone of deeper, northwest-trending ridges and basins (south of the offshore Santo Tomás fault); 3) an outer zone of northwest-trending faults; 4) an inner zone of northwest-trending faults; and 5) a central zone of large-scale folding. Because of the region's complex geology and submarine character, understanding of the tectonic evolution of the California Continental Borderland has been slow to develop. In fact, more than half of the entire region, the part that lies west of Baja California, remains unexplored in detail.

This paper is composed of two major parts: an historical overview of borderland research, and a consideration of current and future tectonic models for the region. After a brief description of earlier developments in the understanding of borderland tectonics, I will discuss some of the more recent tectonic models proposed for the region. More complete bibliographies of borderland research may be found elsewhere (Kennedy and others, 1987). A summary of major elements that define the borderland tectonic evolution follows, including suggested refinements to a composite model of borderland tectonic evolution. I will finish with a discussion of important constraints that some of the most recent data acquired in the region hold for these and future tectonic models. Readers already well versed in borderland research may wish to jump ahead to the "Uncertain and Controversial Issues . . ." section.

Understanding the tectonic evolution of the California Continental Borderland, an integral part of the Pacific-North America (PAC-NOAM) tectonic-plate boundary provides crucial information about the larger plate boundary evolution. With its complex tectonic history, involving all three of the major plate-boundary types—subduction, rifting, and transform faulting—this region continues to provide an exciting research frontier in the earth sciences.

HISTORICAL DEVELOPMENTS

Earlier Studies of the California Continental Borderland

Pioneer Explorations and the Search for Oil in Coastal California.— Geologic investigations in the southern California coastal and offshore region have been occurring for more than a century (Blake, 1855, 1857; Davidson, 1869, 1887; Lawson, 1897, 1924; Fairbanks 1897), and yet, the tectonic evolution of this important offshore region remains poorly understood. Largely because of the relative inaccessibility of the submerged part of the continental margin, most of the detailed geologic studies focused on the landward areas including the Channel Islands. Earliest studies relied upon the observations of geologists and other naturalists traveling with expeditions to explore the California frontier. Some submarine information was available in the soundings taken by the Coast and Geodetic Survey. Striking variations in lithology, structure, and physiography could be seen in the subaerially exposed parts of the continental borderland and adjacent coastal regions.

The early Spanish explorers noted that the offshore islands were roughly parallel to the mountain ranges along

From Shoreline to Abyss, SEPM Special Publication No. 46

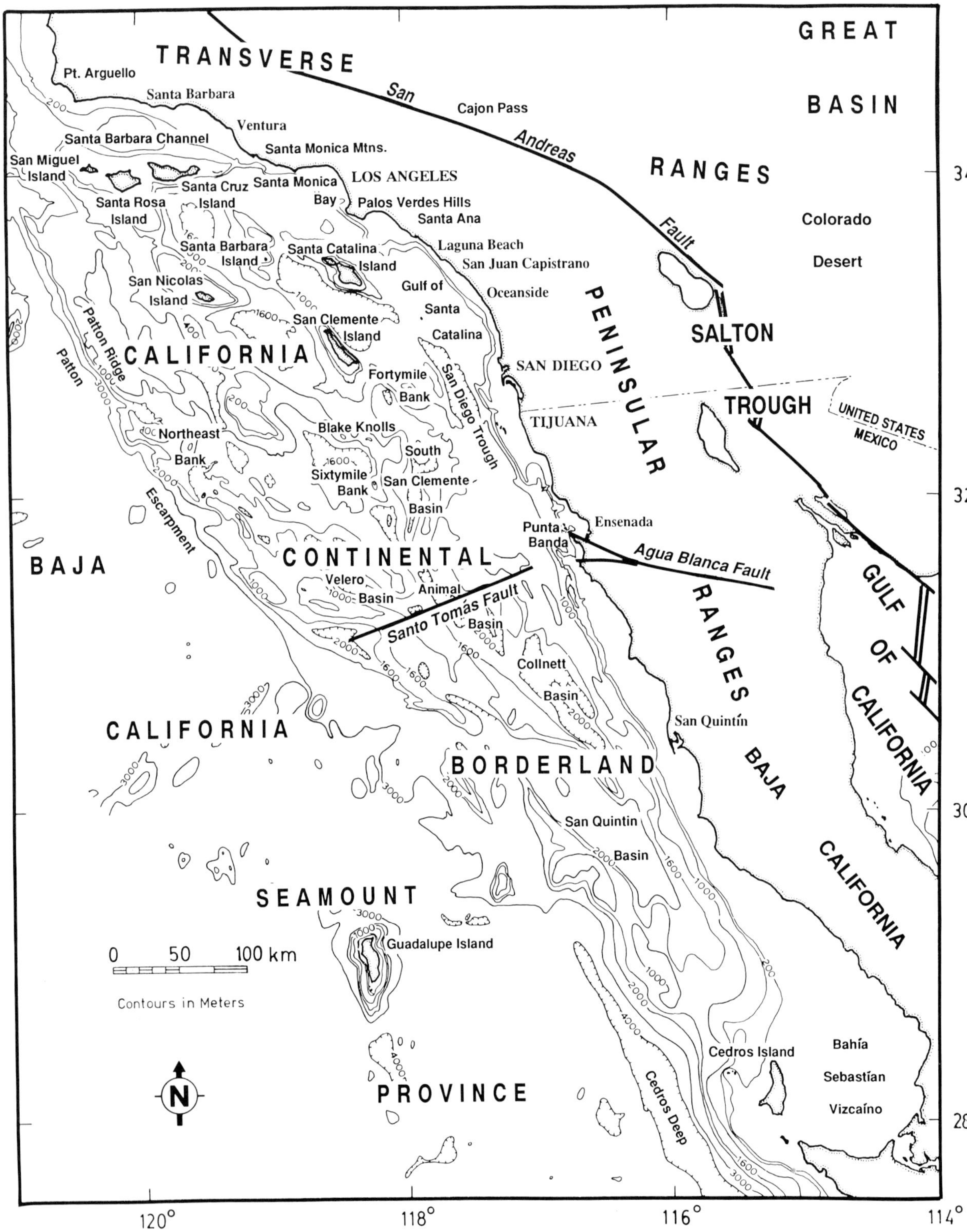

FIG. 1.—Location map showing major geomorphic provinces around the California Continental Borderland and geographic features referred to in the text.

the nearby coast (Fig. 1; Vedder, 1976). W. P. Blake's (1855, 1857) geologic map of the region, compiled during the explorations for a Pacific railroad route, recognized major geomorphic provinces and named the "Coast Mountains," "Bernardino Sierra" (Transverse Ranges), "Peninsular Sierra," and "Oceanic Ranges." Later California geologists (Lawson, 1897; Smith, 1898) proposed fault origins for the steep curvilinear slopes recognized as bounding the major islands and submarine ridges. The escarpment along the eastern side of San Clemente Island is a classic example (Fig. 1).

Early settlers in California experienced numerous strong earthquakes along the coast (Bancroft, 1888; Holden, 1898; Townley and Allen, 1939; Agnew and others, 1979; Toppozada and others, 1979). Perhaps because of this, Blake (1857), in his map of southern California geology, labeled the offshore region between San Diego and San Juan Capistrano: "Earthquake Bay." The area is presently known as the Gulf of Santa Catalina. The earthquake of 8 December 1812, which destroyed the mission at San Juan Capistrano, was the likely origin of this name. Although recent studies have relocated the source of that shock from near the coast inland to the San Andreas fault near Cajon Pass (Fig. 1; Jacoby and others, 1987), many of the strong, and potentially damaging, earthquakes felt by the early settlers along the southern California coast could have had offshore epicenters. Later studies correlated earthquake activity with the seafloor physiography (Clements and Emery, 1947).

Pioneer geologists, including the early oil explorationists, found abundant evidence of substantial vertical movements in the borderland geologic history. The discovery of submarine canyons led to the suggestion that large areas of the offshore region had subsided (Davidson, 1887, 1897; Shepard and Dill, 1966). Numerous prominent, uplifted, marine terraces on the offshore islands and coastal hills showed that some areas had experienced geologically recent tectonic uplift (Lawson, 1897; Fairbanks, 1897; Smith, 1897, 1900, 1933; Vickery, 1927; Grant and Shepard, 1939; Woodring and others, 1946; Orme, 1974). The relative lack of marine terraces on Santa Catalina Island compared to the numerous prominent terraces on the nearby Palos Verdes Hills, Santa Barbara, and San Clemente Islands showed that these vertical movements could be quite variable across the borderland (Lawson, 1897; Shepard and others, 1939). Attempts to correlate terraces based solely on elevation failed because of the numerous faults that separated the region into blocks with different tectonic movements (Vedder and Norris, 1963; Orme, 1974; LaJoie and others, 1979; Vedder and Howell, 1980).

Studies of sedimentary facies in the Transverse Ranges and Los Angeles Basin (Bailey and Jahns, 1954; Woodford and others, 1954) showed sequences of subsidence and basin filling. Associated diastrophism caused uplift, erosion, and renewed subsidence that eventually produced angular discordance of later basin-filling sequences. Later studies of benthic fossils showed that some offshore banks and ridges had gone from bathyal to subtidal and back down to bathyal depths throughout Neogene time (Bandy, 1953, 1963; Doyle and Bandy, 1972; Arnal, 1976; Arnal and Vedder, 1976; Ingle, 1980; Vedder and Howell, 1980). This alternating "up-and-down" tectonic style, colloquially referred to as "porpoising" (Crowell, 1981), provides an interesting clue to understanding borderland tectonics. Rather than a steady uplift or subsidence associated with uniform contraction or extension, a more complex tectonic mechanism, such as wrench tectonics, is required to explain the "porpoising."

The large magnitude of borderland vertical tectonics is impressively illustrated by the uplifted blueschist-facies rocks of the western basement complex (Catalina Schist) exposed on Santa Catalina Island and the Palos Verdes Hills (Fig. 2). These rocks are present as abundant clasts, up to boulder size, within the San Onofre Breccia along the coast (Woodford, 1925). Studies of metamorphic processes show that the lawsonite and glaucophane minerals formed at pressures found at subterranean depths of 10 to 40 km (Ernst, 1973). These rocks were subsequently uplifted and exposed to erosion. Conversely, the great thickness of Tertiary sediments in the Los Angeles and Ventura Basins show that subsidence of 10 to 15 km or more had occurred there (Yerkes and others, 1965).

Geologic mapping showed basement rocks juxtaposed against youthful and relatively undeformed sedimentary rocks and provided additional evidence of substantial vertical movements. For example, on Santa Catalina Island, the Catalina Schist basement is in fault contact with Miocene sedimentary rocks (Woodford, 1924; Bailey, 1941; Platt, 1973, 1975, 1976), and on Santa Cruz Island, the Alamos Pluton, Willows Plutonic Complex and Santa Cruz Island schist are in fault contact with Miocene Blanca Formation (Rand, 1933; Hill, 1976; McLean and others, 1976). Along the Newport-Inglewood fault zone (Fig. 2) in the Los Angeles Basin, the juxtaposition of the western basement terrane of Catalina Schist against the eastern basement terrane of continental plutonic rocks of the southern California batholith suggests that major lateral movements also have occurred (Schoellhamer and Woodford, 1951; Woodford and others, 1954; Hill, 1971). The timing for some of these early borderland vertical movements is provided by the late-early to middle Miocene age of the San Onofre Breccia (Woodford, 1925; Kleinpell, 1938; Ingle, 1972; Stuart, 1975). The Mesozoic age of metamorphism for the blueschist-facies basement rocks (70 to 90 Ma; Suppe and Armstrong, 1972) provides a maximum age for initiation of their uplift.

Regional geologic investigations of the basement rocks showed a major "diastrophism, plutonic intrusion, and metamorphism" (Moore, 1969), during late Mesozoic time (Woodford and others, 1954; Jahns, 1954; Sorensen, 1985). This produced the major basement unconformity that separates pre- and post-batholithic rocks of the Peninsular Ranges batholith. Although some plutonic rocks are exposed on Santa Cruz and Santa Catalina Islands, the Catalina Schist basement was suspected to underlie much of the California Continental Borderland to the west (Woodford, 1924; Rand, 1933; Bailey, 1941; Schoellhamer and Woodford, 1951; Yeats, 1973; Vedder and others, 1974).

Regional geologic investigations of the Transverse Ranges and Los Angeles Basin also noted the abundance of thick sedimentary sequences of Late Cretaceous to Quaternary

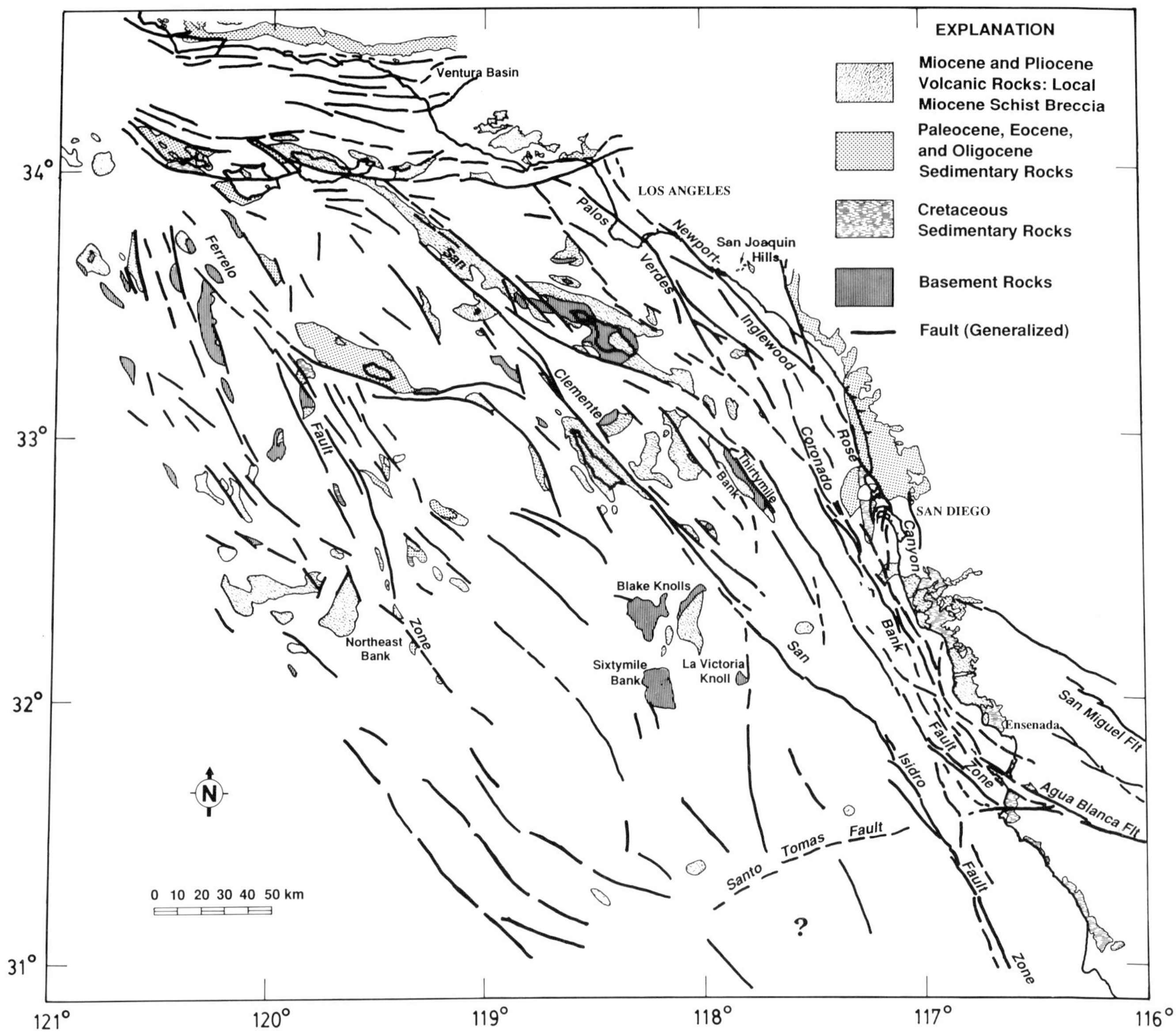

FIG. 2.—Generalized geologic map showing distribution of pre-Miocene sedimentary and acoustic basement-rock outcrops as well as the major faults in the California Continental Borderland (modified from Vedder and others, 1974; Gastil and others, 1975; Doyle and Gorsline, 1977; Legg, 1985; and Vedder, 1987).

basin fill, including turbidites (Reed and Hollister, 1936; Jahns, 1954; Yerkes and others, 1965). Late Cretaceous to early Miocene strata are found in the central borderland, but are conspicuously absent in the inner borderland, except for isolated locations along the coast near San Diego, west of the Newport-Inglewood and Coronado Bank fault zones, and possibly on the southeast end of Santa Catalina Island (Howell and Vedder, 1981; Vedder, 1987). Middle Miocene sedimentary rocks are widespread throughout the borderland. The incomplete record of late Miocene strata in the area with most of these and younger deposits concentrated in the modern basins suggests that much of the borderland basin and range physiography formed during middle Miocene time (Vedder and others, 1974; Howell and Vedder, 1981; Vedder, 1987), although significant basin development has continued into Quaternary time (Teng, 1985). In the central borderland, post-middle Miocene strata overlie the Late Cretaceous to Paleogene sequences, but in the inner borderland, they are found to directly overlie Catalina Schist basement. The observed disparate stratigraphic sequences found both at large scale, such as the tectonostratigraphic terranes of Howell and Vedder (1981), and at small scale, such as the diverse subsurface sections observed on Santa Cruz Island (McLean and others, 1976), require a complex tectonic history to explain their distribution (Howell and Vedder, 1981; Vedder, 1987).

Volcanic rocks of Miocene age, mostly basalts and andesites in composition, are widespread throughout the northern part of the California Continental Borderland (Fig. 2). Flows and volcaniclastic rocks are found interbedded with other fossiliferous sedimentary rocks of middle Miocene age (Vedder and others, 1981; Howell and Vedder, 1981; Vedder, 1987). In many places, the Miocene and older strata were intruded by volcanic rocks, which were dated isotopically as middle Miocene (12 to 16 Ma; Turner, 1970; Merifield and others, 1971; Hawkins and Divis, 1975). Other investigators found abundant volcanic rocks, but of progressively younger age to the south, including late Miocene to Pliocene at Northeast Bank (4 to 6 Ma; Hawkins and Divis, 1975) and Pliocene to Recent west of Baja California and along the coast at Punta Collnett and San Quintín (<1 to 4 Ma; Fig. 1; Allison, 1964; Krause, 1965; Gastil and others, 1975; Doyle and Gorsline, 1977). A major middle Miocene thermal event was recognized west of southern California, which occurred at progressively later times southeastward across the Baja California borderland.

TECHNOLOGICAL DEVELOPMENTS

Advances in Geology and Geophysics

Echo Sounding and the Growth of Submarine Geology.— Based upon numerous hydrographic surveys conducted by the U.S. Coast and Geodetic Survey, Shepard and Emery (1941) provided the first comprehensive maps of the physiography of the sea floor off California. Recognizing that the borderland physiography resembled the Great Basin, the borderland structure was inferred to be dominated by normal faulting. In addition, Shepard and Emery (1941) proposed 40 km of right-slip faulting along the San Clemente fault based upon a realignment of San Clemente Island with Fortymile Bank (Fig. 1). On the basis of seabottom sampling from dredges and cores, and extrapolation of coastal and island outcrops, subsequent investigations correlated seafloor lithology with the borderland bathymetry (e.g., Emery and Shepard, 1945; Emery and others, 1952; Emery, 1960; Vedder and others, 1974).

Exploration Geophysics.—Additional echo sounding, marine geophysical studies using towed magnetometer and shipboard gravimeter, and scattered seafloor sampling provided broad-scale reconnaissance geologic and geophysical mapping of the California Continental Borderland, including the part located west of Baja California (Krause, 1965; Harrison and others, 1966; Theberge, 1971; Pisciotto and others, 1974; Henyey, 1976; Plawman, 1978; Beyer, 1980). These studies noted that the southern borderland has geomorphic character similar to, but is narrower and generally deeper than, the borderland to the north (Fig. 1). A major transverse structure, the southwest-trending Santo Tomás fault, originally considered an extension of the transpeninsular Agua Blanca fault (Allen and others, 1960; Krause, 1965), was identified as the structural boundary between the northern and southern borderland (Fig. 1).

A major program of seismic-refraction studies was undertaken by scientists at the Scripps Institution of Oceanography throughout the borderland and into the adjacent Pacific Ocean Basin (Shor and Raitt, 1958; Shor and others, 1976). These studies showed that the Moho deepens progressively across the borderland, from about 9 to 11 km at the Patton Escarpment to about 24 km at the coast and down to 32 km in the Peninsular Ranges. Because most of the crustal thickening occurs in a layer having seismic velocity of about 6.7 km/s, typical for oceanic crust and faster than continental crust, Shor and Raitt (1958) suggested that borderland crust is intermediate between continental and oceanic crust. These studies also showed that some basins, particularly those near the coast, have thick sedimentary fills (up to 4-5 km), whereas other basins, mainly in the outer and southern borderland, have thin (<1 km) sedimentary fills. High-velocity rocks under several ridges were inferred to be volcanic intrusions or uplifted blocks of basement rocks.

Continuous Seismic-Reflection Profiling and Subbottom Imaging.— Based upon more than 5,000 line-km of seismic-reflection profiles, D. G. Moore (1969) described the submarine geology of the entire California Continental Borderland, from Transverse Ranges to Bahía Sebastian Vizcaíno. Following this important reconnaissance study, numerous detailed local studies using seismic-reflection profiles have been accomplished (e.g., Ridlon, 1969; Nardin and Henyey, 1978; Crouch, 1979, 1981; Kennedy and Welday, 1980; Kennedy and others, 1980; Legg, 1985); most of these were focused on the northern borderland. In particular, more systematic study of the United States part of the borderland was accomplished by the U.S. Geological Survey during the 1970s in preparation for lease sales with regard to offshore petroleum exploration efforts (e.g., Vedder and others, 1974; Taylor, 1976; summarized by Vedder, 1987). Likewise, oil companies and geophysical contractors acquired many thousands of miles of marine seismic data in the region and prepared analyses of the regional geology and tectonic evolution for exploration purposes (Parker, 1971), but much of these data remain proprietary at this time. Large areas still remain unknown or poorly understood because of complex structural relations, poor seismic data, and inadequate or a complete lack of samples (Fig. 2). Even though broad-scale reconnaissance mapping of bathymetry, total magnetic-field anomalies, and free-air gravity anomalies have been accomplished throughout the borderland west of Baja California, it is the least explored area of the California borderland, has the sparsest of seafloor and subbottom sampling, and consequently has the poorest control for geologic mapping.

Tectonic Models

Plate-Tectonics Paradigm.— In the framework of the plate-tectonics paradigm, Atwater (1970) described detailed models of the Cenozoic tectonic evolution of the Pacific Coast of North America that included the California Continental Borderland. Later studies updated and revised her original models (Atwater and Molnar, 1973; Engebretson and others, 1985; Atwater, 1989), but the major events she originally described still provide the framework for all subsequent models of borderland tectonic evolution. In particular, this evolution involves (Fig. 3): 1) the collision of the East Pacific Rise (EPR) spreading center with North America along a Mesozoic-Paleogene subduction zone; 2) subsequent southward migration of the Rivera ridge-trench-transform-fault (RTF) triple junction, northward migration of the Mendocino trench-transform fault-transform fault (TFF) tri-

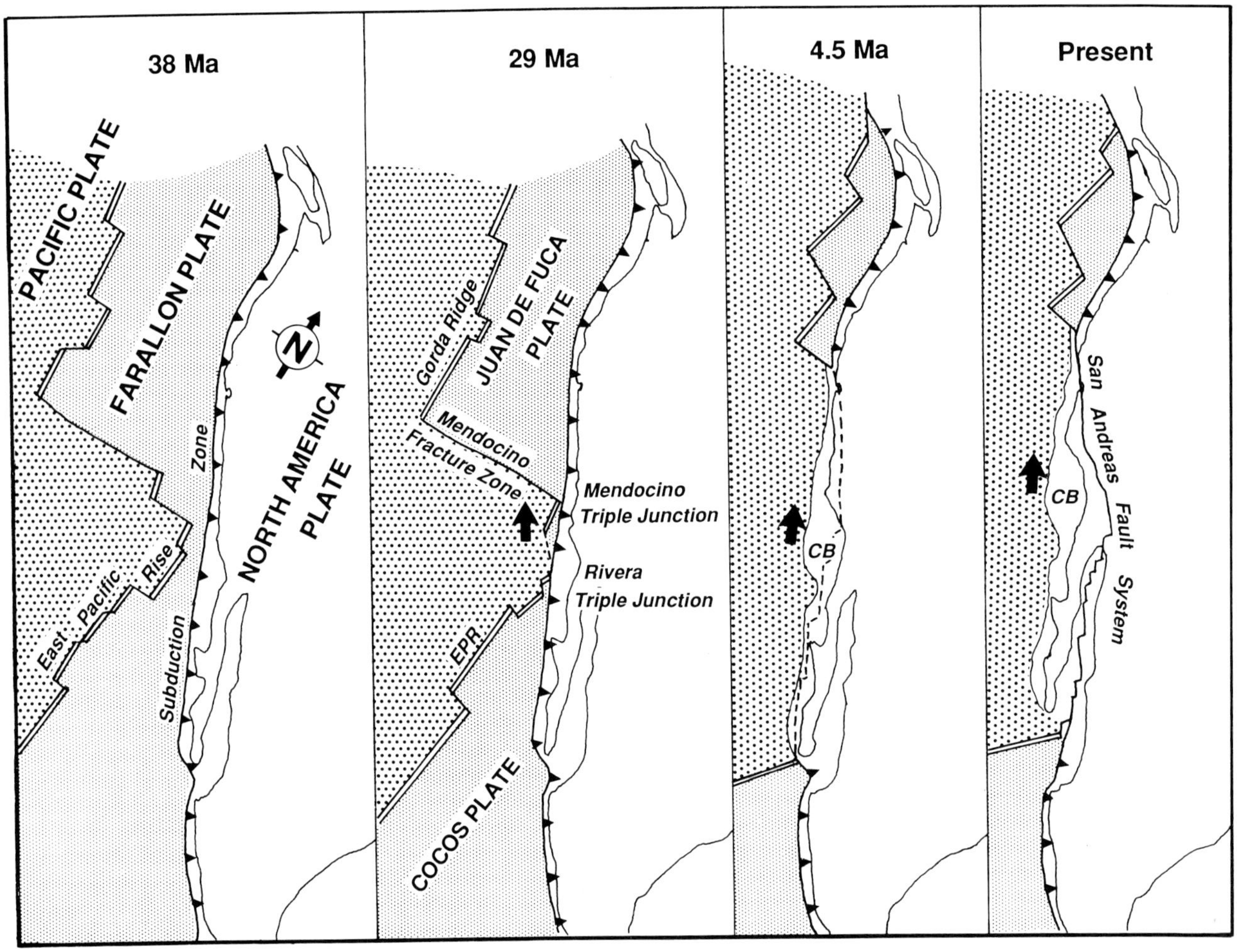

FIG. 3.—Simplified diagrams showing major features of the Cenozoic plate-tectonic evolution of western North America (after Atwater, 1970; Atwater and Molnar, 1973; Blake and others, 1978; Hamilton, 1978).

ple junction away from the borderland region, and the lengthening of the San Andreas transform-fault system between the two triple junctions; and 3) the eventual rifting of the Baja California peninsula from the North American continent as the transform fault jumped eastward of the former volcanic arc to form the southern San Andreas-Gulf of California transform-fault system. The proximity of the California Continental Borderland to the active-plate boundary during this time was responsible for the complex tectonic evolution of the region.

The timing of these major tectonic events is constrained both by the age of magnetic-anomaly patterns in the deep Pacific to the west of the borderland (Fig. 4) and near the mouth of the Gulf of California, and by the various diastrophic events documented by the various geologic investigations in the borderland region. Deep-sea drilling cores at the edge of the borderland show the age of the oceanic crust (about 15-17 Ma) formed when subduction ceased and the EPR (at the Rivera triple junction) passed by the area (Yeats and Haq, 1981). In addition, these cores show the sedimentation history for borderland basins; major hiatuses or slowing of terrestrial-sediment input record the formation of borderland "ridge and basin" physiography and isolation of the outer basins from the continental-sediment source areas. Regional basement rocks are tied to the Mesozoic subduction zone (Hamilton, 1969, 1978; Jones and others, 1976): the Catalina Schist, inferred equivalent to the Franciscan basement of the southern Coast Ranges and related to a subduction complex; the Upper Cretaceous and Paleogene sedimentary sequences, inferred equivalent to the Great Valley sequence and related to the forearc-basin deposits; and the western continental (granitic) basement of the Peninsular Ranges batholith, inferred equivalent to the Sierra Nevada and related to the arc volcanism and plutonism. Widespread middle Miocene volcanism and progressively younger basin formation from north to south across the borderland, including the tectonic evolution of the Los Angeles and Ventura Basins, is correlated with the southward migration of the EPR and the Rivera triple junction (Yerkes and others, 1965; Crowell, 1976; Blake and others, 1978; Dickinson and Snyder, 1979).

Recognition of the right-slip character of the San Andreas transform-fault system in southern California led to further hypotheses of significant dextral shear, or wrench faulting, in the California Continental Borderland (Wilcox and others, 1973; Harding, 1973; Howell and others, 1974;

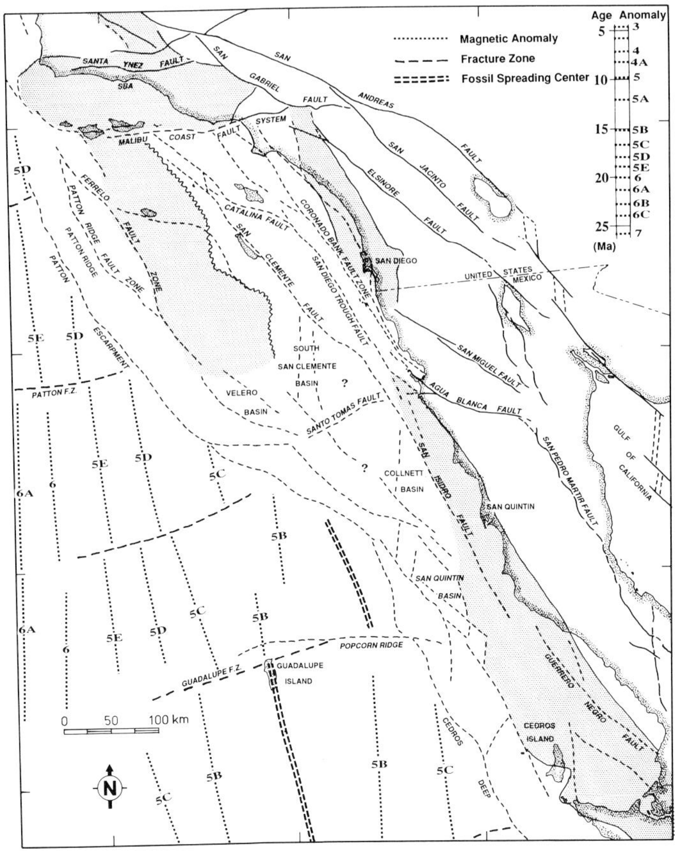

FIG. 4.—Map showing major tectonic features in the present California Continental Borderland and adjacent region. Stippled areas show distribution of early Cenozoic and Late Cretaceous basin deposits (Eocene basin in the north after Blake and others, 1978; and Kamerling and Luyendyk, 1985; and Late Cretaceous forearc basin in the south after Hamilton, 1978; and Crouch, 1981). Magnetic anomalies and fracture zones associated with seafloor spreading in the eastern Pacific are from Atwater and Severinghaus (1989).

Crouch, 1979; 1981; Vedder and others, 1974; Greene and others 1979; Legg and Kennedy, 1979; Howell and Vedder, 1981). In these studies, seismic-reflection profiles were used to interpret fault character, but major lateral offsets were inferred based upon proposed correlations of widely separated distinctive lithologic units (e.g., Eocene Poway and other conglomeratic suites, Howell and others, 1974; Kies and Abbott, 1983). Still, the overall ridge and basin physiography of the borderland convinced others that a rifting origin for the major structural and geomorphic features is most likely (Yeats, 1968, 1976; Yeats and others, 1974; Moore, 1976). Yet others, citing abundant west-to-northwest-trending folds in association with major dextral-shear zones, proposed convergent wrench faulting for the late Cenozoic evolution of borderland structure (Junger, 1976; Nardin and Henyey, 1978; Field and Richmond, 1980).

Paleomagnetism and Block Rotations.—The recent application of paleomagnetism to infer tectonic rotations and translations of rocks containing remanent magnetism from their time of cooling (volcanic) or deposition (sedimentary) provides significant new understanding of the PAC-NOAM plate-boundary evolution in the southern California region. In particular, the work of Luyendyk and others (1980, 1985) showed that elongate crustal blocks in the western Transverse Ranges and northern Channel Islands area had undergone substantial clockwise rotations (about 70° to 90°) since Oligocene time (Kamerling and Luyendyk, 1979, 1985; Luyendyk and others, 1980, 1985; Hornafius, 1985). Lateral translations of these blocks, inferred from inclination anomalies, are also significant; in some cases as much as 10° to 20° northward transport is suggested (Marshall and others, 1979; Champion and others, 1981; Luyendyk and others, 1985). The relative contribution to these inclination anomalies of tectonic transport *versus* non-tectonic causes, such as initial stratal dip or non-dipole geomagnetic-field effects, is uncertain at present. The rotation (declination) anomalies, however, are interpreted to show that the western Transverse Ranges, Santa Monica Mountains, and northern Channel Islands blocks formed the northwest part of the borderland during early to middle Miocene time (Fig. 5a; Crouch, 1979, 1981; Kamerling and Luyendyk, 1979, 1985; Luyendyk and others, 1980, 1985).

UNCERTAIN AND CONTROVERSIAL ISSUES POSED BY CURRENT MODELS OF BORDERLAND TECTONIC EVOLUTION

Relative Importance of Strike-Slip Faulting, Extension, and Contraction

Numerous models of the tectonic evolution of the California Continental Borderland and the southern California region have been proposed in the past two decades. In general, all are based on the plate-tectonics model described earlier. Details of the borderland tectonic evolution vary, in each model's interpretation, by the relative importance, following impingement of the EPR at the continental margin at about 29 Ma, of strike-slip faulting, extension, and contraction during the development of the transform margin. The current distribution of basement rocks associated with the Mesozoic to mid-Cenozoic convergent margin provides important clues to describe the broad pattern of tectonic movements during the late Cenozoic disruption of the former subduction zone and accretionary prism (forearc basin). Most borderland tectonic models attempt to explain this distribution.

Extension versus Strike-Slip Faulting.—All of the recent models of borderland tectonic evolution involve significant lateral slip. Models involving major strike-slip faulting are based upon the palinspastic reconstruction of the region that places Eocene conglomeratic suites found on San Miguel and Santa Rosa Islands next to the Eocene San Diego embayment and the Ballenas River Poway conglomerate locale (Fig. 5b; Howell and others, 1974; Kies and Abbott, 1983). Early versions of these models inferred counterclockwise rotation of the northern Channel Islands block, which has since been refuted by the paleomagnetic data (Kamerling and Luyendyk, 1979, 1985). Yeats (1968, 1976) preferred an extension model during the interaction of the borderland with the EPR, when the Rivera triple junction passed through the region, tying the Eocene conglomerates of the Channel Islands to the Sespe Formation of the Santa Monica and Santa Ana Mountains. In addition, both the strike-slip faulting and the extensional models placed offshore, inner-bor-

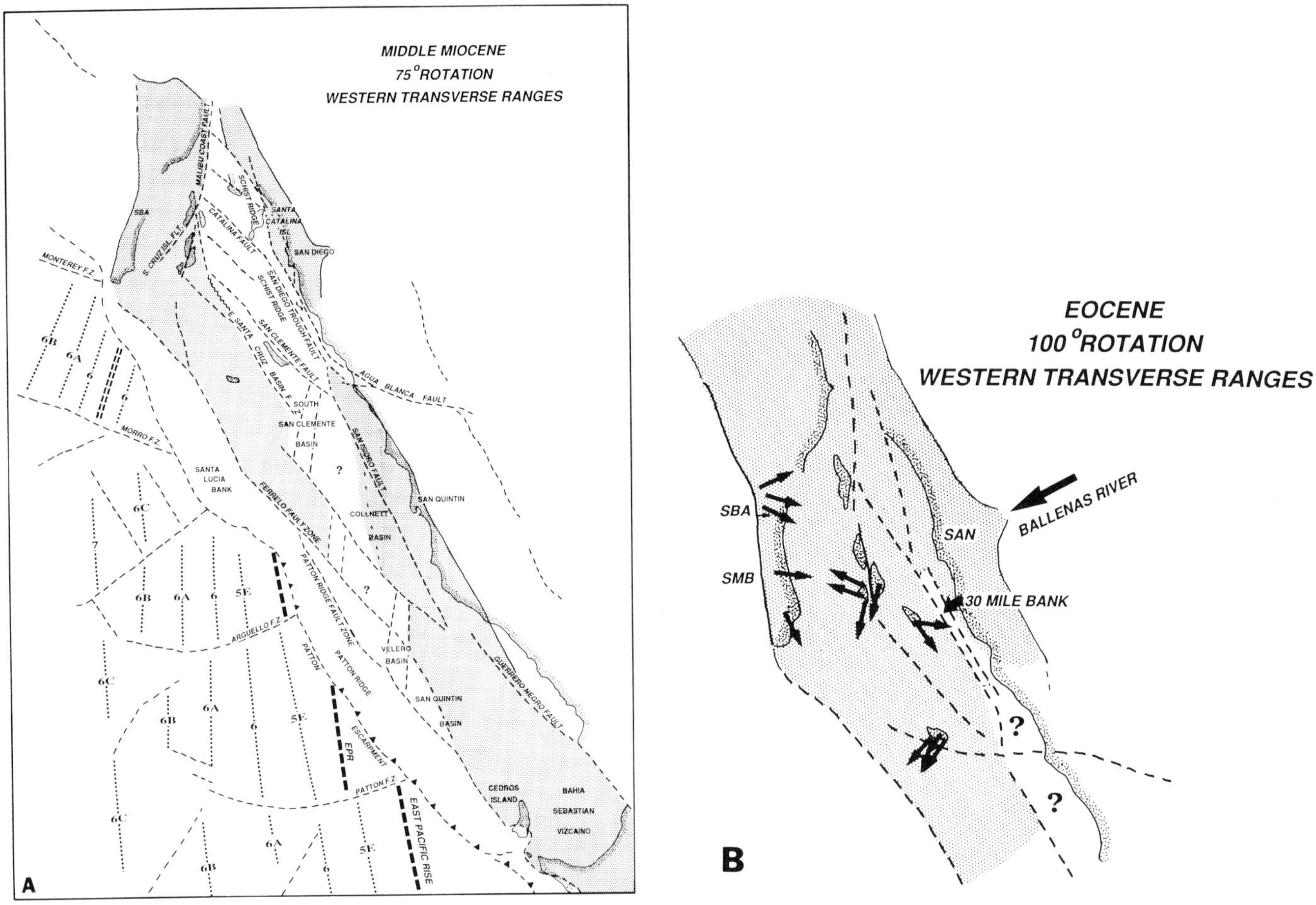

FIG. 5.—Simplified maps showing inferred palinspastic reconstructions of the California Continental Borderland (modified from Crouch, 1981; and Kamerling and Luyendyk, 1985). Magnetic anomalies and fracture zones associated with seafloor spreading in the eastern Pacific are from Atwater and Severinghaus (1989). (A) Middle Miocene reconstruction involving about 590 km of restored PAC-BAJA right-slip faulting and 75° clockwise rotation in the Western Transverse Ranges block. Note how the inferred inner-borderland rift mimics the Gulf of California in this model with the exception of the major uplifted Catalina Schist ridges that are absent in the Gulf. The schist ridge near the coast includes the Palos Verdes Hills and Santa Catalina Island (caught between the Catalina-San Diego Trough fault zone and the Palos Verdes-Coronado Bank fault zone). The location of this rift is roughly aligned with the EPR at the middle Miocene time shown (about 17 Ma). Patton Ridge is also restored to a position near the EPR, as proposed by Crouch (1981). Also note that the northeast trend of magnetic anomalies and fossil spreading center between the Morro Fracture Zone and the Monterey Fracture Zone is roughly aligned with the Western Transverse Ranges, possibly suggesting that a reorganization of the plate boundary shortly after anomaly 6 time (about 20 Ma) may be related to initiation of clockwise rotation of the Western Transverse Ranges. (B) Closure of the inner-borderland rift is almost complete when a northwest-trending Eocene basin, with paleocurrents as shown by the arrows, existed in the borderland region. Forearc rifting concurrent with subduction, as suggested by Platt (1986), is inferred to create the Thirtymile Bank uplift, which provides a laterally restricted source for the Poway fan. This source might be further restricted in the vicinity of the Palos Verdes Hills and Santa Catalina Island schist-ridge uplifts.

derland, Catalina Schist basement ridges adjacent to the coast to provide the requisite source terrane for the San Onofre Breccia (Fig. 5a). Strike-slip faulting allows Santa Catalina Island, a major known outcrop of Catalina Schist that includes the blueschist-facies basement rocks, to be the source for the type San Onofre Breccia between Laguna Beach and Oceanside (Stuart, 1975, 1979a, b). In contrast, the extension model required a now subsided and buried "schist ridge" located west of this coastal area (Junger, 1974).

Advancement of wrench-fault tectonics (Wilcox and others, 1973) led to many models involving wrench-fault origins for the overall borderland structure and physiography (e.g., Crowell, 1974; Blake and others, 1978). Numerous more detailed and localized borderland studies concluded that wrench faults are common in the borderland (Harding, 1973; Junger, 1976, 1979; Junger and Wagner, 1977; Fischer and others, 1977; Nardin and Henyey, 1978; Crouch, 1979, 1981; Howell and others, 1980; Howell and Vedder, 1981); many remain seismically active today (Greene and others, 1979; Legg and Kennedy, 1979; Legg, 1980, 1986; Kennedy and others, 1980; Corbett, 1984; Legg and others, 1989; Hauksson, 1987, 1990). Wrench-fault tectonics provides a convenient way to explain the presence of both extensional and contractional structure within the same fault system (Fig. 6) and is consistent with the right-slip-fault character of the PAC-NOAM transform-fault plate boundary.

More recent models incorporate the middle Miocene clockwise rotations that occurred in the northern borderland

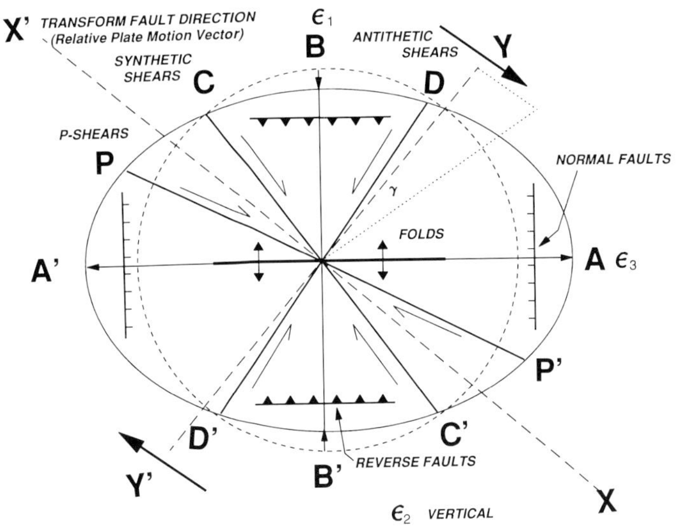

FIG. 6.—Strain ellipse and predicted structural trends for simple shear in wrench-fault tectonics (after Harding, 1974).

(Crouch, 1979; 1981; Kamerling and Luyendyk, 1979; 1985; Luyendyk and others, 1980, 1985). An important feature of these models is that both major right-slip faulting and east-west extension are required. The models vary in the mechanism by which the clockwise rotations, and likewise, inner-borderland extension occur. Luyendyk and others (1980, 1985) involve rotation of a system of elongate, formerly north-to-northeast-trending crustal blocks, whereas, Crouch (1979, 1981) postulates the wholesale northwest translation, clockwise rotation, and scissoring of a large block consisting of the Santa Monica Mountains and western Transverse Ranges. The Luyendyk model creates several "triangular holes," which form localized basins at the corners of the rotating blocks. The Crouch model suggests a major zone of extension in the southern borderland, explaining that the generally deeper basins and ridges and the abundance of Miocene or younger basalts found in that area may have formed in a rhombochasm or seafloor-spreading center. Updates to these models tend to converge toward a common model involving major inner-borderland extension or transtension as the outer borderland and western Transverse Ranges blocks swing away from the coast in a scissoring fashion (Fig. 5a; Kamerling and Luyendyk, 1985; Crouch, pers. commun. 1989). This model explains most of the basement and pre-middle Miocene rock distributions including: the distribution of the Eocene conglomerates, placing San Miguel and Santa Rosa Islands near the Eocene San Diego embayment; the Catalina Schist ridge adjacent to the coast where San Onofre Breccia occurs; and reconstruction of the major lithotectonic belts associated with the Mesozoic subduction zone by placing the outer-borderland forearc basin (Late Cretaceous to Paleogene sedimentary sequence) against equivalent rocks at the coast.

Oblique rifting in the southern California borderland slowed dramatically at the end of the Miocene epoch for two reasons. First, the main transform-plate boundary shifted eastward to the San Andreas-Gulf of California fault system. Second, the PAC-NOAM relative-motion vector changed to a more northerly azimuth (Cox and Engebretson, 1985; Engebretson and others, 1985). Because inner-borderland extension unroofed and elevated the Catalina western basement-source terrane by middle Miocene time, significant uplift of blueschist-facies rocks probably occurred earlier (Fig. 5b), concurrently with subduction, through some form of diapirism (Minch and others, 1976; Cloos, 1982) or tectonic denudation (Platt, 1986).

Post-Miocene Contraction versus Transtension.—Most recently, models involving Plio-Pleistocene convergence or transpression along the south-central California margin have been proposed (Crouch and others, 1984; Minster and Jordan, 1987; Mount and Suppe, 1987; Zoback and others, 1987; Namson and Davis, 1988; Davis and others, 1989). Abundant evidence of late Cenozoic shortening has been described elsewhere for the Transverse Ranges, including the Santa Barbara Channel and Ventura Basin (Vedder and others, 1969, 1974; Yerkes and Lee, 1979; Yeats, 1983). Post-Miocene contraction is evident in the central and outer borderland as broad folds that underlie the ridges and basins (Vedder and others, 1974; Vedder, 1987; Junger 1976, 1979; Field and Richmond, 1980; Crouch, 1981; Howell and Vedder, 1981). In the inner borderland, folding and reverse faulting has been recognized on northwest-trending structures, such as the Palos Verdes Hills fault zone (Fischer and others, 1977; Nardin and Henyey, 1978), although west-trending contractional structures are prominent at the northern edge of the borderland and along the margins of the Los Angeles Basin (e.g., the Malibu Coast fault system). En-echelon folding along the Newport-Inglewood fault zone has created some of the most productive oil fields in California (Harding, 1973; Yeats, 1973). In seismic profiles, transpressional "palm tree" structures have been recognized along the offshore Newport-Inglewood fault zone south of Newport Beach, and youthful thrusts are interpreted at the base of the adjacent slope to the west (J. Crouch, pers. commun., 1989; Nicholson and Crouch, 1989). Continued shortening in the northeastern part of the borderland is manifested by the numerous earthquakes with reverse-faulting mechanisms (Hauksson and Saldivar, 1986, 1989; Hauksson and Jones, 1988; Nicholson and Crouch, 1989; Hauksson, 1990).

Whereas evidence of increasing north-south and northeast-southwest shortening since the end of Miocene time is widespread in the northern borderland, Legg (1985) shows abundant evidence for continued extension or transtension on inner-borderland faults from the latitude of San Diego southward. High-resolution seismic-reflection profiles across the Coronado Bank and San Isidro fault zones (Fig. 7; Kennedy and others, 1980; Legg, 1985) show monoclinal downwarping ("chevron-shaped dilational synclines" according to Kennedy and Welday, 1980) of Quaternary sediments into the widening fault zone. These features are probably the high-resolution equivalents of "negative flower structure," which is diagnostic of transtensional fault zones (Harding, 1985), although the convergence of multiple, near-surface fault traces into a single "master fault" at depth cannot be determined from these data.

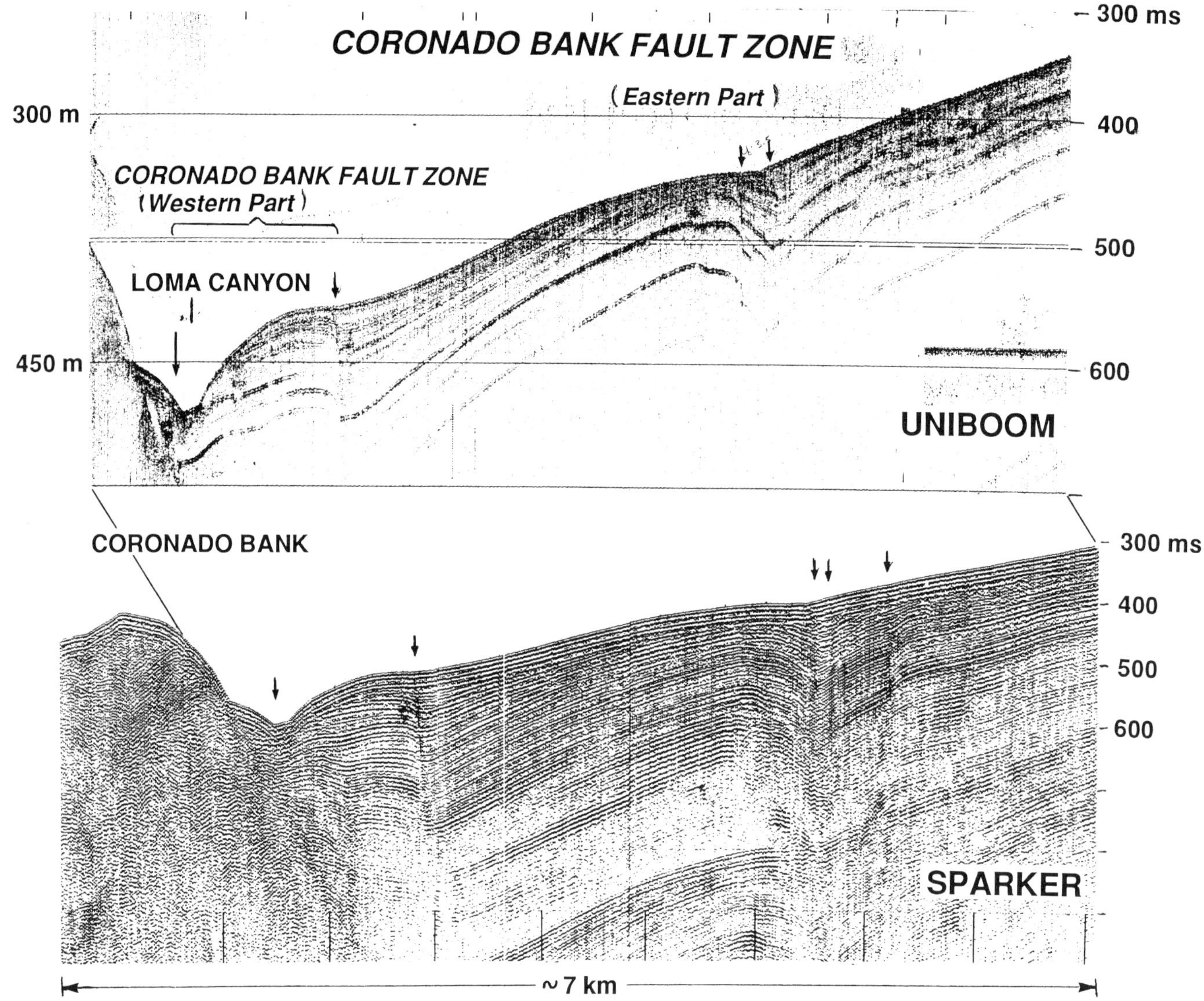

FIG. 7.—High-resolution seismic-reflection profile showing evidence of transtension in the inner California Continental Borderland (Kennedy and others, 1980).

Based upon the systematic change, from transtension in the south to transpression in the north across the inner borderland, Legg (1985, 1989) suggests that the southern California area is a broad shear zone, formed by the splintering of the northern end of a long, narrow, Baja California microplate (Fig. 8). The western boundary of this shear zone is the San Clemente fault, which shows mostly right-slip faulting along a N40°W trend, parallel to PAC-NOAM relative plate motion. The southern boundary is the Agua Blanca fault, which also shows mostly right-slip faulting along its trace that strikes from N50° to 70°W as it curves across the Baja California peninsula. The eastern boundary is the San Andreas and related faults of the Gulf of California transform-fault system. Even though the northern boundary is formed by western Transverse Ranges structure, the big bend in the San Andreas fault forms a kinematic boundary along which major right-slip faulting is transformed into counterclockwise rotation of the southern California area (Weldon and Humphreys, 1986). This rotation is manifest in the borderland as the northeast-southwest-directed contraction in the northern part of the inner borderland (e.g., Palos Verdes fault zone) and as the transtension west of Ensenada and the Agua Blanca fault (San Isidro fault zone). This rotation causes the Agua Blanca fault to act as an intracontinental transform fault, like the Garlock fault in central California (Davis and Burchfiel, 1973), linking inner-borderland extension with Gulf of California rifting. Because the southern California area lacks sufficient strength to rotate as a rigid block without internal deformation, it is breaking up into several subparallel crustal blocks bounded by northwest-trending right-slip fault zones. Furthermore, being constrained by north-south directed Transverse Ranges convergence to the north, and the Pacific Plate to the west, the southern California shear zone may be buckling, in an

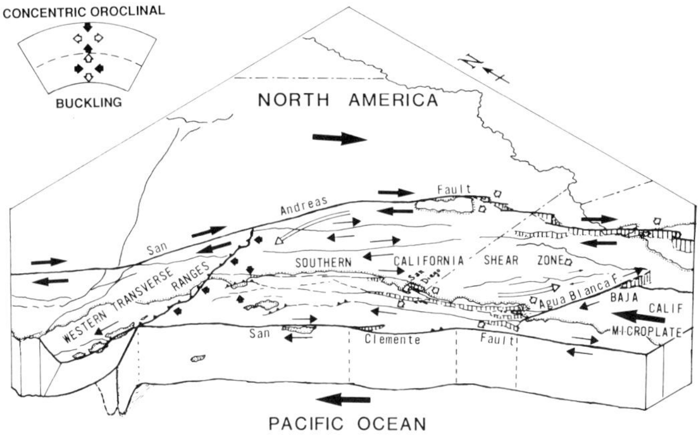

FIG. 8.—Map showing the "Southern California Shear Zone," which represents the splintered northern end of a long, narrow, "Baja California Microplate" caught between the San Andreas-Gulf of California transform-fault system and the San Clemente fault system in the California Continental Borderland (after Legg, 1985, 1989). Buckling of the Southern California Shear Zone caused by curvature in the San Andreas fault and impingement against the western Transverse Ranges may account for complex deformation in the vicinity of San Clemente Island and Fortymile Bank. Inset shows a model of oroclinal buckling as concentric folding that creates local zones of northeast-directed compression on the outer edge of the fold and northwest-directed compression in the inner "hinge" zone of the fold. The numerous subparallel fault zones would represent "bedding plane" slip within the orocline; San Clemente Island and Fortymile Bank would be at the innermost core of the orocline where deformation would be most complex; and the San Clemente fault zone would act as the décollement at the "base" of the orocline. Complex earthquake focal mechanisms (Legg, 1980) and large escarpments formed by oblique(?) extension along the San Clemente fault (Legg and others, 1989) may result from this décollement.

oroclinal fashion, with a hinge zone near San Clemente Island and Fortymile Bank (Fig. 8). This buckling might explain complex, "backwards" earthquake focal mechanisms (left-slip on northwest-trending fault planes) observed there (Legg, 1980).

Palinspastic Reconstruction of San Diego Trough.—The location of blueschist-facies outcrops of Catalina Schist basement lie mostly in the eastern part of the modern borderland, i.e., the inner borderland. Their occurrence as clasts in the San Onofre Breccia and related lithologic units surrounding the inner borderland show the unroofing and erosion of these formerly deeply buried rocks. Because outcrops of San Onofre Breccia with blueschist clasts occur in limited areas, and because paleocurrent trends in these units indicate an offshore, inner-borderland, basement high during their deposition, identification of the basement outcrops in the modern borderland helps to reconstruct the deformation history of the region since the middle Miocene time of San Onofre Breccia deposition (Vedder and Howell, 1976; Stuart, 1979a, b).

A large outcrop of Catalina Schist containing blueschist-facies rocks has been mapped on the east flank of Thirtymile Bank (Figs. 2 and 9; Vedder and others, 1974), whereas San Onofre Breccia containing blueschist clasts outcrops on South Coronado Island and onshore near Tijuana (Minch, 1967; Stuart, 1979a, b). Palinspastic reconstruction of the inner borderland, by oblique closing of the San Diego Trough such that the Thirtymile Bank escarpment meshes with the Coronado Bank escarpment, places the outcrop of Catalina Schist basement near San Onofre Breccia outcrops on South Coronado Island (Fig. 10). Outcrops of Miocene volcanic rocks on Thirtymile Bank also are restored to a position close to the location of the Miocene Rosarito Beach basalt flows located south of Tijuana (Minch, 1967). Furthermore, additional right-oblique extension across the Coronado Bank fault zone (about 10–15 km) would place the coastal basalt flows and breccia units closer to their inferred offshore-source regions by bringing the Coronado Bank-Descanso Ridge alongside the coast, and thereby would align the Miocene volcanic rocks that occur between Tijuana and Punta Salsipuedes (Minch, 1970; Gastil and others, 1975; Ashby, 1989) with the Miocene volcanic rocks along the Thirtymile Bank-Navy Bank ridge. The proposed middle Miocene rifting of the San Diego Trough region would provide a localized source for these volcanics.

The mesh of the two escarpments is very good; note especially how the slight change in strike near the middle of each is aligned and how the protuding block at the southern end of the Coronado Escarpment fits into the reentrant at the southern end of the Thirtymile Bank escarpment, near Boundary and Navy Banks (Fig. 10). The peak in the center of the San Diego Trough, southwest of Los Coronados, may be a volcanic intrusion (note its north trend) or a block of older basement left behind, and rotated clockwise(?) by dextral shear, during the rifting. No bottom samples have been recorded from this peak, although magnetic anomalies recorded across it may be consistent with either hypothesis (Legg, 1985).

The reconstruction shown (Fig. 10) involves about 40 km of right separation on a trend (N47°W) more westerly than the margins of the San Diego Trough or the San Diego Trough fault (N30°W). Because the late Quaternary San Diego Trough fault is interpreted as a simple, parallel, dextral wrench fault (Legg, 1985), the evolution of the San Diego Trough probably involved both right slip faulting and extension. The displacement vector for this reconstruction, however, is parallel to the roughly linear trend of earthquake epicenters across Coronado Bank (Fig. 9), implying that a deep crustal shear zone, perhaps underlying a mid-to-upper crustal detachment, has accommodated the seismogenic and possibly the lithospheric simple shear. Furthermore, this deep crustal shear zone located at the northwest end of the displaced ridges may have exerted tectonic control over the middle Miocene rifting.

The relative timing and detailed character of the oblique rifting in the San Diego Trough remain to be unraveled, although the suggested reconstruction looks reasonably accurate. The complete reconstruction precludes more than about 10 to 15 km of right-slip faulting along the Coronado Bank fault zone but, if confirmed by additional data, right-oblique rifting of 40 km in the San Diego Trough would be one of the larger known fault offsets in the borderland. Because this amount is greater than the total offset inferred for the transpeninsular Agua Blanca fault zone (about 22 km; Allen and others, 1960; Gastil and others, 1975), some

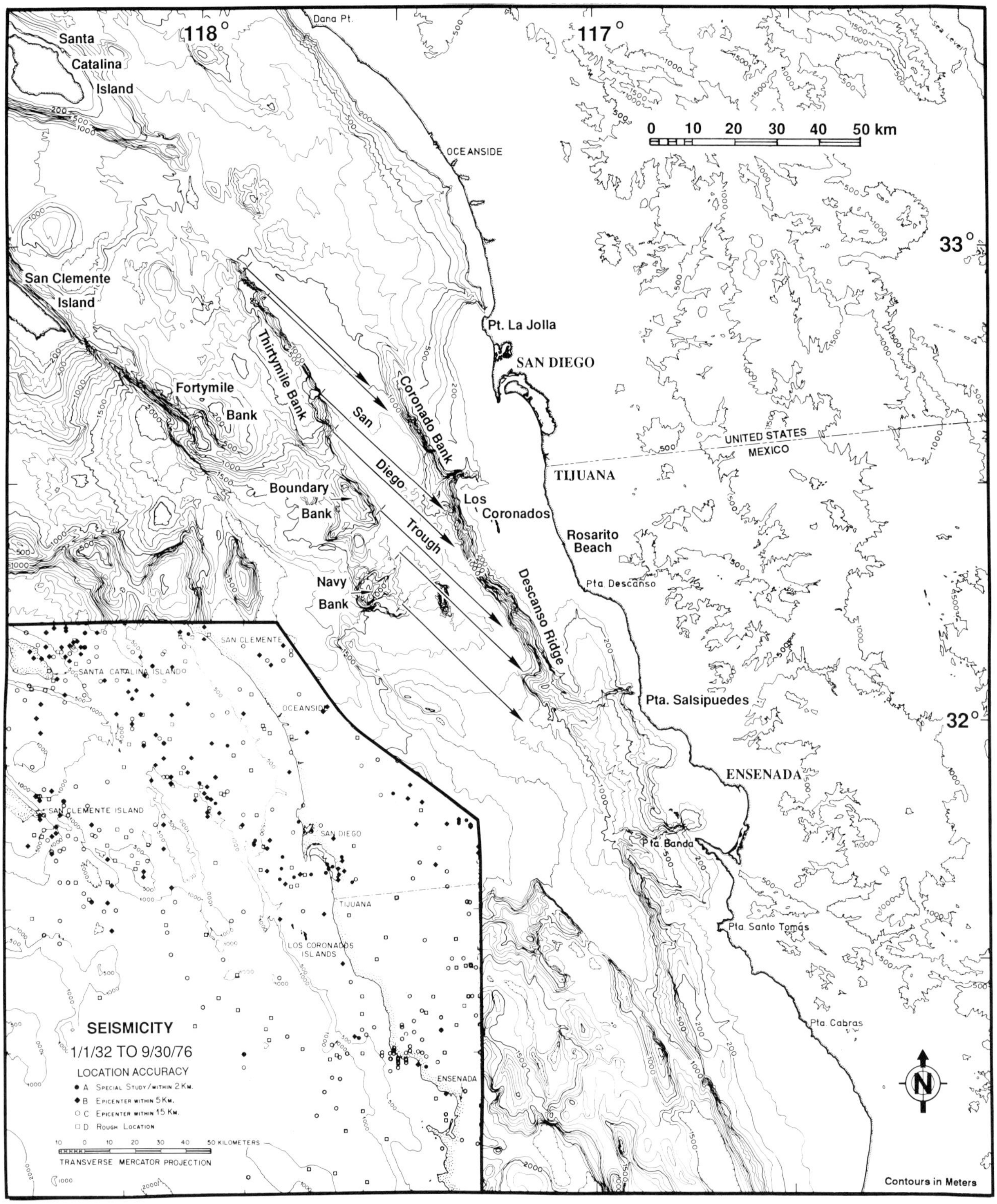

FIG. 9.—Detailed bathymetric map of the inner borderland (Legg, 1985) showing postulated oblique rifting in the San Diego Trough that correlates the escarpment east of Thirtymile Bank with that west of Coronado Bank. Arrows represent vectors for middle Miocene reconstruction of the region (Fig. 10) made by closure of the San Diego Trough that results in a "best-fit" of corresponding contours at the base of the two escarpments. These vectors are aligned parallel to a roughly linear trend of earthquake epicenters across Coronado Bank (Legg, 1980).

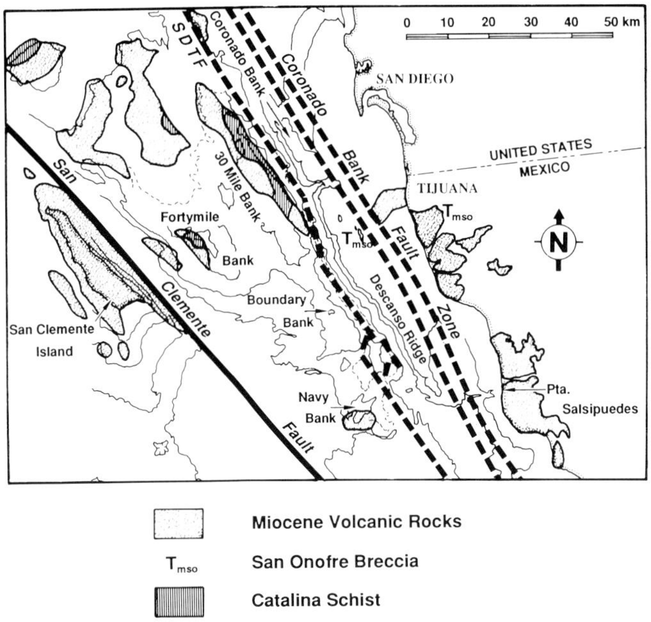

FIG. 10.—Palinspastic reconstruction of the San Diego Trough region for middle Miocene time (approximately 15 Ma). About 40 km of right-oblique extension within the San Diego Trough have been removed along a trend N47°W. In addition, the 40-km right slip along the San Clemente fault, proposed by Shepard and Emery (1941), has been removed, placing San Clemente Island adjacent to Fortymile Bank. An outcrop of Catalina Schist basement along Thirtymile Bank is placed near Los Coronados, where San Onofre Breccia, from a western source (Stuart, 1979a, b), is located. Also, the outcrop of Miocene volcanic rocks on Thirtymile and Navy Banks are located closer to similar rocks on the mainland coast, from Tijuana to Punta Salsipuedes. A better fit of the lateral extent of these volcanic rocks occurs if about 10 to 15 km of right-oblique extension on the Coronado Bank fault zone also is removed.

of the San Diego Trough fault-zone displacement must be transferred to other southern borderland faults, such as the San Isidro fault, west of Punta Banda. This rifting probably began prior to deposition of the San Onofre Breccia, forming the elevated schist ridge-source terrain along Thirtymile Bank, but the proposed 40 to 50 km of right-oblique extension presumably followed the middle Miocene deposition of the San Onofre Breccia and eruption of the Rosarito Beach volcanics (14 ± 0.7 Ma, Hawkins, 1970).

Southern Borderland Rifting.—As the EPR approached the trench, increased heat flow in the forearc area associated with subduction of the thin, buoyant, young oceanic crust may have accelerated extensional processes in the borderland. Possibly, the region domed upward over the hot asthenospheric welt along the edge of the EPR. The Farallon Plate north of the Mendocino transform fault was older and, therefore, colder and thicker than that to the south where the Rivera triple junction eventually formed. Furthermore, while subduction continued to the north of the Mendocino transform fault after intersection of the EPR, hot and thin lithosphere associated with the Rivera triple junction interacted with the NOAM plate margin to the south. The subsequent evolution of the continental margin on each side of the initial PAC-NOAM intersection proceeded in different ways. Possibly due to these differences, a kink in the edge of the North America Plate was created that affected the later margin evolution, perhaps eventually resulting in the formation of the Transverse Ranges (von Huene, 1969; B. Keller, pers. commun., 1985; Luyendyk and others, 1980, 1985).

Regardless of the details, the fundamental split in the plate-boundary evolution created significant differences in the tectonic evolution and geologic character of the California continental margin on opposite sides of the Transverse Ranges. To the north, subduction of older, thicker, and colder lithosphere associated with propagation of the Mendocino triple junction was less favorable for the widespread rift-related volcanism such as occurred in the southern California borderland during middle Tertiary time. As right-slip faulting continues, the San Andreas fault system slices off the western edge of the North American continent but lacks the large-scale rifting found to the south. In fact, at present, the northern area appears to involve a significant component of northeast-southwest convergence (Page, 1966; Gawthrop, 1978; Crouch and others, 1984; Minster and Jordan, 1987; Mount and Suppe, 1987; Zoback and others, 1987).

The rifting to the south combined with the clockwise rotation of the western Transverse Ranges served to broaden the width of the continental margin in the northern borderland to more than double its width elsewhere along the modern transform boundary. Likewise, the continued interaction with the EPR eventually caused the wholesale rifting of Baja California away from the North American continent.

The middle Miocene timing of the passage of the Rivera triple junction, when the EPR intersected the subduction zone and transform faulting ensued, appears to match the timing of widespread volcanism within the California Continental Borderland. In particular, after removing the block rotations (Luyendyk and others, 1985) and lateral slip on borderland fault zones (Crouch, 1981), the late Oligocene to early Miocene borderland is roughly aligned with the north- to northeast-trending Basin and Range Province (Luyendyk and others, 1985). Furthermore, progressively younger volcanism found across the southern borderland corresponds to the southward migration of the Rivera triple junction (Dickinson and Snyder, 1979). The age of the last oceanic crust to form along the margin has been inferred from magnetic-anomaly patterns (Fig. 4; Atwater and Molnar, 1973; Atwater and Severinghaus, 1989) and confirmed by deep-sea drilling (Yeats and Haq, 1981); it ranges from 17 Ma near the Patton Ridge area to 15 Ma near Popcorn Ridge.

The middle Miocene episode of extension and volcanism is more complicated than simple models would suggest (Blake and others, 1978; Dickinson and Snyder, 1979), because greater extension seems to have occurred in the inner borderland, probably related to the rotation of the western Transverse Ranges block away from the coastline (Crouch, 1979, 1981; Kamerling and Luyendyk, 1979, 1985; Luyendyk and others, 1980, 1985) and, perhaps, to possible Paleogene or earlier extension within the accretionary wedge, such as that postulated by Platt (1986). The Neogene extension seems to have occurred along an east-west trend, similar to that of the Basin and Range (Luyendyk and others, 1985). A remnant of this east-west extension may be present as a north-south rift, with central horst, shown by the ridge and basin topography of the south San Clemente Basin region (Figs. 1 and 4). Rocks dredged from the flanks of Sixtymile Bank, Blake Knolls, and La Victoria Knoll (Fig. 2) in this area include blueschist-facies rocks of the Catalina Schist (Vedder and Howell, 1976), which imply that substantial uplift and unroofing of the deep-subduction complex occurred here, forming a horst (another Catalina Schist ridge?). The large magnetic anomaly also found in this area (Theberge, 1971; Pisciotto and others, 1974) may be associated with serpentinized ultramafic rocks of the uplifted upper mantle that once underlay the forearc basin of the North America margin. Alternatively, intrusion of highly magnetic volcanic rocks may have occurred there. The postulated rift is bounded by the San Clemente fault on the northeast and a similar, northwest-trending fault zone roughly delineated by Animal and Collnett Basins on the southwest. These two bounding faults could be ancient transform faults connecting the South San Clemente Basin "continental" spreading center (rhombochasm) with other rift zones or the trench at the outer edge of the borderland. The north-south trend of the rift and 45° angle between transform fault and rift zone are similar to trends in the Basin and Range (e.g., the Las Vegas shear zone). The reconstruction shown in Figure 5A suggests that other similar rift-transform zones were involved in the middle Miocene borderland extension.

I suggest that this borderland extension involved rifting of continental crust (borderland crust) that was abandoned before reaching full-scale seafloor spreading, where new oceanic crust is formed and transform faults are orthogonal to the spreading center-rift valley. Lonsdale (1985) reports older rift margins, on the flanks of the Guaymas Basin in the Gulf of California, that have a north trend, or is about 45° oblique to the modern seafloor-spreading axis. A similar pattern is shown by the Brawley fault where it intersects the Imperial fault in southern California (Sharp and others, 1982), and possibly by other rift segments in the northern Gulf of California (CONMAR, 1989). An analog for Neogene borderland basin formation would be "continental rifts" such as the older, pre-seafloor-spreading, rifts of the Gulf of California or perhaps even the East African Rifts linked by accommodation zones (Rosendahl, 1987). During the middle Miocene rifting of the inner borderland, a similar pattern of rifts may have appeared to the north in the Gulf of Santa Catalina area (Capistrano embayment?). This rifting led to further uplift of the Catalina Schist ridge, the deposition of the San Onofre Breccia, and basaltic volcanism along north-trending dikes, such as those near Pt. La Jolla (Fig. 9; Kennedy, 1975) and in the San Joaquin Hills (Fig. 2; Vedder and others, 1957).

East Santa Cruz Basin Fault Zone.—Along the borders of the inner-borderland rift, major transcurrent faults took up the greatest lateral slip, primarily the East Santa Cruz Basin fault zone (Howell and others, 1974). Because this fault zone appears to have served as a conduit for the voluminous extrusion of volcanic rocks, I suggest that the middle Miocene East Santa Cruz Basin fault system was a leaky transform fault. The modern expression of this transform fault is the zone bounded by the San Clemente fault system to the east and the East Santa Cruz Basin and other faults to the west. Because the San Clemente fault system is most well defined today and appears to mark a major lithostratigraphic discontinuity in pre-Miocene rocks (Crouch, 1981; Howell and Vedder, 1981), I propose that it was the locus of the lateral slip, acting as the principal transform fault of the Miocene plate boundary. It is also feasible that other inner-borderland faults, such as the San Diego Trough and Coronado Bank fault zones, were active parts of this system, but with lesser displacements as previously discussed. Just as relative motion today is spread across a zone more than 300 km wide, with the locus being along the San Andreas fault zone, numerous other subparallel faults in the offshore region probably shared some of the Miocene plate motion. These include major Neogene faults in the outer borderland, such as the Ferrelo fault zone along the Santa Rosa-Cortes Ridge (Fig. 2; Junger, 1979; Field and Richmond, 1980; Couch, 1981; Vedder, 1987).

Post-Miocene Offsets and Slip Rates

In general, post-Miocene fault offsets and slip rates have been considered minor for the California Continental Borderland because the relative plate motion was considered to have shifted eastward to the San Andreas and Gulf of California transform-fault systems (Blake and others, 1978; Crowell, 1979, 1981; Crouch, 1979, 1981; Luyendyk and others, 1980, 1985). Attempts to sum the relative contributions of right-slip across the numerous onshore faults in southern California and reach the total expected PAC-NOAM relative-motion value (50–56 mm/yr) have found a significant discrepancy: "the San Andreas discrepancy" (Anderson, 1979; Weldon and Humphreys, 1986; Minster and Jordan, 1987). This led to models suggesting that a significant portion, as much as 20 to 35 percent, of the relative plate motion is occurring in the offshore region. The high rate of convergence inferred across the Ventura Basin (Yeats, 1983) requires that most of this "missing" relative plate motion occurs within the inner borderland (Weldon and Humphreys, 1986). A novel reconstruction of the San Diego Trough, previously described, may indicate that a significant rate (>2 mm/yr) of right-slip faulting could be associated with some inner-borderland faults. This value may still be short, by as much as 8 to 13 mm/yr, of the total required slip.

Some studies have argued that all of the relative motion, within the error bounds of the data (about ±2 mm/yr), can be accommodated on the onshore-fault zones, including the San Andreas, San Jacinto, and Whittier-Elsinore fault zones (Bird and Rosenstock, 1984; Jordan and Minster, pers. commun., 1984). Their arguments are supported by a downward revision of the total PAC-NOAM relative motion from 56 ± 2 mm/yr (Minster and Jordan, 1978) to about 49 ± 2 mm/yr for the NUVEL-1 model (DeMets and others, 1987). The latter value derives from revised estimates of seafloor-spreading rates based upon magnetic-anomaly patterns at the mouth of the Gulf of California. I consider two problems associated with this revised relative-motion estimate. First, major active right-slip faults, such as the San Clemente-San Isidro and Tosco-Abreojos fault zones, are located entirely west of the Baja California peninsula, and therefore, could shunt significant PAC-NOAM plate motion beyond the mouth of the Gulf of California to the Rivera transform (Spencer and Normark, 1979; Legg, 1985; Legg and others, 1989). Second, as mentioned by DeMets and others (1987), the magnetic-anomaly data for the most recent epoch (central and Jaramillo anomalies, <1 Ma) is better fit by a spreading rate of 51 mm/yr (corresponding to a PAC-NOAM relative motion of about 53 mm/yr in the vicinity of San Diego). Other investigators also suggest faster spreading rates at the mouth of the Gulf of California (65 mm/yr, Ness and others, 1985). Resolution of the slip-rate issue holds great significance to residents of coastal southern California, considering their exposure to offshore earthquake hazards.

NEW DATA CONSTRAINING FUTURE MODEL DEVELOPMENT

Detailed Fault Mapping (3-D Structural Imaging)

The character of offshore faulting now can be mapped in far greater detail than previously possible. In particular, faults can be mapped in three dimensions using closely spaced seismic-reflection profiles so that the character and deformation history of a specific fault zone can be inferred using the techniques of seismic interpretation and sequence stratigraphy (Mitchum and others, 1977a, b; Vail and others, 1977). Right-slip has been demonstrated for most of the major fault zones in the California Continental Borderland (Vedder and others, 1974; Legg and Kennedy, 1979; Greene and others, 1979; Crouch, 1979; 1981; Legg, 1985). The total offset or rate of lateral slip for the offshore faults remains to be determined.

In addition to strike-slip fault zones, most of the major offshore-fault zones show segments with substantial dip-slip faulting, including seafloor uplift. This vertical tectonic component has long been recognized; the timing, however, has become more accurately constrained by the seismic-reflection profiles, seafloor sampling, and deep-sea drilling including deep stratigraphic-test holes and exploratory drilling for petroleum (Yeats and Haq, 1981; Kennedy and others, 1985; Vedder, 1987, 1990). It is recognized that large-scale vertical movements occurred during the Miocene epoch, accounting for the modern basin and range physiography of the borderland (Vedder and others, 1974; Howell and others, 1980; Vedder, 1987). Post-Miocene vertical movements also are recognized, with the greatest uplift (and subsidence) in the vicinity of the Transverse Ranges (Yeats, 1983; Rockwell and others, 1984). Farther south, vertical tectonics associated with divergent, right bends or fault offsets, and convergent, left bends or fault offsets, result in active pull-apart basins or push-ups along the major right-lateral fault zones (Crowell, 1974; Legg, 1985, 1989). Other major faults, such as the Palos Verdes Hills fault, are recognized to have significant thrust or oblique-dextral reverse-slip components (Fischer and others, 1977; Davis and others, 1989). In addition, transpressional structure along the offshore Newport-Inglewood fault zone and prominent northwest-trending thrust faults at the base of the continental slope west of Newport and San Juan Capistrano suggest northeast-southwest convergence in this area (Nicholson and Crouch, 1989; J. Crouch, pers. commun., 1989). All of these features are Quaternary, and, in places, most affect Holocene strata.

Deep crustal imaging, with multichannel seismic-reflection profiling techniques (MCS) provides new data regarding the relation of surficial structure with mid-to-lower crustal structure. The discovery of numerous sub-horizontal, mid-crustal reflections within basement rocks suggests that detachment faulting, such as is found in the Transverse Ranges to the north (Campbell and others, 1966; Yeats, 1983), may be regionally significant. The significance of detachment faulting to borderland tectonics remains to be determined; however, new deep crustal seismic data are being acquired (Bohannon and others, 1990). These data may be interpreted with reference to new models of continental extension involving mid-crustal detachments and upper crustal-block faulting (Wernicke and Burchfiel, 1982; Rosendahl, 1987).

Satellite Geodesy

Satellite geodesy (GPS) provides a measure of the overall direction and rate of slip for the offshore region (Larson, 1989) but is unable to determine the distribution of this slip between the several major submarine-fault zones. To date, preliminary data of the relative motion between San Diego and San Clemente Island suggest high rates of slip, 10 ± 8 mm/yr (Larson, 1989), which represent a significant portion of PAC-NOAM relative motion. Because the San Clemente fault system remains offshore, west of Baja California (Legg, 1985), a significant component of slip, beyond the NUVEL-1 amount (49 mm/yr; DeMets and others, 1987), may bypass the Gulf of California where data for PAC-NOAM motions were obtained for the NUVEL-1 model.

Earthquake-Source Studies

Source studies of the significant earthquakes that occur in the California Continental Borderland have found a mixture of strike-slip faulting, both dextral and sinistral, along with both normal and thrust faulting. Most events along the San Clemente fault show predominately right-slip faulting, with minor components of extension in some cases (Legg, 1980; Corbett, 1984). The Santa Barbara Channel and Transverse Ranges region is dominated by left-slip and left-oblique thrusting (Sylvester and others, 1968; Ellsworth and others, 1973; Lee and Vedder, 1973; Lee and others, 1979; Yerkes and Lee, 1979; Corbett, 1984). To the south, seismotectonics of the inner borderland region are more complex. Studies in the Los Angeles Basin and offshore Santa Monica Bay region show distinct domains of right-slip earthquakes separated from thrust-faulting earthquakes (Hauksson and Saldivar, 1989; Nicholson and Crouch, 1989; Hauksson, 1990). Thrust-fault earthquake mechanisms have been observed as far south as the northern end of the San Diego Trough (Hauksson and Jones, 1988; Pacheco and Nabelek, 1988; Malcolm Ross, pers. commun., 1989). The significance of "blind thrusting" with regard to earthquake hazards has been well demonstrated during the past two decades of California earthquake activity. We are only beginning to understand its significance regarding the regional tectonic framework (Namson and Davis, 1988; Davis and others, 1989; Hauksson, 1990). Still farther south, the occurrence of earthquake swarms west of San Quintín, Baja California (Rebollar and others, 1982), and right-slip earthquakes with extensional components (Legg, 1980) add to the tectonic complexity of the inner borderland.

Other Advanced Technology

Other advances in science and technology, notably marine geophysics through interpretation and modeling of geomagnetic and gravity data, provide additional tools and techniques to infer seafloor and subbottom lithology. Limited heat flow, and almost non-existent magnetotelluric or electromagnetic data in the region may provide additional constraints on the subbottom geology and tectonic evolution, but require more research effort than has occurred to date. Most recently, major advances in computer technology have allowed important improvements in seafloor mapping and subbottom-imaging capability. Multibeam echo sounding or seafloor swath-mapping systems allow bathymetric mapping at greater resolution than previously possible. Although yet to equal aerial photography, the resolution of these data, especially when combined with side-scan sonar images, permits more accurate and detailed geomorphic interpretation of seafloor structure and processes. By reducing the effects of water, bottom multiple reflections, and other coherent noise that interfere with interpretation of seismic reflection profiles, MCS allows better imaging of the subbottom structure at higher resolutions and greater depths of penetration. Deep-sea drilling provides groundtruth geology for the seismic profiles and confirms the complex tectonic history of the borderland (Yeats and Haq, 1981). Increased availability of drillhole data from deep stratigraphic-test holes and exploratory drilling provides additional important subsurface information (Vedder and others, 1981; Vedder, 1987). Access to the sea floor via submersibles or remotely controlled vehicles (ROVs) also permits more accurate sampling programs to determine lithology. Lastly, supercomputers and the general availability of high-powered microcomputers also provide tremendous tools for interpretation, modeling, and graphic display of the diverse forms of data acquired in the region.

CONCLUSIONS

More than a century of research has shown that the North America-Pacific tectonic plate boundary in southern California is a broad, geologically complex zone of dextral shear that includes several major fault zones. An important part of the Cenozoic evolution of this continental-margin plate boundary is recorded in the geology of the California Continental Borderland. This evolution includes the collision of the EPR with North America, subsequent southward migration of the Rivera RTF triple junction, northward migration of the Mendocino TFF triple junction away from the region, and the eventual rifting of the Baja California peninsula from the North American continent. A major middle Miocene episode of oblique rifting resulted in widespread volcanism and the basin and range physiography of the borderland. The disruption of the Mesozoic subduction zone by late Cenozoic transform faulting has led to a distinctive distribution of basement rocks or lithostratigrahic terranes that provide important clues to our attempts at palinspastic reconstruction of the region.

Combinations of westward extension and northwest-directed simple shear of the borderland, away from the continent, produced major rifts in the inner and southern borderland and led to clockwise rotation of large crustal blocks to form the western Transverse Ranges and Santa Monica Mountains. Later north-south convergence in the Transverse Ranges combined with northeast-southwest-directed convergence as a result of the big bend in the San Andreas fault have produced important late Quaternary contractional structure in the northernmost parts of the borderland, including the Los Angeles Basin. Holocene faulting and modern earthquake activity in the region demonstrate that the California Continental Borderland remains a tectonically active part of the present PAC-NOAM plate boundary. Although the rates of deformation, particularly the strike-slip components, are unresolved, preliminary data suggest that the modern borderland is still accommodating a significant fraction of the PAC-NOAM relative plate motion.

Much exciting research is yet to be undertaken in order to understand the evolution and present tectonics of this continental transform-margin plate boundary. Because the California Continental Borderland has endured all three major plate-boundary conditions—subduction, rifting, and transform faulting—an improved understanding of the region's tectonic evolution produces wonderful insight into the nature of global continental-margin and plate-boundary processes.

ACKNOWLEDGMENTS

I have benefitted greatly from numerous discussions about borderland geology and tectonics with Victor Wong, Michael Kennedy, Jack Vedder, John Crowell, Art Sylvester, Bruce Luyendyk, Jacqueline Mammerickx, Bob Tyce, Chris de Moustier, Jim Crouch, Gary Greene, Kim Bailey, Calvin Lee, Tom Henyey, and Craig Nicholson during the past decades of my research career. I am particularly grateful to Bob Bohannon, Steve Eittreim, and Jon Childs for inviting me to join them on the recent MCS cruise aboard the *S. P. Lee*, and to Dave Gallo, Ken Stewart, and the officers and crew of the U.S. Navy *DSV Turtle* for allowing me to see the borderland geology directly. I gratefully acknowledge the helpful reviews by Bob Osborne and Jack Vedder. I am most grateful to Joyce, Geoffrey, and Justin for their patient suffering while I struggled to finish this paper on numerous evenings, weekends and holidays.

REFERENCES

Agnew, D. C., Legg, Mark, and Strand, Carl, 1979, Earthquake history of San Diego, *in* Abbott, P. L., and Elliott, W. J., eds. Earthquakes and Other Perils—San Diego Region: San Diego Association of Geologists Guidebook, p. 123–138.

ALLEN, C. R., SILVER, L. T., AND STEHLI, F. G., 1960, The Agua Blanca fault: A major transverse structure of northern Baja California, Mexico: Geological Society of America Bulletin, v. 71, p. 357–482.

ALLISON, E. C., 1964, Geology of areas bordering Gulf of California: Marine geology of the Gulf of California—A symposium: American Association of Petroleum Geologists Memoir 3, p. 3–29.

ANDERSON, J. G., 1979, Estimating the seismicity from geologic structure for seismic risk studies: Seismological Society of America Bulletin, v. 55, p. 753–797.

ARNAL, R. E., 1976, Miocene paleobathymetric changes of the Santa Rosa-Cortes Ridge area, California Continental Borderland, *in* Howell, D. G., ed., Aspects of the Geological History of the California Continental Borderland: Pacific Section, American Association of Petroleum Geologists, Miscellaneous Publication 24, p. 60–79.

ARNAL, R. E., AND VEDDER, J. G., 1976, Late Miocene paleobathymetry of the California Continental Borderland north of 32°, *in* Proceedings of the Neogene Symposium: Pacific Section, Society of Economic Paleontologists and Mineralogists, San Francisco, California, p. 1–12.

ASHBY, J. R., Jr., 1989, Miocene tectonostratigraphic history of the Rosarito Beach basin, northwestern Baja California: Implications for the early tectonic development of the Southern California Continental Borderland: Unpublished M.S. Thesis, University of California at Los Angeles, Los Angeles, California, 150 p.

ATWATER, TANYA, 1970, Implications of plate tectonics for the Cenozoic evolution of western North America: Geological Society of America Bulletin, v. 81, p. 3515–3536.

ATWATER, TANYA, 1989, Plate tectonic history of the northeast Pacific and western North America, *in* Winterer, E. L., Hussong, D. M., and Decker, R. W., eds., The Eastern Pacific Ocean and Hawaii: Geological Society of America, Decade of North American Geology, v. N, p. 21–72.

ATWATER, TANYA, AND MOLNAR, PETER, 1973, Relative motion of the Pacific and North American plates deduced from sea-floor spreading in the Atlantic, Indian and South Pacific Oceans, *in* Kovach, R. L., and Nur, Amos, eds., Proceedings, Conference on the Tectonic Problems of the San Andreas Fault System: Stanford University Publication, v. 13, p. 136–148.

ATWATER, TANYA, AND SEVERINGHAUS, JEFF, 1989, Tectonic maps of the northeast Pacific, *in* Winterer, E. L., Hussong, D. M., and Decker, R. W., eds., The Eastern Pacific Ocean and Hawaii: Geological Society of America, Decade of North American Geology, v. N, p. 14–20, plates 3A-C.

BAILEY, E. H., 1941, Mineralogy, petrology, and geology of Santa Catalina Island, California: Unpublished Ph.D. Dissertation, Stanford University, Palo Alto, California, 193 p.

BAILEY, E. H., AND JAHNS, R. H., 1954, Geology of the Transverse Ranges province, southern California, *in* Jahns, R. H., ed., Geology of Southern California: California Division of Mines Bulletin 170, p. 83–106.

BANCROFT, H. H., 1988, The History of California: The History Company, San Francisco, 7 vol.

BANDY, O. L., 1953, Ecology and paleoecology of some California Foraminifers: Part II Foraminiferal evidence of subsidence rates in the Ventura Basin: Journal of Paleontology, v. 27, p. 200–203.

BANDY, O. L., 1963, Larger living foraminifera of the continental borderland of southern California: Cushman Foundation for Foraminiferal Research Contributions, v. 14, p. 121–126.

BEYER, L. A., 1980, Offshore southern California, *in* Oliver, H. W., ed., Interpretation of the Gravity Map of California and its Continental Margin: California Division of Mines and Geology Bulletin 205, p. 8–15.

BIRD, PETER, AND ROSENSTOCK, R. W., 1984, Kinematics of present crust and mantle flow in southern California: Geological Society of America Bulletin, v. 95, p. 946–957.

BLAKE, M. C., JR., CAMPBELL, R. H., DIBBLEE, T. W., JR., HOWELL, D. G., NILSEN, T. H., NORMARK, W. R., VEDDER, J. G., AND SILVER, E. A., 1978, Neogene basin formation in relation to plate-tectonic evolution of San Andreas fault system, California: American Association of Petroleum Geologists Bulletin, v. 62, p. 344–372.

BLAKE, W. P., 1855, Observations on the physical geography and geology of the Coast of California, from Bodega Bay to San Diego: U.S. Coast Survey Annual Report, Appendix No. 65, p. 376–398.

BLAKE, W. P., 1857, Geological Report (Williamson's reconnaissance of California), *in* U.S. War Department, exploration and Surveys for a Railroad Route from the Mississippi River to the Pacific Ocean, v. 5, pt. 2, 320 p.

BOHANNON, ROBERT, EITTREIM, STEVE, CHILDS, JON, GEIST, ERIC, LEGG, MARK, LEE, CALVIN, SORLIEN, CHRIS, AND BUSCH, LARRY, 1990, A seismic-reflection study of the California Continental Borderland (abst.): Transactions, American Geophysical Union, v. 71, p. 1631.

CAMPBELL, R. H., YERKES, R. F., AND WENTWORTH, C. M., 1966, Detachment Faults in the Central Santa Monica Mountains, California: U.S. Geological Survey Professional Paper 550-C, p. C1-C11.

CHAMPION, D. E., HOWELL, D. G., AND MARSHALL, M. C., 1981, Paleomagnetic evidence for 3800 km of northwestward translation of San Miguel Island, southern California borderland (abst.): Transactions, American Geophysical Union, v. 62, p. 855.

CLEMENTS, THOMAS, AND EMERY, K. O., 1947, Seismic activity and topography of the sea floor off southern California: Seismological Society of America Bulletin, v. 37, p. 351–354.

CLOOS, MARK, 1982, Flow melanges: Numerical modelling and geologic constraints on their origin in the Franciscan subduction complex, California: Geological Society of America Bulletin, v. 93, p. 421–433.

CONMAR, CONTINENTAL MARGINS STUDY GROUP, 1989, Bathymetric, geomagnetic, and gravity maps of the Gulf and Peninsular Province of the Californias: American Association of Petroleum Geologists, Tulsa, Oklahoma, 6 plates.

CORBETT, E. J., 1984, Seismicity and crustal structure studies of southern California—Tectonic implications from improved earthquake locations: Unpublished Ph.D. Dissertation, California Institute of Technology, Pasadena, California, 231 p.

COX, ALAN, AND ENGEBRETSON, DAVID, 1985, Change in motion of Pacific plate at 5 myr BP: Nature, v. 313, p. 472–474.

CROUCH, J. K., 1979, Neogene tectonic evolution of the western Transverse Ranges and the California Continental Borderland: Geological Society of America Bulletin, v. 90, p. 338–345.

CROUCH, J. K., 1981, Northwest margin of California Continental Borderland; Marine geology and tectonic evolution: American Association of Petroleum Geologists Bulletin, v. 65, p. 191–218.

CROUCH, J. K., BACHMAN, S. B., AND SHAY, J. T., 1984, Post-Miocene compressional tectonics along the central California margin, *in* Crouch, J. K., and Bachman, S. B., eds., Tectonics and Sedimentation Along the California Margin: Pacific Section, Society of Economic Paleontologists and Mineralogists, v. 38, p. 37–54.

CROWELL, J. C., 1974, Origin of late Cenozoic basins in southern California, *in* Dickinson, W. R., ed., Tectonics and Sedimentation: Society of Economic Paleontologists and Mineralogists Special Publication 22, p. 190–204.

CROWELL, J. C., 1976, Implications of crustal stretching and shortening of coastal Ventura Basin, California, *in* Howell, D. G., ed., Aspects of the Geological History of the California Continental Borderland: Pacific Section, American Association of Petroleum Geologists, Miscellaneous Publication 24, p. 365–379

CROWELL, J. C., 1979, The San Andreas fault system through time: Journal of the Geological Society of London, v. 136, p. 293–302.

CROWELL, J. C., 1981, The juncture of the San Andreas transform system and the Gulf of California rift: Oceanologica Acta, Proceedings, 26th International Geological Congress, Geology of Continental Margins Symposium, Paris, p. 137–141.

DAVIDSON, GEORGE, 1869, Pacific Coast—Coast Pilot of California, Oregon, and Washington Territory: U.S. Coast and Geodetic Survey (3rd ed.): U.S. Government Printing Office, Washington, D.C., 262 p.

DAVIDSON, GEORGE, 1887, Submarine valleys on the Pacific Coast of the United States: California Academy of Sciences Bulletin, v. 2, p. 265–268.

DAVIDSON, GEORGE, 1897, The submerged valleys of the coast of California, U.S.A. and of Lower California, Mexico: Proceedings, California Academy of Sciences, 3rd ser., v. 1, (Geology), p. 74–102.

DAVIS, G. A., AND BURCHFIEL, B. C., 1973, Garlock fault of southern California—An intracontinental transform structure: Geological Society of America Bulletin, v. 84, p. 1407–1422.

DAVIS, T. L., NAMSON, J. S., AND YERKES, R. F., 1989, A cross section of the Los Angeles area: Seismically active fold and thrust belt, the

1987 Whittier Narrows earthquake, and earthquake hazard: Journal of Geophysical Research, v. 94, p. 9644–9664.

DeMets, Charles, Gordon, R. G., Stein, Seth, and Argus, D. F., 1987, A revised estimate of Pacific-North America motion and implications for western North America plate boundary zone tectonics: Geophysical Research Letters, v. 14, p. 911–914.

Dickinson, W. R., and Snyder, W. S., 1979, Geometry of triple junctions related to San Andreas transform: Journal of Geophysical Research, v. 84, p. 561–572.

Doyle, L. J., and Bandy, O. L., 1972, Southern continental borderland, Baja California: Its tectonic and environmental development: Geological Society of America Bulletin, v. 83, p. 3785–3794.

Doyle, L. J., and Gorsline, D. S., 1977, Marine geology of Baja California continental borderland, Mexico: American Association of Petroleum Geologists Bulletin, v. 61, p. 903–917.

Ellsworth, W. L., Campbell, R. H., Hill, D. P., Page, R. A., Alewine, R. W., III, Hanks, T. C., Heaton, T. H., Hileman, J. A., Kanamori, Hiroo, Minster, Bernard, and Whitcomb, J. H., 1973, Point Mugu, California, earthquake of 21 February 1973 and its aftershocks: Science, v. 182, p. 1127–1129.

Emery, K. O., 1960, The Sea Off Southern California, A Modern Habitat of Petroleum: John Wiley & Sons, Inc., New York, 366 p.

Emery, K. O., Butcher, W. S., Gould, H. R., and Shepard, F. P., 1952, Submarine geology off San Diego, California: Journal of Geology, v. 60, p. 511–548.

Emery, K. O., and Shepard, F. P., 1945, Lithology of the seafloor off southern California: Geological Society of America Bulletin, v. 56, p. 431–478.

Engebretson, D. C., Cox, Allan, and Gordon, R. G., 1985, Relative Motions between Oceanic and Continental Plates in the Pacific Basin: Geological Society of America Special Paper 206, 59 p.

Ernst, W. G., 1973, Blueschist metamorphism and P-T regimes in active subduction zones: Tectonophysics, v. 17, p. 255–272.

Fairbanks, H. W., 1897, Oscillations of the coast of California during the Pliocene and Pleistocene: American Geologist, v. 20, p. 213–245.

Field, M. E., and Richmond, W. C., 1980, Sedimentary and structural patterns on the northern Santa Rosa-Cortes Ridge, southern California: Marine Geology, v. 24, p. 79–98.

Fischer, P. J., Parker, J., and Farnsworth, R., 1977, Beta Platform Site Evaluations: California State University, Northridge, Marine Studies 77-2, Northridge, California, 60 p.

Gastil, R. G., Phillips, R. P., and Allison, E. C., 1975, Reconnaissance geology of the state of Baja California: Geological Society of America Memoir No. 140, 170 p.

Gawthrop, W. J., 1978, Seismicity and tectonics of the central California coastal region, *in* Silver, E. A., and Normark, W. R., eds., San Gregorio-Hosgri Fault Zone, California: California Division of Mines and Geology Special Report 137, p. 45–56.

Grant, U.S., IV, and Shepard, F. P., 1939, Some recent changes of elevation in the Los Angeles Basin of southern California and their possible significance: Seismological Society of America Bulletin, v. 29, p. 292–326.

Greene, H. G., Bailey, K. A., Clarke, S. H., Ziony, J. I., and Kennedy, M. P., 1979, Implications of fault patterns of the inner California Continental Borderland between San Pedro and San Diego, *in* Abbott, P. L., and Elliott, W. J., eds., Earthquakes and Other Perils—San Diego Region: San Diego Association of Geologists Guidebook, p. 21–27.

Hamilton, Warren, 1969, Mesozoic California and the underflow of the Pacific mantle: Geological Society of America Bulletin, v. 80, p. 2409–2429.

Hamilton, Warren, 1978, Mesozoic tectonics of the western United States, *in* Howell, D. G., and McDougall, K. A., eds., Mesozoic Paleogeography of the Western United States: Society of Economic Paleontologists and Mineralogists, Pacific Coast Paleogeography Symposium 2, p. 33–71.

Harding, T. P., 1973, Newport-Inglewood trend, California—An example of wrench style deformation: American Association of Petroleum Geologists Bulletin, v. 57, p. 97–116.

Harding, T. P., 1974, Petroleum traps associated with wrench faults: American Association of Petroleum Geologists Bulletin, v. 58, p. 1290–1304.

Harding, T. P., 1985, Seismic characteristics and identification of negative flower structures, positive flower structures, and positive structural inversion: American Association of Petroleum Geologists Bulletin, v. 69, p. 585–600.

Harrison, J. C., Von Huene, R. E., and Corbato, C. E., 1966, Bouguer gravity anomalies and magnetic anomalies off the coast of southern California: Journal of Geophysical Research, v. 71, p. 4921–4941.

Hauksson, Egill, 1987, Seismotectonics of the Newport-Inglewood fault zone in the Los Angeles Basin, southern California: Seismological Society of America Bulletin, v. 77, p. 539–561.

Hauksson, Egill, 1990, Earthquakes, faulting, and stress in Los Angeles Basin: Journal of Geophysical Research, v. 95, p. 15,365–54,394.

Hauksson, Egill, and Jones, L. M., 1988, The July 1986 Oceanside (M_L = 5.3) earthquake sequence in the Continental Borderland, southern California: Seismological Society of America Bulletin, v. 78, p. 1885–1906.

Hauksson, Egill, and Saldivar, G. V., 1986, The 1930 Santa Monica and the 1979 Malibu, California, earthquakes: Seismological Society of America Bulletin, v. 76, p. 1542–1559.

Hauksson, Egill, and Saldivar, G. V., 1989, Seismicity and active compressional tectonics in Santa Monica Bay, southern California: Journal of Geophysical Research, v. 94, p. 9591–9606.

Hawkins, J. L., 1970, Petrology and possible tectonic significance of late Cenozoic volcanic rocks, southern California and Baja California: Geological Society of America Bulletin, v. 81, p. 3323–3338.

Hawkins, J. L., and Divis, A. F., 1975, Petrology and geochemistry of mid-Miocene volcanism on San Clemente and Santa Catalina Islands and adjacent areas of the southern California borderland: Geological Society of America Bulletin, Abstracts with Programs, v. 7, p. 323–324.

Henyey, T. L., 1976, Heat flow and tectonic patterns on the southern California borderland, *in* Howell, D. G., ed., Aspects of the Geological History of California Continental Borderland: Pacific Section, American Association of Petroleum Geologists, Miscellaneous Publication 24, p. 427–448.

Hill, D. J., 1976, Geology of the Jurassic basement rocks, Santa Cruz Island, California, and correlation with other Mesozoic basement terranes in California, *in* Howell, D. G., ed., Aspects of the Geological History of California Continental Borderland: Pacific Section, American Association of Petroleum Geologists, Miscellaneous Publication 24, p. 16–46.

Hill, M. L., 1971, Newport-Inglewood zone and Mesozoic subduction, California: Geological Society of America Bulletin, v. 82, p. 2957–2962.

Holden, E. S., 1898, A Catalogue of Earthquakes on the Pacific Coast: Smithsonian Miscellaneous Collections 1087, 253 p.

Hornafius, J. S., 1985, Neogene tectonic rotation of the Santa Ynez Range, western Transverse Ranges, California, suggested by paleomagnetic investigation of the Monterey Formation: Journal of Geophysical Research, v. 90, p. 12,503–12,522.

Howell, D. G., and Vedder, J. G., 1981, Structural implications of stratigraphic discontinuities across the southern California borderland, *in* Ernst, W. G., ed., The Geotectonic Development of California, Rubey Volume 1: Prentice-Hall, Inc., Englewood Cliffs, New Jersey, p. 535–558.

Howell, D. G., Crouch, J. K., Greene, H. G., McCulloch, D. S., and Vedder, J. G., 1980, Basin development along the late Mesozoic and Cainozoic California margin—A plate tectonic margin of subduction, oblique subduction and transform tectonics, *in* Ballance, P. F., and Reading, H. G., eds., Sedimentation in Oblique-Slip Mobile Zones: International Association of Sedimentologists Special Publication No. 4, p. 43–62.

Howell, D. G., Stuart, C. J., Platt, J. P., and Hill, D. J., 1974, Possible strike-slip faulting in the southern California borderland: Geology, v. 2, p. 93–98.

Ingle, J. C., Jr., 1972, Biostratigraphy and paleoecology of early Miocene through early Pleistocene benthonic and planktonic foraminifera, San Joaquin Hills, Newport Bay, Orange County, California, *in* Miocene Biostratigraphy Symposium, Society of Economic Paleontologists and Mineralogists, p. 255–283.

Ingle, J. C., 1980, Cenozoic paleobathymetry and depositional history of selected sequences within the southern California continental borderland: Special Publication, Cushman Foundation for Foraminiferal Research, v. 19, p. 163–195.

JACOBY, G. C., SHEPPARD, P. R., AND SIEH, K. E., 1987, Was the 8 December 1812 earthquake produced by the San Andreas fault? Evidence from trees near Wrightwood: Seismological Research Letters, v. 58, p. 14.

JAHNS, R. H., 1954, Investigations and problems of southern California geology, *in* Jahns, R. H., ed., Geology of Southern California: California Division of Mines and Geology Bulletin 1970, Part 1, p. 5–29.

JONES, D. L., BLAKE, M. C., JR., AND RANGIN, CLAUDE, 1976, The four Jurassic belts of northern California and their significance to the geology of the southern California borderland, *in* Howell, D. G., ed., Aspects of the Geological History of the California Continental Borderland: Pacific Section, American Association of Petroleum Geologists, Miscellaneous Publication 24, p. 343–362.

JUNGER, ARNE, 1974, Source of the San Onofre Breccia: Geological Society of America, Abstracts with Programs, v. 6, p. 199–200.

JUNGER, ARNE, 1976, Tectonics of the southern California Borderland, *in* Howell, D. G., ed., Aspects of the Geological History of the California Continental Borderland: Pacific Section, American Association of Petroleum Geologists, Miscellaneous Publication 24, p. 486–498.

JUNGER, ARNE, 1979, Maps and seismic profiles showing geologic structure of the northern Channel Islands platform, California Continental Borderland: U.S. Geological Survey, Miscellaneous Field Studies, Map MF-991, Scale 1:250,000.

JUNGER, ARNE, AND WAGNER, H. C., 1977, Geology of the Santa Monica and San Pedro Basins, California Continental Borderland: U.S. Geological Survey, Miscellaneous Field Studies, Map MF-820, Scale 1:250,000.

KAMERLING, MARK, AND LUYENDYK, B. P., 1979, Tectonic rotations of the Santa Monica Mountains region, western Transverse Ranges, California, suggested by paleomagnetic vectors: Geological Society of America Bulletin, v. 90, p. 331–337.

KAMERLING, MARK, AND LUYENDYK, B. P., 1985, Paleomagnetism and Neogene tectonics of the northern Channel Islands, California: Journal of Geophysical Research, v. 90, p. 12,485–12,502.

KENNEDY, M. P., 1975, Geology of the San Diego metropolitan area, California: California Division of Mines and Geology Bulletin 200, 56 p.

KENNEDY, M. P., AND WELDAY, E. E., 1980, Recency and character of faulting offshore metropolitan San Diego, California: California Division of Mines and Geology, Map Sheet 40, Scale 1:50,000.

KENNEDY, M. P., CLARKE, S. H., GREENE, H. G., AND LEGG, M. R., 1980, Recency and character of faulting offshore metropolitan San Diego, California: California Division of Mines and Geology, Map Sheet 42, Scale 1:50,000.

KENNEDY, M. P., CLARKE, S. H., GREENE, H. G., AND LONSDALE, P. F., 1985, Observations from *DSRV Alvin* of Quaternary Faulting on the Southern California Continental Margin: U.S. Geological Survey Open-File Report 85-39, 26 p.

KENNEDY, M. P., GREENE, H. G., AND CLARKE, S. H., 1987, Geology of the California continental margin: Explanation of the continental margin geologic map series: California Division of Mines and Geology Bulletin 207, 110 p.

KIES, R. P., AND ABBOTT, P. L., 1983, Rhyolite clast populations and tectonics in the California Continental Borderland: Journal of Sedimentary Petrology, v. 53, p. 461–475.

KLEINPELL, R. M., 1983, Miocene Stratigraphy of California: American Association of Petroleum Geologists, Tulsa, Oklahoma, 450 p.

KRAUSE, D. C., 1965, Tectonics, bathymetry, and geomagnetism of the southern Continental Borderland west of Baja California, Mexico: Geological Society of America Bulletin, v. 76, p. 617–650.

LAJOIE, K. R., KERN, J. P., WEHMILLER, J. F., KENNEDY, G. L., MATHIESON, S. A., SARNA-WOJCICKI, A. M., YERKES, R. F., AND MCCRORY, P. F., 1979, Quaternary marine shorelines and crustal deformation, San Diego to Santa Barbara, California, *in* Abbott, P. L., ed., Geological Excursions in the Southern California Area: San Diego State University, Department of Geological Sciences, San Diego, California, p. 3–16.

LARSON, K. M., 1989, Estimates of crustal deformation in southern and central California from GPS: 1986–1989 (abst.): Transactions, American Geophysical Union, v. 70, p. 1353.

LAWSON, A. C., 1897, The post-Pliocene diastrophism of the coast of southern California: California University Publications *in* Geology, v. 1, p. 115–160.

LAWSON, A. C., 1924, The continental shelf off the coast of California: National Research Council Bulletin, v. 8, pt. 2, 23 p.

LEE, W. H. K., AND VEDDER, J. G., 1973, Recent earthquake activity in the Santa Barbara Channel region: Seismological Society of America Bulletin, v. 63, p. 1757–1773.

LEE, W. H. K., YERKES, R. F., AND SIMIRENKO, M., 1979, Recent Earthquake Activity and Focal Mechanisms in the Western Transverse Ranges, California: U.S. Geological Survey Circular 799-A, p. 1–26.

LEGG, M. R., 1980, Seismicity and tectonics of the inner Continental Borderland of southern California and northern Baja California, Mexico: Unpublished M.S. Thesis, University of California at Santa Diego, San Diego, California, 60 p.

LEGG, M. R., 1985, Geologic Structure and tectonics of the inner Continental Borderland offshore northern Baja California, Mexico: Unpublished Ph.D. Dissertation, University of California at Santa Barbara, Santa Barbara, California, 410 p.

LEGG, M. R., 1986, Earthquake epicenters and selected fault plane solutions of the mid-southern California continental margin. Map No. 2B *in* Greene, H. G., and Kennedy, M. P., eds., Geologic Map Series of the California Continental Margin, California Division of Mines and Geology, Area 2 of 7, (NOS 1206N-15), Scale 1:250,000.

LEGG, M. R., 1989, Faulting and seismotectonics of the inner Continental Borderland west of San Diego, *in* Rocquemore, Glenn, ed., Proceedings, Workshop on the Seismic Risk in the San Diego Region: Special Focus on the Rose Canyon Fault System: Southern California Earthquake Preparedness Project, San Diego, California, p. 50–70.

LEGG, M. R., AND KENNEDY, M. P., 1979, Faulting offshore San Diego and northern Baja California, *in* Abbott, P. L., and Elliott, W. J., eds., Earthquakes and Other Perils—San Diego Region: San Diego Association of Geologists Guidebook, p. 29–46.

LEGG, M. R., LUYENDYK, B. P., MAMMERICKX, JACQUELINE, DE MOUSTIER, CHRISTIAN, AND TYCE, R. C., 1989, SeaBeam survey of an active strike-slip fault—The San Clemente fault in the California Continental Borderland: Journal of Geophysical Research, v. 94, p. 1727–1744.

LONSDALE, PETER, 1985, A transform continental margin rich in hydrocarbons, Gulf of California: American Association of Petroleum Geologists Bulletin, v. 69, p. 1160–1180.

LUYENDYK, B. P., KAMERLING, M. J., AND TERRES, R. R., 1980, Geometric model for Neogene crustal rotations in southern California: Geological Society of America Bulletin, v. 91, p. 211–217.

LUYENDYK, B. P., KAMERLING, M. J., TERRES, R. R., AND HORNAFIUS, J. S., 1985, Simple shear of southern California during Neogene time suggested by paleomagnetic declinations: Journal of Geophysical Research, v. 90, p. 12,454–12,466.

MARSHALL, MONTE, PISCHKE, GARY, MACE, N., GASTIL, GORDON, AND LUYENDYK, B. P., 1979, Paleomagnetism and circum-Gulf of California tectonics. Geological Society of America, Abstracts with Programs, v. 11, p. 472.

MCLEAN, HUGH, HOWELL, D. G., AND VEDDER, J. G., 1976, Miocene strata on Santa Cruz and Santa Rosa Islands—A reflection of tectonic events in the southern California borderland, *in* Howell, D. G., ed., Aspects of the Geological History of the California Continental Borderland: Pacific Section, American Association of Petroleum Geologists, Miscellaneous Publication 24, p. 241–253.

MERIFIELD, P. M., LAMAR, D. L., AND STOUT, M. L., 1971, Geology of central San Clemente Island, California: Geological Society of America Bulletin, v. 82, p. 1989–1994.

MINCH, J. C., 1967, Stratigraphy and structure of the Tijuana-Rosarito Beach area, northwestern Baja California, Mexico: Geological Society of America Bulletin, v. 78, p. 1155–1178.

MINCH, J. C., 1970, The Miocene of northwestern Baja California, Mexico, *in* Allison, E. C., Acosta, M. G., Fife, D. L., Minch, J. A., and Hishikawa, K., eds., Pacific Slope Geology of Northwestern Baja California and Adjacent Alta California: Pacific Sections, American Association of Petroleum Geologists, Society of Exploration Geophysicists, and the Society of Economic Paleontologists and Mineralogists Guidebook, p. 100–104.

MINCH, J. C., GASTIL, GORDON, FINK, WILLIAM, ROBINSON, JOHN, AND JAMES, A. H., 1976, Geology of the Vizcaino Peninsula, *in* Howell, D. G., ed., Aspects of the Geological History of California Continental Borderland: Pacific Section, American Association of Petroleum Geologists, Miscellaneous Publication 24, p. 136–195.

MINSTER, J. B., AND JORDAN, T. H., 1978, Present-day plate motions: Journal of Geophysical Research, v. 83, p. 5331–5334.

MINSTER, J. B., AND JORDAN, T. H., 1987, Vector constraints on western U.S. deformation from space geodesy, neotectonics, and plate motions: Journal of Geophysical Research, v. 92, p. 4798–4804.

MITCHUM, R. M., JR., VAIL, P. R., AND THOMPSON, S., III, 1977a, The depositional sequence as a basic unit for stratigraphic analysis, *in* Payton, C. E., ed., Seismic Stratigraphy—Applications to Hydrocarbon Exploration: American Association of Petroleum Geologists Memoir 26, p. 53–62.

MITCHUM, R. M., JR., VAIL, P. R., AND SANGREE, J. B., 1977b, Stratigraphic interpretation of seismic reflection patterns in depositional sequences, *in* Payton, C. E., ed., Seismic Stratigraphy—Applications to Hydrocarbon Exploration: American Association of Petroleum Geologists Memoir 26, p. 117–134.

MOORE, D. G., 1969, Reflection Profiling Studies of the California Continental Borderland—Structure and Quaternary Turbidite Basins: Geological Society of America Special Paper 107, 142 p.

MOORE, G. W., 1976, Basin development in the California Borderland and the Basin and Range Province, *in* Howell, D. G., ed., Aspects of the Geological History of the California Continental Borderland: Pacific Section, American Association of Petroleum Geologists, Miscellaneous Publication 24, p. 383–390.

MOUNT, V. S., AND SUPPE, JOHN, 1987, State of stress near the San Andreas fault: Implications for wrench tectonics: Geology, v. 15, p. 1143–1146.

NAMSON, J. S., AND DAVIS, T. L., 1988, Seismically active fold and thrust belt in the San Joaquin Valley, central California: Geological Society of America Bulletin, v. 100, p. 257–273.

NARDIN, T. R., AND HENYEY, T. L., 1978, Pliocene-Pleistocene diastrophism of the Santa Monica and San Pedro shelves, California Continental Borderland: American Association of Petroleum Geologists Bulletin, v. 62, p. 247–272.

NESS, G. E., LYLE, M. W., AND LONSETH, A. T., 1985, Revised Pacific, North America, Rivera, and Cocos relative motion poles: Implications for strike-slip motion along the Trans-Mexican Volcanic Belt (abst.): Transactions, American Geophysical Union, v. 66, p. 849.

NICHOLSON, CRAIG, AND CROUCH, J. K., 1989, Neotectonic structures along the central and southern California margin (abst.): Seismological Research Letters, v. 60, p. 23–24.

ORME, A. R., 1974, Quaternary deformation of marine terraces between Ensenada and El Rosario, Baja California, Mexico, *in* Geology of Peninsular California: Pacific Sections, American Association of Petroleum Geologists, Society of Economic Paleontologists and Mineralogists and Society of Exploration Geophysicists Guidebook, p. 67–79.

PACHECO, J., AND NABELEK, JOHN, 1988, Source mechanisms of three moderate California earthquakes of July, 1986: Seismological Society of America Bulletin, v. 78, p. 1907–1929.

PAGE, B. M., 1966, Geology of the Coast Ranges of California, *in* Bailey, E. H., ed., Geology of Northern California: California Division of Mines and Geology Bulletin 190, p. 255–276.

PARKER, F. S., 1971, Petroleum potential of southern California offshore, *in* Cram, I. H., ed., Future Petroleum Provinces of the United States—Their Geology and Potential: American Association of Petroleum Geologists Memoir 15, v. 1, p. 178–191.

PISCIOTTO, K. A., SASNETT, N. B., AND BEYER, L. A., 1974, Residual magnetic map of the California Continental Borderland, *in* Vedder, J. G., Beyer, L. A., Junger, Arne, Moore, G. W., Roberts, A. E., Taylor, J. C., and Wagner, H. C., Preliminary report on the Geology of the Continental Borderland of Southern California: U.S. Geological Survey, Miscellaneous Field Studies Map MF-624, Scale 1:500,000.

PLATT, J. P., 1973, The petrology, structure and geologic history of the Catalina Schist terrain, southern California: Unpublished Ph.D. Dissertation, University of California at Santa Barbara, Santa Barbara, California, 250 p.

PLATT, J. P., 1975, Metamorphic and deformational processes in the Franciscan complex—Some insights from the Catalina-Schist terrane: Geological Society of America Bulletin, v. 86, p. 1337–1347.

PLATT, J. P., 1976, The significance of the Catalina Schist in the history of the southern California Borderland, *in* Howell, D. G., ed., Aspects of the Geological History of the California Continental Borderland: Pacific Section, American Association of Petroleum Geologists, Miscellaneous Publication 24, p. 47–52.

PLATT, J. P., 1986, Dynamics of orogenic wedges and the uplift of high-pressure metamorphic rocks: Geological Society of America Bulletin, v. 97, p. 1037–1053.

PLAWMAN, T. L. 1978, Crustal structure of the Continental Borderland and the adjacent portion of Baja California between latitudes 30°N and 33°N: Unpublished M.S. Thesis; Oregon State University, Corvallis, Oregon, 72 p.

RAND, W. W., 1933, The geology of Santa Cruz Island, California: Unpublished Ph.D. Dissertation, University of California, Berkeley, California, 175 p.

REBOLLAR, C. J., REYES, ALFONSO, AND REICHLE, MICHAEL, 1982, Estudio del enjambre de San Quintin, Baja California, Mexico, ocurrido durante 1975: Geofisica Internacional, v. 21, p. 331–358.

REED, R. D., AND HOLLISTER, J. S., 1936, Structural Evolution of Southern California: American Association of Petroleum Geologists, Tulsa, Oklahoma, 157 p.

RIDLON, J. B., 1969, San Clemente Island Rocksite Project: Offshore Geology, Part 2, Reconnaissance Survey around the Island: Naval Undersea Research and Development Center Technical Publication 156, San Diego, California, 130 p.

ROCKWELL, T. K., KELLER, E. A., CLARK, M. N., AND JOHNSON, D. L., 1984, Chronology and rates of faulting of Ventura River terraces, California: Geological Society of America Bulletin, v. 95, p. 1466–1474.

ROSENDAHL, B. R., 1987, Architecture of continental rifts with special reference to East Africa: Annual Reviews of Earth and Planetary Sciences, v. 15, p. 445–503.

SCHOELLHAMER, J. E., AND WOODFORD, A. O., 1951, The floor of the Los Angeles Basin, Los Angeles, Orange, and San Bernardino Counties, California: U.S. Geological Survey Oil and Gas Investigations Map OM-117, scale 1:62,500.

SHARP, R. V., LIENKAEMPER, J. J., BONILLA, M. G., BURKE, D. B., FOX, B. F., HERD, D. G., MILLER, D. M., MORTON, D. M., PONTI, D. J., RYMER, M. J., TINSLEY, J. C., YOUNT, J. C., KAHLE, J. E., HART, E. W., AND SIEH, K. E., 1982, Surface faulting in the central Imperial Valley, *in* The Imperial Valley, California, Earthquake of October 15, 1979: U.S. Geological Survey Professional Paper 1254, p. 119–143.

SHEPARD, F. P., AND DILL, R. F., 1966, Submarine Canyons and Other Sea Valleys: Rand McNally and Company, Chicago, Illinois, 381 p.

SHEPARD, F. P., AND EMERY, K. O., 1941, Submarine topography off the southern California coast—Canyons and tectonic interpretation: Geological Society of America Special Paper 31, 171 p.

SHEPARD, F. P., GRANT, U.S., IV, AND DIETZ, R. S., 1939, The emergence of Santa Catalina Island: American Journal of Science, v. 237, p. 651–655.

SHOR, G. G., JR., AND RAITT, R. W., 1958, Seismic studies in the southern California borderland: Congreso Geologico Internacional, XXa Seccion, 1956, Mexico, Seccion IV, Segundo Tomo, p. 243–259.

SHOR, G. G., JR., RAITT, R. W., AND MCGOWAN, D. D., 1976, Seismic refraction studies in the southern California borderland, 1949–1974: Scripps Institute of Oceanography Reference 76-13, 70 p.

SMITH, W. S. T., 1897, The geology of Santa Catalina Island: California Academy of Sciences Proceedings, 3rd series, v. 1, 71 p.

SMITH, W. S. T., 1898, A Geological Sketch of San Clemente Island: U.S. Geological Survey, 18th Annual Report, pt 2, p. 465–496.

SMITH, W. S. T., 1900, A topographic study of the islands of southern California: University of California Publications in Geology, v. 2, p. 179–230.

SMITH, W. S. T., 1933, Marine terraces on Santa Catalina Island: American Journal of Science, v. 25, p. 123–136.

SORENSEN, S. S., 1985, Petrologic evidence for Jurassic, island arc-like basement rocks in the southwestern Transverse Ranges and California continental Borderland: Geological Society of America Bulletin, v. 96, p. 997–1006.

SPENCER, J. E., AND NORMARK, W. R., 1979, Tosco-Abreojos fault zone: A Neogene transform plate boundary within the Pacific margin of southern Baja California, Mexico: Geology, v. 7, p. 554–557.

STUART, C. J., 1975, The stratigraphy, sedimentology, and tectonic implications of the San Onofre Breccia, southern California: Unpublished

Ph.D. Dissertation, University of California at Santa Barbara, Santa Barbara, California, 309 p.

STUART, C. J., 1979a, Lithofacies and origin of the San Onofre Breccia, coastal southern California, *in* Stuart, C. J., ed., Miocene Lithofacies and Depositional Environments, Coastal Southern California and Northwestern Baja California: Pacific Section, Society of Economic Paleontologists and Mineralogists Guidebook, Los Angeles, California, p. 25–42.

STUART, C. J., 1979b, Middle Miocene paleogeography of coastal southern California and the California borderland; Evidence from schist-bearing sedimentary rocks, *in* Armentrout, J. M., Cole, M. R., and Terbest, Harry, Jr., eds., Pacific Coast Paleogeography Symposium 3: Society of Economic Paleontologists and Mineralogists, Tulsa, Oklahoma, p. 29–44.

SUPPE, JOHN, AND ARMSTRONG, R. L., 1972, Potassium-argon dating of Franciscan metamorphic rocks: American Journal of Science, v. 272, p. 217–233.

SYLVESTER, A. G., SMITH, S. M., AND SCHOLZ, C. H., 1968, Earthquake swarm in the Santa Barbara Channel, California: Seismological Society of America Bulletin, v. 60, p. 1047.

TAYLOR, J. C., 1976, Geologic Appraisal of the Petroleum Potential of Offshore Southern California: The Borderland Compared to Onshore Coastal Basins: U.S. Geological Survey Circular 730, 43 p.

TENG, L. S-Y., 1985, Seismic stratigraphic study of the California Continental Borderland basins—Structure, stratigraphy and sedimentation: Unpublished Ph.D. Dissertation, University of Southern California, Los Angeles, California, 197 p.

THEBERGE, A. E., JR., 1971, Magnetic Survey off Southern California and Baja California: National Oceanic and Atmospheric Administration, National Ocean Survey, Operational Data Report, NOS DR-12, 10 p.

TOWNLEY, S. D., AND ALLEN, M. W., 1939, Descriptive catalog of earthquakes of the Pacific coast of the United States, 1769–1928: Seismological Society of America Bulletin, v. 29, p. 1–297.

TOPPOZADA, T. R., REAL, C. R., BEZORE, S. P., AND PARKE, D. L., 1979, Compilation of Pre-1900 California Earthquake History: California Division of Mines and Geology Open-File Report OFR 79-6, Sacramento, California, 271 p.

TURNER, D. L., 1970, Potassium-argon dating of Pacific Coast Miocene foraminiferal stages, *in* Bandy, O. L., ed., Radiometric Dating and Paleontologic Zonation: Geological of America Special Paper 24, p. 91–129.

VAIL, P. R., TODD, R. G., AND SANGREE, J. B., 1977, Chronostratigraphic significance of seismic reflections, *in* Payton, C. E., ed., Seismic Stratigraphy—Applications to Hydrocarbon Exploration: American Association of Petroleum Geologists Memoir 26, Tulsa, Oklahoma, p. 99–116.

VEDDER, J. G., 1976, Precursors and evolution of the name California Continental Borderland, *in* Howell, D. G., ed., Aspects of the Geological History of the California Continental Borderland: Pacific Section, American Association of Petroleum Geologists, Miscellaneous Publication 24, p. 6–11.

VEDDER, J. G., 1987, Regional geology and petroleum potential of the southern California borderland, *in* Scholl, D. W., Grantz, A., and Vedder, J. G., eds., Geology and Resource Potential of the Continental Margin of Western North America and Adjacent Ocean Basins, Beaufort Sea to Baja California: Circum-Pacific Council for Energy and Mineral Resources, Houston, Texas, Earth Science Series, v. 6, p. 403–447.

VEDDER, J. G., 1990, Maps of California Continental Borderland showing compositions and ages of bottom samples acquired between 1968 and 1979: U.S. Geological Survey Miscellaneous Field Studies Map MF-2122, Scale 1:250,000.

VEDDER, J. G., AND HOWELL, D. G., 1976, Review of the distribution and tectonic implications of Miocene debris from the Catalina Schist, California Continental Borderland and adjacent coastal areas, *in* Howell, D. G., ed., Aspects of the Geological History of the California Continental Borderland: Pacific Section, American Association of Petroleum Geologists, Miscellaneous Publication 24, p. 326–340.

VEDDER, J. G., AND HOWELL, D. G., 1980, Topographic evolution of the southern California borderland during late Cenozoic time, *in* Power, D. M., ed., The California Islands: Proceedings of a Multidisciplinary Symposium, Santa Barbara Museum of Natural History, Santa Barbara, California, p. 7–31.

VEDDER, J. G. AND NORRIS, R. M., 1963, Geology of San Nicolas Island, California: U.S. Geological Survey Professional Paper 369, 65 p.

VEDDER, J. G., BEYER, L. A., JUNGER, ARNE, MOORE, G. W., ROBERTS, A. E., TAYLOR, J. C., AND WAGNER, H. C., 1974, Preliminary report on the geology of the Continental Borderland of southern California: U.S. Geological Survey, Miscellaneous Field Studies Map MF-624, Scale 1:500,000.

VEDDER, J. G., CROUCH, J. K., AND LEE-WONG, FLORENCE, 1981, Comparative study of rocks from Deep Sea Drilling Project holes 467, 468, and 469 and the southern California borderland, *in* Yeats, R. S., Haq, B. U., and others, Initial Reports of the Deep Sea Drilling Project, v. LXIII: U.S. Government Printing Office, Washington, D.C., p. 907–918.

VEDDER, J. G., WAGNER, H. C., AND SCHOELLHAMER, J. E., 1969, Geologic framework of the Santa Barbara Channel Region, *in* Geology, Petroleum Development, and Seismicity of the Santa Barbara Channel Region, California: U.S. Geological Survey Professional Paper 679-A, p. 1–11.

VEDDER, J. G., YERKES, R. F., AND SCHOELLHAMER, J. E., 1957, Geologic map of the San Joaquin Hills-San Juan Capistrano area, Orange County, California: U.S. Geological Survey Oil and Gas Investigations Map OM-193, scale 1:24,000.

VICKERY, F. P., 1927, the interpretation of the physiography of the Los Angeles coastal belt: American Association of Petroleum Geologists Bulletin, v. 11, p. 417–424.

VON HUENE, R. E., 1969, Geologic structure between the Murray fracture zone and transverse ranges: Marine Geology, v. 7, p. 475–499.

WELDON, R. J., II, AND HUMPHREYS, EUGENE, 1986, A kinematic model of southern California: Tectonics, v. 5, p. 33–48.

WERNICKE, BRIAN, AND BURCHFIEL, B. C., 1982, Modes of extensional tectonics: Journal of Structural Geology, v. 4, p. 105–115.

WILCOX, R. E., HARDING, T. P., AND SEELEY, D. R., 1973, Basic wrench tectonics: American Association of Petroleum Geologists Bulletin, v. 57, p. 74–96.

WOODFORD, A. O., 1924, The Catalina metamorphic facies of the Franciscan series: Department of Geological Sciences Bulletin, California University, v. 15, p. 49–68.

WOODFORD, A. O., 1925, The San Onofre Breccia—Its nature and origin: California University Publications *in* Geological Sciences, v. 15, p. 159–280.

WOODFORD, A. O., SCHOELLHAMER, J. E., VEDDER, J. G., AND YERKES, R. F., 1954, Geology of the Los Angeles Basin, *in* Jahns, R. H., ed., Geology of Southern California: California Division of Mines and Geology Bulletin 170, Part 5, p. 65–81.

WOODRING, W. P., BRAMLETT, M. N., AND KEW W. S. W., 1946, Geology and Paleontology of the Palos Verdes Hills, California: U.S. Geological Survey Professional Paper 207, 145 p.

YEATS, R. S., 1968, Rifting and rafting in the southern California borderland, *in* Dickinson, W. R., and Grantz, Arthur, eds., Conference on Geologic Problems of San Andreas Fault System: Stanford University Publications in Geological Sciences, v. 11, p. 307–322.

YEATS, R. S., 1973, Newport-Inglewood fault zone, Los Angeles Basin, California: American Association of Petroleum Geologists Bulletin, v. 57, p. 117–135.

YEATS, R. S., 1976, Extension versus strike-slip origin of the southern California borderland, *in* Howell, D. G., ed., Aspects of the Geological History of the California Continental Borderland: Pacific Section, American Association of Petroleum Geologists, Miscellaneous Publication 24, p. 455–485.

YEATS, R. S., 1983, Large-scale Quaternary detachments in Ventura Basin, southern California: Journal of Geophysical Research, v. 88, p. 569–584.

YEATS, R. S., AND HAQ, B. U., 1981, Deep-sea drilling off the Californias: Implications of Leg 63, *in* Yeats, R. S., Haq, B. U., and others, Initial Reports of the Deep Sea Drilling Project, v. 63, U.S. Government Printing Office, Washington, D.C., p. 949–961.

YEATS, R. S., COLE, M. R., MERSCHAT, W. R., AND PARSLEY, R. M., 1974, Poway fan and submarine cone and rifting of the inner southern California borderland: Geological Society of America Bulletin, v. 85, p. 293–302.

YERKES, R. F., AND LEE, W. H. K., 1979, Late Quaternary Deformation in the Western Transverse Ranges, California: U.S. Geological Survey Circular 799-B, p. 27–37.

YERKES, R. F., MCCULLOH, T. H., SCHOELLHAMER, J. E., AND VEDDER, J. G., 1965, Geology of the Los Angeles Basin, California—An Introduction: U.S. Geological Survey Professional Paper 420-A, 57 p.

ZOBACK, M. D., ZOBACK, M. L., MOUNT, V., SUPPE, JOHN, EATON, HEALY, J., OPPENHEIMER, D., REASENBERG, P., JONES, L., RALEIGH, C., WONG, I., SCOTTI, O., AND WENTWORTH, C., 1987, New evidence on the state of stress of the San Andreas fault system: Science, v. 238, p. 1105–1111.

SUBJECT INDEX

SUBJECT INDEX

H

I

J

K

L

M

N